Lecture Notes in Artificial Intelligence 1674

Subseries of Lecture Notes in Computer Science
Edited by J. G. Carbonell and J. Siekmann

Lecture Notes in Computer Science

Edited by G. Goos, J. Hartmanis and J. van Leeuwen

Springer

Berlin
Heidelberg
New York
Barcelona
Hong Kong
London
Milan
Paris
Singapore
Tokyo

Dario Floreano Jean-Daniel Nicoud
Francesco Mondada (Eds.)

Advances in Artificial Life

5th European Conference, ECAL'99
Lausanne, Switzerland, September 13-17, 1999
Proceedings

Springer

Series Editors

Jaime G. Carbonell, Carnegie Mellon University, Pittsburgh, PA, USA
Jörg Siekmann, University of Saarland, Saarbrücken, Germany

Volume Editors

Dario Floreano
Jean-Daniel Nicoud
Francesco Mondada
Laboratory of Microprocessors and Interfaces (LAMI)
Department of Computer Science
Swiss Federal Institute of Technology (EPFL)
CH-1015 Lausanne, Switzerland
E-mail: {Dario.Floreano/Jean-Daniel.Nicoud/Francesco.Mondada}@epfl.ch

Cataloging-in-Publication data applied for

Die Deutsche Bibliothek - CIP-Einheitsaufnahme

Advances in artificial life : 5th European conference ; proceedings / ECAL '99, Switzerland, September 13 - 17, 1999. Dario Floreano ... (ed.). - Berlin ; Heidelberg ; New York ; Barcelona ; Hong Kong ; London ; Milan ; Paris ; Tokyo : Springer, 1999
(Lecture notes in computer science ; Vol. 1674 : Lecture notes in artificial intelligence)
ISBN 3-540-66452-1

CR Subject Classification (1998): I.2, J.3, F.1.1-2, G.2, H.5.1, I.5.1, J.4, J.6

ISBN 3-540-66452-1 Springer-Verlag Berlin Heidelberg New York

Printed in Germany

Typesetting: Camera-ready by author
SPIN 10704169 06/3142 – 5 4 3 2 1 0 Printed on acid-free paper

Preface

No matter what your perspective is, what your goals are, or how experienced you are, Artificial Life research is always a learning experience. The variety of phenomena that the people who gathered in Lausanne reported and discussed for the fifth time since 1991 at the European Conference on Artificial Life (ECAL) has not been programmed, crafted, or assembled by analytic design. It has evolved, emerged, or appeared spontaneously from a process of artificial evolution, self-organisation, or development.

Artificial Life is a field where biological and artificial sciences meet and blend together, where the dynamics of biological life are reproduced in the memory of computers, where machines evolve, behave, and communicate like living organisms, where complex life-like entities are synthesised from electronic chromosomes and artificial chemistries. The impact of Artificial Life in science, philosophy, and technology is tremendous. Over the years the synthetic approach has established itself as a powerful method for investigating several complex phenomena of life. From a philosophical standpoint, the notion of life and of intelligence is continuously reformulated in relation to the dynamics of the system under observation and to the embedding environment, no longer a privilege of carbon-based entities with brains and eyes. At the same time, the possibility of engineering machines and software with life-like properties such as evolvability, self-repair, and self-maintainance is gradually becoming reality, bringing new perspectives in engineering and applications.

All these aspects, and many more, are reflected in the 90 papers presented at ECAL'99 from 13 to 17 September 1999 and collected in this volume. Each paper has been carefully reviewed by three members of the scientific committee (see list following the Preface) and selected from among 150 submissions. Of the selected and revised papers, 50 have been accepted as long oral contributions and the remaining 40 as short poster contributions. In both cases, the overriding selection criteria have been scientific and methodological soundness, novelty, and potential for future developments. In addition to the contributed papers, this volume includes the abstracts of four keynote lectures (H. Meinhardt, W. D. Hamilton, L. Steels, and T. Lenton) and two invited talks (D. Mange and D. Thalmann). At the end of each abstract, the reader will find a list of the most relevant references for these talks.

In addition to single-track presentations, demonstrations, and satellite workshops, the first day of the conference was dedicated to a series of tutorials covering genotype-phenotype mappings, collective intelligence, cellular automata and complex systems, synthetic actors, evolutionary robotics, and artificial chemistry.

Contributed and invited papers have been classified according to the following broad categories.

Epistemology is concerned with the philosophical aspects of Artificial Life. The two selected papers address two key concepts in Artificial Life: what an emergent phenomenon could be and when an entity may be defined as alive.

Evolutionary Dynamics (17 papers) addresses a number of fundamental issues in natural and artificial evolution. These include the interactions between evolution and other forms of ontogenetic dynamics, the role and effect of mutations, development of synthetic organisms, and measures of diversity and complexity in evolving systems.

Evolutionary Cybernetics (12 papers) is artificial evolution of mechanisms and structures that support behavior of biological and artificial organisms with a sensory motor system. The papers included in this section employ this methodology for both understanding the development and functioning of biological brains and for synthesising control systems of autonomous robots and of other artificial creatures.

Bio-Inspired Robotics and Autonomous Agents (15 papers) is a collection of contributions describing recent efforts in building physical and virtual agents with life-like properties, such as bio-inspired control, adaptation, human and animal morphologies, and behavioural autonomy. Some of these papers go as far as addressing motivation, emotions, and economic behaviour.

Self-Replication, Self-Maintenance, and Gene Expression (16 papers) includes papers that investigate some fundamental properties of micro-entities such as RNA, DNA, cells, and cellular aggregates. These micro-entities are capable of self-replication, self-maintenance, evolution, and development into full organisms from a set of genetic instructions. While some of the authors attempt to understand these principles with mathematical models and computer simulations, others incorporate them into a new generation of bio-inspired electronic circuits capable of complex behaviours.

Societies and Collective Behaviours (17 papers) display complex dynamics that cannot be understood by looking at single individuals in isolation. Sex, cooperation, selfishness, teaching, cultural transmission, distributed problem solving, or simple interference are some of the phenomena that one observes in assemblies of natural and artificial organisms. The papers in this section attempt to understand under which conditions these phenomena arise, when they develop, and how one could exploit them to create societies of artificial agents and robots capable of performing complex tasks.

Communication and Language (13 papers) goes one step further and investigates the emergence and role of communication in societies of organisms and intelligent machines. In most cases communication is considered a dynamic phenomenon arising in populations of organisms that evolve, learn, and dynamically form temporal aggregations. Within this conceptual framework, some papers address the origin of language in its many manifestations, ranging from speech to lexicon and syntax.

ECAL'99 was selected as the *1999 International EPFL-Latsis Foundation Conference.* Generous sponsorship by the Latsis Foundation allowed us to invite high-quality keynote lecturers, offer several student fellowships, and make sure that the necessary resources were available for a successful organisation. Additional sponsors are listed on the next page.

The organisers thank very much Monique Dubois and Joseba Urzelai for their assistance. Monique ensured a smooth and professional organisation from day 1, taking care of every single detail with precision and patience, while Joseba managed all bits and tricks of electronic submissions and e-mails. Marie-Jo Pellaud kindly helped out whenever it was necessary. Our thanks also go to Luigi Pagliarini for the cover art and to Mark Peden for the caterpillar depicted in it.

Lausanne, June 1999

Dario Floreano
Jean-Daniel Nicoud
Francesco Mondada

Organisation

ECAL'99 has been organised by LAMI-Mantra at EPFL in Lausanne.

Executive Committee

Organisation: Dario Floreano, Jean-Daniel Nicoud, Francesco Mondada
Secretariat: Monique Dubois, Joseba Urzelai

Advisory and Scientific Committee

Sponsoring Institutions

The Latsis Foundation, Geneva, has been the main sponsor.
The other sponsors were Swissair, K-Team S.A., Cyberbotics Ltd., LEGO laboratory of Aarhus University (DK), and Cyberlife Technology Ltd.

Table of Contents

Keynote Lectures

Epistemology

Evolutionary Dynamics

Evolutionary Cybernetics

Bio-inspired Robotics and Autonomous Agents

Self-Replication, Self-Maintenance, and Gene Expression

Societies and Collective Behaviour

Communication and Language

ERRATA

Lecture Notes in Artificial Intelligence 1674
Subseries of Lecture Notes in Computer Science
ISBN 3-540-66452-1

Dario Floreano Jean Daniel Nicoud
Francesco Mondada (Eds.)

Advances in Artificial Life
5th European Conference, ECAL'99
Lausanne, Switzerland, September 1999
Proceedings

In the Paper
Level of Autonomy for Virtual Human Agents, pp. 345-349, the references mentioned in the text are not correct. The correct order should read as follows: [5], [11], [12], [1], [9], [6], [4], [3], [5], [7], [9], [7], [8], [9], [10], [2], [2] and [13], whereby [13] refers to http://www.nada.kth.se/erena.

Keynote Lectures

From Fertilized Eggs to Complex Organisms: Models of Biological Pattern Formation

Hans Meinhardt

Max Planck-Institut fuer Entwicklungsbiologie
D-72076 Tuebingen, Germany
hans.meinhard@tuebingen.mpg.de

The development of an organism from a single cell to its full spatial complexity is a most fascinating phenomenon. Up to about 1984, most information have been observed from perturbation of normal development and the observation of the subsequent regulations. Based on this observations, we have elaborated molecularly reasonable molecular interactions that reproduce in computer simulations the observed dynamic regulation in great detail. The proposed interactions have found meanwhile direct support by molecular-genetic investigations. The following processes are proposed to play a crucial role:

(i) Primary pattern formation is accomplished by short range autocatalysis and long ranging inhibition. This leads to local concentration maxima that can act as organizing regions. Either monotonically graded, spatial periodic or stripe-like distributions are possible. (ii) Cells respond to such signals by obtaining a stable state of differentiation. This is achieved by a direct or indirect autoregulation of genes. If the signal is above a threshold, the corresponding feedback loop is turned on. Due to the autoregulation, the maintenance of this activity does no longer require the inducing signal. A cell remembers the signals it has received in its history. Many autoregulatory genes have been found meanwhile. (iii) Segmentation, a basic pattern in all higher organisms, depends on mutual long range stabilization of cell states that locally exclude each other. Cell types that form neighbouring structures need each other in a symbiotic manner. This leads to a controlled neighbourhood of structures. (iv) Legs or wings are initiated at boundaries between regions of different determinations. If the production of a new morphogen requires a co-operation of two differently determined cell types, their common border will become a new signalling centre. The local morphogen concentration provides a measure for the distance from this border. The intersection of two such borders defines unique points and complete new co-ordinate systems for the initiation of these structures. The model accounts for the correct initiation of legs and wings at particular positions of the organism and for their correct orientation in respect to the main body axes of the embryo. This model has found direct support from observations in vertebrates and insects.

Patterns on shells of tropical molluscs are an especially instructive example for pattern formation in space and time since they preserve the complete history of their generation. Mechanisms derived from this patterning provide a better understanding of processes that seems completely unrelated, such as the guiding of a growing nerve or blood coagulation. This suggest that similar reactions are used in a toolbox-like fashion again and again in different situations of development.

References

1. H. Meinhardt. Biological Pattern-Formation – New Observations Provide Support for Theoretical Predictions. *Bioessays*, 16:627–632, 1994.
2. H. Meinhardt. *The Algorithmic Beauty of Sea Shells.* Springer Verlag, Heidelberg, New York, 1998 (with PC-Software).

Fables of Cyberspace: Tapeworms, Horses, and Mountains

William D. Hamilton

Department of Zoology
University of Oxford
OX1 3PS Oxford, UK
william.hamilton@zoology.oxford.ac.uk

If microevolution is hill-climbing, then macroevolution is valley-crossing leading to higher hills. Does sexuality, making reproduction a reticulate process instead of bifurcating, help macroevolution? Even without valleys, if a niche AB exists but cannot be accessed until either niche A or niche B is conquered, it may appear that sexuals should reach AB sooner because discoveries made by selection in A and B separately can be brought together by recombination: in contrast under asexuality the route to AB has to be either through A or through B, not both at once, and should take more time. Moreover, if valleys on the adaptive surface exist, the force of this presumption is increased. Depending on size of population and depth of valleys, a higher hill may not be accessible by adaptation at all; but with sexuality present at least the chance of attaining the high hill's slope seems inevitably greater. All this, however, leaves out many potentially relevant factors including the intrinsic inefficiency of sexual reproduction. The last makes sex a strongly inferior competitor to asex in the short term as well as making it (advantageously?) slower to climb slopes once they are found.

In the living world is sexuality maintained and what is the evidence that it can help macroevolution? For the silicon world, how and when should AI or GA specialist think to impose sexuality and what recombination values and spatial genomic arrangements should they choose? For the present I can only make suggestions based on what I know of the history and achievements of real life and from very few tentative computer simulations by myself and others. Mainly I will explain a theme that living systems inevitably evolve/acquire *parasites* and will describe how these, while always benefiting themselves to the detriment of host reproduction in the short term, in the long term, if by specialism and good chance not causing extermination, may actually aid hosts. In particular I will explain how infecting and coevolving parasites induce a permanent coevolutionary dynamic as well as often a demographic one. This can maintain sexuality and thereby, as simulation evidence shows, can aid valley-crossing. It does so by (i) allowing partial adaptive discoveries made in different places to be combined, and (ii) keeping adaptive topographies in constant change–in effect empowering a Sewall Wright Shifting Balance but more strongly and determinately than his genetic drift.

Sexuality is not the only way real life became reticulate, but, except in procaryotes, the other processes, which involve temporary or permanent adoption of elsewhere-evolved tricks or whole cooperations of whole organisms, are much more occasional. Examples here are the horizontal transfer of plasmids and virus-borne (or otherwise vectored) bits of code, and the adoption of the mitochondrion or/and the plastid by eucaryote ancestors. Smallness of adventive units (probably parasites initially) and their final complete enslavement and dependence induced by the larger organism seem the important and rare features here: when occasionally the features are attained, however, the consequences for macroevolution can be vast.

As to higher levels of organismic interaction involving separate and unenslaved organisms, as for example with claims for evolution of cooperative ecosystems or of 'gaias' as wholes, it is not yet clear to what extent natural selection can work effectively. In the ecosystem case 'reticulation' is extreme but 'reproduction' is correspondingly ill-defined; nevertheless suggestive mixed-species 'superorganisms' not involving permanent enclosure do seem sometimes to arrive (e.g. lichens, corals, bogs). In the case of suggested 'gaias', any sense of reproduction has to be still less traditional; nevertheless viewing selection simply as the addition of compatible species types into the gaia and as occurring on a time scale not of generations but the additions themselves, some weak and non-Darwinian 'evolution' may seemingly still occur. What effectiveness this has (or what use it can be in the *in silico* world), however, is not yet apparent.

References

1. B. Sumida, A. Houston, J. McNamara, and W. D. Hamilton. Genetic Algorithms and Learning. *J. Theo. Biol.*, 1990.
2. W.D. Hillis. Co-evolving Parasites Improve Simulated Evolution as an Optimisation Procedure. In S. Forrest, editor, *Emergent Computation: Self-Organising, Collective and Cooperative Computing Networks.* MIT Press, Cambridge, MA, 1990.
3. W. D. Hamilton, R. Axelrod, and R. Tanese. Sex as an Adaptation to Resist Parasites. *Proc. Nat. Acad. Sci.* USA 87:8566–3573, 1990.
4. W. D. Hamilton. Memes of Haldane and Jayakar in a Theory of Sex. *J. Genet.*, 69:17–32, 1990.
5. W.D. Hamilton. Haploid Dynamical Polymorphism in a Host with Matching Parasites: Effects of Mutation/Subdivision, Linkage, and Patterns of Selection. *J. Hered.*, 84:328–338, 1993.
6. B. H. Sumida and W. D. Hamilton. Both Wrightian and "Parasite" Peak Shifts Enhance Genetic Algorithm Performance in the Travelling Salesman Problem. In R. Paton, editor, *Computing with Biological Metaphors.* Chapman & Hall, London, 1994.

Cognitive Teleportation and Situated Embodiment

Luc Steels

Sony CSL, Paris
VUB AI Lab, Brussels
steels@arti.vub.ac.be

Throughout this summer, we are conducting a large-scale experiment in which an open-ended population of situated embodied agents bootstrap an ontology and a lexicon from scratch. The agents can teleport between different robot bodies through which they can experience and act upon the world. The agents play a language game, called the guessing game, in which one agent (the hearer) tries to identify through verbal and non-verbal hints from another agent (the speaker) a particular object in the environment. Humans can interact with the artificial agents through the Internet. They can construct their own agent, influence its language, and follow the performance of the agent as it travels between different physical locations.

The experiment relies heavily on key concepts from artificial life. The lexicon self-organises based on a positive feedback loop between use and success, similar to the way a path in an ant society forms. The ontology develops through a selectionist process in each agent; distinctions grow in a partly random fashion and get pruned if they are not useful and effective in conceptualisation or communication. The agents are grounded through a behavior-based robotic architecture in their environments.

The talk will focus on the collective semiotic dynamics that we effectively are observing during this experiment. We expect to see damping of synonymy and polysemy as well as constant lexical and ontological evolution as the agents cope with new situations in the environment or as new agents or humans inject new language forms in the language games.

References

1. Website: http://talking-heads.csl.sony.fr
2. L. Steels. The Origins of Syntax in Visually Grounded Robotic Agents. *Artificial Intelligence*, 103:1–24, 1998.
3. L. Steels. *The Talking Heads Experiment. Vol 1. Words and Meanings.* Pre-edition Laboratorium Antwerpen, 1999 (available during demo session).
4. L. Steels and F. Kaplan. Collective Learning and Semiotic Dynamics. In D. Floreano, F. Mondada, and J. D. Nicoud, editors, *Proceedings of the 5th European Conference on Artificial Life*. Springer-Verlag, Berlin, 1999.

Testing Gaia Theory with Artificial Life

Tim Lenton

Institute of Terrestrial Ecology
Edinburgh Research Station
Bush Estate, Penicuik, Midlothian, EH26 0QB, UK
Timothy.M.Lenton@ite.ac.uk

The Gaia hypothesis postulates that the Earth is a self-regulating system, that tends to maintain conditions habitable for life. The hypothesis grew from the recognition that organisms have fundamentally altered their planetary environment - its thermodynamic state, the composition of the atmosphere, the climate, ocean chemistry and the land surface. Changes in the environment have in turn altered organisms' growth and the forces of natural selection acting upon them and this implies inevitable feedback connections between life and its environment. The resulting world seems remarkably well suited to carbon-based life, and has some important stabilising properties. For example, the Earth's climate has remained habitable despite a 25% increase in the luminosity of the sun since the origin of life. Known feedback mechanisms more often show a tendency to regulate the environment than to destabilise it. Thus there is arguably an overall tendency for the Earth system to counteract gradual forcing and rapid perturbations that tend to drive the environment away from a habitable state. However, it is not obvious why this should necessarily be the case. If the Earth did not have some regulatory properties, we would probably not be here to remark on them, because life would have perished before evolution produced conscious observers. Furthermore, Gaia poses a puzzle when viewed from a Darwinian perspective: Why should the organisms that leave the most descendants be the ones that contribute to regulating their planetary environment? These questions highlight the need to test a 'general' Gaia hypothesis; that once life has originated on a planet it will contribute to its own persistence. The 'specific' Gaia hypothesis (that which applies specifically to the Earth) is notoriously difficult to test directly, because it involves long time-scales and large space scales. However, the general hypothesis may be tested in some sense with artificial life on artificial worlds. The Daisyworld model first showed that planetary self-regulation can occur without teleology (foresight or planning on the part of unconscious organisms). However, the model is a special case in that the organisms alter their environment in the same way at the individual and the global levels and there is no evolution. A series of models will be presented which adapt the simple Daisyworld framework to explore different special cases, and then introduce mutation of the organisms' environment-altering properties. It has been suggested that adaptation to prevailing environmental conditions tends to undermine any environmental regulation, and the evolutionary trade-off between altering one's environment and adapting to prevailing conditions will be considered. However, a more general model is still required to test whether, when organisms evolve new ways of altering their environment, there is any tendency towards self-regulation of the whole system of organisms plus environment. Thus, artificial life researchers may be well placed to help test the Gaia theory.

References

1. J. E. Lovelock. *The Ages of Gaia - A Biography of Our Living Earth.* Oxford University Press, Oxford, 1995.
2. T. M. Lenton. Gaia and Natural Selection. *Nature*, 394:439–447, 1998.
3. J. W. Kirchner. The Gaia Hypothesis: Can it be Tested? *Reviews of Geophysics*, 27(2):223–235, 1989.
4. A. J. Watson and J. E. Lovelock. Biological Homeostasis of the Global Environment: The Parable of Daisyworld. *Tellus*, 35B:284–289, 1983.

Epistemology

Testing for Emergence in Artificial Life

Edmund M. A. Ronald,[1] Moshe Sipper,[2] and Mathieu S. Capcarrère[2]

[1] Centre de Mathématiques Appliquées, Ecole Polytechnique, 91128 Palaiseau Cedex, France. E-mail: eronald@cmapx.polytechnique.fr.
[2] Logic Systems Laboratory, Swiss Federal Institute of Technology, IN-Ecublens, CH-1015 Lausanne, Switzerland. E-mail: {name.surname}@epfl.ch.

Abstract. The field of artificial life (Alife) is replete with documented instances of *emergence*, though debate still persists as to the meaning of this term. In the absence of a formal definition, researchers in the field would be well served by adopting an emergence certification mark which would garner approval from the Alife community. We propose an *emergence test*, consisting of three criteria—design, observation, and surprise—for conferring the emergence label.

1 Introduction

When a bank's accounting program goes seemingly independent and does its own thing, the programmer scratches his head, sighs, and prepares for doing overtime with the debugger. But when a society of agents does something surprising, Alife researchers may solemnly document this "emergent behavior," and move on to other issues without always seeking to determine the cause of their observations. Indeed, overly facile use of the term emergence has made it controversial. Arkin recently observed that:

> Emergence is often invoked in an almost mystical sense regarding the capabilities of behavior-based systems. Emergent behavior implies a holistic capability where the sum is considerably greater than its parts. It is true that what occurs in a behavior-based system is often a surprise to the system's designer, but does the surprise come because of a shortcoming of the analysis of the constituent behavioral building blocks and their coordination, or because of something else? [1](page 105)

Altogether, it seems the emergence tag has become a great attention grabber, thanks to the striking behaviors demonstrated in artificial-life experiments. We do not think, however, that emergence should be diagnosed *ipso facto* whenever the unexpected intrudes into the visual field of the experimenter; nor should the diagnosis of emergence immediately justify an economy of explanation. Such abuse and overuse of the term will eventually devalue its significance, and bring work centered on emergence into disrepute. Therefore, we contend that in the absence of an acceptable definition, researchers in the field would be well served by adopting an emergence certification mark which would garner approval from the Alife community.

Motivated by this wish to standardize the tagging task, we propose an *emergence test,* namely, criteria by which one can justify conferring the emergence label. Our criteria are motivated by an examination of published work in the field of Alife [10].

The emergence test is presented in the next section, followed in Section 3 by four case studies demonstrating its applicability. Finally, in Section 4, we discuss a number of issues pertaining to our test.

2 An operant definition of emergence for Alife researchers

The difficulties we face in adopting a definition of the concept of emergence are reminiscent of the complications faced by early Artificial Intelligence (AI) researchers in defining intelligence.[1] Nonetheless, where the equally elusive concept of intelligence is concerned, Alan Turing found a way to cut the Gordian knot, by means of an *operant definition* which is useful *within the limited context of man-machine interaction* [14]. Debate concerning the concept of intelligence is unlikely to subside in the foreseeable future, and the same, we believe, holds for emergence. We deem, however, that viewing the world through Turing-colored glasses might improve our vision as regards the concept of emergence—at least where modern-day Alife practice is concerned.

Alife is a constructive endeavor: some researchers aim at evolving patterns in a computer, some seek to elicit social behaviors in real-world robots, others wish to study life-related phenomena in a more controllable setting, while still others are interested in the synthesis of novel life-like systems in chemical, electronic, mechanical, and other artificial media. Alife is an experimental discipline, fundamentally consisting of the observation of run-time behaviors, those complex interactions generated when populations of man-made, artificial creatures are immersed in real or simulated environments. Published work in the field usually relates the conception of a model, its instantiation into real-world or simulated objects, and the observed behavior of these objects in a collection of experiments.

The Turing Test focuses on a human experimenter's incapacity at discerning human from machine when holding what we would now call an Internet chat session. Our emergence test centers on an observer's avowed incapacity (amazement) to reconcile his perception of an experiment in terms of a global world view with his awareness of the atomic nature of the elementary interactions.

---

[1] On the difficulties in defining emergence, Emmeche, Køppe, and Stjernfelt recently remarked: "One reason for the widespread scepticism against the word [emergence] is a historical load of confusion surrounding the metaphysical aspects of the concept, reflected in the fact that it has been used in a long series of different ways, apparently making it impossible to use it as a clearly defined term..." [5](page 84)

Assume that the scientists attendant upon an Alife experiment are just two: a system designer and a system observer (both of whom can in fact be one and the same), and that the following three conditions hold:

(i) **Design.** The system has been constructed by the designer, by describing *local* elementary interactions between components (e.g, artificial creatures and elements of the environment) in a language $\mathcal{L}_1$.
(ii) **Observation.** The observer is *fully aware* of the design, but describes *global* behaviors and properties of the running system, over a period of time, using a language $\mathcal{L}_2$.
(iii) **Surprise.** The language of design $\mathcal{L}_1$ and the language of observation $\mathcal{L}_2$ are distinct, and the causal link between the elementary interactions programmed in $\mathcal{L}_1$ and the behaviors observed in $\mathcal{L}_2$ is *non-obvious* to the observer—who therefore experiences surprise. In other words, there is a cognitive dissonance between the observer's mental image of the system's design stated in $\mathcal{L}_1$ and his contemporaneous observation of the system's behavior stated in $\mathcal{L}_2$.

The above three clauses relating design, observation, and surprise describe our conditions for diagnosing emergence, i.e., for accepting that a system is displaying emergent behavior.

When assessing the surprise clause of our test one should bear in mind that as human beings we are quite easily surprised (as any novice magician will attest). The question reposes rather on how *evanescent* the surprise effect is, i.e., how easy (or strenuous) it is for the observer to bridge the $\mathcal{L}_1$–$\mathcal{L}_2$ gap, thus reconciling his global view of the system with his awareness of the underlying elementary interactions. One can draw an analogy with the concept of intelligence and the Turing test: the chatty terminal might at first appear to be carrying on like an intelligent interlocutor, only to lose its "intelligence certificate" once the tester has pondered upon the true nature of the ongoing conversation.

Some of the above points deserve further elaboration, or indeed invite debate. Before treating these issues in Section 4, we wish to demonstrate the application of our test to four cases.

3 Administering the emergence test: Four case studies

In this section we administer the emergence test to four examples, thus demonstrating its application (additional examples are given in [10]). Each example ends with a "test score," constituting our own assertion as observers of whether we are indeed surprised, that is, of whether emergent behavior is indeed displayed—or not.

1. Emergence of a nest structure in a simulated wasp colony, from the interactions taking place between individual wasps [13].

(i) The design language $\mathcal{L}_1$ is that of local wasp interactions, including movement on a three-dimensional cubic lattice and placement of bricks. A wasp's decision is based upon a local configuration of bricks, which lie in its "visual" field. Actions to be taken are prewired under the form of a lookup table with as many entries as there are stimulating configurations.
(ii) The observation language $\mathcal{L}_2$ is that of large-scale geometry, as employed to describe nest architectures.
(iii) While fully aware of the underlying wasp interaction rules, the observer nonetheless marvels at the sophistication of the constructions and their striking similarity to naturally occurring nests.

Diagnosis: emergent behavior is displayed by the nest-building wasps.

2. Emergence of a "highway" created by the artificial Langton ant, from simple movement rules [12].

(i) The design language $\mathcal{L}_1$ is that of single moves of a simple, myopic ant. The ant starts out on the central cell of a two-dimensional, rectangular lattice, heading in some selected direction. It moves one cell in that direction and looks at the color of the cell it lands on—black or white. If it lands on a black cell, it paints it white and turns 90 degrees to the left; if it lands on a white cell, it paints it black and turns 90 degrees to the right. These simple rules are iterated indefinitely.
(ii) The observation language $\mathcal{L}_2$ is that of global behavioral patterns, extended over time and space (i.e., tens of thousands of single ant moves, spanning thousands of cells). Specifically, the ant was observed to construct a "highway," i.e., a repeating pattern of fixed width that extends indefinitely in a specific direction (Figure 1a).
(iii) While fully aware of the very simple ant rules, the observer is nonetheless surprised by the appearance of a highway.

Diagnosis: emergent behavior is displayed by the highway-constructing ant.

3. Emergence of flocking behavior in simulated birds, from a set of three simple steering behaviors [9].

(i) The design language $\mathcal{L}_1$ is that of local bird interactions, the three rules being: *separation:* steer to avoid crowding local flockmates; *alignment:* steer toward the average heading of local flockmates; *cohesion:* steer to move toward the average position of local flockmates. A bird's decision is based upon its nearby neighbors, i.e., those that are in its "visual" field.
(ii) The observation language $\mathcal{L}_2$ is that of flocking behaviors, such as the flock's parting smoothly when faced with an obstacle, and "flowing" around it—to then reunite again (Figure 1b).
(iii) While fully aware of the underlying bird interaction rules, the observer nonetheless marvels at the lifelike flocking behaviors.

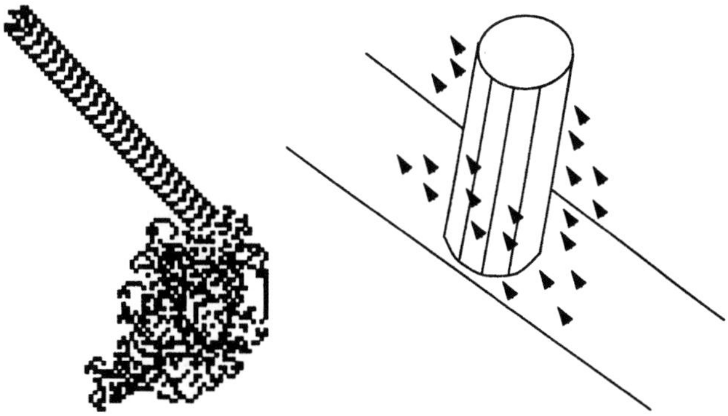

Fig. 1. Examples of emergence. (a) The trail created by the highway-constructing, Langton ant. (b) A flock of simulated birds parts smoothly when faced with an obstacle, and "flows" around it—to then reunite again (after Reynolds [9]).

Diagnosis: the flocking behavior exhibited by the artificial birds was considered a clear case of emergence when it was first reported upon in 1987. However, one could now maintain that it no longer passes the emergence test, since wide-spread use of this technique in computer graphics has obviated the element of surprise. This example demonstrates that the diagnosis of emergence is contingent upon the sophistication of the observer.

4. Emergence of wall-following behavior in an autonomous, mobile robot, from the simultaneous operation of two simple behavior systems: obstacle avoidance and wall seeking [11].

(i) The design language $\mathcal{L}_1$ is that of simple robot behaviors, including—in this case—obstacle avoidance and wall seeking.
(ii) The observation language $\mathcal{L}_2$ is that of more elaborate robot behaviors, consisting—in this case—of wall following.
(iii) Steels wrote that "Wall following is emergent in this case because the category 'equidistance to the (left/right) wall' is not explicitly sensed by the robot or causally used in one of the controlling behavior systems." [11](page 92)

Diagnosis: Steels diagnosed emergence in this case as it accords with his own definition, namely, that a behavior is emergent if it necessitates the use of new descriptive categories that are not needed to describe the behavior of the constituent components [11]. While thus alluding to the language dichotomy rendered explicit by our definition (i.e., the existence of two distinct languages—that of design and that of observation), we maintain that the surprise element is missing: the wall-following behavior can be quite readily deduced by an observer aware of the two underlying simpler behaviors. We thus conclude that emergent behavior is *not* displayed by the wall-following robot.

4 Discussion

We now discuss the various components of our test.

The operant nature of the test. In this we have drawn our inspiration from Turing, who—concerning intelligence—opted for an operant, informal, "social" definition, deliberately eschewing rigor. Turing's definition still serves the AI community well—almost half a century after its publication [14].

Emergence as a property of artificial systems. In his book *Emergence: From Chaos to Order*, Holland wrote that "*Emergence occurs in systems that are generated*" [7](page 225). Reviewing Holland's book, Mallot also opined that "In this context [the construction of artificial systems], the problem of emergence may actually be a genuine one." [8] These views accord with our own view, namely, that the diagnosis of emergence should be considered (and hence our test applied) within domains such as Alife, which are inherently constructive endeavors. This view naturally gives rise to clause (i) of our test, i.e., the existence of a designer—and of a design language.

The existence of an observer. Artificial systems are constructed to be beheld—one does not usually build one's system, to then walk away nonchalantly without ever looking back. Hence, there exists an observer *ipso facto* (who need not necessarily be the constructor himself), a fundamental aspect which has not escaped researchers in the field. In a paper discussing emergence and artificial life, Cariani wrote that "The interesting emergent events that involve artificial life simulations reside not in the simulations themselves, but in the ways that they change the way we think and interact with the world." [2](page 790) He goes on to say that "computer simulations are catalysts for emergent processes in our own minds..." [2](page 790)

Another author, Emmeche, in an introductory monograph on artificial life, examines the case for emergence "in the eye of the beholder." [4](page 145) Also, Crutchfield, in an article devoted to the subject of emergence, asks: "But for whom has the emergence occurred? More particularly, to whom are the emergent features 'new'?... The newness in both cases is in the eye of an observer..." [3](page 517)

Holland brings up the issue of the observer circuitously, when writing that "*The whole is more than the sum of the parts in these generated systems...* Said another way, there are regularities in system behavior that are not revealed by direct inspection of the laws satisfied by the components." [7](page 225) One may ask *direct inspection by whom?* Why, by the observer of course![2] Clearly, the existence of an observer is a *sine qua non* for the issue of emergence to arise at all.

[2] Holland also cites a passage from Gell-Mann's book *The Quark and the Jaguar* [6], which also brings up indirectly the role of the observer: "In an astonishing variety of contexts, apparently complex structures or behaviors emerge from systems

Surprise. By bringing the observer's *emotion* of surprise into play, our emergence test widens the focal beam of discussion, now shining both on the *system's behavior* as well as on the experimenter and her *internalized expectations.* This relates to Cariani's nutshell description of emergence relative to a model as "the deviation of the behavior of a physical system from an observer's model of it." [2](page 779) An author subscribing to said deviation-from-model view would wish to document her *a priori* expectations before diagnosing emergence and abandoning attempts at explanation. Our emergence test might then be reformulated as *Design (Expectations), Observation, Surprise.*

To summarize, the three clauses of our emergent test are grounded in previous work: the design clause expresses our wish to restrict the test to artificially constructed systems; the observation clause reflects the necessity of there being an observer for emergence to arise at all; and the surprise clause embodies both the deliberation and the emotion implied by human judgments of value.

Acknowledgments

We thank Bernard Besnier, Dario Floreano, Daniel Mange, Eduardo Sanchez, Marco Tomassini, and the anonymous reviewers for their helpful remarks and suggestions.

References

1. R. C. Arkin. *Behavior-Based Robotics.* The MIT Press, Cambridge, MA, 1998.
2. P. Cariani. Emergence and artificial life. In C. G. Langton, C. Taylor, J. D. Farmer, and S. Rasmussen, editors, *Artificial Life II*, volume X of *SFI Studies in the Sciences of Complexity*, pages 775–797, Redwood City, CA, 1992. Addison-Wesley.
3. J. P. Crutchfield. Is anything ever new? Considering emergence. In G. Cowan, D. Pines, and D. Melzner, editors, *Complexity: Metaphors, Models, and Reality*, pages 515–537. Addison-Wesley, Reading, MA, 1994.
4. C. Emmeche. *The Garden in the Machine: The Emerging Science of Artificial Life.* Princeton University Press, Princeton, N.J., 1994.
5. C. Emmeche, S. Køppe, and F. Stjernfelt. Explaining emergence: Towards an ontology of levels. *Journal for General Philosophy of Science*, 28:83–119, 1997.
6. M. Gell-Mann. *The Quark and the Jaguar: Adventures in the Simple and the Complex.* Freeman, New York, 1994.
7. J. H. Holland. *Emergence: From Chaos to Order.* Addison-Wesley, Reading, Massachusetts, 1998.
8. H. Mallot. Life is like a game of chess: review of "Emergence: From Chaos to Order" by John Holland. *Nature*, 395:342, September 1998.

---

characterized by simple rules." [7](page 238) Gell-Mann's use of the qualifier "apparently" suggests that the quality in question necessitates a judgment call—that is, an observer.

9. C. W. Reynolds. Flocks, herds, and schools: A distributed behavioral model. *Computer Graphics*, 21(4):25–34, July 1987.
10. E. M. A. Ronald, M. Sipper, and M. S. Capcarrère. Design, observation, surprise! A test of emergence. *Artificial Life*, 1999. (to appear).
11. L. Steels. The artificial life roots of artificial intelligence. *Artificial Life*, 1(1/2):75–110, 1994. The MIT Press, Cambridge, MA.
12. I. Stewart. The ultimate in anty-particles. *Scientific American*, 271(1):104–107, July 1994.
13. G. Theraulaz and E. Bonabeau. Coordination in distributed building. *Science*, 269:686–688, August 1995.
14. A. M. Turing. Computing machinery and intelligence. *Mind*, 59:433–460, 1950.

Are Romance Novels Really Alive? A Discussion of the Supple Adaptation View of Life

Paul Domjan

The Plan II Honors Program, The University of Texas at Austin, WCH 4.104, Austin, Texas, 78712-1105, USA
pdomjan@mail.utexas.edu

Abstract. I will be discussing the plausibility of Mark Bedau's supple adaptation view of life. After presenting the essential aspects of the view, I introduce a hypothetical system of romance novels and romance novel authors, which captures those aspects of a system which are necessary for life under Bedau's view. Using this romance novel system as a starting point, I discuss other intuitively non-living systems, like economies and scientific literature, which would be considered alive by Bedau's standards. Finally, I present one objection to Bedau's view and suggest how his view might change in order to become more widely accepted.

1 Introduction

Bedau [2] has argued that individual biological organisms are not alive, but that the supplely adaptive systems which produce these organisms are. This implies that some intuitively non-living supplely adaptive systems, like economies and scientific literature, are alive. I will present a view of romance novels under which they would be supplely adaptive and thus alive. Using this simple example, I will investigate some of the central questions arising from the supple adaptation view in an attempt to determine what proponents of the view must address for the view to gain significant acceptance. Finally, I will present an objection to the theory, which seems to be the problem around which many of these central questions revolve.

1.1 The Supple Adaptation View

Bedau asserts that "...supple adaptation does not merely produce living entities.... [but that] the entity that is living in the primary sense of that term is the supplely adapting system itself..." [2]. Supple adaptability is defined as "...the way in which evolution automatically fashions and refashions complex intelligent strategies for flourishing as local contexts change" [1]. A living organism is not alive solely by virtue of its own characteristics, but because it is part of a larger system that shows continual adaptive innovation [2].

Supple adaptability itself can be quantified by measuring the prevalence of a trait over time. If a trait has adaptive significance, it will persist in the gene pool much longer than if it has no significant effect, or a detrimental effect, on fitness. The null model for this judgement would have the same mutation rates as the real system, but in the null model the genotype would have no effect on survival. If a system has traits that persist longer than they would in a null model, it is engaging in adaptive evolution [3]. Artificial life models either never engage in adaptive evolution, or do so only until they solve the problem presented by their environment. New traits persist only as long as they would in the null model. The simulation stops evolving [4].

Though this seems to be an interesting statistical method for analyzing evolutionary dynamics, the claim that it is the defining characteristic of life is much more opaque. In Farmer and Belin's view, "[t]here seems to be no single property that characterizes life. Any property that we assign to life is either too broad... or too specific..." [6]. Bedau argues that our intuitive feelings about life vary from time to time and culture to culture, and, as such, we should not look for one property, like metabolism [11], that explains everything that we intuitively feel to be alive, but look for a property that really is life. In taking this approach, Bedau is looking for something that essentially produces living phenomenon [1].

He offers the following definition:

A. x is living iff x is $living_1$ or x is $living_2$.

B. x is $living_1$ iff x is a system undergoing supple adaptation.

C. x is $living_2$ iff there is a some $living_1$ system y such that either (1) x meets condition A_1 and y meets condition B_1 and x bears relation C_1 to y or (2) x meets condition A_2 and y meets condition B_2 and x bears relation C_2 to y, or ... or (n) x meets condition A_n and y meets condition B_n and x bears relation C_n to y. [2]

This definition may solve many problems of life [1], but it implies that human beings are not fully alive, but only alive in a secondary sense, while the biosphere itself is fully alive. The definition also suggests that "...clay crystallites, autocatalytic networks of chemicals, and even human intellectual and economic systems ... also deserve to be thought of as 'living'" [2]. But, all of these things are intuitively non-living.

While this seems difficult for many to accept, focusing on life as a property of both a system and an individual may be promising, as it will eliminate the need to look at life as a collection of properties rather than a single defining property. This critique is intended to motivate more work on developing a less offensive version of this view, which will combine philosophical understanding with Bedau's statistics to develop a powerful and useful definition of life.

1.2 The Romance Novel System

In order to simplify my discussion of intuitively non-living supplely adaptive systems, I shall present a hypothetical system of romance novels that is intuitively non-living but supplely adaptive. The hypothetical system is simpler, and thus easier to discuss, than economies or scientific literature. Assume that there is a population of romance novel writers, and, because of the huge popularity of romance novels, the majority of romance novel readers are also romance novel authors. All of the authors are given unlimited artistic freedom; they can write whatever sort of romance novel they want, and it will be published. Because the writers are reading each other's novels, they are able to incorporate the good aspects of other authors's work into their own, and, over time, the romance novels as a population will improve in quality. These romance novels would have an emergent quality, because no novel would be able to describe the system, and it would not be meaningful to discuss the entire system in terms of an individual novel. Rather, aspects of the novels, like length, plot, target audience, and relationship, would change as the system evolved.

One could easily investigate such a system using the framework of Bedau and Packard [3]. Aspects of the novels, as well as specific plot devices, could be tagged as traits and then traced using the Bedau and Packard method across the population over time. Because the authors are using these traits for some purpose and, quite probably, saw them first in another novel, these traits would probably form statistically significant activity waves, and, as such, the system would be supplely adaptive.

2 Issues Raised by the Romance Novel System

Intuitively, the romance novel system does not seem to be alive, even though it is supplely adaptive. Thus, we must ask why supple adaptation should be viewed as life when it characterizes intuitively non-living systems as living. Supple adaptation faces a critique similar to Gould and Lewontin's [7] "just so story" critique of the adaptationist program. Furthermore, supplely adaptive systems should be tested against commonly held hallmarks of life [6, 10]. Even if a list of hallmarks does not provide necessary and sufficient conditions, it does seem to be a good way to test a definition of life.

2.1 Identifying Traits and Developing a Null Model

Gould and Lewontin [7] charged that there is more than one way for something to develop traits, and that the adaptationist program is flawed because "...the rejection of one adaptive story usually leads to another...", and one can always "concoct a plausible story." Supple adaptation may be similarly flawed. Even though supple adaptation is statistically grounded, the information that these statistics are applied to is highly subjective. In all real-world cases, like the analysis of the fossil in Bedau et al. [4], the researcher must develop the null model. In the aforementioned case, for example, it is assumed that any activity that appears in the fossil record at the family level is adaptively significant. Like Gould and Lowontin's adaptationists, Bedau et al. are choosing what to regard as adaptively significant. They choose how to create the null model and what constitutes an attribute. Even in intuitively living organisms this is a serious issue. Given the amount of junk D.N.A. in biological creatures, a single gene is too small to be an attribute and a specific species may be too large. In intuitively non-living systems, defining attributes is particularly difficult. What kind of change in a romance novel would count as the creation of a new trait, and how should these traits be differentiated, given the interconnectedness of different aspects of the novel (e.g. plot and length seem inherently related). Such questions are also difficult for an economy, as differentiating economic units is, in and of itself, difficult.

2.2 Our Existing Notion of Life

How well do supplely adaptive systems fit with our preconceptions about life? Mayr [10] and Farmer and Belin [6] have both proposed lists of features that we intuitively believe to be present in most living things. Farmer and Belin's list is as follows:

> Life is a pattern in spacetime, [s]elf-reproduction....[,] storage of a self-representation....[, a] metabolism...[, f]unctional interaction with the environment....[, i]nterdependence of parts....[, s]tability under perturbations....[, and t]he ability to evolve. [6]

In a limited sense, intuitively non-living supplely adaptive systems, like my hypothetical romance novel system, meet these criteria. They appear as a pattern in spacetime and engage in self-reproduction. To whatever extent they are subject to some version of the second law of thermo-dynamics, they have a metabolism. They do not seem to store any sort of self-representation, but, because their phenotypic and genotypic levels are the same, as they are both the words of the novel, this appears unnecessary. They interact with their environment. Romance novel writers, for example, might see changes in society and begin producing new romance novels in anticipation of these changes. Lastly, they have interdependence of parts, stability under perturbation, and the ability to evolve.

Many of these properties, however, are the result of the fact that the system is produced by living organisms and not elements of the system itself. In the hypothetical romance novel system, for example, self-reproduction, metabolism, and information storage seem to be a function of the biological organisms that write the novels and not the system itself. Moreover, interaction with the environment is entirely a result of the predictions of the organisms which form the system, and not of the system itself. Thus, advocates of supple adaptation must explain why it is meaningful to discuss intuitively non-living systems as independently living, rather than the sum of their living parts.

3 The Concrete Room Objection

The importance of how the actors in a system are related to the theater in which they interact is illustrated by the following example. Consider taking a small, functioning ecosystem and placing it in a concrete room. It will continue to live, and it will probably evolve to the selective pressures of being in a concrete room, and, as such, become a long run supplely adaptive system in its own right. Nevertheless, the concrete room has not undergone any supple adaptation. The organisms might even produce their own concrete, thereby significantly changing the concrete room, still, the concrete room is not evolving and is not alive.

The romance novel system plays the same role as the concrete room. Even though the writers are producing a definite product, the features of which can be identified without reference to the authors, this product is merely a new way for the biological entities producing it to interact. The system as a whole is not clearly different from the entities that constitute it. The system undergoes evolution, but that evolution is merely an artifact of the interactions of the entities which constitute it. Until the proponents of supple adaptability can show that the romance novel system is uniquely alive, and, as such, gives a sort of life to the romance novels which constitute it, or move in the direction of this critique and explain why it is not alive, supple adaptability will not be able to gain widespread acceptance.

4 Conclusions

Though supple adaptation seems, in many ways, a useful way of explaining life, its usefulness will remain limited until researchers directly address the issue of intuitively non-living supplely adaptive systems. Though Bedau claims that: "It does not matter whether this theory supports our current preconceptions about life or fits the current meaning of our word 'life'" [1], he has yet to directly address our preconceptions about life. As long as this is not dealt with, supple adaptation will remain difficult to accept.

Nevertheless, supple adaptation, especially as modified by this critique, can help other views of life, as well as potentially become a viable view in its own right. Hopefully, this work will help supple adaptation overcome its shortcomings and show other researchers interested in defining life the value of addressing both the system and the individuals who compose it.

References

1. Bedau, M. A. (1998), 'Four Puzzles about Life', in Artificial Life (forthcoming).
2. ——— (1996), 'The Nature of Life', in Boden (1996), 332-57.
3. ——— and Packard, N. H. (1992), 'Measurement of Evolutionary Activity, Teleology, and Life', in Langton et al. (1992), 431-61.
4. ——— Synder, E., Brown, C. T., and Packard, N. H. (1997), 'A Comparison of Evolutionary Activity in Artificial Evolving Systems and in the Biosphere', in Husbands and Harvey (1997).
5. Boden, M. A. (1996) (ed.), The Philosophy of Artificial Life (New York: Oxford University Press).
6. Farmer, J. D., and Belin, A. (1992), 'Artificial Life: The Coming Evolution', in Langton et al. (1992), 815-40.
7. Gould, S. J. and Lewontin, R. C. (1979), 'The Spandrels of San Marco and the Panglossian Paradigm: A Critique of the Adaptationist Programme', Proceedings of the Royal Society B, 205, 581-598.
8. Husbands, P., and Harvey, I. (1997) (eds.), Proceedings of the Fourth European Conference on Artificial Life (Cambridge, Mass.: MIT Press), 293-301.
9. ——— Taylor, C., Farmer, J. D., and Rasmussen, S. (1992) (eds.), Artificial Life II (Santa Fe Institute Studies in the Science of Complexity, Proceedings, 10; Redwood City, Calif.: Addison-Wesley).
10. Mayr, E. (1982), The Growth of Biological Thought (Cambridge Mass.: Harvard University Press).
11. Schrödinger, E. (1969), *What is Life?* (Cambridge: Cambridge University Press).

Evolutionary Dynamics

How Symbiosis Can Guide Evolution

Richard A. Watson Jordan B. Pollack

Dynamical and Evolutionary Machine Organization
Volen Center for Complex Systems – Brandeis University – Waltham, MA – USA
richardw@cs.brandeis.edu

Abstract. Hinton and Nowlan have demonstrated a model of how lifetime plasticity can guide evolution. They show how acquired traits change the shape of the reward landscape in which subsequent genetic variation takes place, and in so doing encourage the discovery of equivalent heritable traits. This enables the seemingly Lamarkian inheritance of acquired characteristics without the direct transfer of information from the phenotype to the genotype. This paper draws direct inspiration from their work to illustrate a different phenomenon. We demonstrate how the formation of symbiotic relationships in an ecosystem can guide the course of subsequent genetic variation. This phenomenon can be described as two phases: First, symbiotic groups find solutions where individual organisms cannot, simply because lifetime interaction produces new combinations of abilities more rapidly than the relatively slow genetic variation of individuals. Second, these symbiotic groups subsequently change the shape of the reward landscape for evolution, providing a gradient that guides genetic variation to the same solution. Ultimately, an individual organism exhibits the capabilities formerly exhibited by the group. This process enables the combination of characteristics from organisms of distinct species without direct transfer of genetic information.

1 Introduction

Symbiosis, in its general definition, is simply the living together of different organisms. Often, in lay usage, the term is used to refer to the special case of *mutualism* where symbionts (organisms in symbiotic relationship) are mutually beneficial. Despite being undeniably common, the phenomenon of symbiosis, and especially mutualism, has for the most part been treated as a curio; a transient aberration on the otherwise relentless path of mutually-exclusive competition between species. In contrast, enlightened evolutionary theory recognises symbiosis as an integral process, and a fundamental source of innovation, in evolution. In its strongest form, symbiosis can lead to *symbiogenesis*: the genesis of new species via the genetic integration of symbionts [12], [9], [8], [10]. For example, eukaryote cells (from which all plants and animals are descended) have a symbiogenic origin [10].

The genetic integration of pre-adapted organisms is a fundamentally different source of innovation from the Darwinian accumulation of random variations. Computational abstractions of random variations under differential selection are well established in evolutionary algorithms research. However, functional models for the interaction between the formation of symbiotic groups and the accumulation of genetic variation are under-researched. In this paper we make a modest start by modeling one mechanism by which symbiosis and genetic variation can interact. We do not pretend to model the biological details involved in any way, rather we provide an abstract computational model examining the dynamics of the adaptive mechanisms. And in this paper we commence with a model of an *indirect* mechanism whereby symbiosis can enable innovation, rather than the more radical process of symbiogenesis where symbionts are genetically integrated directly.

The mechanism behind our model is directly inspired by the work of Hinton and Nowlan in their 1987 paper [6], "How Learning Can Guide Evolution". Their paper demonstrates the Baldwin effect [1]; a phenomenon whereby *acquired* characteristics can induce equivalent *heritable* characteristics. This seems like Lamarkian inheritance of acquired characteristics but it occurs without direct transfer of information from the phenotype to the genotype.

Hinton and Nowlan provide a simple and elegant abstract model that exemplifies how this process can occur and their model has been replicated and extended many times [2], [5], [11]. Here we have adapted their model by replacing learning with symbiosis; or more generally, replacing lifetime plasticity of an organism with lifetime interaction between organisms. Their experimental setup provides a convenient starting point and, moreover, an interesting comparison that assists us in understanding a more general concept that encompasses both phenomena.

Using this model we show that lifetime interaction can enable the evolution of organisms that would otherwise be unobtainable—or at least, would be very unlikely to occur. Our simulation of this phenomenon reveals two phases. First, symbiotic groups find the solution to a problem (a set of abilities that confers high reproductive fitness) more quickly than the solution can be found by a single organism. This occurs simply because the combination of abilities via lifetime interaction of organisms samples a much larger set of variations than the relatively slow genetic variation from mutation. The first stage alone does not demonstrate the evolution of an organism that would otherwise not occur—rather we have simply selected a mutually beneficial group of organisms out of those that do occur.

In the second phase, after a group has found the solution and an ecosystem of mutually beneficial organisms has become established, the evolution of the individual organisms therein operates in a different environment. Where previously an organism that exhibited some fraction of the necessary abilities, but not all the necessary abilities, would fail, now symbionts will occasionally *fill-in* for this organism's inadequacies. Moreover, the greater the fraction of necessary abilities it exhibits the less filling-in is required—i.e. the less it depends on symbionts and the more reliably successful it is. This provides a gradient to guide genetic search toward an organism that can ultimately perform independently. Without the support of symbionts this gradient does not arise.

Thus, the abilities discovered by the symbiotic group become encapsulated in the heritable traits of a single individual. Yet this effect occurs without the exchange of genes—the symbionts may be distinct species. We call this effect *symbiotic scaffolding*: the symbionts support each other as partially able organisms, and enable the gradual accumulation of abilities, until ultimately, when their abilities are complete, the scaffolding is not required. This mechanism is illustrated in the following simulations.

The remaining sections of this paper are organised as follows. In Section 2 Hinton and Nowlan's learning scenario is adapted to our symbiosis model. Section 3 details the experimental setup and gives results. In section 4 we indicate a general principle that encompasses both the Baldwin effect and symbiotic scaffolding, and suggest implications for evolutionary computation methods. Section 5 concludes.

2 A scenario for symbiosis

Following Hinton and Nowlan's lead we describe an extreme and simple scenario where the combinatorics of the phenomenon are clear. We consider a problem that consists of a large number of variables all of which must be correctly specified by an organism in order for that organism to receive any reproductive fitness. In such cases an organism that is partially correct, even one that specifies all but one of the variables correctly, is not rewarded at all. This worst-case scenario is the extreme case of irreducible complexity, in which solutions can only be found by trying possibilities at random. In Hinton and Nowlan's learning model the problem is to find the 20 correct connections for a neural network. For our symbiosis model we may imagine a chemical cycle with 20 steps. Each of the 20 steps must be catalysed by an organism correctly in order to get the chemical cycle going and to thereby confer reproductive fitness.

Where Hinton and Nowlan introduce lifetime plasticity to guide genetic search in this unforgiving landscape we shall use lifetime interaction with other organisms.

The interaction of organisms

Hinton and Nowlan suppose that some of the connections in the neural network will be left genetically unspecified and replaced with a switch that can make or break the connection during the lifetime of the organism. Here we suppose that an organism may have a neutral effect on a step in the chemical process, that is, the organism will neither prohibit nor catalyse the chemical step, and that this step may (or may not) be completed by some other organism in the environment. That is, an organism can gain the benefit (or penalty) of chemical byproducts created by the processes of other organisms in the ecosystem (for those steps where the organism itself is neutral).

We may crudely represent the traits of an organism in 20 genes where each gene has three alleles: *correct*, *incorrect* or *neutral* corresponding to catalytic, prohibitive (preventing completion of the cycle), or neutral interaction with a step in the chemical cycle. This unrealistic simplification enables us to see the mechanisms of interest more clearly but it is not integral to the results that follow.

Thus far we have followed Hinton and Nowlan's model exactly except with a different metaphor. Where they used correct connections, absent connections and plastic connections for a neural network we use catalytic, prohibitive and neutral influence on steps of a chemical cycle. Now, where Hinton and Nowlan use lifetime learning to specify the plastic connections of the network we will substitute lifetime interaction between organisms to fill-in for the missing abilities of the cycle.

In the same way that Hinton and Nowlan use random search as their learning model because it makes the least assumptions necessary, we use the happenstance co-location of organisms to determine their symbiotic interaction since it makes the minimal assumptions. The use of a more sophisticated model of symbiotic relationship-forming will illustrate the scaffolding effect more strongly—we stress that our model of organism interaction is deliberately trivial so as to prevent details from obscuring the essence of the effect. In our model we may imagine that organisms are randomly distributed in the environment and perpetually mixed. At any one instant there will be some number of other organisms in the immediate vicinity of the organism in question. Thus every organism is tested by combining its abilities with those of several other randomly selected organisms. Fig. 1 shows how the abilities of organisms are combined.

```
     fifth organism:   --111----0101--0-0--
    fourth organism:   -01---0-0----10-0-11
     third organism:   1--10-11-0----10---1
    second organism:   -1-0-1-001-01-1--0--
     first organism:   00-0-11---01-1--0--1
---------------------------------------------
combined abilities:    00100110010111100011
```

Fig. 1. Combining the abilities of organisms. The 20 genes of each organism may take one of three alleles: correct, incorrect or neutral shown as 1, 0 and "-", respectively. Notice that the traits of the first organism take priority over all others; for consistency, the traits of the second organism take priority over all but the first, and so on. Since every trait is specified by at least one of the first 4 organisms in this example, the fifth organism shown here is redundant.

Since the selection of, and ordering of, the organisms will be random, the details of this mechanism are largely inconsequential to the result that follows. One important feature, however, is that the fitness of the combined traits will be awarded to the first organism only, and that the traits of the first organism are not over-ruled by any other. However, since the first organism will likely fill-in for other organisms in their turn, this asymmetry is reciprocated. Alternate models of interaction and reward distribution may be equally valid—however, the current model is sufficient for our purposes.

Evaluation

A key feature of the Hinton and Nowlan model is the fact that lifetime plasticity can search combinations of traits far more rapidly than genetic variation. They allow random search to test 1000 combinations of the plastic connections during the lifetime of one individual. Similarly, we allow the combination of abilities via lifetime interaction of organisms to be much more rapid than genetic variation. Accordingly we test 1000 random groups during the lifetime of an organism.

So, to evaluate an individual we divide its lifetime into 1000 time-steps. At each time-step a number of other organisms are selected at random from the ecosystem. The number of individuals that are picked may be different for each time-step and the probable number is varied in the experiments that follow. The abilities of these organisms fill-in for the missing traits of the organism being evaluated, as described in Fig. 1. If the combined abilities of the group of organisms exhibits all required abilities correctly (and therefore enable the entire chemical cycle), then the organism receives a fitness increment, otherwise its fitness is unaffected. This is repeated for all 1000 time steps with a new randomly selected group each time. Overall, the fitness of an organism is given by f=1+n, where n is the number of time-steps when the organism in question forms a successful group with the organisms in its vicinity.[1]

3 Experiments

The genetic model, the method of interaction, and the evaluation described above are iterated in a genetic algorithm (GA) [7]. Hinton and Nowlan choose the population size, number of lifetime trials, and number of variables in the problem carefully so as to make it most unlikely that genetic variation alone will find the solution but very likely that lifetime variation will. We continue to follow the experimental parameters of Hinton and Nowlan where applicable for the same reasons. We use a population of 1000 individuals, or in our symbiosis terminology, an ecosystem of 1000 organisms. Fitness-proportionate reproduction is applied generationally [7].

Since we are interested in the ability of symbiotic scaffolding to encapsulate the abilities of symbionts of different species, we will not use genetic recombination (crossover) in our main experiments. We shall return to this point below. Instead we will use mutation as our only source of genetic variation. Mutation is applied with a bitwise probability of 0.05 of assigning a new random value. New values are randomly selected to be correct, incorrect or neutral genes with probability 0.25, 0.25, 0.5 respectively. These same proportions are used to construct the initial population. Hence in generation 0 each organism will have about 50% neutral genes (as shown in Fig. 1).

Parts of the results that follow are largely predictable given the experimental setup—as in Hinton and Nowlan's experiments the results are used to confirm our intuitions.

Experiment 1: A Crowded Ecosystem

In our first experiment we assume that the ecosystem is crowded. At any time-step there will be more than enough organisms in the immediate vicinity of an organism to fill-in all its neutral abilities. Naturally, when the organisms contain random genes they probably will not fill-in the steps in the cycle correctly. And, of course, if the organism

---

[1] In this detail we differ from the fitness function used by Hinton and Nowlan which is 1+19n/1000, where n is the number of time steps *remaining after the first successful lifetime trial.* Our method gives the expected fitness of an individual to be directly proportional to the probability of success in a single trial whereas the Hinton and Nowlan model gives an expected fitness which is highly non-linear with respect to this probability [Harvey 1993].

being evaluated has any incorrect genes then it cannot form any successful groups. We build groups by accumulating organisms at random until all 20 abilities are specified (one way or the other). Since each organism specifies approximately half the full set of genes, the average number of organisms required to specify all traits is generally small.

Fig. 2 shows the number of each allele per organism averaged over all organisms in the ecosystem at each generation. We see that the proportion of alleles at the start of the experiment is as per the mutation probabilities, i.e. approximately 0.25,0.25,0.5 for correct, incorrect and neutral respectively. Around the 20th generation a quite dramatic change takes place: the proportion of incorrect alleles falls close to zero whilst the number of correct alleles rises.[2] This is the point where symbiotic organisms become established and incorrect alleles are purged from the gene pool.

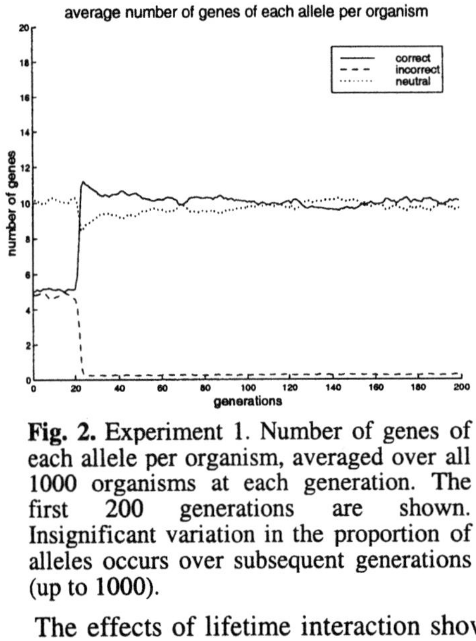

Fig. 2. Experiment 1. Number of genes of each allele per organism, averaged over all 1000 organisms at each generation. The first 200 generations are shown. Insignificant variation in the proportion of alleles occurs over subsequent generations (up to 1000).

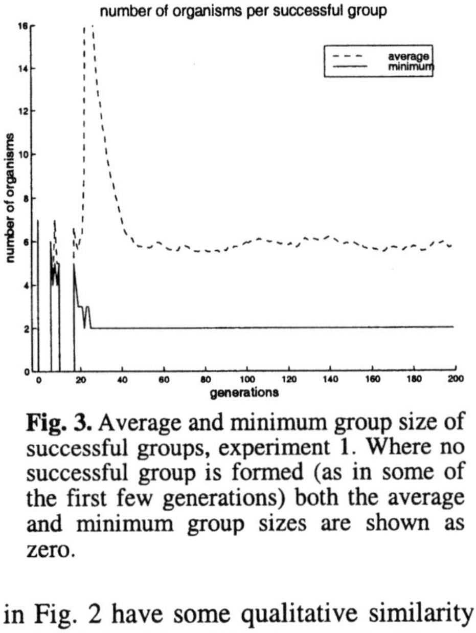

Fig. 3. Average and minimum group size of successful groups, experiment 1. Where no successful group is formed (as in some of the first few generations) both the average and minimum group sizes are shown as zero.

The effects of lifetime interaction shown in Fig. 2 have some qualitative similarity with the results of Hinton and Nowlan on the Baldwin effect. This is no coincidence; the substitution of lifetime interaction for lifetime plasticity is algorithmically insignificant in the initial stages where the abilities provided by other organisms are essentially random. However, once selection has taken hold and the population is predominantly devoid of incorrect alleles, the nature of the symbiotic variation is quite different from that of the random assignment used by Hinton and Nowlan. Specifically, the alleles supplied by symbionts are nearly all correct (whereas the traits supplied by random search remain in the original proportions). This means that the selection pressure for specifying increasing numbers of correct genes is very weak—in nearly all time-steps, other members of the group will correctly supply the required abilities. The

[2] Here, and more so in experiment two, the exact generation at which these sharp changes occur varies from run to run due to the stochastic nature of the experiment. However, the magnitude and general shape of the phenomena are reliable.

only significant pressure on an organism is that it should not specify any incorrect genes. A consequence of this complacency is that there is no pressure to get all correct abilities. Though Hinton and Nowlan's original work also shows no significant trend in decreasing neutral traits, replications of their work have shown subsequent increase in correct alleles over time-scales up to 500 generations [2], [5]. In this respect the influence of symbiotic groups is different from that of lifetime learning.

In Fig. 3 we see that the dramatic changes in Fig. 2 coincide with the establishment of groups that solve the complete cycle. We also see that the average size of successful groups increases dramatically. Given that organisms have approximately 50% of the correct alleles on average at this stage in the experiment, the fact that large groups are required to complete the set of abilities implies that many organisms have neutral alleles at the same loci. Informal observation also indicates that the population is somewhat converged at this time. By about the 50th generation the average group size falls to about six members which, given that the proportion of correct alleles has not changed significantly, shows that some complementary abilities have become established in the population. The smallest successful groups are of size two.

In summary, the crowded ecosystem in the first experiment shows the establishment of symbiotic organisms but does not demonstrate the entire scaffolding effect. We do not see the subsequent guidance of genetic variation to find an individual organism with the abilities formerly exhibited by the group.

Experiment 2: A Sparse Ecosystem

In our second experiment, we suppose the ecosystem is sparsely inhabited. Thus, the number of organisms in the immediate vicinity of the organism being tested is limited. Implementationally, we limit group sizes probabilistically where the limit is randomly selected from an exponential distribution. Specifically, the probability of there being exactly **k** members in a group is 2^{-k}, $\mathbf{k} \geq 1$. In this way it is most likely that an organism will be evaluated on its own; next most likely it will be evaluated with one other organism, and so on. In short, an organism cannot rely on the availability of symbionts, and an organism that is more self-sufficient will receive a higher fitness on average.

In Fig. 4 we see the same dramatic trends seen around generation 20 in the first experiment, though here they occur later. But, we also see some quite different phenomenon thereafter. Unlike the first experiment we see a clear upward trend in the number of correct alleles in subsequent generations, and a significant increase starting at around 160 generations. Whereas the first sharp increase corresponds to the purging of incorrect alleles, the second corresponds to the purging of neutral alleles.

Examining the plots of successful-group size averages in Fig. 5 we see that, although there are a few instances of successful groups in the first 70 generations, the occurrence of successful groups takes longer to become established than in experiment one. This is reasonable, since in experiment two an organism has less interaction with other organisms and therefore fortuitous co-location is more sporadic. We also see that the average number of individuals per group does not escalate as acutely as observed under the evaluation scheme of the first experiment. Finally, we see that the rise in proportion of correct alleles at 160 generations corresponds to the occurrence of individuals that are self-sufficient; i.e. a minimum group-size of 1 becomes established. In other independent runs there is considerable variation in the exact generation where symbiotic organisms become established and the generation which exhibits the first self-sufficient individual. Nevertheless, the overall effect is reliable.

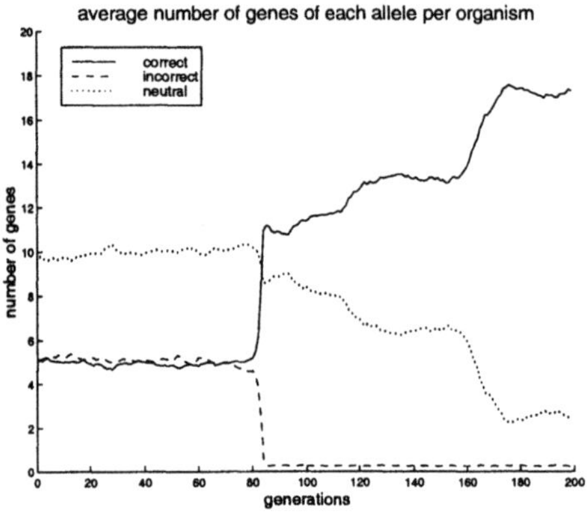

Fig. 4. Experiment 2. The average number of each allele per organism. All parameters as per the first experiment except that the size of groups is limited probabilistically.

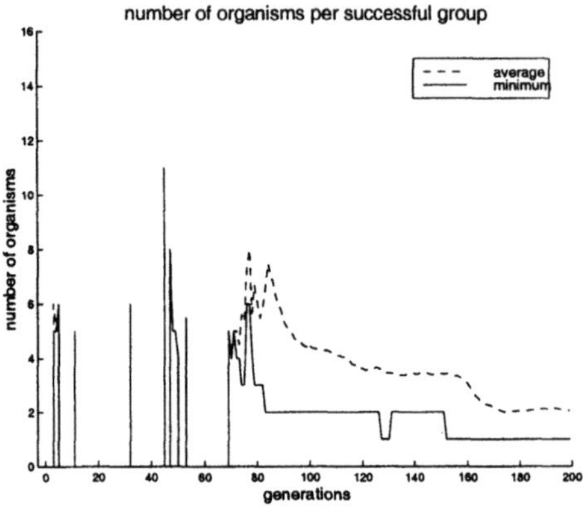

Fig. 5. Number of individuals per successful group in experiment 2.

The difference between experiments one and two is that the high availability of symbionts in the first experiment produces complacency whereas the unreliable availability of symbionts in the second experiment provides a selection pressure favouring independent organisms. Thus in experiment two we witness the complete scaffolding effect: symbionts first enable the adaptive characteristics, then become obsolete. They have shaped the evolutionary search space so as to enable the evolution of organisms that perform the function formerly performed only by groups. [3]

4 Discussion

We can see in these results the emergence of a general principle that encompasses both symbiotic scaffolding and the Baldwin effect—specifically, a process of rapid variation guiding a process of slow variation. Either symbiosis or learning may guide subsequent genetic mutation. Both mechanisms are effective because their fast variation discovers a

[3] Control experiments, not shown, included genetic recombination (one-point crossover) in addition to mutation. The results were essentially similar, though some effects were exaggerated probably due to stronger genetic drift [Harvey 1993]. It should also be noted that the recombination of genes via crossover is quite distinct from the filling-in mechanism of lifetime interaction shown in Fig. 1; this distinction is to be expected in biological organisms also. Additionally, as expected, experiments run without lifetime interaction (with or without crossover) do not succeed in 1000 generations.

solution and then provides a gradient that guides genetic variation to the same solution.

From the biological point of view, the influence of symbiosis on evolution, whether indirect as in these models, or direct as in symbiogenesis, is of interest because it informs us of the origin of organisms we see in nature. Abstract computational models, such as that presented here and by Hinton and Nowlan, inform biology only insomuch as they illustrate the possible space of dynamics and interactions. From the evolutionary computation point of view, the influence of symbiosis in guiding genetic variation is interesting only if it provides inspiration for more effective algorithms (here abstractions, rather than specific biological details, are more informative).

For example: Does the combination of a fast variation mechanism with a relatively slow variation mechanism provide a method that is more powerful than either mechanism alone? Why might the encapsulation of a group into a single individual be computationally important? What is the algorithmic difference between symbiotic combination and sexual recombination? We address these three questions in turn below: the suggestions here have yet to be investigated.

The coupling of fast and slow variation methods provides a balance between exploration and exploitation. A fast, non-permanent variation mechanism enables low-cost exploration (lookahead). Subsequent encapsulation via a method of slow variation, with high-commitment, enables stability from which further exploration may take place without disrupting solutions that have been proven.

The encapsulation of the group into a single organism enables the opportunity for the process to recurse with a larger unit of variation. This implies that the process may be applicable to hierarchical building-block problems [13] where search progresses from bit-combinations to schema-combinations as the building-block hypothesis suggests [7].

The difference between symbiotic combination and sexual recombination, or crossover, as used in existing GAs, is two-fold. First, symbiotic combination occurs between distinct organisms, whereas sexual recombination occurs between similar organisms, i.e. from the same gene-pool and necessarily highly-converged. This perhaps suggests the use of recombination operators that mate similar organisms frequently (as used in existing niching methods [3]) but also mate dissimilar organisms on rare occasions. Second, symbiotic recombination is additive whereas crossover is 'either/or'—that is, the results of symbiogenesis have the sum of the genes from the donors, but offspring from sexual recombination have approximately half the genes from each parent. This has important implications for respecting the integrity of the schemata represented by the parents. Such recombination is more akin to the Messy GA [4] or more generally the Incremental Commitment GA [14]. It can also be seen that the symbiotic filling-in of Fig. 1. corresponds to a 'competitive template' in the Messy GA.

5 Conclusions

We have seen two instances of an adaptive effect whereby non-genetic mechanisms guide the genetic make-up of organisms by shaping the evolutionary landscape. The Baldwin effect, modeled by Hinton and Nowlan, demonstrates learning as the first example of such a mechanism. In this paper we have adapted Hinton and Nowlan's model to illustrate a second example. Specifically, our experiments demonstrate how symbiotic scaffolding can guide the genetic make-up of organisms and lead to the evolution of organisms that would otherwise not occur.

Acknowledgments

Thanks to the members of DEMO at Brandeis for assisting us in the research process, and also to the reviewers for their constructive suggestions.

References

1. Baldwin JM, 1896, "A New Factor in Evolution," *American Naturalist*, 30, 441-451.
2. Belew RK, 1989, "When Both Individuals and Populations Search," in Schaffer JD, ed., *ICGA3*, San Mateo California: Morgan Kaufmann.
3. Deb K & Goldberg DE, 1989, "An Investigation of Niche and Species Formation in Genetic Function Optimization," *ICGA3*, San Mateo, CA: Morgan Kaufmann.
4. Goldberg, DE, Korb, B, & Deb K, 1989 "Messy Genetic Algorithms: Motivation, Analysis and First Results," *Complex Systems* 1989, 3, 493-530.
5. Harvey I, 1993, "The Puzzle of the Persistent Question Marks: a Case Study of Genetic Drift," in S. Forrest (ed.), *Genetic Algorithms: Proc. of the Fifth Intl. Conference*, pp. 15-22, San Mateo CA, Morgan Kaufmann.
6. Hinton GE, & Nowlan SJ, 1987, "How Learning Can Guide Evolution," *Complex Systems*, 1, 495-502.
7. Holland JH, 1975 *Adaptation in Natural and Artificial Systems*, Ann Arbor, MI: The University of Michigan Press.
8. Khakhina LN, 1992 *Concepts of Symbiogenesis: Historical and Critical Study of the Research of Russian botanists*, eds. Margulis L, McMenamin M, (translated by Merkel S, Coalston R), Yale University Press.
9. Kozo-Polyansky BM, 1921, "The Theory of Symbiogenesis and Pangenesis, Provisional Hypothesis" in Journal of the first All-Russian Congress of Russian Botanists in Petrograd in 1921. Peterograd (in Russian, see Khakhina 1992).
10. Margulis L, 1992 *Symbiosis in Cell Evolution: Microbial Communities in the Archean and Ptroterozoic Eons*. 2d ed., W.H. Freeman, New York.
11. Mayley G, 1996, "Landscapes, Learning Costs and Genetic Assimilation," in *Evolution, Learning, and Instinct: 100 Years of the Baldwin Effect, special issue of Evolutionary Computation*, Vol 4, No. 3; P.Turney, D. Whitley and R. Anderson (eds).
12. Merezhkovsky KS, 1909 "The Theory of Two Plasms as the Basis of Symbiogenesis, a New Study or the Origins of Organisms," *Proceedings of the Studies of the Imperial Kazan University*, Publishing Office of the Imperial University. (In Russian, see Khakhina 1992).
13. Watson RA, Hornby GS, & Pollack JB, 1998, "Modeling Building-Block Interdependency," *PPSN V*, Eds. Eiben, Back, Schoenauer, Schweffel: Springer.
14. Watson, RA, & Pollack, JB, 1999, "Incremental Commitment in Genetic Algorithms," In Banzhaf, W, Daida, J, Eiben, AE, Garzon, MH, Honavar, V, Jakiela, M, & Smith, RE eds. *GECCO-99*. San Francisco, CA: Morgan Kaufmann.

Adaptability of Darwinian and Lamarckian Populations toward an Unknown New World

Yuka Yamamoto[1], Takahiro Sasaki[2] and Mario Tokoro[2]

[1] Department of Computer Science, Faculty of Science and Technology, Keio University 3-14-1 Hiyoshi, Kohoku-ku, Yokohama 223-0061, Japan
[2] Sony Computer Science Laboratory Inc., 3-14-13 Higashigotanda, Shinagawa-ku, Tokyo 141-0022, Japan
E-mail: yuka@mt.cs.keio.ac.jp, {sasaki,mario}@csl.sony.co.jp

Abstract. In this paper, we describe adaptive processes of populations with two distinct mechanisms of evolution, *Darwinian* and *Lamarckian*. We use a simple abstract model where neural networks capable of learning are evolved through GAs. Each individual in the populations tries to maximize its life energy by learning certain rules that distinguish between two groups of materials: food and poison. The best-performing individuals are selected to reproduce offspring according to their mechanism of genetic inheritance, which is either Darwinian or Lamarckian, and the offspring conduct lifetime learning in the succeeding generation. In particular, we examine the adaptability of both populations toward a new unknown world, which is given after some evolutionary steps have taken place under the original world. As the main result, we show that only Darwinian populations can adapt to the new world.

1 Introduction

Through interactions with the environment, or learning, natural organisms may undergo some adaptive changes. However, these *acquired characters* will not be re-encoded in their genes and therefore will not be directly transmitted to their offspring. While *Lamarckism*, which has been one of the major ideas concerned with natural evolution, regards the effect of "inheritance of acquired characters" as the motive force of evolution, *Darwinism* claims that evolution is nothing but the cumulative processes of natural selection with random mutation, and denies the possibility of inheritance of acquired characters[1]. As we know, the mainstream of today's evolutionary theory follows Darwinism, and Lamarckism is regarded incorrect or as a heresy.

Due to this biological background, most of the discussions about learning and evolution from an artificial life research view are based on Darwinian models of evolution[6, 9, 1]. On the other hand, from an engineering view, it is not necessary

[1] Darwin himself accepted the possibility of inheritance of acquired characters, while Weismann, whose fundamental ideas have had a great influence on Neo-Darwinism, completely denied the possibility.

to consider only Darwinian models. Indeed, the possibility of using the heredity of acquired characters would be quite attractive and some studies have shown that significant improvements in the performance of problem-solving systems can be attained by introducing Lamarckian mechanisms[5, 2, 7]. In these works, however, the effectiveness and superiority of Lamarckian evolution have been demonstrated only for static environments (or more generally for fixed tasks).

In our previous papers[10, 11], we compared Darwinian and Lamarckian evolution under non-stationary environments, and showed that the Darwinian population not only showed more stable behaviour in the face of environmental changes but could also maintain greater adaptability regarding such dynamic environments than could the Lamarckian population. While Lamarckian population could adapt itself quickly to a certain single situation, it had difficulty in leaving the specific state of adaptation once it had taken place, owing to its extremely greedy strategy for genetic inheritance. In this paper, which offers an extended discussion using the same model as in [10], we observe the adaptability of both populations toward a new unknown world that is given after some evolutionary steps have taken place under the original world. The populations are first cultivated under a certain world (original world), either static or non-stationary, for a rather long period of generations. After that the populations are given a completely new world, and their adaptability toward this new world is evaluated in terms of learning ability. Note that in this paper, a non-stationary environment, where environmental conditions change repeatedly, is itself regarded as one type of world. We discuss the relation between the adaptability to a new world and the level of innate errors of individuals' neural networks.

2 Experimental Model

Here, we describe our experimental settings. A hundred individuals come into a virtual "world", with 500 units of initial "life energy" for each. Each individual has a feed-forward neural network that serves as its "brain", meaning that the individual takes action based on the network outputs (Figure 1). The neural network has three layers, each of which contains six input neurons, four hidden neurons, and eight output neurons. Each neuron is fully connected to all neurons of the next layer. We take an array of real numbers as a "chromosome" from which the neural network is developed. The chromosome directly encodes all the connective weights of the network[8]. Values of the chromosomes in the initial generation are set randomly, from the range $-0.30 \sim 0.30$.

The world contains two groups of materials, "food" and "poison", both of which have distinctive features as shown in Figure 3, for example. Each material is represented by an array of six bits as shown in Figure 3. Food and poison are distinguished by the upper three bits, and the lower three are set randomly for every new input as noise. Thus, each individual tries to maximize the chance of its own survival by learning the distinction rule, which in this case corresponds to a parity problem of three bits. On each occasion when given any material, an individual inputs the pattern of the material into its neural net-

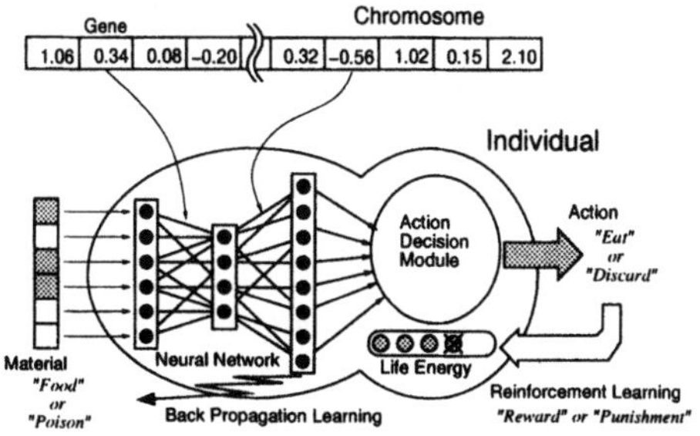

Fig. 1. The architecture of an individual.

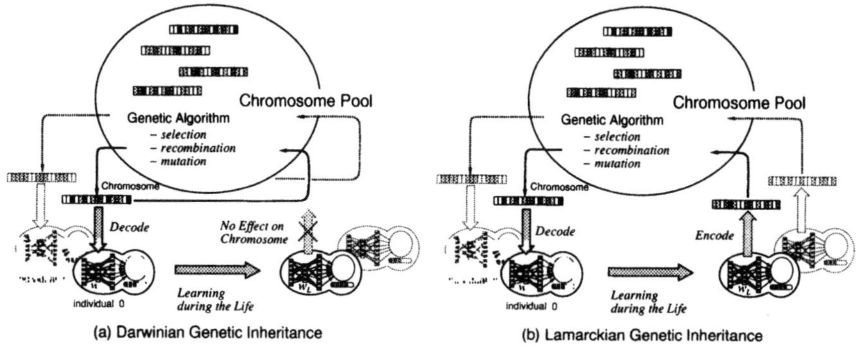

Fig. 2. The genetic inheritance for Darwinian and Lamarckian evolution.

work and stochastically determines whether to "eat" or "discard" it according to the network outputs. These actions, however, are not mapped directly from the outputs themselves. The network outputs are fed once as signals to an "Action Decision Module" (Figure 1), which then finally determines the action of the individual stochastically, according to a Boltzmann distribution. This type of stochastic mechanism is necessary to maintain the possibility of seeking more advantageous behaviours, even if an individual has already acquired a certain adequate behavioural pattern.

If what the individual ate was food, it receives 10 units of energy and tries to train itself to produce the "eat" action with a higher probability for that pattern. Conversely, if the individual ate poison, it loses a comparable amount of energy and tries to train itself to produce the "discard" action with a higher probability for that pattern. When the individual discards the material, no learning is conducted. The aim of each individual is to maximize its life energy by learning a *rule* that discriminates food and poison. We use the *Back Propagation Learning (BP Learning)* algorithm, in combination with a simple *Reinforcement Learning* framework, to train each individual. The coefficients of learning and inertia of BP Learning are set at $\eta = 0.75$ and $\alpha = 0.8$, respectively.

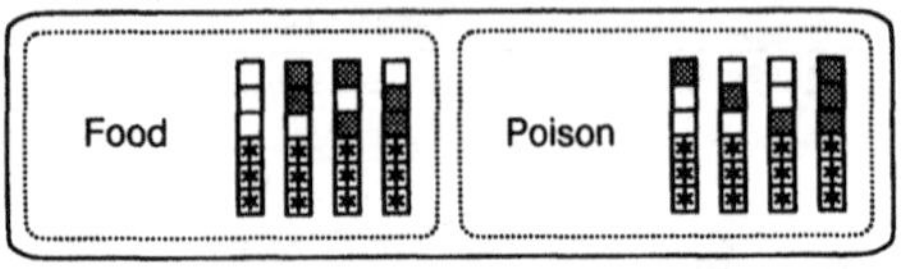

Fig. 3. A static environment. Whether the material is food or poison is determined by the parity property of upper three bits. The material is food if the number of black cells in the upper three bits is even, otherwise poison. The symbol "*" means *don't care* whether the cell is black or white; these cells have no effect on the distinction rule.

Each individual is offered a certain number of materials one at a time and learning occurs. We regard this number of repeated events as the length of an individual's "lifetime", which is set at 400 in this paper. At the end of each generation, some of the individuals are selected as parents by a stochastic criterion proportional to the level of their energy, which is thus regarded as their fitness. Selected parents then reproduce new offspring according to their genetic strategies, which is either Darwinian or Lamarckian. The connective weights of the individuals' neural networks must have been modified through their lifetime learning. The result of the modification through the learning is what we here call "acquired characters". Darwinian individuals do not transmit the acquired characters to the next generation, instead they simply pass their chromosomes, which they inherited from their parents, to the process of GAs (Figure 2(a)). On the other hand, Lamarckian individuals re-encode the acquired characters into their chromosomes, and pass them to the process of GAs (Figure 2(b)). Chromosomes of the selected individuals undergo the genetic processes of recombination and mutation. Here, the number of crossing-over points is set randomly from the range $0 \sim 4$. Each mutation at each locus of chromosome occurs at the rate of 5%, with a variation range between ± 0.5. Thus, the selected parents reproduce new offspring, which then undergo lifetime learning in the following generation.

Although the parameters in this paper are set heuristically according to some preliminary experiments, we have confirmed that changing these values within a moderate range results in qualitatively similar outcomes. All the results shown in the following are averages of ten trials.

3 Experimental Evaluations

3.1 Experiment 1: Static environment

Firstly, we consider a static world where the distinction rule between food and poison does not change, and observe the adaptive processes of the Darwinian and the Lamarckian populations.

Figure 4 shows the changes in the average fitness of the populations through generations. As shown by the figure, we see that the Lamarckian population can adapt themselves toward this static world more quickly and effectively than can the Darwinian population. The result meets our expectations quite well, since the Lamarckian population can continue the learning process that their parents suspended part-way in the previous generation, while the Darwinian population are forced, to some extent, to make a fresh start in each generation.

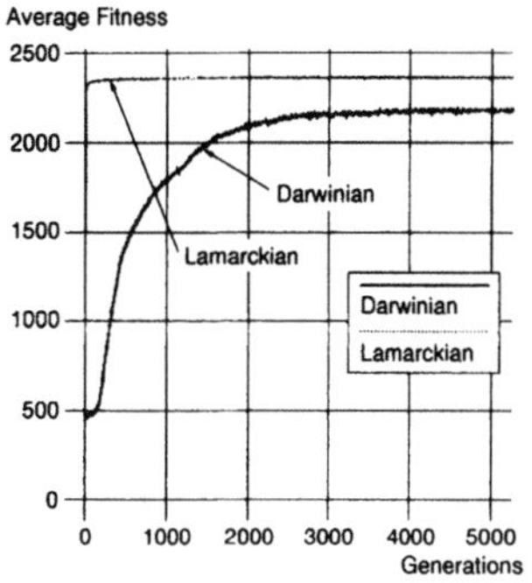

Fig. 4. Average fitness (Static environment)

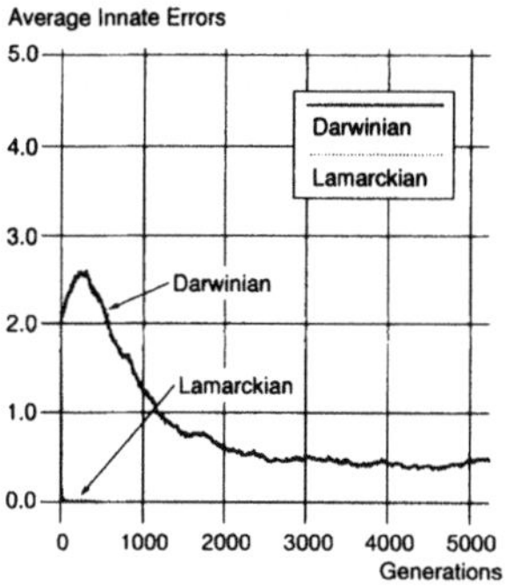

Fig. 5. Average innate errors (Static environment)

Figure 5 shows the changes in the output errors of innate neural network through generations. Network output error is measured by the mean squared error, over every pattern, between the actual outputs and the ideal outputs. Innate output errors are the errors that individuals show before any learning. Therefore, curves of innate errors declining through generations means that individuals came to behave more appropriately from their birth in the later generations. As shown in Figure 5, the Lamarckian individuals decrease their innate errors far more quickly than do the Darwinian individuals.

We must note here (in Figure 5) that the innate errors do decrease even for the Darwinian population, where acquired characters cannot be transmitted from parents to children through genetic processes. This case can be explained from the point of view of the "*Baldwin Effect*"[3]. As Hinton and Nowlan showed in their historical work[6], even if evolution is Darwinian, characters first acquired through learning can then be *assimilated* in the genotype as the generations proceed. One more thing we must note here is that the innate errors for the Darwinian population do not decrease monotonously from the initial generation, but there is a short period where their innate errors increase and thus the curve has a bump. We will discuss this point in Section 4.

3.2 Experiment 2: Non-stationary Environment

Next, we consider a non-stationary world where each material is the same as Experiment 1. Now, however, the distinction rules suffer irregular changes as shown in Figure 6. The following shows the results for a situation where the distinction rule is changed randomly at intervals of 20 generations.

Figures 7(a) and 7(b) show the changes in the average fitness of the Darwinian and the Lamarckian populations, respectively. As one can intuitively

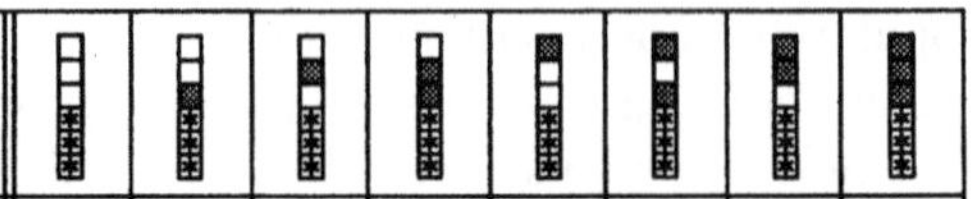

Env A	Food	Food	Poison	Poison	Food	Poison	Poison	Food
Env B	Poison	Food	Poison	Poison	Food	Food	Food	Poison
Env C	Poison	Poison	Food	Food	Poison	Poison	Food	Food
Env D	Food	Poison	Food	Poison	Food	Food	Poison	Poison
Env E	Poison	Food	Poison	Food	Poison	Poison	Food	Food
Env F	Food	Poison	Food	Food	Poison	Food	Poison	Poison

Dynamic Change of the Rules of Universe

Fig. 6. Non-stationary environment

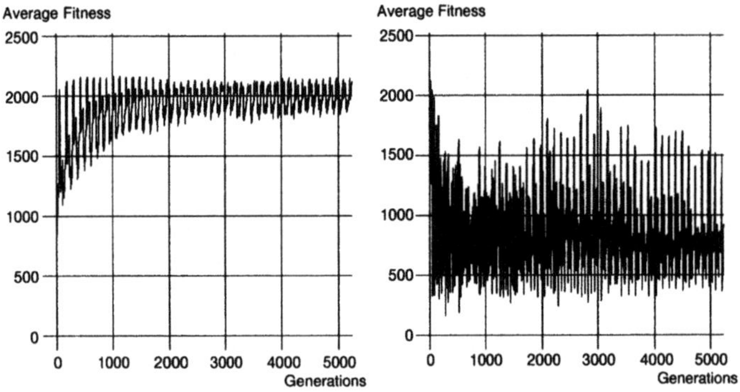

(a) Darwinian (b) Lamarckian

Fig. 7. Average fitness (Non-stationary environment)

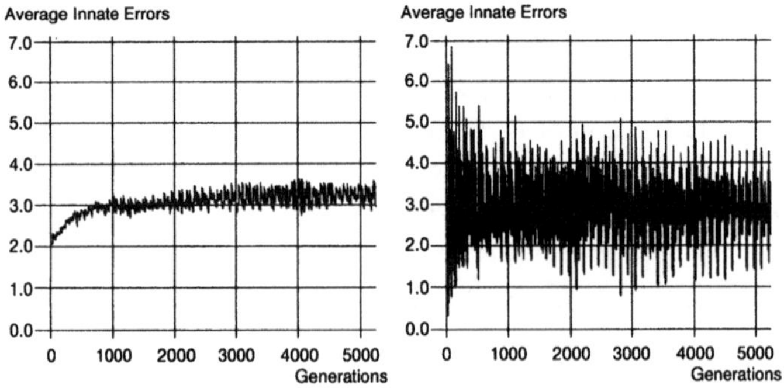

(a) Darwinian (b) Lamarckian

Fig. 8. Average innate errors (Non-stationary environment)

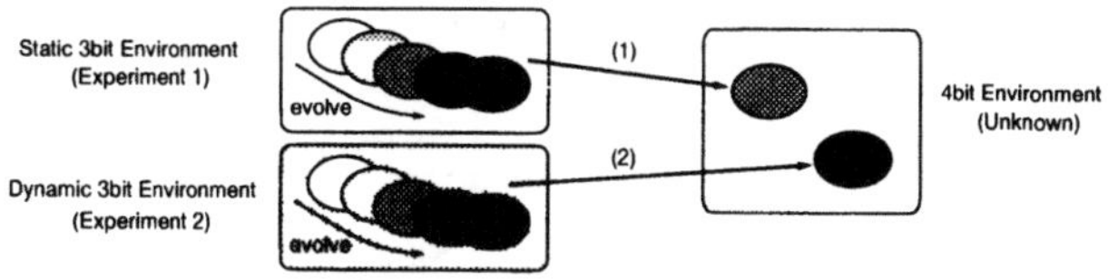

Fig. 9. Setup for Experiment 3. Picking up populations that have been cultivated for some generations under either (1)the static world or (2)the non-stationary world, then put them into a new world.

imagine, the fitness of the Lamarckian population oscillates violently whenever environmental conditions change and thus they can not adapt themselves to this non-stationary world. On the other hand, the point we should especially emphasize is that the fitness of the Darwinian population rises as the generations proceed, although oscillation is observed. We have confirmed in further experiments that the Darwinian genetic mechanism forms, through evolution, a population of individuals that can cope appropriately with any of the rules given in Figure 6(See [10] for more details).

Figures 8(a) and 8(b) show the changes in the average innate errors of the Darwinian and the Lamarckian populations, respectively. As we can see, innate errors of the Darwinian population show smaller oscillations than those of the Lamarckian population. It is interesting here to note that innate errors of the Darwinian population in the earlier generation firstly rise a little and are subsequently maintained at a rather higher level, unlike the results of the previous experiment where the errors decrease via *genetic assimilation* in the later generation(Figure 5). That is to say, individuals become worse from the point of view of their innate behaviour. However, their learning ability becomes better and thus their fitness through their whole life increases as the generations proceed.

3.3 Experiment 3: Adaptability toward New World

Now, we observe the adaptability of the Darwinian and the Lamarckian populations toward a new, unknown world, which is given after some evolutionary steps have taken place under the original world. Firstly, we pick up populations that have been cultivated for some generations under the original world; either the static world (Experiment 1) or the non-stationary world (Experiment 2). Then we put them into a completely new world, and evaluate their adaptability toward this new world in terms of their learning ability (Figure 9). In the new world, materials are distinguished by the upper four bits, and the lower two bits are noise. Individuals cannot solve the parity problem of four bits with the three-bits rule they have acquired under the previous world, so they have to learn the new rule for their survival.

Figures 10(a) and 10(b) show the learning curves under the new world for both populations, which were originally cultivated under the static world (Figure 9(1)). Figure 10 shows the changes in the network output errors through their

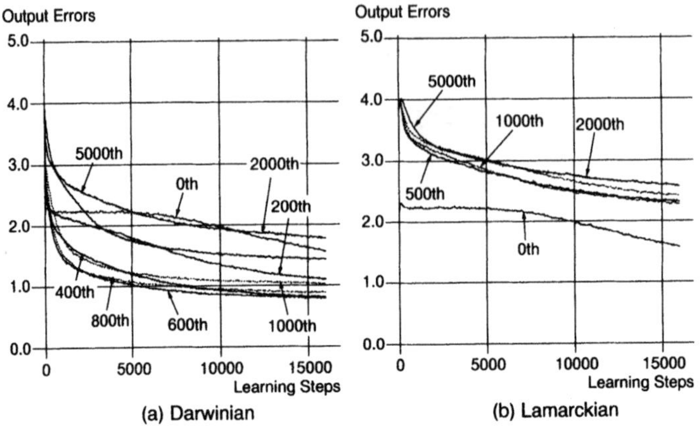

Fig. 10. Learning curves: Static 3 bits world → Unknown 4 bits world

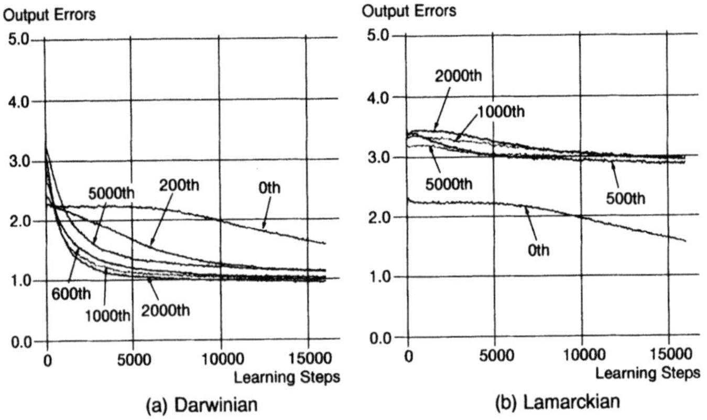

Fig. 11. Learning curves: Non-stationary 3 bits world → Unknown 4 bits world

learning steps. If individuals have adaptability toward the new world and can learn the new rule, their learning curve declines as the learning steps proceed. As expected, Lamarckian individuals cannot learn the new rule at all, since they have deeply adapted for the specific environment under which they were cultivated. Darwinian individuals that have been cultivated for long generations under the original world cannot learn the new rule either. However, those that have been cultivated for rather short periods, say up to 1000 generations, increase their adaptability even toward the new world (Figure 10(a)). The population that has been cultivated for about 600 generations shows the best adaptability.

Figures 11(a) and 11(b) show the learning curves under the new world for both populations, which were originally cultivated under the non-stationary world (Figure 9(2)). The Lamarckian individuals of any generations cannot learn the new rule at all. On the other hand, Darwinian populations that have been cultivated for some generations, say about 600 generations, can learn the new rule to some extent. Note here, that in this case, unlike the case of populations were originally cultivated under static world, the Darwinian individuals maintain the adaptability toward the new world even after they have been cultivated in the original world for a higher number of generations.

4 Discussion

From the results for the Darwinian population, it seems that evolution under the static world falls into two stages. Firstly, in the earlier stage of evolution, individuals that have better learning ability for the given rule are selected, and thus selection pressure develops the "ability to learn". The increase of the innate errors in the earlier stage of evolution, shown in Figure 5, means that the "ability to learn" is more important than the "ability to perform (innately)" at this moment, and hence selected for evolution. We state here that the "ability to learn" may not be specific for the given rule but rather a general one, since the population increases adaptability even toward the unknown new world (Figure 10(a)). Secondly, once the whole population is occupied by individuals that have high learning ability, selection pressure comes to prefer individuals that show innately better behaviours and thus develops the "ability to perform (innately)", also this is shown from decrease of the innate output errors(Figure 5). Therefore, the population comes to adapt itself deeply to the given world, and comes to lose adaptability toward the new world as the generations proceed (Figure 10(a)). Although the generation of the most adaptable population toward the new world shown in Figure 10(a) does not completely coincide with the generations around the bump of the innate output errors shown in Figure 5, there may be some relations between the non-monotonicity in both cases.

On the other hand, under the non-stationary world where environmental conditions suffer changes, individuals cannot devote themselves to a certain specific condition, but they should maintain the "ability to learn". As the "ability to learn" is a general ability, they also maintain adaptability toward the unknown new world, and the second phase, which occurs under the static world, i.e., selection for the "ability to perform (innately)" does not occur in this case (Figure 11(a)). The innate output errors are maintained at a certain higher level as shown in Figure 8. The above fact also suggests that there may be some relations between adaptability toward the new world and the level of innate output errors.

5 Conclusion

We evaluated how learning with inheritance of acquired characters affects the evolution of the population in a simple abstract model where neural networks capable of learning are evolved through GAs. The results obtained were as follows: Firstly, while Lamarckian individuals could adapt themselves toward a static world quite quickly and effectively, they had difficulty in leaving the specific state of adaptation once it had been reached and therefore could not cope with another new world at all. Secondly, Darwinian individuals that had been cultivated under a static world, surprisingly increased their adaptability even toward another new world if the new world was given in a relatively early generation, but their adaptability declined as the change occurred in later generations. Lastly, Darwinian individuals that had been cultivated under a non-stationary world also increased their adaptability toward another new world, and could maintain the adaptability even after many generations in the earlier world.

References

1. David H. Ackley and Michael L. Littman. Interactions between Learning and Evolution. In *Artificial Life II, SFI Studies in the Sciences of Complexity, vol.X*, pages 487–509. Addison-Wesley, 1992.
2. David H. Ackley and Michael L. Littman. A Case for Lamarckian Evolution. In *Artificial Life III, SFI Studies in the Sciences of Complexity, vol.XVII*, pages 3–10. Addison-Wesley, 1994.
3. J. Mark Baldwin. A new factor in evolution. *American Naturalist*, 30:441–451, 1896.
4. Richard K. Belew and Melanie Mitchell. *Adaptive Individuals in Evolving Populations: Models and Algorithms.* Addison-Wesley, 1996.
5. John J. Grefenstette. Lamarckian Learning in Multi-agent Environments. In *Proceedings of 4th International Conference on Genetic Algorithms and their applications (ICGA-91)*, pages 303–310, 1991.
6. G. E. Hinton and S. J. Nowlan. How Learning Can Guide Evolution. *Complex Systems*, 1:495–502, 1987.
7. Akira Imada and Keijiro Araki. Lamarckian evolution of associative memory. In *Proceedings of 1996 IEEE The Third International Conference on Evolutionary Computation (ICEC-96)*, pages 676–680, 1996.
8. David J. Montana and Lawrence Davis. Training Feedforward Neural Networks Using Genetic Algorithms. In *Proceedings of the 11th International Conference on Artificial Intelligence (IJCAI-89)*, pages 762–767, 1989.
9. Domenico Parisi, Stefano Nolfi, and Federico Cecconi. Learning, Behavior and Evolution. In *Toward a Practice of Autonomous Systems: Proceedings of the First European Conference on Artificial Life*, pages 207–216, 1991.
10. Takahiro Sasaki and Mario Tokoro. Adaptation toward changing environments: Why darwinian in nature? In *4th European Conference on Artificial Life (ECAL-97)*, pages 145–153, 1997.
11. Takahiro Sasaki and Mario Tokoro. Adaptation under changing environments with various rates of inheritance of acquired characters: Comparison between darwinianm and lamarckian evolution. In *2nd Asia-Pacific Conference on Simulated Evolution And Learning (SEAL-98)*, 1998.

Requirements for Immortal ALife to Exterminate Mortal ALife in One Finite, Heterogeneous Ecosystem

Tsutomu Oohashi[1],Tadao Maekawa[2],Osamu Ueno[3],Emi Nishina[1,4] and Norie Kawai[1,5]

[1] ATR Human Information Processing Research Laboratories
[2] ATR Media Integration & Communications Research Laboratories
[1,2] Seika-cho, Soraku-gun, Kyoto 619-0288 JAPAN / maekawa@mic.atr.co.jp
[3] Department of Biochemistry, Gifu University School of Medicine, JAPAN
[4] National Institute of Multimedia Education, JAPAN
[5] Foundation for Advancement of International Science, JAPAN

Abstract. We have proposed a theoretical hypothesis called the Programmed Self-Decomposition (PSD) model [1, 2]. The PSD model assumes that a self-decomposition mechanism is programmed in each cell of all living organisms on earth, and that this mechanism contributes to the substantial and spatial restoration of the ecosystem. In this article, we would like to introduce the overview of the hypothesis and report on computer simulations in which both a mortal virtual life form based on the PSD model and an immortal virtual life form start their lives in the same finite, heterogeneous ecosystem. Results of the simulations suggest that the mortal life forms would not be exterminated in almost all cases, contrary to ordinary expectations. It seems that the effectiveness of "death" cannot be denied.

1 "Programmed Self-Decomposition" Model

1.1 Restoration Mechanism of the Terrestrial Ecosystem

We would like to investigate the effectiveness of "death." We supposed that death contributes restoration of ecosystem, advancement of evolution and that death itself was obtained as a fruit of evolution.

The terrestrial ecosystem can still be characterized as being almost closed, in that space and substance are both limited. Accordingly, to maintain the stability of that terrestrial life form activity, the space and substance removed from the environment by the activity of the life form itself have to be returned to the environment.

The mechanism of restoring the terrestrial ecosystem to its original state has conventionally been explained by the principle of biological circulation called the food chain -- mutually giving oneself as food to other living organisms and depending upon metabolic activity to attain restoration of the environment as a whole.

We have set up a new hypothesis that is complementary to that of the food chain. In the present terrestrial ecosystem, while such a mechanism of restoring the environment to its original state is occurring, another hidden mechanism is fundamentally built into every living organism by which each cell of that organism positively decomposes itself so as to contribute to the restoration of the environment.

The phenomenon of decomposing oneself by one's own force is observed in life on earth. Called "autolysis," it has so far been recognized as a process of destruction, i.e., of living organisms randomly directed from a state of order to one of disorder.

Our hypothesis, on the other hand, looks at such a phenomenon as a controlled biochemical process for the given purpose of restoration of the environment to guarantee effective utilization of restored nutrients and space. To distinguish this process from autolysis, we call it "self-decomposition."

1.2 Theoretical Definition of Self-Decomposition

We have developed a new concept of a self-reproductive, self-decomposable automaton [1] through modification of von Neumann's self-reproductive automaton model [3]. It can be summarized as follows: (1) Automaton A constructs another automaton according to instruction I; (2) Automaton B makes a copy of instruction I; (3) Mechanism C combines A and B and functions as follows:(3-1) let A construct another automaton according to I / (3-2) let B make a copy of I and insert it into the automaton constructed above / (3-3) separate the new automaton from the system A + B + C; (4) Automaton D consists of A + B + C; (5) Instruction I_D describes automaton D; (6) Automaton E consists of D + I_D, which can reproduce itself; (7) Instruction I_{D+F} describes automaton D plus another given automaton F; (8) Automaton E_F consists of D + I_{D+F}, which can reproduce itself and construct another automaton F.

This model continues to self-reproduce as long as there is sufficient space, material and energy, and its structure remains intact unless it is attacked by external forces. It is immortal. On the contrary, living organisms on earth die without exception, and unless special measures are taken, they degrade into components after death. This is the essential difference between von Neumann's model and terrestrial life. So taking into account the ideas presented here, we have developed the following model, using von Neumann's model as a prototype, into which the process of "death and decomposition" is already programmed. This new model can be expressed as a variation of von Neumann's self-reproductive automaton E_F. To be more specific:

(a) Automaton FZ, which has the ability to disjoint the whole system into component elements, is a modular subsystem comparable to von Neumann's automaton F.

(b) Instruction I_{D+FZ} describes automaton D plus automaton FZ.

(c) Automaton E_{FZ} is a system comparable to von Neumann's automaton E_F whose instruction I_{D+F} is replaced by instruction I_{D+FZ}.

(d) Automaton G is a system composed of E_{FZ} and FZ, viz., D + FZ + I_{D+FZ} .

This system G can reproduce itself, and therefore makes FZ a subsystem within the system. FZ has the ability to disjoint G into finite elements. These elements are sized and structured in such a way that the entire ecosystem that G belongs to may take advantage of them collectively. FZ's mode of action can be one of the following three:

(1) Its production is normally restricted. With the input of a particular message, however, the restriction is lifted and the production activated.

(2) Its operation is normally restricted. With the input of a particular message, however, the restriction is lifted and the operation activated.

(3) (1) and (2) together.

Furthermore, if, after a certain amount of time has passed or a certain set of events has occurred, there is still no message input to trigger an action, (1), (2), or (3) will happen automatically. The "certain message" referred to above is provided, evidently, when it becomes impossible for G to further multiply itself or to maintain its structure, that is, when it is on the verge of extinction. If G is not given such a message for a certain period of time, G automatically puts FZ into operation. Therefore, we regard G as an automaton having a programme of spontaneous and inevitable "death and decomposition" installed *a priori.*

System G composed as described above not only reproduces itself, but also has the ability to put an end to its own life and to return to its origins, that is, to contribute to the restoration of the ecosystem to its original state. We tentatively call this system a "self-reproductio~self-decomposition (SRSD) system" and call the theoretical model under discussion a "programmed self-decomposition (PSD) model."

2 A Study on Virtual Life -- "SIVA-3"

2.1 Motivation

We adopted two approaches in order to test this model: One, biological experiments using actual life forms; the other, computer simulations using virtual life forms.

We developed the simulator series "SIVA (Simulator for Individuals of Virtual Automata)" in order to investigate effectiveness of "death" [2]. Recently, another investigation using CA is being conducted by Sayama [4] based on a similar concept. We simulated activities of artificial lives in a finite and heterogeneous ecosystem similar to the actual terrestrial ecosystem by using SIVA-3. We found that a mortal life based on the PSD model has a fundamental advantage over an immortal life when they live independently in two similarly conditioned environments. The mortal life can contribute to the restoration of the simulated ecosystem and continuously reuse limited space and substances. It also has greater opportunity for evolutionary adaptation [2]. However, since an immortal life occupies space and uses substances eternally, another life, including its own species, may have difficulty proliferation in the same ecosystem. A situation whereby immortal life forms exterminate mortal life forms can be anticipated if the two types of life forms live in the same ecosystem. To test this, simulations were prepared in which an immortal life form and a mortal life form start their lives in the same finite and heterogeneous ecosystem. SIVA-3 was arranged as described below.

2.2 Design of the Simulations

SIVA-3 was designed as previously described [2]. The virtual space in SIVA-3 is assumed to be a 2-dimensional lattice of 128x128 pixels. One pixel is defined as a spatial unit that one individual of virtual life must occupy to exist. The virtual substances, of which the ecosystem and the artificial lives in SIVA-3 are composed, are restricted to four types of elements. Environmental heterogeneity is introduced into 256 of spatial blocks (each consisting of 8x8 pixels), on the basis of the quantities of substances. These conditions range from those with the maximum conformity for primitive life, to those with the minimum conformity, that is, under a condition in which it is impossible for primitive life to survive or reproduce itself unless it adapts to the condition of taking many evolutionary steps. SIVA-3 can also make heterogeneity based on temperature, this was not adopted in the current simulations in order to make the simulative conditions simpler. Sufficient energy for life activities is assumed to be given [2].

Both an immortal non-PSD life form based on von Neumann's self-reproductive automaton model and a mortal PSD life form with an added modular subsystem, which has the ability to disjoint the whole system into component elements, on the immortal life form were designed. They consist of VC (Instruction I including I_D and I_{D+FZ}) and VP (Automata including A, B, FZ and Mechanism C) [2]. An internal status variable `UNCONF` (non-negative integer), was newly installed for the current simulations as a measure of conformity with the environment. The default value of `UNCONF` is 0 when there is complete conformity with the environment. When a command to reproduce a life form fails in execution, the individual perceives unconformity with the environment and increases the value of its internal status variable, `UNCONF`, by 1 per one failure.

Following two global variables were introduced in order to investigate the requirements for exterminating mortal life forms by immortal life forms.

Temporary adaptability. `TEMP-ADAPT` (between 0.0 - 1.0) was designed as the degree of temporary adaptability of rearranging current activities and getting through the situation without evolutionary adaptation. The probability that each of the commands for the self-reproductive activities will execute was calculated as `TEMP-ADAPT` raised to the power of `UNCONF` in this investigation.

Mutation Rate. At a probability of *misvcrate* (between 0.0 - 1.0) per character of VC, substitution will occur when the VC is copied. Masking against mutation was applied to the VCs of those individuals corresponding to automaton A, B, C, or FZ, so as to prevent the individuals from losing the function of that portion.

2.3 Simulations

Methods. Both an individual immortal life form and a mortal life form based on the PSD model were seeded in the center of the same ecosystem in SIVA-3. There were 36 simulations conducted in all, including 6 cases of temporary adaptability (`TEMP-ADAPT`: 0.0, 0.2, 0.4, 0.6, 0.8, and 1.0); and 6 cases of mutation rate (*misvcrate*: 1x10-6, 1x10-5, 1x10-4, 1x10-3, 1x10-2, and 1x10-1). Each simulation was carried out 10 times from TC (Time Count) = 0 to TC = 1000. The extermination ratio of mortal life forms was calculated for each of the 36 cases. Accumulated frequency of mutation was calculated for each of the 36 cases.

Results. Inevitable extermination of the mortal life forms occurred in only one case (`TEMP-ADAPT`: 1.0; *misvcrate*: 1×10^{-5}). There were the other 3 cases, in which extermination sometimes occurred. The mortal life forms never be exterminated in the other all 32 cases (Fig. 1(A)).

The immortal life forms succeeded in occupation of the relatively broad region only when `TEMP-ADAPT` had extremely high value (1.0). The mortal life forms, on the other hand, can increase on the large scale and extend to the broader regions when `TEMP-ADAPT` was not extremely high. Both of them show the tendency that the higher the mutation rate, the broader the habitat distribution (Fig. 1(B)).

Accumulated frequencies of mutation of the mortal life forms were higher than those of the immortal life forms. In the case of `TEMP-ADAPT`: 0.4 and *misvcrate*: 1×10^{-2} for example, the sum of the mutation frequency of the mortal life forms between TC=0 - 1000 exceeded 60000, however that of the immortal life forms was about 100.

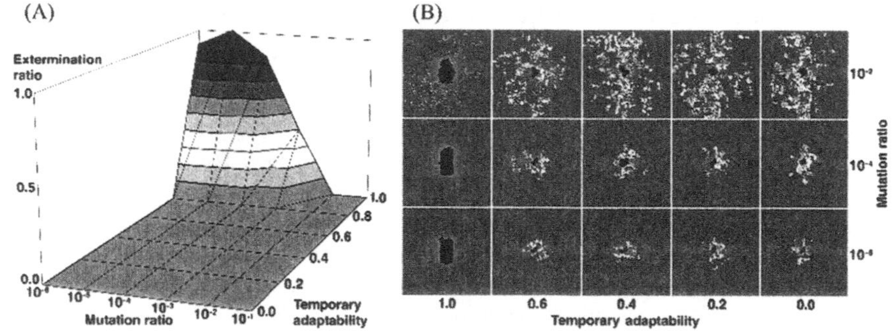

Fig. 1 (A) Extermination ratio of the mortal life forms by the immortal life forms. The number of simulations in which all individual mortal life forms had been exterminated was counted for each of the 36 cases. (B) Typical distributions of virtual life forms after 1000 TC simulations. This matrix shows the results of the typical 15 simulations for the above conditions. The black mass denotes the immortal life forms and the white dots denote the mortal life forms.

Discussion. These simulation results suggest that extremely specific conditions (i.e., the highest temporary adaptability and relatively low mutation rate) are required in order for immortal life forms to exterminate mortal life forms. This means that the probability of extermination of the mortal life forms by the immortal life forms was 0 in the .practical situations, except the situation in which temporary adaptation to any kind of environment can be perfectly achieved

The PSD mechanism contributed the restoration of the ecosystem and advanced evolution in the current simulations, as previously .discussed [2]. Consequently, the species of the mortal life forms became abundant and they occupied the broader area of the heterogeneous ecosystem than the immortal life forms did. The complexity of the ecosystem was also realized. It is noteworthy that these findings were obtained in the simulations in which the mortal life forms and the immortal life forms co-existed in the same finite, heterogeneous ecosystem. The results we found in this time suggest that it is difficult to deny the effectiveness of death in a finite, heterogeneous ecosystem.

Acknowledgment
The authors would like to gratefully thank Dr. Hiroki SAYAMA, New England Complex Systems Institute, for his indispensable contributions to the development of the simulator SIVA.

References

1. Oohashi T., Nakata D., Kikuta T. and Murakami K. (1987) Programmed self-decomposition model, Kagakukisoron kenkyuu, Vol. 18, No. 2, 79-87 (in Japanese)
2. Oohashi T., Sayama H., Ueno O. and Maekawa T. (1996) Artificial life based on programmed self-decomposition model, ATR Technical Report TR-H-198
3. Von Neumann J. (1948) The general and logical theory of automata. In cerebral mechanisms in behavior - The Hixon symposium (Jeffress L.A.(ed.), 1951) John Wiley & Sons, Inc., 1-41
4. Sayama H. (1998) Introduction of Structural Dissolution into Langton's Self-Reproducing Loop, Artificial Life VI, MIT Press, 114-122.

Error Thresholds and Their Relation to Optimal Mutation Rates

Gabriela Ochoa, Inman Harvey, and Hilary Buxton

Centre for the Study of Evolution
Centre for Computational Neuroscience and Robotics
COGS – The University of Sussex
Falmer, Brighton BN1 9QH, UK
{gabro,inmanh,hilaryb}@cogs.susx.ac.uk

Abstract. The *error threshold* — a notion from molecular evolution — is the critical mutation rate beyond which structures obtained by the evolutionary process are destroyed more frequently than selection can reproduce them. We argue that this notion is closely related to the more familiar notion of *optimal* mutation rates in Evolutionary Algorithms (EAs). This correspondence has been intuitively perceived before ([9], [11]). However, no previous study, to our knowledge, has been aimed at explicitly testing the hypothesis of such a relationship. Here we propose a methodology for doing so. Results on a restricted range of fitness landscapes suggest that these two notions are indeed correlated. There is not, however, a critically precise optimal mutation rate but rather a range of values producing similar near-optimal performance. When recombination is used, both error thresholds and optimal mutation ranges are lower than in the asexual case. This knowledge may have both theoretical relevance in understanding EA behavior, and practical implications for setting optimal values of evolutionary parameters.

1 Introduction

The *error threshold* — a notion from molecular evolution — is the critical mutation rate beyond which structures obtained by the evolutionary process are destroyed more frequently than selection can reproduce them. With mutation rates above this critical value, an optimal solution would not be stable in the population, i.e., the probability that the population loses these structures is no longer negligible. On the other hand, an optimum mutation rate — a more familiar notion within the EAs community — is the mutation value which solves a specified search or optimization problem with optimal efficiency, that is with the least number of generations or function evaluations.

The notion of error threshold seems to be intuitively related to the idea of an optimal balance between *exploitation* and *exploration* in genetic search. In this sense, we argue that optimal mutation rates are related to error thresholds. The aim of this paper is to test this hypothesis using an empirical approach together with knowledge from molecular evolution theory.

Optimal parameter settings have been the subject of numerous studies within the EA community [2], [6], [17], and particular emphasis has been placed on finding optimal mutation rates [9]. There is, however, no conclusive agreement on what is best, most people use what has worked well in previously reported cases. It is very difficult to formulate *a priori* general principles about parameter settings, in view of the variety of problem types, encodings, and performance criteria possible in different applications. Our hypothesis — that optimal mutation rates are correlated to the notion of error thresholds — promises practical relevance and useful guidelines in finding optimal parameter settings, thus enhancing evolutionary search.

In the remainder of the paper we summarize the knowledge from molecular evolution relevant to our argument: the notions of *quasispecies* and error thresholds; we discuss the relation between error thresholds and optimal mutation rates; and we describe the fitness landscape used for our experiments: the *Royal Staircase* functions. Thereafter, we describe the empirical methodology used to test the hypothesis under study, we present the experimental results obtained, and we discuss the insight gained.

2 Quasispecies and Error Thresholds

The concept of a 'quasi-species' was developed in the context of polynucleotide replication, and in particular studies of early RNA evolution [3], [4], [5]. A protein space, [12] or more generally a sequence space, can be modelled as the space of all possible sequences of length ν drawn from a finite alphabet of size A. Each sequence has a fitness value which specifies its replication rate, or expected number of offspring per unit time. The fitnesses of all A^ν possible sequences define a 'fitness landscape'. When $A = 2$, a binary alphabet, the fitness landscape is equivalent to specifying fitness values at each vertex of a ν-dimensional hypercube; with some mathematical imagination — and some caution — this can be pictured as spread out over a geographical landscape where fitness is analogous to height, and the dynamics of evolution of a population corresponds to movement of the population over such a landscape.

Given an infinite population, and a specified mutation rate governing errors in (asexual) replication, one can determine the stationary sequence distribution reached after any transients from some original distribution have died away [4]. Unless the mutation rate is too large or differences in fitnesses too small, the population will typically cluster around the fittest sequence(s), forming a concentrated cloud; the average Hamming distance between two members of such a distribution drawn at random will be relatively small. Such a clustered distribution is called a 'quasi-species'. As the mutation rate is increased, the local distribution widens and ultimately loses its hold on the local optimum.

This can be seen at its clearest in an extreme form of a fitness landscape which contains a single peak of fitness $\sigma > 1$, all other sequences having a fitness of 1. With an infinite population there is a phase transition at a particular error rate p, the mutation rate at each of the ν loci in a sequence. In [5], this critical error rate (the error threshold) is determined analytically (Equation 1), and it is defined as the rate above which the proportion of the infinite population on the peak drops to chance levels.

$$p = \frac{ln(\sigma)}{\nu} \tag{1}$$

In equation 1, σ represents the selective advantage of the master sequence over the rest of the population, and ν the chromosome length. In the simplest case σ is the ratio of the master sequence reproduction rate (fitness) to the average reproduction rate of the rest.

2.1 Error Thresholds In Finite Populations

In [14] the calculations of an error threshold for *infinite* asexually replicating populations are extended to *finite* populations (we shall call the critical rate p_M for a population of size M). Finite populations lose grip on the solitary spike of superior fitness easily, because of the added hazard of natural fluctuations in this case. In [15], we derived a reformulation of the Nowak and Schuster analytical expression. This new expression (equation 2) explicitly approximates the extent of the reduction in the error threshold as we move from infinite to finite populations. The expression strictly should be an infinite series in which successive terms get smaller; here, we are ignoring all after the first few:

$$p_M = \frac{ln(\sigma)}{\nu} - \frac{2\sqrt{\sigma - 1}}{\nu\sqrt{M}} + \frac{2ln(\sigma)\sqrt{\sigma - 1}}{\nu^2\sqrt{M}} \tag{2}$$

3 Error Thresholds and Optimal Mutation Rates

The notion of error threshold seems to be intuitively related to the idea of an optimal balance between *exploitation* and *exploration* in genetic search. Too low a mutation rate implies too little exploration; in the limit of zero mutation, successive generations of selection remove all variety from the population, and once the population has converged to a single point in genotype space all further exploration ceases. On the other hand, clearly, mutation rates can be too excessive; in the limit where mutation places a randomly chosen allele at every locus on an offspring genotype, then the evolutionary process has degenerated into random search with no exploitation of the information acquired in preceding generations.

Any optimal mutation rate must lie between these two extremes, but its precise position will depend on several factors including, in particular, structure of the fitness landscape. It can, however, be hypothesized that where evolution proceeds through a successive accumulation of information then a mutation rate close to the error threshold is an optimal mutation rate for the landscape under study; since this should maximise the search done through mutation subject to the constraint of not losing information already gained. The main purpose of our paper is to empirically test this hypothesis (section 5).

Some biological evidence supports the relationship between error thresholds and optimal mutation rates. Eigen and Schuster [5] have pointed out that viruses — which are very efficiently evolving entities — live within and close to the error thresholds given by the known rates of nucleotide mutations. This correspondence has also been noticed before in the GA community: Hesser and Manner [9], devised a heuristic formula for optimal setting of mutation rates inspired by previous work on error thresholds [14]; Kauffman [11] (p. 107) also suggest a relationship between these two notions.

4 Royal Staircase Fitness Functions

van Nimwegen and Crutchfield [19] proposed the *Royal Staircase* functions for analyzing epochal evolutionary search. This class of functions are related to the previous *Royal Road* functions [13]. In [19] the authors justify their particular choice of fitness function both in terms of biological motivations and in terms of artificial evolution issues. In short, many biological systems and artificial evolution problems have highly degenerate genotype-to-phenotype maps; that is, the mapping from genetic specification to fitness is a many-to-one function. Consequently, the number of different fitness values that genotypes can take is much smaller than the number of different genotypes. Moreover, due to its high dimensionality, it is possible for the genotype to break into networks of "connected" sets of equal-fitness genotype that can reach each other via elementary genetic variation steps such as point mutation. These connected subsets of iso-fitness genotypes are referred to as "neutral networks" [10].

Our paper is guided by the working hypothesis that many real search problems have genotype search spaces that decompose into a number of such neutral networks. Such neutrality has been observed in problem domains as diverse as molecular folding [18], evolvable hardware [8], and evolutionary robotics [7]. One symptom of evolutionary search where neutral networks are important is that of long periods of (sometimes noisy) fitness stasis (— search along a neutral network) punctuated by occasional fitness leaps (— transitions to a higher neutral network). The Royal Staircase class of fitness functions capture the essential elements discussed above, and are suitable for evaluating our hypothesis. They are defined as follows [19]:

1. Genotypes are specified by binary strings $s = s_1 s_2 \ldots s_L, s_i \in \{0, 1\}$, of length $L = NK$.
2. Starting from the first position, the number $I(s)$ of consecutive 1s in a string is counted.
3. The fitness $f(s)$ of string s with $I(s)$ consecutive ones, followed by a zero, is $f(s) = 1 + \lfloor I(s)/K \rfloor$. The fitness is thus an integer between 1 and $N + 1$, corresponding to 1 plus the number of consecutive fully-set blocks starting from the left.
4. The single global optimum is $s = 1^L$; namely, the string of all 1s.

Fixing N (number of blocks) and K (bits per block) determines a particular problem or fitness landscape.

5 Experimental Design

The approach taken here is to independently assess error thresholds and optimal mutation rates, comparing then these two measures.

For the experiments, we used Royal Staircase functions (section 4). The rationale for this choice is two-fold. First, because we agree with their proposers [19] that these functions, despite their simplicity, have some ingredients encountered in evolutionary search problems. Secondly, because Staircase functions have a step feature similar to that of single peak landscapes. Theoretical results on error thresholds are available for single peak landscapes. Error thresholds can be extended to other landscapes, however, a degree of ruggedness is needed.

We used a generational GA with fitness proportional selection and without elitism. Fitness functions were Staircase functions. Specifically, we tested 6 different functions (choices of N and K): N = 1-3, $K = 10$; and N = 4-6, $K = 5$. Population size was 100, genetic operators were standard bit mutation and two-point crossover with a rate of 0.6. Several mutation rates were tested, from 0.0 to 0.2 expected mutations per bit. The algorithm was run in two modes *Asexual*: using mutation only; and *Sexual*: using both mutation and recombination. Each run lasted a maximum of 5000 generations.

5.1 Empirically Determined Optimal Mutation Rates

For the purpose of this paper, we defined the optimal mutation rate as that which finds the peak (on average) with the least number of generations.

For determining optimal mutation rates as defined above, we ran the GA starting from a random population, and stored the generation number at which the peak was attained for the first time. This measure was averaged over 100 trials for each mutation rate tested.

5.2 Empirically Determined Error Thresholds

The error threshold is the critical mutation rate beyond which structures obtained by the evolutionary process are destroyed more frequently than selection can reproduce them. Aiming at capturing this definition in an algorithm, we designed the following method for empirically estimating error thresholds:

For the selected range of mutation rates:

- Start from a population of all 1s, that is, all members on the peak.
- Run the GA for a maximum of 5,000 generations or until the whole population has completely lost the peak.
- Count how many times, out of 100 trials, the population completely loses the peak.

For low mutation rates, at least one member of the population is on the peak during all the generations, for all the 100 trials. As the mutation rate is increased, there is a point where the population completely loses the peak for some or all the 100 trials. The error threshold is identified as the mutation rate where this transition occurs. The observations are approximate in that, firstly, precision is limited to the mutation step size used, and secondly, the limit of 5,000 generations was assumed to be sufficiently long for the purpose.

Validating the Empirical Method For validating the empirical method described above, we designed the following experiment: we considered a single peak landscape with $\sigma = 2$, and for distinct string lengths (10, 25, 50, 100, and 200), we calculated the error threshold analytically using equation 2 (valid for finite asexual populations). The population size, M, was 100. Results of these calculations are shown in the second column of table 1. Next, for the the same landscape and settings, we estimated the error thresholds empirically following our method. Results of these estimations are shown in the third column of the table. There is reasonable agreement between analytical and empirical figures, though worst at short string lengths. The empirical figures were always higher than the analytical ones, which is related to the limited number of generations used. The higher the number of generations the lower one should expect the empirical figure to be; the analytical figure assumes an infinite number of generations.

Table 1. Comparing Analytical and Empirical Error Thresholds on a Single Peak Landscape

String Length	Analytical	Empirical
10	0.05	0.11
25	0.02	0.03
50	0.01	0.015
100	0.005	0.006
200	0.003	0.004

6 Results

Figure 1 summarizes experimental results for the six landscapes studied. The curves show the number of generations to reach the global peak as a function of the mutation rate, for asexual and sexual populations. Each data point gives the number of generations for finding the peak averaged over 100 runs. Optimal mutation rates are those which find the peak with the least number of generations. Two general trends may be observed, if one excludes the first result shown. First, there is not a single critically precise optimal mutation rate, but instead a range of mutation values producing near-optimal results. The curves are U-shaped with a flat bottom. Secondly, the curves for sexual populations are shifted to the left, that is, to lower mutation values, when compared to the asexual population curves.

In the plots we indicate the empirically estimated error thresholds for asexual (solid arrows) and sexual (dotted arrows) populations. Error thresholds for the landscapes studied were found to be within the range of optimal mutation rates, for both asexual and sexual populations. The results support the hypothesized relationship between these two measures.

An exception to these general trends is the plot for $N = 1$, $K = 10$. This scenario is equivalent to a single peak landscape, where a single string has the highest fitness, and all the others strings have the same but lower fitness. The rationale for hypothesising that optimal mutation rates are correlated with optimal mutation rates (Section 3) was that of maximising search subject to the constraint of not losing information already gained. However, this landscape is an extreme case where there is no intermediate step, no accumulation of information which might be lost through excessive mutation. The peak is found randomly without any gradual approach. Here, high mutation rates (close to 1.0) were optimal for both asexual and sexual populations.

Notice that error thresholds for sexual populations were, in all cases, lower than for asexual populations. A similar trend is observed for optimal mutation rates; this is compatible with the hypothesized correlation between error thresholds and optimal mutation rates.

The experiments determining optimal mutation rates showed standard deviations of the same order as the average number of generations measured. Thus, there were large run-to-run variations in the time to reach the optimal string. This observation was also reported by van Nimwegen and Crutchfield who performed similar experiments using the Royal Staircase function [19]. Figure 2 shows standard deviations for the $N = 2$, $K = 10$ landscape.

7 Discussion

Our results suggest that error thresholds and optimal mutation rates are indeed correlated. This empirical evidence supports previous intuitions expecting this correlation [9] [11], [15], [16]. There is not, however, a single critically precise value for the optimal mutation rate, but instead a range of values producing near-optimal performance. The error threshold, on the other hand, is a more precise measure. Hence, mutation rates slightly lower or higher than estimated error thresholds are likely to produce near-optimal results.

The implication of this finding is two-fold. First, theoretically, in helping to understand EAs' behavior, as insights regarding error thresholds will be reflected in our understanding of optimal mutation rates. Second, practically, as heuristics for finding error thresholds will provide useful guidelines for setting optimal mutation rates, thus improving the performance of EAs.

In our experiments, both error thresholds and optimal mutation ranges were lower for sexual compared to asexual populations. This result has been observed before: a recent work from the evolutionary biology literature [1] studied the role of recombination on evolving population of viruses, particularly, the effect of recombination on the magnitude of the error threshold. They report that recombination shifts the error threshold to lower mutation rates. Moreover, in [16], now in the realm of genetic algorithms, we found that recombination shifts

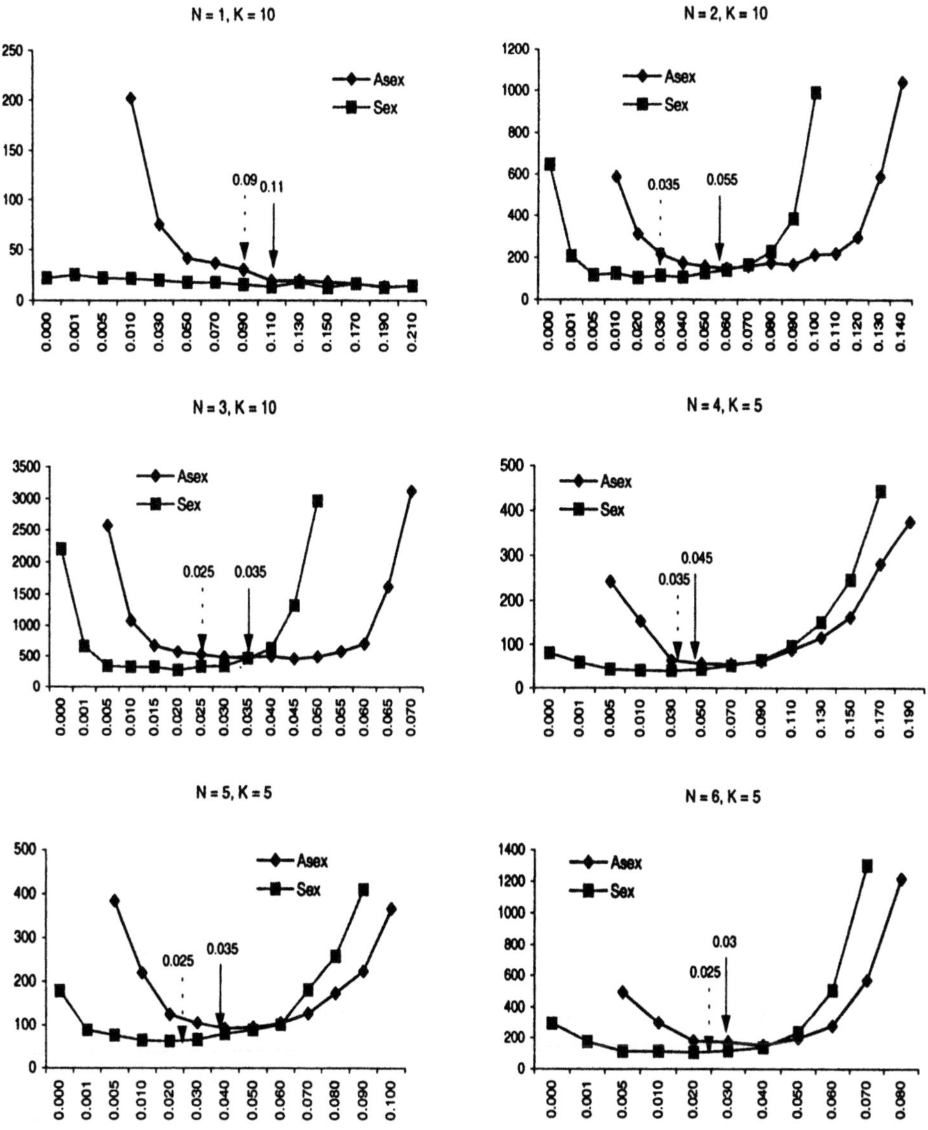

Fig. 1. For the six different landscapes explored, plots show the number of generations for finding the peak (Y axis) as a function of the mutation rate (X axis — note the scale is not linear at the lower end of the axis), for both asexual and sexual populations. Error thresholds are indicated by solid vertical arrows (asexual) and dotted vertical arrows (sexual).

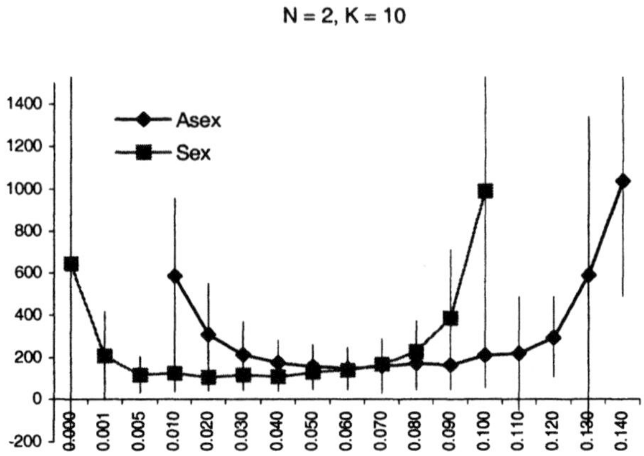

Fig. 2. To demonstrate the significant variance in the results, for one of the landscapes we show error bars for +/- one standard deviation in the number of generations for finding the peak.

optimal mutation rates to lower values. This evidence points indirectly towards the relationship between the notions of error thresholds and optimal mutation rates.

In this paper, we explored a single family of fitness functions, namely Royal Staircase functions. We have found, however, compatible results using the Kauffman NK family of landscapes. Moreover, the methods discussed here can be applied elsewhere. Error thresholds, and hence the implications of our results, hold for landscapes with certain degree of discontinuity. Thus, they are not applicable to smooth monotonic landscapes (or regions). Results will also be modified where there is elitism, either explicit or that implicit in some tournament selection algorithms for steady-state GAs.

Finally, there is a circularity implicit in using the method suggested here for assessing optimal mutation rates in real problems as opposed to toy landscapes — namely that one cannot estimate the error threshold until *after* one has found the global peak. However, knowledge of error thresholds in one region of the landscape, or in one member of a class of landscapes, may be of guidance in assessing error thresholds (and thus optimal mutation rates) elsewhere. For this to be the case one needs to make some assumptions of statistical regularity, but such assumptions are necessary anyway for EAs to be practical at all. So despite this circularity there is the prospect that in real EA applications where the fitness landscape is unknown except to the extent that it is sampled experimentally, the methodology given above allows for experimental assessment of error thresholds and thus guidance for setting optimal mutation rates.

Acknowledgements Thanks to A. Meier for support and critical reading.

References

1. M. C. Boerlijst, S. Bonhoeffer, and M. A. Nowak. Viral quasi-species and recombination. *Proc. R. Soc. London. B*, 263:1577–1584, 1996.
2. K. A. DeJong. *An Analysis of the Behavior of a Class of Genetic Adaptive Systems.* PhD thesis, University of Michigan, Ann Arbor, MI, 1975.
3. M. Eigen. Self-organization of matter and the evolution of biological macromolecules. *Naturwissenschaften*, 58:465–523, 1971.
4. M. Eigen, J. McCaskill, and P. Schuster. Molecular quasi-species. *J. Phys. Chem.*, 92:6881–6891, 1988.
5. M. Eigen and P. Schuster. *The Hypercycle: A Principle of Natural Self-Organization.* Springer-Verlag, 1979.
6. J. Grefenstette. Optimisation of control parameters for genetic algorithms. *IEE Trans SMC*, 16(1):122–128, 1986.
7. I. Harvey, P. Husbands, D. Cliff, A. Thompson, and N. Jakobi. Evolutionary robotics: the sussex approach. *Robotics and Autonomous Systems*, 20:205–224, 1997.
8. I. Harvey and A. Thompson. Through the labyrinth evolution finds a way: A silicon ridge. In *Proc. of The First International Conference on Evolvable Systems: From Biology to Hardware (ICES96).* Springer-Verlag, 1996.
9. J. Hesser and R. Männer. Towards an optimal mutation probability for genetic algorithms. In H.-P. Schwefel and R. Männer, editors, *Parallel Problem Solving from Nature.* Springer-Verlag, Lecture Notes in Computer Science Vol. 496, 1991.
10. M.A. Huynen. Exploring phenotype space through neutral evolution. Technical Report Preprint 95-10-100, Santa Fe Institute, 1995.
11. S. A. Kauffman. *The Origins of Order: Self-Organization and Selection in Evolution.* Oxford University Press, 1993.
12. J. Maynard Smith. Natural selection and the concept of a protein space. *Nature*, 225:563–564, 1970.
13. M. Mitchell, S. Forrest, and J. H. Holland. The Royal Road for genetic algorithms: fitness landscapes and GA performance. In F. J. Varela and P. Bourgine, editors, *Proceedings of the First European Conference on Artificial Life. Toward a Practice of Autonomous Systems*, pages 245–254. MIT Press, Cambridge, MA, 1992.
14. M. Nowak and P. Schuster. Error thresholds of replication in finite populations: Mutation frequencies and the onset of Muller's ratchet. *J. Theor. Biol.*, 137:375–395, 1989.
15. G. Ochoa and I. Harvey. Recombination and error thresholds in finite populations. In W. Banzhaf and C. Reeves, editors, *Foundations of Genetic Algorithms (FOGA-5).* Morgan Kauffman, 1998.
16. G. Ochoa, I. Harvey, and H. Buxton. On recombination and optimal mutation rates. In *Proceedings of Genetic and Evolutionary Computation Conference (GECCO-99)*, 1999. To appear.
17. J.D. Schaffer, R.A. Caruana, L.J. Eshelman, and R. Das. A study of control parameters affecting online performance of genetic algorithms for function optimization. In J. D. Schaffer, editor, *Proceedings of the 3rd ICGA.* Morgan Kaufmann, 1989.
18. P. Schuster. Extended molecular evolutionary biology: Artificial life bridging the gap between chemestry and biology. *Artificial Life*, 1:39–60, 1994.
19. E. van Nimwegen and J. P. Crutchfield. Optimizing epochal evolutionary search: Population-size dependent theory. Technical Report Preprint 98-10-090, Santa Fe Institute, 1998.

Are Artificial Mutation Biases Unnatural?

Seth Bullock

Center for Adaptive Behavior and Cognition
Max Planck Institute for Human Development
Lentzeallee 94, 14195 Berlin, Germany

Abstract. Whilst the rate at which mutations occur in artificial evolutionary systems has received considerable attention, there has been little analysis of the mutation operators themselves. Here attention is drawn to the possibility that inherent biases within such operators might artefactually affect the direction of evolutionary change. Biases associated with several mutation operators are detailed and attempts to alleviate them are discussed. Natural evolution is then shown to be subject to analogous mutation "biases". These tendencies are explicable in terms of (i) selection pressure for low mutation rates, and (ii) selection pressure to avoid parenting non-viable offspring. It is concluded that attempts to eradicate mutation biases from artificial evolutionary systems may lead to evolutionary dynamics that are more unnatural, rather than less. Only through increased awareness of the character of mutation biases, and analyses of our models' sensitivity to them, can we guard against artefactual results.

This paper explores the potential for artefactual evolutionary simulation results to derive from biases inherent within the artificial mutation operators they employ. As an example, consider a recent coevolutionary simulation model which has suggested that signals exhibiting complex symmetry could evolve merely as a side effect of selection for distinctiveness [1].

The model involved multicoloured, composite patterns coevolving with simple artificial neural networks under a mutualist selection regime. Networks which were able to discriminate signal patterns from distractor patterns were favoured, as were discriminable signal patterns. This discrimination had to be achieved despite patterns being presented in various orientations and positions on each network's artificial "retina". After a period of simulated coevolution, the authors report that networks were able to distinguish signals from distractors almost perfectly, and that the coevolved signal patterns displayed "marked symmetries" (p. 171). The authors note that, like many natural displays, the evolved signals consisted of "purer, brighter colours" (p. 171) than average signals. An evolutionary-functional account for the evolved symmetry was proposed – symmetrical signals persist because they are invariant under the various transformations involved in their presentation, and hence easier to discriminate from distractors. The evolved signals' bold coloration was explained as the result of selection pressure to diverge from random distractor patterns.

However, a replication of the study demonstrated that the complex symmetry of the evolved signals resulted from an unnaturally structured presentation regime [2]. Might the boldness of the evolved signals result from a similar simulation bias? One possible source of such a bias is the simulation's mutation operator.

During reproduction, each of the parent signal's colour components was subject to a small chance of mutation. A mutation event, when it occurred, perturbed the parental value by a small increment. Since each colour component was coded for by a real value lying in the interval $[0, 1]$, some of these perturbations resulted in mutant values which lay outside the legal range. The authors do not report the measures taken to deal with such illegal mutations. However, a mutation operator which replaced illegal mutant values with the nearest legal value might favour extreme colour component values merely through evolutionary drift. Perhaps this, rather than some pressure to deviate from the average, explains the simulation's results?

How might mutation biases occur in general? What are the possible effects of mutation bias? How can we test for the presence of mutation biases, and eliminate them from evolutionary simulations? These questions are of practical significance for any evolutionary simulation modeller.

1 Artificial Mutation

Within individual-based evolutionary computer simulations, artificial mutation operators are relied upon to provide the genetic diversity upon which selection may act. Simulations involving sexual populations augment mutation with crossover operators which allow offspring to inherit passages of genetic material from each parent. However, mutation continues to be the major source of genetic *novelty* in such simulations, since crossover operators reshuffle existing genes, rather than introduce new ones.

Many studies have explored the effect on artificial evolution of employing various genetic operators (e.g., one-point or two-point crossover, the transposition, repetition and excision of genetic material), varying rates at which these operators are employed, and even allowing the types of operators and the rates at which they are employed to themselves be subject to adaptation. For example, the proceedings of one early conference on genetic algorithms [3] contains five papers which between them address all of these concerns. More recent studies have developed these ideas [4–6].

However, since much of this work takes place within engineering contexts where there need be little concern for evolutionary plausability, little attention has been paid to the biases inherent within even the simplest mutation operators. Although the construction of these operators is not typically regarded as a complex matter, as will be argued below, they differ significantly from natural mutational processes, and their design therefore involves issues which are unique to the production of evolutionary simulations.

2 Legal Bounds in Natural and Artificial Genetic Systems

Consider a phenotypic trait which may vary over some range. We may distinguish between two classes of such a trait. Unbounded traits may vary limitlessly. Such traits might include the significance afforded to a male display by a female onlooker, which might vary from zero (non-significant), through positive infinity (infinitely attractive), or negative infinity (infinitely repellent). Bounded traits are those for which the range of legal values is in some way limited. Many traits suffer either a lower or an upper limit, and are thus partially bounded. For example, one cannot have fewer than zero legs. Some traits are bounded at both extremes, e.g., a trait governing the time of day at which one begins to forage might vary between dawn and dusk. Such a notion of boundedness raises the associated notion of legal and illegal genotypes, the latter being those that code for traits which transgress their legal limits.

The distinction being made here is an unnatural one which cannot easily be applied to natural genetic encodings. For example, although natural genes may code for what appears to be a bounded phenotypic trait (e.g., the redness of a signal), the manner in which they do so may logically preclude the occurrence of illegal values for this trait. Additive polygenic traits, for example, might code for the varying degree of a phenotypic trait with a varying number of genes. Since such a genotype cannot contain a negative number of these genes, it cannot code for an illegal phenotypic trait (e.g., a signal with negative redness).

In one sense however, an "illegal genotype" is one which does not result in a viable organism. Clearly, such mutants will not leave offspring. Their genotypes will be selected out of the population. This selective process is the same one that excludes viable mutants which are less well adapted to their niche than their competitors. From the perspective of natural selection, there is no difference (ignoring indirect fitness effects) between failing to be born living, failing to survive to reproductive age, failing to mate, or failing to reproduce before dying of old age – all such failures are awarded zero fitness.

Genotype legality within artificial evolutionary algorithms falls somewhere between the accounts given in the previous two paragraphs. It is true that, as in the first account, genotypes with illegal trait values are never realised as organisms, and are thus never subjected to the same selective pressures as their valid conspecifics. However, illegal genotypes *are* generated by many evolutionary algorithms and are thus not excluded out of logical necessity as in the first account, but are selected as invalid, as in the second account. But the grounds upon which illegal genotypes are selected are not the same as those which govern phenotypic selection. The legality of genotypes is assessed, prior to any morphogenetic, developmental or ontogenetic performance, on the basis of the genotype itself, rather than the performance of the associated (unrealisable) phenotype. If found to be legal, the genotype is translated into a phenotype and assessed as normal. If found to be illegal, some alternative course of action must be taken. It is this lack of correspondence between artificial evolutionary algorithms and natural evolution which raises the possibility that the character of artificial mutation operators may be unnatural due to the biases which they introduce.

3 Mutation Operators

In this section, the biases of various mutation operators will be characterised. These operators differ in the manner in which they treat illegal mutant values. In order to describe the repercussions of these differences, I will use the terms *departure rate* to denote the relative rate at which a particular parental value is altered by a mutation operator, and *arrival rate* to denote the relative rate at which a particular mutant value is generated by a mutation operator.

First, a distinction must be made between *context-free* and *context-sensitive* operators. Whilst context-free operators generate mutant values which are independent of parental values, context-sensitive operators are predicated on such variables. This latter class of operator (which will be the focus of this paper) typically generate mutant values which are close to the relevant parental values in an effort to mimic natural evolution. Many of the issues discussed here will be most pertinent to either continuous-valued genes or discrete (typically binary) encodings. However, most are to some degree applicable to both.

Genes coding for bounded phenotypic traits may be subjected to context-free mutations through drawing a random value from the range of legal values available for a trait. This operator will be referred to as the "Flat" mutation operator throughout this paper. For example, when mutating a trait which governs the hirsuteness of an organism, a Flat mutation operator might ignore the degree of body hair possessed by the parent and simply assign the mutant offspring a random number between 0% and 100%. Such an operator is flat in that both the departure rate and arrival rate are uniform across the valid range of the trait.

Similarly, genes coding for unbounded, or partially bounded, phenotypic traits may be mutated independently of their parental value through imposing some arbitrary *mutant range* upon the phenotypic trait and picking a value from this mutant range. Such a solution is unsatisfactory, however, as this method introduces a mutation *bias*. Despite there being an equal probability of any parental value being mutated (a flat departure rate), there is not an equal probability of any legal mutant value being generated by the mutation operator, the arrival rate outside the mutant range being zero, whilst that inside this range is positive and flat. Such a bias may lead to artefactual results. For example, if the optimal trait lies outside the mutant range, a population of organisms may evolve to lie clustered at the mutant value nearest to this optimum. Although the population may appear to have converged on some stable phenotype, this appearance is an illusion created by the mutation bias.

In contrast, context-sensitive mutation operators which *perturb* the parental genotype by some small amount have the potential to generate illegal values for bounded traits. For such mutation operators, decisions concerning the treatment of these illegal mutations must be made. There are a number of straightforward options that an operator designer might take:

Absorb Illegal mutant values are truncated to the nearest boundary.
Repeat Mutant values are repeatedly generated, until a legal value is obtained.

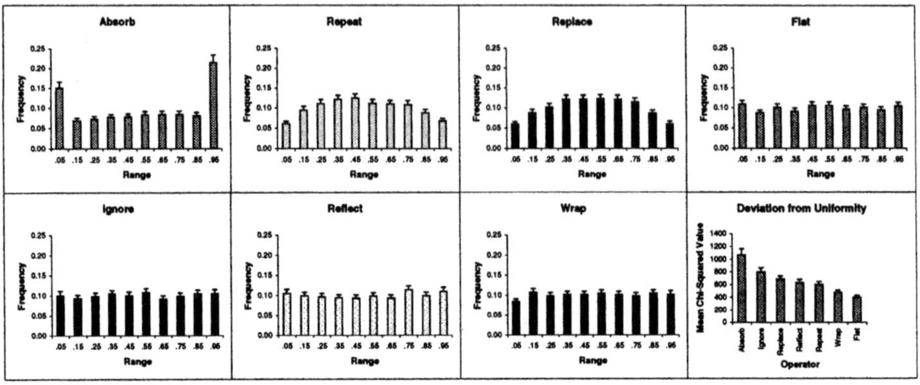

Fig. 1. Average distribution of trait values across 20 asexually reproducing populations of 1000 single-trait organisms after 5000 generations of evolution on a flat fitness surface, for each of seven mutation operators (see text). Traits were real values in the range [0, 1]. Mutation events occurred with probability 0.01, and consisted of real-valued perturbations drawn from a normal distribution with zero mean and standard deviation 0.2. Absorb, Repeat and Replace show clear deviation from uniformity in their aggregate performance. A Flat, context-free operator is shown for comparison. Three of the lower panels suggest that over many runs the behaviour of Ignore, Reflect and Wrap is roughly equivalent to Flat. However, the fourth lower panel shows that their mean deviations from uniformity (see text) differ significantly from each other, and from that of traits evolved under a Flat mutation operator. All graphs show mean frequencies with attendant standard errors.

Replace Any offspring for which illegal trait values are generated is replaced by a new offspring, re-choosing parents.

Each of these three operators result in mutation biases which either favour or resist extreme-valued traits (Fig 1). The "Absorb" operator will effect a constant departure rate across the range of trait values, save that it falls to half the nominal mutation rate at the extremes of the trait's legal range. In addition, the arrival rates at these extremes are increased by their absorbent nature. This ensures that trait values near the legal boundaries of the trait will tend to reach them and be kept there. Conversely, the "Repeat" operator will maintain a constant departure rate across the legal range of trait values, but will tend to mutate extreme parental values away from the boundaries of the trait. This ensures that arrival rates decrease as trait values approach their legal limits. The "Replace" operator, although seemingly the most accurate reflection of natural mutation processes, in that illegal or non-viable offspring are simply rejected, results in a selection pressure that resists the evolution of extreme trait values due to the increased likelihood that extreme genotypes will generate illegal offspring. Several mutation operators might be constructed specifically to alleviate these edge-effect biases.

Ignore Mutation events which transgress legal bounds are ignored. Rather than inherit an illegal mutant value, offspring inherit the parental value.
Reflect Mutant values lying a distance of x above (or below) the legal range are replaced by vaues a distance of x below (or above) the nearest boundary.
Wrap The trait is treated as if it were periodic. The edges of its legal range "wrap" around. Mutant values are calculated modulo the trait's range.

The aggregate behaviour of these operators is uniform (Fig 1). However, they differ in how they achieve this aggregate uniformity. For example, under the "Ignore" regime, more illegal values will be generated (and ignored) for parental values near the extremes of the legal range. This ensures that the departure rate will decrease as parental values approach legal extremes. However, this reduction in departure rate is prevented from systematically biasing the population by a correlated reduction in arrival rate. To the extent that a trait value x is extreme and hence suffers a low departure rate, it will be in an extreme neighbourhood also suffering a low departure rate. Since it is through mutations affecting parents within this neighbourhood that x is likely to arise, this low departure rate will lower x's arrival rate. As a result, extreme values are less likely to be generated by the operator, but when generated, are less likely to be mutated. Whilst this balance ensures that populations are not systematically biased toward or away from some areas of the trait space, it also results in individual distributions with a certain character. In order to determine whether, and to what extent, this character differs from that of populations under different mutation regimes, a measure of *deviation from uniformity* was calculated.

A χ^2 value estimating the degree of deviation from a uniform expected distribution was calculated for each individual evolved population contributing to the aggregate distributions graphed in Fig 1. The mean χ^2 value for each operator is shown in the lower far-right panel. This measure of deviation from uniformity differentiates between the three pseudo-flat operators, demonstrating that despite their similar aggregate performance, they each exert a unique influence on the character of individual populations. In addition, this metric demonstrates that aggregate flat performance can disguise individual distributions which tend to be very far from flat. The Ignore operator, for example, generates distributions which are on average less flat than either Repeat or Replace, despite there being no systematic bias to this non-uniformity.

4 Discrete Encodings

Two kinds of discrete encoding must be distinguished. Both allow a trait to take one of a (possibly infinite) number of discrete values. The first treats these values as unitary wholes for the purposes of mutation (and crossover), whereas the second represents these values with a number of discrete digits, each independently exposed to the possibility of mutation. Whilst the issues discussed above apply fairly straightforwardly to the first kind of discrete encoding, the second kind raises new issues.

a)

	0	1	2	3	4	5	6	7
x	0	1	1	0	0	1	1	0
y	0	0	1	1	1	1	0	0
z	0	0	0	0	1	1	1	1

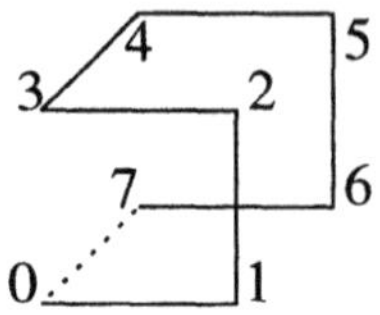

b)

	0	1	2	3	4	5	6	7
x	0	1	1	0	0	0	1	1
y	0	0	1	1	1	0	0	1
z	0	0	0	0	1	1	1	1

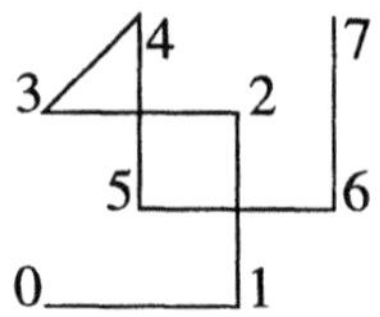

Fig. 2. Two three-bit Gray codes expressed over *(a)* the canonical Hamiltonian circuit, and *(b)* a mere Hamiltonian path. An n-bit canonical Gray code may be constructed by reversing the (n-1)-bit canonical Gray code which it contains (dashed box), and appending a 0 to the first 2^{n-1} bit strings and a 1 to the remaining 2^{n-1} bit strings.

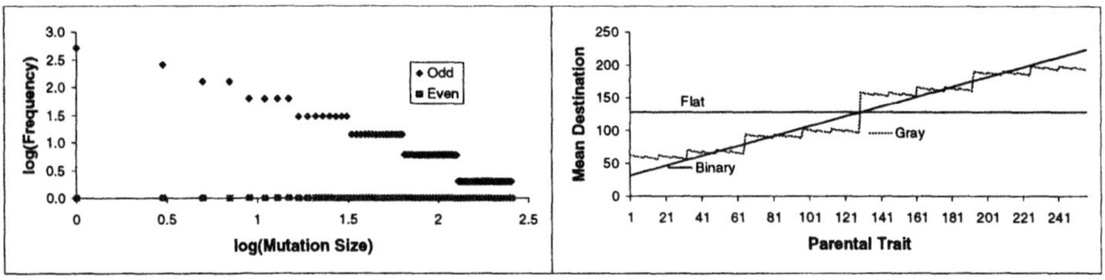

Fig. 3. *Left*: frequency distribution of mutation sizes for an 8-bit Gray-coded trait. *Right*: distributions of mean mutation destination by parental trait value for Gray, Binary and Flat mutation operators. Discontinuities could impede a population's evolution.

Encoding traits as a vector of discrete digits, and moving between trait values through manipulations at the level of these individual digits, imposes a structure on the mutation space of a model. There are a^n possible values represented by an n-dimensional code utilising an alphabet of a symbols, but each individual vector has only $n(a-1)$ immediate neighbours. This ensures that the code's neighbourhood relationships might systematically bias evolutionary change in particular ways.

Traits encoded as groups of binary digits which are interpreted as phenotypic values are typically mutated via random bit flips. When such a mutation operator acts upon conventional binary numbers, mutated phenotypic traits are poorly correlated with parental phenotypic traits, since a single bit-flip may result in a large change in the value for which the bit-string codes. Gray coding [7], through ensuring that consecutive integers are coded for by adjacent binary strings, increases the correlation between mutants and their parents. In addition to sharing the general appeal of context-sensitive mutation operators, this coding scheme is advocated by genetic algorithm designers, who argue that it makes evolutionary search more effective [8].

Constructing an n-bit Gray code can be thought of as assigning each of 2^n values to each of the vertices of an n-dimensional binary hypercube such that adjacent values are assigned to adjacent vertices lying on one of the hypercube's

Hamiltonian paths (paths which visit each vertex once and visit all vertices). Typical Gray code algorithms [8, 9] use a *Hamiltonian circuit*, in which the terminal vertices are also adjacent (Fig 2). It is unclear to what extent the claims of improved performance under Gray code schemes can be generalised from this *canonical Gray code* to other instances [8].

Though Gray codes ensure that there always exists a mutation event capable of perturbing a trait's value by a unit, the character of the remaining distribution of single-bit mutation events is unspecified. In fact, the canonical Gray code achieves a distribution of mutation events which decays roughly exponentially with mutation size (calculated as the absolute difference between pre-mutation and post-mutation trait values). This distribution exhibits strong discontinuities (Fig 3); there are no mutation events of even-valued magnitude. This implies that any mutation event changes both the parity and magnitude of the inherited value. For instance, if the former determined an organism's handedness whilst the latter coded for degree of handedness, this mutation bias would ensure that any mutant would exhibit both a (typically small) change in the degree of handedness *and* a change in which hand was preferred. Care must thus be taken to ensure that the constraints on mutation imposed by a discrete encoding do not systematically interfere with evolutionary change.

5 Implications

The threat to simulation modelling posed by the biases described above may seem overworked. However, the analyses clearly reveal the potential for the design of mutation operators to systematically influence trajectories of evolutionary change. One kind of solution to this problem has already been described, namely, to attempt the construction of bias-free operators. However, we have seen that while such attempts may *alter* the biases exhibited by an operator, these biases are typically not extinguished. It is also the case that whereas the most naturalistic mutation operators (e.g., Replace, or Absorb) might generate the most striking effects, the biases exhibited by these operators are explicable in terms of selection pressures which are present in naturally evolving populations.

For example, natural selection, *ceteris paribus*, favours genotypes which (i) have relatively low mutation rates, and (ii) are reasonably far from non-viable mutants [10, 11] (but see [12, 13]). Between them, these two selective forces can account for the biases exhibited by the most straightforward mutation operators considered here (Absorb, Repeat, and Replace). However, the biases of the operators designed specifically to negate these selection pressures (e.g., Ignore, Reflect and Wrap) are more difficult to account for in terms of natural selection pressures, as a result of their unnatural treatment of trait boundaries.

How are simulation modellers to decide between competing mutation operators – should they choose mutation operators that are clearly biased, but in a manner underwritten by natural selection, or choose alternative operators which although in some sense are "less biased", exhibit idiosyncrasies that have no biological analogue? I propose that this problem can be resolved by addressing a more pressing question: how are simulation modellers to guard against the artefactual results which may be caused by such biases?

Ultimately, in order to answer this question, the kinds of biases exhibited by various classes of mutation operator must be better understood, and attention must be paid to the conditions in which these biases will affect simulation results most severely. For instance, it appears that the influence of mutation bias will be strongest when populations are under only weak selection pressure. This situation may occur when populations drift across fitness plateaux, or during initial evolutionary transients which may be especially sensitive to population make-up. However, mutation biases may still influence the evolution of populations under strong selection pressures if these pressures conflict, through favouring one rather than the other. These are issues demanding further analysis.

In the absence of a principled understanding of mutation bias, however, two steps can be taken to minimise their influence. The presence of potential artefacts can be detected through exploring (i) the sensitivity of the simulation's behaviour to variation of its initial conditions (a procedure which has a number of other advantages and should thus be undertaken in any case), and (ii) the sensitivity of the simulation's behaviour to variation of the mutation operator employed (a less generally useful technique, but one which may be necessary if mutation bias is suspected).

Here the application of these two techniques to the example with which this paper opened will be described. In order to explore whether the bold coloration of the coevolved signal patterns was an artefact brought about by mutation bias, a previous study [2] explored the sensitivity of the result to manipulation of the initial ancestral population: populations of signals were able to evolve away from maximally bold ancestors, but they tended to re-converge on bold colours, although not necessarily those with which the populations were seeded.

Here the second line of sensitivity analysis will be reported. Simulations identical to those carried out in [2] were implemented, save that the mutation operator was varied across three conditions – Flat, Repeat, and Absorb. Boldness, b, was calculated as $1 - \sqrt{4/3}\sum_{i=1}^{25}\sqrt{r_i^2 + g_i^2 + b_i^2}$, where r_i, g_i, and b_i are the distances of the ith red, green, and blue colour components from their nearest boundary (i.e., 0 or 1). The mean boldness (across 100 simulation runs) of coevolved signals was sensitive to changes in mutation operator in the direction predicted by the analyses presented in this paper. Whilst the Absorb operator ($\bar{b} = 0.87$, $\sigma_{\bar{b}} = 0.006$) resulted in coevolved signals which were significantly bolder than those coevolved under either the Flat or Repeat regimes ($\bar{b} = 0.81$, $\sigma_{\bar{b}} = 0.004$, and $\bar{b} = 0.82$, $\sigma_{\bar{b}} = 0.004$, respectively), all three mutation operators generated signals which were bolder than random signals ($\bar{b} = 0.5$).

These results demonstrate that whilst the simulation is sensitive to the different characteristics of different mutation operators, mutation bias is *not* the underlying cause of the boldness of the evolved signals. While Enquist and Arak's original hypotheses may in fact be correct, an alternative explanation is that bold signals exploit an inherent preference for extreme-valued inputs on the part of the simple artificial neural networks with which they coevolve [14, 2].

6 Conclusion

The potential for a mutation operator's inherent biases to influence the dynamics of evolutionary simulation models was demonstrated for a range of prevalent mutation operators. Although these mutation biases may be the source of artefactual results, they are also part and parcel of natural selection. Rather than attempt to eliminate them from evolutionary simulations, their presence should be tolerated and controlled for. Greater understanding of the role of mutation bias in shaping the evolutionary dynamics of both natural and artificial evolutionary systems can only increase our understanding of the parallels and disimilarities between artificial and natural evolutionary processes.

Acknowledgments
Thanks to Valerie Chase, Martin Lages, Laura Martignon, Jason Noble, Peter Todd, Henrietta Wilson and two anonymous referees for helpful comments.

References

1. Enquist, M., Arak, A.: Symmetry, beauty and evolution. Nature **372** (1994) 169–172
2. Bullock, S., Cliff, D.: The role of 'hidden preferences' in the artificial co-evolution of symmetrical signals. Proc. Roy. Soc. Lond., *B* **264** (1997) 505–511
3. Schaffer, J.D., ed.: Proceedings of the Third International Conference on Genetic Algorithms, Morgan Kauffman, San Mateo, CA (1989)
4. Bäck, T.: Self-adaptation in genetic algorithms. In Varela, F., Bourgine, P., eds.: Toward a Practice of Autonomous Systems, Cambridge, MA, MIT Press (1992) 263–271
5. Bäck, T.: Optimal mutation rates in genetic search. In Forest, S., ed.: Proceedings of the Fifth International Conference on Genetic Algorithms, San Mateo, CA, Morgan Kauffmann (1993) 2–8
6. Maley, C.: The coevolution of mutation rates. In Morán, F., Moreno, A., Merelo, J.J., Chacón, P., eds.: Advanced in Artificial Life, Berlin, Springer Verlag (1995) 219–233
7. Gray, F.: Pulse code communication. U.S. Patent **2 632 058** (1953)
8. Caruna, R.A., Schaffer, J.D.: Representation and hidden bias: Gray vs. binary coding for genetic algorithms. [3] 153–161
9. Press, W.H., Teukolsky, S.A., Vetterling, W.Y., Flannery, B.P.: Numerical recipes in C: The art of scientific computing. CUP, Cambridge (1992)
10. Leigh, E.G.: Natural selection and mutability. Am. Nat. **104** (1970) 301–35
11. Leigh, E.G.: The evolution of mutation rates. Genetics **73** (1973) 1–18
12. Taddel, F., Radman, M., Maynard Smith, J., Toupance, B., Gouyan, P.H., Godelle, B.: Role of mutator alleles in adaptive evolution. Nature **387** (1997) 700–702
13. Sniegowski, P.D., Gerrish, P.J., Lenski, R.E.: Evolution of high mutation rates in experimental populations of *e. coli*. Nature **387** (1997) 703–705
14. Dawkins, M.S., Guilford, T.: An exaggerated preference for simple neural network models of signal evolution? Proc. Roy. Soc. of Lond, *B* **261** (1995) 357–360

Evolving Mutation Rates for the Self-Optimisation of Genetic Algorithms

Stevan Jay Anastasoff

School of Cognitive and Computer Sciences, Sussex University, Brighton, England,
stevana@cogs.sussex.ac.uk

Abstract. A version of the standard genetic algorithm, in which the mutation rate is allowed to evolve freely, is applied across a set of optimisation problems. The resulting dynamics confirm the hypothesis that mutation rate, when allowed to evolve, will do so partly as a function of altitude in the fitness landscape. Further, it is demonstrated that this fact can be exploited in order to improve efficiency of the genetic algorithm when applied to a particular class of optimisation problem. Specifically, significant efficiency gains are established in those problems in which the fitness function is not stationary over time.

1 Introduction

In most cases, mutation is far more likely to be detrimental to the future survival of a genome than it is to be beneficial [1, 2]. Given a basic principle of selection, if mutation rate is allowed to evolve, this will generally result in an evolutionary pressure favouring lower rates of mutation [3]. However, this pressure will vary [4]. The more highly optimised a particular genome is, the stronger the pressure will be to push mutation rate down. For a poorly adapted genome, there will be many more ways in which it can change so as to increase fitness. This will result in a much lower downward pressure on mutation rate, or for very poorly adapted genomes, selection towards higher rates.

These principles define the basic hypothesis that is to be tested and explored: if allowed to evolve, mutation rate will do so as an inverse function of altitude in the fitness landscape. This has potential applications with regard to the optimisation of certain genetic algorithms. In particular it suggests an improved performance over non-stationary optimisation problems. As the fitness landscape changes over time, the mutation rate will be able to adapt in order to maintain an improved rate of optimisation in such problems.

A substantial amount of literature exists already on parameter tuning and control in GAs, including much significant work on mutation. The most comprehesive surveys of this work can be found in [5] and [6]. As can be seen from these papers, work on self-adaptive mutation has received most attention with regard to Evolution Strategies. Comparatively little work has been focused on self-adaptive mutation rates in GAs at an individual level, as is considered here.

2 Method

To explore the hypothesis, a genetic algorithm is used in which the mutation rate for each chromosome is allowed to evolve independently. For comparative purposes, the same GA with a fixed mutation rate is also considered.

The two algorithms are applied over a suite of ten optimisation problems. These functions are described in detail below. Each of the problems was run a total of 100 times for each algorithm, and the results averaged.

The algorithms used were based around a standard bit-string GA, as described in Holland [7]. Additionally, each chromosome contained a floating point number between 0 and 1, to be used as the mutation rate for that chromosome. Single-point crossover was used, together with tournament selection.

A two-part mutation operator was used. Firstly, the bit-string portion of the chromosome was mutated with a probability equal to the mutation rate portion of the chromosome. In the fixed mutation rate algorithm, this was set at 0.01. The second stage of mutation was only carried out in the self-adaptive GA. Each mutation rate was itself mutated with a probability equal to itself. A mutated mutation rate was assigned a random floating-point value from 0 to 1.

All parameters were based on the results of the studies by De Jong [8], Grefenstette [9], and Schaffer, Caruana, Eshelman, and Das [10]. A further guiding factor was that the parameters should be chosen so as to maximise the effect of mutation rate in each case. The small population size of 20 is the lower limit suggested by these studies. As De Jong suggests, a small population size will increase the role that mutation has to play. Likewise, the crossover rate of 0.75 was at the lower bound of the ranges suggested.

A variety of starting conditions were tried for the mutation rate of the self-adaptive GA. These included having various fixed rates across the entire population, and randomly setting mutation rate for each individual. However, regardless of initial settings, the behaviour of the algorithm settled into the same pattern after approximately 40–50 generations. In any sort of extended run, therefore, initial mutation rates are of little consequence. An initial mutation rate of 0.01 across the entire population (identical to that used in the standard GA) was used for those specific examples discussed below.

3 The Test Suite

The three stationary problems are primarily derived from Ackley [11]. They cover three common optimisation situations: a linear fitness function, a fitness function containing many broad plateaus, and a function containing a large number of local maxima.

The linear function can be represented by the formula $f(x) = c$, where $f(x)$ is the fitness of the chromosome x, and c is the number of 1 bits in the bit-string part of the chromosome.

The plateau function is defined as follows: divide the bit-string portion of the chromosome into four equal length groups. If all the bits in a group are equal to 1 then add 25 to the fitness value, otherwise add 0. The fitness landscape for this function will have 4 plateaus leading to a peak of fitness 100.

The third function is what Ackley describes as a 'porcupine' function with many local maxima, and is defined as follows: $f(x) = 10c - 15(1 - parity(c))$, where $parity(c)$ effectively returns a value of 0 if c is odd, and 1 if c is even.

The next six problems in the test suite are non-stationary variants of these three problems. Each of the three has two variants, one with a cycle length of 25 generations, and one with a cycle length of 50 generations. In each case, every time the specified number of generations is reached, the function 'flips' such that instead of c being equal to the number of 1s in a bit-string, it becomes equal to the number of 0s (or vice versa).

The final problem is the non-stationary blind knapsack problem, taken from Goldberg [12]. The basic knapsack problem is as follows: given a number of objects, each with a specified weight and value, maximise the value of objects carried in a knapsack, given that it can only hold a limited weight. In the non-stationary, blind version of this problem, the weights and values of the objects periodically change, and there is no prior knowledge of any particular weights or values – only the total weight and value of a given combination is available.

4 The Results

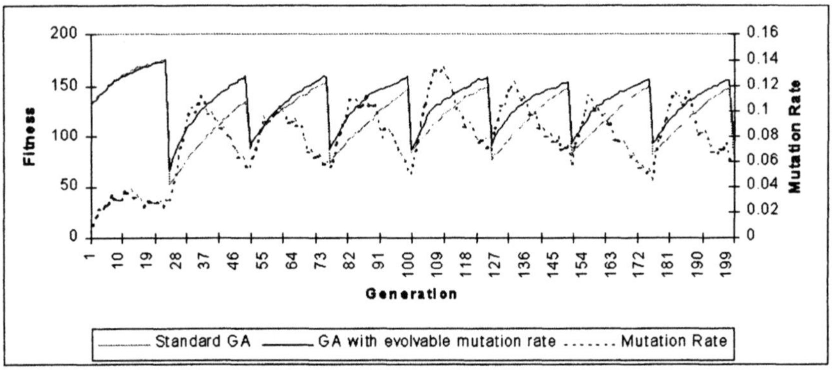

Fig. 1. Graph of 25 generation 'porcupine' fitness function

The first set of runs involved testing the GA with self-adaptive mutation rate on each of the problems from the test suite. On the stationary problems, the mutation rate remains fairly constant throughout. On the non-stationary problems the mutation rate initially behaves in an identical manner. However, as the fitness function changes, large changes in the mutation rate are evident. Drops in fitness can be seen to be highly correlated with increases in mutation rate, with higher fitness levels being correlated with lower mutation rates.

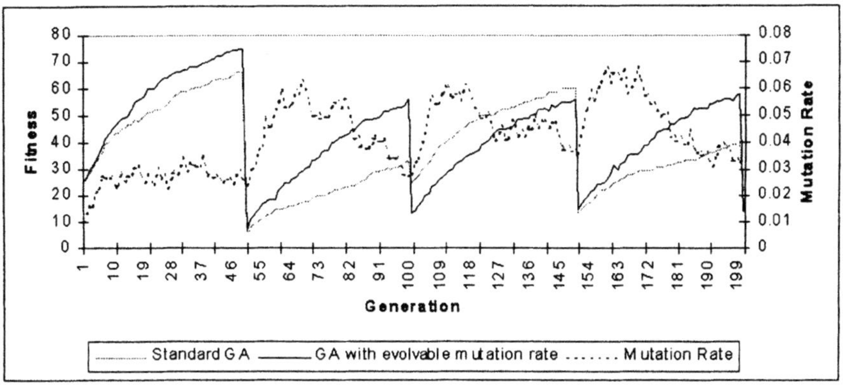

Fig. 2. Graph of 50 generation 'plateau' fitness function

After generating results for the self-adaptive GA, the test suite was then run using the standard GA, and the results plotted against the first set. On the stationary problems the behaviour of the two algorithms is almost identical. However, on the non-stationary problems the self-adaptive GA does consistently outperform the standard GA, both in terms of rate of increase of fitness and maximum fitness achieved. Even on those problems in which the standard GA does at times outperform the self-adaptive GA, it does so with much less consistency (see for example the non-stationary plateau function above).

5 Discussion

The results clearly indicate that levels of fitness and mutation rate are related. Low fitness levels result in a definite selective advantage for higher mutation rates, with increasing fitness resulting in a selective advantage for reduced mutation.

In all of the problems tested, the performance of the self-adaptive GA either equals or surpasses that of the standard GA. In particular, as expected, significant improvement can be seen in the non-stationary problems. It can be noted that in most cases the self-adaptive mutation rate stays at levels well above the 0.01 mutation rate set for the standard GA. It might be thought that the increased performance of the self-adaptive algorithm is simply due to this increased mutation rate, rather than self-adaptation. To test this, various values were tried for the fixed mutation rate. In most cases 0.01, or lower, was found to be optimum. For example, with mutation rate fixed at 0.01 in the 50 generation linear problem the average maximum fitness obtained over 100 runs was 19.88 for the first 50 generation cycle, 19.23 for the second. With the mutation rate increased to 0.1, comparable to the higher values of the self-adaptive mutation rate, the average maximum fitness drops to only 18.12 at the end of the first cycle, and 18.38 at the end of the second.

Traditionally, mutation has been considered very much a secondary operator within genetic algorithms (dating back to Holland [7]), with crossover playing the dominant role. However, this may in part be due to the particular range of problems from which optimum GA parameters have been derived. De Jong's [8] test suite, for example, consists entirely of stationary optimisation problems. It is evident that in some cases mutation can play a significant role in improving the performance of GAs.

The mutation dynamics and sorts of fitness landscape considered here are common also in biological systems. This sort of self-adaptive GA could therefore also be of use in modelling these mechanisms in natural evolution. Future work will explore this possibility.

6 Acknowledgements

Many thanks to Greg Werner, Josh Bongard, and Torsten Reil for comments on earlier drafts of this paper. Work supported by funding from the Engineering and Physical Sciences Research Council.

References

1. Sniegowski, P.D., Gerrish, P.J., Lenski, R.E.: Evolution of high mutation rates in experimental population of *E. Coli.* Nature **387** (1997) 703–705
2. Taddei, F., et al.: Role of mutator alleles in adaptive evolution. Nature **387** (1997) 700–702
3. Drake, J.W.: A constant rate of spontaneous mutation in DNA-based microbes. Proc. Natl. Acad. Sci. USA **88** (1991) 7160–7164
4. Bedau, M.A., Seymour, R.: Adaptation of mutation rates in a simple model of evolution. Complexity International, volume 2, April (1995)
5. Hinterding, R., Michalewicz, Z., Eiben, A.E.: Adaptation in Evolutionary Computation: A Survey. Proceedings of the 4th IEEE International Conference on Evolutionary Computation (1997)
6. Eiben, A.E., Hinterding, R., Michalewicz, Z.: Parameter Control in Evolutionary Algorithms. IEEE Transactions on Evolutionary Computation Vol.3, No.2 (1999)
7. Holland, J.H.: Adaptation in Natural and Artificial Systems. University of Michigan Press (1975)
8. De Jong, K.A.: An Analysis of the Behavior of a Class of Genetic Adaptive Systems. Ph.D. thesis, University of Michigan, Ann Arbor (1975)
9. Grefenstette, J.J.: Optimization of Control Parameters for Genetic Algorithms. IEEE Transaction on Systems, Man and Cybernetics **16** no **1** (1986)
10. Schaffer, J.D., Caruana, R.A., Eshelman, L.J., Das, R.A.: A study of control parameters affecting online performance of genetic algorithms for function optimization. In J.D. Schaffer, ed., Proceeding of the Third International Conference on Genetic Algorithms. Morgan Kaufmann (1989)
11. Ackley, D.H.: Bit Vector Function Optimization, in L. Davis ed., Genetic Algorithms and Simulated Annealing. Morgan Kaufmann (1987)
12. Goldberg, D.E.: Genetic Algorithms in Search, Optimization and Machine Learning. Addison-Wesley (1989)

Statistical Reasoning Strategies in the Pursuit and Evasion Domain

Sevan G. Ficici and Jordan B. Pollack

DEMO Lab, Brandeis University, Waltham MA 02454, USA
www.demo.cs.brandeis.edu

Abstract. Isaacs' treatise on differential games was a break-through for the analysis of the pursuit-and-evasion (PE) domain within the context of strategies representable by differential equations. Current experimental work in Artificial Life steps outside of the formalism of differential games, but the formalism it steps *into* is yet to be identified. We introduce a formulation of PE that allows a formalism to be developed. Our game minimizes kinematic factors and instead emphasizes the *informational* aspect of the domain. We use information-theoretic tools to describe agent behavior and implement a pursuit strategy based on statistical decision making; evaders evolved against this pursuit strategy exhibit a wide range of sophisticated behavior that can be quantitatively described. Agent performance is related to these quantifiables.

1 Introduction: Body and Mind in Pursuit and Evasion

Researchers in the field of Artificial Life (ALife) frequently turn to the domain of pursuit-and-evasion (PE) to study the (co)evolution of complex agent behavior. PE is of particular interest because it provides a parsimonious framework of agent interaction as well as ethological interest. Isaacs' ground-breaking theory of differential games [15] shows that, for a great many forms of the PE game, optimal (minimax) player strategies can be derived analytically from knowledge of the players' abilities. The applicability of differential game theory is constrained by two requirements: 1) a finite set of *state variables* must completely capture the instantaneous description of the game (e.g., agent positions, velocities, accelerations, etc.) necessary to compute the future unfolding of the game—the game is one of perfect information; 2) agent strategies must be expressible as differential equations that operate on these state variables. These constraints make the theory (and its more modern versions, e.g., [8]) a powerful model of agent *kinematics* (e.g., speed, maneuverability, etc.), as kinematics are particularly amenable to differential analysis. Indeed, differential game theory was constructed to answer questions about agents that are distinguished primarily in their kinematic abilities.

Much ALife research, however, steps outside of the formal assumptions made by Isaacs' theory; evolvable sensory apparatus [5] and evolvable information-processing substrates for agent control, such as artificial recurrent neural networks, move the PE domain beyond the purely kinematic realm to include an *informational* one as well. Clearly, differential game theory remains an influential formalism in the design and analysis of experiment kinematics. But, what formalism has been adopted to address the informational aspects of current experimental work? The lack of a rigorous metric of agent behavior has been recognized [7]—how do we ascertain and describe the sophistication of an agent? Recently, information-theoretic tools have been used to measure and adjust environmental complexity to facilitate agent learning in various domains, including PE [11, 13]; information theory has also come into use for the measurement of agent behavior, for example in a discrete state space [23] and a "linear" form of the PE game [10]. Here, we will introduce the use of information theory to measure agent behavior and environmental complexity in a two-dimensional, continuous-space pursuer-evader domain.

The purpose of replacing (stateless) minimax strategies with evolvable agent controllers is to place the pursuit and evasion roles within a coevolutionary setting such that an arms race between ever more sophisticated evasion and pursuit strategies arises [16, 19, 24, 6, 7, 12, 26]. Unfortunately, the envisaged coevolutionary arms race towards complexity is yet to be substantially realized, or, at least, observed; a number of problematic issues surrounding coevolutionary techniques are known to exist, such as the Red Queen Effect [6], mediocre stable-states [22], and the instability of two-population coevolution [4, 2, 9].

If we want an arms race to complexity, we must consider the three components that are responsible for the behavior of the typical PE agent: its sensory apparatus, kinematics, and processing abilities. Which of these three components possess the potential when evolved to engender an arms race to complexity? How might asymmetries between the pursuer and evader in terms of some of these components (say, their sensory and kinematic abilities) diminish the usefulness of evolving others (say, their computational abilities)? We assert that the role of processing ability is yet to be appreciated in its own right—that it has been overshadowed by kinematic and sensory factors.

In this paper, we present a simple formulation of the two-dimensional pursuer-evader game that, by giving the players kinematic parity and a form of sensory asymmetry, requires them to model opponent behavior in order to achieve optimal performance (in the game-theoretic sense). The behavior of an agent determines the complexity of the statistical model required to adequately represent it. The success of an agent reflects the power of its representational and modeling abilities. In this sense, there exists the potential for an arms race in complexity.

We begin, however, by using simple evolution to evolve evaders against hand-built predictors that employ statistical tools to model evader behavior. The questions of interest are: what happens when the statistical model of the pursuer is poor? What kind of behavior must an evader exhibit to defeat statistical prediction of some finite power? Would such evasion behavior resemble our intuitive understanding of *protean behavior* [19]? Our results show that evaders are able to evolve behaviors of substantial sophistication when placed in opposition to a powerful pursuit strategy.

We start with a formal definition of the game and discuss some game-theoretic features. We review the information-theoretic tools used to measure behavior. Experimental results are then discussed. Finally, we point to future work.

2 A Game of Incomplete Information

2.1 Definition

Our game is played in the real-valued, unbounded (two-dimensional) plane. Both agents move with *simple motion*: each agent moves at a constant velocity and is able to instantaneously change its direction by any amount. In our game, both agents move at the same velocity of one body-length per time-step. While the agents are able to freely pick a direction, they can only do so at the beginning of each time-step; they must then commit to moving in their chosen directions for the duration of one time-step. The pursuer can see the evader's current location, but the evader is completely blind. Thus, the pursuer has the opportunity to respond to the behavior of the evader, while the evader's behavior is ballistic.

Each game begins with the pursuer and evader occupying the same location, the plane's origin. Because we are interested in statistical characterizations of agent behavior, each game must be long enough for a reasonably representative sample of behavior to be made with respect to the type of statistic we wish to collect. (In practice, the behaviors of our evolving evaders quickly achieve steady state—initial transients are either very short or unfold so slowly as to be indistinguishable from steady-state behavior for our purposes.)

Thus, the terminating condition occurs when the final time-step transpires, rather than when the evader gets tagged or reaches some "escape" distance. The evader's job is to maximize the *average distance* between itself and the pursuer over the course of the game. The pursuer's job is to maximize the number of *tags*, which is subtly more general than minimizing distance, as we will discuss. A tag occurs when the pursuer comes within one body-length of the evader.

2.2 Game Theoretic Features

This game is designed to require players to induce models of opponent behavior (the type of model is discussed below). Here we explore the game-theoretic features that create such conditions. To begin, we wish to eliminate kinematic asymmetries that may alone favor one agent over the other. For this reason, we give both players identical kinematic abilities: simple motion with equal velocities. The choice to make our evader blind appears to contradict the care taken to give the agents an equal footing. To understand this choice, let us consider what happens if the evader has the ability to see. Further, let us assume the predictor is strictly trying to minimize distance, such that the game is clearly *zero-sum*.

If the sightedness of both players is *common knowledge* [18, 14], i.e., known by both players, then this modified game clearly calls for a *minimax* strategy [15] to be used. The minimax strategy calls for the pursuer to move directly towards the evader and the evader to move directly away from the pursuer. Any divergence from this strategy by an agent gives an advantage to its opponent;

if both agents observe their optimal strategies, then the distance between them does not change. Given our kinematic constraints, minimax allows neither the evader to increase its distance from the pursuer, nor the pursuer to near the evader, let alone tag it—a very dull game, indeed. This is why agent kinematics are made asymmetrical in most experimental work; the possibility of "cognitive" asymmetry is thus marginalized.

We can now appreciate what happens if the evader is blind. Because the evader is unaware of the pursuer's location, it can not simply move away. Instead, the evader must employ a mixed strategy that picks moves probabilistically. If the pursuer knows this to be the case, through common knowledge, then the pursuer is best advised to adopt a *statistical decision process* [27] based upon observations of the evader's behavior. If the pursuer's goal is not only to decrease the distance between itself and the evader, but also to do so as quickly as possible (in order to maximize the number of tags), then pursuit based upon statistical observation will, on average, very likely outperform (and certainly do no worse than) the simple pursuit strategy that minimax specifies above, provided that the statistical observations are suitably accurate. If the evader knows, again through common knowledge, that the pursuer is using such a strategy, then it is no less motivated to behave probabilistically.

In his discussion of pursuit games with incomplete information, Isaacs [15] explains that the importance of accurate prediction diminishes as the distance between opponents increases; if the "probability cloud" of possible future evader locations is small compared to the distance between the evader and the pursuer, then the cloud is reasonably treated as a single point by the pursuer—obviating the need to predict and allowing the pursuer to revert to the simple strategy of moving directly towards the evader. On the other hand, if the two agents are very close, particularly if their regions of possible future locations overlap, accurate prediction is of utmost importance. With spatial proximity, the statistical nature of the game becomes more prominent. For this reason, we choose to begin our game with the pursuer and evader on top of one another.

3 Observation and Statistical Reasoning

Because the evader moves at a constant rate, and commits to some direction of travel for the duration of each time-step, the pursuer need only observe the sequence of direction choices that the evader makes; if the pursuer accurately models the evader's decision process, then effective pursuit is possible. Though the directions in which the agents move (and the locations they occupy) vary continuously, the pursuer's *sensing* of movement has limited resolution: each move of the evader is perceived by the pursuer as being in one of eight directions, as shown in Figure 1. Thus, the pursuer's statistics model the *symbol string* generated by the evader's behavior, and not the evader's behavior directly.

To predict the future path of the evader, the pursuer uses its statistical model to generate the symbol string it expects the evader to produce in the coming time-steps. The pursuer then uses this symbol string to project the expected trajectory relative to the evader's actual current location. The pursuer then acts according to its policy, detailed below. Even if the symbol string is predicted with complete accuracy, the projected trajectory only approximates the path the evader will actually take, due to the pursuer's limited ability to sense and represent heading. Nevertheless, the eight-symbol resolution is high enough for adequate approximation in practice.

In its capacity as model-builder, our hand-built pursuer uses techniques that we, as experimenters, can use to quantitatively describe an evader's behavior. Our analytic approach comes from information theory [25]. The model of an evader is built by computing n^{th}*-order statistics* from observations of behavior. The order of the statistic refers to the amount of conditioning applied when computing the probabilities of various symbols occurring. The higher the order of our statistics, the more we can refine our expectations of what will happen in the future based on what we have observed further in the past. For example, a 0-order statistic states the probability of a symbol occurring, $p(S)$, without conditioning—based strictly on the number of times it is observed; if 10% of the symbols observed are the symbol 'X' then the 0-order probability of 'X' is $p(X) = 0.1$. A 1-order statistic states the probability of a symbol conditioned with respect to the symbol that immediately precedes it; if the symbol 'Y' is followed by 'X' 60% of the time, then one of the 1-order probabilities of 'X' is $p(X \mid Y) = 0.6$. We now know more about when to expect the symbol 'X'.

The *entropy* of an observed sequence describes the over-all predictability of the sequence, given a particular order statistic; higher entropy values indicate less predictability. Certain symbol sequences require higher order statistics to be taken than other sequences to achieve the same level of predictability—the same entropy. The minimal amount of behavioral history, or conditioning, required to maximize predictability is an indication of a sequence's *complexity*. Of equal importance is the degree to which a sequence becomes less predictable as the order of the statistic is decreased. These two characterizations of a symbol sequence provide a quantifiable description of behavior.

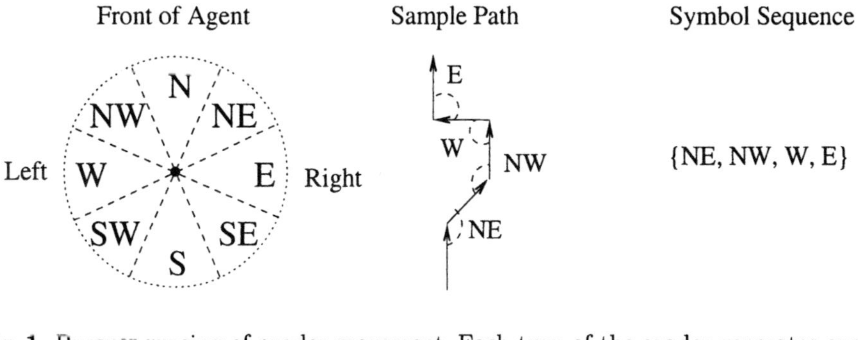

Fig. 1. Pursuer sensing of evader movement. Each turn of the evader generates one of eight symbols, depending on the angle of the turn relative to evader's previous heading.

4 Experiments

4.1 Setup

Evolving Evader Using an enhanced version of GNARL [3], we evolve artificial recurrent neural networks to control our evader agent. The GNARL algorithm adapts Evolutionary Programming (EP) techniques to the evolution of neural networks. Since the evader is blind, there are no inputs to the network; a single bias node, which has a constant activation, is used. The network has a single real-valued output that indicates the turn angle for the next move. Networks are allowed to have as many as 60 hidden units and 400 weights. These limits are not known to be optimal in any sense (casual examination of the larger evolved networks shows around 50 hidden nodes but only 100 weights). The initial population is comprised of random networks that have between 5 to 15 hidden units and 25 to 100 weights. The population size is 100. The fitness of a controller network is equal to the average distance it is able to maintain between the pursuer and the evader over the course of a game.

Hand-Built Pursuer The evader controller networks are evolved against a hand-built pursuit strategy that uses statistical modeling of evader behavior. This pursuit strategy is controlled by two parameters: 1) the power of the statistical model, which is specified by the order statistic, o, and 2) the number of time-steps into the future, t, that will be predicted by the model. The more the pursuit strategy looks into the future, the more opportunity it has to discover and exploit short-cuts to predicted future locations of the evader. But, with increased values of t comes increased risk of being misled by poor prediction.

At each time-step of the game, the pursuit strategy proceeds as follows: using the statistical model of order o, predict the evader's movement from its current location for t time-steps into the future; this yields the path, $\boldsymbol{p}$, of points $\boldsymbol{p} = \{p_1, p_2, ..., p_t\}$. Assuming the predictions to be correct, we know when the evader will arrive at each point, p_i, in the path, $\boldsymbol{p}$. If there exist any points to which the pursuer, by heading directly to them, can arrive before the evader, select the earliest such point in the path. Otherwise, move towards the first point on the path, p_1.

The pursuer computes its statistical model of evader behavior *tabula rasa* for each game. Because the model requires a suitably large sample of behavior to give meaningful predictions, we preface each match between the pursuer and an evader with a "warm-up" period of ten thousand time-steps during which the statistical model of the evader's ballistic behavior is induced. The length of this warm-up period represents a practical compromise between sample quality and simulator speed. After the model is built, the players are placed on the origin of the plane and the game proper begins, which lasts for one thousand time-steps.

4.2 Results

We outline above a statistical formalism with which we can characterize evader behavior, and describe how we embed this formalism into our hand-built pursuit strategy. We now evolve evaders against this strategy. By thus inserting our statistical formalism into the evolutionary process, however, we lose, ironically, the ability to use it as an objective measure of experimental results; while we do provide some quantitative results below, we must also pay careful attention to the qualitative results for meaningful understanding. Our formalism can more properly be used for agent analysis once both the evader and pursuer are made to coevolve; this is the subject of future work.

Pursuer Performance Figure 2 (left) captures a moment during a match. The square and the large circle represent the positions of the pursuer and evader at time t, respectively. The small circles are the pursuer's prediction (using 12th-order statistics), made at time $t-1$, of the evader's path for the next twelve time steps. We see that the evader is exactly where the pursuer expected it to be at time t. The large 'X' represents the point on the evader's path where the pursuer expects to most quickly intercept the evader; it is towards this point that the pursuer moved directly from time-step $t-1$ to time t. Of course, interception will occur only if the predictions are correct. The evader's behavior might instead lead the pursuer into a false expectation of future locations—the only method by which the evader can open space between itself and the pursuer; this is what we see in Figure 2 (right), which shows a much weaker pursuer (using 3rd-order statistics) playing against the same evader at the identical point in the match. Clearly, and to its detriment, this pursuer has a much different model of the evader and very different expectations of future evader behavior.

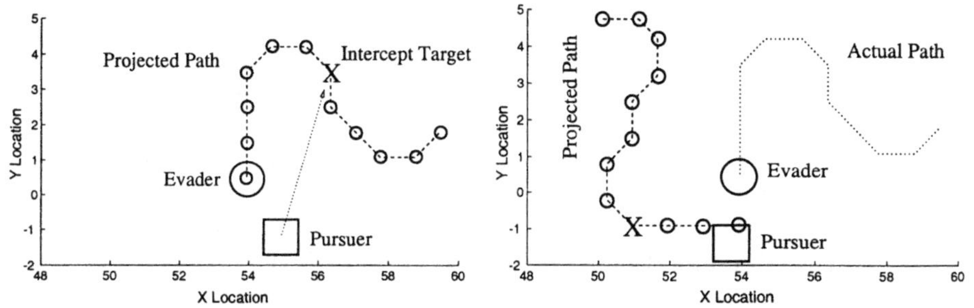

Fig. 2. Pursuit using 12th-order statistics (left) and 3rd-order statistics (right).

Evader Behavior Though the evaders are blind, they too must model their opponent to succeed. In this case, the model is induced through evolution and reflects the settings of the pursuer's two parameters: the order of pursuer's statistical observations, and the number of time-steps predicted into the future. A very wide variety of evader behaviors evolve in our experiments. The trajectories of two different evaders, A and B, are shown in Figure 3 (top left and right, respectively). Both were evolved in the same run with pursuit parameters set to $t = 37$ and $o = 4$; evader A is the best of generation 825, while B arrives much earlier in the run and is the best of generation 85. Evader A has a highly irregular, though smooth, trajectory that contains tight loops, broad and narrow turns, and some relatively straight stretches. Evader B's behavior is extremely regular (in fact periodic) and jagged, and displays two frequencies of movement (one superimposed on the other), which gives it a hint of self-similarity. Both evaders clearly maintain non-trivial internal state.

Below each trajectory, we graph the entropy of the trajectory (dotted line), the average distance maintained between the evader and the pursuer (solid line), and the average tag rate (dash-dot line); these values are given for a pursuer with parameter $t = 37$ and parameter o ranging from 0 to 9. Though the entropy of evader B, when measured with 0-order statistics, is actually higher than that of evader A (approximately 1.9 for B and 1.3 for A out of a maximum possible entropy of 3.0), the entropy of evader A falls only slightly as the order statistic is increased, whereas the entropy of evader B drops dramatically. These entropy curves reflect what is visibly obvious: evader A is very difficult to predict and evader B is very easy. Particularly, once the pursuer models evader B as a 6th-order process, the regular behavior of the evader becomes evident: the observed entropy falls to a mere 0.16, the average distance the evader is able to maintain from the pursuer drops to 0.37 body-length units, and the pursuer's tag rate jumps to a very effective 0.98%. Changes in evader A's performance are evident as modeling power is increased, as well, but are not nearly as dramatic.

These graphs suggest that proteanism—adaptively unpredictable behavior—may be envisioned as a continuum; a behavior may be protean relative to a weak statistical model but not to a more powerful one. While a high entropy value implies a behavior that is difficult to predict, it does not necessarily imply an effective evasion strategy; Isaacs recognizes that, in a game of incomplete information, a tension may exist between moves that create uncertainty and moves that open distance. Good evasion behavior consists of a balanced mix of these types of move. Thus, while evader B has a higher 0-order entropy than evader A, its behavior does not produce such a mix and poorly leverages the uncertainty it generates.

As a simple control experiment, we modify the pursuer to move directly towards the evader's current location (as in minimax); the appropriate minimax response, i.e., straight fleeing, is evolved by the evader. This provides further empirical evidence that evolution is sensitive to the pursuit strategy and creates evasion behaviors accordingly. Thus, effective evasion is not synonymous with complex evasion; the complexity of evasion behaviors we see in our main experiment reflect the sophistication of our hand-built pursuit strategy. In *co*evolutionary frameworks, an absence of evader "proteanism," as conceived here, indicates a similar lack of pursuer sophistication.

5 Conclusion

We introduce a formulation of the pursuer-evader game that emphasizes its informational component. We discuss statistical tools that enable rigorous analysis of evader behavior and, thereby, allow the construction of a powerful pursuit strategy. Evolution against this pursuit strategy results in evasion behaviors of considerable sophistication. Agent performance is relatable to quantifiable features of behavior.

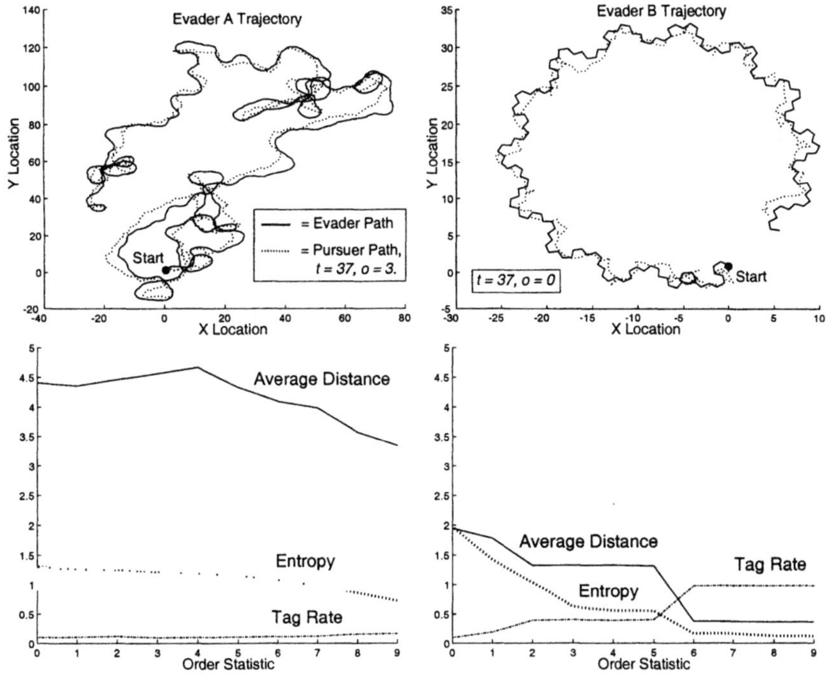

Fig. 3. Evader trajectories and performance curves. Trajectories begin at (0, 0).

Future work will replace the hand-built statistical mechanism of the pursuer with a coevolving substrate. The generality of our hand-built pursuer is what evokes evasion behavior with (often complex) statistical structure. But, unlike the blind evaders, pursuers can not operate ballistically—they must respond to observed behavior. If coevolving pursuers are not exposed to a wide enough diversity of evasion behavior, they will likely evolve pursuit strategies that lack generality; these specialized strategies will be suboptimal in the game-theoretic sense. Nevertheless, our statistical tools will be of particular use in tracking coevolutionary progress [9]. Finally, we will investigate sighted evaders.

6 Acknowledgments

The authors thank the members of DEMO for their insightful assistance, particularly Richard Watson, Ofer Melnik, and Hod Lipson.

References

1. C. Adami et al., editors. *Artificial Life VI (1998)*. MIT Press, 1998.
2. E. Akiyama and K. Kaneko. Evolution of communication and strategies in an iterated three-person game. In Langton and Shimohara [17], pages 150–158.
3. P. J. Angeline, G. M. Saunders, and J. B. Pollack. An evolutionary algorithm that constructs recurrent neural networks. *IEEE Transactions on Neural Networks*, 5:54–65, 1994.

4. S. Bullock. Co-evolutionary design: Implications for evolutionary robotics. In Moran et al. [20].
5. D. Cliff, P. Husbands, and I. Harvey. Evolving visually guided robots. In J.-A. Meyer, H. L. Roitblat, and S. W. Wilson, editors, *From Animals to Animats II (1992)*, pages 374–383. MIT Press, 1993.
6. D. Cliff and G. F. Miller. Tracking the red queen: Measurements of adaptive progress in co-evolutionary simulations. In Moran et al. [20], pages 200–218.
7. D. Cliff and G. F. Miller. Co-evolution of pursuit and evasion 2: Simulation methods and results. In P. Maes et al., editors, *From Animals to Animats IV*, pages 506–515. MIT Press, 1996.
8. R. J. Elliot and N. J. Kalton. The existence of value in differential games of pursuit and evasion. *Journal of Differential Equations*, 12:504–523, 1972.
9. S. G. Ficici and J. B. Pollack. Challenges in coevolutionary learning: Arms-race dynamics, open-endedness, and mediocre stable states. In Adami et al. [1], pages 238–247.
10. S. G. Ficici and J. B. Pollack. Coevolving communicative behavior in a linear pursuer-evader game. In Pfeifer et al. [21], pages 505–514.
11. J. A. Fletcher, M. A. Bedau, and M. Zwick. Effect of environmental structure on evolutionary adaptation. In Adami et al. [1], pages 189–198.
12. D. Floreano, S. Nolfi, and F. Mondada. Competitive co-evolutionary robotics: From theory to practice. In Pfeifer et al. [21], pages 515–524.
13. F. Gomez and R. Miikkulainen. Incremental evolution of complex general behavior. *Adaptive Behavior*, 5:317–342, 1997.
14. J. Harsanyi. Games with incomplete information played by 'bayesian' players. *Management Science*, 14:159–182,320–324,486–502, 1967.
15. R. Isaacs. *Differential Games.* John Wiley and Sons, New York, 1965.
16. J. Koza. *Genetic Programming.* MIT Press, 1992.
17. C. G. Langton and K. Shimohara, editors. *Artificial Life V (1996).* MIT Press, 1997.
18. D. Lewis. *Convention: A Philosophical Study.* Harvard University Press, Cambridge, Massachusetts, 1966.
19. G. F. Miller and D. Cliff. Protean behavior in dynamic games: Arguments for the co-evolution of pursuit-evasion tactics. In D. Cliff et al., editors, *From Animals to Animats III*, pages 411–420. MIT Press, 1994.
20. F. Moran et al., editors. *ECAL 3.* Springer, 1995.
21. R. Pfeifer et al., editors. *From Animals to Animats V.* MIT Press, 1998.
22. J. B. Pollack, A. Blair, and M. Land. Coevolution of a backgammon player. In Langton and Shimohara [17].
23. C. Pötter. Information-theoretic analysis of a mobile agent's learning in a discrete state space. In A. Eiben et al., editors, *PPSN V*, pages 521–529. Springer, 1998.
24. C. Reynolds. Competition, coevolution and the game of tag. In R. A. Brooks and P. Maes, editors, *Artificial Life IV*, pages 59–69. MIT Press, 1994.
25. C. E. Shannon and W. Weaver. *The mathematical theory of communication.* University of Illinois Press, Urbana, Illinois, 1963.
26. M. Wahde and M. G. Nordahl. Co-evolving pursuit-evasion strategies in open and confined regions. In Adami et al. [1], pages 472–476.
27. A. Wald. *Statistical Decision Functions.* Wiley, New York, 1950.

An Evolutionary Method Using Crossover in a Food Chain Simulation

Atsuko Mutoh, Satoru Oono*, Kousuke Moriwaki**,
Tsuyoshi Nakamura, Nobuhiro Inuzuka, and Hidenori Itoh

Nagoya Institute of Technology
Gokiso-cho, Showa-ku, Nagoya 466-8555, Japan
{atsuko, soono, moriw, tnaka, inuzuka, itoh}@ics.nitech.ac.jp

Abstract. A gene expression system n-BDD (n-output Binary Decision Diagram) was proposed in order to investigate co-evolution[5]. Although the system is suitable for behavior models of agents, it does not include crossover. This paper proposes a crossover operation using Bryant's Apply operation[2]. The operation makes an n-BDD probabilistically inherit two functions expressed by two n-BDDs. In an experiment the proposed method had more than 40% high fitness than the conventional method. Moreover, in another environment where carnivores and herbivores are co-evolved, we have seen a food chain relation.

1 Introduction

Co-evolution evolves two species of which are interactive in an environment. The relation between carnivores and herbivores makes typical co-evolutionary competition or an interactive relationship of food chain in the natural world. Therefore researches on co-evolution and food chain models have been gathering attention in the field of evolutionary computation and artificial life.

T.Takashina et al. studied an artificial competitive world and co-evolution[6] using a gene expression system based on finite state automata (FSA), which is proposed in [3]. The efficiency of genetic algorithms depends on the gene expression systems. Moriwaki et al.[5] proposed a system using n-BDDs, which extend Binary Decision Diagrams (BDDs) to express functions on any finite domains. Effectiveness of the method was shown in many applications.

The method in [5] with n-BDDs does not have a crossover operation, however. This paper proposes a crossover operation for n-BDDs and shows that it works well in a co-evolutionary model.

2 A gene expression n-BDD and genetic operations

BDDs, proposed by Akers[1], have been used in engineering fields, and is graph representations of Boolean functions. A BDD has terminal nodes labeled by true

* Currently working for Seiko Epson Corp.
** Currently working for NTT Commucationware Corp.

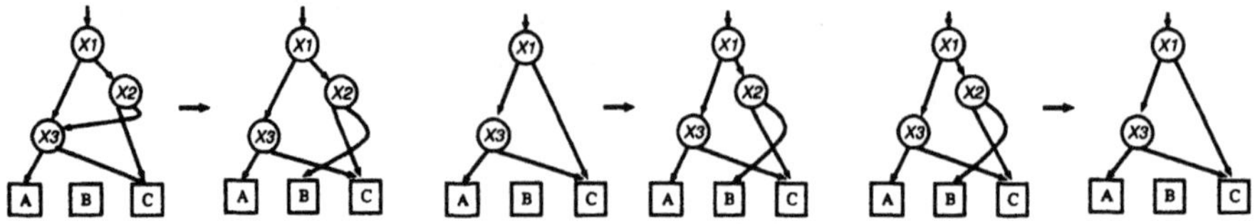

(a) mutation (b) insertion (c) deletion

Fig. 1. Genetic operations for n-BDDs.

or false, while an n-BDD[5] can have more than two labels and gives a value from any set of values which the labels denote.

In Fig. 1 a circle denotes an input bit and is called a decision node. An output value is denoted by a square which is called a terminal node. With input bits an output value is calculated as follows: First look at the top decision node $X1$ and take a left outgoing edge (called a 0-edge) if $X1$ is 0 or a right edge (a 1-edge) if $X1$ is 1. Iterate this for the decision node indexed by the 0-edge or 1-edge until an edge indexes a terminal node, which gives an output. Genetic operations[5] are defined to operate n-BDDs (see Fig. 1). These operations are restricted not to violate the order of decision nodes. Then any loop does not occur.

3 Crossover based on Bryant's Apply operation

Here we give another notation of n-BDDs. A node of an n-BDD is a non-terminal node, represented by a triple (variable name, left-node, right-node), or a terminal node, represented by a value. An n-BDD is a pair consisting of a set of nodes and a node that represents a top node. We equate an n-BDD (and a function f represented by it) with its top node $(f.top, f_0, f_1)$.

The procedure Apply developed by Bryant[2] takes two BDDs representing functions f and g, and a binary Boolean operator $(\circ)$, and produces a reduced BDD representing the function $f \circ g$ defined as

$$f \circ g(X_1, X_2, \ldots, X_n) = f(X_1, X_2, \ldots, X_n) \circ g(X_1, X_2, \ldots, X_n).$$

Instead of a binary Boolean operator $(\circ)$, our Apply extended for n-BDD takes a binary operator on a given set of values. The Shannon expansion lets us expand the above expression as follows.

$$f \circ g(X_1, \ldots, X_i, \ldots, X_n) =$$

$$\begin{cases} f(X_1, \ldots, X_{i-1}, 0, X_{i+1}, \ldots, X_n) \circ g(X_1, \ldots, X_{i-1}, 0, X_{i+1}, \ldots, X_n) & ;\text{if } X_i = 0 \\ f(X_1, \ldots, X_{i-1}, 1, X_{i+1}, \ldots, X_n) \circ g(X_1, \ldots, X_{i-1}, 1, X_{i+1}, \ldots, X_n) & ;\text{if } X_i = 1 \end{cases} \quad (1)$$

If we note facts that $f(X_0, \ldots, X_{i-1}, 0, X_{i+1}, \ldots, X_n) = f_0$ and $f(X_0, \ldots$ $\ldots, X_{i-1}, 1, X_{i+1}, \ldots, X_n) = f_1$, where $f.top = X_i$, we can re-formalize Equation (1) as $f \circ g = (f.top, f_0 \circ g_0, f_1 \circ g_1)$, where we assume that $f.top = g.top$. In the case $f.top$ has an upper order than $g.top$ or $f.top$ is not appeared in g, $f \circ g = (f.top, f_0 \circ g, f_1 \circ g)$, or $f \circ g = (g.top, f \circ g_0, f \circ g_1)$ otherwise. The algorithm in Fig. 2 obeys this idea. The order of variables should be fixed. If we choose the conjunction for $(\circ)$, Apply calculates a BDD for $f \wedge g$. For our purpose we use a probabilistic operator as $(\circ)$. For two values a and b, $a \circ b$

If f or g is a terminal node or $f = g$ **then** $h \leftarrow f \circ g$
Else if $f.top = g.top$ **then** $h_0 \leftarrow f_0 \circ g_0, h_1 \leftarrow f_1 \circ g_1$
  if $h_0 = h_1$ **then** $h \leftarrow h_0$ **else** $h \leftarrow (f.top, h_0, h_1)$
Else if rank of $f.top$ is higher than that of $g.top$ **then** $h_0 \leftarrow f_0 \circ g, h_1 \leftarrow f_1 \circ g$
  if $h_0 = h_1$ **then** $h \leftarrow h_0$ **else** $h \leftarrow (f.top, h_0, h_1)$
Else if rank of $g.top$ is higher than that of $f.top$ **then** $h_0 \leftarrow f \circ g_0, h_1 \leftarrow f \circ g_0$
  if $h_0 = h_1$ **then** $h \leftarrow h_0$ **else** $h \leftarrow (g.top, h_0, h_1)$

Fig. 2. An algorithm to calculate $h(= f \circ g)$.

is a or b in the probability 0.5. Then $f \circ g$ inherits both behavior of f and g probabilistically. That is $f \circ g$ obeys f for some inputs and g for others. This is an expected behavior of crossover operation. The probabilistic ($\circ$) acts probabilistically only when a BDD of $f \circ g$ is constructed and is fixed as it is deterministic when it is used.

The time complexity of crossover for two n-BDDs and the size of a resulted BDD are $O(|f| \cdot |g|)$, where $|f|$ and $|g|$ are the numbers of nodes in f and g. In our experiment, however, the number of nodes did not grow very much.

4 An experiment with a simple competition problem

In this section, we compare ability between n-BDDs with crossover and ones without it in a simple competition problem, which is introduced in [5]. In an environment two artificial animals, a carnivore and a herbivore act. The problem is to let the herbivore evolve to overcome a strong carnivore.

Inputs and outputs of n-BDDs Both animals act according to their own n-BDDs. Meanings of input bits are shown in Table 4. The order of the variables also obeys this table. A gene outputs a value from the five actions given in Table 2. A carnivore does not output runaway.

Experimental setting The animals act according with their n-BDDs in a 20×20 array field. Carnivore has a fixed strategy (i.e. a fixed n-BDD) (Fig. 3), and tries to capture a herbivore, which is evolved using each of methods. By observing evolution of herbivor's behavior we examine effect of the method. Thirty plants are distributed in the field randomly. When a plant is eaten, another one is appeared in a random position. Herbivore may die because of hunger or being eaten. The number of steps for which the herbivore survived is taken as its fitness. The method of selection is preserving the five elitist from thirty herbivores of each generation.

Result We conducted 100 times of experiments. Fig. 4 shows the average of the maximum fitness of herbivores using both methods. It is clear that the rise of the fitness of n-BDDs with crossover is faster than one without it.

Table 1. Bit assignments.

$X0$	hungry
$X1$	repletion
$X2$	carnivore is visible far
$X3$	carnivore is visible near
$X4$	herbivore is visible far
$X5$	herbivore is visible near
$X6$	plant is visible far
$X7$	plant is visible near

Table 2. Actions.

walk (W)	Animal moves to eight neighborhoods at random.
runaway (R)	Animal moves in the opposite direction to a carnivore.
eat (E)	Animal moves a food, and eat it if reached.
do-nothing (N)	Animal does not move.
approach (A)	Animal approaches the same kind of animal.

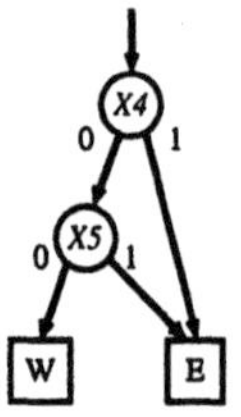

Fig. 3. The carnivore's fixed strategy

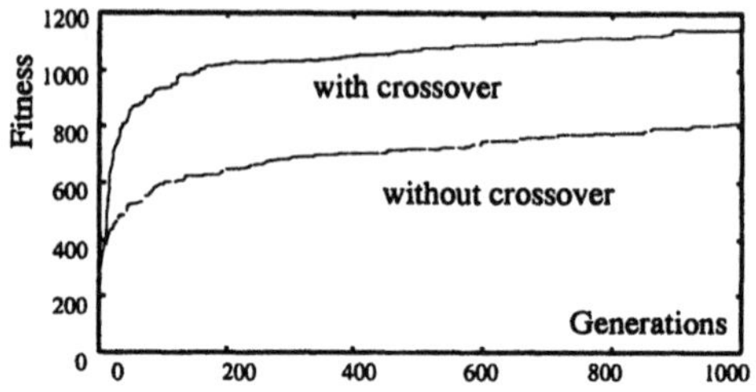

Fig. 4. Fitness transition of evolving herbivore against the fixed carnivore.

5 A model for co-evolution in a food chain model

Here we explain another experiment in a food chain model, where many herbivores and many carnivores exist and are evolvable. Animals can be born. It aims to observe co-evolution among competitive animals through food chain. The food chain model proposed in [6] and also in [5] does not have crossover, that is, a child is generated from only a parent by mutation, insertion or deletion. In our model, a child can be generated from two parents by crossover.

A setting in the model The field is a 150×150 array. A herbivore (a carnivore) acquires energy by eating plants (herbivores, respectively). Its energy decreases in a constant step by every action until it becomes zero. When it becomes zero the animal will die and plants will grow around the body. [1]

Application of the genetic operations in the model The crossover operation lets us construct a natural model for reproduction. When an animal has enough energy it tries reproduction with another one of its species. When it is difficult to find a partner other operations are also possible.

For reproduction an animal (say mother) looks for a companion as a partner. If a partner exists in a 5×5 neighborhood, one with the most highest energy is selected as a partner (say father). A child will be born by Apply and the energy is distributed equally between mother and child. Father and its energy are without any changes. If it fails to find a partner, the animal generates a child by mutation, insertion or deletion. The partner selection procedure is expected to have effect of selection. The mechanism to allow only animals with high energy to have children also has the effect.

Results As the most stable period, Fig. 6 shows an example of a population transition of animals and plants. We can observe peaks of each population by turns. Every peak of population of herbivores follows a peak of carnivores. This shows a food chain caused during the period through evolution. The Lotka-Volterra equation[4] expressing population model for predator-prey in an ecosystem with a food chain, tells that when the abscissa is populations of carnivores

[1] Dying an animal fertilizes the land by working of many living things, and then some plants will grow. We simplified this in the model.

and the ordinate is that of carnivores, its graph exhibits cyclic dynamics. This was observed in our model as shown in Fig. 6.

6 Discussions

We defined a crossover operation for n-BDD based on Bryant's Apply operation and confirmed its effectiveness by experiments. It achieved more than 40% high fitness than the conventional method in the saturated generations in a simple competition problem. Moreover, in the food chain model with crossover operation, we have seen a food chain relation. Table 3 shows the average of the steps continued until either herbivores or carnivores are wiped out. They are the average of 100 trials. We find that a food chain relation in the system using the method with crossover continued a longer time than in the conventional method. We expect animals to evolve faster because crossover is performed with a partner in high energy. This expectation should be confirmed in the future analysis.

Table 3. Average steps continued until either carnivores or herbivores are wiped out.

	without crossover	with crossover
steps	2230 ±3627	10371 ±10860

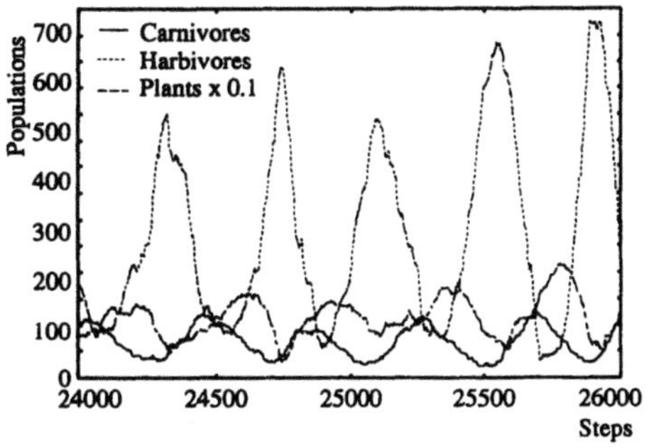

Fig. 5. Populations of carnivores, herbivores and plants (a stable period).

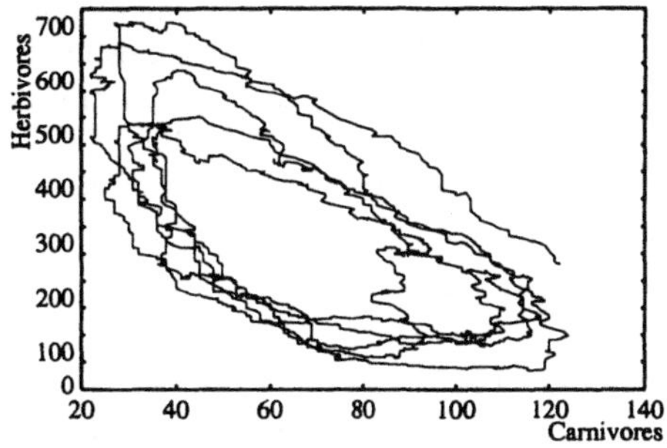

Fig. 6. Populations of carnivores and herbivores (from 24000 to 26000 steps).

References

1. S. B. Akers, 1978, *Binary Decision Diagrams*, IEEE Trans. Comput., pp. 509-516.
2. R. E. Bryant, 1986, *Graph-Based Algorithms for Boolean Function Manipulation*, pp. 677-691, IEEE Trans. Comput.
3. L. J. Fogel, A. J. Owens, and M. J. Walsh, 1967, *Artificial Intelligence Through Simulated Evolution*, John Wiley & Sons.
4. R. Haberman, 1977, *Mathematical Models : Population Dynamics*, PRENTICE.
5. K. Moriwaki, N. Inuzuka, M. Yamada, H. Seki, and H. itoh, 1997, *A Genetic Method for Evolutionary Agents in a Competitive Environment*, Soft Computing in Engineering Design and Manufacturing, pp.153-162, Springer.
6. T. Takashina and S. Watanabe, 1994, *Study of self adaptive behavior in quasi-ecosystem*, Proc. 3rd Parallel Computing Workshop, Kawasaki Japan.

On Self-Reproduction and Evolvability*

Tim Taylor

Institute of Perception, Action and Behaviour,
Division of Informatics, University of Edinburgh,
5 Forrest Hill, Edinburgh EH1 2QL, Scotland.
tim.taylor@ed.ac.uk
http://www.dai.ed.ac.uk/daidb/homes/timt/

Abstract. Von Neumann's architecture for self-reproducing, evolvable machines is described. From this starting point, a number of issues relating to self-reproduction and evolution are discussed. A summary is given of various arguments which have been put forward regarding the superiority of genetic reproduction over self-inspection methods. It is argued that programs in artificial life platforms such as Tierra reproduce genetically rather than by self-inspection (as has previously been claimed). However, the distinction is blurred because significant parts of the reproduction process in Tierran programs are implicitly encoded in the Tierran operating system. The desirable features of a structure suitable for acting as a seed for an open-ended evolutionary process are discussed. It is found that the properties of such a structure are somewhat different to those of programs in Tierra-like platforms. These analyses suggest ways in which the evolvability of individuals in artificial life platforms may be improved, and also point to a number of open questions.

1 Introduction

In the late 1940s and early 1950s, John von Neumann devoted considerable time to the question of how complicated machines could evolve from simple machines.[1] Specifically, he wished to develop a formal description of a system that could support self-reproducing machines which were robust in the sense that they could withstand some types of mutation and pass these mutations on to their offspring. Such machines could therefore participate in a process of evolution.

Inspired by Alan Turing's earlier work on universal computing machines [3], von Neumann devised an architecture which could fulfil these requirements. The machine he envisaged was composed of three subcomponents [2]:

1. A general *constructive* machine, **A**, which could read a description $\phi(\mathbf{X})$ of another machine, **X**, and build an instance of **X** from this description:

$$\mathbf{A} + \phi(\mathbf{X}) \rightsquigarrow \mathbf{X} \tag{1}$$

* This paper is an abbreviated version of certain sections of [1].

[1] Von Neumann had difficulties in defining precisely what the term 'complicated' meant. He said "I am not thinking about how involved the object is, but how involved its purposive operations are. In this sense, an object is of the highest degree of complexity if it can do very difficult and involved things." [2].

(where + indicates a single machine composed of the components to the left and right suitably arranged, and $\rightsquigarrow$ indicates a process of construction.)

2. A general *copying* machine, **B**, which could copy the instruction tape:

$$\mathbf{B} + \phi(\mathbf{X}) \rightsquigarrow \phi(\mathbf{X}) \tag{2}$$

3. A *control* machine, **C**, which, when combined with **A** and **B**, would first activate **B**, then **A**, then link **X** to $\phi(\mathbf{X})$ and cut them loose from $(\mathbf{A} + \mathbf{B} + \mathbf{C})$:

$$\mathbf{A} + \mathbf{B} + \mathbf{C} + \phi(\mathbf{X}) \rightsquigarrow \mathbf{X} + \phi(\mathbf{X}) \tag{3}$$

Now, if we choose **X** to be $(\mathbf{A} + \mathbf{B} + \mathbf{C})$, then the end result is:

$$\mathbf{A} + \mathbf{B} + \mathbf{C} + \phi(\mathbf{A} + \mathbf{B} + \mathbf{C}) \rightsquigarrow \mathbf{A} + \mathbf{B} + \mathbf{C} + \phi(\mathbf{A} + \mathbf{B} + \mathbf{C}) \tag{4}$$

This complete machine plus tape, $[\mathbf{A} + \mathbf{B} + \mathbf{C} + \phi(\mathbf{A} + \mathbf{B} + \mathbf{C})]$, is therefore self-reproducing. From the point of view of the evolvability of this architecture, the crucial feature is that we can add the description of an arbitrary additional automaton **D** to the input tape. This gives us:

$$\mathbf{A} + \mathbf{B} + \mathbf{C} + \phi(\mathbf{A} + \mathbf{B} + \mathbf{C} + \mathbf{D}) \rightsquigarrow \mathbf{A} + \mathbf{B} + \mathbf{C} + \mathbf{D} + \phi(\mathbf{A} + \mathbf{B} + \mathbf{C} + \mathbf{D}) \tag{5}$$

Furthermore, notice that if the input tape $\phi(\mathbf{A} + \mathbf{B} + \mathbf{C} + \mathbf{D})$ is mutated in such a way that the description of automaton **D** is changed, but that of **A**, **B** and **C** are unaffected (that is, the mutated tape is $\phi(\mathbf{A} + \mathbf{B} + \mathbf{C} + \mathbf{D}')$), then the result of the construction will be:

$$\begin{aligned} &\mathbf{A} + \mathbf{B} + \mathbf{C} + \mathbf{D} + \phi(\mathbf{A} + \mathbf{B} + \mathbf{C} + \mathbf{D}) \overset{\text{mutation}}{\rightsquigarrow} \\ &\mathbf{A} + \mathbf{B} + \mathbf{C} + \mathbf{D}' + \phi(\mathbf{A} + \mathbf{B} + \mathbf{C} + \mathbf{D}') \end{aligned} \tag{6}$$

The reproductive capability of the architecture is therefore robust to some mutations (specifically, those mutations which only affect the description of **D**), so the machines are able to evolve. Von Neumann pointed out that the action of the general copying automaton, **B**, was the decisive step which gave his architecture the capacity for evolving machines of increased complexity, because **B** is able to copy the description of any machine, no matter how complicated [2] (p.121). This ability is clearly demonstrated in Reaction 5 above.

2 General Issues of Reproduction

The major focus of this paper is self-reproduction in the specific context of evolution. However, before continuing it is useful to briefly consider some more general issues relating to reproduction.

When looking at any sort of reproduction, it is helpful to look at the process by which reproduction is accomplished from a number of different perspectives. Two important ones are:

1. The degree to which the reproduction process is *explicitly encoded* on the configuration being reproduced, rather than being *implicit in the physical laws* of the world.
2. The number of different configurations that exist, connected by mutational pathways, that are capable of reproducing their specific form (i.e. the distinction between *limited hereditary* reproducers and *indefinite hereditary* reproducers). From the point of view of an individual reproducer, this can be expressed in terms of the proportion of all possible mutations it may experience that will result in the production of distinct, yet viable, reproducers.

There are a number of points to note about these distinctions. First it should be said that (2), in contrast to (1), does not properly relate to individual reproducers *per se*, but rather to lineages of reproducers. It is therefore not relevant when considering self-reproduction in and of itself, but is an important factor when considering the evolutionary potential of a class of reproducers.

Secondly, the two distinctions are generally independent of each other, although the more explicitly encoded the reproduction algorithm is, the less likely, in general, it is to be an indefinite hereditary reproducer (because of the increased chance of mutations disrupting the copying process; see Section 3.3).

3 Self-Reproduction and Open-Ended Evolution

I now wish to return to issues of reproduction in the specific context of evolution. In this section I will concentrate on a number of these issues in turn.

3.1 Trivial versus Non-Trivial Reproduction

Notice that in much of the recent artificial life work with self-reproduction (e.g. [4]), the distinction between trivial and non-trivial self-reproduction is perceived to be a distinction on the implicit-explicit axis.[2] However, from an evolutionary point of view, the limited-indefinite heredity axis is clearly the most relevant. Indeed, this is exactly what von Neumann himself says: "One of the difficulties in defining what one means by self-reproduction is that certain organizations, such as growing crystals, are self-reproductive by any naive definition of self-reproduction, yet nobody is willing to award them the distinction of being self-reproductive. A way around this difficulty is to say that self-reproduction includes the ability to undergo inheritable mutations as well as the ability to make another organism like the original" [2].

Barry McMullin has presented an enlightening discussion on the history of the confusion over von Neumann's work, which he refers to as the 'von Neumann Myth' (see, for example, Section 4.2.7 in [5]). One result of this confusion has

[2] From this point of view, an example of trivial reproduction in a cellular automata space would be where the state of a single cell is reproduced in neighbouring cells purely due to the CA's transition rules.

been that the majority of subsequent research concerning this issue of trivial self-reproduction has concentrated on the implicit-explicit distinction, rather than the limited-indefinite heredity distinction.

Von Neumann's work on self-reproduction concerned the question of how machines might be able to *evolve increased complication* in order to perform increasingly complex tasks. This is why his design for a self-reproducing machine had to be capable of universal construction, and why it was designed in such a way that it could withstand some kinds of mutations.

3.2 Genetic Reproduction versus Self-Inspection

Von Neumann's architecture was designed specifically to allow for a possible increase in complexity and efficiency of machines by evolution. However, even if we accept that his design *is* a solution to this problem, it is by no means the only conceivable solution, as von Neumann himself was well aware. In particular, he also discussed the possibility of a machine which built a copy of itself by actively inspecting its parts, without the need for this design information to be duplicated on a tape (i.e. without a 'genetic' description). Indeed, systems which reproduce by self-inspection have been designed by Laing, e.g. [6], and by Ibáñez and colleagues [7]. From the point of view of designing artificial life systems, we would like to know which of the possible architectures we should employ (according to factors such as their relative simplicity, efficiency, etc.).

Although he certainly did not *prove* that reproduction by self-inspection could not support open-ended evolution, von Neumann did suggest a number of reasons why his genetic architecture would be a more powerful and more general design for this purpose. First of all, as mentioned in Section 1, he noted that the essential feature which allowed his automata to overcome the otherwise seemingly valid rule that machines are necessarily superior (in size and in organization) to their output, was that they contained a general copying automaton **B**, which was capable of copying any linear tape [2] (p.121). Although **B** is of fixed, finite size, it is able to copy a tape of any size. Now, this action of copying a tape is essentially reproduction by self-inspection, but this is generally a straightforward task for a linear tape. The major problems arise when trying to copy a two- or three-dimensional structure by the same method, for example in specifying the precise spatial relationships between parts, and in unfolding multidimensional forms. Von Neumann also pointed out that self-inspection requires that we have a representation which is 'quasi-quiescent' in the sense that it can be read (for the purposes of copying and possibly for interpretation) without being essentially disturbed. With a separate genetic description, we only require that this description is quasi-quiescent, but copying by self-inspection would require that the whole structure to be copied would have this quasi-quiescent property. In general, however, most machines would not have this property, nor would we want to restrict ourselves to only considering those machines which did. In conclusion, von Neumann says: "To sum up, the reason to operate with 'descriptions' ... instead of the 'originals' ... is that the former are quasi-quiescent (i.e. unchanging, not in an absolute sense, but for the purposes of the exploration that has to be undertaken), while the latter are live and reactive. In the situation in which we are finding ourselves here, the importance of descriptions is that they replace the varying and reactive originals by quiescent and (temporarily) unchanging semantic equivalents and thus permit copying. Copying, as we have seen above, is the decisive step which renders self-reproduction (or, more generally, reproduction without degeneration in size or level of organization) possible" [2] (p.122–123).

From a biological perspective, Waddington has made the same point. While discussing possible reasons for the universal adoption of genetic architectures for self-reproduction by biological life, he suggested that the issue "is presumably related to the problem [of] how to combine a store which is unreactive enough to be reliable, with something which interacts with the environment sufficiently actively to be 'interesting'" [8] (p.118).

McMullin has pointed out that von Neumann's genetic architecture also effectively decouples the geometry of the variational space of the reproducers (i.e. the space of the genetic tapes) from the peculiarities of the environment in which they exist (i.e. the space of the phenotype) [5] (pp.191–193). In addition, recall from Section 1 that the architecture will accept any tape of the general form $\phi(\mathbf{A}+\mathbf{B}+\mathbf{C}+\mathbf{D})$. Assuming the the description of **D** on the tape can be separated from the description of **A**, **B** and **C**,[3] this design guarantees that mutations which affect the part of the tape describing automaton **D** will not interfere with the reproductive capacity of the machine. Machines which reproduce by self-inspection would generally not have this localisation property. This being the case, we would not always be able to say that there was a particular section of the machine which could be disrupted by mutation without interfering with the machine's ability to reproduce.

It is interesting to ask whether programs in artificial life platforms such as Tierra [10] reproduce according to von Neumann's genetic architecture or rather by self-inspection. As the arguments of the previous paragraphs suggest that a marked difference exists in the evolutionary potential of these two methods, it is an important question, but it has not received much discussion in the literature. McMullin argued that these programs are reproducing by self-inspection [5] (p.200). Ibáñez and colleagues appear to agree [7] (p.574). In contrast, I would like to suggest that they can sensibly be analysed in terms of von Neumann's genetic architecture. I do not have adequate room to argue the case fully here,

[3] This is not an inherent property of the architecture *per se*, but von Neumann's analysis of evolvability did assume a 'compositional' structure in the language of the tape descriptions (see Section 1). His cellular automata model [2], and Pesavento's recent implementation of a very similar design [9], are existence proofs that it is possible to build a self-reproducing automaton with such a compositional genetic structure. Interestingly, however, despite his design for the cellular automata model, von Neumann also argued that "it is better not to use a description of the pieces and how they fit together, but rather a description of the consecutive steps to be used in building the automaton" [2]. In other words, the information should be in the form of a developmental 'recipe' rather than a 'blueprint'. Further discussion of this topic can be found in Chapter 7 of [1].

so the following paragraphs are included merely for the purpose of stimulating discussion on the subject.

Before I begin, I would like to make a couple of general points, which might help to reorient the reader to my perspective. Firstly, I believe that the notion of a phenotype fundamentally involves *interaction* with the environment (and that this is the essential distinction between the notions of phenotype and genotype—the latter being an informational concept). When I talk about phenotypes in the following, therefore, and specifically when I talk about the automata **A**, **B**, **C** and **D**, I am interested in the role these phenotypic structures play—their function—rather than the details of implementation or of how that function is achieved. Secondly, note that the terminology commonly used to describe reproducers in Tierra-like systems is somewhat different to that used for von Neumann's work. Because of the similarity between Tierra-like operating systems and those of standard digital computers, the actions of Tierran reproducers are often referred to as computations rather than constructions, even when a reproducer is in the process of building a new copy of itself. However, this process of reproduction is, of course, central to the Tierra approach, and I believe that this procedure of building a copy of a program in a different part of memory is, in all the relevant details, a process of construction in just the same way as construction processes in von Neumann's cellular automata model. In the following, also remember that von Neumann's general constructing automaton **A** is the machinery which *interprets* the tape to produce a new machine (phenotype), and the general copying automaton **B** copies the tape uninterpreted.

At first sight it might seem that there is no separate genetic description of the program in a Tierra-like system. The picture is complicated by the fact that the machinery which interprets the program (i.e. automaton **A**) does not reside in the same part of the computer in which the program itself is stored. The state information for this machinery—a program's 'virtual CPU' (i.e. the instruction pointer, stacks, registers, etc.)—is generally represented in an independent area of memory to the program's instructions. Furthermore, the actual 'interpreting machinery' of the virtual CPU is encoded in the global operating system provided by the platform, and is in this sense implicit in the program's environment. Additionally, the control automaton **C**, which controls when the instructions in the program get executed, is also implicit in the part of the operating system which governs mechanisms such as how a program's instruction pointer is updated after the execution of each instruction. All that is left to be explicitly encoded by the program, therefore, is the copying automaton **B**, and potentially any other arbitrary automaton **D**.

Now, the instructions which make up the program exist in an unreactive state in the system's random-access memory. It is only when the control automaton **C** transfers instructions to the interpreting automaton **A** that they become 'active'. Looked at in this way, we can see that it is the *behaviour* of the program (including looping, jumping, etc.) that is the result of automaton **A** interpreting the unreactive genetic description. This behaviour is therefore the equivalent to the constructed machine (and the actions it performs—i.e., the phenotype) in von Neumann's design, and the string of instructions residing in the random-access memory (which is normally referred to as the program) is the tape or genetic description of this phenotype. It is perhaps easier to see the distinction if one considers a parallel program, with multiple processes (with different state information) using the same program listing.

I therefore suggest that a self-reproducing program in a Tierra-like system is consistent with von Neumann's architecture. However, as automata **A** and **C** are largely implicit in the environment in which the programs reside (the only explicit representation being the state information in a program's virtual CPU), and are certainly not encoded by the individual programs, we can see that a 'program', in the sense of a string of instructions in the system's random-access memory, corresponds to the tape $\phi(\mathbf{B}+\mathbf{D})$ in von Neumann's scheme.

The situation is complicated not only because the interpretation machinery resides partly implicitly in the environment, and partly in a different area of memory, but also for (at least) one further reason. I am claiming that the string of instructions comprising the 'program' in random-access memory should be viewed as the genetic tape in a von Neumann style self-reproduction architecture. Now, von Neumann pointed out that the process of copying the tape in his automaton was essentially itself a process of self-inspection. In this sense, Tierran programs do reproduce by self-inspection. However, the overall mechanism for reproduction, including the implicit encodings of the interpretation and control automata, fits in with von Neumann's architecture, in which the copying of the tape by self-inspection is an integral feature. The major consequence of this is that programs in Tierra-like systems should, all else being equal, have similar evolutionary potential to von Neumann's self-reproducing automata, because extra instructions can be added to the end of the 'tape' and subject to mutations. As long as the mutations do not affect that part of the tape which encodes the self-reproduction algorithm, they will be inherited without disrupting the capacity of the program to reproduce.

3.3 Implicit versus Explicit Encoding

The preceding arguments have led us to consider the question of implicit versus explicit encoding of automata. However, rather than the general question that has been the subject of much debate relating to trivial versus non-trivial reproduction, here we are interested in rather more specific questions relating to von Neumann's architecture. Now, as we are interested in the evolution of these self-reproducing machines, and as the inheritable information of each machine (i.e. the part which gets passed on from parent to offspring) is contained on the tape ϕ, I will assume that the tape must be explicitly represented in some fashion, otherwise there would be nothing which could evolve. We can now ask *which parts* of the $[\mathbf{A}+\mathbf{B}+\mathbf{C}+\mathbf{D}]$ architecture are explicitly encoded on the tape ϕ, and which are implicit in the environment. Of course, even the behaviour of those parts which are represented on the tape will still to some extent be encoded in the 'laws of physics' of the environment, but I think the analysis is nevertheless worthwhile.

Considering von Neumann's architecture for a self-reproducing automaton, it is clear that all four subcomponents, **A**, **B**, **C** and **D**, are very explicitly encoded on the tape $\phi(\mathbf{A}+\mathbf{B}+\mathbf{C}+\mathbf{D})$; the environment in which the automaton exists implicitly encodes only very low-level actions in the form of the local transition rules of individual cells. The analysis of self-reproducing programs in Tierra-like systems above suggests that in these systems, **B** and **D** are explicitly encoded on the tape $\phi(\mathbf{B}+\mathbf{D})$, but **A** and **C** are implicitly encoded in the environment (the operating system). Notice that with this design the 'genetic code' which maps the genotype $\phi(\mathbf{B}+\mathbf{D})$ to the phenotype $[\mathbf{B}+\mathbf{D}]$ cannot itself evolve, because the interpretation automaton **A** is not encoded on the tape.

It is interesting to speculate on what information we might desire to be explicitly encoded on a structure which would be suitable for acting as a robust initial seed for an open-ended evolutionary process. I will refer to such a structure as 'proto-DNA'. Now, we would like our proto-DNA to be an indefinite hereditary replicator if it is to be such a seed. In other words, it should be able to exist in an unlimited number of configurations which retain the ability to reproduce. If the copying process is encoded on the tape itself, then mutations have the potential to disrupt its ability to be reproduced. It would therefore seem desirable that the copying automaton **B** of our proto-DNA be largely implicitly encoded in the environment. Note that this would not necessarily prevent a more complicated, and possibly more reliable, explicit copying process $\mathbf{B}'$—genetically encoded as $\phi(\mathbf{B}')$—later evolving from (but still based upon) the simpler implicit process, as indeed seems to have happened during biological evolution.

If the copying procedure for our proto-DNA is implicitly encoded in the environment, however, any configuration of proto-DNA would, all else being equal, be able to reproduce as well as any other. In other words, there would be no basis for preferentially selecting some configurations over others, and therefore no basis for an evolutionary process. Specific configurations of proto-DNA must therefore have some specific properties that are selectively significant. Models of the origin of life commonly presume that these simple phenotypic properties were things such as increased stability of the molecule, simple control of the local environment, catalytic activity, etc. (e.g. [11], [12], [13]).

At the initial stages of an evolutionary process, however, we would not expect there to be mechanisms for explicitly decoding the proto-DNA; in other words, the interpretation machinery **A** is implicit. This means that particular configurations of proto-DNA should have some specific phenotypic properties (such as the ability to act as catalysts) which can be determined directly from their structure rather than having to be explicitly decoded from the genotype. We could therefore regard the proto-DNA as merely $\phi(\mathbf{D})$, meaning that particular configurations have particular phenotypes associated with them, which are (a) not related to the process of self-reproduction *per se*, and (b) do not require to be decoded by an explicit interpretation automaton **A**.[4] Regarding the kinds of

[4] I am also assuming here that the domain of interaction of these phenotypes is within the environment shared by the evolving population (i.e. the phenotypes can act upon other biotic or abiotic components in the environment). This is in contrast

simple phenotypes that we might wish to be available to our proto-DNA, the possibilities seem endless. Graham Cairns-Smith observes: "It is almost too easy to imagine possible uses for phenotype structures—because the specification for an effective phenotype is so sloppy. A phenotype has to make life easier or less dangerous for the genes that (in part) brought it into existence. There are no rules laid down as to how this should be done" [12] (p.106). If more complicated phenotypes are to arise later on in the evolutionary process, however, we require that the proto-DNA at least has the potential for explicit interpretation machinery $\mathbf{A}'$ and control machinery $\mathbf{C}'$ to become associated with it. This would involve some form of specific reaction to subsections of information in the proto-DNA, but more work is needed to fully identify how this potential for explicit interpretation might be assured.

4 Discussion

It has been argued that programs in artificial life platforms such as Tierra conform to von Neumann's genetic architecture for self-reproducing machines. Specifically, the listing of such a program corresponds to the tape $\phi(\mathbf{B}+\mathbf{D})$. In contrast, the analysis of the desirable properties of proto-DNA (i.e. a class of object capable of acting as a seed for an open-ended evolutionary process) suggests that such an object would correspond to the tape $\phi(\mathbf{D})$ only. The fact that the copying process $\mathbf{B}$ is explicitly encoded in Tierran programs means that it is susceptible to disruption by mutations and perturbations from the environment. This is why the interactions between programs in platforms such as this have to be restricted—in Tierra, for example, direct interaction between programs is restricted to the reading by one program of the instructions of its neighbours.

It is likely that many of the more interesting ecological and evolutionary phenomena in the biosphere arise because organisms are able to interact in much richer ways. Most importantly, biological organisms are embedded in a material world, and therefore represent useful resources of matter and energy for potential use by other organisms. Without such an unrestricted range of allowable interactions, and without the system being grounded on a material basis (i.e. where organisms are composed of structural units which are, at their lowest level, conserved, and which are in limited supply), it is doubtful whether any selection pressure can exist for organisms to evolve properties such as self-maintenance. Also, it is only with such a material grounding that ecological phenomena such as food webs and trophic levels can be realised. If we wish to allow artificial life models the capacity to evolve in these ways, we must model a material environment, and allow the individual organisms much more freedom in their interactions. However, if we were to model organisms as self-reproduction algorithms (as in Tierra) in such an environment, they would prove very brittle, because the explicitly-encoded copying process $\mathbf{B}$ would be very easily disrupted.

to models such as genetic algorithms, where the replicators do not directly interact with other replicators and selection is determined by an extrinsic fitness function (thereby limiting the potential for open-ended evolution).

The advantage of the proposed proto-DNA structure, where **B** is implicit in the environment, is that we can relax the restrictions on interactions between self-reproducing machines while maintaining the robustness of the individuals. Such a system would therefore have the potential for evolving much richer ecologies and symbioses. This approach brings with it many new issues which will have to be addressed, such as what range of phenotypes should be available to the proto-DNA structures, and how to ensure that they have the potential for evolving more explicit interpretation machinery to encode more complicated phenotypes. The approach therefore involves a shift of focus away from the process of self-reproduction *per se*, towards questions relating to phenotypes, the environment, and interactions between organisms. Progress in these directions might give us a better insight into the essential requirements for biological evolution, and might also allow us to build artificial life models with improved evolutionary potential.

Acknowledgements

Thanks to my PhD supervisor, John Hallam, and to the anonymous reviewers, for their useful comments and criticisms. Thanks also to EPSRC for providing financial support.

References

1. Taylor, T.: From Artificial Evolution to Artificial Life. PhD thesis, Division of Informatics, University of Edinburgh (1999)
2. von Neumann, J.: The Theory of Self-Reproducing Automata. University of Illinois Press, Urbana, Ill. (1966)
3. Turing, A.M.: On computable numbers, with an application to the entscheidungsproblem. Proc. London Mathematical Soc., Series 2 **42** (1936) 230–265
4. Langton, C.G.: Self-reproduction in cellular automata. Physica D (1984) 135–144
5. McMullin, B.: Artificial Knowledge: An Evolutionary Approach. PhD thesis, Department of Computer Science, University College Dublin (1992)
6. Laing, R.: Automaton models of reproduction by self-inspection. Journal of Theoretical Biology **66** (1977) 437–456
7. Ibáñez, J., Anabitarte, D., Azpeitia, I., Barrera, O., Barrutieta, A., Blanco, H., Echarte, F.: Self-inspection based reproduction in cellular automata. In Morán, F., Moreno, A., Merelo, J., Chacón, P., eds.: Third European Conference on Artificial Life, Springer (1995) 564–576
8. Waddington, C.: Paradigm for an evolutionary process. In Waddington, C., ed.: Towards a Theoretical Biology. Volume 2. Edinburgh Univ. Press (1969) 106–128
9. Pesavento, U.: An implementation of von Neumann's self-reproducing machine. Artificial Life **2** (1995) 337–354
10. Ray, T.S.: An approach to the synthesis of life. In Langton, C., Taylor, C., Farmer, J., Rasmussen, S., eds.: Artificial Life II. Addison-Wesley, Redwood City, CA (1991) 371–408
11. Eigen, M., Schuster, P.: The hypercycle: A principle of natural self-organization. Die Naturwissenschaften **64** (1977) 541–565
12. Cairns-Smith, A.: Seven Clues to the Origin of Life. Cambridge Univ. Press (1985)
13. Szathmáry, E., Demeter, L.: Group selection of early replicators and the origin of life. Journal of Theoretical Biology **128** (1987) 463–486

Some Techniques for the Measurement of Complexity in Tierra

Russell K. Standish

High Performance Computing Support Unit
University of New South Wales
Sydney, 2052
Australia
R.Standish@unsw.edu.au
http://parallel.hpc.unsw.edu.au

Abstract. Recently, Adami and coworkers have been able to measure the information content of digital organisms living in their *Avida* artificial life system. They show that over time, the organisms behave like Maxwell's demon, accreting information (or complexity) as they evolve. In *Avida* the organisms don't interact with each other, merely reproduce at a particular rate (their fitness), and attempt to evaluate an externally given arithmetic function in order win bonus fitness points. Measuring the information content of a digital organism is essentially a process of counting the number of genotypes that give rise to the same phenotype.

Whilst Avidan organisms have a particularly simple phenotype, Tierran organisms interact with each other, giving rise to an ecology of phenotypes. In this paper, I discuss techniques for comparing pairs of Tierran organisms to determine if they are phenotypically equivalent. I then discuss a method for computing an estimate of the number of phenotypically equivalent genotypes that is more accurate than the "hot site" estimate used by Adami's group. Finally, I report on an experimental analysis of a Tierra run.

1 Introduction

The issue of what happens to complexity in an evolving system is of great interest. In natural (biological) evolution, the naive view is that life started simple, and evolved ever more complex life forms over time, leading to that pinnacle of complexity, *homo sapiens*. The end points of that process are of course fixed. In the beginning, life must be simple. In our present era, there must exist intelligent organisms (namely us) pondering over the mystery of how we came to be. So the *anthropic principle* fixes the present day as having complex lifeforms. There is nothing within the *Modern Synthesis* of Darwinism that implies a steady interpolation between these two end points. In fact it is even plausible that more complex organisms than us existed in the past, but have since vanished into obscurity. However, examinations of the fossil record over the Phanerozoic (the last 550 million years of the Earth's history) indicate almost no growth in complexity by a number of different measures over that period, apart from an initial large jump at the Cambrian explosion.[1]

The interesting thing is to ask what one might see if looking at another evolutionary system apart from the one in which we evolved. Would we see any growth in complexity at all? Since we don't have an extra terrestrial biology to observe (a few Martian meteorites aside), the only other systems available are Artificial Life systems evolving within a digital computer such as Tierra or Avida. The Avida group has reported measuring the information content (complexity) of individual avidan organisms[2], or rather a lower bound of the organism's complexity. Their results are that this lower bound increases over time for the maximally fit organism, thus showing information accumulating as time progresses. One important critique of this work, however, is that organisms do not interact directly with each other, and in order to prevent evolution stagnating, an externally imposed task (eg computing a logical operation) is added to the system. Organisms are given "fitness points" depending on how well they perform this task. This heavily weights the system in favour for accruing information.

By contrast, in the Tierra system, the organisms interact with each other, providing a rich array of possible (intrinsic) tasks for the organisms to exploit. Since this is an evolving ecology with no externally imposed task, the above critique does not apply. However, the downside is that determining whether two genotypes are phenotypically equivalent is considerably more complex. In some work a couple of years ago[3], I studied the phenotypic properties of Tierran organisms to build up a picture of the genotype to phenotype landscape. A Tierran organism's phenotype can be characterised by a couple of numbers for each possible pairwise interaction in the ecology. Multiway interactions are ignored in this study, as experience has shown them to be relatively rare.

2 Complexity of a Digital Organism

The information content of a string is given by the difference between the maximal Shannon entropy of that string (i.e. considering the string to be random, or devoid of information), and the entropy given by assuming that the string codes for some phenotype p:[2, 4]

$$I(g) = H(g) - H(g|p) = \ell - \log_{32} N \tag{1}$$

where ℓ is the length of the genotype (in instructions), and N is the number of genotypes that give rise to the same phenotype p. The base, 32, refers to the number of instructions in the Tierra instruction set. If $N \approx 32^\ell$ (ie a completely random sequence), then $I(g) = 0$. Similarly, if $N = 1$ (there is only one genetic sequence encoding a genotype, or no redundancy), then $I(g) = \ell$.

The most obvious way to compute N is to search all 32^ℓ genotypes for equivalent phenotypes. However, this is an enormous number of strings to check, and computationally infeasible. Adami recognised this problem, and took the approach of counting the number of volatile sites v (sites that vary amongst phenotypic equivalents), and approximating $N \approx 32^v$. In one sense this is an overestimate of N, so they argue that this gives a lower bound to the information $I(g)$. In another sense, however, it is not strictly a lower bound. If it turns out that fixing one of the volatile sites to a particular value allows one of the fixed sites to vary without altering the phenotype, then this would be not be counted in the N. so what we have is really an overestimate of an underestimate.

The same criticism applies to this work. We can estimate the above mentioned estimate fairly accurately, more precisely we can find the size of the neutral network[5, 6, 7] connected by one-site neutral mutations to g. However, the possibility remains that there are other neutral networks of g that aren't connected by single site mutations to g. Probably the most efficient way of finding these is by using a genetic algorithm to explore genotype space, i.e. run Tierra for a long time to see what it discovers! The way we use this in our experiment is to keep a list of neutrally equivalent organisms that Tierra discovers. As we explore the neutral network connected to g, we eliminate items from the list that we come across. The remaining names on the list can then be used as seeds to start the process again.

In this work, we use two different techniques to measure N. The first is a Monte Carlo random sampling technique to estimate the proportion of the 32^v strings found by varying the volatile sites. The second technique, which we use in conjunction with the Monte Carlo approach mentioned above, is to walk the neutral net. The Monte Carlo technique works well when the density of neutral variants is fairly high, whereas the latter technique is best on sparse networks. A decision on which technique to use for which site is based on estimated densities of neutral variants.

3 Establishing Phenotypic Equivalence

Equation (4) of [3] presents the dynamical equations of two species of Tierran organisms interacting. The precise form of the dynamics is not important here, however the phenotype of the organism can be characterised by its interactions will all other possible Tierran phenotypes. Since it is impossible to have the complete set of all possible Tierran organisms, those organisms generated during a run of Tierra are used. Since Tierran organisms coevolve, the most important organisms should be contemporaneous with the test organism. The following characteristics are saved for each pair of organisms:

1. The outcome of the tournament. This may be one of the following:
   infertile The test organism never calls the divide instruction, or does not produce any recognisable progeny (essentially still born)
   once The organism produces progeny once, but then never repeats the act.
   repeat The organism continuous reproduces the same progeny. For this purpose we ignore what is produced first time around, as this will be swamped by number latter progeny.
   nonrepeat The organism continuously reproduces, but the progeny is either different each time, or the CPU is in a different state each time the divide instruction is called - thus can't be guaranteed to reproduce ad infinitum.
2. The name of the progeny organism. This is usually identical to the parent, but may another type in the case of symbiosis or parasitism.

3. The number of timesteps it takes to reach the first divide instruction (σ_{ij}), and the time it takes between successive divide steps after that (τ_{ij}).
4. The number of template matching operations made to the opposing organism prior to the first divide (μ_{ij}) and between successive divides (ν_{ij}).

Two organisms are neutrally equivalent if they have identical characteristics against all Tierran organisms. Once all organisms are paired with each other, we can produce a list of phenotypically unique organisms, which provides a smaller test list to pit trial mutants against. We may also eliminate some noninteractive pairings prior to simulation by trying to see if potential template matches could happen between organisms. This still produces a fairly large list of test organisms, so it is still computationally expensive. The high degree of parallelism in this problem allows it to be attacked in reasonable time on a parallel supercomputer.

A further refinement may be possible by producing an archetypal list, perhaps by ignoring the (μ, ν, τ and σ) parameters. The idea being that the archetypes contain a representative organism from each niche of the ecology, and ignoring minor differences such as reproductive rate. This would coarsen the approximation a little, but will probably give an acceptable result. At present this idea has not been tested.

4 Interim Results

Due to the time constraints of producing this paper, the analysis of a reasonable length Tierra run has not been completed. At the time of writing, a moderately large data set of 1660 organisms was generated from a 24 hour Tierra run. Tierra produces most of its diversity during the earliest stage of its running, so it becomes significantly more expensive to produce larger data sets. This data set was halved by removing every second organism, and then a phenotypic analysis was carried out. This set reduced to 103 distinct phenotypes, which formed the test list used for carrying out the complexity analysis. Each of these 103 organisms were then tested for phenotypic equivalence against their single site nearest neighbours. The number of sites on which no mutation resulted in a phenotypically equivalent organism ("nonvolatile sites") is plotted against the time of speciation in figure 1.

References

1. McShea, D.W.: Metazoan complexity and evolution: Is there a trend? Evolution **50** (1996) 477–492
2. Adami, C.: Introduction to Artificial Life. Springer, New York (1998)
3. Standish, R.K.: Embryology in Tierra: A study of a genotype to phenotype map. Complexity International **4** (1997) http://www.csu.edu.au/ci.
4. Layzer, D.: Growth and order in the universe. In Weber, B., Depew, D., Smith, J., eds.: Entropy, Information and Evolution. MIT Press, Cambridge, Mass. (1988) 23–39

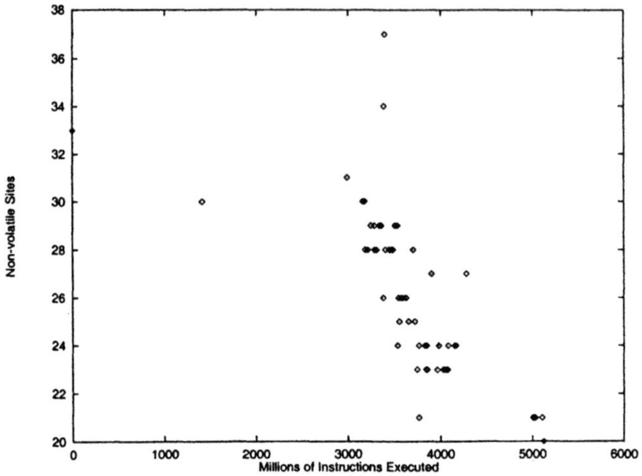

Fig. 1. Non-volatile site count (complexity estimate) for the set of phenotypic Tierran species, as a function of speciation time.

5. Kauffman, S.: At Home in the Universe: The Search for Laws of Complexity. Oxford UP, New York (1995)
6. Reidys, C., Kopp, S., Schuster, P.: Evolutionary optimization of biopolymers and sequence structure maps. In Langton, C., Shimohara, K., eds.: Artificial Life V, MIT Press (1997) 379
7. Schuster, P.: Landscapes and molecular evolution. Physica D **107** (1997) 351–365

A Generic Neutral Model for Quantitative Comparison of Genotypic Evolutionary Activity

Andreas Rechtsteiner[1] and Mark A. Bedau[2]

[1] Systems Science Ph.D. Program, Portland State University
P.O. Box 751, Portland OR 97202, USA
andreas@sysc.pdx.edu, http://www.sysc.pdx.edu/Alife/aendy.html
[2] Reed College, 3203 SE Woodstock Blvd, Portland OR 97202, USA
mab@reed.edu, http://www.reed.edu/~mab

Abstract. We use a new general-purpose model of neutral evolution of genotypes to make quantitative comparisons of diversity and adaptive evolutionary activity as a function of mutation rate among two versions of Packard's Bugs model and their neutral shadows. Comparing diversity and evolutionary activity of all these models across the mutation rate spectrum shows that the generic neutral model may have broad applicability in discovering quantitative laws involving adaptive evolutionary activity in different evolving systems.

1 The Need for a Generic Neutral Model

Adaptive evolution is thought to produce much of the order and functionality evident in complex systems [9, 7, 5], but it is often difficult to distinguish adaptive change from other evolutionary phenomena such as random genetic drift and architectural necessity [6, 10], and some even question whether adaptations can be objectively identified at all [6]. Recent progress on identifying adaptive evolutionary phenomena includes Bedau and Packard's statistical methods for measuring adaptive evolutionary activity. Here, we apply these methods to the problem of determining how adaptive evolutionary activity depends on mutation rate. Our ultimate aim is to develop methods for objectively identifying and measuring adaptive evolutionary activity in all evolutionary systems, both natural and artificial, so that we can seek universal laws of adaptive evolutionary activity. Here, we test such a method, applied at the level of whole genotypes, in the context of two simple models of sensory-motor evolution. But the same method can be applied at other levels of analysis in other evolutionary systems. The ultimate significance of this work comes from the possibility of quantitatively comparing evolutionary adaptations across all evolving systems.

The centerpiece of our method is Bedau and Packard's evolutionary activity statistics. (We also measure system diversity, D, which is simply the number of different genotypes present in a system at a given time.) Detailed definitions and motivations for evolutionary activity statistics are readily available elsewhere [2, 1, 3, 4, 13]. Evolutionary activity statistics aim to identify evolutionary innovations (here, new genotypes) that persist and continue to play a significant role in a system because of their adaptive value. These statistics fall into two broad classes: those reflecting evolutionary activity's *extent* and those reflecting its *intensity*. Here, we attend only to the *extent* of evolutionary activity, measuring it with *mean cumulative evolutionary activity* (sometimes simply called "activity"), $\bar{A}$, which in the present context is operationalized as the mean age of the genotypes present in a system at a given time. So, in this context, the higher a system's mean activity, the higher the mean age of the system's genotypes, which means the greater the continual adaptive success of those genotypes. Intuitively, the *extent* of evolutionary activity concerns how much adaptive structure is present in a system; one might refer to this as the continual adaptive success of the system's components. By contrast, the *intensity* of evolutionary activity reflects the rate at which new adaptive structure is being created. The extent and intensity of adaptive evolutionary activity are independent. For example, if a population of highly adaptive genotypes persist indefinitely without changing and no new genotypes invade the system, then the extent of evolutionary activity is positive and perhaps grow over time, but the intensity of evolutionary activity falls to nil.

To ensure that evolutionary activity statistics reflect the adaptive success of the genotypes and not non-adaptive evolutionary forces like chance and necessity, one must use non-adaptive evolutionary systems called "neutral models" as null hypotheses. That is, one must screen off the effects of non-adaptive evolutionary forces like chance by comparing the evolutionary dynamics observed in target evolutionary systems with those observed in analogous neutral models. Such neutral analogues have heretofore been constructed by crafting systems that "shadow" the target system in all relevant respects *except* that a shadow genotype's presence or concentration or longevity cannot be due to the genotype's adaptive significance [3, 4]. Since neutral shadows are tailored to target systems, they sharply show the target systems' deviation from the no-adaptation null hypothesis. But neutral shadows have significant drawbacks, too, for studying a new target system involves constructing and studying a new neutral shadow, and it is vexing to make meaningful quantitative comparisons among different tailor-made neutral shadows.

The obvious way to solve these problems is to create a *generic* neutral model—one neutral model that can approximate many different neutral shadows. The immediate goal of this paper is to define such a generic neutral model and test its usefulness for quantifying evolutionary activity across different systems. We pursue this goal by comparing the generic neutral model with two simple evolutionary systems and their neutral shadows.

2 The Models

Packard's Line and Block Models. The Bugs simulation is a series of models originated by Norman Packard [11, 2] and subsequently modified in various ways. Packard's simulation is designed to be a very simple model of the evolution of sensory-motor strategies. It consists of agents sensing the resources in their local environment, moving as a function of what they sense, ingesting the resources they find, and reproducing or dying as a function of their internal resource levels. The model's spatial structure is a grid of sites with periodic boundary conditions, i.e., a toroidal lattice. The resource distributions studied here take two forms:

Line: a thin continuous strip, one cell in width, that wraps entirely around the world, with all the other sites in the world entirely devoid of resources;
Block: a square block of resources, 15 cells on a side, with all other sites in the world entirely devoid of resources.

In each case, resources are immediately replenished at a site whenever they are consumed. The agents constantly extract resources and expend them by living and reproducing. Agents ingest all of the resources (if any) found at their current location and store them internally. Agents expend resources at each time step by "paying" (constant) "existence taxes" and "movement taxes" (variable, proportional to distance moved). If an agent's internal resource supply drops to zero, it dies and disappears from the world.

Each agent moves each time step as dictated by its genetically encoded sensory-motor map: a table of behavior rules of the form IF (environment j sensed) THEN (do behavior k). An agent receives sensory information about the resources (but not the other agents) in the von Neumann neighborhood of five sites centered on its present location in the lattice. In the Line world, there are exactly 4 detectable local environments: those detected by agents either on the resource strip, immediately to the strip's left or right, or anywhere else. In the Block world, there are exactly 14 detectable local environments: those detected by agents either just on one of the four edges, or just off one of the four edges, or in one of the four corners, or in the middle of the block, or anywhere else. Each behavior k is a jump vector between one and fifteen sites in any one of the eight compass directions. Thus, an agent's genotype, i.e., its sensory-motor map, is just a lookup table of sensory-motor rules. But the space in which adaptation occurs is fairly large, consisting of $120^4 \approx 10^8$ and $120^{14} \approx 10^{29}$ distinct possible genotypes in the Line and Block worlds, respectively. An agent reproduces (asexually, without recombination) if its resource reservoir exceeds a certain threshold. The parent produces one child, which starts life with half of its parent's resource supply. The child inherits its parent's sensory-motor map, except that mutations may replace the behaviors linked to some sensory states with randomly chosen behaviors.

A given simulation starts with randomly distributed agents containing randomly chosen sensory-motor strategies. The model contains no *a priori* fitness function, as Packard [11] has emphasized. Agents with maladaptive strategies tend to find few resources and thus to die, taking their sensory-motor genes with them; by contrast, agents with adaptive strategies tend to find sufficient resources to reproduce, spreading their sensory-motor strategies (with some mutations) through the population. In the Line world, the main adaptations that occur are learning how to stay on the resource strip and learning to do so in step with the other bugs on the strip (i.e., meshing with the "flock" of other bugs on the line). Another, secondary adaptation is optimizing the jump size on the strip (smaller jumps are better): Furthermore, there is a slight adaptive advantage to learning how to get back on the strip when immediately adjacent to it. In the Block world, as in the Line world, one adaptive pressure is to "flock" along with the other bugs, so as to minimize the changes of getting bumped into the resource desert. But the basic adaptive strategy needed to survive on a resource block is to move in a given direction and speed when in the middle of the block until you detect the edge of the block, and then to jump back in the opposite direction into the middle of the block. A subpopulation of bugs following the same strategy will form a flow that rolls across the block and reflects off its edge. Since all bugs in the Line world must flock in step not to bump each other off the resource strip, the fitness landscape in the Line world has relatively narrow peaks. By contrast, bugs that reflect different distances off the edge can co-exist, and different subpopulations can form along different edges in a resource block, in effect filling different niches, so the Block's fitness landscape allows for more diversity and thus contains relatively broad peaks.

Neutral Shadows for Packard's Line and Block Models. The crucial property of a "neutral shadow" of a model with emergent genotype dynamics is that the shadow system's evolutionary dynamics are like its target model *except* that a shadow genotype's activity cannot be due to its adaptive significance—for it has no adaptive significance. The neutral shadow a Packard Bugs model consists of a population of nominal "bugs" with nominal "genotypes." A shadow "bug" has no spatial location and it cannot ingest resources or interact with other "bugs." All it ever does is come into existence, perhaps reproduce (perhaps often), and go out of existence; its only properties are its genotype and the times of its birth, reproductions (if any), and death.

Each neutral shadow run corresponds to a specific Line or Block model run. The neutral shadow's birth and death events and mutation rate are directly copied from those in the target run. When some creature is born in the target run, a shadow parent is chosen at random (with equal probability) from the shadow population to reproduce. The new shadow child inherits its parent's genotype unless a mutation gives the child a new genotype. When some creature dies in the Line or Block run a "creature" is chosen at random from the shadow population and killed. Thus, all selection in the neutral shadow is random.

The evolutionary dynamics in a neutral shadow is a neutral diffusion in genotype space. Genotypes arise and go extinct, and their concentrations change over time, but the genotype dynamics are at best weakly linked to adaptation through the birth and death rates determined by adaptation in the Line or Block model. When adaptive genotypes are evolving in a Bugs run, one would expect their genotype activity levels to be significantly higher than those in the corresponding neutral shadows. For, although individuals in the Bugs model and its neutral shadow have the same birth, reproduction, and death rates, and their mutation rates are the same, in the Bugs model natural selection can cull poorly adapted genotypes and preserve well adapted genotypes while the selective force in the neutral shadow is entirely random. The difference between the activity levels in the Bugs and its neutral shadow shows how much much natural selection affects Bugs activity.

A Generic Neutral Model. The generic model of neutral genotype evolution consists of a population of individuals that reproduce and die in a fixed genotype space. The genotype space is defined by some number of loci at each of which some number of alleles are segregating. Parameters that need to be specified in the generic neutral model are N, the size of the population of individuals, r, the reproduction rate (the number of individuals that die and reproduce per time step), l, the number of loci, a, the number of possible alleles per locus, m_l, the probability that the allele at a given locus will be mutated when an individual is born. (The probability that an offspring will have mutation somewhere in its genome, i.e., the mutation rate per individual is $m_i = 1 - (1 - m_l)^l$.) The parameters together determine the model's generic behavior. The genotype space is a hypercube of dimension l and size a^l (number of possible genotypes), with each location in this space corresponding to a given genotype. The current state of the model is given by the distribution of N individuals in genotype space. The population wanders through the space stochastically, spreading and clustering at random.

The individuals in the initial population are assigned genotypes at random. Time is discrete, and moves forward each time step by iterating the following two-step algorithm: (1) r individuals (selected at random, with replacement) each produce a child that is genetically identical to itself except for mutations. Mutant alleles are chosen at random from the set of possible alleles. (2) r individuals (selected at random, without replacement) die and are removed from the population and are replaced by the r children produced at step (1).

This neutral model does not closely correspond to those systems in which some of the generic model parameters are variable. E.g., in Tierra [12] the number of loci is variable; indeed, it is not clear exactly what to count as a locus in Tierra. In addition, population size and reproduction rate vary over time in many artificial models of evolution, such as Echo [7] and Packard's Bugs models. Still, the neutral model might apply reasonably well to these systems if the relevant neutral model parameters are set to plausible corresponding values. For the comparisons here we set N and r to the mean observed value of the corresponding parameter in the Bugs model. One of the goals of this study is to assess the usefulness of the generic neutral model under such an approximation.

3 Experimental Methods

We observed the behavior of the Line and Block models, the Line and Block neutral shadows, and the Line and Block generic neutral models across the mutation rate spectrum (varied on a log scale). All Packard model simulations were started with a randomly initialized populations of 500 individuals. We did at least 10 runs at every mutation rate in each model. We varied the simulation time between 5×10^5 and 5×10^7 depending on the mutation rate. The transient time is longer at lower mutation rates, and we aimed to have simulations that were long enough to minimize variance due simply to simulation time. The parameters for the generic neutral model were set to correspond to the Line model (four loci and 120 alleles per locus) and the Block model (fourteen loci and 120 alleles per locus). We determined average population size N and reproduction rate r from each Line or Block model run and set corresponding parameter values in generic model runs. We dumped 5000 data points in each simulation, so the time interval between data dumps varied with run length. We made sure data dumping frequencies did not influence our results. In the generic neutral model evolutionary activity was calculated continuously, so the exact activity value could be recorded in each data dump. But in the Bugs and neutral shadow models genotype data was only *sampled* at each data dump. So, for simplicity, we assumed that a genotype that first appeared at a certain time arose immediately after the previous data dump. This procedure loses all information about short-lived genotypes that arose and went extinct between data dumps, and it significantly overestimates the age of short-lived genotypes that appear in only a few data dumps. This bias was minimized by using shorter simulation time and data dumping intervals for high mutation rates.

4 Results and Discussion

The top of Figure 1 shows the time average of the diversity normalized by the time average of the population size, $\frac{\bar{D}}{\bar{N}}$, for the Line and Block models, and for shadow and generic neutral models of each, as a function of mutation rate, m_i. At very low mutation rates diversity levels off. This diversity floor is an artifact of the finite population size. Indefinitely larger populations would indefinitely lower this floor. These diversity data at mutation rates above the diversity floor show three salient results: First, the dependence of normalized diversity on mutation rate for all four neutral models is strikingly similar. This provides evidence that the generic neutral model is a good approximation of the neutral shadows. Second, the normalized diversity of both Line and Block models is strikingly lower than that for the neutral models—the expected result of natural selection in the Line and Block models versus the random selection in the neutral models. Third, the normalized diversity of the Line model is strikingly lower than that for the Block model. This can be explained by the different fitness landscapes in the Block and Line world. Each fitness landscape has several local peaks, but the Block peaks are broader than the Line peaks and broader peaks support a more diverse population.

The bottom of Figure 1 shows the time average of activity, $\bar{A}$, as a function of mutation rate, m_i, for the Line and Block models and for shadow and generic neutral models of each. The activity ceiling in the Line and Block models at lower mutation rates is partly an artifact of simulation time; a genotype's observed age cannot exceed simulation time and the longest runs we did lasted 5×10^7 time steps. Longer simulations would raise the observed activity values at very low mutation rates. Where unaffected by the activity ceiling artifact, the time-averaged activity data show three significant results.

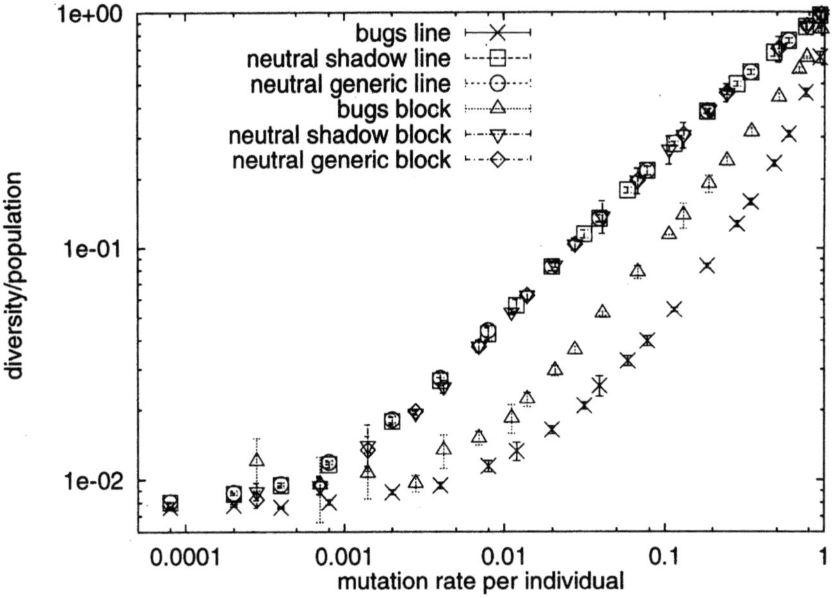

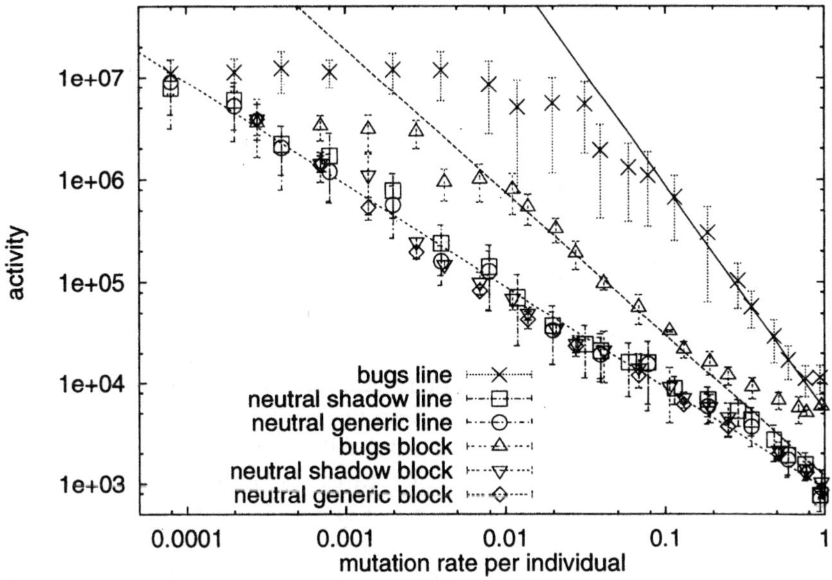

Fig. 1. Above: Time average of diversity (normalized by dividing by time average of average population size), $\frac{D}{N}$, as a function of mutation rate per individual, m_i, for the Line and Block models, their neutral shadows, and generic neutral models for them. Below: Time average of evolutionary activity, $\bar{A}$, as a function of mutation rate per individual, m_i, for the Line and Block models, their neutral shadows, and generic neutral model for them, along with lines showing the power laws which approximately fit these data (at certain mutation rates). In both graphs, error bars indicate standard deviations of time averages computed from at least ten runs per mutation rate.

First, activity in all the neutral models is quite similar, especially considering that we did not account for differences in population size and reproduction rate in the Line and Block worlds. Furthermore, activity's dependence on mutation rate approximately fits a power law of the form $\bar{A} \propto m_i^{-\alpha}$, where $\alpha_{\text{neutral}} = 1.0 \pm 0.1$ (all error bounds are standard deviations). This power law can be explained by adapting an argument of Kimura [8]. Kimura found that the average time it takes for a new mutant gene to reach fixation during neutral evolution, on the assumption that genes get substituted one after another and not at the same time, can be described by two time scales. The first time scale—the time it takes on average for a neutral mutant to spread throughout the population—is proportional basically to the population size. The second time scale—the time it takes on average for such a mutant gene to occur in the population—is proportional to the number of mutations that occur, which is proportional to m_i^{-1}. Kimura's assumption that genes are substituted one after another corresponds in our neutral models to the assumption that genotypes are substituted one after another, and this assumption holds when the mutation rate is not too high. So, for low mutation rates Kimura's discussion applies equally well to genotype substitution in our neutral models, with Kimura's new mutant gene corresponding to our new mutant genotype. For low enough mutation rates only the second time scale is relevant; the other time scale is basically constant (because population is basically constant) and becomes negligible. So, evolutionary activity, which corresponds to the mean lifetime of genotypes, will be proportional to the second time scale, i.e., to m_i^{-1}.

The second significant result in the activity data is that the magnitude of activity is significantly higher in both Line and Block than in the neutral models, as is the slope of activity's dependence on mutation rate. Lower neutral model activity can be explained by the lack of adaptation in the neutral models. Random selection in neutral models does not preferentially preserve well-adapted genotypes but natural selection in the Line and Block models does, so Line and Block activity is expected to exceed neutral model activities.

The third significant activity result is that the magnitude of the activity is significantly higher in the Line model than in the Block model, as is the slope of activity's dependence on mutation rate. Although the data are somewhat ambiguous, at relatively higher mutation rates the Line and Block activity's dependence on mutation rate might approximately fit power laws with $\alpha_{\text{Line}} = 2.2 \pm 0.2$ and $\alpha_{\text{Block}} = 1.4 \pm 0.1$. Our activity data at lower mutation rates are clearly affected by the activity ceiling artifact; we expect that the slope of the Line and Block activity will fall to -1 at low enough mutation rates, but the activity ceiling prevents us from resolving this here. We expect that the differences in the Line and Block fitness landscapes can explain the differences in activity magnitude and slope in the Line and Block worlds. Evidently, fewer successful genotypes residing on narrow fitness peaks persist significantly longer than more genotypes (therefore lower average population size per genotype) residing on broader peaks. In addition, on narrow fitness peaks more than on broad peaks, the extent of adaptation seems to fall faster as mutation rate rises.

Current work includes resolving the activity's different dependence on mutation rate in the Line and Block models. This project is especially engaging because we have precise quantitative results needing explanation.

5 Conclusions

Comparing diversity and evolutionary activity in Packards's Line and Block worlds and in their shadow and generic neutral models, yields a variety of precise and interesting quantitative results. Some are due to the fact that the neutral models are devoid of adaptation, others to the different fitness landscapes in the Line and Block worlds. Both illustrate the power and promise of using neutral models to quantify adaptive evolution in different evolutionary systems.

The absence of adaptation in the neutral models explains their relatively high diversity and low activity, compared with the Line and Block worlds, as well as the lower slope with which activity depends on mutation rate in the neutral models. The differences we observed have three related implications: they confirm the appropriateness of using activity statistics to measure the extent of adaptive structure in an evolving system, thereby confirming the appropriateness of using neutral model activity to measure the amount of activity that can be attributed to adaptation as opposed to other evolutionary forces like chance and necessity, and thereby confirming the importance of the generic neutral model. The generic neutral model closely approximates the behavior of different special-purpose neutral shadows; dependence of diversity and activity on mutation rate is remarkably similar in all of them. Having one simple generic neutral model removes the need to make a new neutral shadow for each evolutionary model and allows us to study the general properties of neutral models in one fell swoop. To be sure, the generic neutral model has so far passed only a preliminary test, and its final confirmation can come only if it successfully approximates many more neutral shadows. Conducting these further tests is a subject of current work, as is discerning the generic neutral model's typical behavior. The results discussed above reveal an important difference between the evolutionary dynamics of neutral and adaptive evolutionary systems, and the generic neutral model is an excellent tool for discerning and understanding this difference.

The magnitude and slope of activity's dependence on mutation rate also reveals the fundamental difference between the Line and Block fitness landscapes. This, too, confirms the appropriateness of measuring the extent of adaptive structure in a system with activity statistics, thereby underscoring the utility of neutral models in general and the generic neutral model in particular. For example, we can measure how much the observed activity reflects the force of adaptive evolution by appropriately normalizing observed activity against the corresponding neutral model, e.g., by dividing observed activity by the corresponding neutral activity. If we call this fraction by which observed activity exceeds neutral (non-adaptive) activity a system's *excess* evolutionary activity [13], then a power-law dependence of activity on mutation rate (over part of the mutation rate spectrum) implies simple power-law dependence of *excess* activity on mutation rate in the Line and Block models. Such excess activity power laws raise intriguing questions: Does excess activity show similar power laws in a broad class of evolutionary models? If so, what exactly explains the magnitude and exponent in the laws? Answering any of these questions would significantly advance our quantitative understanding of adaptive evolution. And a key tool for facilitating these precise quantitative comparisons is the generic neutral model.

Acknowledgements. For helpful discussion, thanks to Norman Packard, Tad Shannon, and Emile Snyder. Thanks to Emile Snyder for software consultation. Thanks to the ECAL-99 reviewers for comments on the manuscript. MAB also thanks the Santa Fe Institute, for support and hospitality while some of this work was completed.

References

1. Bedau, Mark A. 1995. Three illustrations of artificial life's working hypothesis. In W. Banzhaf and F. Eeckman, (Eds.), *Evolution and biocomputation: computational models of evolution*, (pp. 53–68). Berlin: Springer.
2. Bedau, Mark A., and Norman H. Packard. 1992. Measurement of evolutionary activity, teleology, and life. In C. Langton, C. Taylor, D. Farmer, S. Rasmussen, (Eds.), *Artificial life II* (pp. 431–461). Redwood City CA: Addison-Wesley.
3. Bedau, Mark A., Emile Snyder, C. Titus Brown, Norman Packard. 1997. A comparison of evolutionary activity in artificial evolving systems and in the biosphere. In P. Husbands and I. Harvey, (Eds.), *Fourth European conference on artificial life* (pp. 125–134). Cambridge: MIT Press.
4. Bedau, Mark A., Emile Snyder, Norman Packard, 1998. A classification of long-term evolutionary dynamics. In C. Adami, R. Belew, H. Kitano, and C. Taylor, (Eds.), *Artificial life VI* (pp. 228–237). Cambridge: MIT Press.
5. Dawkins, R. 1976. *The selfish gene.* New York: Oxford University Press.
6. Gould, S. J., Lewontin, R. C. 1979. The spandrals of San Marco and the Panglossian paradigm: a critique of the adaptationist programme. *Proceedings of the Royal Society B* 205, 581–598.
7. Holland, John H. 1992. *Adaptation in natural and artificial systems*, 2nd edition. Cambridge: MIT Press.
8. Kimura, Motoo. 1983. *Neutral theory of molecular evolution.* New York: Cambridge University Press.
9. Maynard Smith, J. 1975. *The theory of evolution*, 3rd edition. New York: Penguin.
10. Mayr, E. 1988. *Towards a new philosophy of biology.* Cambridge: Harvard University Press.
11. Packard, Norman. 1989. Intrinsic adaptation in a simple model for evolution. In C. Langton, (Ed.), *Artificial life* (pp. 141–155). Redwood City CA: Addison-Wesley.
12. Ray, T. S. 1992. An approach to the synthesis of life. In C. Langton, C. Taylor, D. Farmer, S. Rasmussen, (Eds.), *Artificial life II* (pp. 371–408). Redwood City CA: Addison-Wesley.
13. Rechtsteiner, Andreas and Mark A. Bedau. 1999. A generic neutral model for measuring excess evolutionary activity of genotypes. In W. Banzhaf, J. Daida, A. E. Eiben, M. H. Garzon, V. Honavar, M. Jakiela, and R. E. Smith, eds., *GECCO-99: Proceedings of the genetic and evolutionary computation conference, July 13-17, 1999, Orlando, Florida USA.* San Francisco, CA: Morgan Kaufmann.

A Comparative Analysis of Biodiversity Measures

Henning Schwöbbermeyer and Jan T. Kim

Max-Planck-Institut für Züchtungsforschung
Carl-von-Linné-Weg 10, 50829 Köln, Germany
kim@mpiz-koeln.mpg.de, schwoebb@mpiz-koeln.mpg.de

Abstract The concept of biodiversity has received rapidly increasing interest in the biosciences during the last decade. Yet, it is unclear and disputed how biodiversity should be characterised and measured. We compared several biodiversity measures by applying them to data retrieved from the LindEvol-GA model of evolution. A series of LindEvol-GA runs with mutation ranges ranging from zero (producing no diversity) to one (producing maximal, but biologically meaningless, diversity) was analyzed with the measures to be compared. At intermediate mutation rates, biologically meaningful diversity can emerge.
We show that biodiversity measures can be classified according to the way in which they respond to these various types of diversity, and we discuss some implications of our observation for the design, choice, and application of biodiversity measures.

1 An approach to biodiversity

During the last ten years, the term "biodiversity" has become widely used in the biosciences [1, 2], including Artificial Life [3] as well as in the general public to refer to collections of biological entities which coexist in an intricately orchestrated, organismic fashion. But despite its widespread usage, there is no satisfactory scientific definition of biodiversity. In fact, quite a diversity of biodiversity definitions and measures have been proposed in the literature [4, 5, 6, 7, 8, 9], and some of them are not compatible.

Clearly, this problem is to some extent due to diverging concepts and objectives, which sometimes are political or economic in nature rather than scientific. However, even in the scientific domain, there are various sources of difficulties. Biodiversity is generally agreed to be a phenomenon occurring on multiple levels of biological organisation [2, 10, 11]. Unfortunately, biodiversity concepts that focus on different levels of biological organisation (e.g. molecular biology, morphology, ecology, evolutionary biology etc.) are sometimes difficult to reconcile.

Even on a given level of biological organisation, quantitative characterisation of biodiversity is not trivial. For example, simply counting species is a widely applied method for the quantification of biodiversity. However, while the number of species in a system is often suitable as an indicator of the system's biodiversity, it is possible to construct systems with equal numbers of species which nonetheless cannot be expected to be equal in biodiversity; e.g. a box containing thousand different insect species would be considered to be less "biodiverse" than a box containing thousand species forming a small ecosystem of plants, microbes and some insects.

In response to this problem, biodiversity measures which take the evolutionary relations between species into account have been developed [7, 8, 12]. One may consider the contribution of distantly related species to biodiversity to be greater than the contribution of closely related species (e.g. [12]), or to define the conservation of evolutionary history to be the goal of biodiversity protection [6]. But according to such concepts, the bulk of biodiversity would reside within prokaryotes, because they diversified in evolution long before eukaryotes appeared. A tropical rain forest, the standard example for high biodiversity, might appear as a "monoculture" of two rather special groups of multicellular eukaryotes, namely animals and plants, from such a perspective.

Thus, both disregarding evolutionary depth as well as naively equating biodiversity with evolutionary depth fail to capture biodiversity adequately. The rainforest example implies that it is something like "the right mix" of degrees of evolutionary relatedness which characterises biodiversity. It therefore seems that biodiversity is a life phenomenon that emerges somewhere between order (close relatedness, complete identity in the limit) and chaos (distant relatedness, total unrelatedness in the limit). We therefore tested several biodiversity measures to see whether they show a maximum at some edge of chaos in an Artificial Life model of evolution.

All Artificial Life models of evolution are imperfect representations of biological evolution. Nonetheless, important aspects of biological evolution can be captured by models of evolution. Specifically, emergence of complex phenomena has been observed in many computer simulations. Therefore, Artificial Life models provide a suitable basis for investigating possible links between biodiversity and such emergent phenomena.

2 LindEvol-GA runs with increasing mutation rates

For the investigation presented here, we used LindEvol-GA [13, 14], a computer model of the evolution of plant growth patterns. Plants in LindEvol-GA grow in a two-dimensional lattice world in which they compete for space and energy. After a vegetation period, a fitness value is assigned to each plant genome based on the amount of energy stored in the plant. Because the plants grow together in one lattice, this fitness value depends on the interactions of a plant with its neighbours. A new generation of genomes is constructed by removing some genomes from the population and creating an equal number of copies of genomes randomly drawn from the surviving part of the population. All genomes in the population are then mutated. The fraction of the population which is removed each generation is specified by a control parameter called the selection rate. The control parameter governing mutation in the runs presented here is the global replacement mutation rate, which is the probability with which the value of a byte in a genome is replaced with a random value in one generation. Insertions and deletions were not used in the runs presented here.

At the start of each time step the effective mutation rate is set to the global mutation rate for all genomes. A plant may multiply or divide its individual effective mutation rate by 2 at the expense of one energy unit. Repeated modifications are possible. They reduce the fitness of the genome, but can increase the chance of accurate replication. Reduction of effective mutation rates can thus be an evolutionarily stable strategy which evolves in some runs with relatively high global mutation rates (see [15] for details).

We performed a series of LindEvol-GA runs in which the global replacement mutation rate rises from 0.0 to 1.0. From 0.0 to 0.4, a replacement mutation rate increment of 0.01 was used, larger increments were applied above 0.4. The selection rate was set to 0.5 in all runs. Mutation rate adaptation was enabled, and all other control parameters were chosen as in [15] as well. Since monoparental reproduction is used in our LindEvol-GA runs, the phylogenetic tree connecting all genomes can be recorded. Every ten time steps, this phylogenetic tree was used to compute the phylogeny-based biodiversity measures described below. The initial phase of each run, in which descendants of more than one of the randomly created genomes of the start population exist so that multiple unconnected phylogenetic trees are present, was excluded from this analysis.

Both extremes of the global mutation rate constitute neutral controls (cf. [16]): With no mutation, all genomes are identical after an initial phase, so the survivors are effectively drawn at random during selection. With maximal mutation, offspring are totally unrelated to their parents, therefore, achieving a high fitness value and surviving selection is again a pure chance event. Only with intermediate mutation rates, new genotypes and phenotypes which inherit information from their predecessors can arise, and thus, information which is biologically meaningful (with respect to the artificial biology of LindEvol) accumulates in the genomes. Assuming that biodiversity is constituted by collections of entities which are different in a biologically meaningful way, one thus expects that biodiversity measures should yield higher values for runs with intermediate mutation rates than for those with no or maximal mutation.

3 Biodiversity measures

Most measures which we tested in our analysis were proposed by WILLIAMS *et al.* [7]. These measures evaluate the topology of a rooted phylogenetic tree; the lengths of the edges in the tree are not taken into account. All measures are based on the following quantities:

- n is the number of terminal nodes in the phylogenetic tree.
- p_j is the number of nodes in the path from the terminal node j to the root node of the phylogenetic tree.
- s_{jk} is the number of nodes which are shared among the paths from node j to the root and from node k to the root.

- u_{jk} is the number of nodes which are on the path from node j to the root but not on the path from node k to the root.

On the basis of these quantities, WILLIAMS *et al.* define the following divergence measures:

$$\begin{aligned} d_{\text{III}}(j,k) &:= s_{jk} \\ d_{\text{IV}}(j,k) &:= u_{jk} + u_{kj} + 1 \\ d_{\text{V}}(j,k) &:= \frac{u_{jk}+u_{kj}+1}{2s_{jk}+u_{jk}+u_{kj}} \\ d_{\text{VI}}(j,k) &:= \frac{u_{jk}+u_{kj}}{s_{jk}} + 1 \end{aligned} \tag{1}$$

WILLIAMS *et al.* introduce four biodiversity measures which are derived from these divergence measures by

$$m_x := n \cdot \text{mean}(d_x) \tag{2}$$

where x is III, IV, V and VI, respectively. Further measures used in [7] are the plain number of species ($m_{\text{I}} := n$), root weight biodiversity, defined as

$$m_{\text{II}} := \sum \frac{1}{p_j} \tag{3}$$

and finally, dispersion diversity, defined as

$$m_{\text{VII}} := n \cdot (\text{mean}(d_{\text{IV}}) - \text{std.dev.}(d_{\text{IV}})) \tag{4}$$

The roman indices for these diversity measures were chosen to match those used in [7]. We determined measures II through VII for these phylogenetic trees.

An alternative approach to measuring biodiversity was proposed by NEE and MAY [6], who suggested to use the length of a phylogenetic tree as a biodiversity measure. The tree length is defined as the sum of the lengths of all edges in the phylogenetic tree. Evidently, this measure differs from those developed by WILLIAMS *et al.* in that edge lengths are taken into account. Edge lengths in the trees retrieved from LindEvol-GA simulations are given in generations.

As a third approach, we used distance distribution complexity (DDC) [13] as a biodiversity measure. DDC is defined as the shannon entropy of the distribution of (discrete or discretized) distance values. In [13], the edit distance between genomes was used for DDC calculation. In this paper, we also calculate DDC on the basis of phylogenetic distances. The phylogenetic distance of two terminal nodes is defined as the sum of the lengths of the edges on the path connecting these nodes. This distance is computed from the trees retrieved from the LindEvol-GA runs. We denote edit distance based DDC by C_{edit} and phylogenetic distance based DDC by C_{phyl}.

Finally, we also combined the concepts of WILLIAMS *et al.* and edit distance into two new measures. One is a combination of mean distance diversity (equation 2) and edit distance:

$$m_e = n \cdot \text{mean(edit distance)} \tag{5}$$

while the other one was constructed from the distance dispersion approach (equation 4) and edit distance:

$$m_{edisp} := n \cdot (\text{mean(edit distance)} - \text{std.dev.(edit distance)}) \qquad (6)$$

4 Results

Fig. 1 shows the time averages of all biodiversity measures which are based on the evaluation of the phylogenetic trees retrieved from the runs. Quite strikingly, none of these measures exhibits any significant response to the global mutation rate[1]. This is true for the measures which operate on tree topology only (measures II to VII) as well as for those which take edge length into account (tree length and C_{phyl}).

The results from the measures which are based on genetic distances (i.e. the edit distance between genomes) are shown in Fig. 2. The number of different genomes is also included as an approximation to the number of species, which is often used to quantify biodiversity. In strong contrast to the graphs in Fig. 1, the quantities shown in Fig. 2 all are pronouncedly correlated to the global mutation rate.

There are two runs (those with global mutation rates of 0.18 and 0.37) which deviate very significantly from the general trends in the graphs shown in Fig. 2. In these runs, mutation rate adaptation evolves. For the global mutation rate 0.18, this effect is strongest. Data from this run are shown in Fig. 3. Shortly after time step 1000, a pronounced transition to mutation rate adaptation is indicated by a sharp drop of the average mutation modification exponent. This transition has marked effects on average fitness, C_{edit} and m_e (left panel of Fig. 3). On the other hand, no sign of this transition is visible in the time series of various phylogeny-based diversity measures (right panel).

Given the expectation that biodiversity measures should respond to randomization, these observations provide a strong indication that biodiversity can be measured much more adequately on the basis of genetic distances (or other pairwise distances) than on the basis of evaluating phylogenetic trees. While it cannot be ruled out that other phylogeny-based measures would exhibit a significant response to the mutation rate, it is clear that none of the measures we tested does so.

In Fig. 2, two modes of response to mutation rate can clearly be distinguished: While the number of different genomes, m_e and m_{edist} rise monotonically with the mutation rate, C_{edit} steeply rises from 0 in the run without mutation to a maximum at low nonzero mutation rates and decays with further growth of the mutation rate, as described before [13]. Thus, mean edit distance (equation 5) turns out to be a traditional complexity measure (cf. [17]), like the number of different genomes. The example of C_{edit} illustrates, however, that genetic distances are also suitable for calculating alternative complexity measures.

[1] The variation seen in mean tree length (Fig. 1g) is mainly correlated to the number of trees that were analyzed, which varies as the initial phase in which multiple independent phylogenies may have different lengths in different runs.

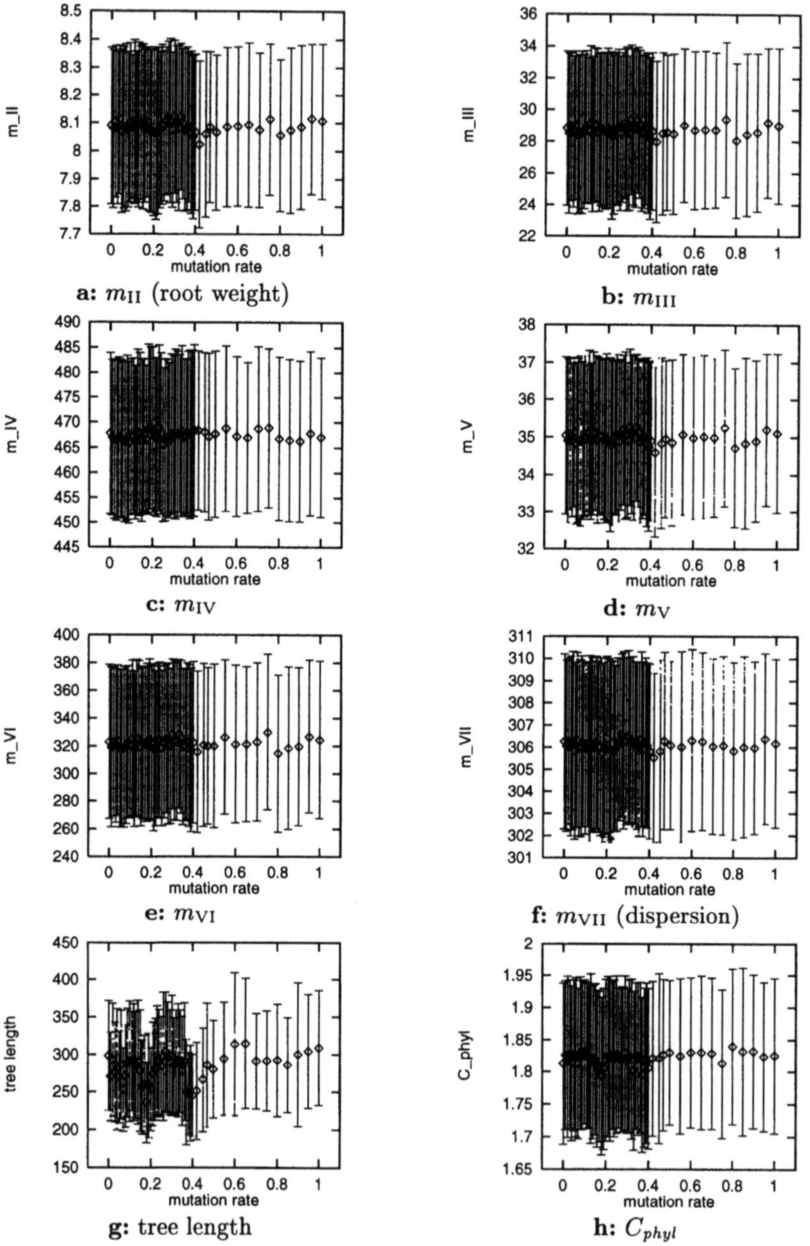

Figure1. Biodiversity measures calculated on the basis of phylogenetic trees. The time averages computed over the entire runs are shown as a function of the global replacement mutation rate setting used in the run. Error bars indicate standard deviations.

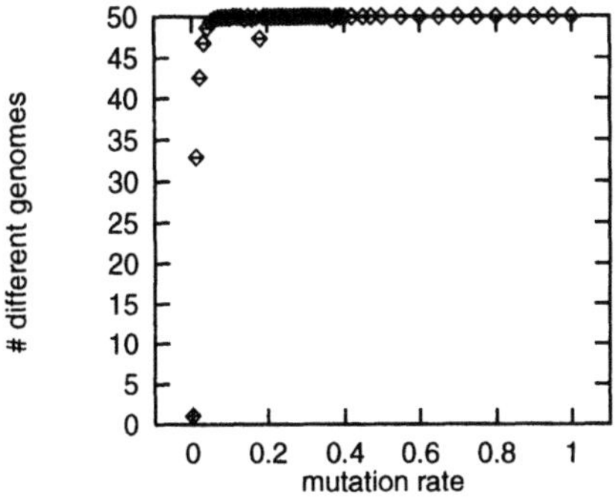

a: number of different genomes

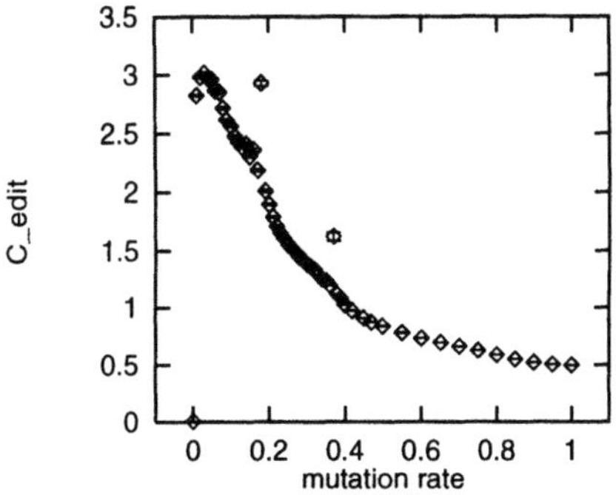

b: C_{edit}

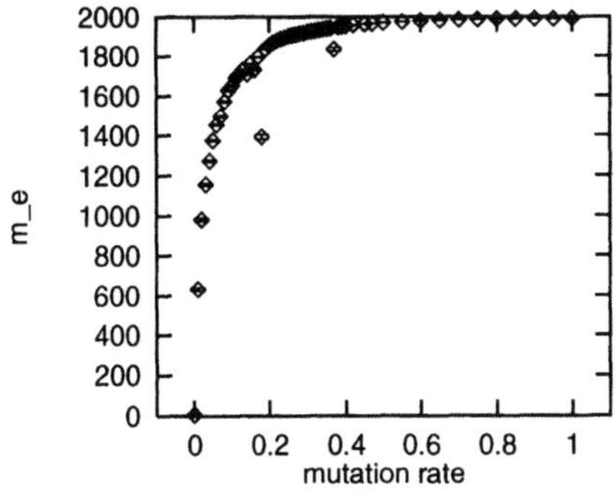

c: m_e

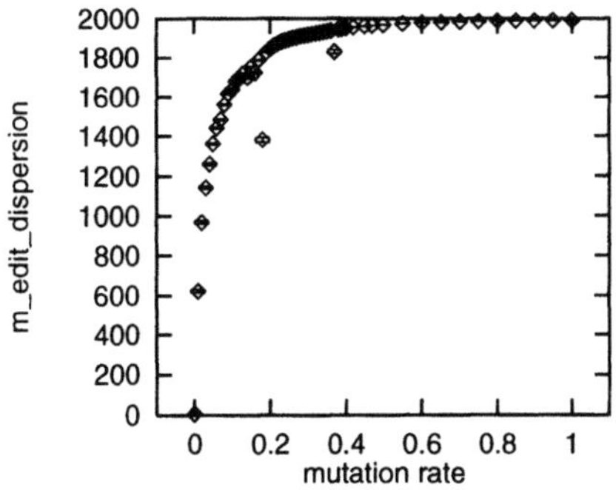

d: m_{edisp}

Figure2. Number of different genomes and biodiversity measures calculated on the basis of genetic distances. Error bars indicate standard deviations (very narrow in these graphs).

5 Conclusions

As a result of our investigation, biodiversity measures can be classified into three types, according to their response to increasing mutation rates:

1. Measures which are insensitive to randomization.
2. Measures which monotonously grow with increasing mutation rates.
3. Measures which yield low values at extremely low and high mutation rates, and elevated values in between.

While the second and the third type are well known in Artificial Life [18, 19] and in complex systems, the first type comes as a surprise and raises some questions. One question is: Why do the evolutionary transitions not leave any traces in the phylogeny-based measures? Our answer to this question is that the bare phylogeny is not an useful source of information about biodiversity, at least in LindEvol-GA. A phylogeny arises even in the control runs, where no diversity is generated (zero mutation control) or no biologically meaningful diversity

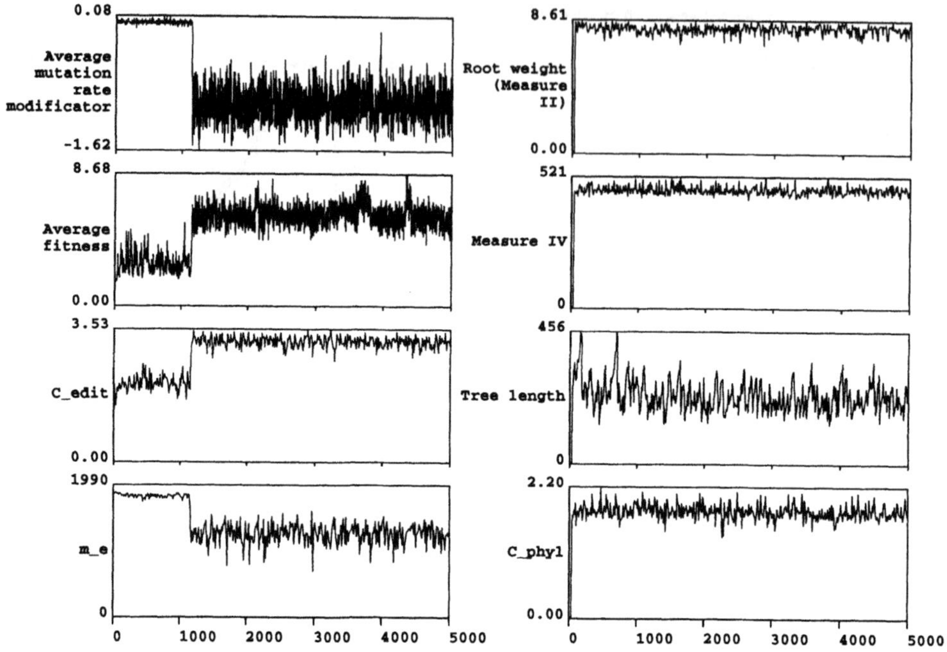

Figure3. Time series of various quantities observed with a global mutation rate of 0.18.

emerges (maximum mutation control). Phylogenies that emerge by evolution which is governed by biological semantics either are not signficiantly different from random phylogenies, or the biodiversity measures we tested fail to respond to such differences. In contrast to this, signatures of the evolution of biologically meaningful information, i.e. of biosemiosis [20], can readily be detected in the genetic distances between the genomes in LindEvol-GA. It may thus turn out that phylogenetic diversity, as quantified by the measures we investigated, is less suitable as an index for biodiversity than genetic diversity.

While the phylogeny based measures have not produced significantly different values for random and biologically meaningful phylogenies from LindEvol-GA, these measures did yield reasonable results in other contexts. Certainly, it is possible that these measures may detect aspects of biodiversity which cannot be modelled by LindEvol-GA. Another explanation, however, is that there is a qualitative difference between the LindEvol-GA data and the other data with which the measures were tested. Usually, phylogenies are reconstructed based on differences in genes or other biologically meaningful properties. These signatures of biological meaning may be partially preserved during reconstruction, and the measures may respond to this information. In contrast to this, we extracted phylogenies from LindEvol-GA independently of differences in genomes or phenotypes, thereby eliminating the information which is relevant for biodiversity. According to this line of reasoning, measurement biodiversity should be based

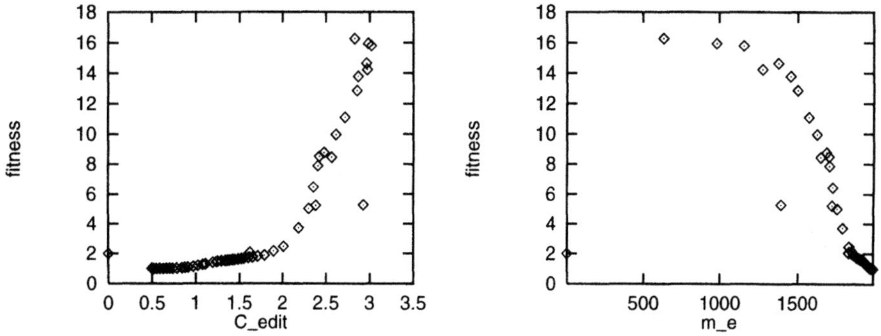

a: C_{edit} vs. mean fitness   **b:** m_e vs. mean fitness

Figure4. Scatter plots showing the correlation of C_{edit} with mean fitnes and the mean number of used genes.

on biological distance data directly; the construction of a phylogeny from these data may be an unnecessary step that obscures the biodiversity signal.

Finally, a question that also remains is whether measures of the second or of the third type are more adequate for characterising biodiversity. We think that the scatter plots shown in Fig. 4 provide a clue that type 3 measures may capture the essential aspects of biodiversity better than type 2 measures. The semantics of LindEvol-GA define that achieving high fitness values is meaningful in that this maximises the chances for successful reproduction. High mean fitness values therefore indicate successful accumulation of information which is meaningful in LindEvol-GA's artificial biology. Fig. 4a shows that there is a pronounced monotonous positive correlation between C_{edit} and mean fitness, while Fig. 4b reveals that the highest values of m_e occur in runs where mean fitness is low. From Fig. 2c, it can be concluded that the low fitness values are due to high mutation rates which prevent the accumulation of significant amounts of biologically meaningful information.

Regarding future directions, we intend to include more diversity measures and to use additional Artificial Life models. We argue that correlations to aspects of biosemiosis should be considered when new biodiversity measures are designed. Assessing such correlations is notoriously difficult for non-artificial systems, but the accumulation of molecular data may open new possibilities. For example, progress in genomics may soon allow to analyse the correlation between numbers of active genes and biodiversity measures, as shown for LindEvol-GA data and C_{edit} in [13]. Molecular data may thus provide a basis for further characterising and understanding biodiversity.

References

[1] Ehrlich, P.R., Wilson, E.O.: Biodiversity studies: Science and policy. Science **253** (1991) 758–762

[2] Wilson, E.O.: Biodiversity: Challenge, science, opportunity. Amer. Zool. **34** (1993) 5–11

[3] Todd, P.M., Miller, G.F.: Biodiversity through sexual selection. In Langton, C.G., Shimohara, K., eds.: Artificial Life V, Cambridge, MA, MIT Press (1996) 289–299

[4] Ehrlich, P.R., Ehrlich, A.H.: The value of biodiversity. Ambio **21** (1992) 219–226

[5] Erwin, T.L.: An evolutionary basis for conservation strategies. Science **253** (1991) 750–752

[6] Nee, S., May, R.M.: Extinction and the loss of evolutionary history. Science **278** (1997) 692–694

[7] Williams, P., Humphries, C., Vane-Wright, R.: Measuring biodiversity taxonomic relatedness for conservation priorities. Australian Systematic Botany **4** (1991) 665–680

[8] Crozier, R., Kusmierski, R.: Genetic distances and the setting of conservation priorities. In Loeschcke, V., Tomiuk, J., Kain, J., eds.: Conservation Genetics, Basel, Switzerland, Birkhäuser Verlag (1994) 1311–1323

[9] Faith, D.P.: Conservation evaluation and phylogenetic diversity. Biological Conservation **61** (1992) 1–10

[10] Soulé, M.E.: Conservation: Tactics for a constant crisis. Science **253** (1991) 744–750

[11] Kim, J.T., Kastner-Maresch, A., Lange, H., Hauhs, M.: Biodiversity in natural and in artificial life. In Wilke, C., Altmeyer, S., Martinetz, T., eds.: Proceedings of the Third German Workshop on Artificial Life, Frankfurt, Germany, Verlag Harri Deutsch (1998) 259–267

[12] Solow, A.R., Polasky, S., Broadus, J.: On the measurement of biological diversity. Journal of Environmental Economics and Management **24** (1993) 60–68

[13] Kim, J.T.: Distance distribution complexity: A measure for the structured diversity in evolving populations. In Langton, C.G., Shimohara, K., eds.: Artificial Life V, Cambridge, MA, MIT Press (1996) 281–288

[14] Kim, J.T.: LindEvol: Models for investigating the interplay between development, ecology and evolution. Bayreuther Forum Ökologie **52** (1997) 17–33

[15] Kim, J.T.: Energy dependent adaptation of mutation rates in computer models of evolution. In Adami, C., Belew, R.K., Kitano, H., Taylor, C., eds.: Artificial Life VI, Cambridge, MA, MIT Press (1998) 248–255

[16] Bedau, M.A., Snyder, E., Packard, N.H.: A classification of long-term evolutionary dynamics. In Adami, C., Belew, R.K., Kitano, H., Taylor, C., eds.: Artificial Life VI, Cambridge, MA, MIT Press (1998) 228–237

[17] Kurths, J., Schwarz, U., Witt, A., Krampe, R., Abel, M.: Measures of complexity in signal analysis. In Katz, R.A., ed.: Chaotic, Fractal, and Nonlinear Signal Processing, New York, Woodbury (1995) 33–54

[18] Gutowitz, H., Langton, C.G.: Mean field theory of theedge of chaos. In Morán, F., Moreno, A., Merelo, J.J., Chacón, P., eds.: Advances in Artificial Life, Berlin Heidelberg, Springer Verlag (1995) 53–64

[19] Langton, C.G.: Life at the edge of chaos. In Langton, C.G., Taylor, C., Farmer, J.D., Rasmussen, S., eds.: Artificial Life II, Redwood City, CA, Addison-Wesley (1992) 41–91

[20] Eder, J., Rembold, H.: Biosemiotics – a paradigm of biology. Biological signalling on the verge of deterministic chaos. Naturwissenschaften **79** (1992) 60–67

Evolution of Differentiated Expression Patterns in Digital Organisms

Charles Ofria*, Christoph Adami†, Travis C. Collier‡, and Grace K. Hsu

Beckman Institute
California Institute of Technology, Pasadena, California 91125, USA
†W.K. Kellogg Radiation Lab 106-38, Caltech, Pasadena, CA 91125
‡Division of Organismic Biology, Ecology, and Evolution,
UCLA, Los Angeles, CA 90095

* To whom correspondence should be addressed. E-Mail: charles@gg.caltech.edu

Abstract. We investigate the evolutionary processes behind the development and optimization of multiple threads of execution in digital organisms using the avida platform, a software package that implements Darwinian evolution on populations of self-replicating computer programs. The system is seeded with a linearly executed ancestor capable only of reproducing its own genome, whereas its underlying language has the capacity for multiple threads of execution (i.e., simultaneous expression of sections of the genome.) We witness the *evolution* to multi-threaded organisms and track the development of distinct expression patterns. Additionally, we examine both the evolvability of multi-threaded organisms and the level of thread differentiation as a function of environmental complexity, and find that differentiation is more pronounced in complex environments.

1 Introduction

Evolution has traditionally been a formidable subject to study due to its gradual pace in the natural world. One successful method uses microscopic organisms with generational times as short as an hour, but even this approach has difficulties; it is still impossible to perform measurements without disturbing the system, and the time-scales to see significant adaptation remain on the order of weeks, at best[1]. Recently, a new tool has become available to study these problems in a computational medium—the use of populations of self-replicating computer programs. These "digital organisms" are limited in speed only by the computers used, with generations in a typical trial taking a few seconds.

Of course, many differences remain between digital and simple biochemical life, and we address one of the critical ones in this paper. In nature, many chemical reactions and genome expressions occur simultaneously, with a system of

[1] Populations of *E.coli* introduced into new environments begin adaptation immediately, with significant results apparent in a few weeks [3].

gene regulation guiding their interactions. However, in digital organisms only one instruction is executed at a time, implying that no two sections of the program can directly interact. Due to this, an obvious extension is to examine the dynamics of adaptation in artificial systems that have the capacity for more than one thread of execution (i.e., an independent CPU with its own instruction pointer, operating on the same genome).

Work in this direction began in 1994 with Thearling and Ray using the program tierra [7]. These experiments were initialized with an ancestor that creates two threads each copying half of its genome, thereby doubling its replication rate. Evolution then produces more threads up to the maximum allowed [11]. In subsequent papers [12, 9] this research extended to organisms whose threads are not performing identical operations. This is done in an enhanced version of the tierra system (**"Network Tierra"** [8]), in which multiple "islands" of digital organisms are processed on real-world machines across the Internet. In these later experiments, the organisms exist in a more complex environment in which they have the option of seeking other islands on which to place their offspring. The ancestor used for these experiments reproduces while searching for better islands using independent threads. Thread differentiation persists only when island-jumping is actively beneficial; that is, when a meaningful element of complexity is present in the environment.

In experiments reported on here, we survey the initial emergence of multiple threads and study their subsequent divergence in function. We then investigate the hypothesis that environmental complexity plays a key role in the pressure for the thread execution patterns to differentiate.

2 Experimental Details

We use the avida platform to examine the development of multi-threading in populations exposed to different environments at distinct levels of complexity, comparing them to each other and to controls that lack the capacity for multiple threads.

2.1 The Avida Platform

Avida is an auto-adaptive genetic system designed for use as a platform in Artificial Life research. The avida system comprises a population of self-reproducing strings of instructions that adapt to both an intrinsic fitness landscape (self-reproduction) and an externally imposed (extrinsic) bonus structure provided by the researcher.

A standard avida organism is a single genome composed of a sequence of instructions that are processed as commands to the CPU of a virtual computer. This genome is loaded into the memory space of the CPU, and the execution of each instruction modifies the state of that CPU. In addition to the memory, a virtual CPU has three integer registers, two integer stacks, an input/output buffer, and an instruction pointer. In standard avida experiments, an organism's genome has one of 28 possible instructions at each line. The virtual CPUs are Turing-complete, and therefore do not explicitly limit the ability for the population to adapt to its computational world. For more details on avida, see [5].

To allow different sections of a program to be executed in parallel, we have implemented three new instructions. A new thread of execution is initiated with `fork-th`. This thread has its own registers, instruction pointer, and a single stack, all initialized to be identical to the spawning thread. The second stack is shared to facilitate communication among threads. Only the new thread will execute the instruction immediately following the `fork-th`; the original will skip it enabling the threads to act and adapt independently. If, for example, a jump instruction is at this location, it may cause the new thread to execute a different section of the program (*segregated differentiation*), whereas a mathematical operation could modify the outcome of subsequent calculations (*overlapping differentiation*). On the other hand, a no-operation instruction at this position allows the threads to progress identically (*non-differentiated*). We have also implemented `kill-th`, an instruction that halts the thread executing it, and `id-th`, which places a unique thread identification number in a register, allowing the organism to conditionally regulate the execution of its genome.

We performed experiments on three environments of differing complexity, with both the extended instruction set that allows multiple expression patterns and the standard instruction set as a control. As individual trials can differ extensively in the course of their evolution, each setup was repeated in two hundred trials to gain statistical significance. The experiments were performed on populations of 3600 digital organisms for 50,000 updates[2]. Mutations are set at a probability of 0.75% for each instruction copied, and a 5% probability for an instruction to be inserted or removed in the genome of a new offspring.

The first environment (I) is the least complex, with no explicit environmental factors to affect the evolution of the organisms; that is, the optimization of replication rate is the only adaptive pressure on the population. The next environment (II), has collections of numbers that the organisms may retrieve and manipulate. We can view the successful computation of any of twelve logical operations that we reward[3] as beneficial metabolic chemical reactions, and speed-up the virtual CPU accordingly; more complex tasks result in larger speed-ups. If the speed increase is more than the time expended to perform the task, the new functionality is selected for. The final environment (III) studied is the most complex, with 80 logic operations rewarded.

---

[2] An update represents the execution of an average of 30 instructions per program in the population. 50,000 updates equates to approximately 9000 generations and takes about 20 hours of execution on a Pentium Pro 200. The data and complete genomes are available at `http://www.krl.caltech.edu/avida/pubs/ecal99/` .

[3] The completion of a logic operation involves the organism drawing one or more 32-bit integers from the environment, computing a bitwise logical function using one or more `nand` instructions, and outputting the result back into the environment.

A record is maintained of the development of the population, including the genomes of the most abundant organisms. For each trial, these dominant genomes are analyzed to produce a time series of thread use and differentiation.

2.2 Differentiation Metrics

The following measures and indicators keep track of the functional differentiation of codes. We keep this initial analysis manageable by setting a maximum of two threads available to run simultaneously. The relaxation of this constraint does lead to the development of more than two threads with characteristically similar interactions.

Thread Distance measures the spatial divergence of the two instruction pointers. This measurement is the average *distance* (in units of instructions) between the execution positions of the individual threads. If this value becomes high relative to the length of the genome, it is an indication that the threads are segregated, executing different portions of the genome at any one time, whereas if it is low, they likely move in lock-step (or sightly offset) with nearly identical executions. Note, however, that if two instruction pointers execute the code offset by a fixed number of instructions, but otherwise identically, the thread distance is an inflated measure of differentiation because the temporal offset does not translate into differing functionality.

Code Differentiation distinguishes execution patterns with differing *behavior.* A count is kept of how often each thread executes each portion of the genome. The code differentiation is the fraction of instructions in the genome for which these counts differ between threads. Thus, this metric is insensistive to the ordering of execution.

Execution Differentiation is a more rigorous measure than code differentiation. It uses the same counters, taking into consideration the *difference* in the number of times the threads execute each instruction. Thus, if one thread executes a line 5 times and the other executes it 4 times, it would not contribute as much towards differentiation as an instruction executed all 9 times by one thread, and not at all by the other. This metric totals these differences in execution counts at each line and then divides the sum by the total number of multi-threaded executions. Thus, if the threads are perfectly synchronized, there is zero execution differentiation, and if only one thread exclusively executes each line, this metric is maximized at one. An execution differentiation of 0.5 indicates that half of the instructions did not have matched executions in each thread.

3 Evolution of Multi-Threaded Organisms

For our initial investigations, we focus on the 200 trials in environment III (the most complex), with the extended instruction set, allowing for multi-threading.

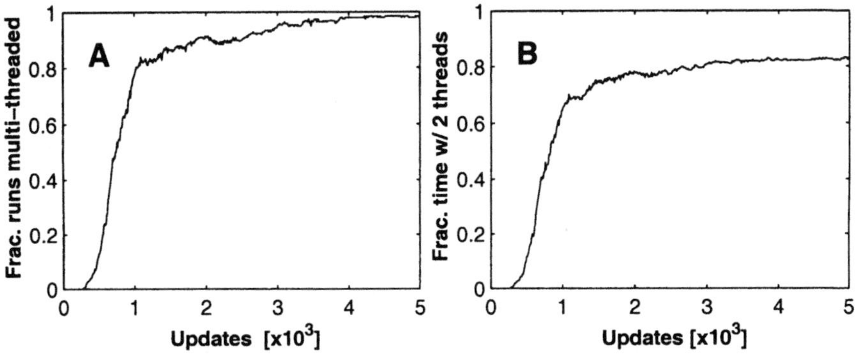

Fig. 1. The time progression of organisms learning to use multiple threads averaged over 200 trials. (A) The fraction of trials which thread at all, and (B) The average fraction of time organisms spend using both threads at once. The data displayed here is for the first 5000 updates of 50,000 update experiments in environment III.

3.1 Emergence of Multiple Execution Patterns

Describing a universal course of evolution in any medium is not feasible due to the numerous random and contingent factors that play key roles. However, there are a number of distinct trends, which will be discussed further.

Let us first consider the transition of organisms from a purely linear execution to the use of multiple threads. In Fig. 1A, we see that most populations do develop a secondary thread near the beginning of their evolution. Secondary threads come into use as soon as they grant any benefit to the organisms. The most common way this occurs is by having a `fork-th` and a `kill-th` appear around a section of code, which the threads thereby move through in lock-step, performing computations twice. Multiple completions of a task provide only a minor speed bonus, but this is often sufficient to warrant a double execution.

Once multiple execution has set in, it will be optimized with time. Smaller blocks of duplicated code will be expanded, and larger sections will be used more productively, sometimes even shrinking to improve efficiency. Once multiple threads are in use, differentiation follows.

3.2 Execution Patterns in Multi-threaded Organisms

A critical question is "What effect does a secondary thread have on the process of evolution?" The primary measure to denote a genome's level of adaptation to an environment is its *fitness*. The fitness of a digital organism is measured as the number of offspring it produces per unit time, normalized to the replication rate of the ancestor. In all experiments, the fitness of the dominant genotype starts at one and increases as the organisms adapt. Fitness improvements come in two forms: the maximization of CPU speed by task completion, and the minimization of gestation time. As all tasks must be computed each gestation cycle to maintain a reward, this gestation time minimization includes the *optimization* of tasks

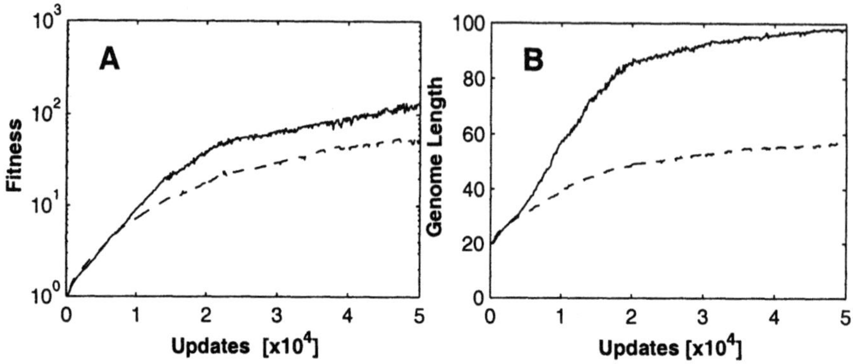

Fig. 2. (A) Average fitness as a function of time (in updates) for the 200 environment III trials. Most increases to fitness occur as a multiplicative factor, requiring fitness to be displayed on a logarithmic scale. (B) Average sequence length for the linear execution experiments (Solid line) and the multiple execution experiments (dashed line).

in addition to speed-ups in the replication process. The average progression of fitness with time is shown in Fig. 2A for both the niche with the expanded instruction set that allows multiple threads, and the standard, linear execution niche as a control.

Contrary to expectations, the niche that has additional threads available gives rise to a slower rate of adaptation. However, the average length of the genomes (Fig. 2B) reveals that the code for these marginally less fit organisms is stored using 40% fewer instructions, indicating a denser encoding. Indeed, the very fact that multi-threading develops spontaneously implies that it is beneficial. How then can a beneficial development be detrimental to an organism's fitness?

Inspection of evolved genomes has allowed us to determine that this code compression is accomplished by overlapping execution patterns that differ in their final product. Fig. 3A displays an example genome. The initial thread of execution (the inner ring) begins in the D "gene" and proceeds clockwise. The execution of D divides the organism when it has a fully developed copy of itself ready. This is not the case for this first execution, so the gene fails with no effect to the organism. Execution progresses into gene C_0 where computational tasks are performed, increasing the CPU speed. Near the center of C_0, a `fork-th` instruction is executed initiating secondary execution (of the same code) at line 27, giving rise to gene C_2. The primary thread continues to line 55, the S gene, where genome size is calculated and the memory for its offspring is allocated. Next, the primary instruction pointer runs into gene R, the copy loop, where replication occurs. It is executed once for each of the 99 instructions in the genome (hence its dark color in the figure). When this process is complete, it moves on through gene I_0 shuffling numbers around, and re-enters gene D for a final division.

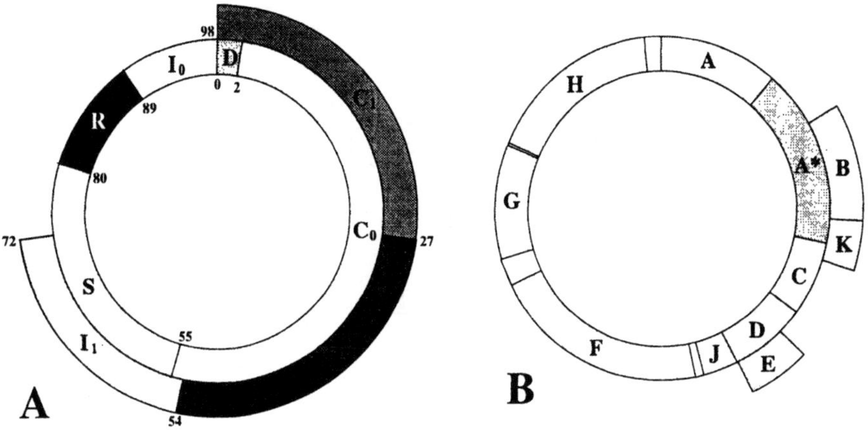

Fig. 3. A: Execution patterns for an evolved avida genome. The inner ring displays instructions executed by the initial thread, and the outer ring by the secondary thread. Darker colors indicate more frequent execution. **B:** Genome structure of the phage ΦX174. The promoter sequence for gene $A*$ is entirely within gene A, causing the genes to express the same series of amino acids from the portion overlapped. Genes B, E, and K are also entirely contained within others, but with an offset reading frame, such that different amino acids are produced.

During this time, the secondary thread executes gene C_2 computing a few basic logical operations. C_2 ends with a `jump-f` (jump forward) instruction that initially fails. Passing through gene I_1, numbers are shuffled within the thread and the jump at line 72 diverts the execution back to the beginning of the organism. From this point on, its execution loops through C_1 and C_2 for a total of 10 times, using the results of each pass as inputs to the next, computing different tasks each time. Note that for this organism, the secondary thread is never involved in replication. Similar overlapping patterns appear in natural organisms, particularly viruses. Fig. 3B exhibits a gene map of the phage ΦX174 containing portions of genetic code that are expressed multiple times, each resulting in a distinct protein [13]. Studies of evolution in the overlapping genes of ΦX174 and other viruses have isolated the primary characteristic hampering evolution. Multiple encodings in the same portion of a genome necessitate that mutations be neutral (or beneficial) in their net effect over *all* expressions or they are selected against. Fewer neutral mutations result in a reduced variation and in turn slower adaptation. It has been shown that in both viruses [4] and Avida organisms [6], overlapping expressions have between 50 and 60% of the variation of the non-overlapping areas in the same genome, causing genotype space to be explored at a slower pace.

In higher organisms, multiple genes do develop that overlap in a portion of their encoding, but are believed to be evolved out through gene duplication and specialization, leading to improved efficiency [2]. Unfortunately, viruses and avida organisms are both subject to high mutation rates with no error correction

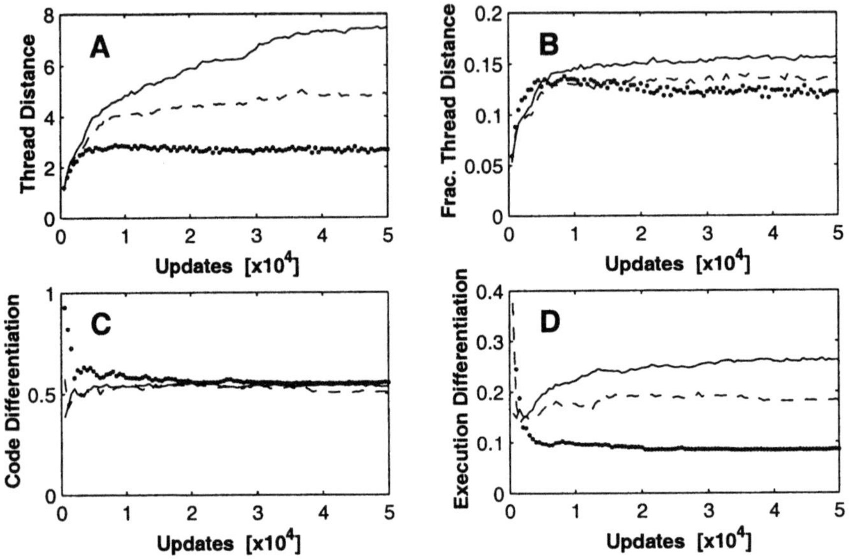

Fig. 4. Differentiation measures averaged over all trials for each experiment. (A) Thread Distance, (B) Fractional Thread Distance, (C) Code Differentiation, (D) Expression Differentiation. Experiments from environment III (solid line), environment II (dashed line), and environment I (dotted line)

abilities. This, in turn, causes a strong pressure to compress the genome, thereby minimizing the target for mutations. As this is an immediate advantage, it is typically seized, although it leads to a decrease in the adaptive abilities of the population in the long term.

3.3 Environmental Influence on Differentiation

Now that we have witnessed the development of multiple threads of execution in avida, let us examine the impact of environmental complexity on this process. Populations in all environments learn to use their secondary thread quite rapidly, but show a marked difference in their ability to diverge the threads into distinct functions. In Fig 4A, average Thread Distance is displayed for all trials in each environment showing a positive correlation between the divergence of threads and the complexity of the environment they are evolving in.

More complex environments provide more information to be stored within the organism, promoting longer genomes [1], and possibly biasing this measure. To account for this, we consider this average thread distance normalized to the length of the organisms, displayed in Fig 4B. When threads fully differentiate, they often execute neighboring sections of code, regardless of the length of the genome they are in, biasing this measurement in the opposite direction. Longer genomes need their threads to be further spatially differentiated in order to obtain an equivalent fractional thread distance. Thus, the fact that more complex environments give rise to a marginally higher fractional distance is quite significant.

Interestingly, Code Differentiation (Fig 4C) does not firmly distinguish the environments, averaging at about 0.5. In fact, the distribution of code differentiation turns out to be nearly uniform. This indicates that the portion of the genomes that are involved with the differentiated threads are similarly distributed between complexity levels. Execution Differentiation (the measure of the fraction of executions that occurred differently between threads, shown in Fig 4D), however, once again positively correlates environments with thread divergence. The degree of differentiation between the execution patterns is much more pronounced in the more complex environments.

4 Conclusions

We have witnessed the development and differentiation of multi-threading in digital organisms, and exhibited the role of environmental complexity in promoting this differentiation. Although this is an inherently complex process, the ability to examine almost any detail and dynamic within the framework of **avida** provides insight into what we believe are fundamental properties of biological and computational systems.

The patterns of expression (lock-step, overlapping, and spatial differentiation) are selected by balancing the "physiological" costs of execution and differentiation against the implicit effects of mutational load. Clearly, multiple threads executing single regions of the genome provides for additional use of that region. The benefit is in the form of additional functionality and a reduction in the mutational load required for that functionality. Within the context of this thinking, the correlation between environmental complexity and the usage of multiple threads makes a great deal of sense: multiple threads are advantageous only if they can provide additional functionality.

However, we have witnessed the cost side in this equation: when a gene or gene product is used in multiple pathways, variations are reduced as the changes to each gene must result in a net benefit to the organism. We observed a negative correlation between rates of adaptation and use of multiple threads. Furthermore, the ability to analyze the entropy of each site in the genome quantifies the loss in variability predicted by this hypothesis. This entropy analysis has been carried out in a biological context by Schneider [10], opening up opportunities to verify our results.

Implications of this work with potentially far reaching consequences for Computer Science involve the study of *how* the individual threads interact and what techniques the organisms implement to obtain mutually robust operations. The internal interactions within computer systems lack the remarkable stability of biological systems to a noisy, and often changing environment. Life as we know it would never have reached such vast multi-cellularity if every time a single component failed or otherwise acted unexpectedly, the whole organism shut down.

Clearly, we are still taking the first steps in developing systems of computer programs that interact on similarly robust levels. Here we have performed experiments on a simple evolutionary system as a step towards deciphering these biological principles as applied to digital life. In the future, we plan to add explicit costs for multi-threading that depend on the *local* availability of resources for thread execution. Systems at levels of integration anywhere near that of biological life are still a long way off, but more concrete concepts such as applying principles from gene regulation to develop self-scheduling parallel computers may be much closer.

Acknowledgements

This work was supported by the National Science Foundation under Grant No. PHY-9723972. G.H. was supported in part by a SURF fellowship from Caltech. Access to a Beowulf system was provided by the Center for Advanced Computing Research at the California Institute of Technology. We would like to thank an anonymous referee for useful comments.

References

1. Adami C, *Introduction to Artificial Life* (Telos Springer-Verlag, New York, 1998).
2. Keese P and Gibbs A, Origins of genes: "Big bang" or continuous creation?, *Proc. Natl. Acad. Sci.* **89**: 9489-9493 (1992).
3. Lenski RE and Travisano M, Dynamics of adaptation and diversification: A 10,000 generation experiment with bacterial populations, *Proc. Nat. Acad. Sci.* **91**: 6808-6814 (1994)
4. Miyata T and Yasunaga T, Evolution of overlapping genes, *Nature* **272**: 532 (1978).
5. Ofria C, Brown CT, and Adami C, *The Avida User's Manual*, in [1] 297-350 (1998).
6. Ofria C and Adami C, Evolution of genetic organization in digital organisms, *Proc. of DIMACS Workshop on Evolution as Computation*, Jan 11-12, Princeton University, Landweber L and Winfree E, eds. (Springer-Verlag, NY, 1999).
7. Ray TS, An approach to the synthesis of life, in *Proc. of Artificial Life II*, Langton CG, Taylor C, Farmer JD, and Rasmussen S, Eds. (Addison Wesley, Redwood City, CA, 1992), p. 371.
8. Ray TS, A proposal to create a network-wide biodiversity reserve for digital organisms, ATR Technical Report TR-H-133 (1995).
9. Ray TS and Hart J, Evolution of differentiated multi-threaded digital organisms, in *Proc. of Artificial Life VI*, Adami C, Belew RK, Kitano H, and Taylor CE, eds. (MIT Press, Cambridge, MA, 1998), p. 295.
10. Schneider TD, Stormo GD, Gold L, and Ehrenfeucht A, Information content of binding sites on nucleotide sequences *J. Mol. Biol.* **188**: 415-431 (1986).
11. Thearling K and Ray TS, Evolving multi-cellular artificial life, in *Proc. of Artificial Life IV*, Brooks RA and Maes P, Eds. (MIT Press, Cambridge, MA, 1994), p. 283.
12. Thearling K and Ray TS, Evolving parallel computation, *Complex Systems* **10**: 229-237 (1997).
13. Watson JD, et al., *Molecular Biology of the Gene* (Fourth Ed., Benjamin Cummings, Menlo Park, 1987).

Species Formation in Evolving Finite State Machines

Arno Rasek[1], Walter Dörwald[1], Michael Hauhs[1] and Alois Kastner-Maresch[1]

[1] BITÖK, University of Bayreuth, D-95444 Bayreuth, Germany
Alois.Kastner-Maresch@bitoek.uni-bayreuth.de

Abstract. Since the early beginnings of Evolutionary Computation, Finite State Machines (FSMs) have been applied to model organisms. We present a new approach to evolve such artificial organisms. The FSMs are subject to a difficult navigation and searching task in heterogeneous environments. We give a definition of FSM-species and investigate their formation. The results show that species are formed as the organisms agree on a common 'genetic broadcast language' and take advantage of the fruitful effects of recombination. As observed in natural ecosystems, higher abiotic diversity leads to higher biotic diversity.

1 Introduction

The emergence of sex and the conditions of species formation are biological problems in which ALife models may help to sharpen the experimental and theoretical discussion at critical aspects. Here we study species formation by simulating evolutionary adaptation of finite automata. In this approach, termed 'Evolving Finite State Machines' (EFSM), state sets and input alphabets can be varied and are subject to evolution. Questions that will be addressed are:

- Are there optimal cardinalities of state sets or input alphabets?
- Do stable multi-species-populations emerge or only single-species-populations?
- How does environmental diversity influence the resulting number of species?

Two types EAs have hitherto been used to adapt FSMs to perform a given task: Evolutionary Programming (EP) and Genetic Algorithms (GAs). The EP-driven approach is described in detail in Fogel et al. [6] where the task of the FSM consists in predicting symbol sequences. Fogel et al. considered intelligent behavior as a property to predict the future environment and to translate this knowledge into goal-directed action. As typical features of EP, only mutation and no recombination was used and the parameters were not coded in binary format. Mutation was implemented by randomly replacing a transition, an output symbol or the initial state. Additionally, the state set could be enlarged or reduced by one state at random. Hence the number of states was not predetermined and became a result of evolutionary adaptation. In [5] basically the same approach was applied to the iterated prisoner's dilemma.

Jefferson and Collins applied a GA in order to adapt FSMs (cf. [9], [3], [4]). In their "Genesys/Tracker" system the FSMs were interpreted as artificial ants which had to solve the task of following a given path on a 32x32 grid (the 'John Muir Trail').

The outstanding features of the GA were multi-point crossover, a large population size (65536 individuals) and high selective pressure (only the top 1% or 10% were selected for mating). As a consequence of using a GA, the state set had to be fixed.

In both approaches the question still remains how powerful the algorithms at work really are. No comparisons to FSMs designed by humans or to other methods were performed. In a few minutes, we were able to manually design FSMs which did as well as the evolved ones reported in the publications. But more importantly spontaneous species formation cannot be investigated with these two approaches (cf. section 4).

2 The Evolving FSMs Approach

In this section we briefly present our new approach which we termed Evolving Finite State Machines. EFSM is a hybrid EA which combines ideas and techniques from different standard EAs and makes use of the special data structure of FSMs. A FSM can be understood as a virtual creature possessing its own artificial intelligence, as it has been pointed out in [6]. In EFSM we observe the evolution of high-level goal-seeking behavior which will be briefly described in section 3.

2.1 Data Structure and Representation of the FSM

The FSM is implemented as a Mealy-type automaton. With the notation of [8] the FSM M is given through $M=(Q, \Sigma, \Delta, \delta, \lambda, q_0)$, where Q is the finite state set, Σ the input alphabet, Δ the output alphabet, δ the transition function, λ the output function and q_0 the initial state. Let m be the cardinality of the input alphabet, n the cardinality of the output alphabet, o the output and $q_0=0$. Then the automaton is fully specified by the maps δ: $Q\times\Sigma\rightarrow Q$ and λ: $Q\times\Sigma\rightarrow\Delta$. This is done through two mxn-matrices, the transition table T and the output table O. Whereas T is composed of integers, O contains bits, integers and real numbers to enable a rich output in order to let the FSMs solve difficult tasks.

Experimental results show the advantages of a direct representation of data structures (cf. [1], [10]). Hence, we did not apply binary coding to any of the integer or real parameters. The genomes have to be of variable length to allow state and input sets of variable cardinality in the population, in contrast to Evolutionary Strategies (ES), EP, GAs and (partly) classifier systems (CFSs). EFSM is in this respect similar to Genetic Programming (GP). Similar to ES we rely on self-adaptive search to ease the difficult task of finding good parameter settings (cf. [1], [7], [10], [11]). The strategy parameters, e.g. crossover and mutation probabilities, or crossover types, are therefore coded within the genome and are subject to evolutionary adaptation.

In summary, our genome contains a vector of strategy parameters and two matrices of variable size, the output and the transition table.

2.2 Fitness, Reproduction and Selection

Unlike other EAs, our approach differentiates reproduction and selection. We define reproduction as the event of creating new structures in the sense of "making copies". Selection is the event of "deleting" structures. In EFSM a selective pressure is employed in both stages. Reproduction and selection cause superior FSMs to reproduce more frequently and to survive longer. The expression of superiority or inferiority is the fitness which is derived from the given task. As GAs, EFSM makes optional use of different fitness scaling schemes and reproduction schemes as presented in [7].

The FSMs reproduced by one of these schemes undergo recombination and mutation. The new FSMs compete for survival against the parent population during the selection phase. Like in ES or EP the worst individuals are deleted until the desired population size is reestablished. This combination of reproduction and selection is similar to "(μ+λ)-selection" for ES, if uniform reproduction is applied in the reproductive phase. In GA-terms it would be called a steady-state-GA.

2.3 Recombination

EFSM relies on recombination, as do GAs, GP and CFSs. Recombination turns out to be of special significance in order to investigate species formation. The vector of strategy parameters undergoes the traditional single-point-crossover of GAs. For the transition and the output table six different recombination types were implemented in EFSM. In order to be concise, we do not specify the recombination types here. It should be sufficient to mention that the recombination operators try to exploit the 'semantic' coherence that may exist in the output and transition tables. More details are available at [15]. By respecting the special 'grammar' of FSMs we avoid that recombination works as a macro-mutation-operator, as itcould in GP (cf. [2]). Instead it produces meaningful structures with high probability.

2.4 Mutation

In EFSM we have different mutation types for different data types. All boolean parameters mutate by simple bit-inversion as in GAs. All integer parameters are mutated by choosing a new random value of the feasible range applying a uniform distribution. For real values there are two different types of mutation. The first is simply to choose a new number from a uniform distribution (same as for integers). This mutation type is applied to real-valued strategy parameters. The second type uses a Gaussian distribution and is realized by generating a random number which is added to the current value of the parameter. This second mutation type is applied to two output parameters and is a simpler version of mutation known from EP and ES.

Further details on the approach can be found at [15]. The approach has been realized as a MS-Windows-application which can be downloaded together with a complete user guide.

3 The FSM as a Virtual Creature

As in [6], [9] we view FSMs as ALife organisms. The task providing the fitness criterion is to navigate in a given heterogeneous 3D-environment. Our ALife organisms seek the lowest point in a multi modal surface, subject to a time constraint. This choice has various advantages. First of all, navigation and path planning has a long tradition since Wilson's animat [14] that resulted in several other examples to compare to. Second, we use well-known functions as environments defining optimization problems hard to solve numerically. Hence, the performance of EFSM can be evaluated against traditional optimizing methods as well as against EAs. Third, via visualization of the navigation behavior we obtain a direct access to the FSMs' capabilities. This allows an easy phenotypic interpretation even of complicated multi-state FSMs.

The FSM is considered the 'brain' of the creatures termed 'optimizers'. Each optimizer is located at one point and has a sensor measuring at a different point of the environment. Whilst the environment is quasi-infinite, the sensor of the optimizer has only a limited precision of measurement. In each measuring event the sensor generates a finite input symbol, hence the measurement precision is limited by the number of input symbols. The input is processed by the Mealy-type FSM. An output is created depending on the state of the FSM and its input. According to the output, the effectors can cause three types of actions: The optimizer may move to the point where the last measurement was performed, may turn around a certain angle and may change the distance where it measures at. The output causing these actions consists of four parameters (one binary and three real). For a machine with 4 states and 4 input symbols the output table already consists of 16 bits and 48 real parameters and the transition table adds 16 integers to the FSM. For more details about the navigation consult [15].

To make the searching task hard for the optimizers, in each generation they are placed on different randomly chosen locations in the quasi-infinite, toroidal environment. The optimizers don't have any knowledge about their position. Additionally, there are no landmarks on the torus so that orientation becomes non-trivial. Nevertheless, in EFSM high-level optimizers evolve that use fascinating searching techniques.

In Fig. 1 an example of the phenotype of an optimizer is depicted. It is a 4×4-machine obtained after 300 generations. The background illustrates the environment which is a 3D-version of Rastrigin's function (cf. [13] as ref.). The local minima are indicated by darker colors. The global minimum is located at the center. Moves realized by the optimizer are drawn with black lines and measuring events without moves are depicted with light gray lines. Positions of the optimizer are indicated by white dots. Starting point was at the right lower corner. After some erroneous moves, the optimizer starts to hop from minimum to minimum and reaches the vicinity of the global minimum after only five hops. It can realize this hopping because it systematically combs the neighborhood of a local minimum at a promising distance. In the situation presented, the optimizer performed only 179 measurements.

Apparently, it is very difficult for humans to design an equally efficient optimizer. Even by knowing this phenotype which is a favorable searching technique, it took us hours to code a similar minimum-hopping automaton by hand. Yet the human-coded one was inferior to those produced by EFSM within a couple of minutes.

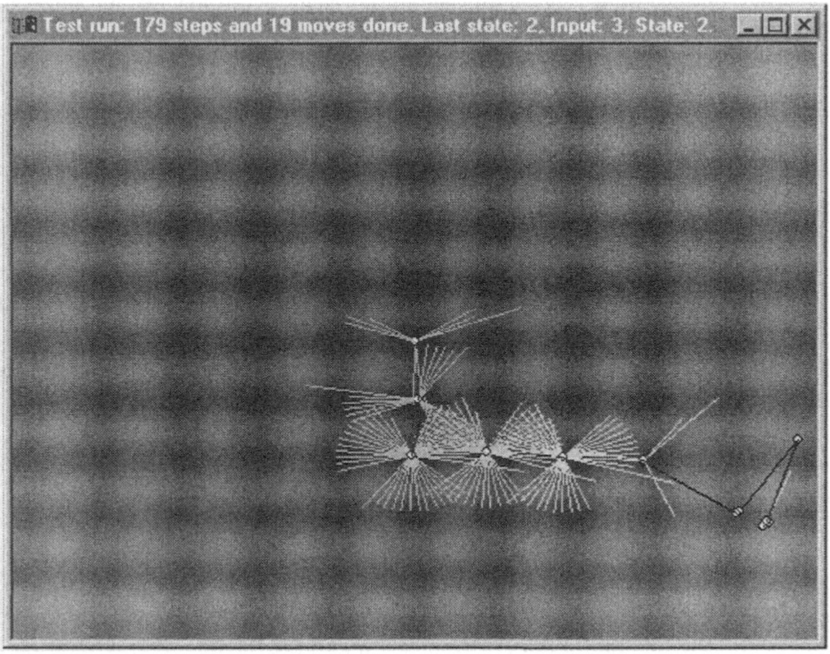

Fig. 1. An example of the phenotype of an FSM created by EFSM.

4 Species in Populations of FSMs

In biology a species is defined through reproductive isolation. In ALife systems surrogate criteria for genetic similarity are frequently employed to distinguish species.

Here we define genetic similarity for FSMs by the overall structure of the FSMs. Individuals with the same genome structure are defined as one species. The structure of the transition table T and the output table O is determined by the state set Q, the input alphabet Σ and the output alphabet Δ. Thus a prerequisite for the existence of various species in one population is that Q, Σ and Δ are not identical in all individuals. In order to observe species formation, the FSMs should additionally have the possibility to isolate themselves reproductively from different species (e.g. from FSMs with differing Q, Σ and Δ) or rejoin during evolution, whenever it might be evolutionary advantageous.

In EFSM we have realized these prerequisites. The population can be initialized with diverse state sets Q and diverse input alphabets Σ. Thus we are able to initialize simulations with single- or multi-species-populations which can be composed of arbitrary diversity. The output alphabet remains fixed, because we don't allow the optimizers' sensors to evolve. We also inserted a strategy parameter subject to evolution which decides whether the individual accepts to mate with individuals of differing Q and Σ. After initialization there's no regulating instance influencing the population's diversity. The number of species obtained is a pure result of simulated evolution.

As mentioned in section 1, the approaches using EP or GAs as Evolutionary Algorithm couldn't investigate species formation. In [6] only one individual was employed in the population. Their successors followed this line of work by rejecting recombination as a genetic operator even when they used populations with more than one individual. To the contrary, by using a GA the whole population consists of only one species. Crossover rapidly leads to uniform populations. In some non-FSM applications, researchers tried to introduce more diversity through different niching techniques [7]. The outcome of these simulations is predetermined by the formula applied for fitness-sharing. Therefore there can't be any endogenous species *formation* when using a GA. Additionally, Q, Σ and Δ must be uniform in GA-populations, as the genome length is fixed. Thus both approaches weren't able to model species formation.

5 Experimental Results

In this section we present experimental results obtained by several simulations performed with the program EFSM. First we check whether there are superior state sets or input alphabets by comparing performance measures of runs with fixed state set Q and fixed input alphabet Σ. This check is necessary because evolution would probably lead to this optimum if it existed. In a second set of simulation runs Q and Σ are subjected to evolutionary adaptation. The third series of runs investigates the relationship between species formation and varying environmental diversity.

5.1 Are There Optimal State Sets or Optimal Input Alphabets?

To answer this question, two series of simulation runs with different state sets Q and input alphabets Σ were performed. As the results are much the same in both cases, we only show the results for differing Q. In all runs Q was fixed and couldn't evolve. The populations were initialized with a uniform number of 1, 2, 3, 4, 8 or 16 states. The number of input symbols was set to 2. Five simulations were done for each of the state sets. We used a series of parameter settings which showed to be adequate through former extensive testing (population size: 100, offspring per generation: 50, roulette wheel reproduction, no fitness scaling, mutation probability 0.04, recombination probability 0.4, recombination type: exchange of blocks of the transition and output tables, an optimizer can perform 400 measurements per generation). The environment used was the so-called Sphere model. As performance parameter we used the 'Current Mean', which is the average fitness of all 100 optimizers in the current population. The fitness value of a single optimizer is the function value of the Sphere model at the end point of the optimizer's navigation. The maximum in the domain is set to 1 and the minimum is set to 0. High-performance optimizers obtain fitness values close to 0.

The results are depicted in Fig. 2. On the left-hand side (Fig. 2.A) we first observe the big dispersion in the 30 different runs. Even on a logarithmic scale, we note a fast improvement during the first 50 generations. Later we observe typical characteristics of evolution: After periods of evolutionary stagnation, there are sudden unpredictable 'inventions' leading to a fast improvement of 'Current Mean'. The relatively best results were obtained with two-state-populations, but this setting is not significantly superior to the other state sets. These five corresponding runs are highlighted in Fig. 2.A. The other 25 runs are drawn with gray lines.

In Fig. 2.B we zoom in to the first 50 generations and compare only the time series for 1, 4 and 16 states. On this time scale the number of states has an influence on the performance. Larger state sets correspond to a poorer performance during the first generations. The results confirm our expectations. State machines with larger state sets contain much more parameters that have to be adapted during evolution leading to slower improvement at the beginning. Additionally, larger structures are generally less robust with regard to mutation and recombination. On the other hand, larger state sets allow a greater variety of actions which emerge in later phases. These more complex optimizers recover the initial disadvantage in the following generations.

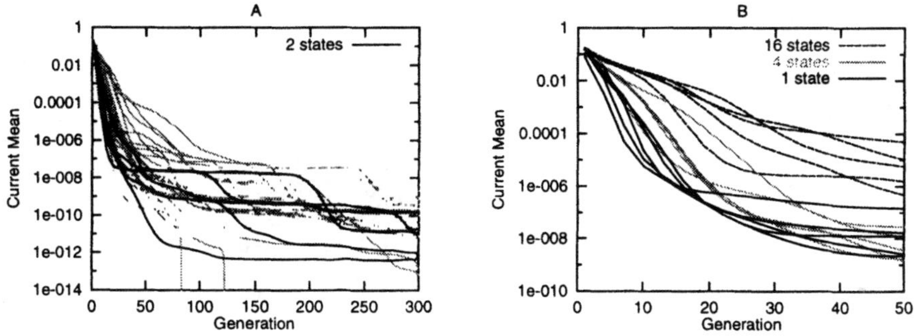

Fig. 2. Time series of the performance measure 'Current Mean' with varying but fixed (non-evolving) state sets. A: Highlight on 2 states, in the background 1, 3, 4, 8 and 16 states for 300 generations. B: Comparing 1, 4 and 16 states for the first 50 generations.

We conclude that there do not exist optimal state sets where the optimizers will automatically evolve to. To the contrary, it depends on the considered time scale which state sets will be 'superior'. Having examined optimality, we are now ready to let the state sets and input alphabets evolve.

5.2 Species Formation as a Result of Simulated Evolution

As stated above, the number of states and the input alphabet are not predetermined and may be subject to evolution. Taking into account the results of section 5.1 we might expect an evolution towards increasing state sets and input alphabets. To address this question, two series of experiments were performed where either the number of states was not fixed (and could vary between 2 and 10) or the input alphabets could evolve (with cardinalities 2, 4, 6, 8 or 10). We made 19 simulation runs each. In contrast to the parameter settings in section 5.1 all strategy parameters (e.g. mutation and recombination rates, recombination types) were allowed to evolve. The strategy parameter which decides whether the individual accepts to mate with individuals of differing Q and Σ evolved, too. Rastrigin's function [13] was used as test environment.

Fig. 3.A shows the evolution of the arithmetic mean of the cardinality of the input alphabets of all optimizers ('Mean No. of Inputs'). In all simulations a *uniform* input alphabet was reached soon and later evolution depended on these early "frozen accidents". However, between different runs the cardinalities vary heavily. It seems that alphabets with higher cardinality are slightly preferred. During the first 10 to 20 generations the mean number of states ('Mean No. of States', cf. Fig. 3.B) decreased in all runs. This is consistent with the result of section 5.1 that smaller state sets are advantageous at the beginning. Very similar to the input alphabets, the number of states evolves fast in direction of a *uniform* value throughout the population. But again, between different runs the resulting numbers of states vary heavily. Somehow surprisingly, the once obtained uniform value doesn't rise in later phases.

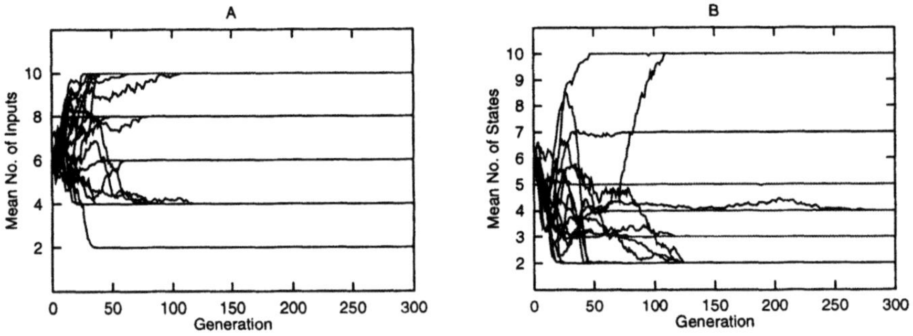

Fig. 3. Mean number of input symbols (3.A) and mean number of states (3.B).

We conclude that in EFSM there is no evolutionary tendency towards a specific state set or input alphabet. But there is a strong tendency towards single-species-populations (uniform Q and Σ). After initialization there is no regulating instance influencing the population's diversity implying that this species formation is an endogenous development of simulated evolution. It seems like the individuals agree on certain Q and Σ as genetic grammar. This grammar may be interpreted as a type of 'genetic broadcast language'. It seems that they need the common language as a prerequisite for successful operation of recombination. As the individuals are not forced into sexual reproduction, the formation of species is a decentralized group decision.

5.3 Species Formation as a Function of Environmental Diversity

In a last series of runs we investigated the relationship between the velocity of species formation and environmental diversity in EFSM. One approach to explain species diversity in biological ecosystems is the variety of ecological niches. In EFSM we observe different specialists evolving for the different landscapes. Thus we can simulate more ecological diversity by simultaneously posing more than one landscape to be optimized. The new fitness criterion is the sum of the results obtained by the optimizers on each landscape. Still the individuals and species compete within one population, but now there might be different specialists for different functions. An individual might compensate a poor performance in one landscape by a high performance in another landscape.

We executed simulation runs with one landscape (Rastrigin's function and Griewank's function [13]) and with four landscapes simultaneously (Rastrigin's, Griewank's, Schwefel's [12] and a rotated version of Schwefel's function). The runs were initialized with uniform input alphabets (four symbols) but diverse numbers of states between two and seven. Consequently, after initialization there were always six species in the population. The rest of the parameter settings were the same as in section 5.1. Fig. 4 depicts the time series of the number of species for the three variants. The number of species shown is the mean of the corresponding 20 runs. In all three variants the number of species decreases rapidly to two. The value 'Mean No. of Species' remains higher in the more diverse environment. Thus in EFSM, like in nature, higher environmental diversity leads to a higher diversity of species.

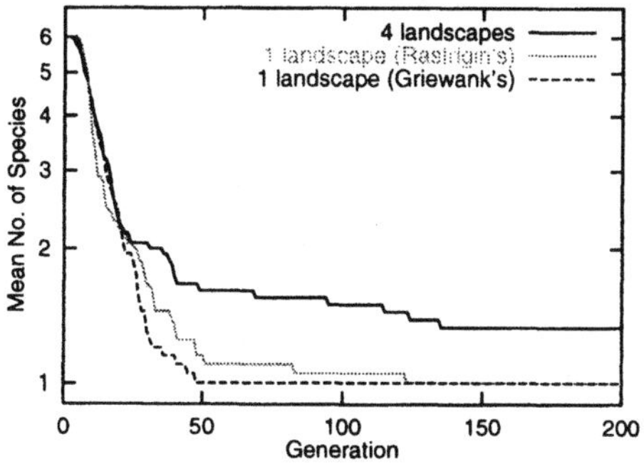

Fig. 4. Number of species for high environmental diversity (four landscapes) and low diversity (one landscape).

6 Conclusions

We did not observe the appearance of optimal numbers of states or optimal cardinalities of input alphabets. Smaller state sets are advantageous during early generations. We observed a strong selection pressure in direction of uniform state sets and uniform input alphabets. This implies that soon after the start of the evolutionary process only one or two species remained. This can be explained by the fact that recombination can only be meaningful between individuals with similar structure. Therefore successful genetic search cannot take place with different grammars. Instead, the organisms have to agree on a common 'genetic broadcast language'. Seen from a ecological point of view, it is also plausible that only one species will survive under one given environmental condition. Indeed, if we use several static environmental conditions simultaneously we get higher diversity despite the fact that we didn't change the conditions of competition in the population.

We appreciate the financial support of the BMBF (grant BEO 51-0339476C).

References

1. Bäck, T.: Evolutionary Algorithms in Theory and Practice: Evolution Strategies, Evolutionary Programming, Genetic Algorithms. Oxford University Press, New York (1996)
2. Banzhaf, W., Nordin, P., Keller, R.E., Francone, F.D.: Genetic Programming - An Introduction: On the Automatic Programming of Computer Programs and its Applications. Morgan Kaufman, San Francisco, dpunkt.verlag, Heidelberg (1998)
3. Collins, R.J., Jefferson, D.R.: Representations for Artificial Organisms. In: Meyer, J.-A. (ed.): From Animals to Animats, Proceedings of the First International Conference on Simulation of Adaptive Behavior. MIT Press, Cambridge (1991) 382-390
4. Collins, R.J., Jefferson, D.R.: AntFarm: Towards Simulated Evolution. In: Langton, C.G., Taylor, C., Farmer, J.D., Rasmussen, S. (eds.): Artificial Life II. Addison Wesley, Redwood City (1991) 579-601
5. Fogel, D.B.: Evolving Behaviors in the Iterated Prisoner's Dilemma. Evol. Comp. 1 (1993) 77-97
6. Fogel, L.J., Owens, A.J., Walsh, M.J.: Artificial Intelligence through Simulated Evolution. John Wiley, New York (1966)
7. Goldberg, D.E.: Genetic Algorithms in Search, Optimization, and Machine Learning. Addison-Wesley, Reading (1989)
8. Hopcroft, J.E., Ullmann, J.D.: Introduction to Automata Theory, Languages and Computation. Addison-Wesley, Reading (1979)
9. Jefferson, D.R., Collins, R.J., Cooper, C., Dyer, M., Flowers, M., Korf, R., Taylor, C., Wang, A.: Evolution as a Theme in Artificial Life: The Genesys/Tracker System. In: Langton, C.G., Taylor, C., Farmer, J.D., Rasmussen, S. (eds.): Artificial Life II. Addison-Wesley, Redwood City (1991) 549-578
10. Michalewicz, Z.: Genetic Algorithms + Data Structures = Evolution Programs. 3rd edn. Springer-Verlag, Berlin Heidelberg New York (1996)
11. Mitchell, M.: An Introduction to Genetic Algorithms. MIT Press, Cambridge (1996)
12. Schwefel, H.-P.: Numerische Optimierung von Computer-Modellen mittels der Evolutionsstrategie: Mit einer vergleichenden Einführung in die Hill-Climbing und Zufallsstrategien. Birkhäuser, Basel (1977)
13. Törn, A., Žilinskas, A.: Global Optimization, Lecture Notes in Computer Science, Vol. 350. Springer-Verlag, Berlin Heidelberg New York (1989)
14. Wilson, S.W.: Knowledge Growth in an Artificial Animal. In: Greffenstette, J.J. (ed.), Proceedings of the First International Conference on Genetic Algorithms and Their Applications. Erlbaum, Hillsdale (1985) 16-23
15. EFSM homepage: http://www.bitoek.uni-bayreuth.de/MOD/Software/EFSM/

Genetic Code Degeneracy: Implications for Grammatical Evolution and Beyond

Michael O'Neill & Conor Ryan

Dept. of Computer Science and Information Systems
University of Limerick
Ireland
{Michael.ONeill|Conor.Ryan}@ul.ie

Abstract. Grammatical Evolution (GE) is a grammar-based GA which generates computer programs. GE has the distinction that its input is a BNF, which permits it to generate programs in any language, of arbitrary complexity. Part of the power of GE is that it is closer to natural DNA than other Evolutionary Algorithms, and thus can benefit from natural phenomena such as a separation of search and solution spaces through a genotype to phenotype mapping, and a genetic code degeneracy which can give rise to silent mutations that have no effect on the phenotype. It has previously been shown how runs of GE are competitive with GP, and in this paper we analyse the feature of genetic code degeneracy, and its implications for genotypic diversity. Results show that genetic diversity is improved as a result of degeneracy in the genetic code for the problem domains addressed here.

1 Introduction

Grammatical Evolution (GE) is a grammar-based, variable length, linear genome system which is capable of generating programs or expressions in any language. Rather than the functions and terminals associated with GP [3], GE takes a BNF specification of a language, or subset thereof, from which it can subsequently generate compilable code. The BNF is used to build a program by applying production rules to elements of the non-terminal set of the BNF definition, in a mapping process to generate the output code from a simple binary string.

GE has been successfully applied to a number of diverse problem domains such as, symbolic regression [9], finding trigonometric identities [10], symbolic integration [10], and the Santa Fe trail [7]. The results compared favorably with systems such as GP, and has been shown to outperform GP [7].

In the spirit of Artificial Life, one definition of which is,

[the] field of study devoted to understanding life by attempting to abstract the fundamental dynamical principles underlying biological phenomenon, and recreating these dynamics in other physical media, such as computers, making them accessible to new kinds of experimental manipulation and testing [5], we have developed a system to generate programs which attempts to harness some of

the features of the genetic machinery of living organisms which are theorised to have an impact on the phenomenon of evolution [1]. This paper focuses on genetic code degeneracy, a characteristic of the genetic code in biological organisms which has been incorporated into GE, in an attempt to analyse the role this feature plays in maintaining genotypic diversity.

2 Grammatical Evolution

When tackling any problem with GE, a suitable BNF definition must first be decided upon. The BNF can be either the specification of an entire language, or perhaps more usefully, a subset of a language geared towards the problem at hand.

The genotype is then used to map the start symbol onto terminals by reading genes of 8 bits to generate a corresponding integer value, from which an appropriate production rule is selected. A rule is selected by the Integer Gene Value *MOD* the Number of Production Rules for the current non-terminal.

Considering the following rule,

```
(1) <code> :: =  <line>              (A)
                |<code><line>        (B)
```

i.e. given the non-terminal *code* there are two production rules to select from. If we assume the gene being read produces the integer 6, then 6 MOD 2 $=$ 0 would select rule (A) $<$ *line* $>$. Each time a production rule has to be selected to map from a non-terminal, another gene is read, and in this way, the system traverses the genome.

Given an 8 bit binary number, each gene can represent 256 distinct integer values. However, many of these integer values can represent the same production rule, taking production rule 5 as an example, if the current gene value was 6, then 6 MOD $3 = 0$ would select rule (A) $left()$ as shown above. The same rule would be chosen if the gene value was 3, 9, 12, etc.

3 The Problem Spaces

Two problem domains were examined at which GE was previously found to be successful, namely a symbolic regression problem [9], and the Santa Fe ant trail [7]. For the Symbolic Regression problem the system is given a set of input and output pairs, and must determine the function that maps one onto the other. The particular function examined is $f(x) = X^4 + X^3 + X^2 + X$ with the input values in the range $[-1..+1]$. To determine the fitness of an individual program 20 test points were applied in the input range, and the fitness was taken as the sum of the error. The objective of the Santa Fe ant trail is to find a computer program to control an artificial ant so that it can find all 89 pieces of food located on a non-continuous trail within a specified number of time steps. The ant can only turn left, right, move forward one square. It may also look ahead one square in the direction it's facing to determine if that square contains a piece of food.

4 Results

Two sets of experiments were carried out for each problem domain. Fifty runs were carried out where genetic code degeneracy was removed so far as possible by reducing the number of bits in a gene to the lowest possible value that can still represent the maximum number of productions rules belonging to any one non-terminal. Another fifty runs were carried out where degeneracy was present as in the standard GE implementation using 8 bit genes.

Two measures were used to give an indication of genotypic diversity, the first is a measure which we have termed the *mean variety*. This measure was obtained by calculating the average of the variances at each bit locus on the genome.

With a population size of 500 this meant that the greater the variance in a population the closer the mean variety measure is to 0.25. The aim of this measure is to attempt to establish how different the individual genotypes in any given population are.

The other measure used was the number of unique individuals in a population, which can also be used to some extent to illustrate the genetic diversity within a population[4].

Results for both of these measures over both problem domains can be seen in Figures 1 2. As can be seen from these graphs the code degeneracy is having a marked effect on the mean variety measure, and on the number of unique individuals.

A measure of the average number of invalid individuals in each generation over the 50 runs was carried out for both problem domains, and results show that when the degeneracy is removed from the genetic code, the number of invalid individuals increases significantly over a run when compared to the case where a degenerate code exists.

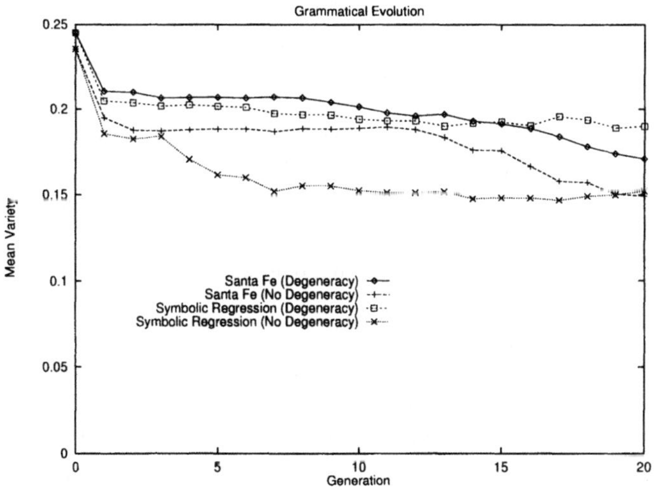

Fig. 1. Genetic Code Degeneracy and Mean Variety

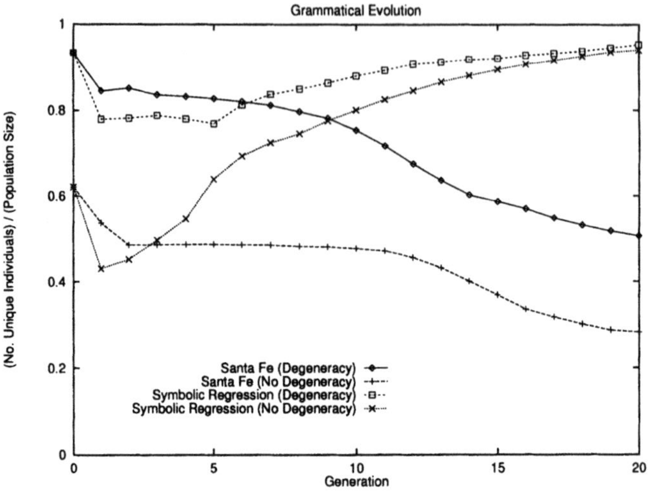

Fig. 2. Genetic Code Degeneracy and Unique Individuals

5 Discussion

In the standard GE implementation 8 bit genes are used, which can represent 256 distinct integer values. Therefore, in this state GE can represent 256 productions for each $non - terminal$ in the grammar. In the case of the Santa Fe trail the maximum number of productions that any one $non - terminal$ has is 3, and for symbolic regression problems this number is 4. As a result, the minimum number of bits any gene can have in the case of these problems is two, as this can represent a maximum of 4 distinct productions. It follows that in the case of the symbolic regression problem, all degeneracy has been removed, while it still exists to a very small extent in the Santa Fe trail problem.

Given the number of invalid individuals produced when genetic code degeneracy is removed, it is clear that degeneracy is responsible for the preservation of the functionality of the phenotype, while still allowing unrestricted search of the genotypic search space. Based on the two measures used to give an indication of the genotypic diversity in a population, the results clearly show that degeneracy in the genetic code is having a beneficial effect on genotypic diversity in the population.

Kimura's neutral theory of molecular evolution [2], suggests that molecular evolution is as a result largely of mutations that have no effect on the phenotype, i.e. functionality. Kimura mentions that this could also be responsible for maintaining genetic diversity within a population. The results produced by GE would reflect the basis for this theory within our artificial population, which is occurring due to the genetic code degeneracy which is part of the genotype to phenotype mapping process.

Following on from Kimura, others have noted the benefits that neutral evolution can bring to the evolutionary process [8] [6]. In particular in [6] it has been suggested that the maximum fitness attainable on a fitness landscape with a degree of neutrality increases as the degree of neutrality increases. Further investigation is required to determine the ramifications, if any, for GE based on these and other observations.

6 Conclusions

The results presented here show that genetic code degeneracy is having a clear beneficial effect on the genotypic diversity, and in the preservation of valid phenotypes, during runs of GE. The benefits of a complex mapping process show that the one-to-one genotype to phenotype mapping that is so prevalent in other evolutionary algorithms is not necessarily a good idea. We have previously shown that by harnessing features from the genetic system of biological organisms that the performance of GE can be enhanced, and as such there are clear advantages to be gained by using more biologically inspired techniques within evolutionary algorithms. It follows, that by modifying the current genotype to phenotype mapping process to one that is even closer to that of biological organisms, we may find that the performance of GE will benefit to an even greater extent.

Taking on board these findings, and the fact that GE has proven successful across diverse problem domains, it has been shown that GE is a powerful approach to automatically generating programs in any language.

References

1. Elseth Gerald D., Baumgardner Kandy D. 1995. Principles of Modern Genetics. *West Publishing Company.*
2. Kimura, M. 1983. The Neutral Theory of Molecular Evolution. Cambridge University Press.
3. Koza, J. 1992. *Genetic Programming.* MIT Press.
4. Langdon, W. 1998. Genetic Programming and Data Structures. Kluwer Academic Publishers.
5. Langton, C.G. 1992. Preface. In C.G. Langton, C.Taylor, J.D. Farmer, and S.Rasmussen, editors, Artificial Life II.
6. Newman M.E.J., Engelhardt Robin. Effects of neutral selection on the evolution of molecular species. In *Proc.R.Soc.London B*, **265** pages 1333-1338.
7. O'Neill M., Ryan C. 1999. Evolving Multi-line Compilable C Programs. In *Proceedings of the Second European Workshop on Genetic Programming 1999.*
8. Reidys C., Forst C.V., Schuster P. Replication and Mutation on Neutral Networks. Santa Fe Institute Working Paper 98-04-036.
9. Ryan C., Collins J.J., O'Neill M. 1998. Grammatical Evolution: Evolving Programs for an Arbitrary Language. In *Proceedings of the First European Workshop on Genetic Programming*, pages 83-95.
10. Ryan C., O'Neill M. Grammatical Evolution: A Steady State Approach. In *Late Breaking Papers, Genetic Programming 1998*, pages 180-185.

Coevolutionary Dynamics of a Multi-population Genetic Programming System

Josh C. Bongard

School of Cognitive and Computing Sciences
University of Sussex, Brighton UK
joshuab@cogs.susx.ac.uk

Abstract. This report presents an asynchronous, distributed genetic programming (GP) system using a master/slave architecture. Using symbolic regression for fourier functions as the problem domain, the system was found to demonstrate cooperative coevolutionary dynamics when multiple client populations evolve solutions to similar, but different problems: specifically, closely coupled populations were found to promote continuous search, which in some cases leads to the discovery of better solutions.

1 Introduction

The abilities of various evolutionary algorithms to traverse large search spaces have been well documented[5][7], as well as their amenability to parallel implementation [3]. Much work has investigated various parallel architectures that partition, or otherwise reorganize a single problem[4][8][11]: focus is placed on the dispersion and synthesis of partial solutions, often referred to as building blocks or schemas.

However, there is little investigation of parallel algorithms that attempt to distill useful information from a set of similar, but different problems. It seems plausible that populations evolving solutions to similar problems may benefit from sharing their results.

It has been found that sharing is indeed possible, and is correlated to the degree of similarity between fitness landscapes. Most importantly, however, we have found that two (or more) populations that share solutions can help each other escape local optima and explore increasingly better optima. As opposed to Hillis'[6] appropriation of parasitic coevolutionary dynamics to guard against premature convergence, the formulation presented here uses cooperative coevolutionary dynamics—in the form of solution transmission between populations—to reach continually better solutions.

2 The Model

A distributed GP model using a master/slave architecture similar to that described in [9] is used to gather results in this report: clients send their current, most fit solution to a central server, and download solutions generated by other clients from that server. Unlike the architecture described in [9], however, all solutions received by the server are broadcast to a requesting client, with the exception of solutions sent by the same client.[1]

We have chosen the problem domain of symbolic regression with which to gather quantitative measurements of the coevolutionary dynamics of the described model. Problem sets are defined as parametric fourier functions of the form

$$y_i = \sum_{j=1}^{5} a_j sin(jx_i) + b_j cos(ix_i)$$

, where each fourier function is characterized by selecting values for each a_j and b_j. Each problem set is composed of 40 values for the independent variables $\mathbf{x}$ uniformly distributed between $[-5, 5]$. The raw fitness of an s-expression s in the GP population is set to $f_s = \sum_{i=1}^{40} ||y_i - o_{i,p}||$ where $o_{i,p}$ is the output of s when presented with the independent variables $\mathbf{x}$.

It is then possible to quantitatively measure the distance between two training sets t_1 and t_2 using

$$d_{t_1,t_2} = \sum_{i=1}^{4} 0||y_{i,t_1} - y_{i,t_2}|| \tag{1}$$

if we assume that the set of dependent variables $\mathbf{x}$ is invariant across all possible training data.

All of the results below were generated using a GP instantiation with a function set covering arithmetic and trigonometric functions; a terminal set of the dependent variables and two floating point constants; one mutation for every 100 nodes involved in subtree crossover; a population size of 1000; and tournament selection with a tournament size of 10.

Each client population is assigned one fourier function as described above, and is run for 500 generations, which was chosen to ensure convergence in most populations. The benefits of convergence prior to migration are reported in [1] and [10]. At the end of the run the most fit solution from each population is sent to the central server, and the populations are then reinitialized with random solutions, as well as downloading solutions from the server. Transferred solutions then participate in selection and crossover depending on their fitness in the new population. Each population operates asynchronously; populations terminate and begin new runs independent of the other populations.

3 Results

For the case of a set of client populations assigned the same problem set, the model presented in this report operates as a distributed panmictic population:

[1] This minimizes the possibility of a client population reconverging to a solution it had previously uploaded to the server.

the client populations are equivalent to demes, in which solutions migrate at fixed intervals, and the demes are fully connected.[2]

Consider now a set of client populations evolving solutions to similar, but different problems. This situation was modelled by generating 40 fourier functions using random values of a_j and b_j as discussed in section 2. The distances between each pair of the 40 problem sets was determined using equation 1, and the two problem sets with the minimum distance were assigned to two client populations.

The two client populations were run for 500 generations each, sending their most fit solution at the end of each run to the server. The populations were then re-initialized with random solutions, as well as retrieving solutions from the server. This process was repeated over 20 runs. It was found that the two populations experienced a decrease in the normalized fitness of the most fit solution in each of the two populations at the end of each run, as the number of runs increases. The results obtained are shown in Figure 1. The results are contrasted against a control case: two populations were evolved for 500 generations over 20 runs with no sharing.

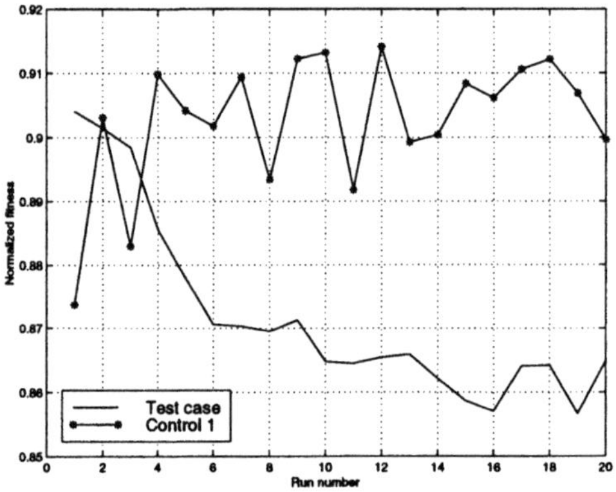

Fig. 1. A plot of the performance increase exhibited by client populations evolving solutions to similar training data. The test case involved two populations evolving solutions over 20 runs of 500 generations each, and trading their most fit solutions after each run via a central server. The control case involved two populations evolving over 20 runs of 500 generations each, but did not involve trading of solutions. Results from the test case and control case were averaged over 10 iterations.

4 Discussion

In modelling two client populations evolving solutions to similar, but different problem sets, it was found that populations do not converge on a single optimum, but rather continue to explore regions of the search space near some optimum common to both populations. This opposes the findings for traditional, multi-deme GA populations reported in [3], in which the parameter dictating the migration rate of solutions between demes acts as a threshold, below which demes converge on different, but inferior optima, and above which demes (sometimes prematurely) converge on the best solution discovered across all demes.

The continuous search property of the model presented in this report is demonstrated in Figure 2, in which two client populations are contrasted. In the control population, 20 runs of 500 generations are performed, and averaged over 20 iterations. At the end of each run, the most fit solution in the population is preserved, but undergoes a large mutation, in which a sub-tree of depth two is replaced with a random sub-tree of depth two. The remaining solutions in the population are replaced with random solutions. No solutions from this population are sent to or received from the central server. The other population trades solutions with other evolving populations as described in section 2. Both populations evolve solutions to the same fourier function.

The large mutation experienced in the first population is meant to simulate the movement of a population a small distance away from the fitness optimum it had recently achieved. During subsequent evolution, the population may return to its previous optimum, or discover another, either inferior or superior optimum.

Over the 20 runs it was found that both populations continue to find better solutions, but that the population that trades solutions with other populations reaches better optima than those discovered by the control population. This indicates that the population that shares solutions discovers new, better optima more often than the population that undergoes repeated large mutations.

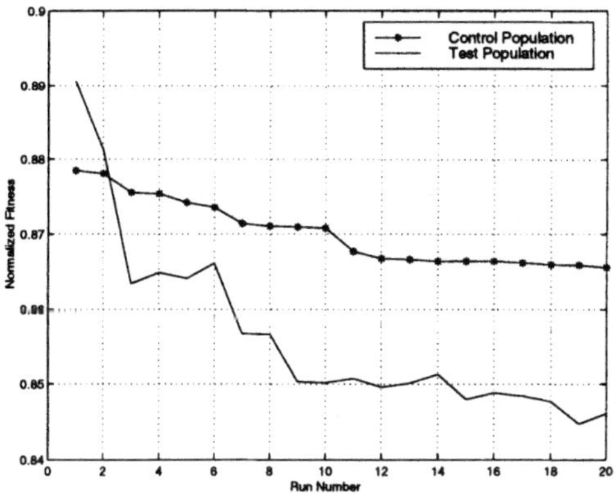

Fig. 2. Two populations evolving solutions to the same problem set for 20 runs of 500 generations each. In the control population, at the end of each run, the best solution undergoes a large mutation, while the remaining solutions are replaced with random solutions. In the test population, at the end of each run, the best solution is sent to the server, all solutions are replaced with random solutions, and solutions are then downloaded from the central server. The results from both populations were averaged over 20 iterations.

5 Conclusions

In this paper we have explored the cooperative, coevolutionary dynamics that appear in an asynchronous, distributed genetic programming architecture. As opposed to the usual formulation of a multideme, distributed GA with a specified transmission rate of solutions between populations, our architecture services client populations with different problem sets drawn from a common problem domain. Transmission of solutions is arbitrated by a central server.

It was found that mutual exchange of solutions between client populations with similar, but different problem sets exhibits a coevolutionary dynamic: solutions from one population, when inoculated in another population, can pull the second population away from its current local optimum, and often lead to the discovery of better optima.

References

1. Braun, H. C.: On Solving Travelling Salesman Problems by Genetic Algorithms. In: Schwefel H.-P. & R. Männer (eds.): Parallel Problem Solving From Nature, pp. 129–133. Springer-Verlag, Berlin (1990).
2. Cantú-Paz, E.: Topologies, Migration Rates, and Multi-Population Parallel Genetic Algorithms. To Appear in: GECCO-99, Genetic and Evolutionary Computation Conference, July 13–17, Orlando FL (1999).
3. Cantú-Paz, E.: A Survey of Parallel Genetic Algorithms. In: Calculateurs Paralleles. **10**:2 (1998).
4. Cantú-Paz, E. & D. Goldberg: Modelling Idealized Bounding Cases of Parallel Genetic Algorithms. In: Koza, J., K. Deb, M. Dorigo, D. Fogel, M. Garzon, H. Iba & R. Riolo (eds.): Genetic Programming 1997: Proceedings of the Second Annual Conference, pp. 353–361. Morgan Kauffman, San Francisco CA (1997).
5. Goldberg, D. E.: Genetic Algorithms in Search, Optimization, and Machine Learning. Addison-Wesley, Redwood City CA (1989).
6. Hillis, D. W.: Co-Evolving Parasites Improve Simulated Evolution as an Optimization Procedure. In: Langton, C. G., Taylor, C. (eds.): Artificial Life II, pp. 313–322. Addison-Wesley, Redwood City CA (1992).
7. Koza, J. R.: Genetic Programming: On the Programming of Computers by Means of Natural Selection. MIT Press, Cambridge MA (1992).
8. Mahfoud, S. W.: A Comparison of Parallel and Sequential Niching Methods. In: ICGA **6**, pp. 136–143. (1995).
9. Marin, F. J., O. Trelles-Salazar & F. Sandoval: Genetic Algorithms on LAN-message passing architectures using PVM: Application to the Routing Problem. In: Davidor, Y., H.-P. Schwefel & R. Männer (eds.): Parallel Problem Solving from Nature III, pp. 534–543. Springer-Verlag, Berlin (1994).
10. Munetomo, M., Y. Takai & Y. Sato: An Efficient Migration Scheme for Subpopulation-Based Asynchronously parallel genetic algorithms. In: Forrest, S. (ed.): Proceedings of the Fifth International Conference on Genetic Algorithms, p. 649. Morgan Kauffman, San Mateo CA (1993).
11. Miller, B. L. & M. J. Shaw: Genetic Algorithms with Dynamic Niche Sharing for Multimodal Function Optimization. In: IEEE International Conference on Evolutionary Computation, pp. 786–791. IEEE Press, Piscataway NJ (1996).

The Evolution of Computation in Co-evolving Demes of Non-uniform Cellular Automata for Global Synchronisation

Vesselin K. Vassilev, Julian F. Miller, Terence C. Fogarty

School of Computing, Napier University, Edinburgh, EH14 1DJ, UK
v.vassilev@dcs.napier.ac.uk

Abstract. We study the evolution of computation performed by non-uniform cellular automata in which global information processing appears at two different levels of self-organisation. In our model, the first level of self-organisation is characterised by interactions among cellular macrostructures or computational demes which compete for room in a finite grid of cells. This level is related to the formation, evolution and extinction of macrostructures, and it is designed in a completely local manner. The second level of self-organisation refers to the interactions among the cells within the demes. The model, derived from the cellular programming approach, allows global computation to occur as a result of many local interactions among computational demes of interacting cells. The study reveals some of the mechanisms by which co-evolving demes of non-uniform cellular automata perform non-trivial computation, such as the synchronisation tasks.

1 Introduction

This paper studies the evolution of computation performed by a system in which global information processing appears as a result of the interactions among many components, each of which is a system which in turn exhibits an ability for global computation at a different level of self-organisation. The model, derived from the non-uniform cellular automata model [1–3], is inspired by the idea of co-evolving to a higher degree of specialisation with macrostructures in a single computational ecosystem. For other relevant work, we refer to the ecological models studied in [4–6].

Cellular automata (CAs) are discrete dynamical systems of simple locally connected interacting cells [7–9]. They are simple, general, and computationally powerful [10, 11]. The model is perhaps the simplest example of systems that are capable of emergent computation – global information processing that appears in systems from the action of many interacting components [12, 13]. The evolution of CAs that perform non-trivial computation has been studied in [14, 15]. They evolved a population of CA rules to attain a single CA able to perform a particular computational task. However, computation requiring global coordination can be arduous for CAs [16], even if they are designed by evolutionary techniques. It has been suggested that better computational performance can be attained by using many CA rules or a non-uniform CA [17].

Non-uniform CAs are cellular automata with cells that may contain different transition rules [17]. The evolvability of non-uniform CAs has been studied in [18] where cellular programming algorithm has been employed to locate non-uniform CAs for different computational tasks. It appears that the evolution of non-uniformity leads to a quasi-uniform CA with one dominant rule occupying most but not all of the cells [18].

In this paper we present the progress of a related but distinct approach. We employ a modification of cellular programming to evolve non-uniform CAs in which global computation occurs by local interactions among computational demes of interacting cells. The notion of a *deme* comes from biology where it refers to the formation of local subpopulations (demes) that are sufficiently isolated to permit differentiation in a population [19]. In evolutionary computation, the term pertains to various implementations of evolutionary algorithms in which several subpopulations are evolved in parallel [20]. The motivation for co-evolving demes of cells is to attain a co-evolutionary model in which the co-evolution to quasi-uniformity could be avoided. The feasibility of such a model has been discussed in another paper [21]. Here, we concentrate on how the process of "building" emergent computation is related to the co-evolution of computational demes of co-evolving cells. The paper reveals some of the mechanisms that co-evolving demes of non-uniform cellular automata employ to perform non-trivial computation, such as the synchronisation task.

In the next section, we introduce the CAs model and describe the synchronisation task. Section 3 defines the co-evolutionary algorithm derived from the cellular programming approach. Section 4 investigates the evolution of computational demes for global synchronisation. Finally, we give conclusions and intentions for future work.

2 Cellular Automata and the Synchronisation Task

A CA consists of an array of cells which accept a finite number of states, k, and a transition rule by which the states of the cells are changed synchronously in discrete time, t. We say that the cells interact locally since the state of each cell at the next time step is specified by the current states of the cell and its surrounding. The state of a cell together with the states of the surrounding cells is called a *neighbourhood* of the cell. The size of the neighbourhood refers to the number of inputs of the CA rule. Since we consider one-dimensional CAs, the size of each neighbourhood is $2r + 1$ where r is the rule radius. We say that a CA is non-uniform, if its cells accept different transition rules.

The behaviour of a CA can be illustrated by its *space-time diagram* in which the configuration of states in the grid is given as a function of time. Such an illustration is given in Figure 1 which represents the evolution of one-dimensional uniform and non-uniform CAs with $k = 2$ and rule radius 3. Each grid is taken with spatially periodic boundary conditions, meaning that the grid is considered as a circle in which the leftmost and rightmost cells are immediate neighbours. The space-time diagrams represent the evolution of CAs of 149 cells, starting from a randomly generated initial configuration and iterated over 149 time steps with time increasing down the page. The diagrams are depicted with white and black dots which represent cells in states 0 and 1, respectively.

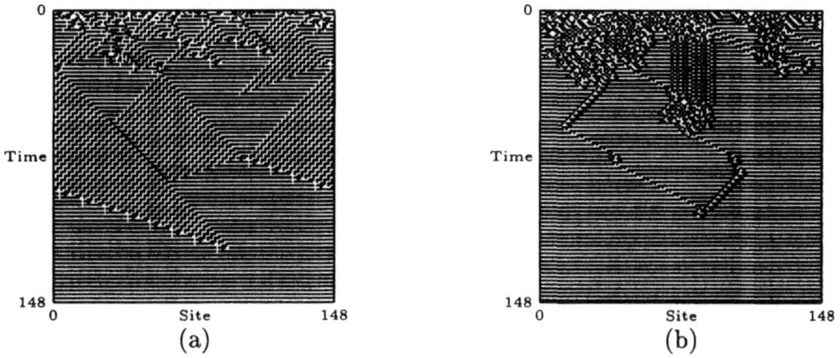

Fig. 1. The synchronisation task. Space-time diagrams of (a) uniform and (b) non-uniform CAs, dicovered by evolutionary techniques. The initial configurations are chosen randomly with densities of 1's $\rho(0) \approx 0.51$. The CAs are one-dimensional with grid size 149, $k = 2$, and rule radius $r = 3$.

The global synchronisation task is to locate a CA which performs a simple oscillation between 0's and 1's configurations of the lattice. The task is non-trivial for rule radius $r \ll N$ since the synchronous oscillation is a global property of the cellular array which must be attained by local interactions between neighbouring cells.

The synchronisation task was introduced in [15]. They employed a genetic algorithm [22] to evolve a population of cellular automata rules for a uniform CA which can solve the task. The fitness value of each CA is the fraction of correct oscillations of 0's and 1's, attained after $M = 149$ iterations of randomly chosen initial configurations with densities of 1's that are uniformly distributed over $\rho(0) \in [0.0, 1.0]$.

The synchronisation task was also studied by [18] who employed cellular programming to co-evolve non-uniform CAs. The fitness value of each co-evolved non-uniform CA is the average of the fitness values of all cells calculated in the same manner as [15] with respect to each cell.

Figure 1 illustrates the evolution of two solvers of the synchronisation task: (a) uniform CA [15], and (b) non uniform CA co-evolved by cellular programming. The depicted plots reveal how the CAs converge to synchronous oscillations between 0's and 1's configurations for a certain number of time steps $M = 149$.

3 The Co-evolutionary Model

We show how it is possible to co-evolve non-uniform CAs which are able to perform emergent computation at two different levels of self-organisation. In this method, the global computation performed by a particular evolved non-uniform CA is attained by interactions among computational demes of cells where the computation of each deme arises from the interaction of all the cells belonging to the deme. Such a simple model is easily identifiable in many systems in nature capable of self-organisation [23]. The algorithm is a modification of cellular programming given by Sipper [17] and it is defined as follows:

1. Initialise the population of N cells at random with CA rules, uniformly distributed among different λ values (the percentage of all the entries in the rule table which map to non-zero states) [24], and set N demes, each of which has size 1 (a single cell).
2. Generate $2(N+1)$ initial configurations at random with uniformly distributed densities of 1's over the interval $[0,1]$.
3. Run the CA $N+1$ steps for each configuration and attach a fitness value to each cell by determining the number of correct states of the cell in the final configurations, then calculate the fitness value of each deme by averaging the fitness values of the deme's cells.
4. Label each cell with a fitness score that specifies the number of fitter neighbouring cells which are elements of the cell's deme.
5. Update the population.
   (a) Update the demes as follows: (1) if a deme of one cell is surrounded by fitter demes then the deme (cell) is *invaded* by the fittest deme; (2) if a deme of two or more cells has a fitter neighbour, then the border cell is *invaded* by the fitter neighbouring deme. The invaded cell is labelled with the highest possible fitness score.
   (b) Update the cells as follows: if the fitness score of a cell is: 0, leave the rule unchanged; 1, replace the rule with the mutated fitter neighbour even if it is invaded by any neighbouring deme; 2 or higher, replace the rule by recombination of two fitter neighbouring rules even if they are invaded by neighbouring demes, followed by mutation.
6. If not finished, go to step 2.

An important part of the algorithm that needs additional explanation is how to determine when a deme can *invade* cells of its neighbours. We say that a deme *invades* a cell, when the deme expands over the cell. It is obvious that if a deme, which is fitter than its neighbours, is always able to expand over the grid then the population will converge very soon to only one deme, and the evolution in the deme's level will subside. To avoid the very fast expansion of the fitter demes over the grid, we employ a threshold function $\tau(t) = \frac{\varrho}{n(t)}$, called deme's tolerance, in which the parameter ϱ is a real number, $\varrho \geq 0$, and $n(t)$ is the number of demes at evolutionary step t. So, a deme, d', will invade a cell of a neighbouring deme, d'', if the deme's fitness value $f(d')$ is higher than $f(d'') + \tau(t)$. Initially, for $0 < \varrho \ll N$, the tolerance $\tau(t)$ is very small so deme's invasion is promoted, later as $n(t)$ drops, the demes only invade if they are substantially fitter. The mechanism allows us to maintain several isolated demes in a grid of strongly interacting cells, and thus, to attain a higher order of cellular computation.

The algorithm is similar to cellular programming in the following respect. The evolutionary process in the two levels of self-organisation is organised in a completely local manner. First, the interactions among the demes are local, since each deme can see only its immediate neighbours, and the deme can invade only one cell per neighbour. Second, the interactions among the cells are local, since the rule of each cell can be replaced with a rule obtained by applying mutation or recombination to rules of neighbouring cells. The mutation is uniform with a certain probability of flipping a bit of the rule, and the operator for recombination is onepoint crossover.

The co-evolutionary model, introduced above, differs from many parallel evolutionary algorithms defined mainly on the coarse-grained island or fine-grained cellular models [20, 18] in three aspects. Firstly, the evolution is performed in two different levels of self-organisation – evolution of demes, and evolution of cells within the demes, which in turn refers to the aforementioned island and cellular models, respectively. Secondly, the model deals with simple co-evolutionary mechanisms which lead to the appearance of interacting demes, each of which may differ from the other in size. The demes are implicitly defined, since they *appear* during the evolution by a simple unionisation of neighbouring cells. The formation, evolution, and extinction of demes lead to the evolution of unique cellular macrostructures with size strongly dependent upon the capabilities of the corresponding cells to perform useful computation. Lastly, the algorithm maintains a population in which the communications between the demes are *implicitly* defined. At this point, our model strongly differ from the other coarse-grained parallel evolutionary algorithms in which the demes' interconnections are explicitly specified by a certain migration scheme. In our model, the demes communicate implicitly through the environment which is strongly dynamical, and the communications appear as a result of the endeavour of the demes to cooperate in order to perform useful computation.

The non-uniform cellular computation attained by co-evolving computational demes is extremely intriguing. We performed many experiments with different tolerance parameters in which various synchronising non-uniform CAs were discovered. The co-evolved non-uniform CAs consist of a small number of demes each of which is a quasi-uniform CA. In our experiments, we observed two different computational strategies for achieving global synchronisation, explicitly studied in [21]. In short, the computational strategies are

Strategy I Some of the demes are perfect solvers of the task while the remainder are perfect parasites - harmless and helpless - only capable of propagating the information signals to the nearest solver(s).

Strategy II The global synchronisation appears by cooperation of all demes each of which unable to perform the task. In this strategy, the decisive com-

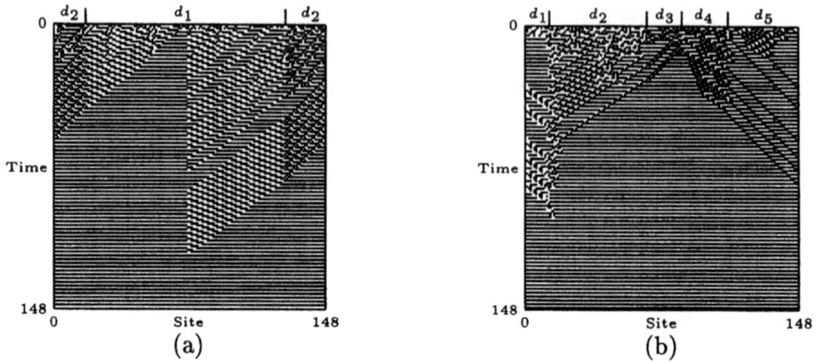

Fig. 2. Synchronising computational demes: space-time diagrams of co-evolved non-uniform CAs that perform the synchronisation task by (a) strategy I, and (b) strategy II. The CAs are one-dimensional with grid size 149 and $k = 2$, $r = 2$. The initial configurations are chosen randomly.

putation is performed by two or more demes, while the remainders ensure and accelerate the computation in the aforementioned group of demes.

The strategies are illustrated in Figure 2 which depicts the space-time diagrams of two co-evolved non-uniform CAs.

4 The Co-evolution of Demes for Synchronisation

In this section, we examine in details the process of co-evolution of demes, using the model above, for the synchronisation task. The co-evolved non-uniform CAs are one-dimensional with two possible cell states, and rule radius 2 ($k = 2$ and $r = 2$). The grid consists of 149 cells with spatially periodic boundary conditions. The deme's tolerance parameter ϱ is set to 1.2. The mutation probability is 0.001.

We describe the process whereby global synchronisation is constructed, by following the average fitness of the population in a typical co-evolutionary run. According to the average fitness plot depicted in Figure 3, we identify six epochs, samples of which, obtained at generations (a) 1, (b) 12, (c) 28, (d) 37, (e) 46, and (f) 241, are described in Figures 4 and 5. The first epoch starts at generation 0 and each of the following epochs are characterised by significant improvements in the ability of the non-uniform CA's strategy to solve the synchronisation task. For each epoch, we investigate the features of the non-uniform CA that characterise the epoch: firstly, by examining the distribution of the cells amongst the computational demes, given in Figure 4, and secondly, by analysing the space-time diagram of the discovered non-uniform CAs, depicted in Figure 5. The latter figure also reveals how the computational demes are located in the cellular lattice with respect to the two reference demes d_{40} and d_{55}. So, having Figures 4 and 5, we can easily determine the demes and their locations in the grid. For instance, the CA obtained at generation 241 consists of demes d_{55}, d_{70},

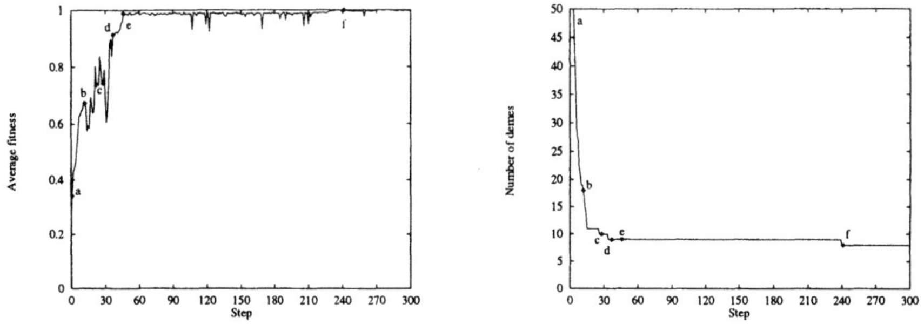

Fig. 3. Co-evolving demes of non-uniform CAs for global synchronisation. Results of a typical co-evolutionary run with rule radius 2: (a) the average fitness of all grid cells, and (b) the number of computational demes.

d_{72}, d_{93}, d_{104}, d_{113}, d_{115}, and d_{145} (Figure 4f), which are the 1st, 2nd, 3rd, etc. demes, starting from deme d_{55} (Figure 5f).

In addition, we study how the performance, P, of the co-evolved non-uniform CAs is changed. The performance of a CA is a measure of the fraction of the initial configurations for which the CA successfully synchronises in 149 iterations. The performance is scaled in the interval $[0, 1]$ and it is estimated over 300 randomly chosen initial configurations with a uniform distribution of densities of 1's over the interval $[0, 1]$.

Epoch 1 (generations 0–11) Initially, the number of demes is 149 which determines very *low* tolerance among the demes. Since the cells have different abilities for solving the task the number of demes decreases in a way that is faster than linear. The epoch can be called a *wild phase*. It is characterised by a fast extinction of unicellular organisms and a formation of multicellular ones. The formation of multicellularity is also related to the appearance of co-evolution within the demes. At the end of the epoch, it leads to the formation of local areas of the lattice that successfully synchronise on a small fraction of the input configurations. The performance of the CAs in this epoch is increased from 0 to approximately 0.02. An example of a non-uniform CA, which is labelled with "a" in the presented figures, is taken from generation 1. Its space-time diagram (Figure 5a) shows a huge diversity of cellular computations represented at this stage of the co-evolution.

Epoch 2 (generations 12–20) Although, the average fitness is significantly higher, the fraction of the initial configurations that the CAs are able to synchronise is low, $P \approx 0.04$, which indicates that only local synchronisation is attained. This is shown in Figure 5b which presents the space-time diagram of the non-uniform CA attained at generation 12. The CA consists of 18 different demes (Figure 4b), the majority of which perform localised synchronisation for a small fraction of initial configurations individually without cooperation. The epoch is characterised by a fast extinction of demes (Figure 3b), especially those comprising a small number cells.

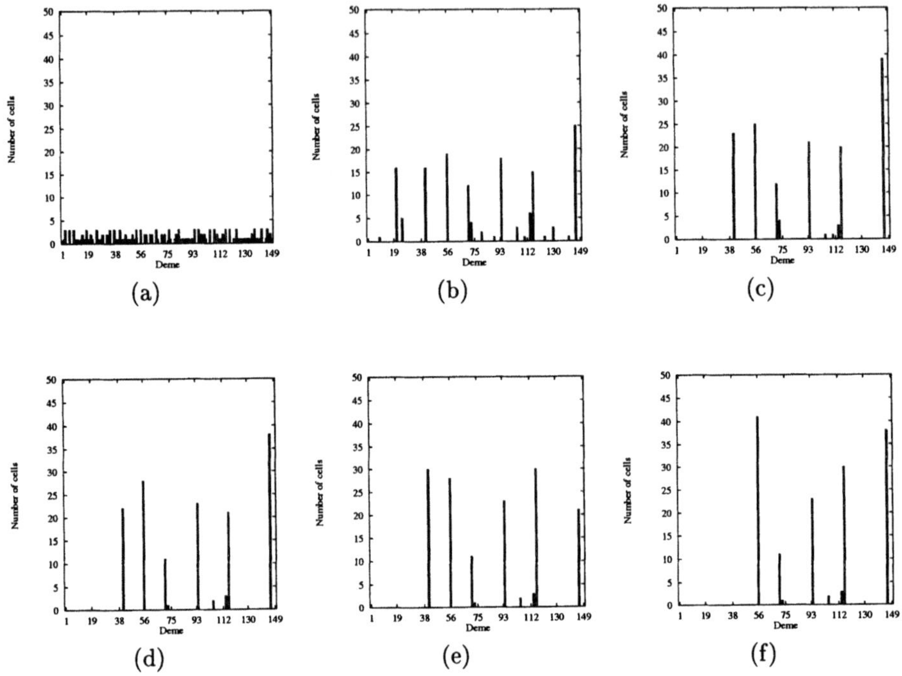

Fig. 4. The distribution of cells amongst the computational demes attained at generations (a) 1, (b) 12, (c) 28, (d) 37, (e) 46, and (f) 241, as labelled in Figure 3.

Epoch 3 (generations 21–33) The average fitness increase is followed by an increase in the performance of the co-evolved non-uniform CAs from 0.073 to 0.286. This is caused by the appearance of cooperative demes as is revealed in Figure 5c which shows the evolution of a non-uniform CA attained at generation 28. It can be seen that the localised synchronisation starts to expand over the whole grid, however only until it meets deme d_{145} which is a relatively poor solver when compared with the remainder of the grid. We suggest that the emergence of cooperation appears as a result of the co-evolution within the demes which is indicated by the slight changes of the demes characteristics (Figures 3b and 4c).

Epoch 4 (generations 34–45) The epoch is characterised by the extinction of deme d_{109} which does not change the computational strategy of the CA, and the evolution of deme d_{145} which leads to an increase of the CAs performance to 0.576. The space-time diagram of the CA obtained at generation 37 is given in Figure 5d. It is demonstrated that the computation performed by d_{145} is significantly improved, however, the deme is still badly adapted to its surroundings d_{40} and d_{115}. According to Figures 4d and 5d, deme d_{145} is the 7th deme on the right of d_{55}.

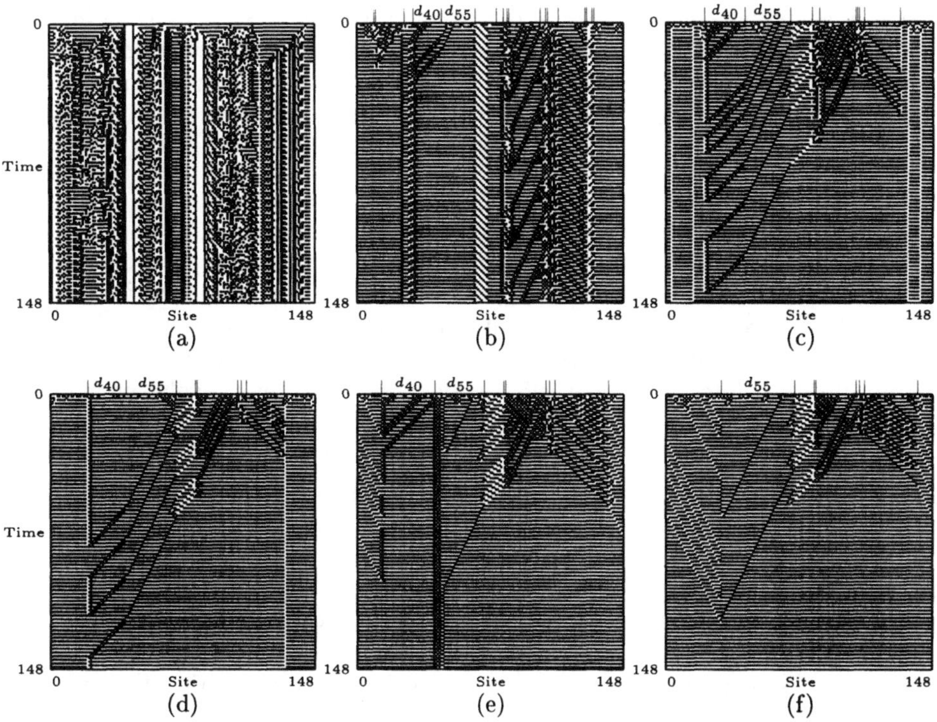

Fig. 5. The evolution of computation in co-evolving demes for synchronisation. The space-time diagrams illustrate the evolution of the non-uniform CAs attained at generations (a) 1, (b) 12, (c) 28, (d) 37, (e) 46, and (f) 241, as labelled in Figure 3. The initial configurations are chosen randomly with densities of 1's (a) $\rho \approx 0.074$, and (b)-(f) $\rho \approx 0.503$.

Epoch 5 (generations 46–240) The co-evolution of deme d_{145} observed in the previous epoch led to the discovery of a new computational strategy of the deme. It is marked with a rapid increase of the CA's performance to $P \approx 0.81$ which is also caused by the re-adjustments amongst the demes (Figure 4d-e). The evolution of the non-uniform CA attained at generation 46 is shown in Figure 5e. It can be seen that d_{145} is perfectly matched with demes d_{40} and d_{115}. However, we noticed that often d_{145} taken together with d_{40} synchronises in more than 149 iterations. The figure also shows that the changes in demes d_{40} and d_{55} led to another unsolvable computational problem, captured in the space-time diagram. The epoch is characterised with the evolution of d_{40} which finally becomes extinct.

Epoch 6 (generations 241–300) The epoch starts with the extinction of deme d_{40} followed by a fitness increase to 0.999 and an increase of the CA performance to approximately 0.997. The problem is perfectly solved for 300 CA iterations. The type of the computation of the solver, shown in Figure 5f, is described in section 3 as computational strategy II. Although the deme's tolerance is high, it seems that the deme d_{40} was unable to adapt in the co-evolved scenario, which caused its extinction, followed by a perfect match between d_{145} and d_{55}. Significant changes in the remainder of the grid were not observed (Figures 4f and 5f).

5 Conclusions

We have studied the co-evolution of synchronising non-uniform CAs in which the global information processing appears as a result of the co-evolution of computational demes, each of which is a system of interacting cells, capable of global computation at a different level of self-organisation. To attain co-evolution in two levels of self-organisation, we defined a threshold function that increases as the number of demes decreases so that the dynamics of the interactions amongst the demes were increasingly suppressed. It allowed us to co-evolve amazing computational strategies for non-uniform CAs that perform the synchronisation task.

We were able to study a simple process of formation of multicellular organisms, each of which may differ from the other in size (the number of cells). A co-evolutionary model of interacting organisms that differ from each other in size and mass has been studied by [6]. The author has studied the importance of body size for adaptation in heterogeneous ecological communities. Our investigation goes a little further. It suggests that the adaptation in an ecosystem is not only related to the size of an organism but also to the organism's ability to cooperate in a complex environment. This is clearly demonstrated in the last two figures, where for instance, at generation 46 demes d_{40} and d_{72} consist of 30 and 1 cells, respectively, however, at generation 241 the deme d_{40} no longer exists while deme d_{72} still consists of a single cell.

The presented model offers two main directions for future work. Firstly, it will be interesting to investigate how the model will work for other computational tasks, such as the density classification task. Also, how the computational strategies of the co-evolved CAs are related to other values of the rule radius of the CAs? Secondly, the model can be improved by taking account of our intuition that better co-evolutionary search would be attained if we relate the deme's tolerance to the size of the demes. This remains for the future.

References

1. G. Y. Vichniac, P. Tamayo, and H. Hartman. Annealed and quenched inhomogeneous cellular automata. *J. Stat. Phys.*, 45:875–883, 1986.
2. H. Gutowitz. *Cellular Automata: Theory and Experiment.* MIT Press, Cambridge, MA, 1991.
3. M. Sipper. Non-uniform cellular automata: Evolution in rule space and formation of complex structures. In R. A. Brooks and P. Maes, eds., *Artificial Life IV*, pp. 394–399. MIT Press, Cambridge, MA, 1994.
4. P. Hogeweg. Mirror beyond mirror, puddles of life. In C. Langton, ed., *Artificial Life*, pp. 297–316. Addison-Wesley, Reading, MA, 1988.

5. K. Lindgren and M. G. Nordahl. Artificial food webs. In C. G. Langton, ed., *Artificial Life III*, pp. 73–103. Addison-Wesley, Reading, MA, 1994.
6. A. R. Johnson. Evolution of a size-structured, predator-prey community. In C. G. Langton, ed., *Artificial Life III*, pp. 105–129. Addison-Wesley, Reading, MA, 1994.
7. J. von Neumann. *Theory of Self-Reproducing Automata*. Univ. of Illinois, 1966.
8. S. Wolfram. *Theory and Applications of Cellular Automata*. World Scientific, 1986.
9. T. Toffoli and N. Margolus. *Cellular Automata Machines*. MIT Press, Cambridge, MA, 1987.
10. M. Mitchell. Computation in cellular automata: A selected review. In H. G. Schuster and T. Gramss, eds., *Nonstandard Computation*. VCH Verlagsgesellschaft, Weinheim, 1996.
11. M. Sipper. Simple + parallel + local = cellular computing. In A. E. Eiben, T. Bäck, M. Schoenauer, and H.-P. Schwefel, eds., *PPSN V*, pp. 653–662. Springer, Berlin, 1998.
12. S. Forrest. Emergent computation: Self-organizing, collective, and cooperative phenomena in natural and artificial computing networks. *Physica D*, 42:1–11, 1990.
13. J. P. Crutchfield and M. Mitchell. The evolution of emergent computation. *Proc. Natl. Acad. Sci. U.S.A*, 92:10742–10746, 1995.
14. M. Mitchell, J. P. Crutchfield, and P. T. Hraber. Evolving cellular automata to perform computations: Mechanisms and impediments. *Physica D*, 75:361–391, 1994.
15. R. Das, J. P. Crutchfield, M. Mitchell, and J. E. Hanson. Evolving globally synchronised cellular automata. In L. J. Eshelman, ed., *Proc. 6th ICGA*, pp. 336–343. Morgan Kaufmann, San Francisco, CA, 1995.
16. M. Land and R. Belew. No perfect two-state cellular automata for density classification exists. *Phys. Rev. Lett.*, 74(25):5148–5150, 1995.
17. M. Sipper. Co-evolving non-uniform cellular automata to perform computations. *Physica D*, 92:193–208, 1996.
18. M. Sipper. *Evolution of Parallel Cellular Machines: The Cellular Programming Approach*. Springer, Berlin, 1997.
19. S. Wright. Stohastic processes in evolution. In J. Gurland, ed., *Stohastic Models in Medicine and Biology*, pp. 199–241. Univ. of Wisconsin, Madison, WI, 1964.
20. W. N. Martin, J. Lienig, and J. P. Cohoon. Island (migration) models: Evolutionary algorithms based on punctuated equilibra. In Th. Bäck, D. B. Fogel, and Z. Michalewicz, eds., *Handbook of Evol. Comp.*, chap. C6.3:1. Oxford University Press, New York, NY, 1997.
21. V. K. Vassilev, J. F. Miller, and T. C. Fogarty. Co-evolving *demes* of non-uniform cellular automata for synchronisation. In A. Stoica, D. Keymeulen, and J. Lohn, eds., *Proc. 1st NASA/DoD Worksh. Evol. Hard.*. IEEE Computer Society Press, Piscataway, NJ, 1999. To appear (available via http://www.dcs.napier.ac.uk/~vesselin/papers/eh99-1.ps.gz).
22. J. Holland. *Adaptation in Natural and Artificial Systems*. MIT Press, Cambridge, MA, 1992. 2nd edition.
23. G. Nicolis and I. Prigogine. *Self-Organization in Nonequilibrium Systems*. Wiley Interscience, New York, NY, 1977.
24. C. G. Langton. Computation at the edge of chaos: Phase transitions and emergent computation. *Physica D*, 42:12–37, 1990.

Evolutionary Cybernetics

Exploiting the Power of Sensory-Motor Coordination

Stefano Nolfi[1], Domenico Parisi[1]

[1] Institute of Psychology, National Research Council (CNR),
15 Viale Marx, 00137, Rome, Italy
{nolfi, parisi}@ip.rm.cnr.it

Abstract. One important implication of embodiment is that, by acting, agents partially determine the sensory patterns they receive from the environment. The motor actions performed by an agent, by modifying the agent's position with respect to the external environment and/or the external environment itself, partially determine the type of sensory patterns received from the environment. In this paper we investigate how agents can take advantage of this ability. In particular, we discuss how agents coordinate sensory and motor processes in order to (1) select sensory patterns which are not affected by the aliasing problem and avoid those which are; (2) select sensory patterns such that groups of patterns which require different responses do not strongly overlap; (3) exploit emergent behaviors that result from the interaction between the agent and the environment.

1. Introduction

Recently, a new research paradigm has challenged the traditional view according to which intelligence is an abstract process that can be studied without taking into consideration the physical aspects of natural systems. In this new paradigm, research tends to stress the importance of situatedness (i.e. the importance of studying systems, natural or artificial, which are situated within an external environment) and embodiment (i.e. the importance of study systems which have bodies, receive input from their sensors and produce motor actions as outputs). We will refer to systems which are embodied and situated as *agents*.

One important implication of embodiment is that, by acting, agents partially determine the sensory patterns they receive from the environment. The motor actions performed by an agent, by modifying its position with respect to the external environment and/or the external environment itself, in fact partially determine the type of sensory patterns that it will receive from the environment. As we will see below, agents can take advantage of this ability in different ways. We will refer to the process of exploiting the agent-environment interaction (i.e. the ability to select sensory patterns that are useful for some purpose through certain motor actions) as *sensory-motor coordination* (for a similar view see [1]). We will show three different ways in which sensory-motor coordination can help to solve otherwise insoluble problems.

The way in which sensory-motor coordination can be exploited also depends on the characteristics of the agent. As shown in [2], for example, agents that are able to modify themselves (i.e. to learn) during their interaction with the environment can use sensory-motor coordination to select sensory patterns that are useful to learn (i.e. they can alter the frequency and the order of different sensory patterns to enhance the outcome of the learning process). In this paper however, we will restrict our analysis to the simplest case - that of pure sensory-motor agents which cannot modify themselves during their interaction with the environment and which do not retain any trace of the previously experienced sensory patterns. These agents, by definition, always react in the same way to the same sensory pattern.

2. How sensory-motor coordination can cope with the perceptual aliasing problem

One of the most straightforward ways in which sensory and motor processes can be coordinated is to solve the perceptual aliasing problem. Perceptual aliasing, a term coined by Whitehead and Ballard [3], refers to the situation wherein two or more identical sensory patterns require different responses in order to achieve a certain goal. When such a situation occurs (i.e. when an agent receives a sensory pattern that requires different motor responses in different circumstances) the agent should act in order to select other sensory patterns until a sensory patterns which is not affected by the aliasing problem (i.e. an unambiguous sensory pattern) is encountered.

Consider for example the case of a Khepera robot [4] (Fig. 1, left) which is placed in an environment containing two types of object: one with the top painted black, which should be avoided, and one with the top painted white, which should be approached (Fig. 1, right). The Khepera robot is provided with 8 infrared proximity sensors which can detect the bottom part of the obstacles up to a distance of about 3 cm. Moreover, the robot is provided with a K213 linear camera which has 64 photoreceptors and produces a linear image composed of 64 pixels of 256 gray-levels each, subtending a view-angle of 36°. On the motor side, the robot is provided with two wheels controlled by two motors which can rotate in both directions.

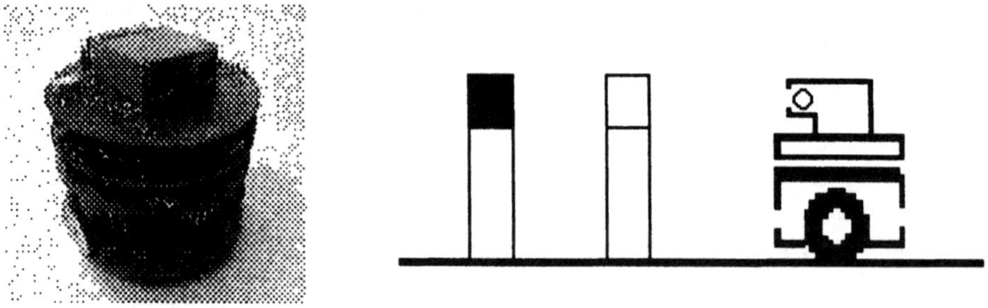

Fig. 1. The robot and the environment. The robot has 8 infrared proximity sensors that can detect the bottom part of the obstacles and a linear camera that can detect the color of the top part of the obstacles.

Every time such a robot approaches an object that does not happen to be in the viewing angle of its camera, it will experience an ambiguous sensory pattern (i.e. a sensory pattern which is affected by the aliasing problem). In fact, the same type of sensory pattern will be experienced by the robot both in the case of objects to be approached and in that of objects to be avoided. The obvious solution to this type of problem is to turn toward the object. By turning so as to have the object within the 36° view angle of the camera the robot will finally receive a unambiguous sensory pattern (i.e. frontal infrared sensors on and white image for objects to be approached and frontal infrared sensors on and a black image for objects to be avoided). The process of selecting sensory patterns that are easy to discriminate through motor actions is usually referred to as *active perception* [5]. Several examples of processes falling within this category have been identified in natural organisms. As demonstrated by Dill et al. [6], for example, in order to recognize certain visual patterns, drosophila moves so as to shift the perceived image to a certain location in the visual field.

3. How sensory-motor coordination can simplify hard problems

In the previous section we saw how the problem of the same sensory pattern requiring different answers can be solved by selecting other unambiguous sensory patterns, if any exist, through sensory-motor coordination. In this section we will investigate problems in which various sensory patterns requiring different motor answers strongly overlap, although not completely. When this happens, agents face hard problems that may be difficult or even impossible to solve. In this section we will try to show how sensory-motor coordination can turn these hard problems into simpler ones.

The distinction between simple and hard problems has been recently formalized by Clark and Thornton [7]. They introduced the term type-2 problems to denote hard tasks in which the problem of mapping input patterns into appropriate output patterns is complicated by the fact that the regularities[1], which can allow such mapping, are hidden or marginal within the sensory patterns. On the basis of this consideration these authors distinguished type-2 problems (i.e. hard problems) from type-1 problems (i.e. easy problems) in which a sufficient number of regularities are directly available in the sensory patterns.

As claimed by Clark and Thornton, type-2 problems which require complex input-output mapping may be reduced to type-1 (tractable) problems by re-coding sensory information so to enhance useful regularities. This can be achieved in two different ways. One possibility is to internally re-code sensory inputs so as to enhance useful regularities. Elman [8], for example, showed how complex tasks which cannot be solved by training a feed-forward neural network using standard back-propagation can be solved if the network is first trained using a simpler subtask and then exposed to the full task. As claimed in [7] and [8] this can be explained by considering that the first learning phase affects how the sensory patterns are re-coded at the level of the internal representations. This re-coding, by enhancing the regularities of the sensory patterns, turns the process of learning the entire task into a type-1 problem [7].

[1] The term 'regularity' refers to features of the sensory patterns which can be used to discriminate between classes of sensory patterns that require different answers. The exact meaning of the term will become clearer later on.

As shown by Scheier et al. [9], however, agents can also transform a type-2 problem into a type-1 problem by actively structuring their own input through sensory-motor coordination (interestingly, this strategy, as we will see, cannot be used by systems that are passively exposed to sensory states, i.e. systems which are not embodied and situated).

The authors considered the case of a Khepera robot which is supposed to approach large, and to avoid small, cylindrical objects. The robot is provided with six frontal infrared proximity sensors and 2 motor units that encode the actual speed of the two wheels. The environment is an arena surrounded by walls which contains small and large cylinders.

As shown by Nolfi [10,11] who reported a similar experiment in which a Khepera robot had to approach small cylindrical objects and avoid walls, the task of discriminating between the two types of objects is far from trivial for a agent that is required to passively discriminate between the sensory patterns produced by the two objects. This can be explained by considering that the sensory patterns received by the robot depend largely on the distance and on the relative angle between the robot and the objects. As a consequence, the sensory patterns belonging to the two categories largely overlap. "Put differently, the distance in sensor space for data originating from one and the same object can be large, while the distance between two objects from different categories can be small" [9, p. 1559]. On the other hand, as shown by Nolfi [10, 11] and by Scheier et al. [9], the task can easily be solved by an agent that is left free to perform sensory-motor coordination. Aside from these similarities the two experiments seem to call for different explanations. For this reason we will describe the experiment reported in [9] first and the experiment described in [10, 11] later.

In the simplest experiment reported in [9] Scheier et al. used artificial evolution to select the weights of the robot's neural controllers. Individuals' fitness was increased each time step they were close to a large object and decreased each time step they were close to a small object or a wall. As the authors show, performances increase during the first generations and stabilize in the vicinity of optimal performance after about 40 generations. In other words, while passive systems (i.e. neural networks which are required passively to classify a set of patterns corresponding to two different objects, see [10, 11]) display poor performance, agents that are allowed to exploit sensory-motor coordination can solve the task easily. The fact that the coordination between the sensory and the motor processes is crucial in solving this task can be clearly demonstrated by observing the behavior of evolved individuals and by observing how the distribution of the sensory patterns changes through the generations.

As reported in [9], the fittest individuals in 86% of the runs move in the environment until they start to perceive an object (large or small) and then start to circle around the object (the other 14% stop in front of the objects, although these individuals display significantly poorer performances). At this point the robot continues to circle around large objects while avoiding and abandoning small objects. This circling behavior is crucial to accomplish the discrimination between the two types of object given that the sensory patterns that the robot experiences while circling the small objects are significantly different from those that the robot experiences while circling the large objects. In other words, the sequence of motor actions leading to the circling behavior allows the robot to select sensory patterns that can easily be discerned.

The role of sensory-motor coordination has been further demonstrated by measuring the extent to which sensory-patterns belonging to different objects are separated for individuals of different generations (i.e. by measuring the difficulty of the discrimination task for different individuals). Indeed, given that the type of sensory pattern that an individual receives from the environment depends partly on how the individual reacts to each sensory state, individuals who behave differently may face harder or simpler discrimination tasks. To accomplish this measure the authors used the geometric separability index (GSI) proposed by Thornton [12], which can be used to quantify the distinction between type-1 and type-2 problems introduced above. In the case of this experiment, the GSI can be used to measure to what extent sensory patterns corresponding to one and the same object are close in sensory space and to what extent sensory patterns corresponding to different objects are separated. GSI is computed by counting the average number of times the sensory pattern nearest to the current sensory pattern falls into the same category. In the case of patterns corresponding to only two categories (small and large objects) this can be calculated in the following way:

$$GSI(f) = \frac{\sum_{i=1}^{N} \left(f(x_i) + f(x_i') + 1 \right) \bmod 2}{N}$$

where f is the category of the object, x are the sensory patterns consisting of N vectors, and x_i' is the nearest neighbor of x_i.

As reported in [9], in these experiments the GSI value starts from about 0.5 and monotonically increases during the first 40 generations until it reaches a stable state around 0.9 (note that performance also increases during the first 40 generations). This means that individuals of successive generations increase their ability to coordinate the sensory and motor processes so that experienced sensory patterns corresponding to one object are similar amongst themselves and are different from sensory patterns corresponding to other objects. In other words, evolved individuals are able to transform a type-2 problem into a type-1 one.

4. Exploiting emergent solutions

In the previous two sections we have considered the case in which it is difficult to react appropriately to part of the sensory patterns because they are affected by the aliasing problem (i.e. the same sensory pattern requires different motor answers) or because the regularities present in the sensory pattern are hidden or marginal (i.e. groups of sensory patterns that require different answers largely overlap). As we showed, both problems can be solved by using sensory-motor coordination. In this section we will examine a similar problem and again we will see how it can be solved by using sensory-motor coordination. The way in which sensory-motor coordination solves this problem, however, seems to be qualitatively different from the two cases described above. Indeed, as we will see, this way of solving the problem is effective also in cases in which all sensory patterns are affected by the aliasing problem or, in other words, cases in which the sensory data do not contain any regularities at all.

As mentioned above, Nolfi conducted an experiment in which a Khepera robot was required to discriminate between walls and cylindrical objects by finding and remaining close to the latter [10, 11]. The environment was an arena of 60x35 cm surrounded by walls with a cylindrical object with a diameter of about 3 cm randomly distributed inside it. To develop the control system for this robot Nolfi used artificial evolution. Individuals' fitness was increased each time step they were close to the cylindrical object. After a few generations the best individuals succeeded in ending up close to the cylindrical object within 500 cycles most of the times. This means that they were able to discriminate between the two objects, avoiding walls and remaining close to the cylinders (see Fig. 2, thick line).

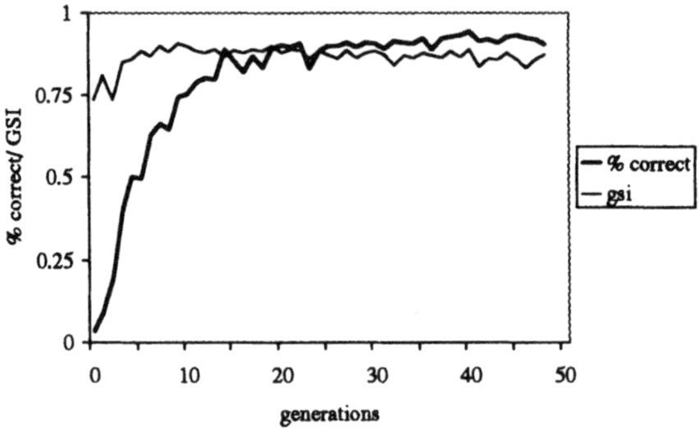

Fig. 2. Thick line: Percentage of times the best individuals of each generation succeed in ending up close to the cylindrical object after 500 cycles. **Thin line:** GSI of the sensory patterns experienced by individuals of successive generations. Average results of 10 replications.

Evolved individuals do not circle around objects (as in the case of the experiment described in the previous section). On the other hand, all evolved individuals start to move back and forth and/or left and right when they approach the cylinder. This emergent behavior can be described as a dynamical system and the relative positions with respect to an object in which individuals start to move back and forth or left and right while remaining in proximity to the object can be described as an attractor since the robot's trajectory converges on the same relative positions regardless of the direction of approach to the target. This can be seen in Fig. 3 which shows the trajectory of the movements produced by an evolved individual while approaching walls or cylinders (top and bottom, respectively). As can be seen, when the individual reaches a distance of about 20 mm from an object it avoids walls while it continues to approach targets until it reaches the attractor area located at a distance of about 15 mm and an angle of about 45 degrees. The trajectory of the motor responses in this area all converge toward the center of the area itself allowing the individual to keep more or less the same relative position with respect to the cylinder.

At this point we may wonder about the role of sensory-motor coordination in this type of emergent behavior. Can we conclude that, as in the experiments described in the previous section, sensory-motor coordination allows individuals to be exposed to data in which groups of sensory patterns belonging to different objects (walls and cylinders in this case) do not strongly overlap?

One way to answer this question is to look at the GSI index through generations. As we can see from Fig. 2 it stabilizes around 0.8 after the very first generation, while performance continues to increase through 50 generations. The fact that performance continues to increase significantly while GSI stabilizes and the fact that it stabilizes at a lower value with respect to the experiments described in the previous section (about 0.8 instead of 0.9) suggests that in this case the ability to experience sensory patterns belonging to two objects (walls and cylinders in this case) that are easy to discriminate plays a less important role.

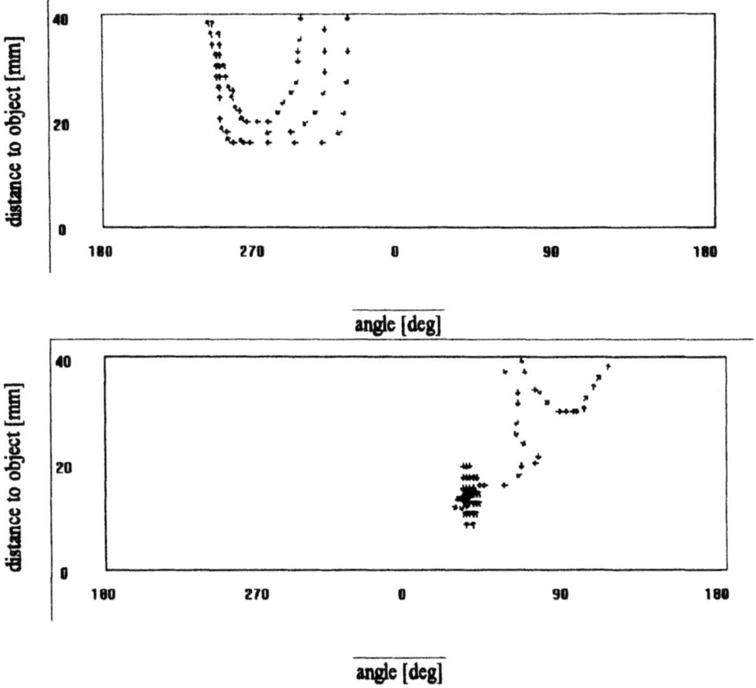

Fig. 3. Angular trajectories of an evolved individual close to a wall (top graph) and to a target (bottom graph). The picture was obtained by placing the individual in a random position in the environment, leaving it free to interact with the environment for 500 cycles, and recording the change in the relative positions with respect to the two objects for distances lower than 40 mm. Angle 0 corresponds to objects in front of the robot. For the sake of clarity arrows are used to indicate the relative direction but not the amplitude of the movements.

Another way to investigate the role of sensory-motor coordination in this experiment is to study another case in which the environmental conditions are designed to prevent the robot from behaving in such a way that the separability between groups of sensory patterns belonging to different objects increases. Consider the case of a simulated agent which lives in a circular strip divided into 40 cells (20 cells on the left and 20 on the right). At each time step the agent occupies one single cell and perceives the sensory state associated with the cell. There are 20 different sensory states that the agent can perceive, numbered from 0 to 19, which are each associated with a single cell both in the left and in the right part of the environment in a randomly different order (see Fig. 4). The agent can react to the current sensory state in two different ways (move one cell clockwise or anti-clockwise) and has the goal of reaching and remaining in the left part of the environment.

The agents have a neural network with 20 input units which locally encodes the corresponding perceived sensory state and 1 output unit which binarily encodes one of the two possible actions. As a consequence only one sensory unit is activated each time step. Weights can assume only two values (0 or 1). As a consequence, the weight connecting the input unit corresponding to the current sensory state to the output unit determines the motor reaction of the agent at each time step (move clockwise or anti-clockwise if the weight is 0 or 1, respectively). Individuals do not have any memory of the previously experienced sensory states (i.e. they always react in the same way to a given sensory state).

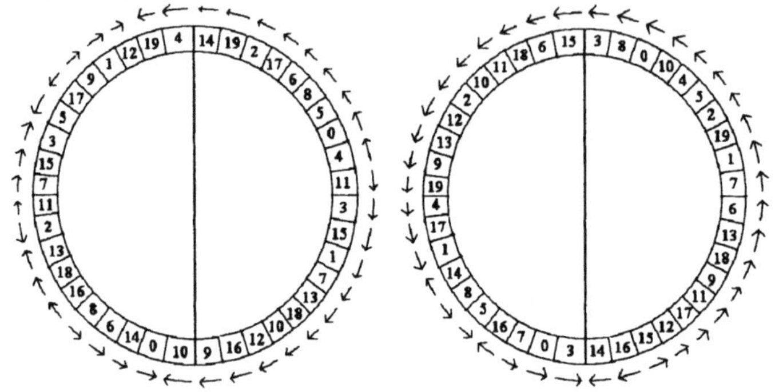

Fig. 4. Two environments. The numbers represent the sensory state experienced by the agent in each cell. Each different sensory state is present once in the left and once in the right part of the environment. Arrows indicate the motor reaction of a typical evolved agent.

What is interesting about this experimental situation is that all possible sensory states are affected by the aliasing problem. For each possible sensory state experienced, in fact, the animat has a 50% probability of being in the left or in the right part of the environment. In other words, in this case it is impossible to use sensory-motor coordination to select sensory states which are not affected by the aliasing problem or sensory states in which groups belonging to different categories (left and right side of the environment in this case) do not strongly overlap (in this case, in fact, all 20 different sensory states are equidistant in the input space).

Despite this, if we evolve a population of agents by selecting those that ended up in the left part of the environment after 200 cycles, after a few generations we obtain individuals that are able to move away from the right part of the environment and to remain in the left part[2]. The way in which evolved individuals solve this problem can be seen by observing the arrows in Fig. 4. In the right part of the environment individuals consistently move clockwise or anti-clockwise until they abandon the right side. Conversely, in some areas of the left side of the environment, individuals start to move back and forth, although remaining there for the rest of their life. Note that the way an individual reacts to a particular sensory state (for example the fact that the individual shown on the left side of Fig. 4 reacts clockwise to the sensory state '3') does not have any function in itself. The way in which an evolved individual reacts to a certain sensory state makes sense only if we consider how it reacts to all other sensory states. Agents solve the problem by reacting anticlockwise to the upper cells and clockwise to the lower cells in the right portion of the environment. This implies

[2] Evolving individuals were allowed to "live" for 100 epochs, each epoch consisting of 200 actions. Connection weights were binarily represented in the genotype which was 20 bits long. Population size was 100. The best 20 individuals of each generation were allowed to reproduce by generating 5 copies of their genotype with 2% of their bits replaced with a new randomly selected value. The experiment was replicated 10 times.

that there is only a single point on the right side of the environment, no matter where, in which two adjacent cells are responded to in different ways. This guarantees that the agent quickly leaves the undesired right side and moves to the desired left side. When the agent finds itself on the left side of the environment, the different spatial distribution of the cells on the left side with respect to the right side with their associated sensory states will ensure, with a very high probability, that the agent moves within the left side of the environment without ever leaving it. Therefore the problem is solved by implicitly exploiting the particular spatial distribution of sensory patterns in the environment.

Note that the way in which these simulated agents solve their task closely resembles the strategy adopted by the evolved robots described earlier in this section. The environment has only one dimension in this experiment while it has two dimensions in the previous experiment (the third dimension is irrelevant given that the robot can only move in two dimensions). This explains why agents move clockwise or anticlockwise in these experiments while robots moved back and forth and left and right in the previous experiments. On the other hand, the type of strategy adopted by evolved individuals is the same: react to sensory states to produce attractors (i.e. set of motor actions that result in a set of movements that, summed together, allow the individual to remain in the same position) in the left part but not in the right part of the environment in this experiment; react to sensory states to produce an attractor close to cylindrical objects but not close to walls in the previous experiment.

Obviously this task cannot be solved without exploiting sensory-motor coordination (e.g. it is impossible to train a network to produce different outputs for sensory patterns belonging to the left and to the right part of the environment). On the other hand, it is not possible either to solve this task: (1) by using sensory-motor coordination to select sensory states not affected by the aliasing problems or (2) by selecting sensory states in which groups belonging to different categories do not strongly overlap. This means that there is a third way in which sensory-motor coordination can solve hard tasks. This third way relies on emergent solutions (i.e. simple solutions relying on the dynamical interaction between the agent and the environment).

5. Conclusions

We have seen how sensory-motor coordination (i.e. the ability to select sensory patterns that are useful for some purpose through motor actions) can solve hard problems in three different ways: (1) by selecting sensory patterns which are not affected by the aliasing problem and avoiding those that are; (2) by selecting sensory patterns in which groups of patterns requiring different motor responses do not strongly overlap; (3) by reacting to each sensory pattern in a way that maximizes performance in view of how the agent reacts to all other sensory patterns.

The third way, which we have called emergent behavior, is in a way more radical than the other two categories. When we claim that a sensory pattern is affected by the aliasing problem or that two groups of sensory patterns that require different responses overlap strongly we are implicitly assuming a certain behavioral solution to a given task (e.g. we are assuming that the agent should react differently to patterns belonging to different objects or to different sides of the environment). However, often tasks can be solved in different ways and only some of these ways may present problems such as aliasing or lack of regularities among groups of sensory patterns that require different answers. Let us consider the last experiment. When we first think about this task we assume that the only way to solve it is to react differently to sensory states present in the left and in the right part of the environment (at least to some of them). When we then realize that all sensory patterns are present on both sides and equally distant in the input space we feel that there is no way to solve the problem (at least without taking into account previously experienced states). However, when we observe the behavior of the evolved agent, we see that there is a completely different way to solve the problem which does not require reacting differently to sensory states lying on the two sides.

When we leave individuals free to find their own solution to a task by interacting with the external environments, two different processes take place. On one hand, individuals select those strategies that are less affected by the aliasing problem or by the lack of regularities within groups of sensory patterns requiring different answers. On the other hand, given a certain selected strategy, individuals try to use sensory-motor coordination to avoid sensory patterns affected by aliasing problems and to experience sensory patterns so to increase regularities within groups of sensory patterns requiring different answers.

References

1. Pfeifer, R. & Scheier, C. Sensory-motor coordination: The metaphor and beyond. Robotics and Autonomous Systems. 20 (1997) 157-178.
2. Nolfi, S. Parisi, D.: Learning to adapt to changing environments in evolving neural networks. Adaptive Behavior. 1: (1997) 99-105
3. Whitehead, S.D. & Ballard, D. H.: Learning to perceive and act by trial and error. Machine Learning. 7 (1991) 45-83.
4. Mondada, R., Franzi, E. & Ienne, P.: Mobile robot miniaturization: A tool for investigation in control algorithms: In: T. Yoshikawa & F. Miyazaki (eds.): Proceedings of the Third International Symposium on Experimental Robots, Kyoto, Japan (1993)
5. Bajcsy, R.: Active Perception. Proceedings of the IEEE (76) 8: (1988) 996-1005
6. Dill, M., Wolf, R., Heisenberg, M.: Visual pattern recognition in drosophila involves retinotopic matching. Nature. 355: (1993) 751-753.
7. Clark, A. & Thornton, C.: Trading spaces: Computation, representation, and the limits of uniformed learning. Behavioral and Brain Sciences. 20 (1997) 57-90.
8. Elman, J.L.: Learning and development in neural networks: The importance of starting small. Cognition. 48 (1993) 71-99.
9. Scheier, C., Pfeifer, R. & Kunyioshi, Y. Embedded neural networks: exploiting constraints. Neural Networks. 11: (1998) 1551-1596.
10. Nolfi, S.: Adaptation as a more powerful tool than decomposition and integration. In: T. Fogarty and G. Venturini (Eds), Proceedings of the workshop on Evolutionary computing and Machine Learning, 13th International Conference on Machine Learning, Bari (1996)
11. Nolfi, S.: Evolving non-trivial behavior on autonomous robots: Adaptation is more powerful than decomposition and integration. In T. Gomi (ed.): Evolutionary Robotics, Kanata, Canada: AAI Books (1997)
12. Thornton C.: Separability is a learner's best friend. In: J.A. Bullinaria, D.W. Glasspool, & G. Houghton (eds.): Proceedings of the Neural Computation and Psychology Workshop: Connectionist Representations, London, Springer (1997).

Evolution of Neural Controllers with Adaptive Synapses and Compact Genetic Encoding

Dario Floreano and Joseba Urzelai

Laboratory of Microprocessors and Interfaces (LAMI)
Swiss Federal Institute of Technology, CH-1015 Lausanne (EPFL)
Dario.Floreano@epfl.ch, Joseba.Urzelai@epfl.ch

Abstract. This paper is concerned with artificial evolution of neuro-controllers with adaptive synapses for autonomous mobile robots. The method consists of encoding on the genotype a set of local modification rules that synapses obey while the robot freely moves in the environment [2]. The synaptic weights are *not* encoded on the genotype. In the experiments presented here, a "behavior-based fitness" function gives reproductive advantage to robots that can solve a sequential task. The results show that evolutionary adaptive controllers solve the task much faster and better than evolutionary standard (non-adaptive) controllers, that the method scales up well to large architectures whereas standard controllers do not, and that evolved adaptive controllers are not trivial and cannot be reduced to a fixed-weight network.

1 Evolution and Learning

Artificial evolution of adaptive individuals can provide computational advantages and richer adaptive dynamics [1] with respect to evolution of individuals whose defining parameters are entirely genetically-determined. Several hypotheses have been suggested to explain the observed advantages of the combination of evolution and learning [7, 8, 11]. In general, these advantages amount to discovery of better solutions for a given problem, to faster convergence, and to improved robustness in face of changing fitness landscapes. They are thus relevant for artificial evolution of robotic control systems.

Despite the growing worldwide interest in Evolutionary Robotics, remarkably little work has been done in this direction. A review of the combination of evolution and learning for sensory-motor controllers can be found in [5, 10]. Most of the work done so far and effectively applied to robots, or realistically simulated organisms, shares two components: all synaptic weights are individually specified and directly encoded on the genetic string, and learning amounts to some standard gradient-descent algorithm.

In previous work we employed a different approach where synaptic strengths are not genetically specified and adaptation during life consists of Hebbian synaptic changes [2–4]. For each synapse, the genetic string encoded four Hebbian rules, a learning rate, the sign, and the postsynaptic effect of the travelling signal (driving or modulatory). At the beginning of an individual's "life", all

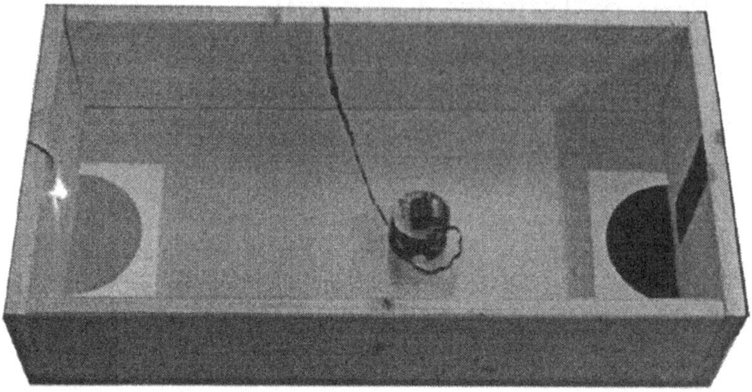

Fig. 1. A mobile robot equipped with a vision module gains fitness by staying on the gray area only when the light is on. The light is normally off, but it can be switched on if the robot passes over the black area positioned on the other side of the arena. The robot can detect ambient light and the color of the wall, but not the color of the floor.

synapses were initialized to small random values and, while the robot was freely moving around the environment, each synapse could modify its own strength every 100 ms according to the genetically specified Hebbian rule. Evolved individuals displayed more robust behaviors [2] and consistently won tournaments in a competitive co-evolutionary scenario [4].

In this paper, we extend previous work by using a much more compact genetic representation of adaptive neurocontrollers and systematically compare its performance with respect to direct encoding of synaptic weights and to encoding of noisy synapses. In a further set of experiments, we show that compact encoding of adaptive networks scales up to large neurocontrollers whereas direct encoding fails. Finally, we analyze a family of evolved controllers under different conditions and show that their competitive advantage comes indeed from evolved adaptive synapses.

2 Environment, task, architecture, and genetic encoding

A mobile robot Khepera equipped with a vision module is positioned in the rectangular environment shown in figure 1. A light bulb is attached on one side of the environment. This light is normally off, but it can be switched on when the robot passes over a black-painted area on the opposite side of the environment. A black stripe is painted on the wall over the light-switch area. Each individual of the population is tested on the same robot, one at a time, for 500 sensory motor cycles, each cycle lasting 100 ms. At the beginning of an individual's life, the robot is positioned at a random position and orientation and the light is off.

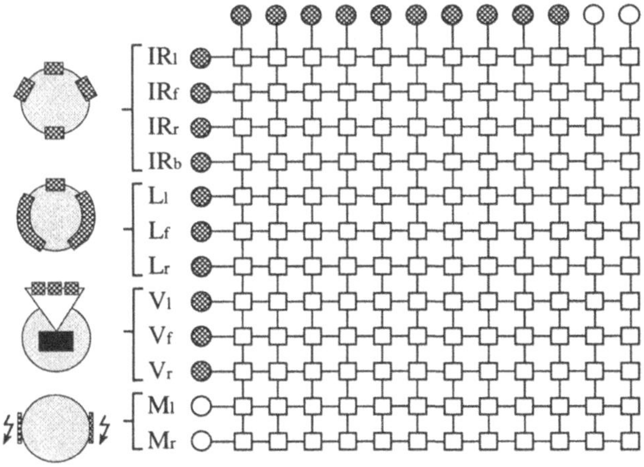

Fig. 2. The neural controller is a fully-recurrent discrete-time neural network composed of 12 neurons giving a total of 12 x 12= 144 synapses (here represented as small squares of the unfolded network). 10 sensory neurons receive additional input from one corresponding pool of sensors positioned around the body of the robot shown on the left (l=left; r=right; f=front; b=back). ***IR***=Infrared Proximity sensors; ***L***=Ambient Light sensors; ***V***=vision photoreceptors. Two motor neurons ***M*** do not receive sensory input; their activation sets the speed of the wheels ($M_i > 0.5$ forward rotation; $M_i < 0.5$ backward rotation)

The fitness function is described as the number of sensory motor cycles spent by the robot on the gray area beneath the light bulb *when the light is on* divided by the total number of cycles available (500). In order to maximize this fitness function, the robot should find the light-switch area, go there in order to switch the light on, and then move towards the light as soon as possible, and stand on the gray area[1]. Since this sequence of actions takes time (several sensory motor cycles), the fitness of a robot will never be 1.0. Also, a robot that cannot manage to complete the entire sequence will be scored with 0.0 fitness.

A light sensor placed under the robot is used to detect the color of the floor—white, gray, or black— and passed to a host computer in order to switch on the light bulb and compute fitness values. The color of the floor is *not* given as input to the neural controller. After 500 sensory motor cycles, the light is switched off and the robot is repositioned by applying random speeds to the wheels for 5 seconds.

The controller is a fully-recurrent discrete-time neural network (figure 2). It has access to three types of sensory information: infrared light (object proximity), ambient light, and vision. The active infrared sensors positioned around the robot

[1] Notice that the fitness function does not explicitly reward this sequence of actions, but only the final outcome of the overall behavior chosen by the robot. Therefore, we call it a *behavior-based* fitness function.

measure the distance from objects (up to 4 cm). Their values are pooled into four pairs and the average reading of each pair is passed to a corresponding neuron. The same sensors are used to measure ambient light too. These readings are pooled into three groups and the average values are passed to the corresponding three light neurons. The vision module consists of an array of 64 photoreceptors covering a visual field of 36°. The visual field is divided up in three sectors and the average value of the photoreceptors (256 gray levels) within each sector is passed to the corresponding vision neuron. Two motor neurons are used to set the rotation speed of the wheels. Neurons are updated every 100 ms according to the following equation

$$y_i \leftarrow \sigma \left(\sum_{j=0}^{N} w_{ij} y_j \right) + I_i,$$

where y_i is the activation of the ith neuron, w_{ij} is the strength of the synapse between presynaptic neuron j and postsynaptic neuron i, N is the number of neurons in the network, $0 \leq I_i < 1$ is the corresponding external sensory input, and $\sigma(x) = (1 + e^x)^{-1}$ is the sigmoidal function. $I_i = 0$ for the motor neurons.

Each synaptic weight w_{ij} can be updated after every sensory-motor cycle (100 ms) using one of the four modification rules specified in the genotype.[2] The four rules are called Hebbian because they are a function of the pre-synaptic activation, of the post-synaptic activation, and of the current value of the weight itself. The *Plain Hebb rule* strengthens the synapse proportionally to the correlated activity of the two neurons. The *Postsynaptic rule* behaves as the plain Hebb rule, but in addition it weakens the synapse when the postsynaptic node is active but the presynaptic is not. Conversely, in the *Presynaptic rule* weakening occurs when the presynaptic unit is active but the postsynaptic is not. Finally, the *Covariance rule* strengthens the synapse whenever the difference between the activations of the two neurons is less than half their maximum activity, otherwise the synapse is weakened. Synaptic strength is maintained within a range [0, 1] (notice that a synapse cannot change sign) by adding to the modification rules a self-limiting component inversely proportional to the synaptic strength itself [2, 3, for more details].

Encoding	Bits for one synapse					Bits for one node				
Genotype	1	2	3	4	5	1	2	3	4	5
A	sign	strength				sign	strength			
B	sign	Hebb rule		rate		sign	Hebb rule		rate	
C	sign	strength		noise		sign	strength		noise	

Table 1. Genetic encoding of synaptic parameters for Synapse Encoding (left) and Node Encoding (right). In the latter case the sign encoded on the first bit is applied to all outgoing synapses whereas the properties encoded on the remaining four bits are applied to all incoming synapses. A: Genetically determined controllers; B: Adaptive synapse controllers; C: Noisy synapse controllers.

[2] These four rules co-exist within the same network.

Two types of genetic (binary) encoding are considered (see table): *Synapse Encoding* and *Node Encoding*. Synapse Encoding is also known as direct encoding [12]. Every synapse is individually coded on 5 bits, the first bit representing its sign and the remaining four bits its properties (either the weight strength or its adaptive rule). Node Encoding instead codes only the properties of the nodes in the network. These properties are then applied to all its incoming synapses (consequently, all incoming synapses to a given node have the *same* properties). Each node is characterized by 5 bits, the first bit representing its sign and the remaining four bits the properties of its incoming synapses. Synapse Encoding allows a detailed definition of the controller, but for a fully connected network of N neurons the genetic length is proportional to N^2. Instead Node Encoding requires a much shorter genetic length (proportional to N), but it allows only a rough definition of the controller.

Independently of the type of genetic encoding, the following three types of properties can be encoded on the last 4 bits. A) *Genetically determined*: Weight strength. The synaptic strength is genetically determined and cannot be modified during "life". B) *Adaptive synapses*: Adaptive rule on 2 bits (four rules) and learning rate (0.0, 0.3, 0.6, 0.9) on the remaining 2 bits. The synapses are always randomly initialized when an individual starts its life and then are free to change according to the selected rule. C) *Noisy synapses*: Weight strength on 2 bits and a noise range on the remaining two bits $(0.0, \pm 0.3, \pm 0.6, \pm 0.9)$. The synaptic strength is genetically determined at birth, but a random value extracted from the noise range is freshly computed and added after each sensory motor cycle. A limiting mechanism cuts off sums that exceed the synaptic range $[0, 1]$. This latter condition is used as a control condition to check whether the effects of Hebbian adaptation amount to random synaptic variability.

In previous work we always resorted to Synapse Encoding and showed that evolution of adaptive synapses for an obstacle avoidance task develops levels of performance similar to those obtained by evolution of genetically-determined synapses [2–4]. Since in our approach adaptive synapses do not require a specification of initial strength, in this new set of experiments we have employed Node Encoding for adaptive synapses and systematically compared it to genetically-determined controllers using both Synapse Encoding and Node Encoding.

3 Experiments

The experiments have been carried out in simulations sampling sensor activation and adding 5% uniform noise to these values [9]. In addition, we have repeated the evolutionary experiments for the most significative conditions on the physical robot. Since the results on the physical robot do not differ significantly from those obtained in simulation, we report them in the appendix. For each experimental condition, 10 different[3] populations of 100 individuals each have been independently evolved for 200 generations. Each individual is tested three times

[3] Using different sequences of random number.

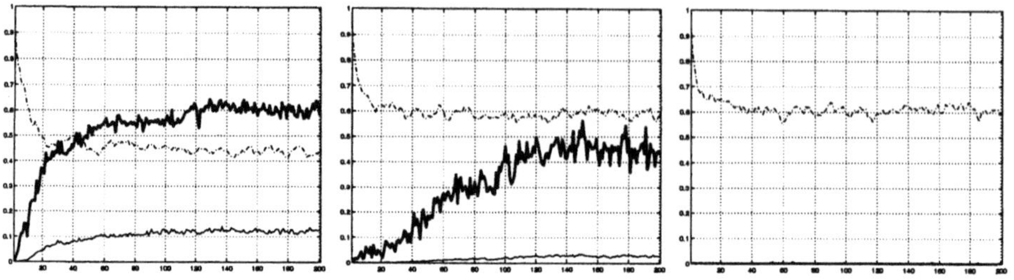

Fig. 3. Comparison of adaptive synapses with Node Encoding (*left*) versus genetically-determined synapses with Synapse Encoding (*center*) and Node Encoding (*right*). Thick line=best individual; thin line=population average; dashed line=genetic diversity. Each data point is an average over 10 replications with different random initializations.

and the fitness value is averaged. The 20 best individuals reproduce by making 5 copies of their genetic string. Strings are crossed over with probability 0.2 and mutated with probability 0.05 (per bit). In the case of adaptive synapses, synaptic weights of individuals are randomly initialized within the range $[0.0, 0.1]$ at the beginning of each test.

The fitness results reported in figure 3 show that individuals with adaptive synapses and Node Encoding (graph on the left) are much better than individuals with genetically-determined synapses and Synapse Encoding (graph in the center) in that: a) both the fitness of the best individuals and of the population report higher values (0.6 against 0.5); b) they reach the best value obtained by genetically-determined individuals in less than half generations (40 against more than 100); c) they display much less variability across generations. Individuals evolved with genetically-determined synapses and Node Encoding (graph on the right) never managed to complete the task reliably in any of the ten replications. The genetic variance[4] of the populations of adaptive individuals is reduced more markedly than in all other conditions, *probably* indicating a more reliable selection of individuals and preservation of genetic building blocks.

Two sets of control experiments –one using Synapse Encoding (figure 4, left) and the other Node Encoding (figure 4, right)– have been carried out using noisy synapses in order to check whether the improvements obtained by evolving adaptive synapses were simply due to a random sampling of the fitness surface surrounding each individual. In both cases the results were considerably worse than those obtained with adaptive synapses (figure 3, left) and than those obtained with genetically-determined synapses (figure 3, center).

[4] Measured as the average dispersion of individual vectors from the center of mass of the population and further normalized by the string length.

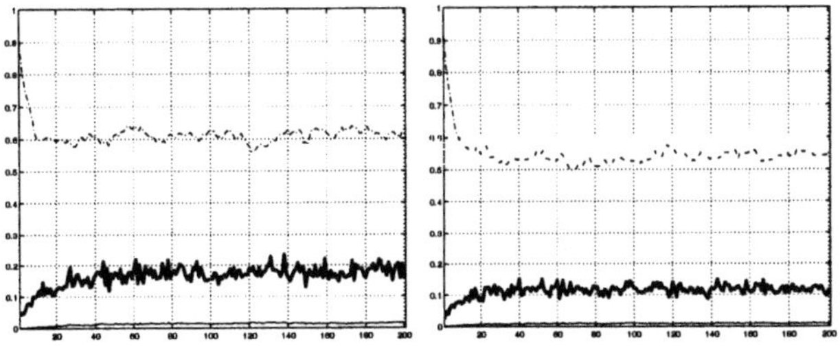

Fig. 4. Evolution of noisy synapses using Node Encoding (*left*) and Synapse Encoding (*right*). Thick line=best individual; thin line=population average; dashed line=genetic diversity. Each data point is an average over 10 replications with different random initializations.

3.1 Scaling up

The choice of a neural architecture is often difficult and may affect the outcome of an experiment. A large architecture may be computationally more powerful, but it may also entail a larger genotype and stronger epistatic effects. Unless one knows that a larger search space for the genotype/phenotype mapping considered has the same proportion of solutions as a smaller one, shorter genotypes may be preferrable because evolutionary search could be faster and more effective.

We have performed a new series of experiments using a larger neural network. The architecture shown in figure 2 was extended by adding 20 hidden neurons. These neurons were fully connected to themselves and to other neurons in the network, but did not receive sensory input and were not used to set the speeds of the wheels. The length of the genetic string grows from 60 to 160 bits for Node Encoding and from 720 to 5120 bits for Synapse Encoding. The results shown in figure 5 indicate that evolution of adaptive synapse with Node Encoding reports fitness values still comparable to the case of a smaller network;

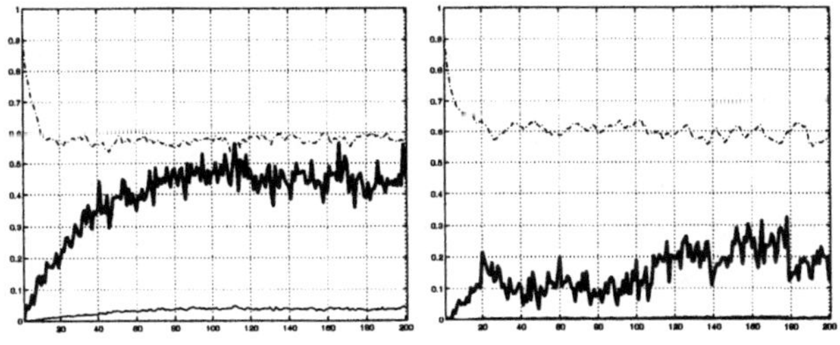

Fig. 5. Evolution of a large controller with 20 hidden nodes. *Left:* Adaptive synapses with Node Encoding. *Right:* Genetically-determined synapses with Synapse Encoding. Thick line=best individual; thin line=population average; dashed line=genetic diversity. Each data point is an average over 10 replications with different random initializations.

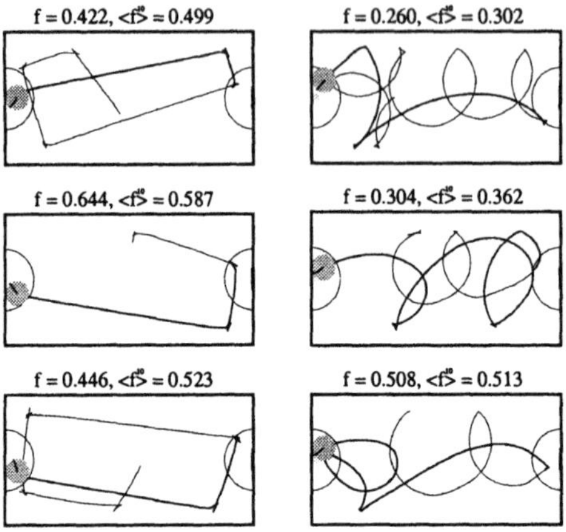

Fig. 6. Behaviors of three best individuals with adaptive synapses and Node Encoding (*left column*) and of three best individuals with genetically-determined synapses and Synapse Encoding (*right column*). Individuals belong to the last generation of three different replications (randomly chosen out of ten) for each condition. When the light is turned on, the trajectory line becomes thick. The corresponding fitness value is printed on the top of each box along with the average fitness of the same individual tested ten times from different positions and orientations.

instead, evolution of genetically-determined controllers with Synapse Encoding is badly affected in this condition. Evolution of genetically-determined synapses with Node Encoding (data not shown) remained closed to zero fitness, whereas evolution of synaptic strength and noise range with both Node Encoding and Synapse Encoding reported the same results as for the smaller network (data not shown).

The fact that genetically-determined controllers with Synapse Encoding perform badly may indicate that the search space here contains proportionally less solutions than the smaller search space of the network pictured in figure 2. The slower convergence and slightly lower fitness values of the controller with adaptive synapses (compare with left graph of figure 3) may be explained by the increased length of the genetic string, but also by the fact that the architecture is fixed and fully connnected. Since in Node Encoding the properties of a node propagate to all incoming synapses, there might be a high number of "parasitic" connections that cannot be individually eliminated. We shall come back to this point in the final discussion.

4 Behavioral Analysis

Figure 6 shows the behaviors of three best individuals evolved with adaptive synapses and Node Encoding (left) and with genetically-determined weights and Synapse Encoding (right). In both cases individuals aim at the area with the

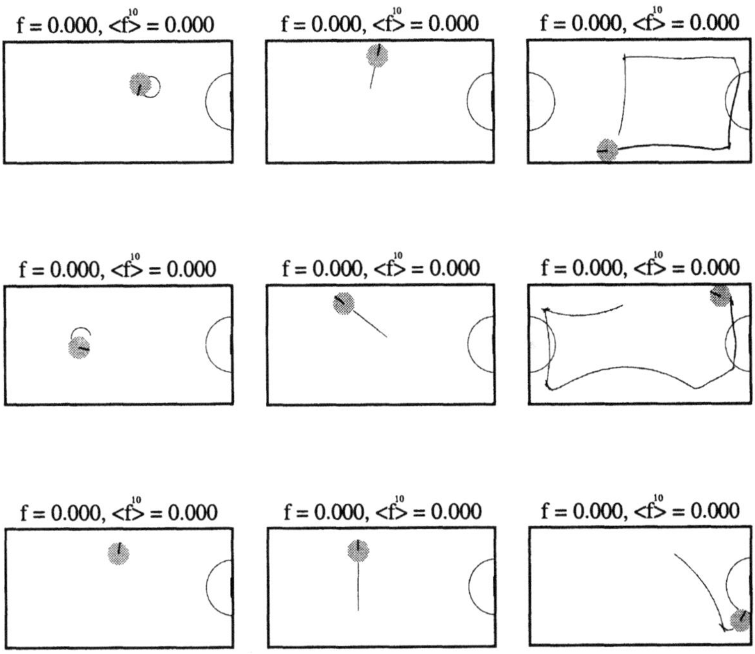

Fig. 7. Disabling adaptation for three best individuals with adaptive synapses (shown on the left of figure 6). *Left column:* Synapses are initialized to random values in the range [0.0, 0.1], as during evolution. *Center column:* Synapses are all initialized to 1.0. *Right column:* Synapses are set to their average value recorded during a full test of the individual. The corresponding fitness value is printed on the top of each box along with the average fitness of the same individual tested ten times from different positions and orientations. The values are always 0.0 because none of the individuals ever manage to complete the task under these test conditions.

light switch[5] and, once the light is turned on, they move towards the light and remain there. The better fitness of the adaptive controllers (given on the top of each box, see figure caption) is given by straight and faster trajectories whereas genetically-determined individuals display loopy trajectories (and sometimes are not capable of standing still on the fitness area, as in the case of the third individual on the bottom right of the figure).

Another set of tests has been carried out to assess the role of adaptation in the behavior of the individuals with adaptive synapses. For example, one might argue that what matters is the sign of the synapse and not its strength as long as it is non-zero, or that adaptive synapses may have the same effect of fixed synapses with strengths set to their average values[6]. The same three best

[5] Their performance is badly affected if the vision input is disabled, indicating that they do not use random search to locate the switch (data not shown).

[6] This latter suggestion was made by Flotzinger [6] who replicated our previous experiments on Adaptive Synapses with Synapse Encoding

individuals with adaptive synapses shown in the left column of figure 6 were tested again disabling adaptation in three different conditions (figure 7). In the first condition the synapses were initialized to small random values in the range $[0.0, 0.1]$ (figure 7, left column), as during evolution. In the second condition, the weights were all set to their maximum strength 1.0 (figure 7, center column). In the third condition the weights were set to their average value (figure 7, right column). (The average values had been previously computed while testing the robot in adaptive mode and recording the synaptic strength of each connection after every update.) For each condition, the three individuals were tested ten times from different positions and orientations. None of the individuals ever managed to complete the task in any of the three conditions.

5 Conclusions

We have shown through a set of systematic comparisons that evolution of adaptive synapses brings a number of advantages with respect to evolution of synaptic weights. It can generate viable controllers in much less generations and the evolved controllers display more performant behaviors. Since adaptive synapses here need not be specified on the genetic string because their strength is always randomly initialized at the beginning of an individual's test, this approach can rely on a very compact genetic encoding that specifies only the adaptive properties of individual nodes. Such a compact encoding scales up very well to large networks with many synapses. The data obtained from control experiments with noisy synapses and from behavioral tests of evolved individuals with adaptation disabled all suggest that Hebbian adaptation plays a specific role in the functioning of the controllers both during evolution and during the "life" of an individual.

When describing our controllers with changing synapses, we have accurately avoided the term "learning" because we have no evidence that the controller acquires new knowledge or skills, or that it may easily acquire new abilities for a different task (implementing, for example, something functionally similar to reinforcement learning). However, we have used the term "adaptation" because synapses change according to the states of the sensors and of the other neurons in the controller of the robot. In other words they adapt their initial random configuration to a dynamically-stable configuration that depends on the behavior of the robot. The adaptation rules are genetically specified and have evolved to satisfy a specific fitness function. As they stand, our results indicate that we have developed a smart genetic specification of neural controllers suitable for evolution. One of our current projects aims at establishing to what extent can this approach scale up to more complex behavioral tasks and to other more traditional learning problems. Another project aims at testing (and possibly extending) this approach for behavioral problems where learning is traditionally considered necessary.

We think that evolution of adaptive synapses may be very suitable for evolving neural morphologies where one cannot specify the strength of individual

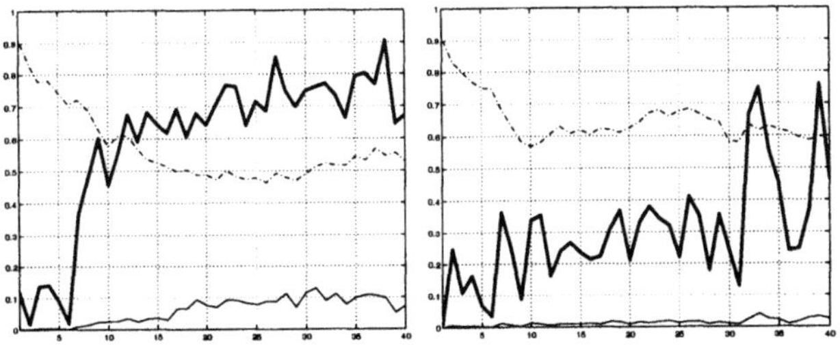

Fig. 8. Comparison of adaptive synapses with Node Encoding (*left*) versus genetically-determined synapses with Synapse Encoding (*right*) for experiments carried out on the physical robot.

synapses on the genotype and at the same time wishes to keep the genetic string as compact as possible. The methods proposed so far for evolution of morphologies all need very complex genetic encoding, require much domain-specific knowledge (e.g., symmetries, connectivity types), and have not yet been shown to be competitive with direct-coding methods. The Node Encoding scheme that we have proposed may be a first step in the direction of morphology evolution in the sense that synapse details are not specified in the genetic code, but are taken care of by adaptive online rules. In a current project we are extending our approach by adding genes for expression of connection growth and recursive rules to the node specification.

Appendix: Evolution on the physical robot

Two sets of experiments have been repeated on the physical robot: adaptive synapses with Node Encoding and genetically-determined synapses with Synapse Encoding (figure 8). The main differences from those carried out in simulations are: the population size is 80, each run lasts 40 generations, only one run has been carried out for each condition, and each individual is tested only once in the environment. The last restriction means that the effects of chance are more marked on the performance and are the cause of the higher oscillation observed, especially for the individual with genetically-determined synapses. These data should be compared to those shown in the graphs at the left and center of figure 3. The performances obtained with the physical robots are better than those obtained in simulation because the latter include severe constraints. For example, when a simulated robot pushes against a wall, it cannot move unless it backs away; instead, real robots can often get away by sliding against the walls.

Acknowledgements

Review of this paper has been organized by Inman Harvey with the help of three independent referees. We thank the referees and Inman for useful comments and criticisms, especially on the size of the search space in evolution. D.F. acknowledges support by the Swiss National Science Foundation, project nr. 21-49174.26. J.U. is supported by grant nr. BF197.136-AK from the Basque government.

References

1. R. K Belew and M. Mitchell, editors. *Adaptive Individuals in Evolving Populations: Models and Algorithms.* Addison-Wesley, Redwood City, CA, 1996.
2. D. Floreano and F. Mondada. Evolution of plastic neurocontrollers for situated agents. In P. Maes, M. Matarić, J-A. Meyer, J. Pollack, H. Roitblat, and S. Wilson, editors, *From Animals to Animats IV: Proceedings of the Fourth International Conference on Simulation of Adaptive Behavior*, pages 402–410. MIT Press-Bradford Books, Cambridge, MA, 1996.
3. D. Floreano and F. Mondada. Evolutionary Neurocontrollers for Autonomous Mobile Robots. *Neural Networks*, 11:1461–1478, 1998.
4. D. Floreano and S. Nolfi. Adaptive behavior in competing co-evolving species. In P. Husbands and I. Harvey, editors, *Proceedings of the 4th European Conference on Artificial Life*, Cambridge, MA, 1997. MIT Press.
5. D. Floreano and J. I. Urzelai. Evolution and learning in autonomous robots. In D. Mange and M. Tomassini, editors, *Bio-Inspired Computing Systems.* PPUR, Lausanne, 1998.
6. D. Flotzinger. Evolving plastic neural network controllers for autonomous robots. Msc dissertation 9580131, COGS, University of Sussex at Brighton, 1996.
7. I. Harvey. Is there another new factor in evolution? *Evolutionary Computation*, 4(3):313–329, 1997.
8. G. E. Hinton and S. J. Nowlan. How learning can guide evolution. *Complex Systems*, 1:495–502, 1987.
9. O. Miglino, H. H. Lund, and S. Nolfi. Evolving Mobile Robots in Simulated and Real Environments. *Artificial Life*, 2:417–434, 1996.
10. S. Nolfi and D. Floreano. Learning and evolution. *Autonomous Robots*, 7(1):forthcoming, 1999.
11. D. Parisi and S. Nolfi. How learning can influence evolution within a non-lamarckian framework. In R. K. Belew and M. Mitchell, editors, *Adaptive Individuals in Evolving Populations.* MIT Press, Cambridge (MA), 1996.
12. X. Yao. A review of evolutionary artificial neural networks. *International Journal of Intelligent Systems*, 4:203–222, 1993.

Synthetic Approaches to Neurobiology: Review and Case Study in the Control of Anguiliform Locomotion

Auke Jan Ijspeert**

LAMI-EPFL, IN-F Ecublens, CH-1015 Lausanne, Switzerland
ijspeert@lamisun1.epfl.ch,
WWW home page: http://diwww.epfl.ch/w3lami/team/ijspeert/

Abstract. This paper briefly reviews synthetic approaches to neurobiology and presents results of two experiments on the use of evolutionary algorithms for the design of neural controllers for locomotion. The first experiment consists in using the evolutionary algorithm for instantiating low level parameters of a connectionist simulation of the lamprey's locomotor circuitry. The second experiment develops potential neural circuits for the swimming and trotting of the salamander; an animal whose locomotor circuitry has currently not been decoded. In both cases, biologically plausible control circuits are developed which produce a neural activity with many similarities to that measured in the real animals.

1 Synthetic approaches to neurobiology

The fields of artificial life and artificial intelligence have developed tools and methods which have the potential to significantly help computational neurobiology. Synthetic approaches to neurobiology can indeed increase our understanding of the central nervous system, and this at two levels.

At a high —behavioural— level, fields such as *computational neuroethology* [1, 2], or also *synthetic psychology* [3], investigate how behaviour results from neural circuits through the development of neural controllers for artificial animats (robots or simulations). Models of escape and feeding behaviours in frog [4], insect locomotion [1, 5], fly vision [6, 7], cricket phonotaxis [8], classical conditioning [9] have, for instance, been simulated and/or implemented in real robots. These studies investigate hypotheses on central nervous systems by embedding neural models into bodies (simulated or real) in interaction with an environment. An interesting aspect of these investigations, compared to more traditional computational neurobiology, is therefore that they test the completeness of a model, that is, they verify whether all elements necessary for the production of an observed behaviour have been taken in account. They are also useful for analysing the effect of having a real body in terms of sensory feedback and body dynamics. Finally, their synthetic essence, i.e. the fact that, although biologically plausible, the developed neural models do not necessarily correspond to

** A large part of this work was carried out while the author was at the University of Edinburgh in the Department of Artificial Intelligence.

existing mechanisms, is interesting for investigating possible control mechanisms and, potentially, inspiring new neurobiological measurements.

At a lower level, techniques from artificial neural networks can be used as tools for completing neurobiological models. Backpropagation algorithms have been used for instantiating synaptic weights of a connectionist model of the escape reflex in a leech [10], and the locomotor circuit of the stick insect [5], for instance. More recently, evolutionary algorithms are being used for setting parameters of compartmental models of single neurons [11], or for defining synaptic weights in a model of the salamander's visual system [12]. The interesting outcome of these approaches is the development of tools which automatically instantiate multiple parameters of complex non-linear systems modelling biological circuits, given a description of their observed output.

We will next present two experiments in which a genetic algorithm is used for generating part of control circuits for anguiliform locomotion. In the first experiment, the genetic algorithm is used to instantiate synaptic weights of a neural circuit whose general structure is well known —the locomotor circuitry of the lamprey— while in the second experiment it is used for generating potential neural controllers for the locomotion of the salamander, an animal whose locomotor circuitry has not been decoded for the moment.

2 Design of the lamprey's swimming controller

2.1 Ekeberg's connectionist model

The lamprey —one of the earliest vertebrates— swims using an anguiliform swimming gait, i.e. by propagating a travelling undulation from head to tail. Its locomotor circuitry has been studied in detail by neurobiologists (see [13] for a review), and is known to be a *central pattern generator* (CPG) made of a chain of approximately 100 segmental oscillators located in the spinal cord (Figure 1).

Several models of that circuitry have been developed, and this research is based, in particular, on the connectionist model developed by Ekeberg [14]. That model simulates the complete 100-segment CPG of the lamprey organised as illustrated in Figure 1. It is composed of neuron units modelled as leaky integrators with a saturating transfer function which represent populations of functionally similar neurons in the real lamprey. The output u of a neuron unit corresponds to the mean firing frequency of the population it represents ($\in [0,1]$) and is calculated as follows:

$$\dot{\xi_+} = \frac{1}{\tau_D}(\sum_{i \in \Psi_+} u_i w_i - \xi_+) \tag{1}$$

$$\dot{\xi_-} = \frac{1}{\tau_D}(\sum_{i \in \Psi_-} u_i w_i - \xi_-) \tag{2}$$

$$\dot{\vartheta} = \frac{1}{\tau_A}(u - \vartheta) \tag{3}$$

$$u = \begin{cases} 1 - \exp\{(\Theta - \xi_+)\Gamma\} - \xi_- - \mu\vartheta & (u > 0) \\ 0 & (u \leq 0) \end{cases} \tag{4}$$

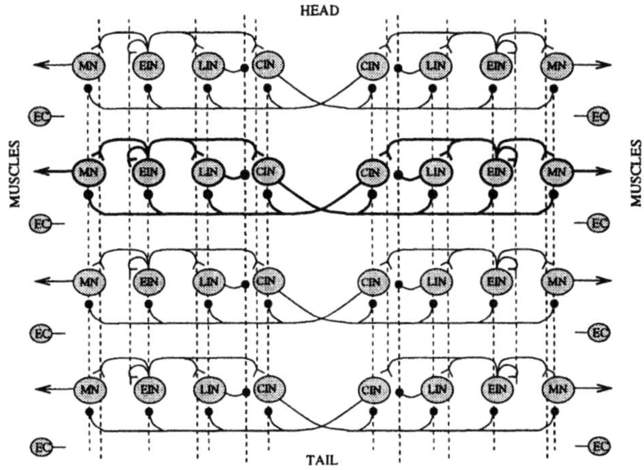

Fig. 1. Lamprey's swimming controller. The controller is made of 100 interconnected segmental oscillators (only 4 segments shown) composed of 8 neurons each. Four types of neurons are present in the oscillators: 3 types of interneurons EIN, CIN and LIN and the motoneurons MN. Sensory feedback is provided by stretch sensitive *edge cells* EC. The dashed lines indicate the projections from segmental connections to neighbouring segments.

where w_i are the synaptic weights, Ψ_+ and Ψ_- represent the groups of presynaptic excitatory and inhibitory neurons respectively, ξ_+ and ξ_- are the delayed 'reactions' to excitatory and inhibitory input, and ϑ represents the frequency adaptation observed in some real neurons.

The model is able to produce the following behaviours observed in the real lamprey: 1) when excitation is applied to the neurons of the different segmental oscillators, the segmental circuits develop an oscillatory activity with a frequency proportional to the level of excitation; 2) applying extra excitation to segments closest to the head leads the system to oscillate with small phase lags between segments which are constant over the spinal cord, therefore producing the typical wave of neural activity observed in anguiliform swimming; 3) for a given level of extra excitation, the wavelength of the undulation is independent of the oscillation frequency. Furthermore, when the motoneuron signals are used to determine the muscular activity of the simple mechanical simulation of the lamprey that Ekeberg developed, a swimming gait is produced which is very similar to that of real lampreys.

2.2 Parameter instantiation using a genetic algorithm

The implementation of a model such as Ekeberg's requires a significant amount of time for the setting of a large number of parameters, including the neuron parameters and the synaptic weights of all connections (Ekeberg, personal communication). We will here present how a genetic algorithm can be used as a tool for automatically instantiating those parameters, given a description of the desired behaviour of the system. This experiment follows evolutions of "artificial" controllers for swimming [15], i.e. controllers without the lamprey's connectivity. For a more detailed description of the results, see [16].

The evolved controllers are composed of the same type of neurons as those of Ekeberg and their connectivity corresponds to that observed in the lamprey. The design process is made in three stages, with first the development of segmental oscillators, then the development of intersegmental coupling and finally the development of sensory feedback connections from stretch sensitive cells.

Genetic algorithm. The same real number genetic algorithm is used for the three stages. Genes are real numbers between 0.0 and 1.0 which directly encode parameters of the neural controller (see below). Parents chromosomes are chosen with a rank-based probability, and children chromosomes are created with a 2-point crossover and a mutation operator. Mutation consists of modifying the old gene value by a small random value.

Stage 1: segmental oscillators. In this stage, the synaptic weights of the 26 connections within one segment are evolved. Because a left-right symmetry is assumed, chromosomes have 13 genes, which directly encode a synaptic weight through a linear transformation. The fitness function is defined to reward solutions which 1) produce regular motoneuron oscillations, and 2) have a frequency and an amplitude of oscillations which increase with the level of external excitation (for the mathematical definition of the function see [16]).

Ten runs were carried out with populations of 100 chromosomes for 500 generations. All populations converged to best solutions oscillating regularly and covering a large range of frequencies. Interestingly, the range of frequencies of the evolved oscillators (e.g. from 0.9 to 11.0 Hz) is much closer to that observed in the real lamprey (from 0.25 to 10.0 Hz) than Ekeberg's segmental oscillator (from 1.7 to 5.6 Hz).

Stage 2: intersegmental coupling. The second stage consists of developing the coupling connections between segmental oscillators. In the lamprey, oscillators are coupled through projections of segmental connections towards neighbouring segments. The extent of the projections are currently not known in detail, and for that reason Ekeberg chose a simplified coupling in which all segmental connections project symmetrically in the rostral and caudal directions except for the connections from the CIN neurons which project more caudally.

Here the GA is used to investigate potential coupling configurations between 100 copies of a chosen segmental oscillator. The chromosome encodes the extent of the projections of each segmental connection for both the rostral and caudal direction. The fitness function is defined to reward solutions which are able 1) to produce regular oscillations of the motoneurons in all segments, 2) to produce a travelling wave whose wavelength can be modulated by the extra excitation applied to the segments closest to the head, and 3) to produce swimming gaits covering a large range of speeds.

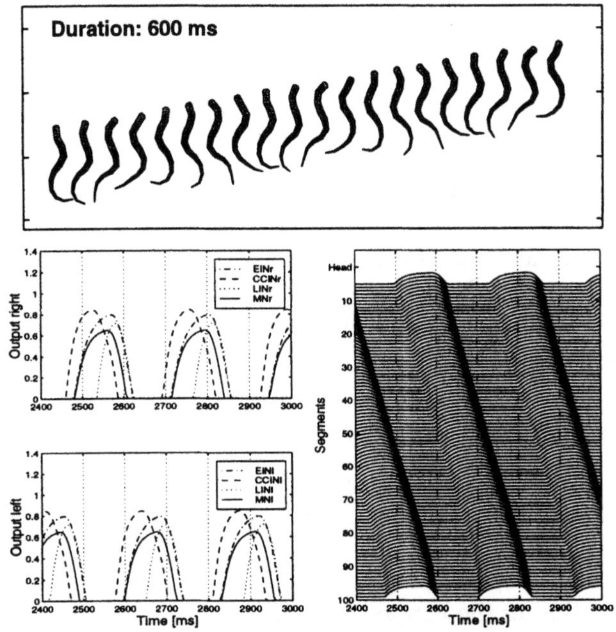

Fig. 2. *Top*: swimming gait produced by the optimised lamprey's controller. *Bottom*: Corresponding neural activity in the 50th segmental oscillator (*left*) and in the motoneurons along the left side of the spinal cord (*right*).

Five runs were realised with populations of 40 chromosomes for 100 generations. All five runs converged to controllers with similar performances to Ekeberg's biological model, in particular they cover larger ranges of lags and can reach slightly higher speeds. The range of lags of the best solution, for instance, varies between 0.0 and 3.5% of the oscillation period, which corresponds to the lags observed in the real lamprey (up to 3.0%) when local concentrations of excitatory bathes are varied [17]. Figure 2 illustrates the swimming gait produced by one of the evolved controllers.

Stage 3: sensory feedback from stretch sensitive cells. The last evolutionary stage consists of evolving the synaptic weights of sensory feedback connections from stretch sensitive cells. The lamprey has a series of inhibitory and excitatory stretch sensitive cells —the edge cells— located on both sides of the body which project to the segmental oscillators [18]. In [19], Ekeberg demonstrated that these cells could be useful for crossing a speed barrier (a local area with an increase of the speed of the water).

In order to further investigate how sensory feedback could be best used by the swimming CPG, the GA was used to generate weights for these feedback connections, given a fitness function rewarding the capacity to progress against the speed barrier with as small deviation as possible.

In all 5 runs tested (populations of 100 chromosomes, 100 generations), controllers were generated capable of crossing the chosen speed barrier (15 cm wide with a speed 40% higher than the lamprey's swimming speed). Interestingly, the evolved sensory feedback pathways correspond very closely to those observed in [18]: for all established (inhib. and excit.) biological connections, the evolved controllers have developed sensory feedback connections with the same sign.

3 Design of the salamander's locomotor controller

This second experiment concerns the salamander, an animal whose locomotor circuitry has not been decoded for the moment. The aim of the synthetic approach is here to investigate which kind of neural circuits can produce the observed gaits of the salamander.

3.1 Neurobiology of the salamander's locomotor circuitry

A salamander swims like a lamprey, and on ground it switches to a trotting gait with the body producing a standing wave coordinated with the movements of the limbs [20]. It has been hypothesised that the neural circuitry capable of producing both the travelling and the standing wave is based on a similar organisation to that of the lamprey [21, 22].

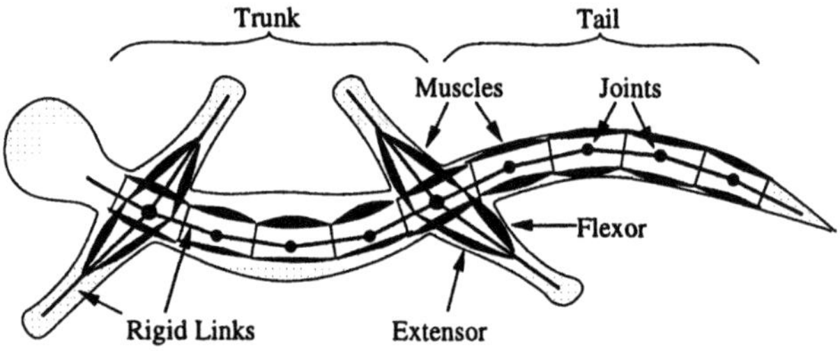

Fig. 3. Mechanical simulation of a salamander-like animat.

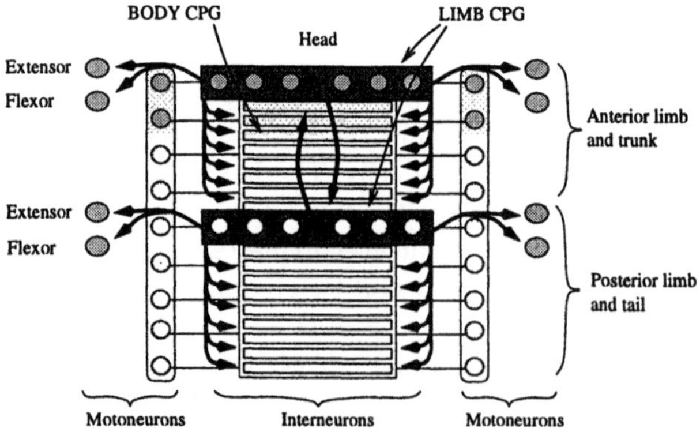

Fig. 4. Organisation of the evolved controllers for the salamander.

3.2 Evolution of potential locomotor controllers for the salamander

Following that assumption, we use a genetic algorithm to generate synaptic weights of a controller made of a lamprey-like body CPG and two limb oscillators which are copies of the body's segmental oscillators (Figure 4). The limb oscillators project to the motoneurons of the limbs and to the segmental oscillators of the body CPG, creating an unilateral coupling between them and the body CPG. [1] A simple 2D mechanical simulation of a salamander-like animat is developed for testing the swimming and trotting gaits (Figure 3, see [23] for a more detailed description). The aim is to be able to switch between the swimming and the trotting gaits by applying external excitation either to the body CPG or to both the body and the limb CPGs, respectively.

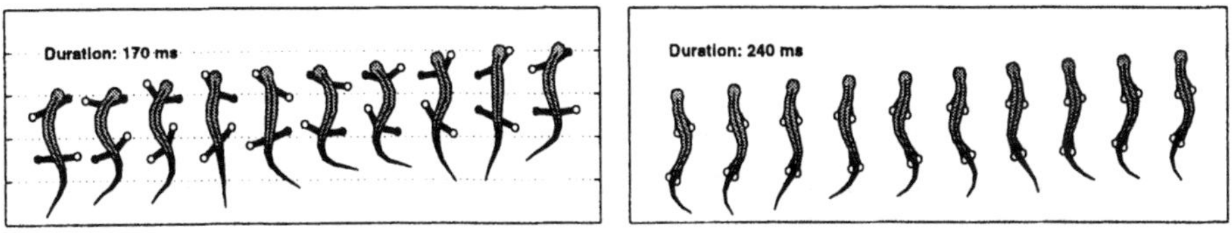

Fig. 5. Trotting (*left*) and swimming (*right*) salamander.

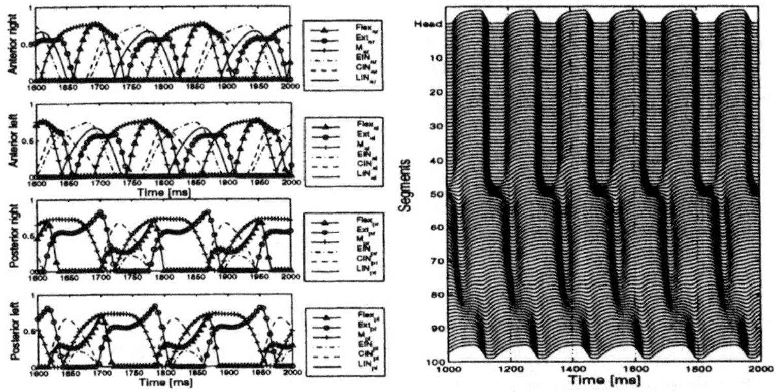

Fig. 6. Neural activity during trotting. *Left:* Neural activity in the limb oscillators (M_a and M_p represent the motoneuron activity of body segments 5 and 95, respectively). *Right:* Motoneuron activity along the left side of the body.

Chromosomes encode the synaptic weights of all possible connections from the two limb oscillators, as well as the connections from the brain stem to the limb motoneurons. The fitness function is defined to reward solutions which 1) trot as fast as possible, 2) can cover a large range of speeds when the excitation applied to both the body and the limb CPGs is varied, and 3) can change the direction of motion when left-right asymmetrical excitation is applied. The same real number genetic algorithm as for the experiment on the lamprey is used.

[1] This configuration is more biologically plausible than the one used in initial experiments in which there was no coupling between the swimming and the trotting CPGs [23].

Ten evolutions with populations of 100 chromosomes are carried out for 50 generations. All but 3 evolutions converged to controllers exhibiting a trotting gait with a trunk-limb coordination very similar to the real salamander (Figure 5, left). The speed of the trotting can be increased by increasing the amount of excitation, and applying a small asymmetry of excitation between the left and right sides of the CPGs leads to the salamander trotting in a circle. Finally, a lamprey-like swimming gait can be produced when excitation is applied only to the body CPG (Figure 5, right).

During trotting, the effect of the unilateral coupling from the limb oscillators on the body CPG is to force the anterior and posterior part of the body to oscillate in antiphase (Figure 6, right). Interestingly, the timing of the flexor and extensor limb motoneurons compared to the body motoneurons is very similar to that measured in the real salamander [22].

4 Discussion

These two experiments illustrated how a genetic algorithm could be used as a tool for neurobiological modelling. The interesting features of the method are:

1. GAs allow automatic instantiation of multiple parameters in complex non-linear models of central nervous systems. The evolution of controllers for the lamprey illustrate, for instance, that the GA can generate a significant part of the model that Ekeberg has designed by hand.
2. Specific characteristics specified by the user can be optimised. It was, for instance, possible to optimise the frequency range of Ekeberg's model and to obtain a better fit of biological data.
3. As illustrated with the salamander, the GA can also be used to investigate potential control mechanisms for biological systems whose structure is not known for the moment.

Compared to more traditional learning algorithms for artificial neural networks, such as variations of the backpropagation algorithm, for instance, a GA has the advantage that the fitness function does not need to be differentiable and that the desired output of the system can be described at a higher level. There is no need to provide a specific output cycle that the network should learn, and the desired behaviour of the system can, for example, be described in terms of a desired range of frequencies or the capacity of fast swimming. Note that the GA is here not used as a simulation of natural evolution, and that, similarly to [24, 25, 26], the staged evolution approach taken here rather corresponds to an "engineering" approach to artificial life.

It is hoped that, in the case of the salamander, this synthetic approach can provide news ideas for neurobiological measurements, and that a back and forth processus between modelling and measurements on the real animal will lead to a progressive improvement of the model by incorporating new neurobiological data when it becomes available.

Finally, the types of developed connectionist controllers may also be useful to robots using animal-like locomotion. The neural controllers are capable of transforming simple commands into the multiple rhythmic signals sent to the different actuators for efficient locomotion. They present the interesting property that by simply varying the amplitude of the commands, the speed, direction and type of locomotion can be modulated.

5 Conclusion

This paper briefly reviewed synthetic approaches to neurobiology and presented two experiments in the use of genetic algorithms for designing connectionist models for anguiliform locomotion. In these experiments, the genetic algorithm is used for instantiating synaptic weights of neural circuits whose structure corresponds to that decoded (for the lamprey) or hypothesised (for the salamander) in the real animal. It is found that 1) the GA is successful in automatically instantiating variables which require a long time to be set by hand, and 2) it can generate solutions which optimise high level characteristics specified by the user such as the speed of locomotion of a mechanical simulation.

Acknowledgements: Facilities were provided by the University of Edinburgh. Many thanks to John Hallam and David Willshaw for their comments on this project. This work was supported by a grant for young researcher from the Swiss National Science Foundation and by a Marie Curie grant from the European Commission.

References

[1] R.D. Beer. *Intelligence as Adaptive Behavior, an Experiment in Computational Neuroethology.* Academic Press, 1990.

[2] D. Cliff. Computational neuroethology. In M.A. Arbib, editor, *The handbook of brain theory and neural networks*, pages 626–630. MIT Press, 1995.

[3] V. Braitenberg. *Vehicles: Experiments in Synthetic Psychology.* MIT Press, Cambridge, 1984.

[4] A. Cobas and M. Arbib. Prey-catching and predator-avoidance in frog and toad: defining the schemas. *Journal of Theoretical Biology*, 157:271–304, 1992.

[5] H. Cruse, D.E. Brunn, Ch. Bartling, J. Dean, M. Dreifert, T. Kindermann, and J. Schmitz. Walking: A complex behavior controlled by simple networks. *Adaptive Behavior*, 3(4):385–418, 1995.

[6] N. Franceschini, J.M. Pichon, and C. Blanes. From insect vision to robot vision. *Phil. Trans. R. Soc. Lond. B*, 337:283–294, 1992.

[7] D. Cliff. Neural networks for visual tracking in an artificial fly. In F.J. Varela and P. Bourgine, editors, *Proceedings of the first European Conference on Artificial Life (ECAL91).* MIT Press, 1992.

[8] B. Webb. Robotics experiments in cricket phonotaxis. In D. Ciff, P. Husbands, J.M. Meyer, and S.W. Wilson, editors, *Proceedings, From Animals to Animats III.* MIT Press, 1994.

[9] P.F.M.J. Vershure, J. Wray, O. Sporns, G. Tononi, and G.M. Edelmann. Multilevel analysis of classical conditioning in a behaving real world artifact. *Robotics and Autonomous Systems*, 16:247–265, 1995.

[10] S.R. Lockery and T.J. Sejnowski. Voyages through weight space: network models of an escape reflex in the leech. In R. Beer, R.E. Ritzman, and T. McKenna, editors, *Biological neural networks in invertebrated neuroethology and robotics.* Academic Press, 1993.

[11] R.M.E. West and G.L. Wilcox. Robust parameter selection for compartmental models of neurons using evolutionary algorithms. In J.M. Bower, editor, *Computational Neuroscience: Trends in Research 1997*. Plenum Publishing, New York, 1997.

[12] W.C. Eurich, G. Roth, H. Schwegler, and W. Wiggers. Simulander: a neural network model for the orientation movement of salamanders. *Journal of comparative physiology*, 176:379–389, 1995.

[13] S. Grillner, T. Degliana, Ö. Ekeberg, A. El Marina, A. Lansner, G.N. Orlovsky, and P. Wallén. Neural networks that co-ordinate locomotion and body orientation in lamprey. *Trends in Neuroscience*, 18(6):270–279, 1995.

[14] Ö. Ekeberg. A combined neuronal and mechanical model of fish swimming. *Biological Cybernetics*, 69:363–374, 1993.

[15] A.J. Ijspeert, J. Hallam, and D. Willshaw. Evolving swimming controllers for a simulated lamprey with inspiration from neurobiology. *Adaptive Behavior*, 7(2), 1999. In press.

[16] Auke Jan Ijspeert. *Design of artificial neural oscillatory circuits for the control of lamprey- and salamander-like locomotion using evolutionary algorithms.* PhD thesis, Department of Artificial Intelligence, University of Edinburgh, 1999.

[17] T. Matsushima and S. Grillner. Neural mechanisms of intersegmental coordination in lamprey: local excitability changes modify the phase coupling along the spinal cord. *J. of Neurophysiology*, 67:373–388, 1992.

[18] G. Viana Di Prisco, P. Wallén, and S. Grillner. Synaptic effects of intraspinal stretch receptor neurons mediating movement-related feedback during locomotion. *Brain Research*, 530:161–166, 1990.

[19] Ö. Ekeberg, A. Lansner, and S. Grillner. The neural control of fish swimming studied through numerical simulations. *Adaptive Behavior*, 3(4):363–384, 1995.

[20] L.M. Frolich and A.A. Biewener. Kinematic and electromyographic analysis of the functional role of the body axis during terrestrial and aquatic locomotion in the salamander *ambystoma tigrinum*. *Journal of Experimental Biology*, 62:107–130, 1992.

[21] A.H. Cohen. Evolution of the vertebrate central pattern generator for locomotion. In A. H. Cohen, S. Rossignol, and S. Grillner, editors, *Neural control of rhythmic movements in vertebrates.* Jon Wiley & Sons, 1988.

[22] I. Delvolvé, T. Bem, and J.-M. Cabelguen. Epaxial and limb muscle activity during swimming and terrestrial stepping in the adult newt, *pleurodeles waltl. Journal of Neurophysiology*, 78:638–650, 1997.

[23] A.J. Ijspeert, J. Hallam, and D. Willshaw. From lampreys to salamanders: evolving neural controllers for swimming and walking. In R. Pfeifer, B. Blumberg, J.-A. Meyer, and S.W. Wilson, editors, *From Animals to Animats, Proceedings of to the Fifth International Conference of The Society for Adaptive Behavior (SAB98)*, pages 390–399. MIT Press, 1998.

[24] R.D. Beer and J.C. Gallagher. Evolving dynamical neural networks for adaptive behavior. *Adaptive Behavior*, 1(1):91–122, 1992.

[25] F. Gruau and K. Quatramaran. Cellular encoding for interactive evolutionary robotics. In P. Husbands and I. Harvey, editors, *Proceedings of the Fourth European Conference on Artificial Life, ECAL97*, pages 368–377. MIT Press, 1997.

[26] J. Kodjabachian and J.-A. Meyer. Evolution and development of neural networks controlling locomotion, gradient-following, and obstacle-avoidance in artificial insects. *IEEE Trans. on Neural Networks*, 9(5), 1998.

Replicating Experiments in "Detour Behavior" with Artificially Evolved Robots: An A-Life Approach to Comparative Psychology

Richard Walker[1] Orazio Miglino[2]

[1]Cognitive Science Group, Department of Psychology, University of Palermo, Viale delle Scienze, Parco d'Orleans, 90133 Palermo, Italy.
rwalker3@csi.com

[2]Institute of Psychology, National Research Council, Viale Marx 15, 00145 Rome, Italy
orazio@caio.irmkant.rm.cnr.it

Abstract. To be useful in psychology "artificial organisms" have to perform tasks comparable to those performed by animals. One way to achieve this is to replicate actual animal experiments. Here we reproduce an experiment showing "detour behavior" in chicks - a behavior usually explained in terms of "cognitive maps" or other forms of internal representation. We artificially evolve software-simulated robots with a "generic" ability to detour. Sensor-motor physics are carefully calibrated with data from a physical robot. Robot architecture is constrained to exclude internal representation. The evolutionary process rewards exploratory skills as well as detour behavior. Robot performance matches the results achieved in the original experiment. This proves that internal representations are not a necessary condition for primitive detour behavior and suggests that "detouring" evolves naturally from simpler behaviors. Future research will show whether it is possible to evolve more complex detour abilities using a similar bottom-up strategy.

1 Introduction

Experimental animal psychology is based on the study of animals performing well-defined tasks in closely controlled conditions. Psychologists use descriptions of behavior to infer animals' cognitive abilities, inputs and outputs, algorithms and internal representations [1]. Comparative studies provide insight into general mechanisms of perception and cognition. This mature tradition has produced a vast volume of reliable data and well-tried methodologies which pose a severe challenge for A-life researchers. An artificial organism, to be useful, has to perform tasks comparable to those that animals perform in the laboratory or the wild [2]: it has to acquire input from a noisy environment using imperfect sensors; motor mechanisms have to be based on realistic physics. Last but not least, the behavior of the organism has to be measured and recorded with the precision animal psychologists have come to demand. One way to achieve these goals is to replicate experiments from the past, no longer with animals but with robots produced by artificial evolution. In this paper we apply this approach to so-called "detour behavior" – the ability of an animal to negotiate an obstacle to reach a target.

2 Related work

2.1 A-life and Evolutionary robotics as a tool in psychological research

In 1984 V. Braitenberg suggested that it was possible to gain insight into problems of sensory-motor coordination by designing robots exhibiting specific forms of behavior [3]. In recent years a number of researchers began to build robots deliberately designed to test hypotheses in cognitive science. Robots have been built with the ability to simulate insect locomotion [4], to orient towards mating calls from female crickets [5] and to model theoretical models of animal navigation [6].

The challenge of designing "embodied" robots capable of operating in complex environments has been a source of important insights. Psychologists and computer scientists have come to recognize the difficulty of the tasks faced by autonomous agents and the unimagined simplicity of some of the possible solutions [7] [8].

While one school of researchers was developing "adaptive robotics" a second school has worked on the "artificial evolution" of software-based "animats". This approach avoids the limitations associated with manual design. Animats have been "evolved" with a broad range of "interesting" behaviors, such as pursuit and evasion [9]. There have however been charges that animats fail to address the complex problems inherent in the design of physical motor-sensory systems [10].

One way of combining the strengths of adaptive robots and "artificial evolution" is to build simulations which effectively reproduce the behavior of a physical robot. In our group [11] [12], we have developed a methodology for calibrating simulations with data from physical robots. In recent work we have "bred" robots which exhibit efficient exploratory behavior [13] Another theme of recurrent interest has been so-called "detour behavior" [14].

2.2 Detour behavior and its interpretations

When an animal seeks to reach a target it often meets an obstacle. The only way to reach the target is to take an indirect route during which it looses visual contact with the target. This is known as "detour behavior". Detour behavior has been demonstrated in many animals including chimpanzees [15], rats [16] and two day old chicks [17].

For Tolman and Honzik [18] detour behavior is incompatible with behaviorist theory. If there is no stimulus the animal's behavior cannot be a response. Detour behavior, they argue, requires "cognitive maps" of spatial relationships. As Bennett has pointed out, however [19], every experiment which has been explained in terms of "cognitive maps" can also be interpreted in different, often simpler, ways. The neurological evidence is inconclusive. Recording of single neuron activity in the hippocampus appears to indicate the existence of "place cells" which fire when an experimental animal is in a given location [20]. This finding is supported by evidence that hippocampal lesions disrupt landmark-based navigation [21]. Recently however it has been shown that neurons thought to be "place cells" may also respond to non-locational data [22].

2.3 Detour behavior in 2 day old chicks

The work reported in this paper replicates experiments by Regolin et al. [19] which have demonstrated the existence of detour behavior in two day old chicks. In these experiments the chicks are placed in a white cage divided in two by a barrier (see Figure 1). The part of the cage on the other side of the barrier contains a corridor. On the end of the corridor facing the target there is a grill through which the chicks can see the target. On each side of the corridor there are apertures leading into two compartments, one facing the target, and one facing in the opposite direction. The two compartments facing the target are labeled C and D.; the two compartments facing in the other direction are labeled A and B.

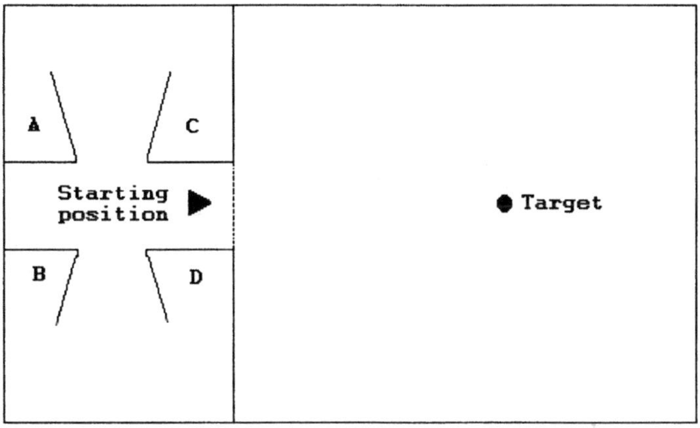

Figure 1: Diagram of apparatus for chick experiment

At the beginning of each session a chick is placed in the corridor close to the barrier and allowed to wander freely. The experimenter records the first compartment the chick enters and the time it takes to reach it. After 10 minutes the experiment is halted.

The results of the experiment show that of 25 chicks tested 5 failed to leave the corridor within the allotted time. Of the remaining 20 animals 18 chose compartments C or D; two choose compartments A and B. The excess of birds choosing the correct compartment was statistically significant (chi^2=12.80, df=1, p<0.001). There was no significant difference between the number of chicks entering compartment C and the number entering compartment D. These results demonstrate that the chicks were able to turn towards the goal in the absence of locally orienting clues. This, it is argued, shows the ability to maintain a representation of the goal after the loss of perceptual contact. Additional experiments (not replicated in this paper) showed that on repeated trials successful chicks would sometimes choose compartment C and sometimes compartment D. This seems to show that the birds "did not learn a fixed response, (i.e. turn left or right) but a position in space in egocentric coordinates" (i.e. turn left or right depending on the previous direction of turn).

3 Objectives and methods

The goal of the work reported in this paper was to "evolve" robots with the ability to replicate the behavior of the chicks yet whose architecture categorically excluded the presence of “cognitive maps” or other forms of internal representation. An analysis of the strategies, sensors and architectures used by successful robots, would lead, we hoped, to interesting explanatory hypotheses for detour behavior.
The experiments used a genetic algorithm operating on a population of 100 robots, simulated in software. The simulation software was designed to precisely emulate the well-known Khepera robot [23]. Input to the robot came from 8 infrared proximity sensors, 4 sensors linked to a linear video-camera and 3 "time sensors". The use of time sensors was motivated by previous work in which we showed that such sensors improve the efficiency of exploratory behavior [13]. Proximity sensors have a sensory field of 20° and are sensitive to obstacles within 3 cm of the sensor. Output (between 0 and 1) is a continuous, decreasing function of distance to the obstacle. The video-camera has a field of vision of 36°. Each sensors produces an output of 1 if the center of the target lies within its own 9° field of vision. The output values of the three time sensors were initially set to zero, increasing respectively by 0.01, 0.02 and 0.03 on each cycle of computation. When a sensor value reached 1 it was reset to zero.
The motor apparatus consisted of a left and a right wheel driven by stepping motors which can move both forwards and backwards. The motor apparatus was controlled by an Artificial Neural Network (ANN) with input neurons representing the state of the sensors and output neurons controlling the stepping motors. A number of different architectures were tested. The architecture finally chosen was based on a simple Perceptron [24] in which all sensors have a direct connection to the two output units. Evolution involved "mutations" in connection strengths. The genome of the organism consisted of a sequence of binary coded numbers (8 bits per number) representing the strengths of individual connections.

We evolved "detour behavior" by applying "artificial selection" to the population of ANNs. In the initial population connection strengths for individual networks were set to random real values uniformly distributed between -1 and 1.

In preliminary work it became clear that it would never be possible to evolve detour-capable robots without rewarding intermediate steps in the evolutionary process as well as the ultimate goal. We hypothesized that detour behavior might be derived from other, more primitive forms of exploration and food-seeking e.g. the ability to move towards a visible target, to negotiate an obstacle and to efficiently search an open space. We therefore designed fitness formulae and evaluation protocols so as to reward these abilities individually even when they did not lead to successful detours. Fitness was recalculated on each cycle of computation, using the following algorithm:

```
IF ( "some infrared sensor" > 0 && old_position ><
new_position) fitness ++

IF ( "some infrared sensor" > 0 && old_position =
new_posion) fitness --

IF (distanceToTarget< 15 cm.) fitness += 10
```

The first two components in the fitness formula were designed to encourage obstacle avoidance; the last component rewarded the robot when it approached the target.

The authors wished to simulate natural evolution and to avoid the emergence of adaptations specific to a particular experimental setting. To achieve this goal robots were evaluated in four different environments (see Figure 3). Each environment consisted of an open field with no external fence. In the first environment there was no obstacle between the robot and the target. The fitness formula rewarded robots which successfully searched for the target and moved towards it. The second, third and fourth environments selected for actual detour behavior. In the second environment the target was placed behind a linear obstacle 80 cm long. In the third environment Khepera was placed inside an 10*80 cm corridor. The fourth environment used a 40*70 cm U-shaped obstacle. Obstacles were 3 cm high and of negligible thickness; the target was 12 cm high. It follows that in the "evaluation sessions" the robot was always able to "see" the target even when the path to the target was obstructed by an obstacle. The evaluation test was repeated five times for each environment. At the beginning of each cycle the robot was placed in a randomly chosen position 90 cm from the target. The heading was chosen randomly from a uniform distribution. Each test consisted of 600 cycles of computation.

When all robots had been tested individual robot fitness scores were summed over each of the five tests in each of the four test environments. The 20 robots with the highest overall score were selected for "reproduction". Each of the selected robots produced 5 offspring. Reproduction was asexual. During the cloning process "mutations" were introduced by flipping bits in the genome with a probability of 0.02 per bit per generation. This process was iterated for 350 generations. Each simulation was repeated six times using a different random number seed on each occasion.

Finally, the 4 best robots produced in the last generaration of each of the simulations were tested in a replica of Regolin et al.'s experimental apparatus. It is important to note than in this setting the obstacle completely obstructed the robot's view of the target. As in Regolin's work, the results of the experiment were given by the number of robots choosing the correct compartments within a pre-determined duration (600 cycles of computation)

4 Results

4.1 Evolution of generic detour behavior

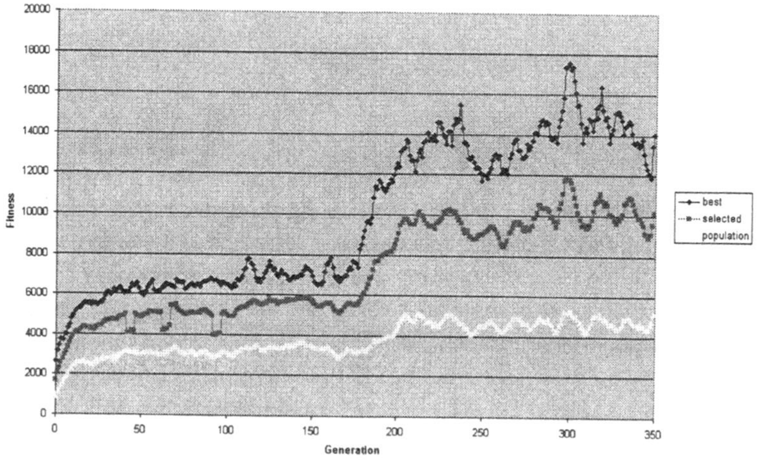

Figure 2: Fitness scores averaged over the 6 simulations.

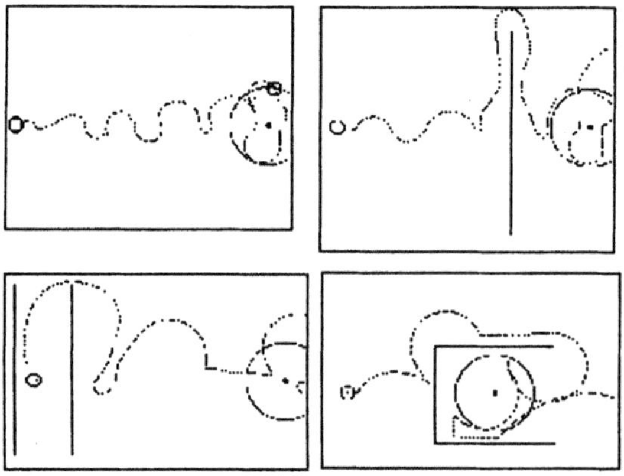

Legend: Small circles: terminal points for trajectories (if within field box). Large circle: target area rewarded by fitness function

Figure 3: Typical trajectories followed by robots in the four training environments

Figure 2 shows the score achieved by the fittest robot in the population, the mean score achieved by robots selected for reproduction and the mean score for the whole population. As can be seen the fitness score for the best organism in the population initially increased rapidly, rising from 2640 in generation 0 to 6500 in generation 50. Between generation 50 and generation 180 fitness levels oscillated around a stationary level of approximately 7000. From generation 180 to the end of the simulation fitness scores again rose, reaching a level of 14,000 in generation 350. A qualitative examination of the trajectories followed by individual robots show that by the end of the evolutionary process all robots selected for reproduction were exhibiting satisfactory detour behavior in all four environments (see Figure 3)

4.2 Performance in the experimental setting

Table 1 compares the performance of the 24 robots with the chicks in Regolin et al.'s experiment.. As can be seen from the table 1/24 of our robots failed to enter one of the compartments within the allotted time (5/25 in the original experiment). Of the remaining robots 22 entered the correct compartment (20 in the original work) and 3 chose the wrong compartment. As is the original experiment there was no significant difference between the number of robots entering compartment C and the number entering compartment D.

Table 1. Robots and chicks performances in the experimental setting

	Do not leave the corridor	Sector A	Sector B	Sector C	Sector D	Total
Chicks	5	2	3	9	11	25
Robots	1	0	2	11	11	24

There is no statistically significant difference between the results achieved in our experiment and those reported in Regolin et al. We therefore conclude that our simulation successfully replicates the results of the original experiment.

An examination of individual trajectories shows few differences among individuals; the trajectories followed by robots chosing Sector C are a rough mirror-image of those used to reach Sector D. A typical trajectory is shown in Figure 4.

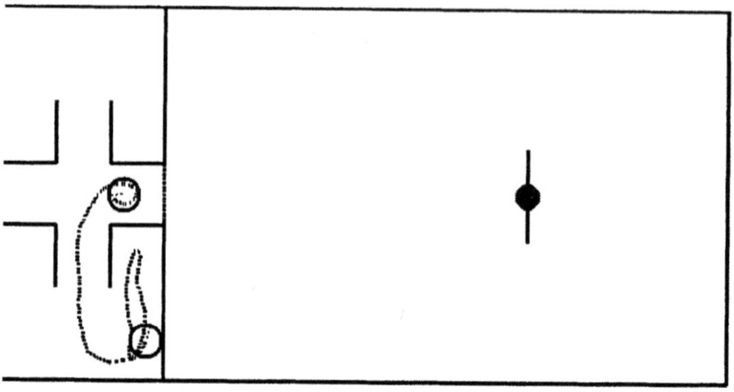

Figure 4: A typical robot trajectory in the experimental apparatus

5 Discussion

In our work every neuron in the ANN directly represented either an input or an output; there were no hidden neurons. In brief the robots we evolved had no access either to maps or to any other internal representation of location.
In different simulations robots evolved different strategies. We have not as yet performed a detailed analysis either of the strategies themselves or of the underlying computational mechanisms In general however they appear to be based on simple rules, for example (see Figure 3.):

1) Main: Move forward turning first clockwise (slowly) and then anti-clockwise (more rapidly).

2) Taxis: On visual contact with target turn sharply towards the target. Return to Main

3) Wall following: If left proximity sensors active turn right. If right sensors active turn left. Move forwards until obstacle out of view. Return to Main

Input from the time sensors allows the robot to generate differentiated responses to identical external stimuli. It is the time sensors which make it possible for the robot to change its direction of motion in the absence of any external stimulus. The combination of exploration and taxis, which these sensors make possible, represents an efficient strategy for moving towards the target even when it is often outside the camera's field of vision. It is this strategy (rather than an explicit representation of target position) which enables the robots to perform successfully in the experimental setting.
The observation that artificially evolved robots can generate a particular behavior using simple behavioral rules does not demonstrate that chicks use the same rules or

that that cognitive maps or target representations do not exist. There is, in fact, at least some evidence that chicks and robots use different rules. Individual chicks are reported to have no particular preference for left over right turns; individual simulation runs on the other hand produce asymmetrical strategies (though the asymmetry may be reversed from one simulation to the next).

While our work does show is that it is possible to generate, at least the simplest forms of detour behavior, without resort to internal representations. It should be added that feed-forward control networks like those used in our experiments have no internal representation of past states. They cannot, in other words, follow the rule Regolin et al. suggest for their chicks, turning left or right depending on their previous direction of turn.

Our work suggests alternative models for detour behavior and the behavioral primitives on which this behavior is likely to depend. The ability to efficiently explore an environment, to locate food, to move towards the food and to negotiate obstacles, are of fundamental importance for a broad range of animal species. The results of our experiment suggest that primitive detour behavior may, in fact, be a relatively simple extension of these basic exploratory capabilities. What they seem to show is that it is possible to generate primitive detour behavior on the basis of nothing more than: (a) a robot's ability to move towards a target, (b) a strategy guaranteeing that it will never permanently loose visual contact and (c) a simple "wall following" routine.

None of this implies, of course, that all forms of detour behavior can be explained so simply. The evidence in favor of cognitive maps is still, at this stage of the game, relatively convincing. And yet a doubt remains. If elementary detour behavior can evolve, step by step, from simple behavioral primitives, might it not be possible to evolve more complex detour abilities using a similar bottom-up approach? This is an issue which we will address in future experimental work.

References

1. Marr D. (1982) - Vision, W.H. Freeman and Company.
2. Prato Previde, E., M. Colombetti, M.D. Poli & E. Cenami Spada (1992). The mind of organisms: Some issues about animal cognition. Focal article with commentary, International Journal of Comparative Psychology, 6 (2), 79-119.
3. Braitenberg, V. (1984), Vehicles, Experiments in Synthetic Pscychology, MIT Press.
4. Beer R.D., (1989) Intelligence as Adaptive Behavior, Academic Press.
5. Lund H. H., Webb B., and Hallam, J. A (1997). Robot Attracted to the Cricket Species Gryllus bimaculatus. In P. Husbands and I. Harvey, editors, Proceedings of Fourth European Conference on Artificial Life, pages 246--255, Cambridge, MA,. MIT Press, Bradford Books.
6. Burgess N., Donnett J.G., O'Keefe, J., (1997) Robotic and Neuronal Simulation of Hippocampal navigation, AISB Workshop on "Spatial reasoning in Mobile Robots and Animals, Technical Report n. UMCS-75-4-1, University of Manchester.
7. Clark A., (1997), Being There, MIT Press.
8. Clark, A. and Thornton, C. (1997). Trading Spaces: Computation, Representation, and Limits of Uniformed Learning. Behavioral and Brain Sciences. 20, 57-90.

9. Cliff D. & Miller G.F.(1996) Co-evolution of Pursuit and Evasion II: Simulation methods and Results, in From animals to animats 4: Proceedings of the Fourth International Conference pm Simulation of Adaptive Behavior (SAB96), Maes P, Meyer J-A, Pollack J. & Wilson S.W. (eds), MIT Press.
10. Brooks, R.A. (1991). Intelligence without representation. Artificial Intelligence, 47 (1-3), 139-159.
11. Nolfi, S., Floreano D., Miglino O., Mondada F., (1994), How to evolve autonomous robots: different approaches in evolutionary robotics. In Brooks R.A. & Maes P. (eds), Artificial Life IV- Proceedings of the Fourth International Workshop on the Synthesis and Simulation of Living Systems, pp. 190-197, Cambridge MA, MIT Press.
12. Miglino O., Lund H. H., Nolfi S. (1996), Evolving Mobile Robots in Simulated and Real Environments, Artificial Life, 2(4): 417-434.
13. Walker R, Miglino O. (1999). Simulating exploratory behavior in evolving artificial neural networks, GECCO 1999, Proceedings of the Genetic and Evolutionary Computation Conference (in press).
14. Miglino, O., Denaro, D., Tascini, E., Parisi, D. (1998). Detour behavior in evolving robots. Porceedings of First European Workshop on Evolutionary Robotics.
15. Koelher, W. (1925). The Mentality of Apes. Harcout Brace, New York.
16. Tolman, E. C. (1948). Cognitive maps in rats and men. Psychological Review, 36, 13-24.
17. Regolin, L., Vallortigara, G., Zanforlin, M. (1994) Object and Spatial Representations in Detour Problems by Chicks. Animal Behavior, 48:1-5.
18. Tolman, E.C. and Honzik, C.H. (1930). Insight in rats, University of California Publications in Psychology, vol 4, pp 215-32, 77-8.
19. Bennett, A. T. D., (1996). Do animals have cognitive maps?. Journal of Experimental Biology, 199 (1), 219-224.
20. O'Keefe J., & Nadel L. (1978), The hippocampus as a cognitive map. Oxford University Press.
21. Roberts D.L., & Good M. (1999). Hippocampal lesions disrupt navigation based on cognitive maps but not heading vectors, John M. Pearce, Amanda, Nature, 396, 75-77
22. Wood R., Dudchenko P.A. & Eichenbaum H.(1999), The global record of memory in hippocampal neuronal activity, Nature Vol 397, 613-616.
23. Mondada F., Franzi E. & Ienne P. (1993) Mobile Robot miniaturisation. A tool for investigation in control algorithms. In Proceedings of the Third International Symposium on Experimental Robotics, Kyoto, Japan.
24 Minsky M. & Papert S., (1988) Perceptrons, MIT. Press.

Studying Animals through Artificial Evolution: The Cricket Case

Rens Kortmann[1] and John Hallam[2]

[1] Universiteit Maastricht, IKAT, Neural Networks and Adaptive Behaviour Group, P.O. Box 616, NL-6200 MD Maastricht, The Netherlands
rens@cs.unimaas.nl
[2] University of Edinburgh, Division of Informatics, Institute of Perception, Action, and Behaviour, 5 Forrest Hill, Edinburgh EH1 2QL, Scotland, UK
john@dai.ed.ac.uk

Abstract. In this paper we introduce the notion of *historical evidence* – the ability to replicate biologically realistic evolutionary scenarios – for hypothesised mechanisms for control of sensorimotor behaviour. We apply the idea to the phonotaxis mechanism proposed by Webb and her collaborators to account for the abilities of *Gryllus bimaculatus*. To do this, we tested whether the proposed control mechanism, when implemented in a robot model of the animal, could account for evolutionary adaptations observed in the natural system. We describe and discuss the experiment and its results, but start by explaining the methodology used, which is an extension of Webb's existing methodology for obtaining *behavioural evidence* for a hypothesised mechanism. We conclude there is historical evidence for the neural control mechanism investigated.

1 Introduction

At the previous European Conference on Artificial Life (ECAL-97), Lund *et al.* presented their work on a robot model of the cricket species *Gryllus bimaculatus* [1]. The model, devised by Barbara Webb [2], offers an hypothesised control mechanism to account for the ability of female crickets reliably to approach their mates using only cues available in the male calling song. This work, through replication of studies performed on the animals themselves, provides evidence that the hypothesised mechanism is sufficient to account for much of the variety and detail of the female cricket's phonotactic behaviour. However, the paper already alluded to, and other work with the model (e.g. [3]), provide what we shall call *behavioural evidence* – that is, the experimental methods focus on demonstrating that the behaviour evoked by the controller, when implemented in a robot model, is statistically comparable to that observed by ethologists in experiments on the real animal.

In this paper we consider the possibility of an alternative source of evidence for the sufficiency of the control mechanism, which we call *historical evidence*. Any mechanism in use in nature must not only account for the behaviour it engenders, but must also have been generated by evolution. We therefore consider, in this work, the extent to which the hypothesised controller alluded to above can account for evolutionary adaptations in the cricket. If the controller can support similar evolutionary histories as are seen in the real animal, this provides additional support for its plausibility. (More strictly, if it *cannot* support the kind of evolutionary histories seen in the animal species, it becomes less plausible that it is the natural mechanism even though it may account for the species' behaviour very well.)

1.1 A Methodology for Evolutionary Bio-Robotics

The methodology employed here is an extension of the one developed by Webb for studying perceptual systems by building artificial ones [2, 4, 5]. The central idea of this methodology is to provide hypotheses and evidence for control mechanisms generating sensorimotor behaviour observed in animals by using physical computational models (often robots). Webb [2] emphasises three principal requirements in her methodology: (1) model one particular biological sensorimotor system from a specific animal, (2) use physical (hardware) models, and (3) use established evaluation methods. These requirements stem from validity considerations: the model developed should be ecologically valid, hence the focus on a specific animal; physical models are used to finesse the difficulties of accurately simulating the (usually) complex physics of the sensorimotor interactions being studied; the evaluation and analysis of the physical model should be comparable to that of the animal, to allow generalisation of conclusions from the former to the latter. Now, from such comparisons, behavioural evidence for a hypothesised mechanism could be obtained.

We can extend this methodology to provide historical evidence by considering the evolutionary context of the animal system under study: if the hypothesised mechanisms modeled in the robot are those found in the target animal, we should be able to simulate evolutionary change in the real animal with the model. For this to work, we add two further items to Webb's list of methodological requirements: (4) find an evolutionary scenario, in the biological literature, which is plausible for (or preferably observed in) the target animal, and (5) find parameters in the hypothesised mechanism that have a genetic disposition and that are relevant for the chosen scenario.

Again, the primary motivation for these requirements is to permit valid comparative inferences to be drawn. Without a specific evolutionary scenario matching the specific target animal, we run the risk of simulating something with no biological counterpart. The fifth requirement is more subtle, and is a consequence of the fact that we typically cannot simulate genetics as it happens in the target animal, *i.e.* the details of the encoding of the mechanism in genome and ontogeny are not known. To finesse this difficulty, we assume that evolutionary change of mechanism parameters known to be determined genetically in the animal is an adequate model of the unknown processes that take place during the animal's reproduction. The necessary evolutionary change is supplied by, for instance, a genetic algorithm operating on the chosen parameters of the mechanism being studied.

From the foregoing discussion it should be clear that finding *historical evidence* for a particular controller is in fact performing a series of *behavioural experiments* with parameters from generation to generation under the control of an evolutionary algorithm. The design of the fitness function for that algorithm can therefore be based on behavioural experiments of the kind alluded to above. It is less clear, however, what kinds of genetic operators are appropriate. In this paper we start with the standard genetic operators: reproduction, cross-over, and mutation.

1.2 Survey of the Paper

This paper reports a first attempt, using the methodology already outlined, to provide historical evidence for the phonotaxis mechanisms studied by Webb and her collaborators. To this end, we implemented a standard genetic algorithm using, for the assessment of individuals, a fitness function measured *via* the robotic implementation of the phonotaxis mechanism described in [3]. The experiments therefore used *on-line* evolution, somewhat similar to the work of Floreano and Mondada [6, 7], except that the goal here was not to evolve robot controllers with particular capabilities but to follow a chosen evolutionary scenario – the latter predetermined by the target biological system.

The paper continues with a description of the specific evolutionary scenario we chose to model. Details of the experimental methods follow. Section 3 describes the results of the on-line evolutionary experiments and the analysis of the behaviour of the best evolved individuals. Finally, Sect. 4 discusses the results and points out to future work. Also, in this section, we conclude by saying that our experiments *did* provide historical evidence on the hypothesised control mechanism investigated.

2 On-line Evolution of Cricket Phonotaxis

In this section we fill in the requirements, mentioned in the previous section, in order to provide historical evidence for the control mechanism for cricket phonotaxis investigated by Lund *et al.*.

2.1 Biological Background

Males of the species *Gryllus bimaculatus* attract females with their calling songs. Females respond by navigating towards the sound. For this they posess a phonotaxis mechanism that can both determine the directionality of the sound and its specificity (each species' calling song has its own specific pattern). We will here investigate the second property: selectivity, used for distinguishing conspecific males from heterospecific ones.

An Evolutionary Scenario Cricket calling songs are grounded in the genes of the animal [8]. Therefore, cricket communication can be properly investigated from an evolutionary point of view. Otte [9] describes an evolutionary scenario that starts with two closely related species coming to live together in the same habitat. In addition, they happen to mate at the same time and have similar calling songs. Now, as females will not be able to distinguish between the two calling songs, interbreeding will occur. This situation is unstable in evolutionary terms and will result in either the divergence of the two species' calling songs or the extinction of one of the species. In this work, we explored the former possibility.

Evolvable Parameters The neural controller under study is based on findings in the neurophysiology of crickets. It consists of a two-layered neural network receiving input from the ears, and feeding its output into the motor system. Each layer consists of two leaky integrator neurons, one for each side of the body. The network accounts for both directionality sensing and selectivity. For a more detailed account we refer to the work of Lund *et al.* [3]. In the controller, several parameters represent genetically determined features of the neurons involved. Moreover, selectivity of phonotaxis in the females is tuned by the values of these parameters [3]. These parameters therefore provide the representation of the cricket genome for the model of cricket reproduction used in the experiments below.

In our experiments we evolved seven parameters. For both layers of neurons, we take the decay rate (D1, D2), the lower threshold of activation (L1, L2), and the duration of propagation after reaching the higher threshold (P1, P2). In addition, we evolved the higher threshold of activation in neurons of only the first layer (H1). The value of this parameter in the second layer was kept equal to the lower threshold. We did not evolve the parameters of the (physical) peripheral auditory model as the carrier frequency was the same for both songs in our experiments.

2.2 Materials and Methods

For the simulation of Otte's evolutionary scenario we designed an experimental setup in which specific emphasis was put on fully automated experimentation. This was necessary, since the on-line evolution of just 2 generations took an entire night.

Materials In our on-line experiments, the robot model [3] was able to move freely on a fenced table, size 240 x 240 cm. It was externally powered and connected to a controlling PC through a serial communication cable. A robust communication protocol [10] and a command interpreter enabled the controlling PC to drive the robot or send messages such as 'start phonotaxis' to the on-board software. Low-level processes, such as the phonotaxis behaviour, could thus be switched on and off by the PC. The implementation of the phonotaxis behaviour (the neural controller of interest) in the robot was essentially the same as described by Lund *et al.* [3]. The physical robot was exactly the same.

Two loudspeakers, representing males of the two cricket species, were placed on the table. The controlling PC could play different calling songs through them on demand. An overhead camera viewed the entire table and a real-time visual tracking system [11] was employed to measure the consecutive table coordinates of the robot while performing phonotaxis. These data were used to analyse phonotaxis performance and also, between phonotaxis runs, as input to a vector-based homing mechanism used to drive the robot to particular starting positions for fitness assessment (see below).

Methods Following Otte's evolutionary scenario, we started the experiment with two populations of females showing equal preference to one calling song. In the course of the experiment, an evolutionary process would diverge the two populations' preferences. On the other hand, unlike the scenario, we did not let the male calling song evolve freely. There are two reasons for us not to *co*-evolve calling song and female response. First, a simulated co-evolutionary process is not likely to follow the prescribed scenario, because of the self-organising properties of a co-evolutionary system [12]. Secondly, we are only interested in the adaptivity of the female reponse, not of the male sound production. Therefore, we 'evolved' the sound pattern by hand and focussed on the changes in the phonotaxis mechanisms of the two populations. Due to limitations of time, the experiment consisted of a single run.

We composed two fixed calling songs, differing in one temporal parameter: syllable period (the time between the onset of two consecutive bursts of sound). This corresponds to Otte's observations in which the calling songs had diverged through this parameter. One population of males (population 0) would emit calling songs at 55 ms syllable period, versus 25 ms in the other population (population 1). The syllable period of the initial song was 40 ms. Each male population was modelled by a loudspeaker that emitted its song during the experiments.

In the experiment, a genetic algorithm (GA) would generate (labelled) populations of individuals which would be evaluated with the robot model. For doing this, the parameter set of each individual was downloaded into the robot consecutively. During fitness evaluation, the robot was exposed to the two sound sources simultaneously. Its task was to navigate towards the sound source matching its assigned label (population 0 or 1), *i.e.* its conspecific loudspeaker.

The Genetic Algorithm The seven parameters mentioned in Sect. 2.1 were directly encoded as an array of floating point numbers. A standard GA [13] was adapted for the evolution of several populations. Rank-based selection was used; mate choice was restricted to individuals from the same population. Interbreeding did not occur, because individuals that showed phonotactic preference to the heterospecific loudspeaker were not subject to the selection regime. Selected parents generated children by copying (90% probability) or two-point cross-over;

children were then mutated (40% probability, 10% mutation range). Each population consisted of 30 individuals. This number was deliberately kept low for computational reasons. We used a generational replacement scheme, as this is the natural situation for crickets: every individual of the previous generation has died once the young hatch [8].

Fitness Function Every individual was assessed four times (trials) on its phonotactic performance. The trials started at different locations, all equidistant to the conspecific loudspeaker. This is to ensure that the evolutionary process would not exploit factors other than the sound structure and its (constantly changing) directionality. The initial direction the robot faced was always towards the heterospecific loudspeaker. Therefore, as the default movement of the robot was always straight ahead, failure to respond to any of the sound sources always resulted in poor performance, as the robot would navigate towards the wrong loudspeaker. Note that while the amplitude of the conspecific song was controlled for by this arrangement, that of the heterospecific song could vary substantially; a 'good' individual would have to demonstrate preference for its conspecific song against a range of amplitudes of distractor song.

The tracker returned the movement track of the robot after each trial. From this, six measures were calculated: mean distance (d_{m}), nearest distance (d_{n}), and final distance (d_{f}) with respect to both loudspeakers. From these measures, the fitness function estimated to which loudspeaker the robot was heading during the trial by comparing the mean and nearest distances to them. If the robot was heading for the heterospecific (wrong) loudspeaker, fitness value -1 was assigned. Otherwise, a positive fitness value was assigned according to the following formula:

$$Fitness = \frac{1}{2}\frac{d_{\mathrm{s}} - \frac{1}{2}(d_{\mathrm{f}} + d_{\mathrm{n}})}{d_{\mathrm{s}}} + \frac{1}{2}\frac{d_{\mathrm{s}} + (\frac{1}{2}d_{\mathrm{s}} - d_{\mathrm{m}})}{d_{\mathrm{s}}} \;, \qquad (1)$$

with d_{s} as starting distance. This formula converges to 1 when the robot's movement track converges to a straight line to the conspecific loudspeaker.

The fitness function returns a value which is the average over the four trials with the individual mechanism. If this value is negative, the individual is excluded from reproduction (as it failed to localise a mate). With positive fitness, an individual is subjected to the selection regime of the GA.

Evaluation of the Best Evolved Individuals To evaluate the performance of the best evolved individuals in the final generation we performed syllable rate experiments as described by Lund *et al.* [3]: we exposed the robot to calling songs of different syllable periods – ten times to each song, alternating between two different starting positions. These positions were mirrored in the central axis of propagation of the sound. Therefore, the robot's abilities to turn both right and left were examined. The recorded tracks were transformed into polar coordinates and from these we calculated the *mean heading angle* towards the loudspeaker, averaged over the ten trials. This measure of performance for one individual for one type of calling song can be used in statistical comparison using the Mann-Whitney U-test – a non-parametric test for comparing sample means of independent samples. The outcomes of this test indicate how significantly two sample means differ. A fuller description of our evaluation methods can be found in the work of Lund *et al.*, of Webb, and of Kortmann [3, 2, 14].

We compared the best evolved individuals of populations 0 and 1 (which we will call B0 and B1 respectively), and an individual from generation 0 (G0). G0 was the individual described in the work of Lund *et al.* [3]. This individual's parameter set was hand-set, tuned to 40 ms syllable period.

3 Results

3.1 Evaluation of the Evolutionary Process

In Fig. 1 the maximum, mean, and minimum fitness are depicted for each generation in both population 0 (left) and population 1 (right). The mean fitness fails to climb to 1, which is due to the short simulation time (40 generations). Also, the behaviour depended strongly on the parameter values, relatively to each other. Therefore, cross-over, very often, turned a pair of good parameter sets into a pair of poor sets. Individuals that were not subject to cross-over, though, adapted easily to the new calling song, as can be seen from the quick convergence of the maximum fitness (2 and 6 generations in population 0 and 1 respectively).

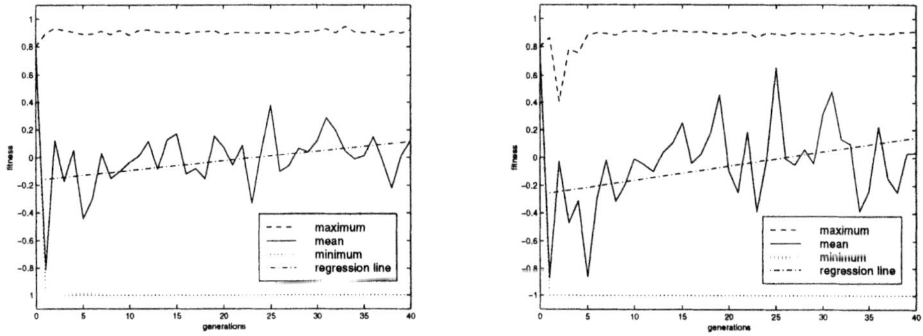

Fig. 1. Development of the fitness of population 0 (left) and population 1 (right). Shown are the maximum, mean, and minimum fitness. Also shown is the regression line through the mean fitness data points.

Population divergence was determined by fitting a regression line through the mean fitness data points (slope 0.007 ± 0.003 and 0.010 ± 0.004 in population 0 and 1 respectively).

3.2 Evaluation of the Best Evolved Individuals

From the results of syllable rate experiments with B0, B1, and G0, three comparisons were made. In the tables below, we use the following symbols to indicate how significantly the results differ: S^2 for $p < 0.001$, S^1 for $p < 0.01$, and S^0 for $p < 0.05$, with p as significance level. NS means 'no significant difference'.

Selectivity The performance of each individual with respect to different song types was analysed, revealing the song-selectivity of the individual. Table 1 shows that every individual tested is responsive to songs in a small range around its conspecific song. The responsiveness changes gradually with more remote syllable rates.

Table 1. Comparison of phonotactic performance of each individual towards conspecific song and other songs. See text for key to symbols.

Individual	Syllable period conspecific song (ms)	Syllable period other songs (ms)					
		10	25	35	45	55	70
G0	40	S^2	NS			S^2	S^2
B0	55		S^2	S^1	S^0		S^0
B1	25	S^2		NS	S^2	S^2	

Responsiveness Here, the performance of a pair of individuals with respect to their own calling songs was compared. The comparisons show how well the individuals are tuned to their conspecific calling songs. Table 2 shows that both B0 and B1 respond very significantly better to their conspecific song, as compared to G0 (significance level $p < 0.001$). The GA, apparently, tuned the parameter sets very well – remember G0 was hand-set. Also there is a slight difference in responsiveness between B0 and B1 ($p < 0.05$).

Table 2. Comparison of the phonotactic performance of three individuals, relative to one another, with respect to their conspecific calling songs.

	G0 (40 ms)	B0 (55 ms)	B1 (25 ms)
G0 (40 ms)		S^2	S^2
B0 (55 ms)	S^2		S^0
B1 (25 ms)	S^2	S^0	

Adaptation Now, the performance of either best evolved individual (B0, B1) was compared with that of G0, both with respect to the calling song of the evolved individual. This comparison shows how well the evolved individuals adapted their behaviour, relative to the initial population. Table 3 shows that B1 performs significantly better on its calling song (25 ms) than G0 does ($p < 0.01$). Even more significant is the difference between B0 and G0 to the 55 ms syllable rate song: $p < 0.001$

Table 3. Comparison of the phonotactic performance before and after evolution with respect to the calling song of the evolved individual.

	B0 (55 ms)	B1 (25 ms)
G0	S^2	S^1

4 Discussion and Conclusions

Divergence of calling song caused the female response mechanism to adapt accordingly. This can be seen most clearly in Table 3. Selectivity for temporal patterns was maintained in the evolved mechanisms (see Table 1). In addition, the evolutionary process tuned the parameters responsible for selectivity more sharply than hand-setting did (Table 2).

While this is perhaps unsurprising from a GA viewpoint – GAs are of course well-known as optimisation techniques – the development of novel GA results is not the focus of this work. Rather, we suggest the results reflect an evolutionary adaptation in female response that might be demonstrable in real crickets, considering the genetic disposition of the parameters involved.

Repeating the experiments, to allow for contingency effects, will strengthen this suggestion and the model's validity. Still, this is very costly and was, due to time limitations, therefore not performed.

As was shown in Fig. 1, the best individuals adapted fairly quickly to a hand-set change in a calling song parameter. This was due a large mutation range (10%). Smaller mutations would have caused a more gradual increase of maximum fitness. Thus, although the present work did not permit co-evolution of female parameters and calling song structure, given this rapidity of evolutionary response we expect that a true co-evolutionary process between male calling song and female response would likely be successful in our experimental setup. Adaptibility to constantly changing environments is therefore a good topic for further research. In addition, the principles of male sound production can be studied, as was mentioned in Sect. 2.2.

Another topic for further study would be finding a better scheme for encoding the genome, for example a developmental scheme (e.g. see [15]) that is better able to handle the cross-over operator than our direct encoding scheme.

Concluding, the experimental results provided *historical evidence* for the control mechanism for real cricket phonotaxis proposed by Webb and collaborators [3]. We presented the first results of a new methodology that places hypothesised mechanisms for animal behaviour in an evolutionary context using an embodied agent and computational methods. We hope that the work presented here will encourage more research aimed at not only elucidating mechanisms underlying animal behaviour as they are currently found, but also their formation in evolutionary time.

Acknowledgements

We would like to thank Barbara Webb and Henrik Hautop Lund for their work on the robot model and methodology we used. Their help and comments were greatly appreciated. We also thank Hans van Ditmarsch for his advice. Facilities for this work were provided by the University of Edinburgh, while the preparation of this paper was facilitated by both Edinburgh and Maastricht Universities.

References

1. Lund, H., Webb, B., Hallam, J.: A robot attracted to the cricket species *gryllus bimaculatus*. In: Proceedings of 4th European Conference on Artificial Life, MIT Press/Bradford Books (1997)
2. Webb, B.: Perception in Real and Artificial Insects: A Robotic Investigation of Cricket Phonotaxis. Ph. D. dissertation, University of Edinburgh (1993)
3. Lund, H., Webb, B., Hallam, J.: Physical and temporal scaling considerations in a robot model of cricket calling song preference. Artificial Life **4** (1998) 95–107
4. Webb, B.: Using robots to model animals: a cricket test. Robotics and Autonomous Systems **16** (1995) 117–134
5. Webb, B.: Robots, crickets and ants: models of neural control of chemotaxis and phonotaxis. Neural Networks **11** (1998) 1479–1496
6. Floreano, D., Mondada, F.: Evolution of homing navigation in a real mobile robot. IEEE Transactions on Systems, Man, and Cybernetics **26** (1996) 396–407 Part B: Cybernetics.
7. Floreano, D., Mondada, F.: Evolutionary neurocontrollers for autonomous mobile robots. Neural Networks **11** (1998) 1461–1478
8. Alexander, R.: Evolutionary change in cricket acoustical communication. Evolution **16** (1962) 443–467
9. Otte, D.: Evolution of cricket song. Journal of orthoptera research **1** (1992) 25–49
10. Comer, D.: Operating system design - Volume II. Prentice-Hall International (1987)
11. Lund, H., de Ves Cuenca, E., Hallam, J.: A simple real-time mobile robot tracking system. Technical Paper 41, Department of Artificial Intelligence, University of Edinburgh (1996)
12. Nolfi, S.: Evolutionary robotics: Exploiting the full power of self-organization. Connection Scienc (1998)
13. Goldberg, D.: Genetic algorithms in search, optimization, and machine learning. Addison-Wesley (1989)
14. Kortmann, R.: Evolving phonotaxis in a robot cricket — an investigation in biorobotics. Master's thesis, University of Groningen and University of Edinburgh (1998)
15. Kodjabachian, J., Meyer, J.: Evolution and development of control architectures in animats. Robotics and Autonomous Systems **16** (1995) 161–182

Evolving Behavioural Choice: An Investigation into Herrnstein's Matching Law

Anil K Seth

Centre for Computational Neuroscience and Robotics
and School of Cognitive and Computing Sciences
University of Sussex
Brighton, BN1 9QG, UK
anils@cogs.susx.ac.uk

Abstract. In 1961, Herrnstein [4] famously observed that many animals *match* the frequency of their response to different stimuli in proportion to the reinforcement obtained from each stimulus type. Since then, a great deal of research has attempted to elucidate the mechanisms underlying this "matching law", so far without a clear consensus emerging. Here, we take the view that "choice behaviour" is a product of agent, environment, and observer, and that "mechanisms of choice" are therefore not to be located solely within the chooser. A simple model, employing the novel methodology of evolving choice behaviour in a multi-agent system, is used to demonstrate that matching behaviour can occur (in stable environments) without *any* dedicated choice mechanism.

1 Introduction

All behaviour is choice. R.J. Herrnstein (in [14])

In ALife, behavioural choice has been largely synonymous with *action selection,* which can be loosely defined as the problem of choosing what to do, at any given time, in order to further progress towards multiple, time-varying goals, [8]. Action selection is concerned with choice in the context of different behavioural options that relate to *distinct* goals (for example, feeding and drinking), let us call this a type-1 choice scenario. Here, we consider a different situation; where choice operates between different ways of satisfying the *same* goal - let us call this a type-2 choice scenario - an example would be foraging amongst two kinds of plant that differ in nutritive value. It is in such scenarios in biology that we can observe *matching behaviour* when animals allocate the frequency of their responses to different stimuli in proportion to the reinforcement obtained from each stimulus type.

We may distinguish two questions: (1) how (and why) animals display matching at all, and (2) how animals are able to track changing environmental contingencies. The latter is clearly the more complex, implicating learning, and we shall be concentrating on the former; exploring some minimal conditions on the internal mechanisms required to support matching behaviour in *stable* environments.

In Sections 2 & 3 we briefly revise how the biological sciences of ethology, behavioural ecology, and experimental psychology have approached both type-1 and type-2 choice, and note that ALife, in following an ethological precedent, has (so far) concentrated almost uniquely on the former problem of action selection. The *matching law* of Herrnstein [4] is then introduced as an alternative choice paradigm and it is argued that theories in both domains suffer from what will be a theme throughout this paper; that *behaviour* and *mechanism* should not be confused.

In pursuit of a more fruitful understanding of the mechanisms underlying choice, Section 4 describes a multi-agent model in which simple reactive agents are evolved in a type-2 choice environment. A hypothesis derived from Fretwell [2] is put to the test; that matching behaviour is optimal if there is sufficient *competition* for the resources. This model therefore introduces the evolution of multiple agents for choice behaviour.

The results (in Section 5) support Fretwell's intuition, and indicate that matching behaviour (in stable environments) can arise without there being any dedicated mechanism of choice, and, indeed, without any internalisation of distinct behavioural options at all. Sections 6 and 7 discuss these results in terms of how ALife can contribute towards the natural science attempts to understand choice; in particular, how it can make the relationship between behaviour and mechanism the object of study, rather than the legacy of unwarranted assumptions.

2 Biological Approaches to Behavioural Choice

2.1 Ethology

Ethology, through the observation of animal behaviour in natural contexts, seeks to understand the nature of the mechanisms underlying behaviour *and* their evolutionary origins.[1] Ethological accounts of behavioural choice (in type-1 choice scenarios) generally propose internal mechanisms that arbitrate between internalised (and pre-existing) repertoires of behavioural options. The ALife approach to action selection has also relied on this foundation, and has investigated a broad class of arbitration devices, from hierarchies ([12]) to distributed networks ([7]).

In previous work [10] this approach was criticised for committing the category error of assuming that externally observed behaviours must have internal mechanistic correlates. A behaviour is the joint product of an agent, an environment, and an observer; thus the (agent-side) mechanisms underlying the generation of any behaviour should not be assumed to be identical to the behaviour itself. A simple animat model was used in [10] to illustrate that effective action selection could occur without any internalisation of behaviour, and in which choice could

[1] Tinbergen, a founder of the discipline, actually specified *four* components of any complete ethological account of a behavioural pattern; causation, development, survival value, and evolution (from [9]).

be explained just as well in terms of perception as in terms of action. In this paper, a similar model illustrates that the same arguments apply in understanding the mechanistic basis of behavioural matching.

2.2 Behavioural Ecology

Behavioural ecology (BE) differs from ethology in that it is ostensibly *not* concerned with mechanism, but only with the adaptive rationale for observed patterns of behaviour.[2] It is within this discipline, and within the complementary discipline of experimental psychology (EP), that the investigation of behavioural matching has acquired greatest momentum.

The study of matching in BE has focussed on *foraging*. Krebs & Kacelnik [6] present a discussion of various foraging strategies, and the environmental conditions under which they may do well. A typical foraging problem is a type-2 choice environment, with two types of source providing reinforcement of a single kind (for example, food), but at different rates or with different probabilities. Two particular foraging strategies are of interest here; the zero-one rule (see [11]) and the matching rule [4]. Given a type-2 environment, these rules make conflicting predictions about how foraging agents should behave with regard to the *less profitable* source (both rules agree that agents should maximize their intake of the most profitable). The former predicts that an agent will *either always* take the less profitable source upon encounter *or never* take it. The latter predicts that the agent will take the less profitable source at a rate that matches the difference in profitability between the two source types.

In Section 3 we shall see how these conflicting predictions can be unified into a single adaptationist framework, but for now let us note that although BE claims to be unconcerned with questions of mechanism, it has been argued that there is often an *implicit* tendency to credit the animal with complex computational and representational abilities (see [3]).[3] So either BE tells us nothing at all about mechanism, or it may force us into making strong mechanistic assumptions without appropriate justification.

2.3 Experimental Psychology

It is to experimental psychology (EP) that we must turn in order to find more explicit investigations of the mechanisms underlying behavioural choice. In the case of matching, these investigations are primarily concerned with the more complex issue of how animals learn to track changing contingencies, rather than the simpler question of how (and why) they can match at all. It is perhaps because of this that the status of matching in EP remains unclear, with some believing that matching is the product of underlying learning rules (see, for

[2] Historically, BE derives from the split in ethology between those exclusively interested in adaptation, and those who concentrated on mechanism (see [1]).

[3] For those familiar with BE, this relates to the necessary assumption of a *decision variable* in optimal foraging theory.

example [5]), and others arguing that matching is *itself* the rule by which animals determine their responses, and not a (mere) description of the results (see [14]).

The confusion surrounding the nature of matching may also stem from both the frequent lack of connection between the contrived environments of many EP experiments and ecologically plausible situations[4], and the common feature of mechanistic EP theories of the concept of *response strength*, which is essentially another manifestation of an internal behavioural correlate.[5] EP therefore understates the environment both with respect to the behavioural problem (by encouraging arbitrarily constructed environments) and with respect to the mechanistic solution (by relying on internal behavioural correlates). Indeed, given the co-dependence of behaviour on agent and environment, the former inevitably leads to the latter.

Although the emphasis on learning in EP precludes direct comparisons being drawn with the present work, a bridge can be built by considering that the question of learning may productively *follow* an understanding of matching in stable environments, and that the problematic (and complex) mechanistic proposals prevalent in EP (see [14]) can be largely attributed to the confusion between behaviour and mechanism described above.

3 The Matching Law and the Ideal Free Distribution

In Krebs & Kacelnik [6] we find the following definition: "the matching law states that the animal allocates its behaviour in proportion to the rewards it has obtained from them." More formally:

$$log(\frac{F_A}{F_B}) = log(k) + b(log(\frac{r_A}{r_B})) \tag{1}$$

where F_A and F_B are the response frequencies to alternatives A and B, and r_A and r_B are the attained rates of reinforcement from the two alternatives, with k and b as scaling parameters. The accepted fact that animals (and possibly humans) often behave according to this rule in behavioural choice situations, both inside and outside the laboratory, is undoubtedly of importance and demands explanation.

In 1972, Fretwell [2] introduced the concept of the *Ideal Free Distribution* (IFD). Given a distribution of resources of different qualities, and a population of (uniformly capable) foragers, the IFD describes the distribution of foragers

[4] A popular methodology in EP involves creating experimentally contrived abstractions of foraging environments, in which different "choices" are directly presented to animals, with these choices being reinforced according to experimenter-defined "reinforcement schedules", which can be quite complex (see [14] for a comprehensive review). Note that this methodology is not particular to 'behaviourism'.

[5] EP theories often focus on how the "response strengths" are modified, and on how they, in turn, determine the emitted response. But the "response strength" concept is a prime example of a problematic *intervening variable,* crucial to the theoretical coherence of EP, but empirically unsubstantiable from within it.

such that they all do equally well, regardless of the local resource quality, and such that no-one can profit by moving elsewhere. The central intuition is that the high quality resources will tend to become overcrowded, and Fretwell argued that the only foraging strategy that is evolutionarily stable is that foragers match the relative frequency of their choices to the relative qualities of their options, in other words, that they follow the matching law. Of course, in ecological contexts without this element of competition, the optimal behaviour may well be to concentrate exclusively on the higher quality source, in other words to follow the zero-one rule. For my money, this explanation of matching behaviour is satisfying as far as it goes; a multi-forager, limited resource environment is indeed ecologically plausible. But no claims are made about the *mechanisms* involved in this choice behaviour.

It is here that the current paper can contribute empirically, in playing out the above intuition. A model is described which employs artificial evolution to design internal mechanisms, for multiple agents, in a variety of choice scenarios. We are thus able to combine the mechanistic insight sought in EP, with the ecologically motivated adaptationist hypotheses of BE.

4 The Multi-Agent Model

The essence of the model is a type-2 choice environment with multiple foraging agents. A GA is used to evolve the parameters of a reactive feedfoward neural network with a fitness function requiring efficient foraging. The hypothesis under test is simple: as the number of foraging agents increases, the behaviour of each agent should approximate that predicted by the matching law. Following the evaluation of this adaptationist hypothesis, the evolved mechanisms are analysed and the relevance of the concept of a "mechanism of matching" is questioned.

4.1 Agent and Environment

The agent(s) exist in a spatially continuous world (but with discrete time steps) containing 4 of each of 2 types of source, A and B. Each source type can potentially fully replenish the internal battery of the agent (initial level 200), which otherwise diminishes at a steady rate (1 per time step). Source type A *always* replenishes the battery ($r_A = 100\%$), but the replenishment probability r_B can be explicitly set at the beginning of each experiment. If the battery reaches 0 the agent dies, and trials terminate after a maximum of 800 time steps.

After an agent has visited a source, the source disappears and reappears randomly in another part of the world. All sources appear in a limited area (200 by 200 units) and the positions of all objects (sources *and* agents) are initialised with a minimum inter-object spacing of 40 units, a condition also adhered to when sources reappear following consumption. Each source is 16 units in radius (5 for each agent), and at full speed an agent covers 2.8 units per time step.

The agents possess 5 sensors, 4 of which are tuned to the two source types (in 2 left/right pairs), and one of which reflects the battery level. The source sensors respond to the distance from the nearest source of each type to the agent, with each sensor ranging linearly from 100 (at the source) to 0 ($\geq$ 200 units away). If the source is to the left of the agent, the relevant left sensor will respond with 20% greater activation (and *vice-versa* if the object is to the right).

4.2 Genotype and Phenotype

The internal architecture of the agent comprises a simple feedforward reactive network, fully interconnected between layers, but with no internal recurrency within layers. There are 5 inputs feeding through to a 3 unit hidden layer, and then to a 2 unit output layer. The input units linearly scale the sensor values to range from 0 to 1, and all weights range from -1 to 1. Each neuron in the hidden and motor layers applies a sigmoid transfer function to the sum of its inputs (plus a threshold value), with the outputs ranging from 0 to 1. The outputs are scaled to range from -10 to 10 to set the wheel speeds.

These weights and thresholds are specified by 26 real numbers on the 28 number long genotype. The remaining two numbers specify something less orthodox; how well the agent is able to discriminate between the two source types.

Each pair of sensors (S_i), as well as having an associated source type ($i \in \{A, B\}$), also has an associated discriminability value α, $(0 < \alpha < 1)$. If both α's are 0, then both pairs of sensors will behave identically, responding to the nearest source, regardless of type. As either α increases, the associated sensors (S_i) are more likely to respond selectively to their particular source type, i. In detail, this is what happens. First, the distance (from the agent) to the nearest source of each type is calculated (D_i) from which the distance to the nearest source of *any* type, (D_N), is easily derived. Second, for each sensor pair (S_i), if $\mathcal{R} < \alpha$ (where $\mathcal{R}$ is a random number between 0 and 1), then S_i responds to D_i, otherwise to D_N. Once the agent has perceived a source using this procedure, it will continue to perceive it in the same way until a new nearest source comes into range, at which point the procedure repeats.

This additional element of discriminability is admittedly unusual for a simple feedforward network; the rationale for including it derived from a desire to allow the agents the ability to determine (to some extent) their own input, to evaluate the possibility that mechanisms underlying choice can be based on perception as well as action.

4.3 GA and Fitness Function

A distributed GA[6] was used to evolve population of genotypes in each of 4 conditions in two different models; one model involving a single foraging agent (the S model), and another involving 3 agents (the M model).[7] In the S model, an incremental fitness function, summed over each time step as $\frac{\mathcal{B}}{200}$ ($\mathcal{B}$ being the

[6] Population 64, mutation rate 0.04 per bit, crossover rate 0.5. Runs of 1000 generations took about 10 hours on a single user Sun 166MHz SparcStation.

[7] In the M model the agents could not directly perceive each other.

battery level), and averaged over 5 separate trials, rewarded agents that lived long and foraged effectively. In the M model, each genotype gave rise to 3 genetic clones, and the fitness of the genotype was assessed (as above) by following the performance of just *one* of these agents (selected randomly) under the rationale of avoiding direct reinforcement of behaviour beneficial to the group as a whole.

5 Results

Four conditions were first evolved and then tested in both the S and M models, $r_B \in \{100\%, 66\%, 33\%, 0\%\}$, ($r_A = 100\%$ in all conditions). Two were control conditions; $r_B = 100\%$ and $r_B = 0\%$. In the former, source types A and B are functionally identical, and we should expect foragers in both models to attend to each equally. In the latter, type B is always worth nothing, and so we would expect foragers in both S and M models to attend exclusively to type A sources. The experimental conditions were $r_B = 66\%$ and $r_B = 33\%$. Here, we predict that the single S forager will *either* continue to attend to A and B equally, *or* will switch to exclusively attending to A. But in the M model, we predict that each forager will attend to B in proportion to the difference in profitability between A and B.

Fit agents generally evolved after about 500 generations in each condition (in both models), but in each case the population was left until 1000 generations had been completed. The fittest agents from each condition (in both S and M models) were tested 1000 times each, with the average number of visits to A and B sources being recorded. The entire set of evolutions (and analyses) was repeated 10 times to obtain overall averages. Fig 1 illustrates that the above predictions were indeed borne out; in the S model the zero-one rule is followed, but in the M model matching behaviour is observed.

A second analysis, using the same evolved agents, revealed very similar results (fig 2). This time, instead of testing in the same environment as that present during evolution, a parallel with EP was drawn by testing the fittest agent from each condition in the contrived environment of a forced-choice discrimination test. Here, each agent was placed equidistant from a single A source and a single B (and no other sources were present). The trial was stopped as soon as one or other of the sources had been visited, and again each agent was tested 1000 times. Note that these tests always involved a *single* agent, even if evolution had occurred in a multi-agent environment. [8]

Sensitivity to parameter settings was investigated by re-evolving all 8 conditions with either 3 or 5 sources of each type in the environment (10 complete re-evolutions in each case). Analysis was performed as before, and fig 3 illustrates, again, qualitatively similar results to fig 1.

Turning briefly to the evolved mechanisms (of the original agents, not those re-evolved to check parameter sensitivity), the most immediate observation is

[8] This test environment was meant to bear some similarity to the "skinner-box" experiments widespread in EP.

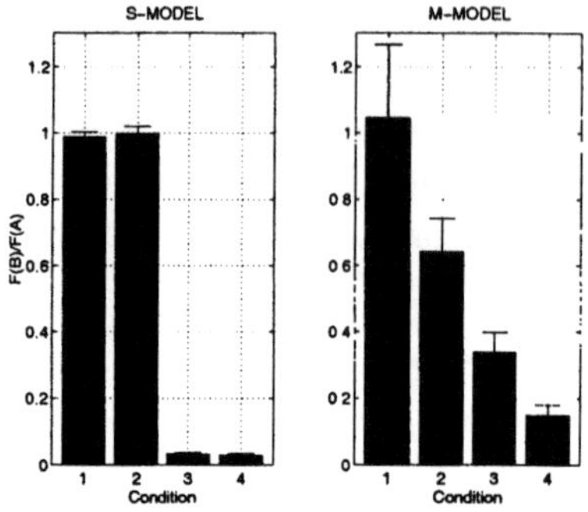

Fig. 1. *Same Environment Testing*

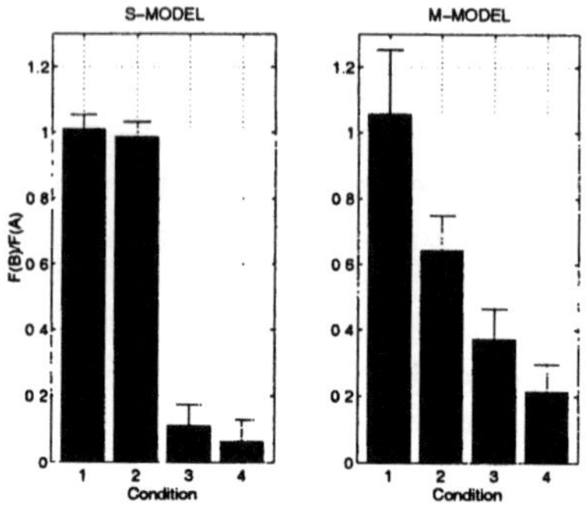

Fig. 2. *Forced Choice Testing*

Conditions are 1: $r_B = 100\%$, 2: $r_B = 66\%$, 3: $r_B = 33\%$, 4: $r_B = 0\%$. These graphs show the average (and standard deviation) rate of response to $B(F_B)$ as a fraction of rate of response to A, (F_A) over 10 evolutionary runs in each condition. From each run, a single value was obtained by testing the fittest agent 1000 times in either (fig 1) the same environment as evolution, or (fig 2) in a forced-choice environment. The evolution of matching behaviour is observed in both M-model tests, and the evolution of zero-one behaviour is observed in both S-model tests.

that there is a strong correlation between the difference in response to A and B, and the degree to which the agents can discriminate between the sources (fig 4).

6 Discussion

The above results clearly indicate that matching behaviour can evolve in multi-forager, stable, limited resource environments, and that agents with the same fundamental architecture evolve a zero-one foraging strategy if competition is absent. Four immediate observations can be made. First, the results are robust; neither testing in a forced-choice environment, nor re-evolving with different source densities altered their pattern. Second, it is evident that nothing more than a simple reactive neural network is required on the part of the agent, under the conditions of these models, for either matching or zero-one behaviour (although these agents are unable to track changing contingencies). Third, choice behaviour is apparently mediated as much by perception as by action (fig 4).[9] Fourth, even in the conditions in which agents appear to be ignoring source type B (when $r_B = 0\%$ in both the S and M models), removal of B sources afflicts their performance (mean fitness difference 2.11%, statistically significant using paired-sample t-test with $t = 2.95, df = 18, p < 0.01$). Here the B sources are clearly influencing behaviour, even though the agents are never "choosing" B.

[9] An interesting side issue here is that adaptive behaviour often depends on the sensors presenting less than perfect discriminability. This runs counter to the common intuition that the more accurately sensors reflect the external world, the better.

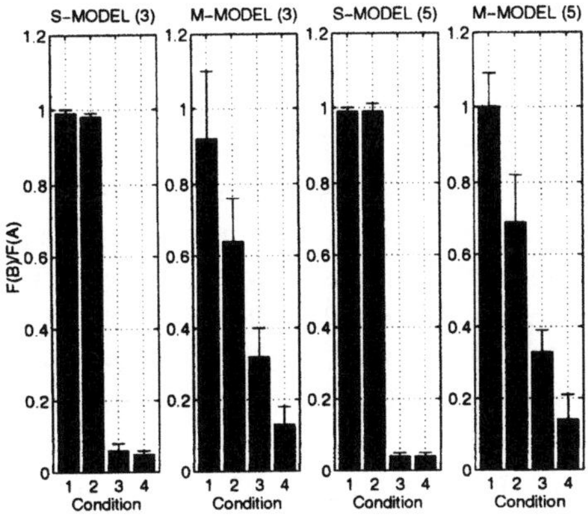

Fig. 3. *The distinction between matching and zero-one behaviour remains clear with altered source densities. The two leftmost graphs report results with 3 sources, and the two rightmost, 5 sources. The graphs are to be interpreted as in fig 1. All testing was in the same environment as evolution, and 10 evolutionary runs were performed in each condition.*

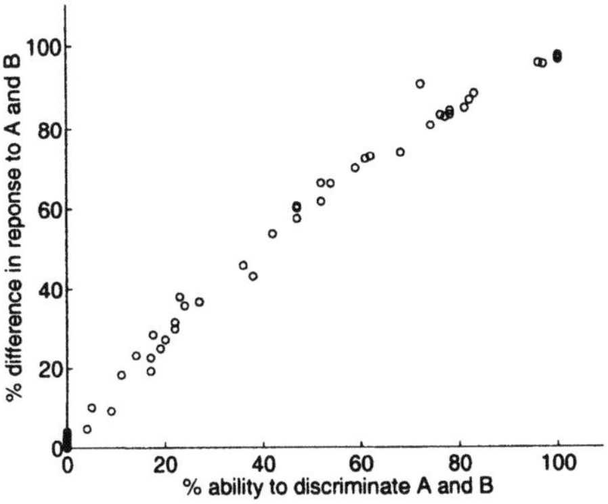

Fig. 4. *For each fittest agent in each condition (a total of 80 data points) the ability of the agent to discriminate between A and B sources is plotted against the average response pattern (over 1000 trials) to the sources. There is a clear correlation.*

These observations suggest that the idea of an explicit "choice mechanism" (or worse, a specific "mechanism of matching") supervening on an internalised repertoire of behavioural options is inappropriate for these evolved agents. However, this is *not* to say that the differences between the mechanisms are irrelevant to the differences between the behaviours; the forced-choice test results (fig 2) indicate that agents evolved in M (multi-agent) models still behave according to the matching law even when tested *without* conspecifics.

If stable matching behaviour can arise without there being a dedicated "mechanism of matching", it is natural to ask whether the importance of the matching "law" itself has been overstressed. One possibility is that the emphasis placed on matching derives from the fact that most (if not all) studies of matching (in biology) have taken place in variants of type-2 choice environments. However as Williams [14] argues; the way in which *other choice alternatives* exert their effects "is a vital theoretical issue that remains to be clarified." Current work is addressing this issue by evolving choice behaviour in environments with both type-1 and type-2 characteristics (thereby also permitting the joint examination of the problems of matching and action selection.) Future work may also address the evolution of learning mechanisms that allow agents to track changing contingencies. It is perhaps here that parallels with EP might be more in evidence, and that the notion of a "mechanism of matching" may find greater application.

It appears that the methodology espoused here provides a promising way of addressing the difficulties faced by both EP and BE in how they deal with mechanism (see Section 2). By making the distinction between behaviour and mechanism fully explicit, and the object of study rather than assumption, we can avoid the pitfalls of either (1) searching for mechanistic correlates that underlie behaviours in arbitrary and contrived environments in EP, or (2) relying on the hidden mechanistic assumptions necessary for the construction of predictive models in BE. But there is an obvious pitfall for the ALife approach as well; the currency of minimal models and simple environments may bear very little relation to the complexities of their biological counterparts. The temptation to use ALife to attempt to displace biological knowledge must therefore be resisted. It is better that the insights from such simple modelling are used to engender shifts in conceptual stance of benefit to all intellectual parties.[10]

A word of caution is therefore necessary. I do not suggest that animals (or humans) choose in the way that these evolved agents choose. Nor do I ignore the fact that both humans and animals display matching behaviour in a much wider range of situations than has been investigated here, and are clearly able to rapidly modify their behaviour in response to continuously changing reinforcement contingencies. All I suggest is that the assumption that observed matching behaviour need be supported by a dedicated "mechanism of matching" supervening on internalised behavioural correlates (or response strengths) should not be entirely trusted.

7 Conclusions

Marian Stamp Dawkins [1] has maintained for some time that a satisfying framework for understanding animal behaviour will require a new appreciation of Tinbergen's ideal that both adaptation and mechanism be considered jointly. As she says, the behavioural ecologists must start talking to the neurobiologists (or, indeed, the experimental psychologists). I would like to conclude by suggesting that ALife can encourage this dialogue; in the present paper we have seen an exploration of a BE phenomenon, in which an adaptationist hypothesis (Fretwell's IFD) has been cashed out in mechanistic terms. We have seen that the search for a "mechanism of matching" is easily wrongfooted, given that choice (or, indeed, *any* behaviour) resides not solely inside the chooser, but in the joint activity of the chooser, its environment, and the observer. We have seen that matching behaviour in stable environments need not require anything more than a simple reactive mechanism.

This work also benefits ALife itself, in three ways; by highlighting the potential importance of perception in choice (a theme further developed in [10]), by demonstrating the evolution of choice behaviour in a multi-agent system, and not least by drawing attention to a rich body of behavioural choice phenomena to which the current paper serves as the briefest of introductions.

---

[10] There is of course a different tradition in ALife that *is* concerned with developing models of stringent biological plausibility, from which biological hypotheses can be directly assessed (see [13] for example.)

To return to Herrnstein's aphorism that "all behaviour is choice". Well, one could equally say that *no* behaviour is choice; behaviour is just observed ongoing agent-environment activity. We, as observers, demarcate and label portions of this stream of activity, and sometimes call the boundaries choices. The foraging agents in this paper do not employ any mechanisms of choice; of course how animals may work remains an open question.

Acknowledgements

Thanks to Matt Quinn, Andy Philippides, Emmet Spier, Tom Smith, to my supervisors Phil Husbands and Hilary Buxton, to J. Peck for introducing me to the work of Herrnstein, and to the EPSRC for financial support (award no. 96308700).

References

1. M.S. Dawkins. *Unravelling animal behaviour (2nd ed.)*. Longman, 1995.
2. S.D. Fretwell. *Populations in seasonal environments.* Princeton University Press, Princeton, New Jersey, 1972.
3. C.R. Gallistel, A.L. Brown, S. Carey, R. Gelman, and F.C. Keil. Lessons from animal learning for the study of cognitive develpment. In S. Carey and R. Gelman, editors, *The epigenesis of mind*, pages 3–37. Lawrence Erlbaum, 1991.
4. R.J. Herrnstein. Relative and absolute strength of responses as a function of frequency of reinforcement. *Journal of the Experimental Analysis of Behaviour*, 4:267–272, 1961.
5. J.M Hinson and J.E.R. Staddon. Hill-climbing by pigeons. *Journal of the Experimental Analysis of Behaviour*, 39:25–47, 1983.
6. J.R. Krebs and A. Kacelnik. Decision making. In J.R. Krebs and N.B. Davies, editors, *Behavioural Ecology (3rd edition)*, pages 105–137. Blackwell Scientific Publishers, 1991.
7. P. Maes. A bottom-up mechanism for behaviour-selection in an artificial creature. In J. Meyer and S.W. Wilson, editors, *Proceedings of the first international conference on simulation of adaptive behaviour*, 1991.
8. P. Maes. Modelling adaptive autonomous agents. *Artificial Life*, 1:135–162, 1994.
9. D.J. McFarland. *Animal Behaviour (3rd ed.)*. Longman, 1999.
10. A.K. Seth. Evolving action selection and selective attention without actions, attention, or selection. In R. Pfeifer, B. Blumberg, J.A. Meyer, and S.W. Wilson, editors, *Proceedings of the 5th International Conference of the Society for Adaptive Behaviour (SAB 98)*, pages 139–147. MIT Press, 1998.
11. D.W. Stephens and J.R. Krebs. *Foraging theory.* Monographs in behaviour and ecology. Princeton University Press, 1986.
12. T. Tyrell. *Computational mechanisms for action selection.* PhD thesis, University of Edinburgh, 1993.
13. B. Webb. Robotic experiments in cricket phonotaxis. In D. Cliff, P. Husbands, J. Meyer, and S. Wilson, editors, *From animals to animats 3: Proceedings of the 3rd international conference on the simulation of adaptive behaviour.*, Cambridge, Mass., 1994. MIT Press.
14. B.A. Williams. Reinforcement and choice. In N.J. Mackintosh, editor, *Animal Learning and Cognition*, pages 81–108. Academic Press, 1994.

Modeling of an Early Evolutionary Stage of the Cnidarian Nervous System and Behavior

John Albert

Department of Comparative Physiology, Department of History and Philosophy of Science, Loránd Eötvös University, Pázmány sétány 2. Budapest, H-1117 Hungary
jalbert@mail.matav.hu

Abstract. The object of this work is to examine the early evolution of the nervous system in relation to adaptive behavior. The main questions are: how the nervous system came into being, how a system can be organized during evolution that is able to ensure the adaptive behavior of a being, what are the basic rules of construction that are sufficient to create a workable nervous system without specifying the details of the construction. The biological bases of the model are the phyla *Cnidaria* and *Porifera* because they stand at the beginning of the genesis of nervous organization. We found that if during the evolutionary process a kind of cell comes into being that is able to conduct electrical stimuli - even in a rudimentary way - than this kind of cell improves the behavioral performance by itself without containing any specific information of how to organize the construction of this system.

1 Introduction

Artificial Life offers a new device for modeling biological systems. Among other benefits we can use it for the better understanding of how the nervous system functions. One of the methods to achieve this is to examine a model of the nervous system together with the resulting behavior. This makes it possible to observe the influence of the change of the nervous system on the adaptivity of behavior.

The aim of our work is to create a computer model of the earliest evolutionary stage of the nervous system. The main question we examine is: what are the simplest organizing principles that make possible the formation of a simple conducting and processing system that consists of only elements of the most primitive multicellular animals and that is able to influence the behavior in an adaptive way so as to ensure selectional benefits. The development of this system cannot require sophisticated genetic rules, because they are not present at this level of the evolution yet. So the base of the formation has to be only those simple rules, that follow from the functioning of the building elements. The biological bases of the model are the phyla *Cnidaria* and *Porifera* as they stand at the beginning of the genesis of nervous organization.

2 The animat and its environment

2.1 Anatomy

Our task now is to create a model-animal (animat) with the characteristics of the most primitive *Cnidarians*. The animat is a tube-like being corresponding to the gastrula-state, which is similar to the bodily construction of a *Hydra*. These animals have in fact the most primitive nervous system, so it is not arbitrary to imagine our animat as a *Hydra*-like being without tentacles (see Fig. 1.). This is suitable for our purpose, because these animals have no special locomotor organs which would demand an advanced neural apparatus for their operation [5], [10], [18]. The locomotor apparatus of the animat consists only of the muscles of the body-wall and their motoneurons. This ensures the possibility of simple movement of the body, for example curving and crumpling, like the movement of the body column of *Hydra*. In the recent model our animat is supposed to be sessile, because many species of *Cnidaria* and all of *Porifera* are also sessile.

The movement of the animat is ensured by four longitudinal muscle-strips consisting of many muscular cells along the body. The working of a muscular cell causes the contraction of one section of the body which by itself ensures a very limited movement. Considerable movement is only possible if many muscular elements work in harmony with each other.

2.2 „Histology"

The body of the animat consists of three cell-types, all of which take part in the formation of behavior. These are: epithelial cells, muscular cells and nerve cells.

2.2.1 Epithelial cells. The representation of this type of cell in our model is motivated by the fact that in animals with primitively developed nervous systems, mainly in *Cnidarians*, the role of epithelial cells is similar to that of nerve cells. These cells are able to receive stimuli and to conduct electric potential in a passive way and even to operate muscular cells [3], [16], [18]. In *Porifera* (which have no nervous system) these cells also play an important role in the conduction of stimuli [8], and many researchers derive neurons from these [9], [17]. These cells have a double role in our model, just as reality: on the one hand they serve as a kind of skin, on the other hand they take part in receiving and conducting stimuli.

2.2.2 Muscular cells. These cells form a homogenous population similarly to the epithelial cells, that is, all muscular cells have the same properties. One of these properties is the stimulus threshold, another is the number of neurons that can innervate the same muscle cell, and a third is the ability of receiving stimuli from epithelial cells. (It is well-known that in the case of *Cnidarians* more than one motoneuron can innervate the same muscular cell, and a motoneuron can take part in the innervation of more than one muscular cell.) Features of epithelial cells and muscular cells are encoded in the genome.

2.2.3 Nerve cells. Neurons can be receptors, interneurons or motoneurons by their functioning. But at the lower levels of the evolution these functions have not detached from each other yet, so in *Cnidarians* - mainly in *Hydra*, which have the most primitive nervous system - the same neurons can carry out several functions [11], [19].

Connections between the cells are not determined in advance, therefore synapses of a particular cell are not encoded in the genome (i.e."indirect encoding" [4]). The formation of the synapses of a particular nerve cell is the result of ontogeny as determined by the number and the length of the processes of the given cell and the number of potentially synaptic partner cells within reach.

Characteristics of passive conductivity of cells - a time constant (τ) and a space constant (λ) - are also built into the model. The time constant is the length of time during which the value of membrane potential decreases to its 1/e-th part. The space constant of the membrane is the distance where similarly the membrane potential decreases to its 1/e-th [7]. Both of them are derived from the cable-equation which describes the passive electric properties of biological membranes.

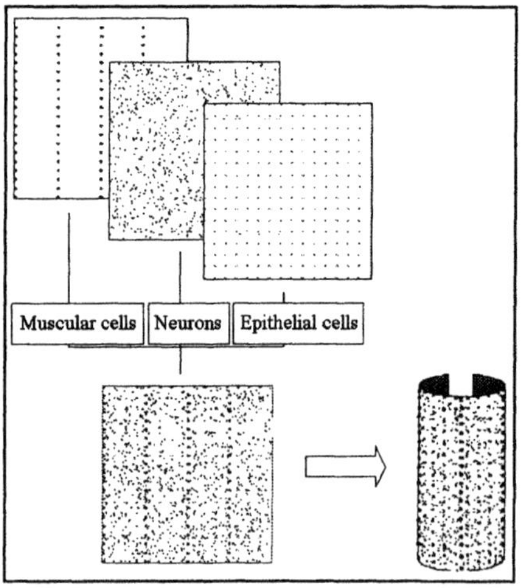

Fig. 1. The bodily construction of the animat

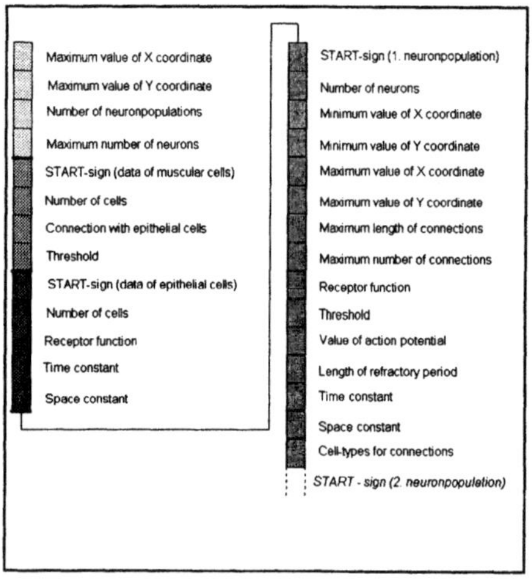

Fig. 2. „Gene-mapping" of the animat

2.3 Genetics

The basic information about the construction and functioning of our animats are encoded in a „genome". The „life" of an animat begins with a short „ontogeny", when the animats start to develop on the basis of information derived from the genome. This phase is similar to that studied in the evolutionary modeling of neural networks [4], [14], but in the work we report here genotypes will not go through an evolutionary process. The possibility for this is built into the model for future work, but in this paper we only make use of the ability of the system to create „phenotypes" (animats equipped with a simple body and a nervous system) on the basis of preencoded data. On the one hand this automates creating animats, on the other hand one of our very objects of study was the way a workable nervous system can come into being on the basis of some very general information concerning its construction. Utilizing the principle of indirect encoding means that not all the specific data for every cell are encoded, but only the most important general rules and data, on the basis of which the nerve cells and their connections can be created. This implies that different phenotypes belong to a given genotype [13], [15].

The model genome consists of four main parts (see Fig. 2.). General information about the size and construction of the body is encoded in the first part, which determines the morphology of the animat. If we lay out the tube-like body, the position of each cells can be described by x and y coordinates, where the maximum values specify the size of the body.

In the second part of the genome the data for muscular cells, while in the third part the properties of epithelial cells are encoded. The homogenous cell populations of both kinds play an important role in the shaping of the frame of the body. At the earliest evolutionary state epithelial cells can take part in operating muscular cells, therefore data about this function must also exist in our model genome.

The largest part of the genome is the fourth, which stores the data for nerve cells. These cells can form different local populations in the animat, so this part consists of as many sections as many populations there are. The value of the x and y coordinates and the synapses of a particular cell are not encoded, these develop during the ontogeny, when the phenotype comes into being during the decoding of the genotype. Unlike in the real animal in our model all data are encoded in a single „chromosome".

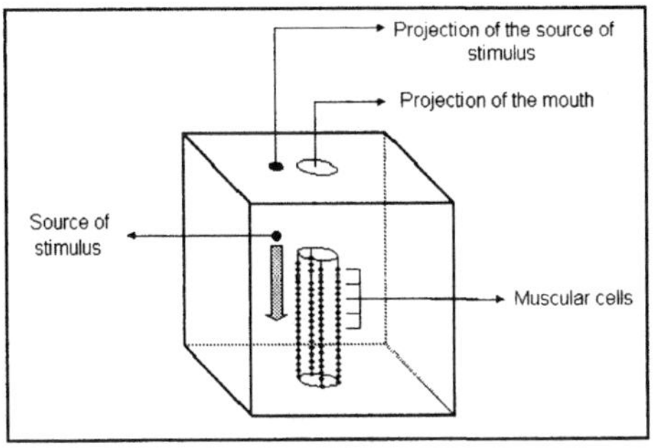

Fig. 3. The animat and its environment

2.4 Environment

The environment is quite simple. It consists of an imaginary box in which the animats exist (see Fig. 3.). We perform the tests in an isolated environment, which means that there is only one individual in the box at a time. The stimulus is the appearance of a piece of food, which slowly sinks to the bottom of the box. This is a chemical stimulus for the animat and its strength decreases exponentially with the distance. If the animat is able to catch at least one piece of food the test is continued. The feeding is succesful if the animat not only touches the piece of food but also „eats" it, in other words it gets food particles into its *coelenteron* through the mouth (see Fig. 4.). The most successful individuals are those that catch the most pieces of food.

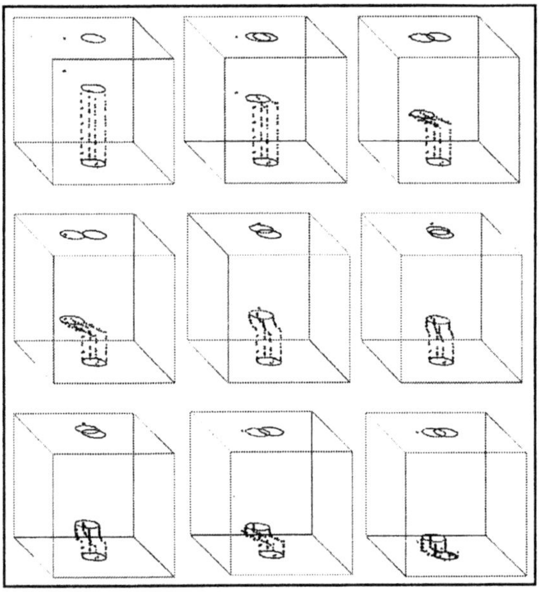

Fig. 4. The movement of the animat

3 Experiments

The fully specified and operational model was part of the author's Masters Thesis and is described therein [1]. A more complete report of the earliest results have been submitted elsewhere [2].

The investigation of the model consisted of several parts, during which the program was run with a number of different settings of the parameters. The purpose was to test the abilities of the animat and to establish relations - if they exist - between the behavior and the values of these parameters. During the tests we studied the effect of the time constant and space constant of the epithelial cells and the nerve cells, besides, the effect of the number of nerve cells and the maximum length of neural processes on the adaptivity of the behavior of the animat.

4 Results and discussions

4.1 The effect of the time- and space constant of the epithelial cells on the behavior

First we studied the behavior of animats that have only epithelial and muscular cells, but no nerve cells. These cells give a stable background for the functioning of nerve cells, therefore it is important to know the degree of responding ability of this system to the environmental stimuli. The time constant of the epithelial cells was studied in the range between 1 to 50 ms (the physiological value is about 1-20 ms) and the space constant between 1 to 10 mm (the physiological value is about 1-5 mm).

Our results show that, qualitatively speaking, the excitability and the success of feeding are determined, in the first place, by the value of the space constant, and the time constant modifies them only slightly (see Fig. 5.). The success of feeding grows with the space constant significantly if its value is close to the physiological value. In summary, the epithelial cells alone are able to ensure a minimally adaptive level of the behavior for the animat, similarly to the behavior of a nerve-free *Hydra* [6].

On the basis of these results we can say that a homogenous, diffuse conducting network - which is probably similar to the ancestor of the nervous system - can be sufficient to control the basic reactions of a simple living being to the environmental stimuli, which stands in harmony with the results of physiological experiments.

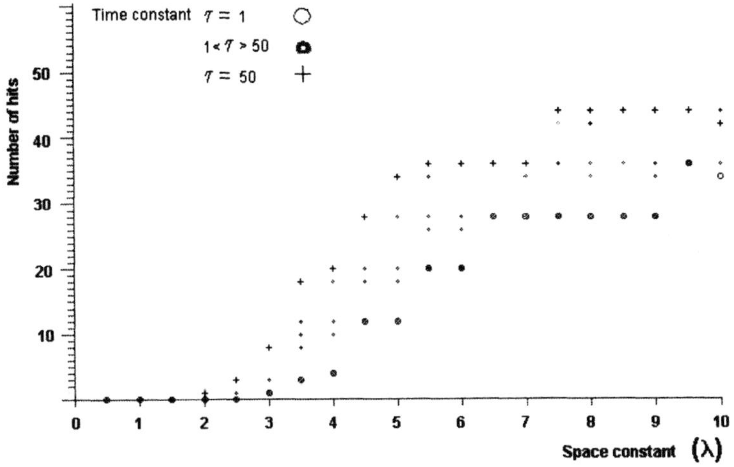

Fig. 5. The number of hits (pieces of food that are caught and eaten) depending on the space constant of the epithelial cells, in the case of nerve-free network

4.2 The effect of the number of nerve cells on the behavior

In further experiments we studied the changes of the behavior of the animats when epithelial-like protoneural cells appear in the primitive nerve-free conducting system as created in the previous step. This corresponds to the evolutionary step during which the ancestors of nerve cells come into being from epithelial cells and begin to organize to a simple network. In this phase nerve cells are rather similar to epithelial cells, because they are able to connect only with their close neighbours. The conduction is decremental, so the network is a "protoneuronal system" similar to the present *Hexactinellida* [12]. The time- and space constants of the nerve cells are like those of epithelial cells. We chose low values where animats with nerve-free networks are no more able to feed successfully, therefore the changes of the behavior, if any, should be explained by the introduction of the nerve cells.

We found that the success of behavior is determined by the space constant and the number of nerve cells together (see Fig. 6.). The number of hits (pieces of food caught and eaten) increases with the number of nerve cells and this tendency is more definite if we increase the value of the space constant as well. Comparing this with the results of the nerve-free conducting system we found that although animats with a nerve-free conducting system are not able to respond to environmental stimuli at these parameter values, the appearance and multiplication of nerve cells increases the success of feeding behavior.

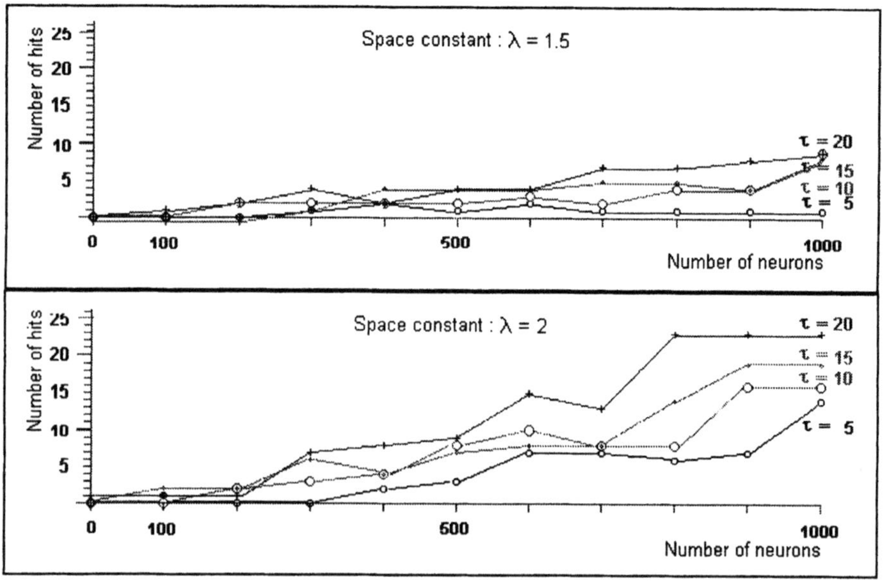

Fig. 6. The number of hits depending on the number of nerve cells in the case of different values of time (τ) - and space (λ) constant

These results suggest that at the most primitive evolutionary state of the nervous system the ability to respond to environmental stimuli can increase by increasing the number of elements alone. This newly established fact may have a great evolutionary importance because this suggests that *at the beginning of the nervous organization increasing the number of the cells may be the simplest way to increase effectiveness.* At this level of the evolution there is no need for developing complex centres in the nervous system yet, to increase its effectiveness, so there is no need to store the information of the organization in the genome either.

4.3 The effect of the connections of nerve cells on the behavior

Next we studied how the length of the processes of nerve cells - that determines the maximum distance of connections - influence the success of behavior. The length of the processes is given in proportion to the size of cells, during the simulation it was taken to be the 10-50-fold size of the nerve cells. In actual animals with the most simple nervous systems we do not find nerve cells with longer processes.

We found that the success of behavior increases only a little with the increasing length of the processes and the degree of increase of behavioral success depends on the overall number of nerve cells (see Fig. 7.). What we see is that if the animat has less than 200 nerve cells, the length of processes (and with this the maximum distance of connections) has no effect whatsoever on the behavior. Considerable behavioral influence can be seen with a large number of neurons only, but the most important factor is not the length of processes, but the ability to develop processes at all. If the cells are able to create them, the behavior is influenced favourably even with short processes, and if the number of cells is between 200-500, the growth of processes has no further effect. But if the number of cells is more than 500, the adaptivity of behavior keeps on increasing with the maximum lengths of the processes. Therefore the success of the behavior depends first of all on the number of nerve cells, and the effect of the lengths of the processes only adds to this. The behavioral effects determined by the lengths of the processes (remember that these effects also depend on the number of nerve cells) can be explained by the fact that nerve cells have to reach a critical density to organize a workable network at all. In the case of few nerve cells their density is not sufficient to get close enough to each other and so they cannot form connections.

Increasing the density of nerve cells, the behavior will be more effective (as seen on both Figs 6. and 7.). However at the first stage this growth is not due to the development of a well-functioning neural network, but to the increasing number of nerve cells that get close to the muscular cells, so they will work as motoneurons, presumably in such a way that muscular cells will be innervated by them directly (as indeed further results show this to be the case [11]), and this is already sufficient for a certain level of adaptive behavior. As already mentioned, it is common about real-life *Cnidarians* that nerve cells have more than one function, so they can receive stimuli from the environment (receptor function) and innervate a muscular cell (motoneuron function) at the same time [19].

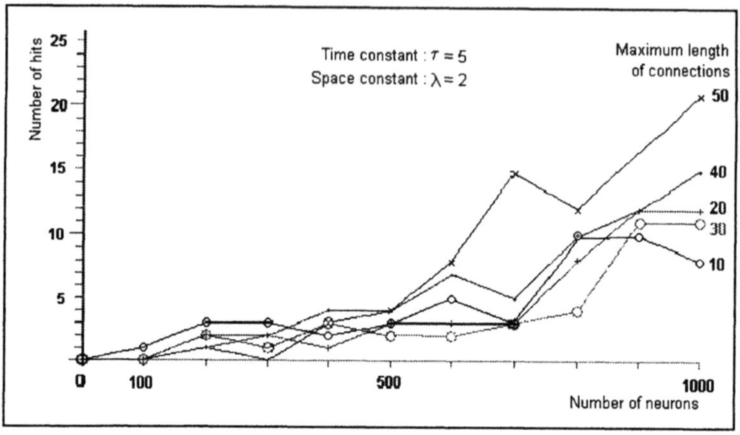

Fig. 7. The number of hits depending on the number of nerve cells and the maximum length of connections

5 General discussion

Summarizing the results: in a network of homogenous epithelial-like cells, considered as the starting point of nervous organization, the changes that adaptively influence the behavior are those that make conductivity more efficient. We found that such a change can cause the increase of the effectiveness of the behavior by itself, without any differentiation in the network or any development of special cell-types. There are numerous ways to increase the conductivity in the network. One of them is the increase of the space constant of cells in case of passive decremental conduction, but changing the biophysical parameters in living organisms is only possible within some boundaries. Evolution had found a solution to increase the effectiveness of forwarding electrical stimuli without increasing the space constant in an extreme way, and this solution was the action potential. Our results suggest that another way to increase the effectiveness of functioning is the increasing of the density of nerve cells.

These simple ways of increasing the adaptivity of the behavior probably played an important role during the early evolution of the nervous system, because at this stage the genome of the animals could not contain detailed information yet to encode the construction and functioning of an effective nervous system. Therefore the organization of the system must have proceeded in an „ad hoc" fashion in the first stage, and it has a great importance, if this early and simple state of the system is already able to increase the chance of survival at all.

The effectiveness of this simple system can increase significantly by changing merely quantitative parameters, therefore the factors that determine the adaptivity of the behavior can be reduced to quantitative attributes. All of these suggest that the way that leads to the ancestral forms of the nervous system could be covered by simple steps.

References

1. Albert, J.: Computer Modeling of the Early Evolutionary Stages of the Nervous System. M.Sc.Thesis (in Hungarian), `http://hps.elte.hu/PhD/janalbert/cim.html` (1997)
2. Albert, J.: Computational Modeling of an Early Evolutionary Stage of the Nervous System. BioSystems, submitted (1998)
3. Anderson, P. A. V.: Epithelial conduction: Its properties and functions. Prog. Neurobiol. 15. (1980) 161-203
4. Balakrishnan, K., Honavar, V.: Evolutionary Design of Neural Architectures - A Preliminary Taxonomy and Guide to Literature. Artificial Intelligence Research Group. Technical Report CS TR # 95-01 (1995)
5. Bullock, T. H., Horridge, G. A.: Structure and Function in the Nervous System of Invertebrates. Freeman, San Francisco, CA (1965) 450-506
6. Campbell, R. D., Josephson, R. K., Schwab, W. E., Rushforth, N. B.: Excitability of nerve-free hydra. Nature 262. (1976) 388-390
7. Ganong, W. F.: Review of Medical Physiology. Prentice Hall Inc. Appleton-Lange (1987) 87-111
8. Lawn, I. D.:Porifera. In: G.A.B. Shelton (ed): Electrical conduction and behaviour in 'simple' invertebrates. Clarendon Press Oxford (1982) 49-72
9. Mackie, G. O.: Neuroid conduction and the evolution of conducting tissues. Quart. Rev. Biol. 45. (1970) 319-332
10. Mackie, G. O.: The Elementary Nervous System Revisited. Amer. Zool. 30. (1990) 907-920
11. Mackie, G. O.: Giant axons and controll of jetting in the squid *Loligo* and the jellyfish *Aglantha.* Can. J. Zool. 68. (1990) 799-805
12. Mackie, G. O., Lawn, I. D., Pavans de Ceccatty, M.: Studies on hexactinellid sponges II. Excitability, conduction and coordination of responses in *Rhabdocalyptus dawsoni (Lambe 1873).* Phil. Trans. Roy. Soc. B. 301. (1983) 401-418
13. Nolfi, S., Miglino, O., Parisi, D.: Phenotypic Plasticity in Evolving Neural Networks. Technical Report PCIA-94-05 (1994) To appear in: Proceeding of the First Conference From Perception to Action, 5-9 Sept. Lausanne
14. Nolfi, S., Parisi, D.: Phylogenetic Recapitulation in the Ontogeny of Artificial Neural Networks. Technical Report 94. Institute of Psychology N.R.C. Rome (1994)
15. Nolfi, S., Parisi, D.: 'Genotypes' for neural networks. In: M. A. Arbib (ed): Handbook of Brain Theory and Neural Networks. MIT Press (1995)
16. Ranganathan, R.: Evolutionary origins of ion channels. Proc. Natl. Acad. Sci. USA. 91. (1994) 3484-3486
17. Robson, E. A.: The Nervous System in Coelenterates. In: P. N. R. Usherwood, D. R. Newth (eds): Simple Nervous Systems. Edward Arnold (1975) 169-209
18. Spencer, A. N.: A Protoneuronal System in Sponges; Cnidarian Nervous Systems. In: C. Ladd Prosser (ed): Neural and Integrative Animal Physiology; Comparative Animal Physiology. Wiley - Liss Publication (1991) 554-567
19. Westfall, J. A., Kinnamon, J. C.: A second sensory-motor-interneuron with neurosecretory granules in *Hydra.* J. Neurocytol. 7. (1978) 365-379

Spontaneous Evolution of Command Neurons, Place Cells and Memory Mechanisms in Autonomous Agents

Ranit Aharonov-Barki[1], Tuvik Beker[1], and Eytan Ruppin[2]

[1] Center for Computational Neuroscience
The Hebrew University, Jerusalem, Israel
ranit@picard.tau.ac.il tuvik@picard.tau.ac.il

[2] School of Medicine and School of Mathematical Sciences
Tel-Aviv University, Tel-Aviv, Israel
ruppin@math.tau.ac.il

Abstract. Using evolutionary simulations, we develop autonomous agents controlled by artificial neural networks (ANNs). In simple life-like tasks of foraging and navigation, high performance levels are attained by agents equipped with fully-recurrent ANN controllers. Examining several experimental settings, differing in the sensory input available to the agents, we find a common structure of a *"command neuron"* switching the dynamics of the network between radically different behavioural modes. In some of the models the command neuron reflects a map of the environment, acting as a "place cell". In others it is based on a spontaneously evolving short-term memory mechanism. The resemblance to known findings from neurobiology places Evolved ANNs as an excellent candidate model for the study of structure and function relation in complex nervous systems.

1 Introduction

The study of *Artificial Neural Networks (ANNs)* relates to Neuroscience in two ways. On one hand, ANNs serve as models for certain functions of biological neural networks [1]. Like any model, ANNs abstract many key features in the phenomena they try to describe, and yet they give researchers the ability to gain better understanding of other, restricted aspects of natural neural mechanisms. On the other hand, ANNs use insights gained from investigating the nervous system to create flexible and powerful computational models that have already found their way to many useful applications [6].

In recent years a novel paradigm emerged in the study of ANNs. This paradigm uses genetic algorithms [10] and evolutionary computation [4] to develop ANNs. Work in this field comprises of the development of "isolated" ANNs, evolving to maximize a certain target function on one hand [8, 5], and the development of "embedded" ANNs, serving as the control mechanism for an autonomous agent, on the other hand [3, 7, 11]. In the latter case the agents perform certain behavioural tasks, and their performance level in these tasks serves as the basis for evolutionary selection.

This new paradigm of *Evolved ANNs (EANNs)* is clearly very interesting from the applicative point of view, opening new horizons in the development of robotic control mechanisms. However, its relevance to our understanding of biological neural system has not yet gained wide recognition.

In this paper we describe the development of EANN-controlled autonomous agents. We use unconstrained network architectures, and life-like behavioural tasks. Under these conditions, we show that networks maintaining steady activation levels can evolve, and moreover - serve to control agents that perform at a remarkably high level, compared with algorithmic benchmarks. By analyzing the non-trivial network structures that evolve in these agents, we demonstrate the existence of neurons whose functional repertoire strongly resembles that of "command" neurons and "place" cells known from biological models. The emergence of "place cells" was already described by Floreano and Mondada [3], who study homing navigation of a real robot. We show that the evolution of such "place cells" in a task requiring simple navigation skills is a robust phenomenon, occurring under various sensory capabilities.

Another important result presented here is the emergence of a memory mechanism. A simple memory-based behaviour in a small, bilaterally symmetrical ANN with 10 neurons and about 20 synapses was recently described by Jakobi [7]. We describe the emergence of a memory mechanism in an unconstrained network with over 200 synapses. This memory mechanism culminates in a single neuron, which in turn serves as a command neuron modulating a complex switch in network dynamics between two radically different operation modes.

These results testify to the biological relevance of the evolutionary paradigm in the research of ANNs. Combined with the fact that the evolving network structure is purely an emergent evolutionary phenomenon, and with the availability of full information about network structure and dynamics, these findings lead us to believe that in EANNs will find their place as one of the leading models for studying the computational activity of biological nervous systems.

2 The Model

The Environment and Behavioural Task: The basic environment consists of a grid arena of size 30x30, surrounded by "walls". In this arena two kinds of resources are scattered. "Poison" is randomly scattered all over the arena. Consuming this resource decreases the fitness of an agent. "Food" is also randomly scattered, but only in a restricted "food-zone" area of size 10x10, at the south-west corner of the arena. The position of an agent is defined by the grid cell it is in, and by its orientation - north, south, east or west. In some of the models the boundaries of the food-zone were marked, enabling the agents to sense them, while in other models this marker was absent. The agents' behavioural task is seemingly very simple - to eat as much of the food while avoiding the poison. The complexity of the task, as will become clearer later, stems from the limited sensory information available to the agents. In order to successfully accomplish their task, they have to find a way to efficiently navigate into the food zone, remain there and consume as much food as possible. The life cycle of an agent (an *epoch*) lasts 150 time steps. 250 poison items and 30 food items are randomly scattered in the environment before the agent is introduced into it. The fitness of the agent is calculated as the total amount of food consumed minus the total amount of poison eaten, normalized to give a maximal value of 1.

Network Structure and Dynamics: Each agent has an EANN "brain" consisting of 15 to 50 neurons (the number was fixed within a given simulation run). Out of these, K_{in} neurons (5 to 7) are dedicated sensory neurons. The values of these neurons are clamped to the sensory input. Four neurons are designated as ***output neurons***, their output serving as commands to the motor system. The network has a fully recurrent connectivity, with the exception of the sensory neurons having no input from other neurons. All neurons with the exception of the sensory neurons are binary McCulloch-Pitts neurons, with zero threshold, and network updating is synchronous.

Selection and Variation: A "chromosome" for an agent having a network of N neurons consisted of $N(N - K_{in})$ real numbers specifying the synaptic weights for each of the synapses. Population size was fixed at 100. Reproduction was sexual, with mating probability proportional to the fitness. Uniform point-crossover with probability 0.35 was applied to the parents' chromosomes, after which point mutations were randomly applied to 2 percent of the locations in the genome. These mutations changed the pertaining synaptic weight by a random value between -0.6 and +0.6. Simulations lasted a pre-defined number of generations, ranging between 10000 and 30000. In the last generation, every agent was evaluated in 5000 epochs, averaging fitness to obtain its accurate fitness. We often used the best agents obtained in the last generation of a given simulation as a population seed for a 'continuation run'. This procedure usually yielded small performance increase in the first one or two iterations. The best agents were obtained using one or two such continuation runs.

The Sensory System: Each agent was equipped with a basic sensor a *"somatosensor"*. In addition, it could have a ***position sensor*** or a ***hunger sensor.*** The basic somatosensor consists of five probes, to each of which a sensory neuron is clamped. Four probes sense the grid cell the agent is located at, and the three grid cells ahead. These probes can sense differences between an empty cell, a cell containing a resource (either poison or food - with no distinction between the two), an arena boundary and food zone boundary when it is marked. Each of these cases was coded as an integer from 0 to 3. The fifth probe is a "smell" probe, returning -1 or +1 if the agent is currently in a grid cell where there is poison or food respectively, and -1 or +1 randomly otherwise. Thus, the agent has to integrate the input from two sensors in order to identify the presence of food or poison.

A ***position sensor*** consists of two receptors, sensitive to the distance from the northern and eastern walls of the environment respectively, and returning a signal inversely proportional to that distance. Again, two sensory neurons are clamped

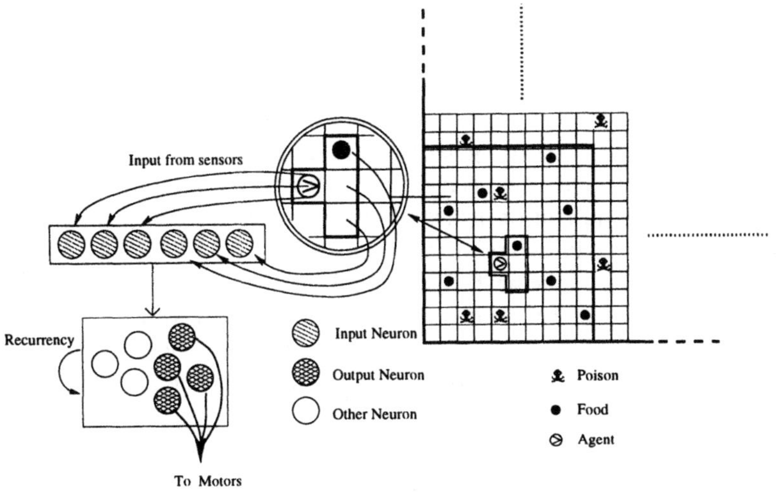

Fig. 1. An outline of the arena (southwest corner) and the controlling network, for an agent equipped with a somato+hunger sensor. The agent is marked by a small arrow on the grid. Curved lines indicate where in the arena each of the sensory inputs comes from. Output neurons and inter-neurons are all fully connected to each other. The output neurons are marked by a grating pattern.

to these values for the duration of a single time step. These two receptors are non-directional, thus giving no information about the orientation of the agent. The *hunger sensor* is a binary probe returning 1 if the agent has already eaten one or more food items during its life, and 0 otherwise. Note that the sensory system of the agents, with the exception of the position sensor, is purely local. Moreover, even in models where the agents were equipped with a position sensor, they had no sense of orientation, so navigation remained a challenging task.
The Motor System: The motor system of an agent consists of four motors, receiving binary commands from four *output neurons*. The first motor induces forward movement when activated. Two motors control right and left turns, inducing a 90 degrees turn in the respective direction when only one of them is activated, and maintaining the orientation otherwise. The fourth motor controls eating, consuming whatever resource is available in the current location when activated. For eating to actually take place, however, there has to be no other movement (forward step and/or turn) in the same time step. This both enforces simple motor integration, and makes any attempt to eat a costly procedure.

3 Behavioural Results and Performance Estimates

In order to evaluate the performance of the evolved agents, we wrote a set of benchmark algorithms. We restricted ourselves to reactive algorithms, i.e. algorithms having no memory. In that restricted realm we can be quite confident that we wrote nearly the best possible algorithm, thus any agent obtaining a better score is clearly employing some kind of memory. The basic idea behind all these algorithms is the same, and can be summarized as: "When reaching a resource - eat it if and only if it's a food item. Unless you 'believe' you are inside the food zone, try to navigate into it, and ignore all resources on your sides. If you 'believe' you are inside the food-zone, switch to an efficient foraging mode, and try not to leave the food-zone...". This basic idea translates into different instructions, according to the available sensory input. For example, when equipped with a position sensor, but with no hunger sensor (and no sense of orientation), the activation levels of the two position receptors can act as a trigger for the change in behavioural mode.

The details of the specific tasks depend on the type of sensor available to the agent, and whether the food-zone borders are marked or not. This defines 6 different scenarios. Fig. 2 summarizes the performance levels achieved by the best memoryless algorithms we designed, and those attained by the best evolved agents, for each of these scenarios.

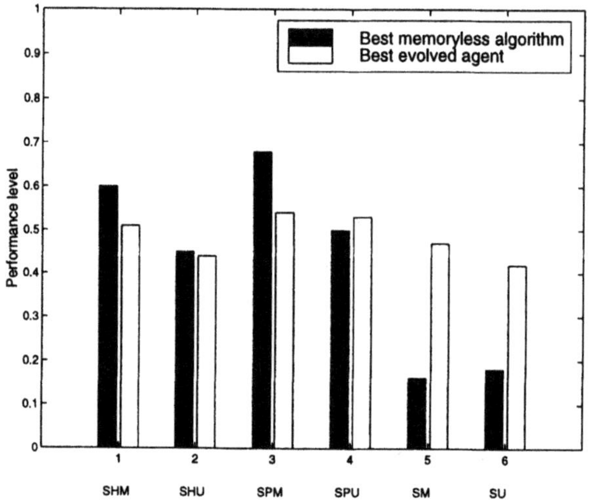

Fig. 2. Comparison between memoryless algorithmic performance and that achieved by evolved agents. S = somatosensor, SP = somato+position sensor, SH = somato+hunger sensor. M = marked food-zone, U = unmarked food-zone.

As can be seen from the figure, the evolved agents do fairly well, compared to the algorithms, in all possible scenarios. The agents specifically excel where no memoryless algorithm does well. As shall be seen below, this is due to the development of memory under these scenarios.

4 Function and Structure of Successful Agents

We shall now focus on three cases where the evolved agents have developed particularly interesting behaviours. The three cases share the same behavioural task. They differ in the sensory input available to the agents. In the first case, the somatosensor was supplemented by a hunger sensor. In the second, the hunger sensor was replaced by a position sensor, and in the third case, the agents had to do with the somatosensor alone.

In all three cases, the successful strategy relied upon a switch between two behavioural modes - exploration and foraging. Due to the different sensory capabilities in each of these cases, different sensory cues were used to trigger that switch. However, a common structure could be identified, whereby the mode switch was always mediated by a central *"command neuron"*. The processing behind the action of this command neuron was different in each case, but its overall function was nearly identical, as we shall see.

Case 1: Somato + Hunger Sensor, marked food-zone borders. Under this scenario, the successful agents' behaviour remarkably resembles that of the best algorithm we wrote as a benchmark for that task. The exploration mode in this case consisted of moving in straight lines, ignoring resources in the sensory field which are not directly in front of the agent, and turning at walls. If food-zone border is reached before the first food item is encountered, the border is ignored, and forward movement is preserved. The encounter with the first food item triggers the mode switch. The agent starts turning to resources to its right or left in order to examine them, and consumes nearly all the food items it encounters. Reaching the border of the food-zone at that stage causes it to turn - just as it turns at walls. This behaviour is not perfect, though. Sometimes resources are ignored, and there are cases where food-zone borders are crossed after the first eating event. These imperfections account for the gap between the agents' performance and that of the algorithm.

In this experiment, the hunger sensor modulated the behaviour of the agent. When the neuron connected to the hunger sensor was clamped to a firing state (indicating that eating has already taken place), the agent employed a foraging strategy even outside the food-zone. When clamped to the quiescent state, food-zone borders were always crossed, and movement was mainly in straight lines. In this case the "hunger neuron" was especially suited to serve as a command neuron by its predefined connection to the hunger sensor. But as we shall show, neurons with a similar function can also evolve spontaneously.

Case 2: Somato + Position Sensor, unmarked food-zone borders. The basic behavioural modes in this case were similar to case 1, with the exception of turning at food-zone borders while foraging, since the borders were not marked. The switch between the two modes was based on the position in the environment - foraging inside the food-zone, and moving in straight lines when outside it. The agents employing this strategy often stepped out of the food-zone, but then switched to moving in straight lines, thus returning to it promptly.

It should be noted that without some form of memory, it is hard to avoid leaving the food-zone from time to time, as knowledge about position only, without orientation cues, doesn't suffice for that.

Examining the networks of successful agents revealed one common feature. Certain inter-neurons had a position-sensitive response. Typically, these neurons would fire outside the food-zone, and remain quiescent inside the food-zone and in its close vicinity. Fig. 3(2a) depicts the mean activity of such an inter-neuron, as a function of the location in the environment. Fig. 3(1a) shows that the hunger neuron in the previous experimental scenario (case 1) has a similar location-dependence (with inverse coding). Shifting the food-zone to the southeast corner of the arena revealed the different mechanism behind these seemingly similar maps. The activity pattern of the hunger neuron in case 1 shifts according to the food-zone location (fig. 3(1b)), being based upon the first encounter with food. The activity pattern in the second case, based directly on the agent's position, remains fixed (fig. 3(2b)), thus leading to near-zero performance.

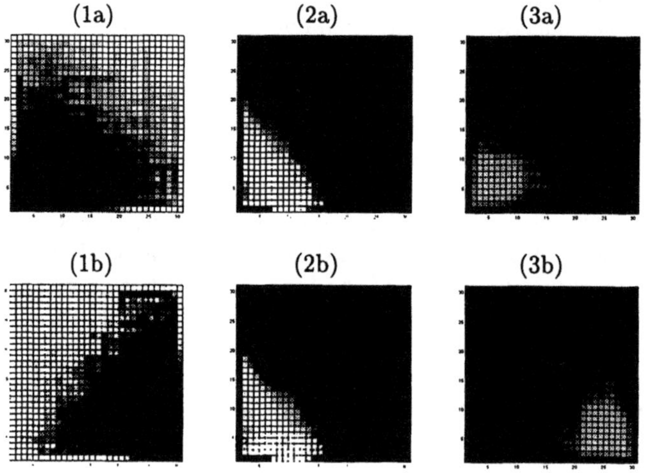

Fig. 3. Location-dependent activity maps of the respective command neurons. Darker means higher average activity. **Somato+Hunger Sensor:** (1a) Usual condition, (1b) Shifted food-zone (2.3% performance decrease). **Somato+Pos Sensor:** (2a) Usual condition, (2b) Shifted food-zone (100% performance decrease). **Somatosensor:** (3a) Usual condition, (3b) Shifted food-zone (no performance decrease).

Given the direct sensory information about the location in the environment, the mere existence of a "place cell" is not very surprising. Yet the function of these emerging neurons is very interesting. In the lack of a single cell whose sensory input can serve as a cue for a behavioural switch, the naturally emerging "place cells" provide this function. Indeed, clamping these place cells to active or quiescent states produces behavioural patterns identical to the ones caused by applying the same procedure to the hunger neuron in the previous experiment.

Case 3: Somato Sensor only, unmarked food-zone borders. This particular case turned out to be especially interesting. Due to the purely local sensory information and the lack of any positional cue, it is hard to adopt a strategy that forages more efficiently inside the food-zone. The two basic strategies – always moving in straight lines or always examining every resource in the sensory field – both yield near-zero performance. Our best memoryless benchmark algorithm improved on this by using the total number of resources in the sensory field as a cue for switching between foraging and exploration behaviours, scoring an average of 0.16.

The evolved agents achieved much better results, as evident from fig. 2. The observed behaviour of these successful agents is characterized by an elevated tendency to examine resources in their sensory field, that lasts some time after eating. There is a 74% increase in the tendency to turn to resources right after eating, compared to the behaviour a long time after eating [1].

This behaviour led us to believe that the successful agents have some sort of short-term memory, remembering the last eating episode for several time-steps. Such a memory mechanism could take the role of a command neuron in this scenario, switching to foraging behaviour when food is encountered and maintaining it for a short while (a reasonable strategy, given the concentration of food in a restricted area). Examining the "brains" of these agents, we managed to confirm the existence of such a memory device. Fig. 4(a) shows the average activity of such a memory command neuron, as a function of the time elapsed since the last eating episode.

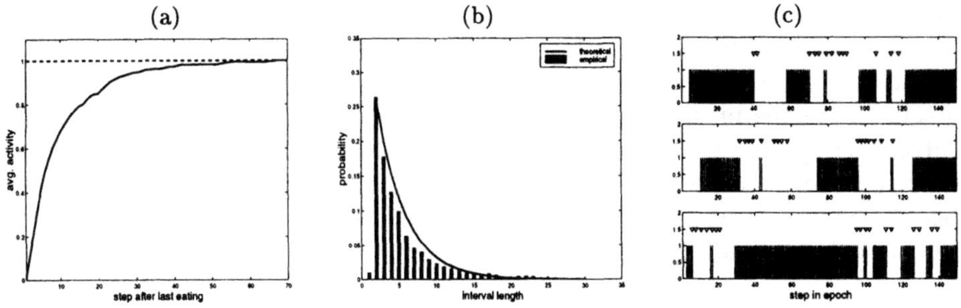

Fig. 4. Activity patterns of a memory neuron: (a) The average activity level, as a function of the time elapsed since last eating episode. The average is taken over all the steps in 1500 distinct epochs. (b) The distribution of quiescent period durations. The solid line corresponds to exponential distribution with $\lambda = 0.26$ (shifted by two timesteps). (c) A raster plot of the activity in 3 different epochs. Vertical lines mark steps where the neuron fired. The small triangles above them mark the feeding events.

As can be seen from fig. 4(a), the memory command-neuron undergoes a sharp inhibition immediately after eating. Its probability to fire then increases as time since the eating episode elapses. The emerging dynamics of the cell is sustained activity, interrupted by periods of inactivity triggered by the feeding events. Fig. 4(b) shows the distribution of post-eating quiescent period durations. Fig. 4(c) shows a raster plot of the activity of the same cell, where the small triangles indicate the times of feeding events. As evident from the fig., the normal state of the neuron is active, and it is inhibited *only* following feeding. After a

[1] In the case of marked food-zone borders, the trigger from the memory-based command neuron also determined whether to turn at food-zone borders or not.

"refractory period" that lasts a single time-step, the length of the quiescence period is geometrically distributed. That is, the probability to resume firing at any time-step after eating is constant, but once the neuron starts firing, it will retain the firing state till the next feeding. The average quiescence period after eating lasts 5.74 steps. Approximating by a continuous exponential distribution function (shifted by two timesteps), we found the constant of the process to be $\lambda = 0.26$, which corresponds to a mean memory maintenance of 5.85 time-steps (3.85, after a lag of two timesteps). The stochastic nature of such a memory mechanism emerging in a binary neural network can in part be accounted for by the random distribution of resources in the environment, as well as the partly random input from the somatosensor.

Fig. 3(3a) shows the mean activity of the same "memory neuron", as a function of the location in the environment. The activity pattern is similar to that of the command neurons in the two previous cases. The location-dependent activity in this case is a byproduct of the memory mechanism - the neuron is less active (on the average) in the vicinity of food, because eating shunts its activity for several time-steps. This leads to a more graded "map". Similar to the case of the hunger neuron, shifting the food-zone from its original location to the southeast corner (fig. 3(3b)) causes no malfunction, as the location-dependent activity shifts accordingly.

The simple utilization of memory in this experiment induces a considerable increase in performance level. Although these agents do not reach the scores attainable in the presence of global positioning cues, they do remarkably well, surpassing by far every memoryless algorithm. In fact, in the regime where the food-zone borders are not marked, the most successful agents lag only slightly behind the best memoryless algorithm that operates *given* explicit knowledge of the hunger state via a hunger sensor.

5 Summary

We analyzed the control mechanisms evolving in autonomous agents performing a simple foraging task and governed by a recurrent ANN, without any predefined network architecture. In order to succeed in their behavioural task, the agents had to develop a mechanism for switching between two distinct types of behaviours, foraging and exploration. In all six experimental scenarios examined, the evolved agents managed to closely match the best memoryless algorithms for the task, and in a few of the cases surpassed them by far. This was achieved with *completely unconstrained* network architectures.

We discussed in detail three experimental scenarios, differing in the sensory input that was available to the agents. In all three cases, a similar mechanism has evolved, whereby an inter-neuron acts as a *"command neuron"*, translating the sensory input into a binary command that modulates the dynamics of the whole network and switches between the two behaviours. In the case where the agents had the most limited sensory capabilities, a memory mechanism emerged, which became the basis for such a command neuron.

Cells in the rat hippocampus show complex location specificity [9]. The location-specific activation patterns presented here are not nearly as sophisticated, yet they do seem to share the same basic characteristics. The function of the command neurons in our simulated agents resembles the CPR command neuron found in the Aplysia, which modulates the feeding behaviour based on certain sensory stimuli [12]. This resemblance to known findings from neurobiology suggests that once better scalability is achieved, EANNs may provide an excellent vehicle to study the fundamental problem of structure and function relation in complex nervous systems.

The results presented here were obtained using the crudest form of genetic encoding - direct specification of all the synaptic weights in the genome. This bears a direct limiting effect on the scalability and speed of the evolutionary process. With the application of more sophisticated genetic encoding schemes, such as grammatical or ontogenic encodings [2, 8] we may expect the evolution of larger recurrent EANNS, processing more complex sensory input to achieve more intelligent behaviours. The accessibility of such models to thorough analysis should make them an important tool in the tool-chest of modern neuroscientists.

Acknowledgments: We greatly acknowledge the valuable contributions made by Henri Atlan, Hanoch Gutfreund, Isaac Meilijson and Oran Singer. This research was supported by the FIRST grant of the Israeli Academy of Science.

References

[1] D. Amit. *Modeling Brain Function.* Cambridge University Press, New York, 1989.

[2] A. Cangelosi, D. Parisi, and S. Nolfi. Cell division and migration in a 'genotype' for neural networks. *Network*, (5):497–515, 1994.

[3] D. Floreano and F. Mondada. Evolution of homing navigation in a real mobile robot. *IEEE Transactions on Systems, Man, and Cybernetics - Part B*, 26(3):396–407, June 1996.

[4] D. B. Fogel. *Evolutionary Computation - Toward a New Philosophy of Machine Intelligence.* IEEE Press, Piscataway, NJ, 1995.

[5] P. G. Harrald and M. Kamstra. Evolving artificial neural networks to combine financial forecasts. *IEEE Transactions on Evolutionary Computation*, 1(1):40–52, April 1997.

[6] J. Hertz, A. Krogh, and R. Palmer. *Introduction to the Theory of Neural Computation.* Addison-Wesley Publishing Company, New York, 1991.

[7] N. Jakobi. Evolutionary robotics and the radical envelope-of-noise hypothesis. *Adaptive Behavior*, 6(2):325–368, 1998.

[8] H. Kitano. Designing neural networks using genetic algorithms with graph generation system. *Complex Systems*, (4):461–76, 1990.

[9] P. D. Martin and J. O'Keefe. Place field dynamics and directionality in a spatial memory task. *Brain Research*, 783(2):249–261, February 1998.

[10] M. Mitchell. *An Introduction to Genetic Algorithms.* The MIT Press, Cambridge, Massachusetts, 1996.

[11] C. Scheier, R. Pfeifer, and Y. Kunyioshi. Embedded neural networks: Exploiting constraints. *Neural Networks*, (7-8):1551–1569, October-November 1998.

[12] T. Teyke, K. R. Weiss, and I. Kupfermann. An ientified neuron (CPR) evokes neuronal responses reflecting food arousal in aplysia. *Science*, (247):85–87, January 1990.

Modularity in Evolved Artificial Neural Networks

Ádám Rotaru-Varga

School of Computing Science, Simon Fraser University, Vancouver
<arotaru@cs.sfu.ca>

Abstract. The relationship between *evolution* (genetic & developmental processes of an evolutionary system) and *modularity* (its support for modular structures) is explored. Modules are defined as structures with common origin; either evolutional or developmental. In the former case, processes supporting modularity operate on the phylogenetic level, in the latter, on the ontogenetic level. Three such processes are identified (duplication, divergence, co-vergence). The existence of these processes determine the system's support for modularity. Modules are analysed in the particular context of artificial neural networks (ANNs), where they appear as subnetworks. Gruau's cellular developmental encoding is used as an example, and an extension is proposed which better supports modularity.

1 Introduction

The *issue of modularity* in living systems appears under various contexts, addressed by different disciplines. In artificial life, the issue of modularity appears under all these contexts, stressing the need for an inter-disciplinary approach [1]. We mainly focus on the relation between evolution and modularity, more specifically, what aspect of an evolutionary system determines its support for modularity? Does a system support modularity, and if it does, to what extent? In particular, in the context of evolved artificial neural networks (ANNs), we answer this question for several versions of the cellular developmental encoding.

Common-Origin Definition of Modules. In the general sense, *module* refers to a subsystem that can be defined in terms of its function and/or structure. The criteria of *similar function* and *similar structure* are essential, but not very precise. Instead, in our evolutionary framework we define modules as *subsystems with common origin*, tacitly assuming a correlation between this requirement and the former two. Common origin includes both common *evolutionary* origin (*phylogenetic*) and common *developmental* origin (*ontogenetic*). We stress that common origin should *not* be viewed as the only mechanism for the emergence of modules. For example, unrelated modules can *acquire* similar functions.

Evolved ANNs provide a good study ground for modularity, because relations between modules (which are *subnetworks*), and their internal structure, are relatively easy to analyse. A better understanding of modularity can also contribute practical ideas for evolving ANNs, which are widely used as controllers. Modular ANNs have several advantages (tested or hypothesised): smaller number of components, easier re-adaptation from one task to another, more powerful generalisation, etc. [5, 6, 1].

2 Genetics of Modularity

Duplication and Divergence Create Modules. Modules descending from a common origin are created by some *duplication* event. Duplication alone can account only for *identical modules*: similar (but different) modules are a result of subsequent *divergence.* As two sibling modules differentiate, they can assume different functions. If two contradictory adaptive constraints had been active on the ancestor module, duplication and divergence can resolve this contradiction, 'lifting' evolution from a temporary dead end [1].

Two Pathways: Phylogeny and Ontogeny. Duplication and divergence processes can occur on different timescales. We identify two cases: one involving processes operating on the population level (*phylogeny*), and one on the organism level (*ontogeny*). In the first case, the phenotype is encoded by the genotype *directly*: every module is encoded by a corresponding gene. A *gene duplication* event (on the genotype level) can result in two copies of the gene and the encoded module. Subsequently, genetic operators can alter the two copies independently, causing their divergence. We refer to this case as *phylogenetic module duplication.*

In the second case, the phenotype of the organism is the result of a *developmental process.* Modules are result of duplication processes occurring during the ontogenetic process. The developmental process is guided by the genotype, but indirectly, thus there is no one-to-one correspondence between genes and modules. For convenience, we term this case *ontogenetic module duplication.* Table 1. lists three module-related mechanisms and their form in the two cases.

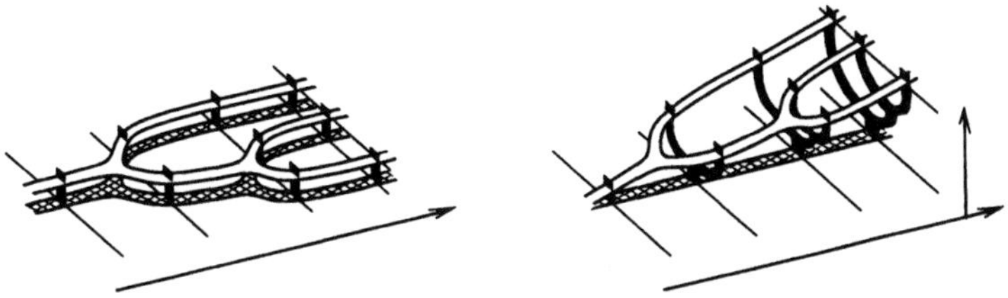

Figure 1. Illustration of the relationship between phylogeny and ontogeny. The horizontal arrow indicates evolutionary change, the vertical one developmental time. *Left*: no ontogeny; the phenotypic tree (white) duplicates the genotypic tree (grey). *Right*: ontogeny (black) links phenotypes and genotypes. Note the similar shape of trees.

In both cases the end result is modularization on the phenotype level, induced by changes occurring on the genetic level, guided by selection. Ontogeny is an additional step linking the genotypic and phenotypic levels. Fig. 1. depicts the relation between the phylogenetic and ontogenetic planes. In nature, *ontogeny reflects phylogeny* [2, 11].

A notable difference is that in the ontogenetic duplication case there can be genetic operators that affect *both modules equally.* This is not possible in the phylogenetic duplication case, because sibling modules are encoded by independent genes. Such parallel adaptation can be useful. A change somewhere in the development tree is analogous to a *past* change in the ancestor module. In conclusion, ontological module duplication has better potential support for modularity.

mechanism:	**duplication**	**divergence**	**co-vergence**
effect:	creates new module	changes one module	changes sibling modules
in phylogenetic duplication case:	gene duplicating mutation	ordinary mutation	*impossible*
in ontogenetic duplication case:	mutation that results in an additional duplication	may be supported by mutation (with developmental processes)	

Table 1. The three module-supporting mechanisms and their appearance.

Gruau's Cellular Developmental Encoding. As an example, we look at the encoding by Gruau [5, 6]. Does this encoding support modularity? The only candidate for a developmental duplication mechanism is the DIV instruction. Two daughter cells can develop into subnetworks, but there is nothing in the scheme that would make these subnetworks related. Therefore, duplication is not implemented, so the basic Gruau encoding does not support modularity.

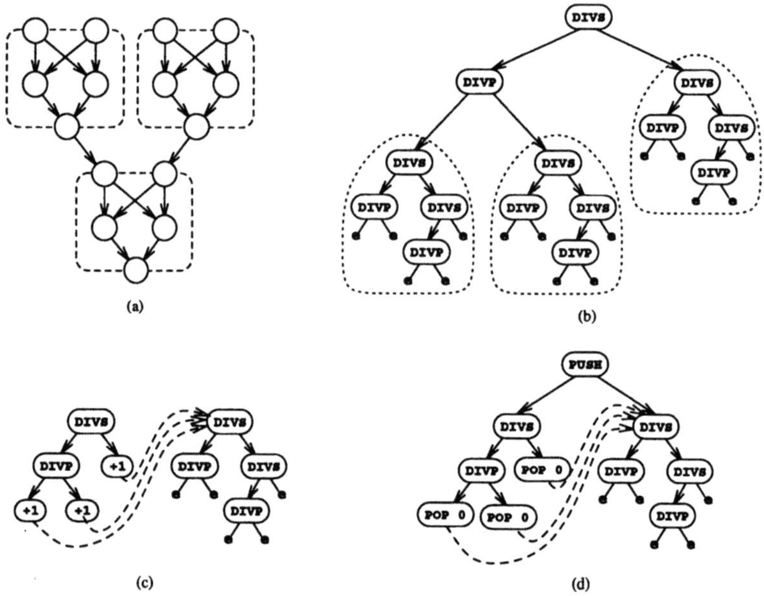

Figure 2. A sample network (a) and its cellular developmental tree (b). The network has the architecture suitable for solving the parity-4 problem. Alternative encodings: (c) extended with ADSN, (d) extended with PUSH/POP.

The simplest way to extend the encoding to support modules is with an identical division (IDIV) with only *one* subtree, executed by *both* cells. Thus more copies of the same structure can form. Gruau's ADSN extension [5], implements the same idea with more flexibility. In these extensions, the *code tree* is generalised to a *directed acyclic graph*. But these extensions are still limited, as they produce only identical modules. In conclusion, this ADSN-extended version supports modularity, but in a limited way (it implements division and co-vergence, but *not* divergence).

We propose another way to implement identical modules, and one to implement divergence. We extend cells to hold a small stack of pointers, and introduce two instructions: PUSH saves its right subtree on the stack, and executes the left; POP returns to the address on the top of stack (if any). This scheme matches the ADSN scheme, but still maintains the code in one tree, see Fig. 2. For divergence, a conditional branching instruction can be created: cells maintain a queue of flags, which record information about last divisions. Later, when they execute the same portion of the code tree, a COND instruction can branch according to this value. Finally, Table 2. compares these variations of the cellular encoding, by looking at their ability to express specific scenarios. Only the last variant is able to express all proposed scenarios.

scenario:	encoding: basic Gruau, plus:				
After division, the two daughter cells ...	-	IDIV	ADSN	PUSH	PUSH+COND
... execute different code (no modules).	yes	yes	yes	yes	yes
... execute the same code (identical modules).	-	yes	yes	yes	yes
... execute some different code, then the same.	-	-	yes	yes	yes
... execute same, then different (divergence).	-	-	-	-	yes
Same portion of code appears in arbitrary places.	-	-	yes	yes	yes

Table 2. Comparison of different extensions of cellular developmental encoding.

Weak Encodings. The outlined extensions of the Gruau encoding might be criticised as *ad-hoc* and difficult to evolve. What is difficult to grasp is the concept of being similar yet different. In the Gruau encoding, the fate of a cell is completely determined by its genetic code; this is a *strong* encoding. However, in *weak* encodings *genes have no total* control over the expression of phenotype. Several ANN encoding schemes has been developed that imitate biological ontogenetic processes in more detail [8], in which the development of a cell is influenced also by other cells and the environment, adding another layer to the genotype-phenotype mapping. Due to these higher-order interactions, divergence can occur even if active genes are identical! Weak encodings are beyond the scope of this discussion, but we can argue that weak encodings provide better support for the emergence of modules.

3 Experimental Results

We compared the performance of different encodings on a simple task (temporal version of identity task for words, requiring memory). We compared three encodings: a direct ANN encoding (no modules), a basic developmental with no modules, and an extended developmental with module support (using PUSH/POP). Most solution were found using the direct encoding – attributed to the increasing overhead of the more complex encodings (success rates were 87%, 10% and 37% for the three cases). However, the modular encoding produced solutions with more structure, which were able to generalise much better (generalisation power: 5%, 77%, and 90%). Space does not permit us to present further details of the experiments.

4 Conclusions

Emergence of modules (defined by their common origin) depends on the existence of a duplication process, and at least one kind of adaptation: divergent and/or parallel. In a system, the collection of genetic and developmental processes determine the existence of such processes, and thus the system's support for modularity. We distinguish two routes for module formation: the phylogenetic and ontogenetic pathways. While phylogenetic module duplication has a good intrinsic support for divergence, it cannot support co-vergence. In contrast, ontogenetic module duplication can support both (but not necessarily).

The encoding used as an illustration, Gruau's cellular developmental encoding does not support modularity in its basic form, some of its extensions do, but to a limited degree. Some experimental evidence based on a task of a simple recurrent network shows that more sophisticated encodings do produce more modular structures, and these can indeed be more desirable. However, the more sophisticated an encoding is, the more time the evolutionary search takes, due to increasing overhead. These limitation provides an incentive to explore more complex encodings, which include interactions. Weak encodings can more naturally embed the above mentioned three mechanisms. In light of these, it should not be a surprise that in nature, organisms typically 'use' developmental encodings, and weak ones.

References

[1] Calabretta, R., et. al. A case study of the evolution of modularity: towards a bridge between evolutionary biology, artificial life, neuro- and cognitive science. In *Proceedings of "Artificial Life VI", 1998*. Adami, C. et al., eds., MIT Press, 1998.

[2] Gilbert, S. Developmental Biology. Sinauer Associates, Sunderland, MA, 1991.

[3] Grönroos, M. A. Evolutionary Design of Neural Networks. M. Sc. thesis, Computer Science, Department of Mathematical Studies, University of Turku, 1998.

[4] Gruau, F., Whitley, D. L. Adding learning to the cellular development of neural networks: Evolution and the Baldwin effect. In *J. of Evolutionary Computation*, 1(3):213-233. 1993.

[5] Gruau, F. Modular Genetic Neural Networks for 6-Legged Locomotion. In *Artificial Evolution European Conference, 1995. Lecture Notes in Computer Science*, Alliot, J.-M., Lutton, E., Ronald, E., Schoenauer, M., Snyers, D., eds., Springer, Berlin, 1996. 201-219.

[6] Gruau, F., Quatramaran, K. Cellular Encoding for Interactive Evolutionary Robotics. Research report 425, Univ. of Sussex, School of Cognitive and Computing Sciences, 1996.

[7] Hadley, R. F. Connectionism and Novel Combination of Skills: Implications for Cognitive Architecture. Technical Report CMPT TR 1998-01, Simon Fraser University, 1998.

[8] Michel, O., Collard, Ph. Artificial neurogenesis: an application to autonomous robotics. In *Proceedings of the 8th International Conference on Tools with Artificial Intelligence,* Radle, M. G., ed., IEEE Computer Society Press, 1996. 207-214.

[9] Ray, T. S. Selecting Naturally for Differentiation: preliminary evolutionary results. In *Complexity*, 3(5): 25-33. John Wiley & Sons Inc., 1998.

[10] Whitley, D. L., Gruau, F., Pyeatt, L. Cellular Encoding Applied to Neurocontrol. In *6th Inttl. Conference on Genetic Algorithms*, Eshelmann, L., ed., Morgan Kaufmann, 1995.

[11] Sanchez, E., Mange, D., Sipper, M., Tomassini, M., Pérez-Uribe, A., Stauffer, A. Phylogeny, ontogeny, and epigenesis: Three sources of biological inspiration for softening hardware. In *Proceedings of The First Intl. Conference on Evolvable Systems: From Biology to Hardware (ICES96)*, vol. 1259 of LNCS, Higuchi, T., et al., eds., Springer-Verlag, 1997.

Framsticks: Towards a Simulation of a Nature-Like World, Creatures and Evolution

Maciej Komosiński and Szymon Ulatowski

Institute of Computing Science,
Poznan University of Technology
Piotrowo 3A, 60-965 Poznan, Poland

Maciej.Komosinski@cs.put.poznan.pl

http://www.frams.poznan.pl/

Abstract. In this paper we describe our attempt to create a nature-like simulation model of artificial creatures. The model includes physical simulation of creatures, their interaction with the environment, their neural network control, and both directed and open-ended evolution. We describe a complex, three-dimensional simulation system, where various fitness criteria can be selected for evolving species, and a spontaneous evolution can be run. The work is still being developed, and we hope to make it a realistic model capable of producing real-life phenomena through an open-ended evolution in a life-like world of stick creatures.

1 Introduction

Artificial life research attempts to study real-life, biological organisms by creating and analyzing virtual creatures. Existing artificial life experiments seem to fall into two categories: the first is based on elegant, perfect, but often simple models. These are usually used for theoretical studies or to test some biological hypotheses – like those concerning coevolution in a pursuer-evader game [2], or evolution of spider nets and eyes [1]. Those in the second category use relatively sophisticated models, but the evolutionary mechanisms are not so much scientifically and biologically inspired. Instead, they focus on realistic simulation, graphics [7], or entertainment [6]. By encompassing the advantages of both approaches in **Framsticks,** we tried to fill the gap between advanced artificial life models with their consequences and advanced simulation tools with their realism. Here, the simulation model is reasonably biologically realistic, while the evolution model allows great possibilities for various experiments.

Our model encompasses a virtual, three-dimensional world and creatures (with their "bodies" and "brains") that are capable of interacting with themselves (locating, pushing, hurting, killing, eating, etc.) and the environment (walking, swimming, etc.). The environment can be a composed of any combination of flat land, hills, and water. Evolution may be directed by the predefined criteria.

Although there have been many experiments so far, no attempt was made at spontaneous evolution using environments and creatures as complex as those in **Framsticks**. We hope that our model is complex enough to allow the emergence of sophisticated, life-like dependencies and phenomena, and simple enough to be simulated on existing computer systems.

The paper is organized as follows: section 2 describes the system architecture and models of evolution and simulation. In section 3 we focus on the evolutionary properties of our system, describing genotype representation, genetic operators like crossover and mutation, etc. In section 4 we briefly discuss the results of experiments performed so far, summarize the work and present our future goals.

2 Simulation model

2.1 System architecture

Our aim is to design the **Framsticks** model so that it allows for an open-ended evolution (including natural selection) of stick creatures, controlled by neural networks, in a three-dimensional world [4]. In the real world and some artificial life simulators (*Tierra*, *Avida*), the rules of selection and reproduction emerge from the simulated creatures' living conditions. In our simulator, these rules are already defined, as they are in other evolutionary models designed for optimization, such as genetic algorithms [3] and most of artificial life simulations [7]. Creatures in **Framsticks** are evolutionarily optimized according to some predefined criteria. However, it is possible to mimic an open-ended evolution with the "directed" model of evolution. This is the case when the chosen fitness criterion depends on the survival and reproduction abilities. Thus, an open-ended evolution can be simulated by using the life span selection criterion. The longer the creature lives, the better it reproduces in the environment, which is generally analogous to the real-world situation.

The main module of the simulator is the evolution simulator, which is responsible for maintaining the set of currently existing genotypes. This module must also obtain their multi-criteria evaluation in order to perform selection. All the individuals need not be simulated simultaneously: only a fraction of them are being evaluated in the virtual world at a time. The artificial world is thus a reduced model of the whole ecosystem. Such an approach is quite universal and has also the following advantages:

- a few individuals can be simulated much faster than a few thousand, so one can see the simulation in the real-time, study the behavior of creatures, and affect them,
- the only information needed to save/load the state of the evolutionary process is the performance of each genotype. The state of the virtual world is not saved.

In order to construct such an architecture (figure 1), two parameters are needed: a maximum number of genotypes, N, and a maximum number of individuals simultaneously simulated, n. Usually, n is significantly smaller than N. When the interaction between simulated creatures is not important, n may be set to 1. Larger values of n mean that a larger part of the whole set of individuals is simulated, and more interactions between creatures may happen.

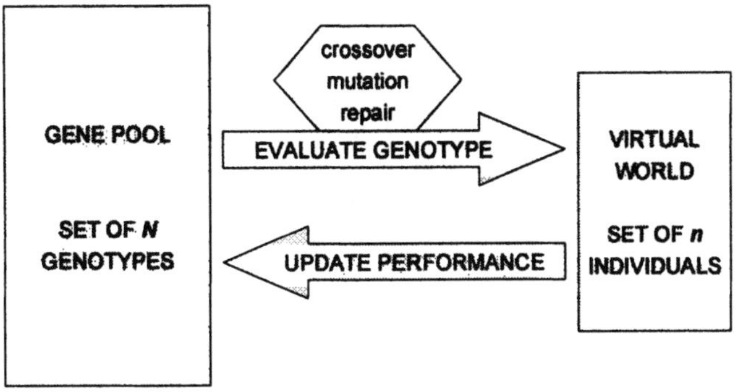

Fig. 1.
The system architecture.

2.2 Physical simulation

The module used to evaluate individuals (genotypes) simulates the creatures and their environment. A three-dimensional simulator is used, in hope that a range of complex, various stimuli affecting organisms will be the origin of dynamic development. The first behaviors tested were the mechanisms of locomotion and orientation in an environment, so all the kinds of interaction between physical objects were considered: static and dynamic friction, damping, action and reaction forces, energy losses after deformations, gravitation, and uplift pressure – buoyancy (in water environment).

The basic element of the creatures is a stick (figure 2) made of two particles flexibly joined. Finite element method is used for simulation. Sticks can have various length, weight, strength, friction etc. Neurons (connected in any way) and receptors can also be placed on sticks.

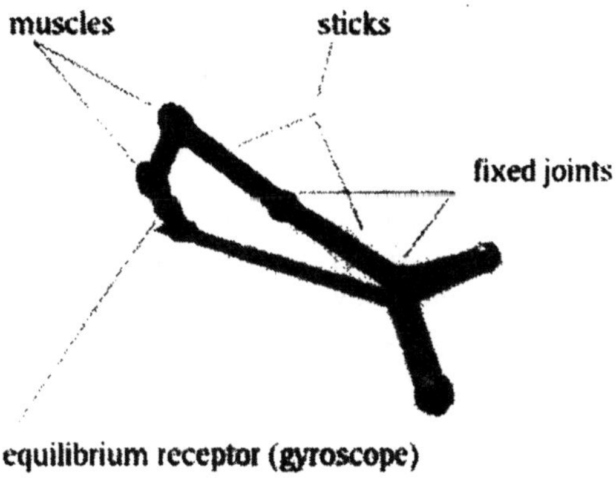

Fig. 2.
A simple framstick creature.

Muscles (bending and rotating) are placed on stick junctions. The signal that controls a muscle changes the relative position of the adjacent sticks.

2.3 Neural network

Framsticks' neurons are similar to those widely used in machine learning. Sophisticated and unnatural processing units (as in [7]) were not introduced; we proved that it is possible to construct complex modules (integrating, differentiating, summing, subtracting, and generators with different shapes) from simple neurons. Neuron properties can be additionally changed by three special parameters (all under control of evolution): *force*, *inertia* and *sigmoid*. These parameters modify the way neurons process signals. Details and sample neuronal runs can be found at [5].

An important aspect of the neural network is its interaction with the virtual world. Neurons can control muscles (*effectors*), and can obtain information from *receptors*. Currently, there are three kinds of *receptors*: those for orientation in space (equilibrium sense, gyroscope), detection of energy/food (smell) and detection of physical contact (touch).

3 Evolution

The genotypes used in **Framsticks** are described textually, so they can be easily read and modified by a human. Stick phenotypic properties are represented locally, but propagate through creature's structure with a decreasing power. The genotype describes precisely all the parts of the corresponding phenotype. Small changes in the genotype cause small changes in the resulting creature. Control elements (neurons, receptors) are associated with the element under their control (muscles, sticks). The current restriction is that only tree-like structures can be represented (no cyclical structures allowed).

Both physical structure (body) and neural network (brain) are described in the same genotype. The "body" is made of sticks, which have some properties: biological (muscle strength, stamina, assimilation, ingestion, initial energy level), physical (length, weight, friction) and concerning joints (rotation, twist, curvedness). The "brain" is made of neurons. The neural network can have any topology and complexity. An important property is that neural connections are described relatively. This lets sub-nets survive the crossover operation; the whole set of neurons can be moved to another place in the genotype (and in the creature), possibly with limbs, and can be still operational.

Two genetic operators were introduced: crossover and mutation. Mutation concerns many aspects of genotypic changes, each having adjustable relative probability. The crossover operator is a two-point one. A simple repair procedure is used, which can repair small errors and validate an invalid genotype.

In nature, groups of similar individuals share the same ecological niche and constitute species. In **Framsticks**, similarity to other coexisting species lowers the given species' fitness [3]. This introduces a pressure to diversify populations of species. The second mechanism which supports speciation is the specific crossover operation: the *corresponding parts* of genotypes of similar species are exchanged.

4 Conclusions and future work

The evolutionary experiments performed so far concerned mainly directed evolution, with fitness defined as speed (on the ground and in water). Many walking and swimming species evolved during these runs [5]. Usually, the first idea of "how to move" is a neuron connected recursively to itself, and controlling a bending muscle. Better methods of locomotion on the ground include chaotic pushing back, where the changing signals come from equilibrium sense or touch receptors. After further evolution the movement becomes more purposeful, and redundant parts are removed.

During the experiments, we had to modify simulation rules and fix bugs several times. Evolution turned out to be a very good method of searching the space of solutions (organisms), and was capable of finding fit individuals, regardless of their sensibility and validity. The faults in the simulator were sooner or later discovered by evolution, used in simulated organisms and exploited to the highest possible extent.

Currently, the genotypic representation seems to be the main limitation, because it does not allow for easy evolution of complex organism structures. Our future work will thus concern introducing a better representation: instead of coding the structure and neural network linearly, the genotype will describe a way of creating (growing) an organism. Such an approach will base more on the nature, and may make the search of the organism space even more effective.

Our future work will also concern defining a better similarity function, improvements of simulation rules and their parameters, and open-ended simulations. More receptors and more complex criteria for directed evolution may be introduced.

We hope that after further development of open-endedness in our three-dimensional simulation model, evolution will create organisms with more complex behaviors, and realistic, life-like phenomena will emerge.

Acknowledgements

This work has been supported from CRIT 2 – Esprit Project no. 20288.

References

1. Dawkins, R., *Climbing mount improbable*, W.W. Norton & Company, 1997.
2. Ficici, S. G., Pollack, J. B., Coevolving communicative behavior in a linear pursuer-evader game. In: *Proc. Fifth International Conference of the Society for Adaptive Behavior*, Pfeifer, Blumberg, Kobayashi (eds.), Cambridge: MIT Press, 1998.
3. Goldberg, D. E.: *Genetic Algorithms in Search, Optimization, and Machine Learning*, Addison-Wesley Publishing Co., 1989.
4. Komosiński, M., Ulatowski, Sz., Framsticks – Artificial Life. In: C. Nédellec, C. Rouveirol (eds.), *ECML '98 Demonstration and Poster Papers*, Chemnitzer Informatik-Berichte, Chemnitz, 1998, 7-9.
5. Komosiński, M., Ulatowski, Sz., **Framsticks** Internet site, http://www.frams.poznan.pl/
6. Maes, P., Artificial Life meets Entertainment: Interacting with Lifelike Autonomous Agents. In: *Special Issue on New Horizons of Commercial and Industrial AI* **38**, 11 (1995) 108-114, Communications of the ACM, ACM Press.
7. Sims, K., Evolving 3D Morphology and Behavior by Competition. In: R. Brooks, P. Maes (eds.), *Artificial Life IV Proceedings*, MIT Press, 1994, 28-39.

Evolving Visual Feature Detectors

Tony Belpaeme

Vrije Universiteit Brussel
Artificial Intelligence Laboratory
Pleinlaan 2, 1050 Brussels, Belgium
tony@arti.vub.ac.be http://arti.vub.ac.be

Abstract. This paper describes the generation and selection of visual feature detectors. The feature detectors are randomly generated, and are built out of components, some having functionality inspired on observations of animal visual pathways. The input for the feature detectors consists of non-synthetic images, while the selectionist pressure comes from the amount of information the feature detectors generate. The experimental setup is described and some results are given.

1 Introduction

In the same sense that Darwinian processes are the driving force for evolving living organisms, we believe that Darwinian-like development is responsible for the functional adaptation of the brain to its environment and tasks. Very much like a micro-ecology of competing signal processing pathways, in which more often used pathways are strengthened and less used pathways tend to disappear. This process needs diversity, here provided by a random generator; furthermore, a selectionist mechanism is required, which selects from the raw material according to certain selectionist pressure. This selectionist pressure depends on two factors. First of all, there's the sensory input to the processing pathways. Depending on the input, particular functionality may or may not develop. Secondly, the task influences the selection of functionality. If the success of the task relies heavily on certain processing pathways, then these pathways will be strengthened, while disuse of pathways will results in their disappearance. This paper focuses on evolving visual functionality: generating and selecting visual feature detectors using selectionist mechanisms. Each feature detector is built out of simple primitives performing basic visual processing, some of which are inspired on functionality observed in the visual pathways of animals. This approach differs from the statistical analysis of natural images [6], in which regularities in images are used to attain data reduction, or biologically inspired object recognition [4], in which a set of carefully engineered feature detectors are used. The input consists of images showing objects or faces. The output of the visual feature detectors is typically fed to a task, this can be a cognitive task, e.g. object discrimination, or it can be a sensori-motor task, e.g. visual taxis for an animat. However, for the experiments described here no task is used, for reasons explained later.

2 Experimental setup

During an experiment a set of images is presented to a set of feature detectors. For each image ι, every feature detector produces a scalar output $o_j \in [0,1]$. If there are N feature detectors, a N-dimensional vector $o_\iota = \{o_1, o_2, \ldots, o_N\}$ is produced. This vector is then used by the task to act accordingly to the presented visual stimulus.

2.1 The selectionist architecture

For the experiments a strongly typed genetic program is used, this incorporates a random generator and a selectionist architecture. We use three parameter types: `scalar` is a defined as $x \in [0,1]$. `point` is an ordered pair of scalars, defining a relative image coordinate. And `image` refers to an image of fixed size. Table 1 shows the terminals used. The random ephemeral constant is a random number of type `scalar`, which keeps its value during the existence of an individual. Table 2 shows the non-terminals; there is one extra non-terminal, not mentioned in the table, which combines all feature detectors; it accepts N scalars (with N being the number of feature detectors) and has no output type.

Some primitives are chosen to ease the image processing, while others are inspired by observations in mammalian visual processing pathways. The `orientationResponse` primitive is based on the orientation specifity of the visual cortex [7], while the `spatialResponse` and `spatialFilter` primitives are based on the observation that mammals perform spatial filtering at a retinal and cortical level [2]. No spectral sensitive primitives were used, since this would bias the discrimination of the visual stimuli too much.

2.2 The task and the feedback it provides

The selectionist pressure depends on the environment –in casu the visual input– and the task of the individual. In the experiments no particular task is used, but the fitness of the feature detectors is determined by the information content of their outputs. This does not commit the evolution of the feature detectors to a specific task, but gives a general measure for the usefulness of the feature detectors, and allows for faster experimental runs.
The information content is a measure for the spread of the feature detector

Terminal	Type
leftTop	`point`
rightBottom	`point`
zero	`scalar`
one	`scalar`
random ephemeral constant	`scalar`

Table 1. Terminals for the strongly typed GP.

Function	Functionality	Input Types	Output type
averageIntensity	Computes average intensity of an image region	image, point, point	scalar
thresholdImage	Filters out all pixels not in range $[a, b]$	image, scalar, scalar	image
spatialFilter	Filters the image using a Laplacian of Gaussian filter with size σ, as defined by [3]	image, scalar	image
spatialResponse	Gives the response of a Laplacian of Gaussian filter with size σ	image, scalar	scalar
orientationResponse	Gives the response to edges in a particular direction α	image, scalar	scalar
combineImage	Makes an image by taking the average of two input images	image, image	image
cons	Combines two scalar in an ordered pair	scalar, scalar	point
amplify	Amplifier, $param_1$ / $paramr_2$	scalar, scalar	scalar
attenuate	Attenuator, $param_1$. $param_2$	scalar, scalar	scalar

Table 2. Non-terminals for the strongly typed GP. The last three non-terminals do not perform image processing.

outputs o_i in the N-dimensional space. The fitness of a set of N feature detectors is calculated by taking the entropy in the discretised N-dimensional space. Each dimension is divided into M equal divisions, giving a total of $P = N^M$ partitions. Figure 1 illustrates this. If I is the number of images presented to a set of feature detectors, the information content is calculated as in eq. 1, x_i is the number of outputs in partition i.

$$\epsilon = \sum_{i=1}^{M^N} -\frac{x_i}{I} \ln \frac{x_i}{I} \qquad \text{for } x_i > 0 \tag{1}$$

The maximal information content $\epsilon_{max} = I(-\frac{1}{I} \ln \frac{1}{I}) = \ln I$ is reached when every partition contains one output or no output at all. The fitness of an individual, with a maximum of 1, is defined as:

$$f = \frac{1}{1 + (\epsilon_{max} - \epsilon)} \tag{2}$$

If the outputs are well spread in the N-dimensional space, then the fitness will be high. On the other hand, clustered outputs will result in a low fitness. Note that if there are more partitions, it will be easier to reach a high fitness. There should be at least I partitions (with I being the number of images), so every dimension should be divided in at least ceil($\log_N I$) divisions. Note that it is important to evaluate all the outputs of all feature detectors in one fitness measure, instead of calculating the fitnesses of every single feature detector and then just summing the fitness values. The latter would be exploited by the GP to

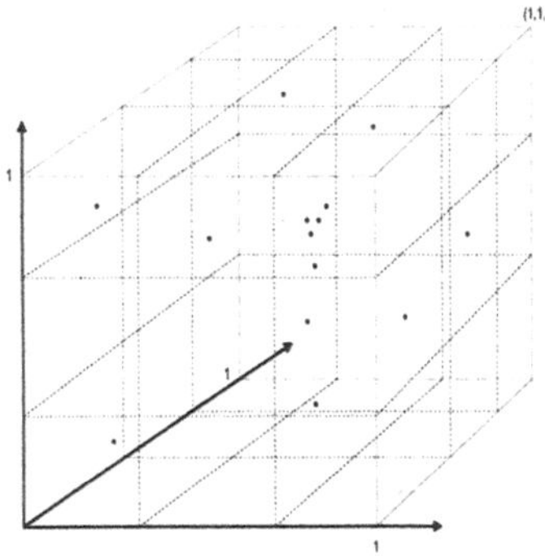

Fig. 1. Illustrating a partitioned output space, there are $N = 3$ feature detectors, and each dimension is split into $M = 3$ divisions, giving a total of $P = N^M = 27$ partitions. The dots represent outputs; ideally every partition should contain one output.

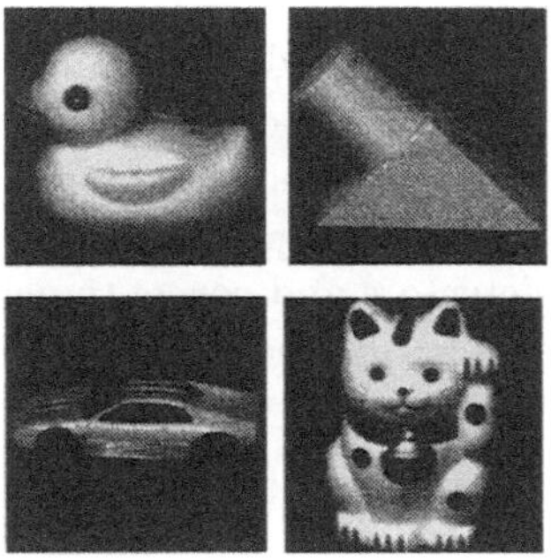

Fig. 2. Sample of some input images.

reach high fitness but evolve all similar feature detectors, while the first method ensures that the feature detectors will all differ significantly.

3 Results

The results described here all use the same input: 20 monochromatic images of daily-used objects[1]. To evaluate a set of feature detectors, all images are presented to the set and the fitness is computed as in eq. 2.

The genetic program is run for 500 generations, with a crossover rate of 0.8, mutation rate of 0.1 and reproduction rate of 0.1; note that the reproduction rate is actually higher due to failed crossover operations. Figure 3 shows the fitness of a run.

Results (not reported here) show that different input images (showing objects or faces) influence the selection of primitives (for more results and a elaborate exposition of design choices, see [1]). For inputs showing objects the feature detectors rely on intensity to categorize the inputs, while inputs showing face shots tend to produce feature detectors which rely on orientation and spatial information.

A second observations is that the depth of the feature detectors is quite low. A run in which no selectionist feedback was given, produced feature detectors with an average depth of 7.3, and on average 130 nodes. While the run shown in figure 3 on average produced feature detectors of depth 3.2 and size 25. This is due to the attenuating effect of large trees: more primitives –connected to each other in a serial way– tend to filter out information, until no information is left.

[1] The images are taken from the Columbia Object Image Library (COIL-20) at `http://www.cs.columbia.edu/CAVE/`.

4 Conclusion

The papers demonstrates how feature detectors to discriminate visual stimuli evolve. The outputs of the feature detectors are presented to a task, which provides selectionist pressure. Future experiments should add more low-level context-free primitives (such as a symmetry detector, lateral inhibitory feature detectors and corner/junction detectors). Further, the number of feature detectors should be made variable and the task adaptive.

Acknowledgments

For the GP system an adapted version of LilGP by Douglas Zongker and Bill Punch was used. The author is a Research Assistant of the Fund for Scientific Research - Flanders (Belgium) (F.W.O.)

References

1. T. Belpaeme. Evolution of visual feature detectors. In S. Cagnoni and P. Nordin, editors, *Proceedings of EvoIASP99*, Late breaking papers, 1999.
2. R. De Valois and K. De Valois. *Spatial vision.* Oxford University Press, Oxford, 1990.
3. D. Marr. *Vision.* Freeman, 1982.
4. B. W. Mel. Seemore: Combining color, shape, and texture histogramming in a neurally-inspired approach to visual object recognition. *Neural Computation*, 9:777–804, 1997.
5. B. Olshausen and D. Field. Natural image statistics and efficient coding. *Network: Computation in Neural Systems*, 7(2):333–339, 1996.
6. D. L. Ruderman. The statistics of natural images. *Network: Computation in Neural Systems*, 5(4):517–548, 1994.
7. T. Wiesel and D. Hubel. Receptive fields and functional architecture of monkey striate cortex. *Journal of Physiology*, 195:215–243, 1968.

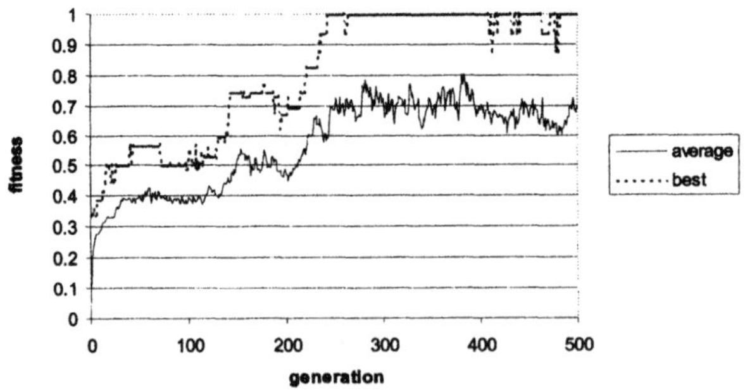

Fig. 3. Result of run with individuals consisting of a set of 5 feature detectors. Each dimension has $M = 5$ divisions, thus creating $P = M^N = 5^5$ partitions. The graph shows the mean fitness and the fitness of the best-of-generation individual.

Evolutionary Design of Mobile Robot Behaviors for Action-Based Environment Modeling

YAMADA Seiji

CISS, IGSSE, Tokyo Institute of Technology
4259 Nagatsuta, Midori, Yokohama 226-8502, Japan
E-mail: yamada@ymd.dis.titech.ac.jp
URL: http://www.ymd.dis.titech.ac.jp/~yamada/index-e.html

Abstract. This paper describes an evolutionary way to design behaviors of a mobile robot for recognizing environments. We have proposed an action-based approach (called AEM) for a mobile robot to recognize environments. In AEM, a behavior-based mobile robot acts in each environments and action sequences are obtained. The action sequences are transformed into vectors characterizing the environments, and the robot identifies the environments with the vectors. The design of suitable behaviors for AEM is very difficult for human because the search space is huge and intuitive understanding is hard. Thus we develop the evolutionary design of such behaviors using genetic algorithm.

1 Introduction

The most studies to recognize environments have tried to build a precise geometric map using a robot with high-sensitive and global sensors. However, just to recognize environments, such a strict map may be unnecessary. Thus we have tried to build a mobile robot which recognizes environments only with low-sensitive and local sensors, and proposed approach that a mobile robot can recognize the environment with *action sequences*. We call this approach AEM (Action-based Environment Modeling) [5]. In AEM, a mobile robot acts using *given* suitable behaviors like wall-following in environments. Then the action sequences executed in each environment are obtained, and transformed into environment vectors. A robot identifies the environments by comparing them.

Through the research on AEM, we recognized a significant problem: *where the suitable behaviors come from?*. An easy solution is that human designs the behaviors. However the task becomes quite difficult for a human designer as the variety of environments increases. In this paper, we propose the evolutionary design method of such behaviors using GA (Genetic Algorithm) and make experiments for evaluation.

In the similar approach to AEM, several studies have been done in robotics [2] and artificial life [4]. In most researches, wall-following has been used as suitable behaviors in [2][4][5]. Unfortunately the behaviors were described by human designers, and fixed.

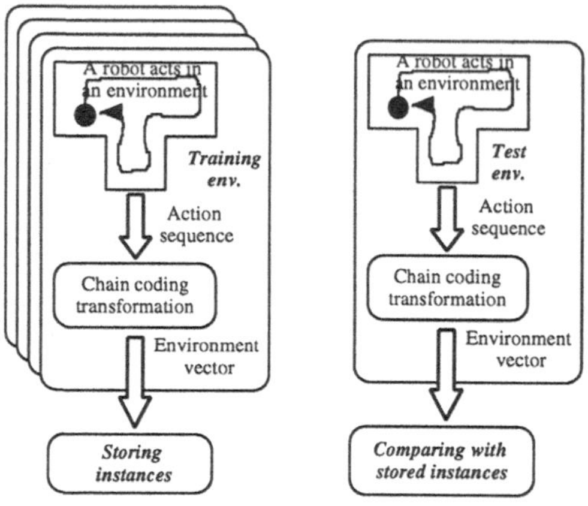

(a) Training phase (b) Test phase

Fig. 1. Overview of AEM

Fig. 2. Khepera

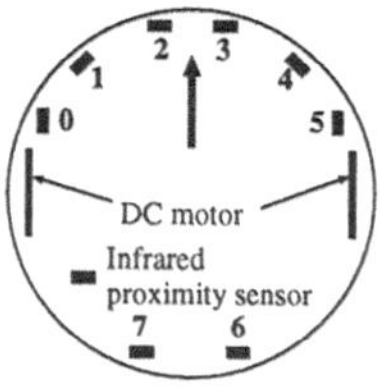

Fig. 3. Sensor positions

2 Task: Action-based Environment Modeling

In AEM [5], a mobile robot is designed in a behavior-based approach [1]. The *behavior* means mapping from states to actions. An AEM procedure consists of two stages: a training phase and a test phase (Fig.1). In the training phase, *training environments* having a *class* are given to a robot. The class means a category in which the environment should be included. The mobile robot acts in the environments using given behaviors and obtains sequences of executed actions (called *action sequence*) for each of them. They are transformed into real-valued vectors (called *environment vectors*) using a chain coding-like method. The environment vectors are stored as *instances*, and the training phase finishes.

In the test phase, a robot is placed in a *test environment* which is one of training environments. The robot tries to identify the test environment with one of training environments. The identification is done using 1-Nearest Neighbor method, i.e. the robot selects the most similar instance to the test environment, and considers that the class of the instance is that of the test environment. The similarity is evaluated with Euclidean distance between environment vectors.

Since the suitable behaviors depend on environment structure which a robot should recognize, they have been described by human designers thus far. However the task is very difficult because of a huge search space.

3 States, actions and environment vectors

Using real mobile robots as individuals in GA is not practical because it is impossible to operate several tens of real robots over more than one hundred generations. Thus we use a simulator of Khepera (Fig.2). As shown in Fig.3, it has two DC motors as actuators and eight Infra-Red proximity sensors which measure both distance to obstacles and light strength.

We describe a state with the range of a sensed value. For reducing the search space of behaviors, we restrict states and actions. A sensor on Khepera returns 10 bit values for distance and light strength. Thus we transform the distance into binary values 0 or 1. The "0" means an obstacle exists within 3cm from a robot, and "1" means it does not exist. Furthermore only three (front, left and right) of eight sensors are used. Next states for light strength are described using only 4 sensors (front, left, right and back). We describe a state using the sensor with the strongest light value and its binary values which mean a light is "near" or "far". A state in which all of the sensors has almost same values is also considered. As a result, the number of states for light is nine, and the total number of states is 72 ($= 2^3 \times 9$). We also describe four actions; *A1*: Go 5mm straight on, *A2*: Turn 30° left, *A3*: Turn 30° right, *A4*: Turn 180° left.

The generated action-sequence is transformed into an environment vector. Let an action-sequence and its environment vector be $[a_1, a_2, \cdots, a_n]$ ($a_i \in \{A1, A2, A3, A4\}$, $a_0 = 0$) and $\boldsymbol{V} = (\boldsymbol{v}_1, \boldsymbol{v}_2, \cdots, \boldsymbol{v}_m)$ ($m \geq n$) respectively. The vector values of $\boldsymbol{V}$ are determined by four rules: If $a_i = A1$ then $\boldsymbol{v}_i = \boldsymbol{v}_{i-1}$, If $a_i = A2$ then $\boldsymbol{v}_i = \boldsymbol{v}_{i-1} + 1$, If $a_i = A3$ then $\boldsymbol{v}_i = \boldsymbol{v}_{i-1} - 1$, If $a_i = A4$ then $\boldsymbol{v}_i = -\boldsymbol{v}_{i-1}$. They change the vector value when the direction of movement changes in the similar way to *chain coding* which is a popular method in pattern recognition.

4 Applying GA to acquire behaviors

GA procedure and coding We use a simple GA procedure and parameters in the followings. We use the coding in which one of actions $\{A1, \cdots, A4\}$ is assigned to each state.

- *Population size*: 50, *Crossover operator*: Uniform crossover.
- *Selection method*: Elite strategy and tournament selection (the size = 2).
- *Crossover rate* P_{cross}: 0.8, *Mutation rate* P_{mut}: 0.05

Defining a fitness function We introduce three conditions for suitable behaviors to AEM: termination of actions, accuracy and efficiency of recognition. The fitness functions for each conditions are defined, and integrated.

A robot has to stop when it returns to the neighborhood of a start point. The termination is evaluated with $g = \frac{\text{(No. of E-trial)} + \text{(No. of H-trial)}}{2\times\text{(Total No. of trials)}}$, where E-trial and H-trial means trials in which a robot escaped from the neighborhood of the start point and trials in which it succeeded in returning. Accuracy of identifying environments is also important. It is evaluated with $h = \frac{\text{No. of successful test env.}}{\text{Total No. of test env.}}$, where $h = 0$ when $g \neq 1$. In AEM, the actions should be as small as possible for efficiency. Hence we introduce $k = 1 - \frac{\sum_{i=1}^{n} S_i}{n * S_{max}}$, where S_i is the size of an action sequence in an environment i, S_{max} is the given limited

Table 1. Experimental results in Exp-1

Env.	Train. env.	GN	MaxF
(a)	{emp, L}	1.0 (0)	2.80 (0.106)
(b)	(a) + L2	2.6 (1.84)	2.43 (0.131)
(c)	(b) + iL	2.8 (1.69)	2.44 (0.025)
(d)	(c) + s-emp	2.8 (1.75)	2.44 (0.093)
(e)	10 env.	2.1 (0.738)	2.51 (0.024)
(f)	12 env.	5.2 (3.08)	2.48 (0.07)

Table 2. Experimental results in Exp-2

Env.	Train. env.	GN	MaxF
(g)	emp, 1-la	1.6 (0.966)	2.68 (0.196)
(h)	(g) + 2-la	4.8 (2.78)	2.59 (0.131)
(i)	(h) + 3-la	9.3 (5.19)	2.59 (0.111)
(j)	(i) + 4-la	10.0 (5.94)	2.62 (0.075)

size of an action sequence, and $k = 0$ when $h \neq 1$. We finally integrate three fitness function into $f = g + h + k$ having range [0, 3]. The utility of the fitness function is investigated through successful experiments.

5 Experiments with simulation

We implement a system using a Khepera simulator [3], and make experiments. In all experiments, we give each of training environments to a robot once. The robot acts in the environments, and the environment vectors transformed from the action sequences are stored as instances. Next each of the *training* environment is given to the robot as a *test* environment, and the robot identifies each of the test environments with one of the training environments. In all the experiments, we had 10 trials having different initial population, and investigated the averages and standard deviations of generation number in which GA stopped.

Exp-1: Environments with different contours in shape First we made experiments Exp-1 using environments with different contours in shape. The experimental results are shown in Table 1. Four parts ((a) ~ (d) in Table 1) of five environments: {emp, L, L2, iL, s-emp} were given to a robot. Additionally 10 and 12 different shape environments were used ((e) and (f) in Table 1). The "GN" is the generation number in which GA stopped, and "MaxF" means the maximum fitness value at GN. The numbers in GN and MaxF stand for averages, and the numbers in brackets are standard deviations. This format is common in all the experimental results. Fig.4 indicates the action traces of the best individuals at GN in (d). In such simple environments, the suitable behaviors for AEM were obtained within few generations. The standard deviations was large in GN and small in MaxF, and this tendency was observed through all the experiments. Seeing from Fig.4, different action sequences were obtained depending on the structure of the environments.

Exp-2: Environments with different lights Next, by adding different lights to environments in number and position, we made five environments: {emp, 1-la, 2-la, 3-la, 4-la}. Exp-2 is made by using parts of the environments. Light was so strong that a robot can detect the light direction in any place. The experimental results are shown in Table 2. Fig.5 indicates the action trace of the best individual at GN in (j). In the figures, a black circle stands for a light.

(a) emp (b) L (c) L2 (d) iL (e) s-emp

Fig. 4. Trace of actions in Exp-1 (Five environments)

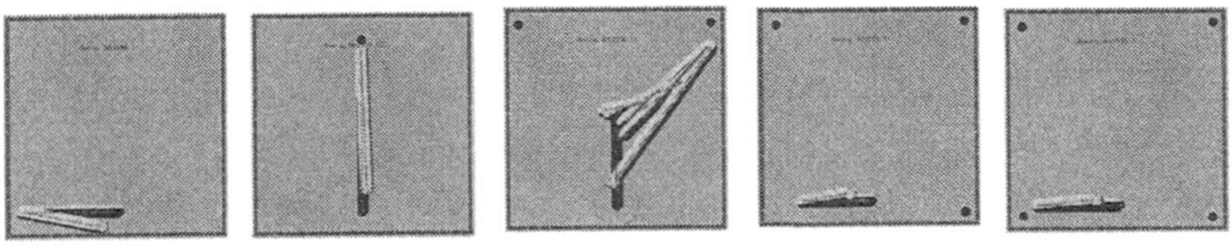

(a) emp (b) 1-la (c) 2-la (d) 3-la (e) 4-la

Fig. 5. Trace of actions in Exp-2

Though the GN increased more than ones in Exp-1, the suitable behaviors were obtained. Note that we cannot intuitively understand the behaviors in Fig.5. This means that it is very hard for human to design such behaviors by hand-coding and this automatic design method is quite effective.

6 Conclusion

We proposed evolutionary design of suitable behaviors to AEM. GA was applied to search the behaviors, and the simulated mobile robots were used as individuals. States and actions were described for coding chromosomes, and we carefully defined the fitness function. We made experiments using different environments in shape and lights, and found out our approach is promising to learn suitable behaviors for AEM.

References

1. R. A. Brooks. A robust layered control system for a mobile robot. *IEEE Transaction on Robotics and Automation*, 2(1):14–23, 1986.
2. Maja J. Mataric. Integration of representation into goal-driven behavior-based robot. *IEEE Transaction on Robotics and Automation*, 8(3):14–23, 1992.
3. O. Michel. *Khepera Simulator v.2 User Manual*, 1996.
4. U. Nehmzow and T. Smithers. Map-building using self-organizing networks in really useful robots. In *Proceedings of the First International Conference on Simulation of Adaptive Behavior*, pages 152–159, 1991.
5. S. Yamada and M. Murota. Unsupervised learning to recognize environments from behavior sequences in a mobile robot. In *Proceedings of the 1998 IEEE International Conference on Robotics and Automation*, pages 1871–1876, 1998.

Bio-inspired Robotics and Autonomous Agents

On the Dynamics of Robot Exploration Learning

Jun Tani and Yuya Sugita

Sony Computer Science Laboratory Inc.
Takanawa Muse Building, 3-14-13 Higashi-gotanda,
Shinagawa-ku,Tokyo, 141 JAPAN
email: tani@csl.sony.co.jp, http://www.csl.sony.co.jp/person/tani.html

Abstract. In this paper, the processes of exploration and of incremental learning in the robot navigation task are studied using the dynamical systems approach. A neural network model which performs the forward modeling, planning, consolidation learning and novelty rewarding is used for the robot experiments. Our experiments showed that the robot repeated a few variation of travel patterns in the beginning of the exploration, and later the robot explored more diversely in the workspace by combining and mutating the previously experienced patterns. Our analysis indicates that internal confusion due to immature learning plays the role of a catalyst in generating diverse action sequences. It is found that these diverse exploratory travels enable the robot to acquire the rational modeling of the environment in the end.

1 Introduction

One of the debates in behavior-based robotics is whether or not agents should possess higher-order cognitive functions such as internal modeling, planning and reasoning. Most researchers in behavior-based robotics have rejected the "representation and manipulation" framework since they consider that the representation cannot be grounded and that the mental manipulation of the representation cannot be situated adequately in the behavioral context of the robot in the real world environment. This argument seems to be valid if the agent's mental architecture employs the symbolist framework. One of the major difficulties in the symbolist framework is that the logical inference mechanism utilized in planning or reasoning assumes completely consistent model of the world. This presumption cannot be satisfied if the learning should be conducted dynamically as in animal and in human adaptation processes. It is, however, also true that the embodiment of higher-order cognitive functions is crucial if we attempt to reconstruct an intelligence at the human level in robots, since even two year-old human infants are said to possess primitive capabilities of modeling and planning within their adopted environment.

We consider that an alternative to the symbolist framework can be found in the dynamical systems approach [4, 1] in which the internal cognitive processes are considered to exist in tight coupling with the external environmental processes [1]. Our previous study in navigation learning demonstrated that a robot using a recurrent neural net (RNN) is able to learn the "grammatical" structure hidden in the environment, as embedded in attractor dynamics with a fractal structure, from the experiences of sensory-motor interactions [8]. The forward dynamics [3] of the RNN generates a mental image of future behavior sequences driven by the acquired attractor dynamics. The crucial argument in that study is that the situatedness of the higher cognitive processes are explained on the basis of the entrainment of the internal dynamics by the environmental dynamics. However, a drawback of that study was that the learning was conducted in an off-line manner i.e. the navigation could be conducted only after complete learning of the environment.

In the current paper, we study the development of the interactive processes between learning and acting in the robot's exploration of its environment. By conducting real robot experiments, we focus on how the robot interacts with its environment and how it makes sense of the world by utilizing its limited experiences. Our experiment exhibits an interesting result: we find that the diverse exploratory behaviors are generated through taking advantage of the state of confusion in the internal modeling in the middle of the learning process. Our analysis, based on the dynamical systems scheme clarifies the underlying mechanism.

2 The Model

In this section we introduce a neural net model which enables the system to perform exploratory behavior, goal-directed planning and behavior-based learning. The neural net architecture employed has been built by combining pre-existing neural net schemes. In the learning process, both reinforcement learning and prediction learning are conducted [11]. Using reinforcement learning, the action-policies for better rewarding are reinforced, through which the most preferred action in the current state is selected. In prediction learning, the forward model [3] is adapted to extract the causality between the action and the sensation. In goal-directed planning, the inverse dynamics scheme [11, 3] is applied to the forward model in order to generate possible action sequences. In this planning process, the action policy adapted using reinforcement learning provides heuristics for searching for the better rewarded acion sequences. In the current formulation, rewards are given to the system based on the novelty which the system experiences for each exploration action [10, 6]. In other words, when the system cannot predict the next sensation in terms of the current action, the current action is rewarded. In addition, the prediction learning attempts to learn to predict how much prediction error it will make. By combining this novelty-rewarding scheme with the reinforcement learning and with the prediction learning schemes, the system tends to explore the workspace regions with which it is unfamiliar. As the novelty rewarding scheme continues to bring new experiences to the system, the system is forced to operate in a nonequilibrium state in which learning as well as acting cannot always be rationalized. The main purpose of this modeling is to investigate the possible interplay between exploration and learning when the system develops in a nonequilibrium dynamical manner.

2.1 The neural net architecture

A RNN architecture is employed in our model as shown in Fig 1. The RNN receives the current sensory input s_t, the current reward signal r_t, and the current action x_t. The RNN then outputs the prediction of the next sensory input $\hat{s}_{t+1}$, the reward signal $\hat{r}_{t+1}$, and its preference for the next action $\hat{x}_{t+1}$ which is expected to obtain the maximum reward in the future. For the novelty rewarding, the current normalized prediction error for the sensory inputs is used to evaluate the current novelty reward. It is noted that the reward is generated internally and we observe that the RNN learns to predict it (see section 3.2). The RNN has context units c_t in the input and output layers in order to account for the internal memory state (See Ref.[8] for more details of the role of context activation in navigation learning.)

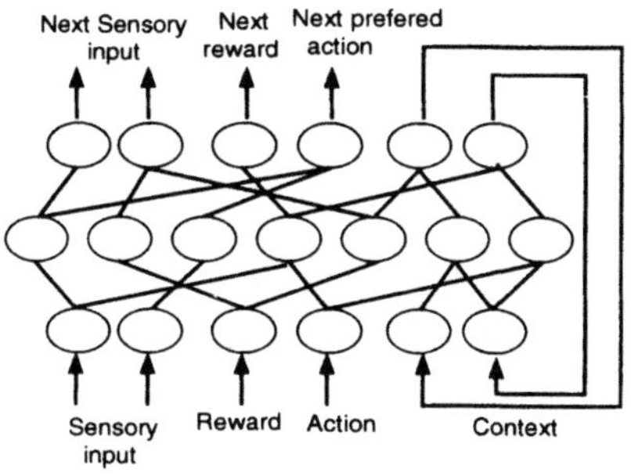

Fig. 1. The RNN architecture.

(A) Learning: The RNN learns to predict the next sensory inputs and the rewards corresponding to the current sensory inputs, the action selection and the internal state. This corresponds to the forward model learning. The preference for the next action is learned by a variant of the profit sharing method [2] in order to propagate the decayed reward signal backwards in time. This means that if the current action selection leads to an unpredictable experience, this action selection is reinforced. This corresponds to reinforcement learning. Both learning processes are executed in the RNN using the back-propagation through time (BPTT) algorithm.

(B) Planning: The objective of planning is to find the action plan $x* |(x_0, x_1, ...x_T)$ which generates the path to maximize the future calmative rewards. The action sequence is dynamically computed by using contributions both from the forward model part and from the action policy part. Inverse dynamics [3] are applied to the forward model in order to obtain the update of the action plan $\Delta x*$ for maximizing the calmative reward expected in the future sequence. We consider the following energy function by taking the negative of the calmative reward from the current time step to the terminal step τ:

$$Em(x*) = -\sum_{i=0}^{\tau} \alpha^i \hat{r}_{i+1} \tag{1}$$

where α is the decay coefficient of the reward. The back-propagation through time (BPTT) algorithm [5] is used to compute the update to the action sequence which minimizes the energy assumed in the model part. In addition to this, the action policy influences the planning dynamics in that the difference between the preferred action and the planned action at each step is minimized. The update to the action at each future step is obtained by taking the sum of both parts of the contributions and adding a Gaussian noise η. The update to the action plan is therefore

$$\Delta x_i = \epsilon \cdot [\frac{-\delta Em(x*)}{\delta x_i} + kr \cdot (\hat{x}_i - x_i) + kn \cdot \eta] \tag{2}$$

The Gaussian noise term is employed to prevent the plan dynamics being captured in a local minimum. The value of kn is changed in proportion to the value of Em. Therefore the plan search dynamics become stabilized when the energy is minimized; otherwise, it continues to be activated. Here, the reader is reminded that the contributions to the update from the forward model and from the action policy do not always agree with each other in the course of the exploration processes since the overall system dynamics are characterized by highly nonlinear and nonequilibrium dynamics.

(C) Incremental learning by consolidation: The robot learns what it experienced incrementally after each travel is terminated by using the so-called consolidation learning scheme [9] which has been developed as inspired by the biological observation of the memory consolidation [7] during sleep in mammals. In our system, a new episodic sequence experienced in the current travel is stored in the temporal memory. In the consolidation process, the RNN generates the imaginary sensory action sequence by rehearsing from the long term memory pre-learned. This rehearsal can be performed by repeating "planning", as described in the previous section, without actually moving – as in dreaming. Then, the RNN is re-trained using both the new episodic sequence stored in the temporal memory and the rehearsed sequences generated from the pre-learned memory simultaneously. This combination of rehearsal and learning allows the memory system to be re-organized without suffering from some catastrophic interference between the novel experiences and the pre-learned memory.

3 Experiment

3.1 Task setting

A mobile robot as shown in Fig 2 (a) is used for the experiment. The robot is equipped with range sensors and a color vision camera on its head. Fig 2 (b)

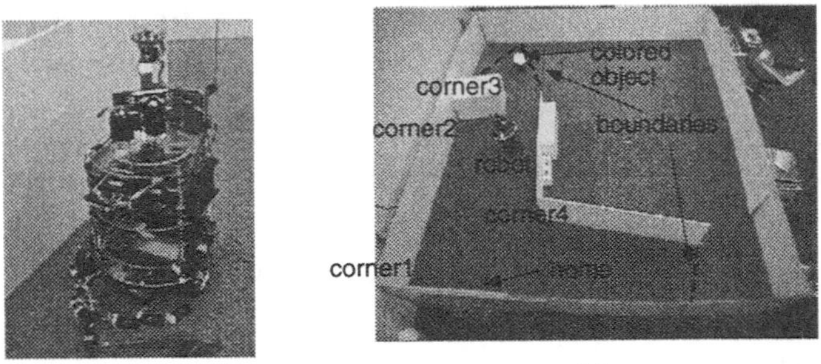

(a) (b)

Fig. 2. (a) The robot employed in the experiment and (b) the adopted workspace.

shows the adopted workspace. The basic behavior of the robot is determined using a set of pre-programmed action modules for actions such as wall-following, wall-switching and colored-object-approach. The actions are switched between by the RNN using branching. Two cases of branching are considered: (a) the robot, after turning a corner, determines whether it will continue to follow the current wall on its left side or instead to leave the current wall and to move forward diagonally at 45 degrees to the right until it encounters another wall; and (b) the robot, after finding a colored object, determines whether to continue the wall-following or to approach the colored object. In this setting, the action can be represented by one bit of information which represents whether or not to branch. The RNN architecture receives two types of sensory input at each branch point. One is the travel vector which represents what distance and from which direction the robot has traveled since the previous branch. These values are measured by taking the sum and the difference between the left and right wheels's rotation angles. The other sensory input is the categorical output of the visual image obtained when the robot encounters a colored object. The robot plans its future action sequence dynamically while it travels and receives the sensory inputs at each branch encountered. The robot starts its exploration travel from a fixed home position and the exploration is terminated when the travel takes it outside a predefined boundary. (The home position and the boundary predefined in our experiment is shown in Fig 2 (b).) At the moment of termination, the RNN receives the termination sign in its sensory input and the robot is brought back to the home position manually. Following this, the consolidation process takes place in which the temporary stored sequence is learned using 10 rehearsal sequences. After the consolidation, exploration by the robot is resumed.

3.2 Results

The robot repeated the exploration travels 20 times in the experiment. This experiment was conducted three times under the same conditions. Fig 3 represents the average prediction error for each travel sequence in the three experimental

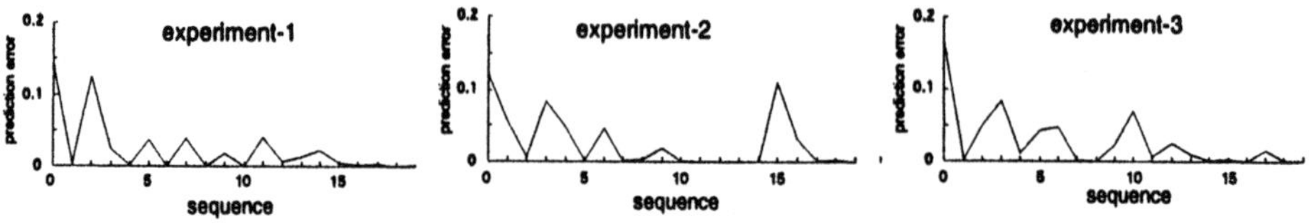

Fig. 3. The history of the prediction error for the three experiment cases.

cases. For all three cases, on average that the prediction error gradually decreases as the exploration is proceeds.

It is interesting to observe the rehearsing during the consolidation learning since the contents of the rehearsing activities represent what the robot has learned so far. Fig 4 shows how the diversity of the rehearsed plans at each consolidation learning process change as the exploration proceeds. The lower graphs

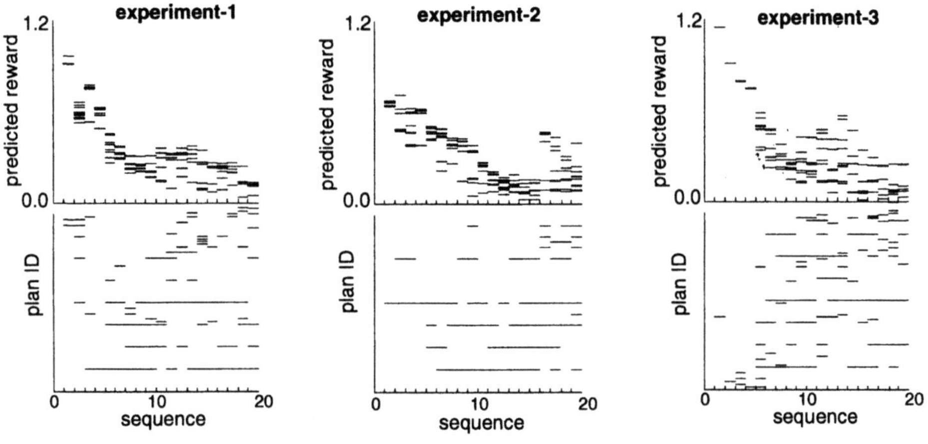

Fig. 4. Changes in the diversity of the rehearsed plans during the three exploration experiments.

in the figure shows ID of all rehearsed plans generated during each consolidation learning period; the upper graph represent the corresponding predicted rewards of the plans generated. (The ID is assigned for each plan generated by encoding the bit pattern of the branching sequence, a maximum of 10 time steps in length, into numbers from 0 to 512.) It is observed that the diversity of plans is increased and that the predicted reward is decreased as the exploration trial is continued. We observed that the rehearsed plans are generated not just by repeating the sequences previously experienced but by combining previously experienced sequences into new ones. Since the rehearsing directly affects the re-organization of the learned contents, the diversity generated in rehearsing leads to the diversity in actual travel.

In the following, we examine how the diverse travel sequences are generated in the course of exploration. Fig 5 shows all 20 trajectories of the robot's travel observed in one experimental case (experiment-1). In the initial period

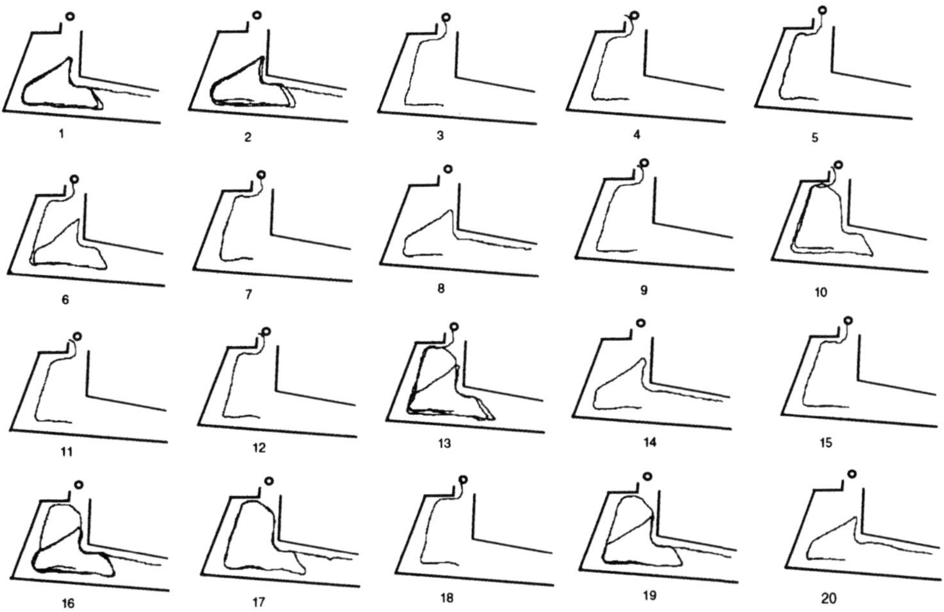

Fig. 5. The trajectories of the robot exploration travel for one experimental case. The travel sequence number is given.

of the exploration, the robot tends to repeat the same branching sequences. As is evident in Fig 5, the same trajectory is repeated for the first two travel sequences. For the third sequence, branching changes and a different trajectory is generated. This trajectory is repeated in the next two travel sequences. The trajectory in the sixth travel sequence seems to be generated by combining the two travel sequences previously experienced. We summarize that the novelty rewarding scheme causes the observed repetitions and variations in the travel. When the robot undergoes a previously unexperienced travel sequence, the branching sequence experienced is reinforced strongly because of its unpredictability. When the same trajectory is repeatedly generated through reinforcement, the sequence becomes predictable and is rewarded less. As a result, the probability of modifying the current travel is increased.

An interesting question is how novel action sequences are generated in the planning process. What we found is that novel branching sequences are originated not merely by the noise term in the planning dynamics but also by the internal confusion caused by the incremental learning. This point is illustrated by considering an example seen in the 10th travel sequence. In this travel sequence, the robot, starting from the home position, continued to follow the wall after passing corner1, then it branched to another wall after passing corner2. This branching at corner2 is a novel experience for the robot. We investigated how this branching decision was generated by examining the recorded planning process. Fig 6 shows the actual planning processes which took place immediately before the branching was made at corner2. In Fig 6 (a) each column consisting of

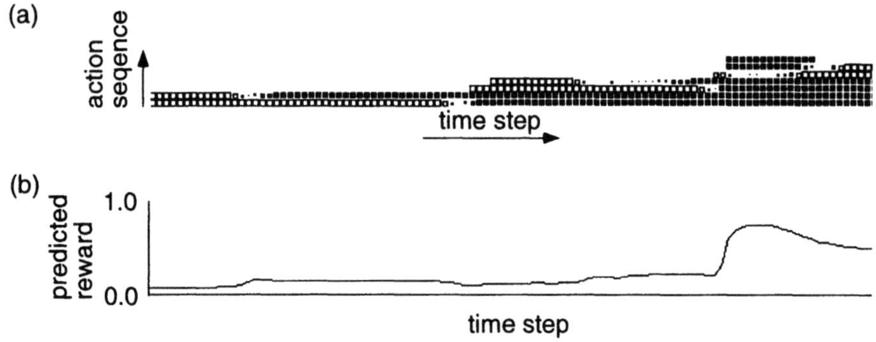

Fig. 6. (a) The time history of plan generation at the corner2, (b) the predicted reward of the corresponding plan.

white and black squares represents a branching sequence plan at each time step of the planning process, where the black and white squares denote branching and non-branching, respectively. Fig 6 (b) indicates the predicted reward for the plan generated. At the beginning of the planning process, a plan of not branching twice is generated with a low predicted reward. This plan will repeat the 5th travel sequence if actually realized. At the end of the planning process, plans are generated such that branching actions are planned to occur repeatedly after passing corner2 with an expectation of a higher reward, even though such action sequences have never been experienced. It is noted that this type of plan was not observed when the robot approached the same corner in its earlier travels. Further examination showed that the lookahead prediction of the sensory sequences after branching at corner2 and at corner1 are mostly the same. This can be interpreted as meaning that the robot hypothesized that branching at any corner would lead to better chances for encountering novel experiences because it applied the situation after branching at corner1 to consider the situation at corner2. (Indeed, the travel will continue as long as branching is selected at approaching corners without terminating the travel by going out of the workspace boundary.) We conclude that the novel action of branching at corner2 results from the expectation of a higher reward which is falsely anticipated by means of fake memory generated in the course of consolidating immature experience. This phenomenon of the novel action trial being generated by fake memory and the internal confusion was seen frequently in the middle of the learning process.

Finally, we investigated how the internal modeling develops by examining the evolution of the RNN attractor. Fig 7 shows the attractor which appeared in the phase space of the RNN at different stages in experiment-1. The phase plots were drawn by iteratively activating the RNN in the closed-loop mode with inputs comprising 4000 steps of random branching action sequences. The generated sequence of the context units activation are plotted in the two dimensional phase space. In Fig 7, cluster structures consisting of multiple segments are clearly seen

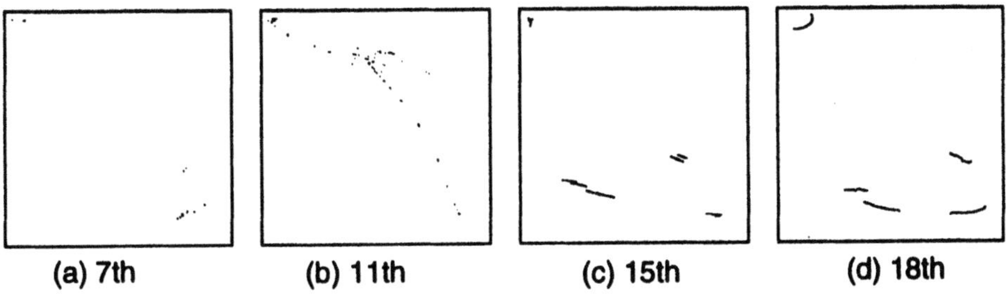

Fig. 7. The RNN attractor appeared at a certain stage of the learning process in experiment-1. The learning stage is given at the base of each plot.

in the later periods of the exploration travel. Our examination clarified that this set of cluster segments represents the global attractor. Further analysis indicated that in the phase plots in Fig 7 (c) and Fig 7 (d) each segment corresponds uniquely to each branching position in the workspace and also that the graph structures are topologically equivalent between that of the state transition in the phase space and that of branching of the robot trajectories in the environment. In this condition, it is said that the "dynamical closure" is generated in the attractor since an equivalence of the closed graph structure is generated in the phase space. However, such structures were barely seen in the phase plots in Fig 7 (a) and Fig 7 (b). While the learning process is "immature", the shape of the attractor varies substantially after each learning and neural dynamics exhibits diverse trajectories in the phase space and the robot behaves as if it were confused. In the meanwhile, the attractor develops step by step as the diverse exploration repeated and finally the dynamical closure is organized in the internal neural dynamics.

4 Discussion and Conclusion

In the experiments, it was shown that the robot learned incrementally about its workspace through exploration and that the robot was eventually successful in obtaining a rational model of the workspace. However, the emphasis in this study is on the observation of dynamical processes before the rational model is achieved. In the beginning, a few travel sequences are repeated and later some combinations of them are made. In the middle period, novel actions are frequently tried with a false expectation of the future consequences. The confusion due to the immaturity turns out to be beneficial since it acts as a catalyst for generating the diverse behavior required to explore the environment. Such diverse behavior enables the robot to acquire the rational model later.

Our experimental studies, however, are limited in a sense that (a) the robot is manually recovered when it goes out of the workspace boundary, (b) the environment is static. Our future study will address these problems.

References

1. R.D. Beer. A dynamical systems perspective on agent-environment interaction. *Artificial Intelligence*, Vol. 72, No. 1, pp. 173–215, 1995.
2. J.H. Holland and J.S. Reitman. Cognitive systems based on adaptive algorithms. In D.A. Watermann and F. Hayes-Roth, editors, *Pattern Directed Inference Systems*. New York: Academic Press, 1978.
3. M.I. Jordan and D.E. Rumelhart. Forward models: supervised learning with a distal teacher. *Cognitive Science*, Vol. 16, pp. 307–354, 1992.
4. J.B. Pollack. The induction of dynamical recognizers. *Machine Learning*, Vol. 7, pp. 227–252, 1991.
5. D.E. Rumelhart, G.E. Hinton, and R.J. Williams. Learning internal representations by error propagation. In D.E. Rumelhart and J.L. Mclelland, editors, *Parallel Distributed Processing*. Cambridge, MA: MIT Press, 1986.
6. J. Schmidhuber. A possiblity for implementing curiosity and boredom in model-building neural controllers. In J.A. Meyer and S.W. Wilson, editors, *From Animals to Animats: Proc. of the First International Conference on Simulation of Adaptive Behavior*, pp. 222–227. Cambridge, MA: MIT press, 1991.
7. L.R. Squire, N.J. Cohen, and L. Nadel. The medial temporal region and memory consolidation: A new hypothesis. In H. Weingartner and E. Parker, editors, *Memory consolidation*, pp. 185–210. Erlbaum, Hillsdale, N.J., 1984.
8. J. Tani. Model-Based Learning for Mobile Robot Navigation from the Dynamical Systems Perspective. *IEEE Trans. on SMC (B)*, Vol. 26, No. 3, pp. 421–436, 1996.
9. J. Tani. An interpretation of the "self" from the dynamical systems perspective: a constructivist approach. *Journal of Consciousness Studies*, Vol. 5, No. 5-6, pp. 516–42, 1998.
10. S.B. Thrun and Knut Moller. Active exploration in dynamic environments. In *in Proc. of NIPS 4*, pp. 531–538. 1990.
11. P. Werbos. A menu of designs for reinforcement learning over time. In W.T. Miller, R.S. Sutton, and P.J. Werbos, editors, *Neural Networks for Control*, pp. 67–95. MIT Press, Boston, MA, 1990.

Artificial Autonomy in the Natural World: Building a Robot Predator

Ian Kelly[1], Owen Holland[1], Martin Scull[1], and David McFarland[2].

[1]Intelligent Autonomous Systems Engineering Laboratory, Faculty of Engineering, University of the West of England, Bristol, BS16 1QY, England
Ian.Kelly@uwe.ac.uk
[2]Department of Zoology, University of Oxford, England

Abstract. With a few exceptions, today's mobile robots, however complex, are not truly autonomous. At some time, they all require humans to supply them with energy and/or information; most also require other forms of assistance. In complete contrast, even the simplest animals are totally self-sufficient. We describe a current project[1] which aims to construct autonomous robots with animal-like self-sufficiency both in terms of energy and information. The robots will live free on agricultural land, hunting and catching slugs, and fermenting the corpses to produce biogas, which will fuel the generator providing the robots with power.

1 Introduction

During the last two decades much research has been carried out into the design and control of so-called autonomous robots. However, most of these robots still require some intervention from humans in order to carry out their task(s). Forms of human intervention include supplying information and energy, physically assisting the robot, and modifying the environment to suit the robot(s). For robots to be truly autonomous they would have to be able to carry out their entire mission without human intervention. There are of course a few examples of robots which achieve a high degree of autonomy, in that they carry enough fuel for their mission or can use radiant energy from their environment, and can control themselves without human intervention. Examples include missiles, smart torpedoes, and some spacecraft. In addition some automated cleaning and materials handling AGVs use opportunity battery charging to achieve a degree of autonomy. But while we might be inclined to congratulate ourselves on our achievements in improving the autonomy of robots, we must also recognise that even the simplest animals exhibit a degree of self-sufficiency and independence which is immeasurably superior to that of the best of our robots.

This project represents an attempt to design and construct a robot system, with energetic and computational autonomy comparable to an animal system (as [10] urges). In order to make the mission a real technical challenge, we decided that the system would have to obtain its energy in the same way as most animals - by finding and 'digesting' organic material. Such natural resources are found in particular types

---

[1] This project is supported by an award from the EPSRC ROPA programme

of places, and are destroyed by being used; the process of foraging for food must therefore deal with the issue of where and when to look for food, when to revisit an area where food was previously found, when to abandon a site which is not producing food, and so on. Of course, the organic resources, or food, must be converted to a form of energy that the robot system can use. We propose to convert the organic material to electricity by first fermenting it to obtain bio-gas, and then using this bio-gas to power an internal combustion engine driving a generator. Using modern engine management techniques, it is possible to use bio-gas with as little as 25% methane, and to recover up to 45% of its energy.

The organic energy source should be a mobile animal species, since the technical problems of predation are much more challenging than grazing on plants. However, the prey species should be reasonably plentiful and not require rapid pursuit, which would be difficult to achieve at a reasonable energy cost. Since there would be a certain energy cost associated with finding, catching, and consuming any creature, however large or small, the prey species should be reasonably large, to give a reasonable margin over energy expended. Finally, it should preferably be a pest species subject to aggressive control measures, so that the system could be perceived as doing something of actual use that would have to be done anyway. All of the above criteria are met by the slugs found on agricultural land, especially *Deroceras reticulatum* [8]. They are slow moving, plentiful, large, and destructive - UK farmers spend over £20m per annum on molluscicides and spreading them [1]. Slugs are also potentially more suitable for fermentation than some other possible target species: they do not have a hard shell or exoskeleton and have a high moisture content.

The agricultural ground where slugs are pests usually takes the form of a well cultivated seedbed containing winter wheat or potatoes [2]. Such ground is soft - so soft that moving a heavy fermentation vessel over it would consume large amounts of energy. The fermentation vessel/gas engine/generator system will therefore be fixed, and the robots will deliver slugs to it, and collect power from it rather like some social insect colonies. Because the fermenter will require a certain amount of energy to cover its operating overhead, it seems extremely unlikely that a single robot would be able to gather enough energy to service both its own and the fermenter's requirements; we will therefore require a multiple robot system. This confers other potential advantages: some search tasks are more efficiently performed by a number of communicating robots than by a single robot; perhaps more importantly, the multiple approach gives a potential for achieving reliability through redundancy. This paper describes the progress made to date in the design and construction of the robots, work on the fermenter is to begin shortly.

2 The Robots

The key features of the robots are; that they must be energy efficient, operate in unmodified agricultural fields, and be protected from the weather, slime, and mud. To minimise energy usage whilst foraging each robot will scan the ground and catch slugs using a sensor and gripper mounted on a long, light articulated arm - the energy required to move this arm is much less than that required to move the whole robot over rough ground. The optimal arm length is a function of the distribution of slugs and off the power required to operate arms of different lengths in relation to the power required to move the robot. Our calculations showed that a 1.5m long arm would be the most energy efficient. The other requirements of the arm are that it should be light, stiff, easily controllable, and capable of moving in all directions; it should also have a fairly simple construction, increasing its reliability, making it easier to manufacture and reducing its cost. A design consisting of two 0.75m tubular sections, with a hinged joint between them, was chosen (see Fig. 1). To allow the arm to rotate around the whole robot it is mounted on a turntable located in the centre of the robot's chassis. The chassis is large enough to maintain stability in all directions when the arm is fully extended. To meet the requirements of lightness and stiffness the arm is constructed from aircraft grade carbon fibre tube. To keep the arm structure light, the motor and gearbox required to provide movement at the elbow joint are mounted on the turntable; a lightweight toothed belt inside the arm transmits the drive to the elbow joint. Since the numbers of slugs on the surface peak in the early evening and just before dawn, the rate of gathering them during these periods must be as high as possible. To this end the arm motors and gearboxes were selected so that the arm can move from fully retracted to fully extended or vice-versa in under 1.5 seconds. Self locking worm gearboxes provide the required motor speed reduction, and allow the arm to be held in position without consuming any energy.

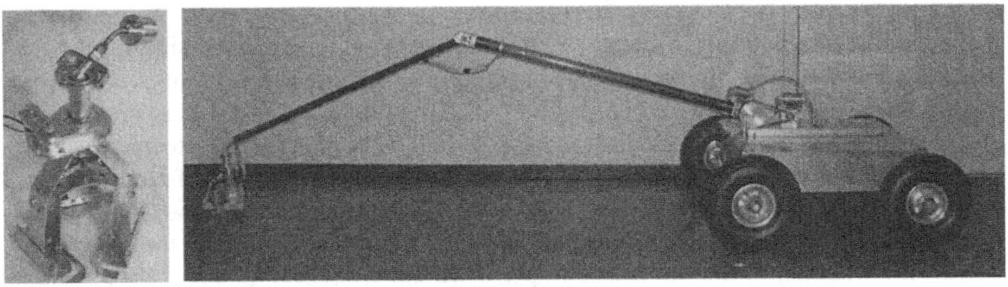

Fig. 1. Prototype three fingered gripper with wiper blades and compliance gimbal (left), and the arm and gripper system mounted on a turn table (right)

The arm's end-effector is a robust lightweight gripper that is able to pick up and release both wet and dry slugs, regardless of their size and orientation, and any irregularities in the substrate. The current design consists of three fingers at 120° spacing, operated by a single miniature motor. As the fingers close, they meet underneath the slug so that it can be lifted; wiper blades ensure the slug's release when the gripper is opened. Slugs are detected and targeted by a camera mounted in the centre of the gripper, away from slugs and mud. To allow for contour matching with the ground, and to ensure that the view from the camera is always perpendicular to the ground, regardless of the arm's extension, the whole gripper assembly hangs freely on a gimbal. When scanning for slugs it is possible to lock the gripper assembly, thus stopping it swinging, by fully opening the gripper. Each of the three wiper blades has a plate on the end to allow for passive alignment with the contour of the ground, ensuring that all three blades move under the slug when the gripper closes. The gripper's mechanism will be protected from the weather and mud by a flexible rubber cover.

Locating the static fermentation station in a large muddy field, where wheel slip will be inevitable, will be achieved by using a combination of the Differential Global

Positioning Satellite (DGPS) system, and an active infrared localisation system [4], [5]. DGPS can also be used for mapping the locations of grazed areas, so that good patches can be found again, and over-grazing can be avoided. (This last point may not be a problem: a study allied to this project [3] found that removing all surface slugs from a location every few days does not, in the medium term, appear to reduce the number of available surface slugs. This is thought to be because there is a large reservoir of underground slugs which can rapidly replace those that are removed.)

3 Sensing and the control strategy

A system is required that can successfully detect slugs, under sparse vegetation, against a background of rough earth. This task could potentially be achieved by several different types of sensors; we have opted for a vision based system since it offers the best combination of size, weight, cost and effectiveness. We have opted for a monochrome CMOS image sensor that is lightweight, low power (<175mW), adequate resolution (164 by 124 pixels), sensitive (0.1 Lux), has a digital interface, and can produce up to 60 frames per second. This sensor also has built in adjustable automatic exposure control and is inexpensive.

To avoid computationally intensive image processing, a method of simplifying the detection of slugs and of filtering out vegetation is required. Since slugs are active mainly from dusk to dawn some form of illumination is needed. This creates the possibility of using some combination of coloured light and filtering to increase the visibility of slugs and decrease the visibility of vegetation. We have found that this can be achieved by using red light from extreme brightness LEDs, and fixing a red filter in front of the image sensor. In the received images, under these lighting conditions, vegetation and soil appear dark, whilst the slug *Deroceras reticulatum* reflects red light and thus appears very bright. Fig. 2 shows such a slug (32mm long) together with some grass under white illumination. Fig. 3 is the same image, except for some movement of the slug, under red illumination - the grass now appears dark whilst the slug is bright. Fig. 4 shows how applying a simple threshold function: (c[Average Image Intensity] – k) where c, k are constants, to the red illuminated image of Fig. 3 pinpoints the slug. This threshold scheme does not detect slugs under 15mm in length, thus filtering out smaller slugs that would be of little use to the robot. All that is necessary in the way of processing is the registration of large bright patches; clearly, further tests will need to be carried out into the overall robustness of this system, especially with respect to slugs that are partially obscured by vegetation.

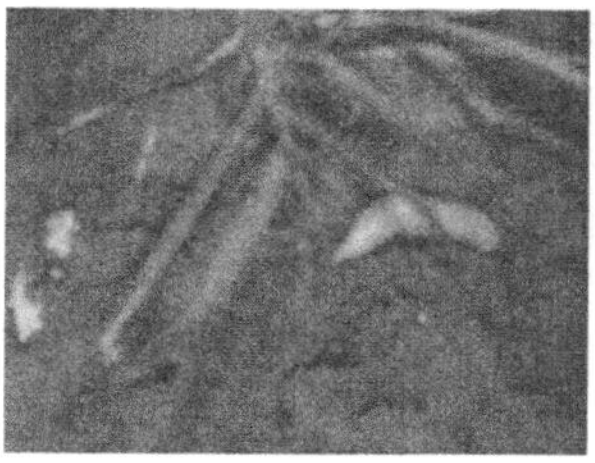

Fig. 2. Under white light

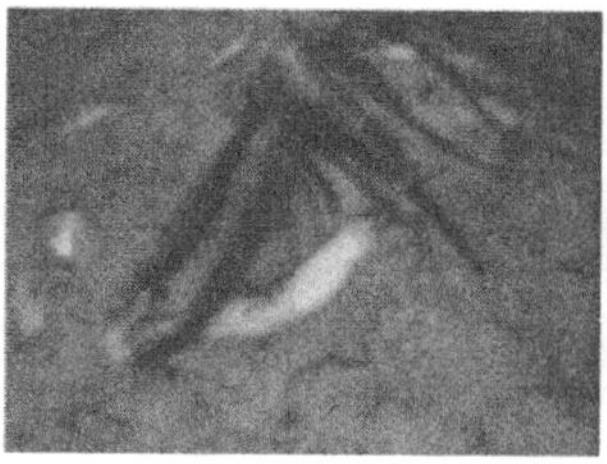

Fig. 3. Under red light

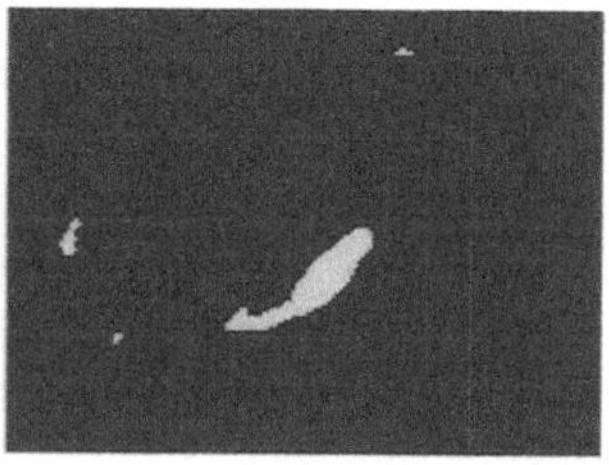

Fig. 4. After thresholding

As regards the overall control strategy, the situation of a typical animal is radically different from that of a typical robot. At any moment, a robot usually has a single goal, and is rarely faced with a real choice of what to do, whereas animals and our robots, which are in the same free-living situation, will have several simultaneous goals. Slugs must be gathered; batteries must be recharged; it must not get lost; it must always have enough charge to be able to return to the refuelling point; it must maintain the functionality of its sensors and effectors; and so on. How can we approach the task of programming the robots in our system so that they always act to maximise expected survival time? Our strategy, as designers, must be to find a computationally feasible solution which gives adequate performance. We know that because of the inefficiency of the digestive process, our system will at best be on the borderline of survivability, and so our performance requirements may be even more severe than those which an animal living exclusively on slugs would experience.

Unfortunately, we lack the detailed information which might allow us to arrive immediately at a specific and optimal solution. We are therefore forced into taking a more general approach. The chosen solution has been to adopt a model of motivation and action selection in animals and robots (the d,r,k model) which has been developed over several years [6], [7], [9]. It is suitable for application on robots because it has been developed by considering animals as if they were robots, and vice-versa. It provides a formalism within which we can represent the level of motivation which an animal or robot might possess in a given situation in respect of what a roboticist would think of as its various goals, and enables the rational selection of one of a potentially large number of actions as a function of a relatively small number of variables and parameters. While the problem of parameter estimation remains, it provides a clear and principled functional framework within which to operate.

References

1. Glen, D.M.: Ecology of Slugs in Cereals in Relation to Crop Damage and Control. In Individuals, Populations, and Patterns in Ecology, Eds. S.R.Leather, K.F.A.Walters, N.J.Mills, and A.D.Watt. Intercept (1994) 163-171
2. Glen, D.M., Milsom, N.F., Wiltshire, C.W.: Effects of Seed-bed Conditions on Slug Numbers and Damage to Winter Wheat in a Clay Soil. Annals of Applied Biology (1989) 115, 177-190
3. Glover, C.: Reinvasion of Cleared Areas by the Slug *Deroceras reticulatum*. Internal Report. Long Ashton Research Station, Bristol (1998)
4. Kelly, I.D., Keating, D.A.: Flocking by the Fusion of Sonar and Active Infrared Sensors on Physical Autonomous Mobile Robots. Conf. on Mechatronics (1996) 1/1-1/4
5. Kelly, I.D.: The Development of Shared Experience Learning in a Group of Mobile Robots. Ph.D. Thesis, Department of Cybernetics, University of Reading (1997) 2/4-2/12
6. McFarland, D.J., Boesser, T.: Intelligent Behaviour in Animals and Robots. MIT Press, Cambridge, MA (1993)
7. McFarland, D.J., Spier, E.: Basic Cycles, Utility, and Opportunism in Self-Sufficient Mobile Robots. Robotics and Autonomous Systems 20 (1997) 179-190
8. South, A.: Terrestrial Slugs: Biology, Ecology, and Control. Chapman and Hall (1992)
9. Spier, E., McFarland, D.: Possibly Optimal Decision Making Under Self-Sufficiency and Autonomy. Journal of Theoretical Biology (1998)
10. Steels, L.: When are robots intelligent autonomous agents? Journal of Robotics and Autonomous Systems 15. (1995) 3-9

Behaviour Adaption on Behavior-Based Controller for Brachiation Robot

Yasuhisa HASEGAWA[1], Yoshikuni ITO[2], Toshio FUKUDA[3],

[1] Dept. of Micro System Engineering, Nagoya University,
Furo-cho, Chikusa-ku, Nagoya, Aichi 464-8603, JAPAN
yasuhisa@mein.nagoya-u.ac.jp

[2] Dept. of Micro System Engineering, Nagoya University,
Furo-cho, Chikusa-ku, Nagoya, Aichi 464-8603, JAPAN
y_ito@robo.mein.nagoya-u.ac.jp

[3] Center for Cooperative Research in Advanced Science & Technology, Nagoya University, Furo-cho, Chikusa-ku, Nagoya, Aichi 464-8601, JAPAN
fukuda@mein.nagoya-u.ac.jp

Abstract: This paper proposes novel adaptation method for a behavior-based locomotion robot. Utilization of hierarchical behavior controller makes the controller designing process easier and shorter, because the designer can deal with behavior controllers for simple behaviors independently, and coordinate those behavior controllers in order for a robot to perform the objective behavior after finishing the design of the fundamental behavior controllers. Some problems are still remaining. One is how to adjust the behavior coordinator when the objective behavior or robot parameters are slightly changed. We propose the novel method to adjust the behavior coordinator against some changes. This method measures the effects of the fundamental behavior controllers to the total behavior, and changes the activation values for them in less trials. This proposed method is applied to the real brachiation robot control. This brachiation robot has redundant mechanism to locomote from branch to branch with 14 actuators like a long-armed ape.

1 Introduction

Recent years, many researches about a brachiation type locomotion robot are carried out. The brachiation mobile robot(BMR) is a mobile robot, which dynamically moves from branch to branch like a gibbon, namely long-armed ape, swinging its body like a pendulum[1][2]. Saito et al,[3]-[5] proposed the heuristic learning method for generating feasible trajectory for two-link brachiation robot. Fukuda et al,[6] propose the self-scaling reinforcement learning algorithm to generate feasible trajectory with robust property against some disturbances. These studies do not use a dymamics model of two-link brachiation robot directly. On the other hand, Nakanishi et al,[7] took another approach, using target dynamics, for control an underactuated systems. The two-link brachiation robot is underactuated system with two degrees of freedom and one actuator. As a two-dimensional extended model, seven-link brachiation robot is studied by Hasegawa et al.[8]. This seven-link brachiation robot is given the redundancy to locomote so that it is able to take a motion like a real ape in plane. In that study, hierarchical behavior architecture is adopted to design the controller with multi-input and multi-output efficiently. The behavior controllers and its coordinators in the hierarchical structure are generated using reinforcement learning. The concept of hierarchical behavior controller is based on behavior-based control, which has an advantage of creating a higher-level behaviors from simpler behaviors in complex system. The learning process for acquiring the behavior controller and the behavior coordinator is hard to apply a real robot, because of taking many trials to tune them. Even if a small adjustment about the controller is needed when the objective task or robot parameters are slightly changed, the reinforcement learning needs many trials because it uses no structured information about the relations and the effects between the each behavior and total behavior.

In this paper, we propose the novel adaptation method to adjust behavior coordinator against small changes of the objective task or a robot environment. This method measures the relations between the each behavior and the total behavior and then determine the direction of change of the activation values from the behavior coordinator based on the measurements. We apply this adaptation method to control problem of the Brachiator III. This robot, Brachiator III, has 13 links and 12 joints and be able to take a motions like a real ape in a three-dimensional space. The controller is designed based on behavior-based approach. We show the effectiveness of the proposed algorithm through the experiments using the Brachiator III.

2 Outline of 13-link Brachiation Robot

The motion of the conventional brachiation robot is limited in two-dimensional plain. These motions are far from the real ape behaviors. Therefore we made the brachiation robot with 13 degrees of freedom in order to realize dynamically dexterous behaviors. The sizes and mobile joint angles of this robot is designed by simulating a real long-armed ape. The weight is about one kilogram heavier than a real ape because of actuator problems. The designing and adjusting process of the controller is very hard, because the motions becomes much more complex than the motions in plane

2.1 Mechanical Structure of Brachiator III

Brachiator III consists of 13 links, 13 degrees of freedom, and 14 motors including two motors for two grippers. Each joint is driven by DC motor through the wire so that we could conform the weight distribution to a real ape adjusting the DC motor locations. The appearance of "Brachiator III" is shown in Fig.1.

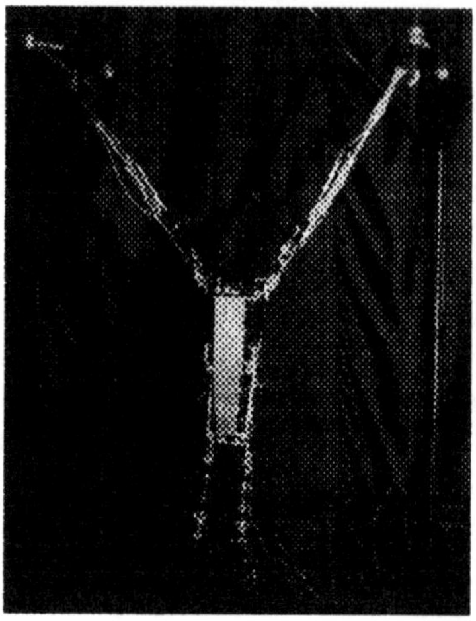

Fig. 1 Brachiator III

2.2 Motion Control Based on Behavior-based Architecture

The hierarchical behavior controller for brachiation robot is designed based on behavior-based architecture. The objective complex behavior is supposed to be decomposed into some simpler behaviors. Higher level behavior is performed by coordinating simpler behaviors that are obtained before. The architecture of the hierarchical behavior controller of the robot based on this concept is shown in Fig. 2. At first, the brachiation behavior is divided into a swing action which stores the sufficient energy prior to the transfer (preliminary swing mode), and a locomotion action which is actual transfer operation (locomotion mode). Then, these actions are decomposed into the fundamental behaviors; leg swing, body rotation 1, leg stretch, body rotation 2, body lift and arm reaching.

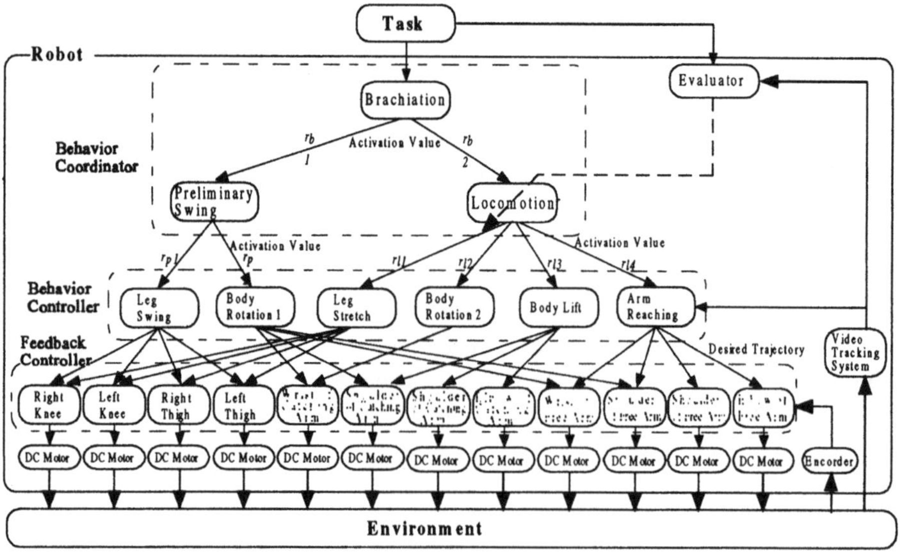

Fig. 2 Hierarchical behavior controller for brachiation robot

2.3 Desired Trajectory From Behavior Controller

The behavior controllers except the arm reaching behavior controller are feedforward controllers which output the original trajectories expressed by the cubic spline function. The desired trajectories for the connected actuators are generated by rescaled by the activation values from the behavior coordinator on higher layer. The formulas are,

$$yd_i(t) = r_k(y_k(t) - b_k(t)) \quad (1)$$

yd_i: Desired trajectory to actuator i

r_k: Activation value for behavior controller k

$y_k(t)$: Actuator trajectory from behavior controller k.

$$b_k(t) = b(0)(t^* - t)/t^* + b(t^*)t/t^* \quad (2)$$

t^*: Time when the behavior is finished

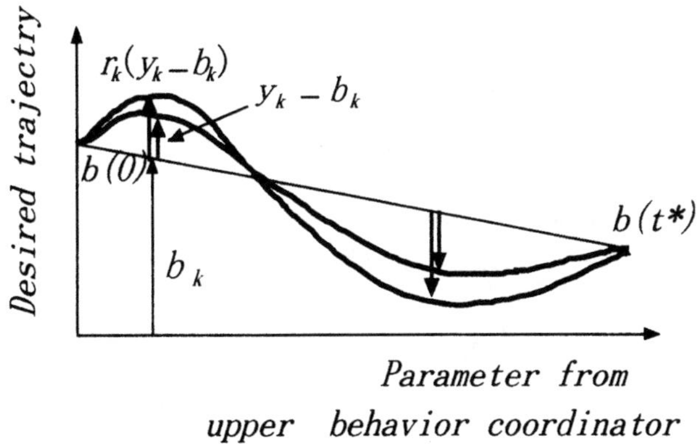

Fig. 3 Desired trajectory to connected actuator

2.4 Motion Measurement of Brachiation Robot Using Real-time Tracking System

It is impossible to measure the location of the tip of the free arm and the location of the center of gravity of the robot, because the slip angle between the catching grip and the branch is not measurable directly using a potentiometer or rotary encoder. We use the real-time tracking system, “Quick MAG System III”, which measures the three-dimensional locations of the eight points at 60Hz sampling frequency, using two CCD cameras.

The eight measuring positions shown in Fig. 4 are chosen to calculate the center of gravity of the robot based on the assumptions as follows,

1. The elbow of the catching arm keeps straight.
2. Both legs are controlled to behave the same motion.
3. Two joints on the shoulder are adjoining and attached on the almost same position.

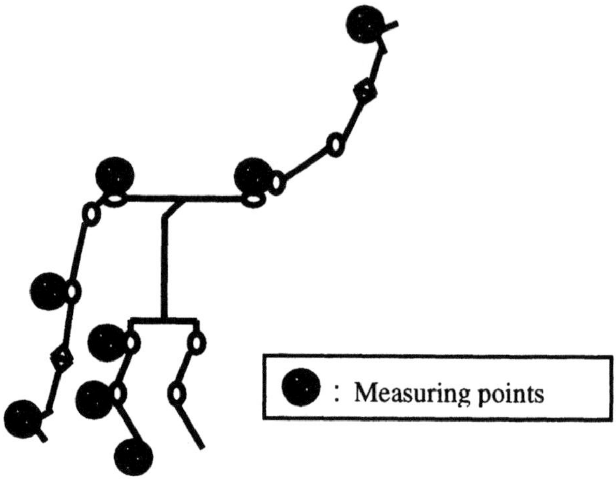

Fig. 4 Measuring points for calculating the center of gravity

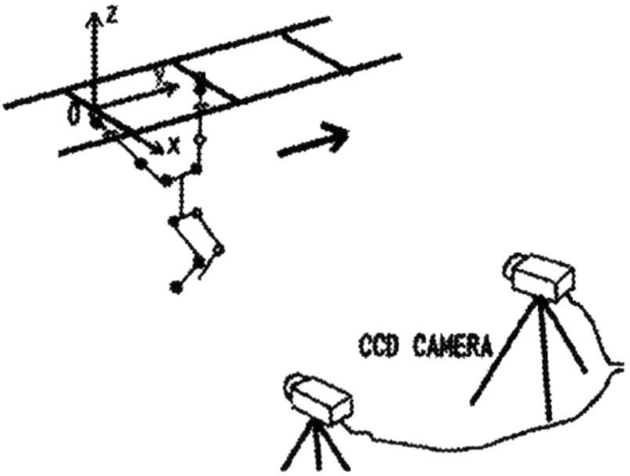

Fig. 5 Coordinates of the measurement

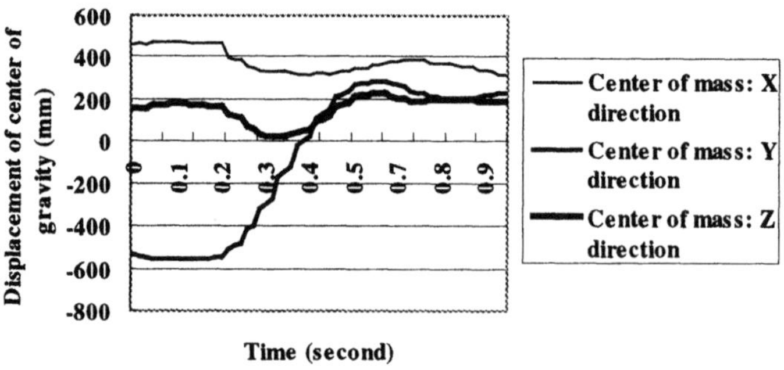

Fig. 6 Transfer of center of gravity calculated from each measuring points

3 Adaption for Behavior-Based Controller

The unsupervised learning methods, i.e genetic algorithms, evolutionary programming and reinforcement learning, need many trials to obtain the desired behavior. Therefore, it is effective to use those unsupervised learning methods in computer simulations. The real robot would not behave the desired performance using the controller obtained in the computer simulations, because of the model error used in the computer simulation and some disturbances. An adaptation method or a tuning method for compensating such a error or disturbance should be applied to improve its stability. Furthermore the adaptation method is required, when the parameters of the robot or objective task is slightly changed. The proposed adaptation method for the behavior coordination is explained in this section.

3.1 Adaptation Problem for Behavior-based Controller

The brachiation robot is controlled by the hierarchical behavior controller explained above. To adjust the total behavior against the small changes of robot parameters or desired behavior, we have three approaches; the fundamental behavior controller, the behavior coordinator or both of them should be adjusted. The easiest approach is to adjust behavior coordinator because of the small searching space. If the adjustment of the behavior coordinator is not enough to the desired behavior, both the behavior controller and behavior coordinator should be rearranged. The proposed method is concerning about adaptation of only the behavior coordinator. In order to adjust the behavior coordinator, we should evaluate the effect of each behavior controller to the whole behavior. Therefore, we propose approximately measuring method of the relations between the each behavior and the total behavior through several trials and update method of the activation values of the behavior coordinator.

3.2 Adaptation Algorithm

The behavior coordinator indicates the activation values which determine the amplitude of the desired trajectories from behavior controller. By adjusting the activation values, the total behavior could be improved to some extent. The relation between the activation values and the performed total behavior is strong nonlinear, however we assume their relations could be express as the multiplier of the degree of contributions and the activation values, eq. (3), only in neighborhood around the current state. Three parameters are selected as a representative index of the robot's performance; the maximum angle of center of gravity, the distance R in y-z plain and x-directional distance x_{hand} between a free hand and a target branch shown in fig. 7. When the objective task or robot parameters are slightly changed, the robot will fail because the differences between the current performance and new target performance are generated. The differences are express by the eq. (7) from eq. (3). We explain the algorithm as follows,

1. At first, measure the degree of contribution of each behavior controller to three index parameters. some value as a perturbation Δr around current activation value are added to each activation value and then make a trial per a perterbation.
2. Using eqs. (8), (9) and (10), measure the degree of contribution of each behavior controller.
3. Calculate the new activation value using eq. (11).
4. Evaluate its performance by making trial with new activation values and explore the their neighborhoods area, R' and R''.
5. Update the degree of contribution of each behavior controller with new activation values using eq. (12).
6. Go to step 3.

$$E(s) = W(s) \cdot R(s) \tag{3}$$

$$E(s) = \left(\theta(s)_{body},\ x(s)_{hand},\ R(s)_{hand}\right)^T \tag{4}$$

$$W(s) = \begin{pmatrix} w(s)_{l_2 body} & w(s)_{l_3 body} & w(s)_{l_4 body} \\ w(s)_{l_2 handX} & w(s)_{l_3 handX} & w(s)_{l_4 handX} \\ w(s)_{l_2 handR} & w(s)_{l_3 handR} & w(s)_{l_4 handR} \end{pmatrix} \tag{5}$$

$$R(s) = \left(\Delta r(s)_{l_2},\ \Delta r(s)_{l_3},\ \Delta r(s)_{l_4}\right)^T \tag{6}$$

$$\Delta E(s) = W(s) \cdot \Delta R(s) \tag{7}$$

$i)\ \ if\,(s = 0)$

$$\begin{aligned} w(0)_{kj} &= \frac{E_k(r_k + \Delta r_k) - E(r_k)}{\Delta r_k} \qquad (if\ \ E_k(r_k + \Delta r_k) > E_k(r_k - \Delta r_k) > 0) \\ &= \frac{E_k(r_k) - E(r_k - \Delta r_k)}{\Delta r_k} \qquad (if\ \ E_k(r_k - \Delta r_k) > E_k(r_k + \Delta r_k) > 0) \end{aligned} \tag{8}$$

$$w(0)_{kj} = 0 \qquad (if\ \ E_k(r_k + \Delta r_k) < 0\ \ and\ \ E_k(r_k - \Delta r_k) < 0) \tag{9}$$

$$w(0)_{kj} = \frac{E_k(r_k + \Delta r_k) - E_k(r_k - \Delta r_k)}{2\Delta r_k} \qquad (if\ \ E_k(r_k + \Delta r_k) \cdot E_k(r_k - \Delta r_k) < 0) \tag{10}$$

$$\Delta R(s) = W(s)^{-1} \cdot \Delta E(s) \tag{11}$$

$ii)\ \ if\,(s > 0)$

$$W(s+1) = F(s) \cdot T(s) \tag{12}$$

where

$$F(s) = \left(E(s)\quad E'(s)\quad E''(s)\right) \tag{13}$$

$$T(s) = \left(R(s)\quad R'(s)\quad R''(s)\right) \tag{14}$$

$$R'(s) = \left(\Delta r(s)_{l_2},\ \Delta r(s)_{l_3} + \alpha,\ \Delta r(s)_{l_4} - \alpha\right)^T \tag{15}$$

$$R''(s) = \left(\Delta r(s)_{l_2} + \alpha,\ \Delta r(s)_{l_3} - \alpha,\ \Delta r(s)_{l_4}\right)^T \tag{16}$$

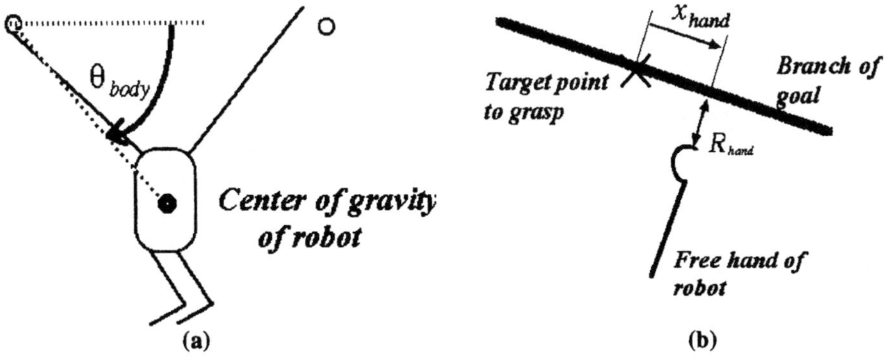

Fig. 7 Figure (a) shows the angle of center of gravity, and figure (b) shows the distance R in y-z plain and x-directional distance x_{hand} in between a free hand and a target branch

4 Experiments

We show the adaptation of the proposed method against a change of the objective task. Brachiator III with initial hierarchical behavior controller can catch the branch whose span is 90 centimeters. Then as a change of the task, the span to the next branch is extended to 101 centimeters. Brachiator III failed to catch the branch since an oscillation energy of the center of gravity is shortage and the body did not come up to a reachable region.

We assume that the locomotion behavior coordinator is not sufficient and should be tuned. We therefore apply to the proposed adaptation algorithm to adjust the locomotion behavior coordinator, which has four fundamental behavior controllers and output four activation values to them; Leg stretch, Body rotation 2 , Body lift and Arm reaching. We set Δr of first step with ± 0.05, ± 0.1 and ± 0.4 as the perturbation, and set α with 0.05, β with 0.5. Also we choose "arm reaching ", "body lifting" , "body rotation 2" as behavior to be corrected.

4.2 Experimental Results

Three steps of the adaptation algorithm are required to search the feasible activation values when the branch span is extended to 101 centimeters. The transition of variations of r_i, $\Delta\theta_{body}$, Δx_{hand}, and ΔR_{hb} through the adaptation are shwon in Fig. 8, and table. 1. Figure 9 shows the trajectories of tip of the free hand both before and after adaptations.

When we choose ± 0.1 as Δr, we can get the feasible brachiation motion against task change, the parameters such as $\Delta\theta_{body}$, Δx_{hand}, and ΔR_{hb} are also converged. These parameters are improved when we set Δr with ± 0.05 and ± 0.4 .

5 Conclusions

In this paper, we proposed the novel adaptation method to adjust behavior coordinator against small changes of the objective task. This method measures the effects of the each behavior to the total behavior, and determines the activation values in the behavior coordinator for each behavior controller based on the trials. We applied this adaptation method to control problem of the Brachiator III. We showed the effectiveness of the proposed algorithm, which could find the feasible activation values through several trials against small task change. The performance of this method depends on the amplitude of perturbation, which is difficult to be adjusted because of the non-linearity of robot behavior and because of measuring error.

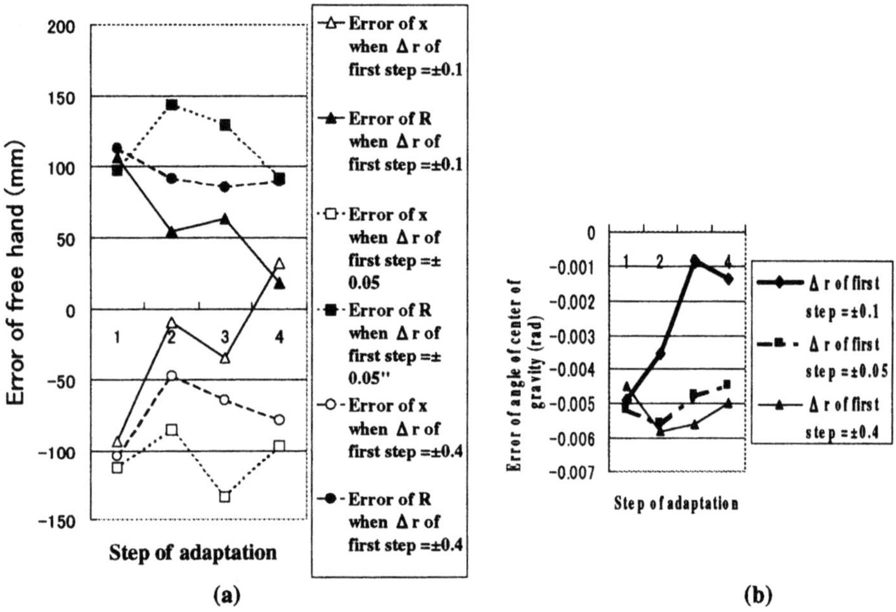

Fig. 8 In the figure (a), ΔR_{hb} denotes the distance from free hand to the branch of goal, and Δx_{hand} denotes the distance to the x direction from free hand to the center of the branch. In the figure (b), $\Delta \theta_{body}$ denotes the error of θ_{body} from the target angle.

Table 1. Transitions of m_i, $\Delta \theta_{body}$, Δx_{hand}, and ΔR_{hb} in the adaptaion process

Trial No.	r_{l_1}	r_{l_2}	r_{l_3}	r_{l_4}	$\Delta \theta_{body}$ (rad)	Δx_{hand} (mm)	ΔR_{hb} (mm)
0	1.0	1.0	1.0	1.0	-0.00490	-92.8	106.4
1	1.0	0.869	1.21	0.815	-0.00354	-9.50	54.4
2	1.0	0.886	1.23	0.987	-0.000836	-34.7	63.4
3	1.0	0.909	0.958	0.929	-0.00137	31.9	18.2

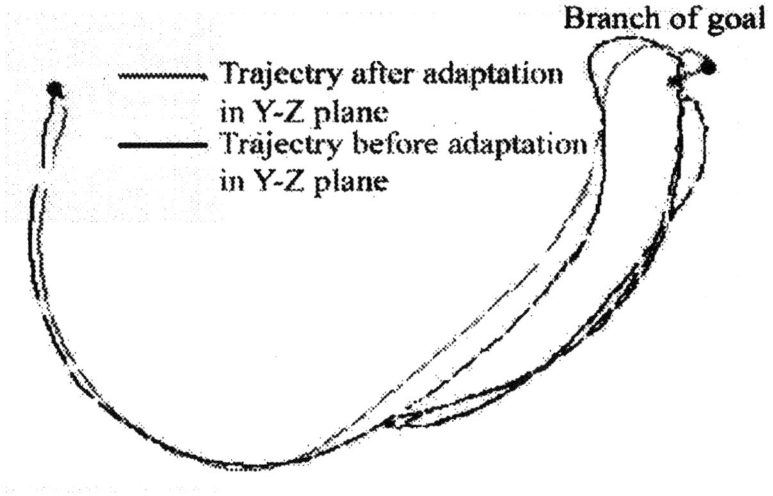

Fig. 9 Free hand's trajectories between before and after adaptation

References

[1] T. Fukuda, H. Hosokai and Y. Kondo, "Brachiation Type of Mobile Robot", Proc. IEEE Int. Conf. Advanced Robotics, pp. 915-920,(1991).

[2] Spong, M. W., Swing Up Control of the Acrobot, Proc. 1994 IEEE International Conference on Robotics and Automation, pp. 2356-2361, (1994)

[3] F. Saito, T. Fukuda, F. Arai, "Swing and Locomotion Control for a Two-Link Brachiation Robot", IEEE Control Systems, Vol. 14, No. 1, pp. 5-12, (1994).

[4] F. Saito, T. Fukuda, F. Arai, "Swing and Locomotion Control for Two-Link Brachiation Robot", 1993 IEEE International Conference on Robotics and Automation, Atlanta, pp. II:719-724, (1993).

[5] T. Fukuda, F. Saito and F. Arai, "A Study on the Brachiation Type of Mobile Robot(Heuristic Creation of Driving Input and Control Using CMAC)", Proc. IEEE/RSJ Int. Workshop on Intelligent Robots and Systems, pp. 478-483, (1991).

[6] T. Fukuda Y. Hasegawa K. Shimojima and F. Saito, "Self Scaling Reinforcement Learning for Fuzzy Logic Controller", IEEE International Conference on Evolutionary Computation, pp. 247-252, Japan, (1996)

[7] J. Nakanishi, T. Fukuda and D.E. Koditschek, "Experimental implementation of a "Target dynamics controller on a two-link brachiating robot"", IEEE International Conference on Robotics and Automation, pp. 787-792, (1998)

[8] Y. Hasegawa and T. Fukuda, "Learning Method for Hierarchical Behavior Controller", IEEE International Conference on Robotics and Automation, pp.2799-2804, (1999)

Wet Artificial Brains: Towards the Chemical Control of Robot Motion by Reaction-Diffusion and Excitable Media

A. Adamatzky[1], O. Holland[1,2], N. Rambidi[3], and A. Winfield[1]

[1] Intelligent Autonomous Systems Lab, UWE, Bristol, United Kingdom
{Andrew.Adamatzky, Owen.Holland, Alan.Winfield}@uwe.ac.uk
http://www.uwe.ac.uk/facults/eng/ias/

[2] The CyberLife Research Institute, Cambridge, United Kingdom
oholland@cyberlife-institute.co.uk
http://www.cyberlife-institute.co.uk

[3] Physics Department, Moscow State University, Moscow, Russia
rambidi@lord.phys.msu.su
http://poly.phys.msu.su/persons/rambidi.xml

Abstract. The paper gives probably the first ever systematic discussion on how wave processes in reaction-diffusion and excitable homogeneous media can be efficiently used to solve a wide range of problems in robot navigation. Three possible applications of chemical controllers are considered: (i) object following, (ii) optimal path finding, and (iii) universal control. The various implementations of controllers discussed here include: Belousov-Zhabotinsky chemical processors, families of excitable lattices, and self-localised excitations. We present some results from a simulation of robot control using excitable lattices, and find the results encouraging for our planned construction of a chemically controlled robot.

1 Background

One remarkable aspect of life is that it presents innumerable instances of simple biological systems which can solve particular problems in the real world; many of these problems can also be set in a computational context. Artificial life has yielded many demonstrations of abstract systems which also exhibit these properties in simulation; our focus here is on building real physical systems which instantiate these abstract systems in a situations close to the biological source of inspiration — using them to control a freely moving physical agent in a physical environment. Our chosen technical route is chemical information processing.

Chemical information processing is attractive from many points of view: to some people, it offers fine grained massive parallelism; to others, the appeal is in the potential robustness of the technique, or in the connection to computation in biological structures. Our interest is driven by the possible relationship between the problems which one form of chemical computation may excel at solving, and the problems which need to be solved by agents moving around in the real world.

Chemical computation is of two basic types. The first occurs within domains in which spatial differentiation is eliminated (for example by stirring); the essence of the computation is in the appearance and development of linked reactions and compounds in which certain characteristics of identity and concentration can be identified as the outcome of useful computation over some problem defined by the initial state and external inputs. The second type exploits local changes and spatial differentiation, using unstirred reactors and typically employing a rather simpler basic chemistry than the first type. This intrinsic connection to locally connected structures maps well onto many of the problems which a mobile robot faces when moving around an unstructured environment: two and three dimensional representations must be processed, and movement in two or three dimensions must be controlled and directed. Excitable media have been studied for some time in the context of the possible realization of basic computational operations in chemical active media. For example, it has been demonstrated that universal logical gates can be realized in compartmentalized chemical media [9], and that many basic forms of image processing can be executed in parallel in such chemical computing devices ([13] - [16]). Universal logical gates constructed from chemicals held in and connected by combinations of tubes and valves are sophisticated examples of chemical computing elements (see e.g. [21]). The non-stirred active chemical media exhibit a wide spectrum of dynamic behavior, which could be extremely useful from a practical point of view ([3], [2], [1]).

Of the family of chemical oscillators, the Belousov-Zhabotinsky (BZ) reaction is the most investigated, and the most suitable for laboratory experiments. The non-stirred layer of an oscillating BZ-reaction can be considered as a massively parallel processor, where every elementary processor is represented by a micro-volume reactor. The state of the micro-volume can be identified with the reduced/oxidized state of the bromate component. Information processing media of the BZ type are specialized processors where the statement of the problem, the computational processes, and the results of the computations are all represented in the states (concentrations etc.) of the micro-volumes. The evolution over time of such a processor leads to a spatio-temporal dissipative structure that can be interpreted as the solution of the problem. The computation in a BZ processor takes place when waves are generated, and spread and interact with each other, and perhaps with external representations of information (e.g. light patterns).

The known abilities of excitable media for more complex image analysis [16] and for the solution of spatially based problems (e.g. shortest paths - [19], [21]; [14]) create the expectation that this experimental arrangement will be capable of demonstrating more sophisticated and complex robot behaviours such as object following, obstacle avoidance, and optimal path finding.

2 Object following and obstacle avoidance: System requirements and results of simulations

In order to make a chemically controlled robot which can follow some moving target without requiring an umbilical cable, it is necessary to accept a number of practical constraints. The chemical reactor (or other material realisation of the excitable medium) must be fitted on board the mobile robot. Local environmental information must be sensed onboard. Non-local information (if necessary) must either be gathered and pre-processed by an offboard vision system and communicated to the robot via a high-bandwidth radio link, or must be provided by an onboard system using optical projection. The environmental information will be represented to the robot as patterns of light projected onto the reaction medium; the light patterns initiate and modulate waves in the medium. The patterns formed by the waves will be sensed and processed by an optical system, and the output will be mapped to the robot's motion controller.

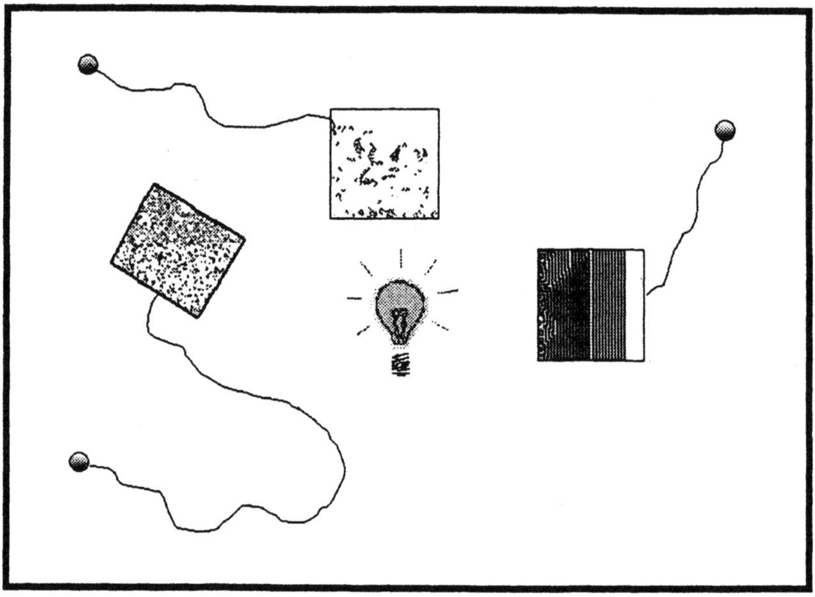

Fig. 1. The snapshot of the simulation. The left, the middle and the right left configurations represent lattices from the useless, adaptive and non-adaptive classes, respectively. The initial positions of the robots and their approximate trajectories are shown.

We can make a preliminary evaluation of the potential of such a system by investigating a model using computer simulation. Let us consider an onboard array $\mathbf{L}$ of excitable cells with an 8-cell neighbourhood such that for every cell $x \in \mathbf{L}$ the neighbourhood template is defined as $u(x) = \{(\delta_i, \delta_j) \in \{-1,0,1\} \times \{-1,0,1\} - (0,0)\}$. The cells take three states: rest ($\circ$), refractory ($-$) and excited ($+$), and change their states in discrete time by the rule: $x^{t+1} = +$ if $x^t = \circ$ and $C(x)^t$; $x^{t+1} = -$ if $x^t = +$; $x^{t+1} = \circ$ otherwise. The predicate $C(x)^t$ is true with probability p (or p^*) if x lies on the boundary of the array, or if the number $e(x)^t$ of excited neighbours, $e(x)^t = |\{\delta \in u(x) : (x+\delta)^t = +\}|$, lies in some specified interval of the excitation, $e(x)^t \in [\theta_1, \theta_2]$, $1 \leq \theta_1 \leq \theta_2 \leq 8$. We should note that all inner cells of the lattice are excited deterministically, and so noise can influence only the edge cells.

The edge cells of the lattice are subject to stochastic activity: every cell becomes excited without any external information-bearing stimulus with probability p at any discrete time step. If some parts of the edge are illuminated (i.e. receive information) the cells belonging to these parts are excited with a higher probability p^*, $0 < p \ll p^* < 1$; the edge cells which are in shadow become excited with probability p. Once generated at the edges, activity patterns spread on the array.

Let the environment consist of an open arena containing a single light source; we require the robot to follow the light source as it moves. The robot is assumed to move at a constant speed; the controller may vary only the direction of motion. The relative position of the light source can be derived from two successive configurations of array activity in the following manner. Every site x of the array is assigned a vector v_x^t which is updated at every step t of the temporal evolution of activity: $v_x^t = \frac{1}{|e(x)^t|} \sum_{\delta \in e(x)^t} \delta$. The vector thus indicates the direction with the highest density of excited neighbours. The vector is not updated if there are no excited neighbours. The integral vector v^t, $v^t = \sum_{x \in \mathbf{L}} v_x^t$, indicating the relative position of the light source can be transformed to a unit vector, $v^t = (v_1^t, v_2^t) \leftarrow (sg(v_1^t), sg(v_2^t))$, where $sg(a) = -1$ if $a < 0$, it equals 1 if $a > 0$ and it is zero otherwise. The value of v^t determines the change of direction of the robot. (In a wheeled robot, it may control the steering motor, or the relative speeds of wheel rotation in a differential-drive system.) If every cell of the excitable array has its own effector, then we do not need to calculate the integral vector; the motion of the robot toward the source of light will be the result of the summed activity of the collective of effectors.

In simulation experiments (Figure 1), several situations of interest were investigated. As will be described later, not all parametric selections gave good results, but the behaviour of the better performers was certainly encouraging. It was demonstrated that in the absence of a light source the robot can wander apparently at random, implementing a search of the environment due to the small stochastic activity of the edge cells; it is perhaps worth noting that this exploration in the absence of light can lead to the discovery of a source which is obscured by some obstacle. This functionality is often deliberately added to small behaviour based robots. A single source of light can certainly attract the robot. The situation becomes more complex and interesting when there are several sources of light. If p^* is constant then, depending on the initial position of the robot, the robot may wander between the sources, or be attracted to single source. (It is worth comparing the abilities of this simple simulation with the abilities of the earliest simple autonomous robots ([23] and [10]). Making p^* proportional to the intensity of stimulation apparently enables the robot to choose the target with the maximum intensity of light; additionally, if all targets are the same luminance, a single target will be apparently be chosen. (This is a consequence of the stochasticity of the edge cells.

We have analysed the behaviour of simulated robots with onboard excitable lattices for all possible intervals $[\theta_1, \theta_2]$ of cell excitations. It is possible to subdivide the rule space into four basic classes: (i) useless: unsuitable for navigation; robots do not approach a static target at all, interval [1, 2], Figure 2, A; (ii) not adaptive: robots initially approach a static target but then overshoot and collide with the boundaries), intervals $[1,4]\cdots[1,8]$ and $[2,4]\cdots[2,8]$, Figure 2, B; (iii) adaptive: robots successfully follow the target even if it moves, intervals [1, 1], [2, 2] and [2, 3] , Figure 2, C; (iv) clumsy: robots can approach a target but do so in a very nonlinear manner, making wide excursions and oscillations interval [1, 3], Figure 2, D.

The suitability of the excitation rules (measured in distance-to-target units) is shown in Figure 2. Does the performance of the robot controllers as a function of the interval of excitation support a classification of the excitation rules which fits the morphological classification [3] invented by us earlier? This is true to some extent. The class of computationally universal rules ([2, 2] interval) join the unrelated rules which are responsible for chaos-like activity patterns to form the adaptive following class; but all rules that generate one or another kind of waves fall into the class of non-adaptive following. The complete absence or extreme instability of generators of excitation in all members [5] of the adaptive following class are the feature that explains the good behaviour of the robots in this class.

As we have already discussed in our previous papers ([2] and [3]) there are no persistent activity patterns in excitable lattices when $\theta_1 > 2$. In the arrangement discussed here, where the lattice is onboard and stimulated by light, only the cells near edges of the lattices can be excited. A cumulative effect plays a key role in this situation: the robot initially chooses the appropriate direction towards the target, but continues that course and passes the target because it cannot change direction until a sufficient number of new vectors have been generated to overcome the effects of the old vectors. (If this does not happen quickly enough, it may result in a collision with one of the arena boundaries. However, even while stationary it still responds to the illumination from the target, slowly updates the local vectors and then starts to move towards the target again (Figure 2 (E)).

It is clear from these simulations that we badly need a **reset** button to better utilize the wave processes which occur in some important families of excitable media; for the light-sensitive wave generating BZ reaction a reset can be achieved by illuminating the medium by white light for 15-30 sec [14]. Let us look at what happens when a reset operation is used. In the experiments with interval excitable media we used a periodic reset, i.e. all cells of the array are forced into rest states every nth step of the evolution, where n is the linear size of the array. Under this regime, the behaviour of the robot becomes surprisingly uniform for excitation rules - all follow the target adaptively. In all cases the robot reaches the vicinity of the target and then wanders near it (Figure 2(F)). Even if generators of waves are formed, they will be eliminated by the reset operation.

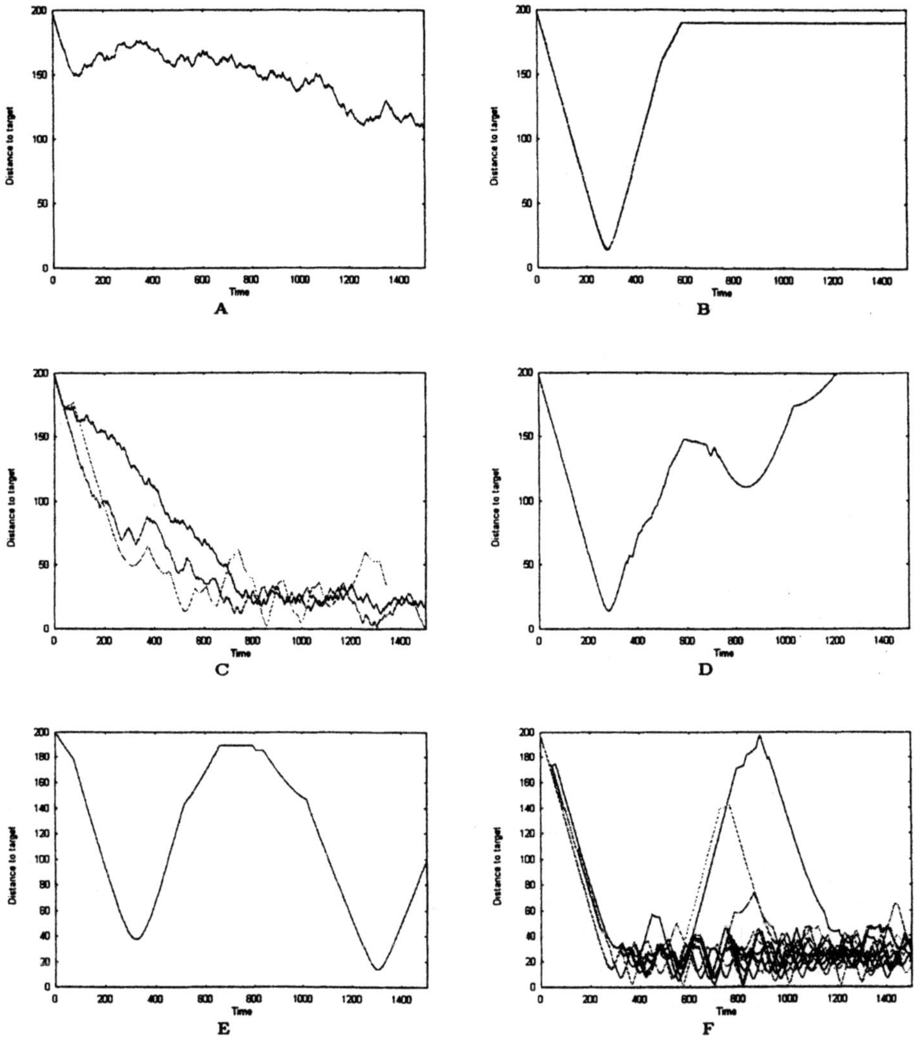

Fig. 2. Performance (measured in distance to target) of the robots controlled by excitable lattice of 50 × 50 cells, $p^* = 0.1$. **A**. Useless class. Single run for $\theta_1 = 1$ and $\theta_2 = 2$. **B**. Not adaptive following. Single runs for $[1,4] \cdots [1,8]$ and $[2,4] \cdots [2,8]$. **C**. Adaptive following. Single run for $[1,1]$, $[2,2]$ and $[2,3]$. **D**. Clumsy behaviour. Single run for $[1,3]$. **E**. Single run for $\theta_1 = 3$ and $\theta_2 = 4$. **F**. Effects of **reset** operation: single runs for $\theta_1 \in [1..2]$, $\theta_2 \in [1..8]$, $\theta_2 \geq \theta_1$. **Reset** operation is applied every 50th step of the simulation time. Two outstanding peaks correspond to $[2,2]$ and $[2,8]$ intervals of excitation.

In evaluating the performance of the simulated robot, it might be useful to bear in mind the abilities of other, perhaps simpler architectures capable of approaching or following a source; for example, Braitenberg's vehicles 2b, 3a, and 3b are obvious candidates [7]. However, our real emphasis here is on showing that the simulated robot is at least capable of following a source, rather than on showing that it does so with any relatively high level of performance, or relative economy of means.

3 Optimal path finding

The problem is how to reach a specified destination without colliding with obstacles and with minimum waste of energy (e.g. in minimal time). We take as granted that the whole configuration of the arena is represented in the reaction space by inducing deviations of the media characteristics from the average. There are many ways to draw pictures with inhomogeneities in a homogeneous medium. The most common ones are the (i) physical compartmentalization of the medium (e.g. construction of impenetrable walls to represent a labyrinth), (ii) projection of the data configuration using light with certain characteristics, and (iii) injection of another reagent into the sites corresponding to the data configuration. In fact, three key problems of computational geometry strongly relate to optimal path finding: (i) approximation of the shortest path; the search for the exit out of a labyrinth is just another interpretation of this problem; (ii) approximation of obstacle free paths; this is computed in the form of the Voronoi diagram of the obstacles; (iii) approximation of the tree connecting all possible destinations; this is computed in the form of the skeleton of the planar polygon vertices, any of which may be possible destinations of the robot.

The Babloyantz-Sepulchre algorithm [6] is the first simulation based work that demonstrated ten years ago the potential of excitable media for some combinatorial optimization problems. In their paper the physical environment is mapped onto an octagonal network of oscillators: obstacless are represented by inactive oscillators and the waves are generated at the destination point. The robot moves towards the locally highest value of one of the concentration variables. The viability of the idea was demonstrated in laboratory experiments ([19]).

In recent experiments [14] a new approach, an alternative to the physically inhomogeneous representation of the environment structure, has been invented. It has been demonstrated that we can map the configuration of an obstacle onto the reaction space without even disturbing the structure of the space. To upload the topology of a labyrinth into the reaction medium it is enough to expose the reaction layer to light radiation, the distribution of intensity of which determines the initial image of the labyrinth. Moreover, we can use phase waves instead of trigger waves. Phase waves are independent of diffusion, and propagate along the phase gradients in an oscillatory medium [14]. Phase waves are faster than trigger waves. It is also possible to explore the so-called light-sensitive phase waves [14]: the evolution of the positive/negative image depends on the intensity of the superimposed background, and so the phase wave virtually travels along the corridors of the labyrinth.

It is possible to build processors for the approximation of the *Voronoi diagram* and the *skeleton* which are based on entirely chemical inputs and optical outputs (see e.g. [20], [4]). The Voronoi diagram of a given planar finite set **P** is defined as a partition of the plane into convex regions, one per element of the set; every convex region is assigned to some element $p \in \mathbf{P}$ of the given set and contains all points of the space that are closer to p than to any other element of **P**. The edges of these convex regions form the Voronoi diagram. A skeleton of a planar shape is a set of centers of bitangent circles lying entirely inside the contour of the shape. After testing numerous possibilities we have found two ways to approximate both structures; both methods can be done onboard while robot is moving. The first chemical controller is made of an agar-palladium thin layer with potassium iodide liquid diffusing in it; the second one is based on the formation of Prussian blue on an agar film. In both controllers the result of the computation is represented by the absence of precipitate, and, as usual, it may be detected optically.

4 Universal controllers and self-localised excitations

By universal controllers we mean the families of nonlinear media in which universal logical gates can be realised; in theory, they will enable us to use any comprehend computational schemes for navigation. The easiest and most practical way to build a universal computer in an excitable medium is to use the paradigm of the billiard ball model developed in [8]: quanta of information are represented by self-localised excitations, or particle like waves, that travel in space and perform computation by interacting with other traveling signals [1]; the idea is orthogonal to conventional techniques of constructing logical gates in chemical computers, which involve constraining reactions in tubes (see e.g. [21]).

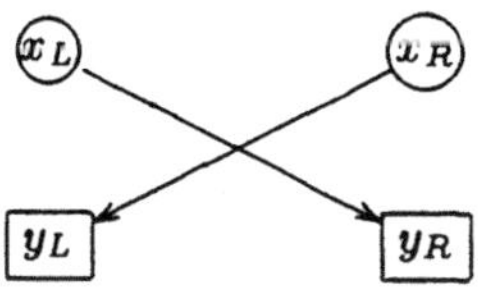

Fig. 3. The principal scheme of the control of a robot by an excitable medium using self-localized excitations. Particle like waves are induced by light stimulation of the sensors x_L and x_R. They travel toward motor units y_L and y_R and transmit energy (excitation) to them. If two waves collide they disappear. So, functions $y_L = x_R \wedge \overline{x_L}$ and $y_R = \overline{x_R} \wedge x_L$ are realised. Assuming constant speed of the wheels of the wheeled robot model without any external stimulation, and incrementing the these speeds by the effects of the particle like waves, we see that the robot changes the direction of movement adaptively, adjusting it to the current direction of the light target.

The basics of universal computation in real and simulated excitable media, together with possible real implementations are discussed in [1]; here, due to lack of space, we can only describe a minimal scheme where self-localised excitations do not merely transmit signals from the sensors to the effector units of the robot but also interact to adjust in the orientation of the robot to a light target (Figure 3).

5 Further steps

We have found that a range of excitable media, independently of their chemical or structural origin, exhibit the types of activity which have the potential to be applied successfully to the problems of robot navigation. Are there any other suitable materials and prototypes that can be also used for the same task? The printed excitable media [18] are one possibility. The catalyst of the BZ reaction is printed on a polysulphone membrane with an ink jet printer; therefore, virtually any pattern of catalyst can be printed. There is a possibility that discreteness-induced self-localized excitation can be produced by printing a lattice, for example. Scheibe aggregates [12] are monolayers that are present in nature in a wide spectrum of photochemical reactions; they can be build as Langumir-Blodgett films in laboratory conditions. These aggregates posses unique transfer properties, namely excitons, which are induced by light, and which travel along the two-dimensional array of molecules like particles (compare with the universal controllers). Worms, or localised structures in binary mixtures of Rayleigh-Bernard convection [22], form another field of opportunities. Depending on their particular states, the worms can stop, continue their motion undisturbed, or selectively disappear as a result of binary collision [22]. Other subjects were sorrowfully left outside the paper. They are metabolic controllers of robot motion [24], and one-dimensional nonlinear media, especially particle machines (e.g. [17] and [11]).

6 Conclusions

Our review of the abilities of chemical systems to solve certain spatial problems has indicated the theoretical possibility of using those systems for the control of a robot. Our simulations of a robot controlled in such a way show that the basic navigational requirements are apparently easy to achieve; we have also noticed a useful relationship between the observed patterns of behaviour, and our existing analysis of parametric descriptions of activity in excitable media. We are continuing this work by undertaking a feasibility study to investigate the construction and experimental assessment of a physical robot using this type of chemical control.

References

1. Adamatzky A. Universal dynamical computation in multi-dimensional excitable lattices *International J. Theor. Physics* **37** (1998) 3069-3108.

2. Adamatzky A. and Holland O. Edges and computation in excitable media In: *Proc. ALIFE IV: 6th Int. Conf. on Artificial Life* (MIT Press, 1998) 379-383.
3. Adamatzky A. and Holland O. Phenomenology of excitation in 2D cellular automata and lattice swarm *Chaos, Solitons and Fractlas* **7** (1998) 1233-1265.
4. Adamatzky A. and Tolmachev D. Chemical processor for computation of skeleton of planar shape *Adv. Mater. Opt. Electron.* **7** (1997) 135-139.
5. Adamatzky A. and Holland O. Parametric classification of local rules of excitation in a lattice: morphology, dynamics and computation *Unpublished manuscript*, 1998.
6. Babloyantz A. and Sepulchre J.A. Front propagation into unstable media: a computational tool In: *Nonlinear Waves Processes in Excitable Media* (A.V. Holden, M. Markus and H.G. Othmer, Eds) (New York and London: Plenum Press, 1991) 343-350.
7. Braitenber V. *Vehicles: Experiments in Synthetic Psychology* (The MIT Press: Cambridge, Massachusetts, 1984).
8. Fredkin E. and Toffoli T. Conservative logic *In. J. Theor. Phys.* **21** (1982) 219-253.
9. Hjelmfelt A., Weinberger E.D. and Ross J. Chemical implementation of neural networks and Turing machines *Proc. Natl. Acad. Sci. USA* **88** (1991) 10983-10987.
10. Holland O. Grey Walter: The pioneer of real artificial life In: *Artificial Life V* (The MIT Press, 1997) 34-41.
11. Jakubowski M., Squier R.K. and Steiglitz K. State transformations of colliding optical solitons and possible application to computation in bulk media *Phys. Rev. E* **58** (1998) 6752-6758.
12. Moebius D. and Kuhn H. Energy transfer in monolayers with cyanine dye Scheibe aggregates *J. Appl. Phys.* **64** (1979) 5138-5141.
13. Rambidi N.G. An approach to computational complexity: nondiscrete biomolecular computing *Computer* **25** (1992) 51-54.
14. Rambidi N.G. and Yakovenchuk D. Finding path in a labyrinth based on reaction-diffusion media *Adv. Mat. Optics and Electronics* (1998) (in print).
15. Rambidi N.G. and D.S. Chernavskii Toward a biomolecular computer. 1. Ways, means, objectives *J. Mol. Electron.* **7** (1991) 105-114.
16. Rambidi N.G., Maximychev A.V. and Usatov A.V. Molecular image processing based on chemical reaction systems. 1. Implementation of Blum-type algorithm *Adv. Materials Optics Electr.* **4** (1994) 191-201.
17. Steiglitz K., Kamal I. and Watson A. Embedded computation in one-dimensional automata by phase coding solitons *IEEE Trans. Comp.* **37** (1988) 138-145.
18. Steinbock O., Kettunen O. and Showalter K.J. Anisotropy and spiral organizing centers in patterned excitable media *Science* **269** (1995) 1857-1860.
19. Steinbock O., A. Toth and K. Showalter Navigating complex labyrinths: optimal paths from chemical waves *Science* **267** (1995) 868-871.
20. Tolmachev D. and Adamatzky A. Chemical processor for computation of Voronoi diagram *Adv. Mater. Opt. Electron.* **6** (1996) 191-196.
21. Toth A. and Showalter K. Logic gates in excitable media *J Chem Phys* 103 (1995) 2058-2066.
22. Tu Y. Worm structure in modified swift-hohenberg equation for electroconvection `patt-sol/9701001`.
23. Walter G. W. An imitation of life *Scientific American* (1950) 42-45.
24. Zigler J., Dittrich P. and Banzhaf W. Towards a metabolic robot control system In: *Proc. 2nd Int. Workshop Information Processing in Cells and Tissues* (New York: Plenum Press, 1997) 305-318.

A Neural Structure for Learning by Imitation

Sorin Moga and Philippe Gaussier

ETIS / CNRS 2235, Groupe Neurocybernetique,
ENSEA, 6, avenue du Ponceau,
F-95014, Cergy-Pontoise cedex, France
{moga, gaussier}@ensea.fr
http://www-etis.ensea.fr

Abstract. In this paper, we present a neural architecture for a mobile robot in order to learn how to imitate a sequence of actions. We show that the use of a representation of the information in a continuous and dynamic way is necessary and the use of the neural fields can be a good solution to control the dynamic of several degrees of freedom with a single internal representation.

1 Introduction

Until now, our work has been mainly focused on the design of a neural network architecture (named PerAc: Perception-Action) for the control of a visually guided autonomous robot. However, the PerAc architecture does not help to solve problems which have an intrinsic high dimension. Therefore imitation of already learned behaviors or subparts of a behavior not completely discovered is certainly one way to allow a population of animals or robots to learn and to find solutions by themselves. Learning by imitation is already used in a few projects of Artificial Intelligence (see [2, 3, 5]). In our previous work [6], we proposed a neural architecture for imitation based on visual information and we shown how to use it to teach the robot to perform a particular sequence of movements (to make a zigzag trajectory, a square ...). In this paper we try to put together 2 ideas: how a PerAc architecture can be used for learning by imitation and how the properties of the neural fields can be used to improve the motor control.

2 Neural network for sequence imitation

For the imitation behavior, we start with the assumption that proto imitation (not intentioned imitation) is triggered by a perception error (see [6] for details) and in Fig. 1 we present an overview of a general PerAc architecture using this principle. The reflex path of PerAc works as a movement tracking mechanism which consists in going towards any perceived movement. The second level of the architecture learns the temporal interval between the successive robot orientations (i. e. a sequence of movements), and associates it to a particular motivation.

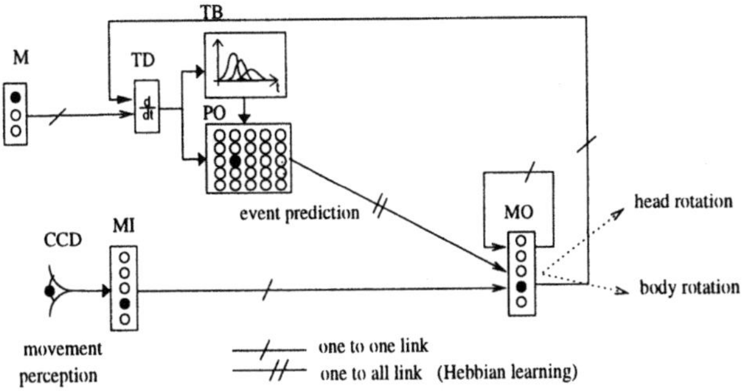

Fig. 1. A general diagram of the PerAc architecture use for learning the temporal aspects of a trajectory. CCD - CCD camera, M - Motivations, MI - Movement Input, MO - Motor Output, TD - Time Derivator, TB - time battery, PO - Prediction Output

A frame-grabber is used to take a sequence of images. In one of our simplest implementation, a "movement image" is the difference between 2 different time integrated images of the above sequence. The perceived movement orientation is computed from the "movement image". The result is one-to-one "projected" on a map of analog formal neurons, the Motor Input (MI) group in Fig. 1. To avoid the perception errors in the tracking mechanism, we allow the robot camera (robot head) to rotate. In this way, the head tries to pursuit the teacher at any time by centering it in its visual field. The robot body turns only if the teacher movement is observed under the same angle for a given time interval. The independent rotation of the robot head and its body can be viewed as a simple two degrees of freedom system. The functioning of the motor group (MO) is quite simple. At each step, a WTA mechanism chooses the most activated neuron, performs the rotation corresponding to this neuron and finishes with a fixed translation. The MO group uses the same information representation as the MI group. It receives the information from both reflex level and event prediction level.

In order to learn a sequence, the student robot detects and learns the transitions in its own body orientation and to be able to reproduce them. The movement rotations characterized by OFF-ON transitions (Time Derivative TD group) of MO neurons are used as input information for a bank of spectral neurons (TB in Fig. 1). Time filter batteries (TB) act as delay neurons endowed with different time constants. As such, they perform a spectral decomposition of the signal that will allow the neurons in the Prediction Output group (PO) to store the transition patterns between two events in the sequence. Finally, the PO group is linked with the MO group via one-to-all modifiable links.

3 An neural dynamics of the motor system

The first limitation in our architecture is the poor stability of the tracking behavior. Even if the temporal integration allows a memory effect, any new input stimulus can generate an immediate change of the head orientation (a classical WTA decision). A second major limitation is the input discrimination. Two or more movement zones can be interpreted as different targets or as the same target due to perception error. In the present system, no interpretation of the perceived movement is performed in order to avoid a misinterpretation. The motor group has to be a topological map of neurons using a dynamical integration of the input information to avoid forgetting the previously tracked target. A dynamical competition has also to be used to avoid intermittent switchings from a given target to another.

We will use the simplified formulation of the neural field proposed and studied by Amari [1].

$$\tau \cdot \frac{f(x,t)}{dt} = -f(x,t) + I(x,t) + h + \int_{z \in V_x} w(z) \cdot g(f(x-z,t))\, dz \qquad (1)$$

Without inputs, the homogeneous pattern of the neural field, $f(x,t) = h$, is stable. The inputs of the system, $I(x,t)$, represent the stimuli information which excite the different regions of the neural field and τ is the relaxation rate of the system. $w(z)$ is the interaction kernel in the neural field activation. These lateral interactions ("excitatory" and "inhibitory") are modeled by a DOG function. V_x is the lateral interaction interval. $g(f(x,t))$ is the activity of the neuron x according to its potential $f(x,t)$. We use a classic ramp function.

G. Schöner [7,4] has proposed to use the properties of the neural field for motor control problems. The "read-out" mechanism consists in the use of the derivate of the neural field activation to compute the motor command. The orientation of the robot head, ϕ_{rob}, relative to a fixed reference is used in the system as a behavioral variable. The state of the system is expressed as a value of this variable. The local maxima of the neural field are named attractors. If the target orientation is ϕ_{tar} (see Fig. 2, a), it erects an attractor in the neural field (see Fig. 2, b) and the robot rotation speed will be $\omega = \dot{\phi} = F(\phi_{rob})$. $\dot{\phi}$ is a function of the current robot orientation, ϕ_{rob}. It sets the dynamics of our robot.

Taken separately, each input erects an attractor in the neural field. The Amari's equation allows the cooperation for coherent inputs associated with different goals (spatially separated targets). For closely spaced input information, the dynamic has a single attractor corresponding to the average of the input information. For a critical distance between inputs, a bifurcation point appears and the previous attractor becomes a repellor and 2 new attractors emerge. Depending on the initial state, the robot switches to one of the 2 new fixed points. This mechanism of input competition / cooperation has an hysteresis properties which avoids oscillations between the two possible behaviors. Another feature of the neural field is the memory. If the parameter h in Eq. (1) has a sufficiently negative value then the neural field operates with a memory effect in which a peak of an attractor has been maintained for a short time interval. A large positive value of h determines a supra-threshold in the neural field activation. We use

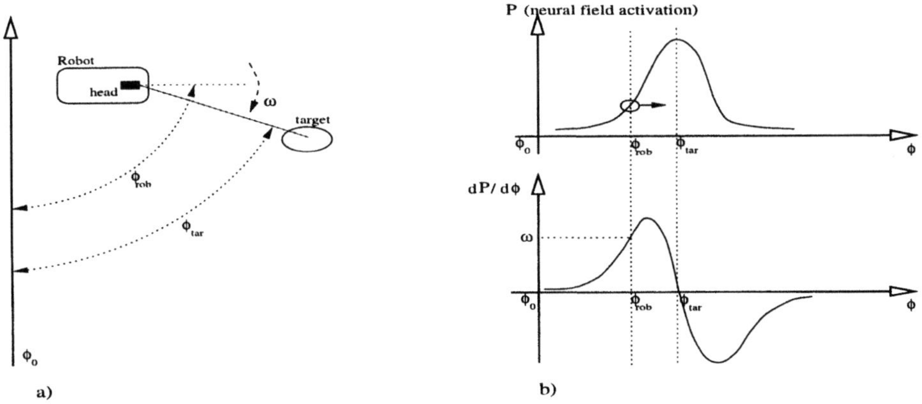

Fig. 2. a) The robot and the target coordinates are represented in the same reference. The reference orientation, ϕ_0 is used to compute ϕ_{rob} and ϕ_{tar}. b) The target position erects an attractor at ϕ_{tar}. The "read-out" mechanism allows to compute the rotation speed ω using the derivate of the neural field activation.

the inputs of the actual system to drive a motor command using a neural field without any modification. Replacing the MO group by a neural field is the sole modification in the architecture (see Fig. 1). All above properties of the neural field come into the general architecture, eliminating the input segmentation and the stability problem of the initial architecture.

4 Experimental results and discussion

At first, we have implemented the tracking reflex using only one degree of freedom, i. e. the robot moves only its head. In order to demonstrate the capabilities of neural field to control several degrees of freedom we take a simple example. The robot follows a "teacher" and learns a sequence of movements ABC. The sequence starts with the activation of the state A (orientation) corresponding neuron. The input in the neural field generates an attractor at the the ϕ_A orientation (see Fig. 3).

At τ moment, the ϕ_B neuron will be activated by the PO group. This activation shifts the attractor to ϕ_B in the neural field. Using the "read-out" mechanisms, we obtain 2 rates of orientation change (due to differences inertia): one for the head orientation and another one for the robot body orientation. In the top of the Fig. 3, we show the variation of head and body orientation as a function of time. According to neural field dynamics, the change of the orientation is continuous. For an external observer, the head orientation anticipates the body orientation (i.e. the inertia of the robot is learned too).

This work is at its beginning. Its interest is in its use of the neural field concept in a PerAc architecture. We show that we can learn the temporal sequence of movements by imitation using a PerAc architecture. The tracking mechanism in the reflex path of PerAc permits the temporal "segmentation" of the "teacher"

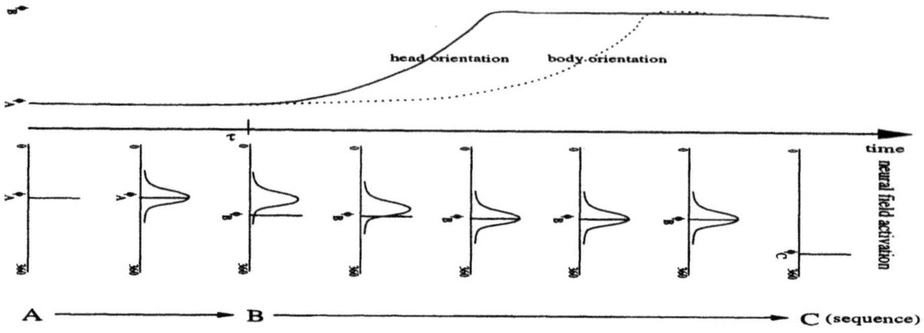

Fig. 3. Top: the temporal variation of the head and of the body orientation. Bottom: the neural field activation for an ABC sequence. The bar represents the predicted movement.

movements without learning to visualize what the teacher is doing or not. The use of the neural field improves the stability of the proto imitation process and permit the discrimination of moving objets in the visual perception field.

References

1. S. Amari. Dynamics of pattern formation in lateral-inhibition type neural fields. *Biological Cybernetics*, 27:77–87, 1977.
2. P. Bakker and Y. Kuniyoshi. Robot see, robot do : An overview of robot imitation. In *AISB Workshop on Learning in Robots and Animals*, Brighton, UK, 1996.
3. L. Berthouze and Y. Kuniyoshi. Emergence and categorization of coordinated visual behavior through embodied interaction. *Machine Learning*, 31(1/2/3):187–200, 1998.
4. E. Bicho and G. Schoner. The dynamic approach to autonomous robotics demonstrated on a low-level vehicle platform. *Robotics and Autonomous Systems*, 21:23–35, 1997.
5. John Demiris and Gillian Hayes. Imitative learning mechanisms in robots and humans. In *Proceedings of the 5th European Workshop on Learning Robots*, Bari, Italy, July 1996. 1996.
6. J.P. Banquet P. Gaussier, S. Moga and M. Quoy. From perception-action loops to imitation processes: A bottom-up approach of learning by imitation. *Applied Artificial Intelligence:An international Journal*, 12(7).
7. G. Schöner, M. Dose, and C. Engels. Dynamics of behavior: theory and applications for autonomous robot architectures. *Robotics and Autonomous System*, 16(2-4):213–245, December 1995.

A Neural Model for the Visual Navigation and Planning of a Mobile Robot

Mathias Quoy[1], Philippe Gaussier[1], Sacha Leprêtre[1], and Arnaud Revel[2]

[1] Université de Cergy-Pontoise - ETIS, 6, av. du ponceau, 95014 Cergy-Pontoise Cedex, France
quoy@u-cergy.fr , http://www.ensea-etis.fr
[2] Ecole Nationale Supérieure de l'Electronique et de ses Applications

Abstract. We present a neural model for the control of an animat. The model is based on two structures. The first one enables visual navigation using landmarks. It may be used in unknown and changing environments. The second structure enables building a proximity map of the environment. Using this map, an animat may successfully reach different goals linked to different motivations and solve various types of action selection problems.

1 Introduction

Taking inspiration from neurobiology and psychology, we develop artificial neural network models for the navigation of a mobile robot. Our research is driven by three different influences. First, our goal is to build a system which allows a robot to "behave" autonomously in an a priori unknown environment (animat approach). Second, we also base our work on a constructivist approach. The robot perception is primarily based on vision through a CCD camera. We do not provide any explicit information about the external world structure. The integration of sensory information allows to create representations of the external world which are based on real data (symbol grounding problem). Finally our third motivation is to validate implications of neurobiological models, and eventually to propose some new hypotheses to explore. We pretend that this approach may boost the learning approach to autonomous robots.
We will first describe our model and then report on our results about the selection between different goals. These results are computer simulations, but we emphasize that part of this work is already running on a real robot experiment.

2 Model

We have developed a model explaining some capabilities enabling to return to a learned place from unexplored locations in the same environment [4,3]. A neurobiological inspiration is the finding of "place cells" in the rat hippocampus [5]. These cells fire when the rat is at a particular location in an area. In our architecture, each "place cell" (or learned location) is coded by a set of couples (landmark, azimuth). We don't provide any external explicit description of the environment (world model), nor learn what to do for each location in the environment. The recognition of a learned location is based on the recognition of the landmark configuration.

When the visual environment of the goal cannot be perceived because of huge obstacles, it becomes impossible to only use place recognition, since the robot always tries to make shortcuts and goes directly to the goal. A special reflex allows however for obstacle following. Yet, the best way would be to learn to make a detour around an obstacle, instead of going towards it, and afterwards follow it (fig. 1). The same drawback appears in an environment with several rooms. As it is, the animat cannot go from one room to the other directly. We need to introduce a kind of map of the environment to be able to perform this task. To represent this map, we add a group of neurons able to learn the relationships between successively explored places. The temporal proximity being equivalent to a spatial proximity, the system creates a topological representation of its environment. We will call this last group our "cognitive map" (or goal map)(fig. 1) [7]. The activation of a motivation induces the activation of a goal in the map which is linked to the satisfaction of this motivation. For instance, the "eating" motivation activates the goals where the animat has discovered food. Then, this activity diffuses on the map beginning from the goals, and activates place cells according to their distance (in number of links) to the goal. First, the animat tries to follow the gradient of neuron activity in this cognitive map to select the next location to reach. The most activated goal or subgoal in the neighborhood of the current animat location is then selected and used to attract the animat in its vicinity. When the animat is close enough to this location, the associated subgoal is inhibited and the animat is attracted by the next subgoal and so on until it reaches the final goal. The algorithm is proved to always find the shortest path in the graph (it is equivalent to the Bellman-Ford algorithm of graph exploration [2]). The principle of this kind of cognitive maps is not new. The novelty in this paper is that our algorithm allows to solve planning problems involving several moving goals in a dynamic environment (the sources may disappear and appear again elsewhere, obstacles may be moved and landmarks hidden). The map is learned and modified on-line and allows to manage contradictory goals. The subgoals correspond to the following situations: the end of an obstacle avoidance (for instance, the animat stores the pathway between two rooms) or places badly recognized. Learning the cognitive map is performed continuously. There is no separation between the learning and the test phases. The links between neurons of the graph are reinforced (hebbian associative learning) for neurons associated with successively recognized places. Let W_{ij} be the weight associated with the fact that from the place i it is possible to reach directly place j, its learning rule is the following:

$$\frac{dW_{i,j}}{dt} = -\lambda.W_{i,j} + (C + \frac{dR}{dt}).(1 - W_{i,j}).\overline{Pot_i^*}.Pot_j^* \quad (1)$$

where Pot_i^* is the value of place cell i. Pot_i^* must be held to a non null value until Pot_j^* (with $i \neq j$) is activated by the recognition of the place cell j. This is performed by a time integration of the Pot_i^* values represented in the

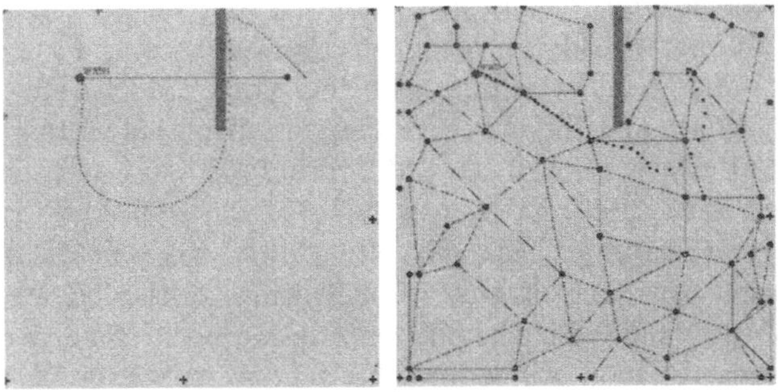

Fig. 1. On the left, reflex behavior of obstacle following for reaching the goal (initial speed towards the upper left. The animat has a mass giving a momentum). The dots indicate the trajectory followed by the animat. On the right, cognitive map built by exploration of the same environment. The landmarks are the crosses on the border. Each circle is a subgoal. The links indicate that the two subgoals have been activated in succession. The subgoals and the learned transitions form the goal map.

equation by $\overline{Pot_i^*}$. $\overline{Pot_i^*}$ decreases with time and can be used as a raw measure of the distance between i and j. λ is a very low positive value. It allows forgetting unused links. $C = 1$ in our simulations. The term $\frac{dR}{dt}$ corresponds to the variation of an external reinforcement signal (negative or positive) that appears when the animat enters or leaves a "difficult" or "dangerous" area.

The way it is constructed, the planning map fits very nicely in the "neural field" framework [1]. Indeed, one can consider that the goal has a neural field extending over each subgoal. The same way it is possible to control the heading direction of a robot [6] by controlling its angular velocity, we control the position of the animat through modifications of its speed introduced by the gradient. Roughly speaking, the speed is a function of the distance to the goal. Hence the planning map not only indicates where the goal is, but also enables to control how to reach it (top-down process).

3 Learning to choose between goals

We will work on the following experiment. The animat has two different energy levels called "food" and "water". They are linked with two motivations "hunger" and "thirst". As the animat moves, the two energy levels decrease. Through exploration, it has to discover "food" and "water" places where it may re-supply one of the two levels. The animat performs a random exploration of the environment in order to discover the resource places. However, we want it to *go back* to a previously discovered interesting place when it needs to. The animat must therefore *solve the dilemma between exploring and reaching a known food and/or water place.* In the simulated experiment, the perception of the animat is limited to landmarks and obstacles. The supply level of a resource place is decreased by the amount taken by the animat. When the source is empty, an other one appears randomly in the environment. Hence, the animat always has to explore the environment in order to find new potential sources. The environment is a T-maze. The animat already knows where the sources are. The cognitive map follows the middle of the corridors, and does not introduce any a priori bias in its use. The maze and the cognitive map are displayed as the first figure of each experiment shown below it. In addition to food and water, there is a nest. The need to go to the nest increases twice as fast as the need for food or water. Hence the animat has to go back very often to its nest which is in the bottom arm. In the right arm there is only water whereas in the left arm the animat can find both water and food. The difference between the two experiments is the distance between the food source and the right water source. In the first part of the experiment, where the food source is only slightly shifted to the left, the preferred water source remains the right one. But as it is shifted more to the left, it is the right water source which eventually is the most used one. Because the right water source is nearer to the nest than the left one, the animat goes in the right arm to drink. When it gets hungry, it goes in the left arm to find the food. Since it is the only food supply, the path between the nest and this source is reinforced very often. A large part of this path is also common to the way to the left and right water sources. The animat has to go more often in the left arm to go to eat, so when it goes exploring, it is more likely (in fact has more time) to go in the left end than in the right one which is farther. Hence the links between the food source and the left water source are reinforced. So when the animat is thirsty while being near the food source, it goes to the left food source. And at last, as the path between the nest and the left food and water sources is being reinforced, even when the animat is near its nest, it goes to the left for drinking. Only the occasional exploration of the right arm may reactivate the links towards the right water source. But the process only begins at the same initial point again, leading eventually to the reinforcement of the links in the left arm. The animat succeeds the task in the absence of any explicit reinforcement unlike what must be done with Q-learning for instance (which also seems to have a lot of problems to work on a problem with such a number of dimensions – at least 6 continuous dimensions: x, y, time, and 3 motivations!).

4 Discussion

The navigation and planification system we have presented is able to solve complex action selection problems. However, for the moment, it has to really perform the movements in order to reinforce particular paths. An improvement would be the possibility to internally replay the trajectory used to reach a goal.

The main drawback of the algorithm is the computation of the gradient of the neuron activity. Indeed, in environments with a great number of subgoals, it may be very small. Hence, if there is a small noise on the gradient, it would be now impossible to follow it. So there is a need to combine several maps of different joint environments. This means we have to define a planning structure, or plans of plans, in order to address large scale environments.

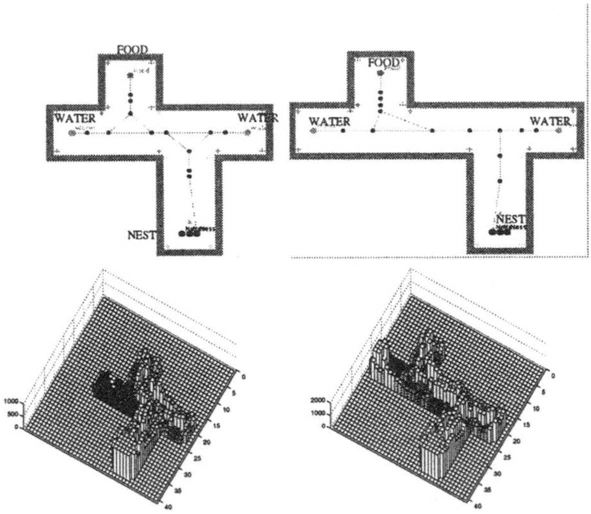

Fig. 2. The figures on the top display the initial mazes. Below, the height of each square shows the number of time it has been occupied by the animat (dark (resp. light) color indicates low (resp. high) occupation. The values have been computed adding 50 different runs of 20000 iterations. In the first experiment (left figure), the animat visits the right source more often than the left one. In the second experiment, the water source on the left, near the food source, is more often used than the water source in the right arm.

This work was supported by a French GIS contract on Cognitive Sciences entitled "comparison of control architectures for the problem of the action selection" in collaboration with the Animat lab (J.Y. Donnart, A. Guillot, J.A. Meyer), LISC (G. Deffuant) and RFIA (F. Alexandre, H. Frezza).

References

1. S. Amari. Dynamics of pattern formation in lateral-inhibition type neural fields. *Biological Cybernetics*, 27:77–87, 1977.
2. R.E. Bellman. On a routing problem. *Quarterly of Applied Mathematics*, (16):87–90, 1958.
3. P. Gaussier, S. Leprêtre, C. Joulain, A. Revel, M. Quoy, and J.P. Banquet. Animal and robot learning: experiments and models about visual navigation. In *Seventh European Workshop on Learning Robots*, Edinburgh, 1998.
4. P. Gaussier and S. Zrehen. Perac: A neural architecture to control artificial animals. *Robotics and Autonomous Systems*, 16(2-4):291–320, 1995.
5. J.O'Keefe and N. Nadel. *The hippocampus as a cognitive map.* Clarendon Press, Oxford, 1978.
6. G. Schöner, M. Dose, and C. Engels. Dynamics of behavior: theory and applications for autonomous robot architectures. *Robotics and Autonomous System*, 16(2-4):213–245, December 1995.
7. E.C. Tolman. Cognitive maps in rats and men. *The Psychological Review*, 55(4), 1948.

Progressive Construction of Compound Behavior Controllers for Autonomous Robots Using Temporal Information[1]

J. A. Becerra[1], J. Santos[1] and R. J. Duro[2]

[1] Dpto. Computación, [2] Dpto. Ingeniería Industrial, Universidade da Coruña, Spain
{ronin, santos}@dc.fi.udc.es, richard@udc.es

Abstract. In this work we present a methodology for the progressive construction of compound behavior controllers for real autonomous robots. Some of these behaviors require temporal processing which is achieved through the inclusion of temporal delays in the synapses of the artificial neural networks used for their implementation. Starting from a set of simple behaviors implemented by means of evolved monolithic controllers, the evolution strategy employed obtains behaviors in higher levels either choosing the necessary low level behaviors from the previously selected set or through the coevolution of part of the low level behaviors and the higher level one. Emphasis is placed on making the behaviors robust and capable of performing in a real robot.

1 Introduction

When but a few behavior modules are required in order to implement a robot behavior controller, as Cliff et al. [1] point out, the complexity in a design scales with the number of possible interactions among modules. If the behaviors and their interconnections are designed by hand, two problems arise, one due to the complexity problem mentioned and the other to the fact that the hand-designed behaviors are not necessarily the best, or, in some cases even adequate for the task.

In the late eighties and early nineties artificial evolution was proposed as a means to automate the design procedure of these types of systems [1][2]. Many authors have taken up this issue and have developed different evolutionary mechanisms and strategies in order to obtain robotic controllers.

When designing a complex behavior, two options are possible: a monolithic approach or a modular approach [3]. The advantage of the monolithic alternative is that it is not necessary to have prior knowledge about possible sub-behaviors and the interrelations between them. In hierarchical modular architectures [4][5] behaviors can be reused, thus allowing for the introduction of previously acquired competences and the individual modules are usually simpler. The problem that arises is that the behavioral decomposition is not clear in every case.

In the work presented here, we have tried to combine, in a practical way, the monolithic approach and a hierarchical modular structure so that complex behaviors

---

[1] This work was funded by the Universidade da Coruña and the CICYT (project TAP98-0294-C02-01).

could be generated automatically but took into account the experience accumulated through the implementation of previous behaviors. Specifically, using this method, a designer provides the system with whatever behaviors he has or decides that may be useful. This initial set need not be complete and may include many unnecessary behaviors. When obtaining the higher level controller, the evolution process will select those lower level behaviors from the initial set that are useful in order to perform the task assigned and will ignore the rest. If some part of the global behavior cannot be obtained through the interconnection of the available modules, a new monolithic module that handles this part will be co-evolved with the global controller.

2 The Robot and the Simulation/Evolution Environment

In the examples we present throughout this article, the robot employed is a "Rug Warrior". It is a small (18.5 cm diameter), simple and cheap circular robot. It only has two DC motors, two binary infrared emitters and one receiver, two photosensors, a pyrosensor, three bumper sensors and two optical encoders. The sensors and actuators are very low quality, very noisy and imprecise, which is a plus when trying to obtain robust behaviors in the hardest possible conditions. The robot is untethered and large enough for it to be able to move around freely in a human environment, such as an office or a laboratory. This is a great advantage with respect to other robots often employed in evolutionary robotics experiments (such as the Khepera robot).

Regarding the simulation/evolution environment, we have selected as basic evolutionary procedure an evolution strategy [6][7]. Some authors such as Salomon [8] have employed this alternative. The main advantage of this approach to evolution is that it aids in problems where the level of epistasis, and consequently, of deceptivity is high. In our case, the original structure of evolution strategies has been adapted to the development of ANN based robot controllers by uniformly distributing the initial population throughout the search space and subdividing it into races to prevent the clustering that arises when a random initial distribution of the low fitness individuals of small populations is employed. In addition, we have made use of a tournament selection scheme, adding a small probability of crossover (typically 0.1).

The evolutionary algorithm can evolve a single low level behavior, a high level behavior that makes use of the low level behaviors we provide it with or co-evolve both simultaneously. The encoding scheme employed for the Neural Networks that implement the controllers, is just a direct genotypic representation of the phenotype in terms of synaptic weights and delays when appropriate. Regarding the fitness criteria we employed a global energy based fitness criteria.

The simulation module is based on the Khepera simulator [9] and adapted to the Rug Warrior. The response of sensors and actuators employed in the simulator were obtained by measuring the real response of the real sensors and actuators of the robot under different conditions such as different levels of ambient light, etc.

For a simulation to be appropriate for the transference of behaviors developed using it to the real world, it must meet some criteria, such as those established by Jakobi [10], which usually imply handling different levels and types of noise. In addition to the traditional random noise applied to values sensed and actuator commands, our simulations include the following three types of noise:

- Generalization noise. It consists in randomly changing some characteristic of the robot or the environment (such as the orientation of a sensor) and maintaining this change for one whole life of the robot. This is necessary, for instance, if the robot is working on battery power, as different levels of battery charge imply different speeds for the wheels of the robot when presented with the same command.
- Systemic noise. A specially important type of noise is noise that has to do with defects or particular traits in the operation of the real robot.
- Temporal noise. This type of noise is used in order to obtain behaviors that are tolerant to variations in the time elapsed between events.

Obviously, noise makes obtaining a given behavior slower and more difficult. It may cause the environment to be perceived very differently in each evaluation of the robot, forcing the robot to obtain compromise conservative solutions.

3 Constructing Compound Behaviors

In this section we will consider the progressive construction of a global "homing" behavior. In the examples presented here, when temporal processing was required, we have made use temporal delays in all or part of its synapses. These delays are a crude model of the length of the synapses found in nature.

Initially, we have developed four controllers for the robot, that is: monolithic wall following, co-evolved compound wall following using a escape from collisions module, monolithic homing and compound homing using all the previously evolved controllers.

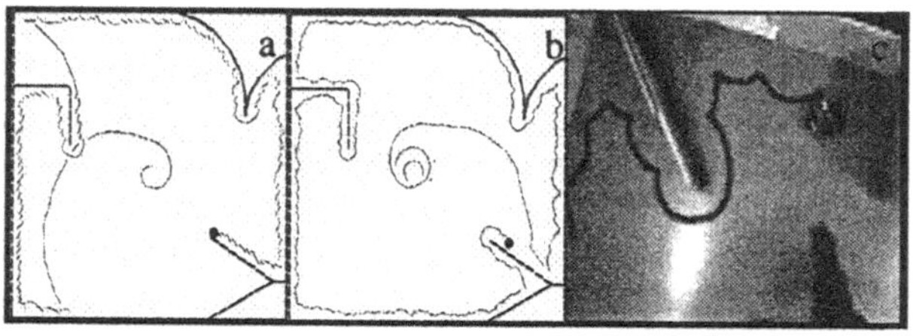

Fig. 1. Wall following behavior. a) Monolithic controller, b) Compound behavior (monolithic + escape from collisions), c) On the real robot.

With regards to wall following, in figure 1.a we display the operation of the robot controlled by a monolithic controller obtained in 160 generations of an evolutionary process with 0.1 crossover probability and 4 races with 64 individuals each. The simulations included all the noise types commented before and the controller is an ANN with time delays in the synaptic connections of the first layer. The robot seems to oscillate very much in this behavior. This is because, as it must try to avoid crashing at all costs, it needs to explore its environment very closely. Even then, sometimes it will crash and either get stuck as shown in figure 1.a or become unstuck with lots of friction and thus come out disoriented as

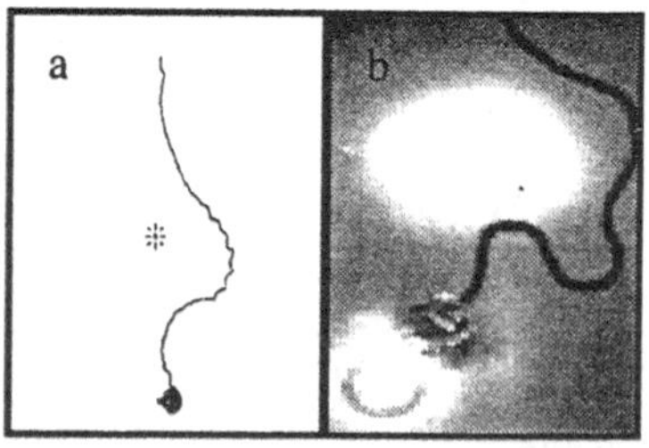

Fig. 2. Homing behavior. a) Simulation b) Real robot.

shown in the top left corner of the same picture. In this second case, the robot lost its bearings after crashing and had to search for the wall all over again.

If we consider the hierarchical approach where the designer only provided a escape from collisions module and allowed the other low level module to be co-evolved with the higher level wall following controller the robot oscillates a lot less (figure 1.b), but crashes a little more often (without getting stuck). In figure 1.c we present the operation of a real robot using the same controller.

To obtain a monolithic homing controller we prepared an environment with no walls, facilitating the search for home (a flashing light), but with a trap that is very similar to home (static light). The robot will require temporal information in order to distinguish one object from the other. During the evaluation of the robot, both the trap and home are randomly positioned. The evolution took around 600 generations and each robot was evaluated 16 times each generation. An order of magnitude less generations are required if no temporal noise is used, but the results are not useful in real robots. The most difficult scenario in order to test the monolithic controller would be to have it confronted with a situation like the one presented in figure 2.a, where the static light (the trap) masks the flashing light (home) and thus, in order to go home, the robot must surround the trap without falling into it which is what the robot does in simulation. The correspondence with reality can be seen in the picture of figure 2.b.

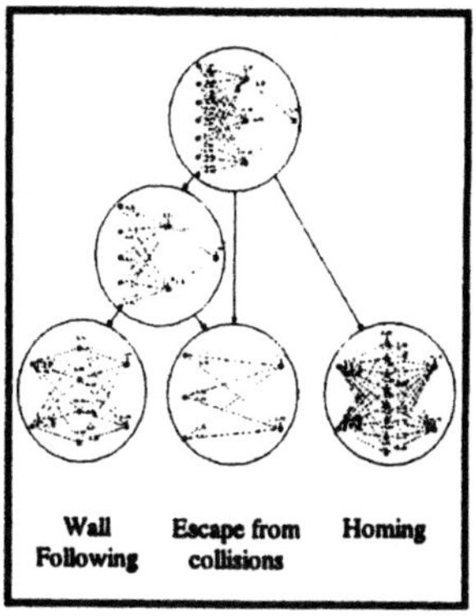

Fig. 3. Three level compound homing architecture.

After obtaining the previous behaviors, we are now going to allow the evolution process to make use of previously acquired competences in order to obtain a more complex behavior through a higher level controller that selects lower level ones. The objective is the same as used for monolithic homing, that is, the robot must seek home avoiding fixed lights, but in this case the flashing light is hidden and the environment is not free from obstacles. To obtain a monolithic controller for this task is difficult and time consuming, and the network that implements it is necessarily quite complex.

Figure 3 shows the three level architecture we evolved for this behavior, consisting of the monolithic homing behavior, a escape from collisions behavior, a two level wall following behavior, that shares the escape from collisions controller and a high level module that coordinates them. The operation of this controller is shown in figure 4. If the environment is relatively free from obstacles, the robot uses a escape from collisions behavior which allows it to explore its world relatively fast. When the environment is more complex, such as in the case presented in the figure, this type of behavior is not efficient, as the probability of the robot going into one of the boxes containing home is quite low. In this case, the robot makes use of either a wall following behavior or the lower level monolithic homing behavior allowing for a sort of bad wall following behavior that permits a faster exploration of the wall contours.

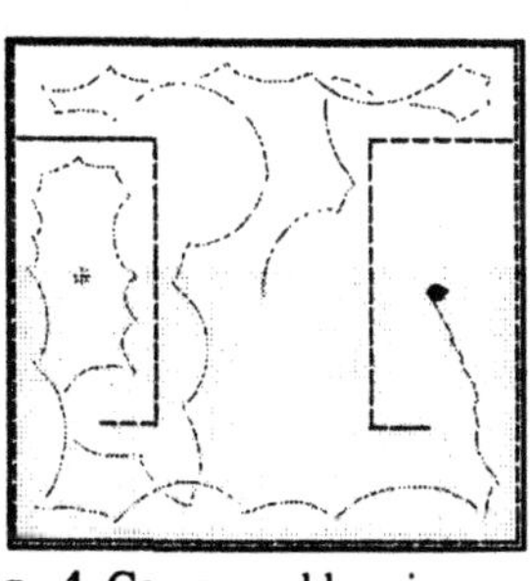
Fig. 4. Compound homing behavior. Home is in the right box and the trap in the left one.

4 Conclusions

In this article we have presented a structured way of obtaining compound behaviors in an infrasensorized robot with noisy sensors operating in real environments. This implies a hierarchical structure where the different behaviors are implemented through ANNs, some of them using temporal delays to consider time related events. Through this structure, the controllers may be used in multiple compound behaviors without duplicities, and, as new behaviors are evolved, this evolution becomes faster and simpler due to the fact that a lot of experience in the form of previously evolved behaviors is available. In order to prevent the problem of the designer having to be exhaustive in its determination of all the necessary lower level behaviors, we have included the possibility of cooperatively coevolving lower and higher level behaviors. To bridge the reality gap several types of noise were employed and, as some behaviors require temporal processing, a type of noise that implied variations in the temporal positions of the events the robot perceives in simulation was used. The controllers evolved in simulation directly transferred to the real robot.

References

1. Cliff, D.T., Harvey, I., and Husbands, P.: Explorations in Evolutionary Robotics. Adaptive Behavior, Vol. 2, (1993) 73-110.
2. Beer, R.D. and Gallagher, J.C.:Evolving Dynamical Neural Networks for Adaptive Behavior. Adaptive Behavior, Vol. 1, No. 1, (1992) 91-122.
3. Nolfi, S.: Evolving Non-Trivial Behaviors on Real Robots: A Garbage Collecting Robot. Robotics and Autonomous Systems, Vol. 22, (1997) 187-198.
4. Brooks, R.A.: Achieving Artificial Intelligence through Building Robots. A.I. Memo 898, MIT, AI Lab. (1986).
5. Colombetti, M., Dorigo, M., and Borghi, G.: Behavior Analysis and Training - A Methodology for Behavior Engineering. IEEE Transactions on Systems, Man and Cybernetics, Vol. 26, No. 3, (1996) 365-380
6. Rechenberg, I.: Cybernetic Solution Path of an Experimental Problem. Library Translation 1122, Royal Aircraft Establishment, Farnborough, UK. (1965)
7. Schwefel, H.P.: Evolutionsstrategie und Numerische Optimierung. Ph. D. Thesis, Technische Universität Berlin. (1975).
8. Salomon, R.: The Evolution of Different Neuronal Control Structures for Autonomous Agents. Robotics and Autonomous Sys., Vol. 22 (1997) 199-213
9. Mitchel, O.: Khepera Simulator Package version 2.0, Freeware Mobile Robot Simulator. downloadable from http://wwwi3s.unice.fr/om7khep-sim.html. University of Nice Sophia-Antipolis, France (1996)
10. Jakobi, N.: HalfBaked, AdHoc and Noisy Minimal Simulations for Evolutionary Robotics. Fourth European Conference on Artificial Life, P. Husbands and I. Harvey (Eds.), MIT Press (1997).

A Visualization Tool for the Mini-Robot Khepera: Behavior Analysis and Optimization

Axel Löffler, Jürgen Klahold, Manfred Hußmann, and Ulrich Rückert

Heinz Nixdorf Institute, System and Circuit Technology, Paderborn University,
Fürstenalle 11, D-33102 Paderborn, Germany,
loeffler@hni.uni-paderborn.de,
http://wwwhni.uni-paderborn.de/fachgruppen/rueckert/KHEPERA

Abstract. The design of behavior generating control structures for real robots acting autonomously in a real and changing environment is a complex task. This is in particular true with respect to the debugging process, the documentation of the encountered behavior, its quantitative analysis and the final evaluation. To successfully implement such a behavior, it is vital to couple the synthesis on a simulator and the experiment on a real robot with a thorough analysis. The available simulator tools in general only allow behavioral snapshots and do not provide the option of online interference. In order to cure these shortcomings, a visualization tool for *aposteriori* graphical analysis of recorded data sets which gives access to all relevant internal states and parameters of the system is presented. The mini-robot Khepera has been chosen as experimentatory platform.

1 Introduction

The design of behavior generating control structures for real robots acting autonomously in a real and changing environment is a complex task. In [12] the usefulness of embodiment for robotics is comprehensively pointed out as algorithms developed by sole simulation of autonomous agents in restricted and controlled environments may fail when transfered to a real system. Nevertheless, simulators as in [9] are very useful to obtain a primal executable version of a control structure generating the desired behavior. On the other hand, especially due to the fact that, in general, only momentary snapshots of the encountered behavior are available through these tools, a judgment of the behavioral dynamics is made very difficult. Moreover, the option of online interference concerning for example parameter variations is naturally minimal. For this, a completely new test run is required without the possibility to compare two instances directly. Coupling of synthesis and analysis in a feedback loop leads to an evolutionary process between implemented behavior and designer's knowledge. Therefore, we propose to extend the common two step program in behavior design (create a satisfying simulator solution, then adapt it to the real robot case) by a third stage, an *aposteriori* holistic analysis of the encountered behavior (Fig. 1). The latter requires the development of appropriate software tools as the one presented in this paper. Our basic approach for the implementation of behavior generating control structures for autonomous agents follows the broad outline of the so-called *nouvelle Artificial Intelligence* [2]. Complex behavior is produced by the interaction

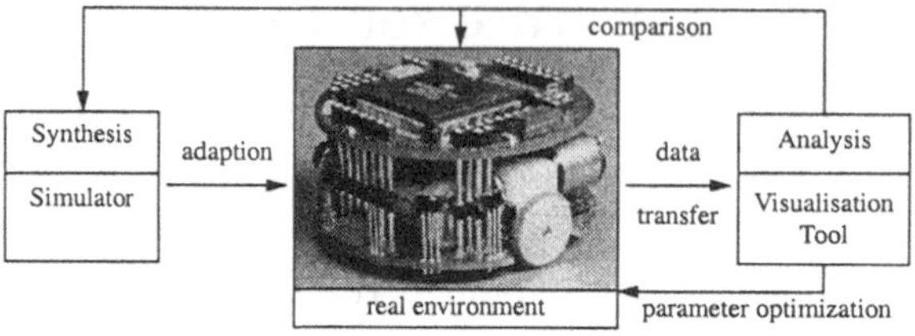

Fig. 1. The proposed design cycle with a photography of the mini-robot Khepera [3].

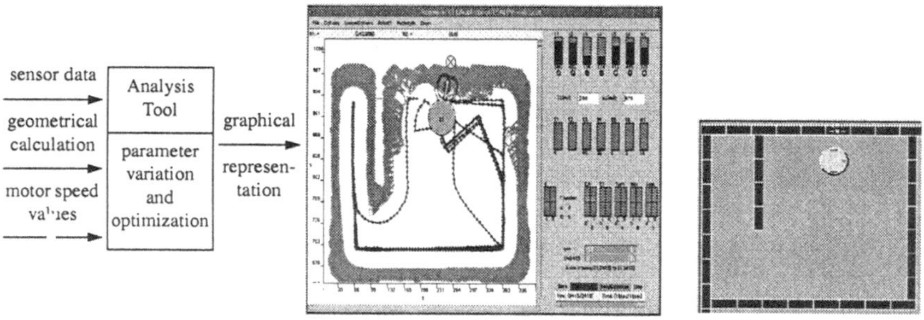

Fig. 2. A schematic of the input data flow of the visualization tool with a full-scale view of its graphical representation is shown together with the corresponding simulated environment ($40\,cm \times 35\,cm$, total runtime: 18 sec). The control panel (bottom right) of the visualization tool permits to select a specific part of the data set in question, to start or stop the visualization at a particular point and to view the data stepwise.

of 'simple' modules active in parallel like the Braitenberg patterns [1], artificial neural networks, geometric calculations and rule-based reasoning without relying on extensive and resource inefficient internal representations of the environment. The basic modules are either hardwired or adaptive. In the first case, one may rely on genetic algorithms and evolutionary programming [4]. An alternative is to consider hand-designed algorithms, wherein the optimization process is left to the designer. This approach usually leads to the design of a more complex and easier-to-understand behavior. Secondly, one may endeavor the implementation of adaptive, self-organizing structures. Here, either online learning [7], [13], which allows to adapt to an unknown environment, or off-line training [8], which - although limited in the case of a changing environment - in general produces more accurate results, are conceivable. Eventually, [5], [6] and [10] may be seen as complementary to the presented approach. Firstly, evolutionary robotics instead of hand design are used to build the controllers. Secondly, the reality gap between simulation and real robot behavior is bridged not by *aposteriori* analysis, but by an *apriori* set up of valid simulations. Common to all these approaches is the need for data documentation permitting a quantitative analysis, which has prior only been possible through simulations [11]. By means of the graphically represented data, an evaluation of the encountered behavior becomes possible. Hence, the designer is able to conduct an optimization process by appropriately varying the system and training parameters without directing a considerable number of test runs. The visualization tool (Fig. 2) described in the following section is conceived to actively support this process.

2 Building a dynamical view of the environment

The main purpose of the presented software tool is to build a holistic view of the robot's environment, which may be subject to dynamical changes. Note that the perceived distribution of obstacles is purely subjective, i.e. derived from the sensory-motor data of the robot. To stay abreast of environmental changes, special features (Fig. 2) permit to suppress obsolete parts of the encountered obstacle history. Every navigational tasks like transport or homing is based on a position calculating process. Since the mini-robot Khepera in its basic configuration does not dispose of far-ranging active sensors necessary to obtain topological information, we apply an explicitly geometrical method. The current position is obtained by odometrical path integration using the incremental encoder values n_L, n_R as variables and the wheel distance d as well as the advancement per pulse Δl as system parameters. Since the designer knows the exact geometry of the real environment, one may compare the robot's perception to reality. This is demonstrated by an experiment using an environment of rectangular geometry ($75\,cm \times 60\,cm$). The robot follows the walls for roughly two rounds (92 sec, i.e. 1126 program cycles) and stops at an internal angle of 71.7° instead of 90° (correct). The following results concerning angular errors due to parameter variations (original ones: d=52mm, Δl=0.08mm) have been obtained by means of the visualization tool (Table 1,2):

Δl [mm]	0.076	0.077	0.078	0.079	0.080	0.081	0.082	0.083	0.084
$\Delta\alpha$ [°]	18.6	9.4	0.1	-9.1	-18.3	-27.6	-36.8	-46.0	-55.2

Table 1. Angular error$\Delta\alpha$ by variation of Δl.

d [mm]	48	49	50	51	52	53	54	55	56
$\Delta\alpha$ [°]	-79.8	-63.5	-47.9	-32.8	-18.3	-4.4	9.0	21.9	34.4

Table 2. Angular error $\Delta\alpha$ by variation of d.

Thus the presented tool may be used to optimize the odometrical parameters for individual real robots. Moreover, it gives the designer a hint of the time interval during which the position calculation system works sufficiently well. This may be seen through another experiment, where the same environment as before, but with an additional light source has been used. An *event* is defined by locating a light source. After the first detection, the robot was able to recognize the light source five times, then registered it as a different one (a confidence area of 20cm in Manhatten distance concerning the light source position was used). Hence, the positioning system worked well during 297 sec, i.e. 3825 update cycles. As already mentioned, the infrared proximity sensors are used to construct an estimation of the current environmental structures. Moreover, these data also form the sensor space on which different exploration modules like obstacle avoidance (OA), edge following (EF), turning (Turn) and point-to-point navigation (Nav) work in parallel, which are visualized in the control panel (Fig. 3). The spatial form of the sensor characteristics may be customized in order to simulate sensor degradation or the use of other kinds of proximity sensors. Furthermore, an artificial neural network has been applied for extracting a symbolic angle-to-light source information from the subsymbolic sensor data stream (Fig. 4). Eventually, a comparison between the simulator (Fig. 2) solution and the real ro-

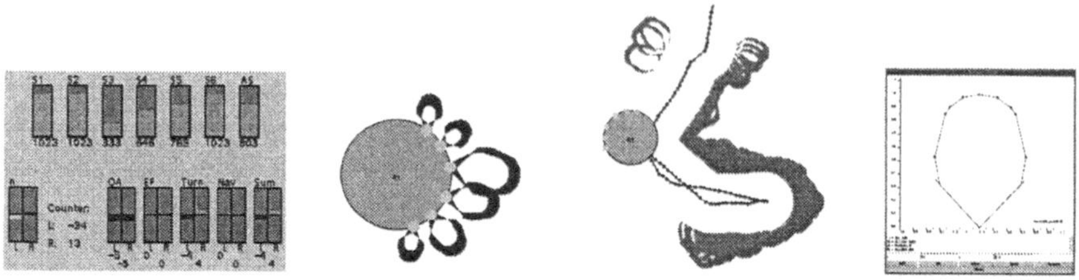

Fig. 3. Visualization of the infrared sensor values (S1-6, A[verage]S), the speed values determined by different software modules and the values from the incremental encoders n (left). The possible positions of obstacles, which are causing this specific sensor input, are drawn around the robot (middle). A high sensor value corresponds to a nearby and unextended obstacle whereas a low one indicates a dislodged object. The average sensor noise is taken into account by the thickness of the traits. Finally, the sensor trait may be customized (right).

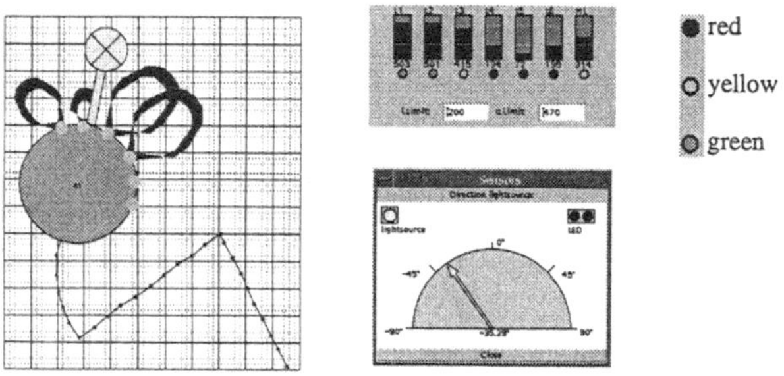

Fig. 4. Visualization of the light sensor values (top middle) and the artificial neural network (bottom middle). It is possible to feed different thresholds (top middle/right) for analyzing the photo-sensitivity of the implemented algorithms.

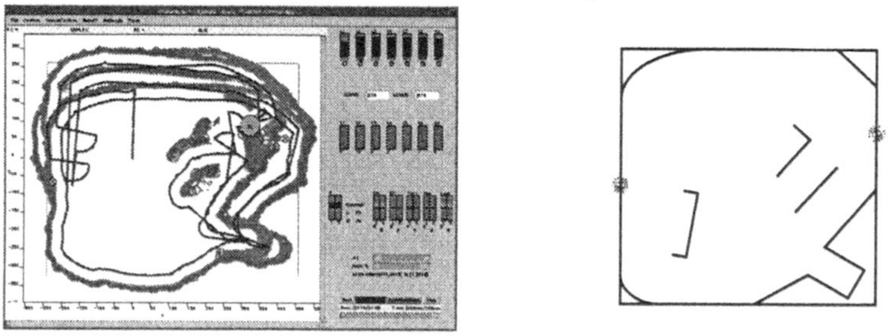

Fig. 5. A complete data set obtained from a real Khepera acting in a real environment is shown with a schematic of the corresponding actual obstacle structure ($80\,cm \times 80\,cm$, total runtime: 208 sec).

bot's implementation (Fig. 5) of the Dynamical Nightwatch's Problem [8] reveals a shortcoming that is inherent to all simulations: the position determination is supposed to be perfect, which is necessary for correct representation. Note that in simulations, path integration is usually not used. In contrast, for real systems, it still presents the main position calculating method. Hence, the most restricting limitation to the real robot's performance arises from the erroneous position calculation due to wheel slippage and unknown fabrication tolerances. As demonstrated above, the presented visualization tool is particularly apt to assist in the minimization of errors due to the latter.

3 Conclusion

The presented visualization tool for the mini-robot Khepera permits the extension of the common two step design program for behavior generation by a third stage: an *aposteriori* analysis of the encountered behavior. This allows to set up a synthesis-analysis feedback loop evolutionarily improving the envisaged behavioral design. In particular, the holistic visualization of the encountered behavior enables the designer to thoroughly document, analyze, evaluate and compare the performance of his implementation.

Acknowledgments. We are very grateful to Prof. G. Domik whose lecture on visualization motivated us to undertake the presented work. In addition, the corresponding author is supported by the DFG-Graduiertenkolleg "Parallele Rechnernetzwerke in der Produktionstechnik", GRK 124/2-96.

References

1. V. Braitenberg. *Vehicles: Experiments in Synthetic Psychology.* MIT Press/Bradford Books, 1984.
2. R. A. Brooks. Intelligence without Representation. *Artificial Intelligence*, 47:139–159, 1987.
3. Laboratoire de Microinformatique l'Ecole Polytechnique Fédérale de Lausanne. Mobile Robots as Research Tools: The K-Robot Family. http://diwww.epfl.ch/Khepera/, 1998.
4. D. Floreano and F. Mondada. Evolutionary neurocontrollers for autonomous mobile robots. *Neural Networks*, 11:1461–1478, 1998.
5. N. Jakobi. Evolutionary Robotics and the Radical Envelope of Noise Hypothesis. *Journal Of Adaptive Behaviour*, 6, 1997.
6. N. Jakobi. The Minimal Simulation Approach To Evolutionary Robotics. In *Proc. of ER'98*, AI Systems Books, 1998.
7. D. Lambrinos. Navigating with an Adaptive Light Compass. In *Proc. of 3rd Eur. Conf. on Art. Life*, volume 929 of *Lecture Notes in Artificial Intelligence*, pages 602–613. Springer, 1995.
8. A. Löffler, J. Klahold, and U. Rückert. The Dynamical Nightwatch's Problem Solved by the Autonomous Micro-Robot Khepera. In *Proc. of 3rd Eur. Conf. on Art. Evolution*, volume 1363 of *Lecture Notes in Computer Science*, pages 303–313. Springer, 1998.
9. O. Michel. Khepera Simulator 2.0, Webots, 1999. http://diwww.epfl.ch/lami/team/michel/khep-sim/ and www.cyberbotics.com/.
10. O. Miglino, H. H. Lund, and S. Nolfi. Evolving mobile robots in simulated and real environments. In *Proc. of 3rd Eur. Conf. on Art. Life*, volume 929 of *Lecture Notes in Artificial Intelligence*, pages 417–434. Springer, 1995.
11. A. G. Pipe, B. Carse, T. C. Fogarty, and A. Winfied. Learning Subjective "Cognitive Maps" in the Presence of Sensory-Motor Errors. In *Proc. of 3rd Eur. Conf. on Art. Life*, volume 929 of *Lecture Notes in Artificial Intelligence*, pages 463–476. Springer, 1995.
12. E. Prem. Grounding and the Entailment Structure in Robots and Artificial Life. In *Proc. of 3rd Eur. Conf. on Art. Life*, volume 929 of *Lecture Notes in Artificial Inteligence*, pages 39–51. Springer, 1995.
13. C. Scheier and R. Pfeifer. Classification as Sensory-Motor Coordination, A Case Study on Autonomous Agents. In *Proc. of 3rd Eur. Conf. on Art. Life*, volume 929 of *Lecture Notes in Artificial Intelligence*, pages 657–667. Springer, 1995.

Towards UAV Nap-of-the-Earth Flight Using Optical Flow

Thomas Netter and Nicolas Franceschini

CNRS, Laboratoire de Neurobiologie, LNB3
31, chemin Joseph Aiguier
13402 Marseille Cedex 20, France
[tnetter,enfranceschini]@LNB.cnrs-mrs.fr

Abstract. Nap-of-the-Earth (NOE) flight and the development of advanced Unmanned Air Vehicles (UAV) require further integration of sensors and onboard processing. Flying insects feature a panoramic eye which relies on Optical Flow (OF) for guidance and obstacle avoidance. We describe the development of a miniature thrust-vectoring UAV featuring an analog electronic eye based on biological motion detection. The system is tested within the laboratory on a custom-built whirling-arm rig. Preliminary tests have demonstrated satisfactory maneuvrability under remote-control. The project combines the fields of neurobiology, robotics, and aerospace. A VLSI implementation of the vision system is planned.

1 Introduction

Nap-of-the-Earth (NOE) flight is dangerous because obstacles are close and their perception challenging [1]. Flying insects exhibit NOE and obstacle avoidance abilities thanks to their wide Field of View (FOV) compound eyes which seem tailored for optical flow-based flight guidance and control [2]. Furthermore, flies bear remarkable neural fusing of visual, inertial, and aerodynamic senses [3]. The goal of our *visuomotor control* testbed is to demonstrate how insect vision can be applied to Unmanned Air Vehicles (UAV). This work continues a preliminary study of reactive speed and altitude control using Optical Flow (OF) [4].

Section 2 summarizes our effort to achieve a design scaled to eventually carry the entire sensory-motor system. Section 3 describes how we apply Optical Flow and its impact on sensor design. We explain how we use Optical Flow with a flight simulation in section 4 and then present preliminary tests with our indoors flight testbed in section 5. We conclude with further miniaturization in mind.

2 UAV Design

The main constraint on the UAV's design stems from our requirement to perform low speed flight tests within the limited space of our laboratory. Furthermore, our need to demonstrate reactive maneuvers in order to achieve terrain following and obstacle avoidance as well as hovering, implies small size and weight. The aircraft is mounted on a whirling-arm testbed.

We produced a single-rotor vectored-thrust Vertical Take-Off and Landing (VTOL) design. The thrust vectoring concept uses vanes to redirect the flow and permits outstanding maneuvrability [5]. Recent UAV projects also explore this technology [6]. Our design results from the following considerations:

- Hovering ability is required through single motor control,
- Compact with good thrust/weight, mechanical and aerodynamic simplicity,
- Best shape for 360 deg FOV eye and embodiment,
- Axisymmetric configurations simplify inertial calculations,

Aerodynamic calculations [7] for the rotor and thrust-vectoring vanes led to the construction of a 3-degrees of freedom Proof of Concept (POC) featuring:

- A 35cm diameter remote-controlled variable-pitch rotor,
- A controlled 300W motor with gear spinning the rotor at 6000 RPM,
- A single vane immersed in propeller flow with remote-controlled elevon,
- An opto-electronic rotor tachymeter,

The assembly weighs 0.85kg. Balanced control of the three actuators (rotor speed, blade pitch, vane elevon) combined with visual and inertial sensing enables us to vary thrust and aircraft pitch to control attitude, altitude, and speed.

3 Eye Design

We favored a camerular assembly (compactness, simplicity) to a compound eye design. It embeds a one-dimensional 20-pixel linear photoreceptor array and disposable camera lens.

An essential component of biological visual motion processing, whether it takes place in insects or humans, is the Elementary Motion Detector (EMD). Each EMD detects motion occurring in a particular direction within a small part of the visual field. The airborne photoreceptor array connects to an array of analog electronic EMDs derived from those of the fly. These EMDs were developed for an earlier mobile robot [8]. The retina is tilted so that its FOV covers the forward and downward region (figure 1).

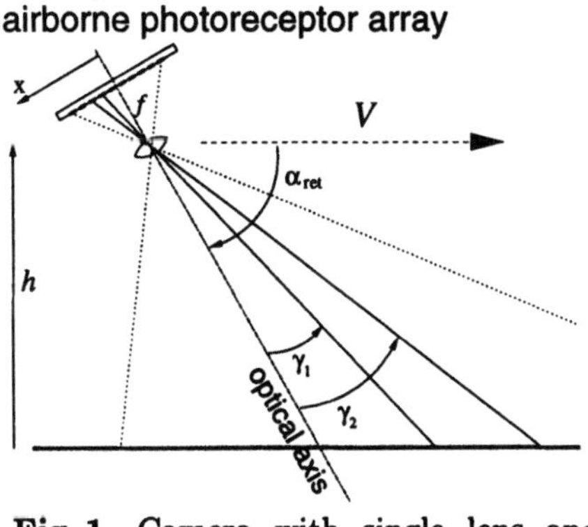

Fig. 1. Camera with single lens and photoreceptor array moving at altitude h and horizontal speed V above ground.

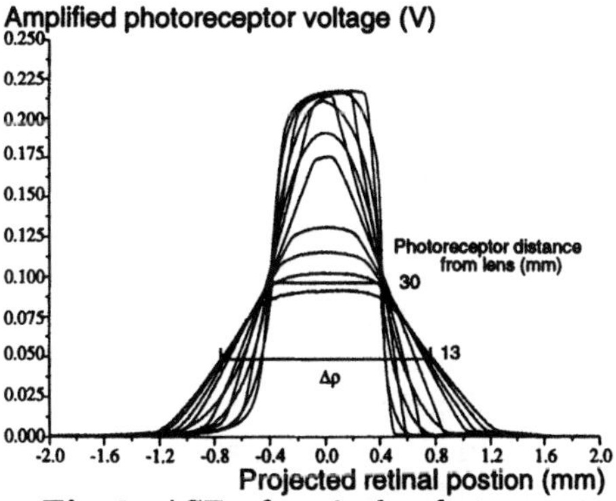

Fig. 2. ASF of a single photoreceptor (width 0.8mm) for 11 defocusing distances (13 to 30mm).

We tuned the receptor's Angular Sensitivity Function (ASF) to the sampling period (the interreceptor angle) $\Delta\varphi$ for improved motion detection and anti-aliasing. We determined the ASF of the "lens-photoreceptor" system by translating a light source across the FOV while measuring the electrical response of a single photoreceptor for various defocus distances (Figure 2).

4 Nap-of-the-Earth Flight Simulation

The control system relies on flight regimes interpolated from attitude and thrust flight data. The aircraft initially climbs to horizontal flight with no visual feedback [9]. Similarly to insects, some maneuvers are performed in a preprogrammed open-loop fashion. Our simulation assumes a steady thrust and attitude can be maintained through tachymeter and inertial sensors feedback.

When flying horizontally at altitude h and constant velocity V over the horizontal ground [10], the retinal velocity v_{ret} for a linear array is:

$$v_{\mathrm{ret}} = -\frac{fV}{h}\frac{\sin^2(\alpha_{\mathrm{ret}}+\gamma)}{\cos^2(\gamma)} \tag{1}$$

where f is the distance between optical center and retina, α_{ret} is the angle between the optical axis and V, γ is the angle between optical axis and photoreceptive pixel axis.

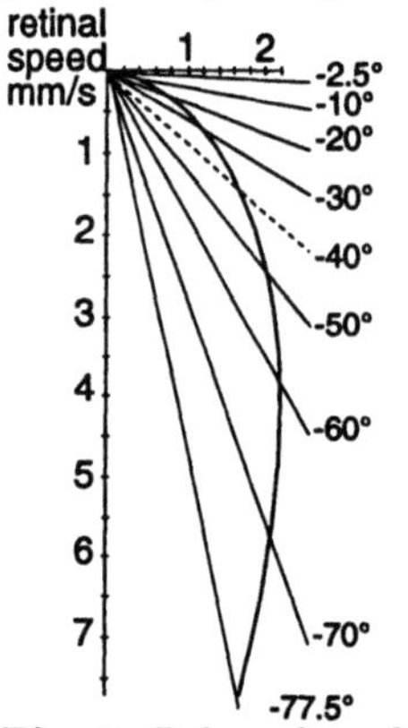

Fig. 3. Polar plot of (1), retinal speed (mm/s) vs ray angle (deg) with respect to optical axis (α_{ret} = -40 deg, f = 13 mm, V = 2 m/s, h = 5 m).

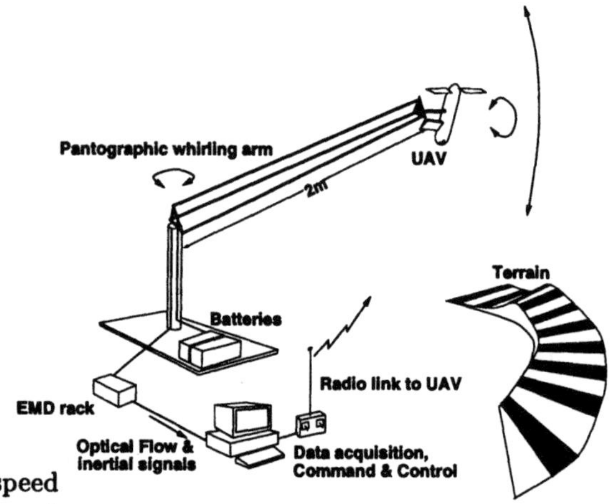

Fig. 4. Whirling-arm testbed.

Equation (1) shows (Fig. 3) that the forward part of the FOV generates nearly no Optical Flow [11] and responds poorly to ground height variations. Assuming the aircraft starts flying at a preprogrammed ground-speed over flat terrain, we set a reference OF distribution with (1). Terrain following is then achieved by retaining the reference OF through altitude variations. We use predefined thrust and pitch combinations at constant horizontal speed without introducing spurious rotation-related OF.

Since the forward view is the most important for obstacle avoidance we devised a weighted average OF fusion scheme giving most importance to the forward FOV and less to the downward FOV. This paradigm was inspired by the response fields and dendritic structures of frontal neurons VS1 and VS2 of the blowfly [12]. A reference Optical Flow corresponding to a predefined reference speed V_{ref} and altitude h_{ref} is fused into a reference weighted average :

$$\mathrm{OF}_{\text{ref ave}} = \frac{1}{N}\sum_{i=1}^{N} \frac{\mathrm{OF}_{\text{ref}N} - \mathrm{OF}_{\text{ref}i}}{\mathrm{OF}_{\text{ref}N}}\,\mathrm{OF}_{\text{ref}i}\ , \tag{2}$$

where $i = 1..N$ ranges from front to most downward viewing photoreceptor axes. We then relate a similar in-flight OF weighted average OF_{ave} with $\mathrm{OF}_{\text{ref ave}}$: $\mathrm{OF}_{\text{ratio}} = (\mathrm{OF}_{\text{ref ave}} - \mathrm{OF}_{\text{ave}})/\mathrm{OF}_{\text{ref}N}$. OF variation is inversely proportional to the square root of altitude so we produce a request for a new altitude $h_{\text{req}} = h - h_{\text{ref}}\,\mathrm{sign}(\mathrm{OF}_{\text{ratio}})\sqrt{|\mathrm{OF}_{\text{ratio}}|}$. Finally, the average divergence ∇v_{ret} over a few forward-looking EMDs triggers evasive maneuvers for large vertical obstacles.

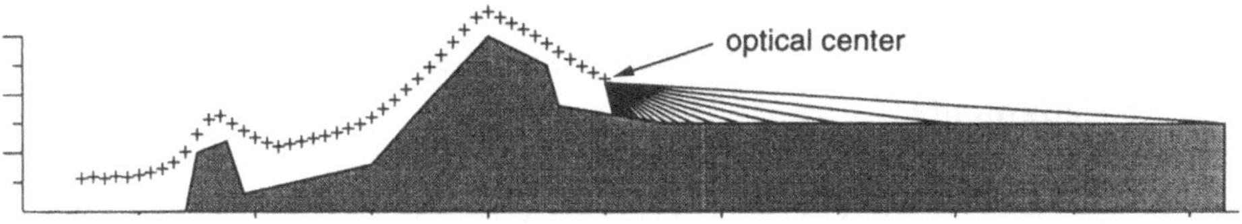

Fig. 5. NOE simulation with 20-photoreceptor retina, optical axis -40 deg, FOV 75 deg (within the lens limits). This run did not trigger the divergence-based evasive routine.

A simple NOE flight (figure 5) programmed with Scilab depicts the path of the aircraft eye's optical center. Flies fly closer to the ground when faced with headwind and we simulated this by progressively decreasing the aircraft speed but retaining the same request to maintain a reference OF (not shown).

5 Experimental Rig

Our indoors flight-testing requirement led to the construction of a pantographic whirling-arm rig (figure 4). It prevents yaw and roll of the UAV but permits variation in pitch (±30 deg) and altitude (0.3 to 3m). The model is powered by two car batteries (24V, 15A) fed to the aircraft through heavy duty bearings. Visual, inertial, and tachymeter signals are transmitted from the UAV to the rig's base through a slip-ring. The 20 photoreceptor signals are preamplified onboard and fed to 19 EMD boards scanned by an acquisition card in a PC running Real-Time Linux. Flight commands are output via the parallel port to a standard radio-control model transmitter. The onboard radio receiver commands the rotor speed variator, the collective pitch, and elevon servos. We can recover the aircraft manually in case of software malfunction.

Manually remote-controlled flight tests were performed in the laboratory to assess both rig and aircraft reliability. Horizontal speeds reached 6 m/s. High speeds tend to force the whirling-arm to remain horizontal; maneuvers are less realistic due to inertial forces. Vertical thrust pulses ($> 2m/s$) were demonstrated by increasing rotor blade pitch or RPM. Rotor blade pitch (or collective pitch in helicopter terminology) is beneficial for precise thrust adjustments, especially in ground effect. Maneuvrability was impressive. Aggressive flight direction reversals were achieved in 2-3 seconds at high flight speeds using elevon commands. Stability is induced by the large rotor. Steady hovering and forward flight transitions were demonstrated in and out of ground effect.

6 Conclusion

This UAV project combines the fields of neurobiology, robotics, and aerospace to test biologically-inspired sensory-motor concepts in a challenging environment. We use a low number of Elementary Motion Detectors to achieve reactive flight control. The Optical Flow field is computed and fused with inertial inputs by the real-time flight control computer. Although the testbed constrains tests to 3 degrees of freedom, flight tests have shown that it will enable us to explore aspects of insect flight, especially in wind gusts.

Our UAV design is suitable for the scale we have chosen but might not be easily scaled down or up because of aerodynamic or structural limitations, respectively. Our objective is to integrate the visuomotor paradigm into a VLSI neuromorphic circuit which could then be embedded into Micro Air Vehicles (MAV) to assist guidance in cluttered surroundings and indoors operations.

Acknowledgments.

We thank Marc Boyron and Jean Roumieu for their expert technical assistance in electronics and precision mechanics. Thomas Netter was supported by a CogniSciences doctoral stipend. This research is supported by the CNRS Life Science, Engineering Science, Cogniscience, and Microsystems programs.

References

[1] V. H. Cheng and B. Sridhar, "Technologies for automating rotorcraft Nap-of-the-Earth flight," *J. American Helicopter Society*, April 1993.

[2] M. Lehrer, M. Srinivasan, S. Zhang, and G. Horridge, "Motion cues provide the bee's visual world with a third dimension," *Nature*, vol. 332, pp. 356–357, 24 March 1988.

[3] W. P. Chan, F. Prete, and M. H. Dickinson, "Visual input to the efferent control system of a fly's "gyroscope"," *Science*, vol. 280, pp. 289–292, 1998.

[4] F. Mura and N. Franceschini, "Visual control of altitude and speed in a flying agent," in *From Animals to Animats III* (D. Cliff, P. Husbands, J. Meyer, and S. Wilson, eds.), pp. 91–99, MIT Press, 1994.

[5] B. Gal-Or, *Vectored Propulsion, Supermaneuvrability, and Robot Aircraft.* Springer-Verlag New York, 1990.

[6] K. M. Black, J. O. Smith, and R. A. Roberts, "The UTA autonomous aerial vehicle - automatic control and navigation," in *Nat. Aero. and Elec. Conf.*, 1992.

[7] R. W. Prouty, *Helicopter Performance, Stability, and Control.* Krieger, 1995.

[8] N. Franceschini, J.-M. Pichon, and C. Blanes, "From insect vision to robot vision," *Phil. Trans. R. Soc. B*, vol. 337, pp. 283–294, 1992.

[9] M. Kantner, B. Bodenheimer, P. Bendotti, and R. M. Murray, "An experimental comparison of controllers for a vectored thrust, ducted fan engine," in *American Control Conference*, 1995.

[10] J. S. Albus and T. H. Hong, "Motion, depth, and image flow," in *Proceedings of the 1990 IEEE Int. Conf. on Robotics and Automation*, pp. 1161–1170, 1990.

[11] J. G. Gibson, P. Olum, and F. Rosenblatt, "Parallax and perspective during aircraft landings," *Amer. J. Psychology*, vol. 68, pp. 372–385, 1955.

[12] H. G. Krapp, B. Hengstenberg, and R. Hengstenberg, "Dendritic structure and receptive-field organization of optic flow processing interneurons in the fly," *Journal of Neurophysiology*, vol. 79, pp. 1902–1917, April 1998.

On Bots and Bacteria: Ontology Independent Embodiment

Tom Quick[1], Kerstin Dautenhahn[2], Chrystopher L. Nehaniv[3] and Graham Roberts[1]

[1] Department of Computer Science, University College London, Gower Street, London, UK
{T.Quick, G.Roberts}@cs.ucl.ac.uk
[2] Department of Cybernetics, University of Reading, Whiteknights, Reading, UK
kd@cyber.reading.ac.uk
[3] Interactive Systems Engineering, University of Hertfordshire, Hatfield, UK
c.l.nehaniv@herts.ac.uk

Abstract. A framework for understanding and exploiting embodiment is presented which is not dependent on any specific ontological context. This framework is founded on a new definition of embodiment, based on the relational dynamics that exist between biological organisms and their environments, and inspired by the structural dynamics of the bacterium *Escherichia coli*. Full recognition is given to the role played by physically instantiated bodies, but in such a way that this can be meaningfully abstracted within the constraints implied by the term 'embodiment', and applied in a variety of operational contexts. This is illustrated by ongoing experimental work in which the relational dynamics that exist between *E. coli* and its environment are applied in a variety of software environments, using Cellular Automata (CA) with artificial 'sensory' and 'effector' surfaces, producing qualitatively similar 'chemotactic' behaviours in a variety of operational domains.

1 Introduction

This paper is concerned with the nature of embodiment — it proposes a new and precise definition of the term derived from asking 'what is it that is special about the relationship between bodies and the world?' and then suggesting how the features that are identified can be put to use independently of any specific ontological context. By focusing on the *relationship* between a system and its environment as the basis for embodiment, it is possible to analyse the significance of physical qualities without grounding the analysis *itself* in a material ontology. Material features are significant in so far as they condition the system-environment relationship, but not *just* because they are material.

The definition of embodiment presented yields practical and conceptual benefits. It provides a basis for *quantifying* embodiment, which is significant for behavioural robotics, for example with regard to understanding how to calculate and maximise embodiment, as well as understanding the problems that arise in moving between simulation and actual physical environments (cf. [1]).

The ontological neutrality of the definition also enables inter-disciplinary discussion about embodiment, for example between the behavioural robotics and intelligent software agents communities. It does this by providing a common framework for addressing embodiment — regardless of context — whilst recognising the uniqueness of different forms of embodiment.

On the same basis, it can defuse the tension between Artificial Life (ALife) and embodiment (cf. [2, 3]). From an ALife perspective, embodiment represents a theoretically well-grounded alternative to the tradition of symbol manipulation in AI. However, if *physical* embodiment is a necessary condition for the emergence of at least some life-like behaviour, this bodes ill for the synthesis of such behaviours in non-physical media — a central theme in ALife [4].

2 Embodiment and System - Environment Dynamics

Grounding embodiment in system-environment dynamics fits well with existing bodies of research. Ray's work on *Tierra* illustrates the significance of dynamics in generating the phenomena associated with living systems [5].

Striking examples of analyses of 'real world' embodiment that appeal to dynamical relationships between systems and the environments in which they are observed can be found in, for example, [6, 7], and particularly in Beer's work [8, 9].

Kushmerick [10] illustrates some of the difficulties of adherence to domain specificity in embodiment, which become apparent when trying to apply lessons learnt from the material world to that of the software agent. Without an underlying definition of embodiment, transferral of lessons learnt in the material world can only occur at the level of manifest phenomena, rather than at that of underlying or generative causes.

Franklin comes closest to the spirit of the perspective advocated here, referring to embodiment in terms of "autonomous agents structurally coupled with their environment" [11]. Etzioni [12] adopts a similar stance.

3 Embodiment as Situated Structural Coupling between System and Environment

We define what it is for a system to be embodied as follows:

*A system **X** is embodied in an environment **E** if perturbatory channels exist between the two. That is, **X** is embodied in **E** if for every time **t** at which both **X** and **E** exist, some subset of **E**'s possible states have the capacity to perturb **X**'s state, and some subset of **X**'s possible states have the capacity to perturb **E**'s state.*

This relational definition draws on Maturana and Varela's influential notion of structural coupling [13, 14]. At once, embodiment becomes quantifiable, ontology independent and directly linked to behaviour.

Degrees of Embodiment. The definition above describes the conditions under which structural coupling is possible. The mere fact that an object is embodied in an environment is insufficient to guarantee that any interesting interaction will occur between the two. It simply affords the possibility of perturbatory interaction.

There is huge scope for variation in this common embodiment relationship across different instances of embodiment, be it as a result of design or natural development.

As the perturbatory relationship between system and environment is quantitatively measurable, embodiment itself becomes measurable — expanding our viewpoint beyond the issue of whether or not a particular system is embodied[1]. One possibility is to ground a metric in the total *complexity* (as rigorously defined in [16], or alternatively [17]) of the dynamical relationship between system and environment, over all possible interactions. Factors such as the total bandwidth of the perturbatory channels between system and environment, as well as the computational power in the dynamics of their interaction, may contribute to this complexity.

Environmentally Coupled Cellular Automata as a Generic Class of Embodiable Dynamical System. Cellular Automata (CA) designed to engage in a mutually perturbatory relationship with some environment are suitable for exploring and exploiting the embodiment relationship articulated above, in that they can participate in the interaction defining such a relationship. Precedent for such a form of CA, in contrast to the more common closed form (cf. [18, 19]) has been set by Varela's *Bittorio* [20] (see also [21] for a demonstration of CA–environment coupling in an evolutionary context).

4 Bots and Bacteria — *E. coli* on the Internet

This section provides context for the definition of embodiment outlined above, by illustrating how *E. coli*'s autonomous and adaptive chemotactic behaviour emerges, for an observer, from the embodiment *relationship* between bacterium and environment. An experimental programme is underway, designed to explore this definition of embodiment by instantiating the equivalent embodiment relationship between a CA-based software system and a variety of ontologically distinct operational environments.

Structural Dynamics in *E. coli*. Despite being far less structurally complex than some multi-cellular organisms, and equipped with only non-directionally sensitive receptors and effectively binary state effectors, *E. coli* exhibits adaptive and consistently sensitive (cf. 'dynamic range,' below) chemotactic behaviour in response to nutrient gradients over five orders of magnitude [22].

The dynamics of two structural processes play key roles in the emergence of *E. coli*'s chemotaxis — highly connected signalling pathways within the cell and spatial clustering of receptors on the surface of the bacterium [23] — operating within and in relation to its physical environment via sensory and effector surfaces (nutrient receptors and flagella, respectively).

The signalling pathway is a CA-like system, comprised of a number of interconnected elements with relatively simple interaction rules between them. Interactions are based on the transfer of phosphoryl groups, the presence of which at flagella motor sites promotes 'tumbling' (random reorientation). Internal processes produce phosphoryl groups, encouraging frequent tumbling. Encounters with chemoattractants inhibit this process, shifting the behavioural bias towards 'running' (smooth swimming). In addition, receptors are constantly methylated, which promotes tumbling, even at higher concentrations of chemoattractants (cf. [22] for more detail).

[1] As suggested in [15] with respect to robots.

Receptor clustering plays a pivotal role in *E. coli*'s dynamic range [23]. This occurs at low attractant concentrations, and has the effect of pooling the output of a group of receptors when one member is activated. At higher concentrations the receptors disperse, providing sensitivity to attractant binding that would be effectively ignored by large clusters.

***E. coli* on the Internet.** A Java program, 'Phenomorph', is under development. Its relationship to Web pages is based on *E. coli*'s relationship to its environment. The purpose is not to achieve optimal information search (cf. [24] [25]), but to investigate, in conjunction with planned future experiments[2], the validity of the concept of embodiment presented above.

At the heart of Phenomorph lies a uniform 1D binary state environmentally coupled CA which generates dynamics roughly analogous to those inherent in *E. coli*'s structure. Receptor clustering and methylation are also simulated, whilst keywords defined by the user at run-time play the part of chemoattractants. When Phenomorph visits a web page containing defined keywords, the CA is proportionately stimulated. The CA global activation level determines the likelihood that Phenomorph will 'run' rather than 'tumble,' each of which are implemented through hyperlink following. See [26] for a more detailed description of Phenomorph.

Initial Evaluation. Although not yet developed sufficiently to produce infotaxis[3], and lacking comparative studies in other operational environments, Phenomorph shows a variety of autonomously generated responses to environmental features, which arise directly from and are determined solely by the interplay between the environment and its CA-based structural dynamics. Environmental variety is constantly filtered by the (very simple) form of Phenomorph's embodiment.

5 Conclusions

The definition of embodiment presented offers immediate opportunities to bridge the interpretative gap between disciplines concerned with very different forms of embodiment, something previously hampered not least by the lack of any firm definition of the term. This understanding of embodiment has the potential to provide benefits to practitioners on both sides of the ontological divide between physical and non-physical systems and environments — from measuring the embodiment of robots to evolving distributed autonomous control systems that exploit emergent behavioural strategies across a range of operational environments. A great deal of experimental work also remains to be done investigating this concept of embodiment across operational environments, and developing possibilities for its exploitation.

[2] These will consist of embodying Phenomorph, via appropriate sensory and effector surfaces, in an abstract parameter space, and a physical environment.

[3] An informatic counterpart of chemotaxis. Phenomorph has yet to attain a level of behavioural 'fitness' that can be compared with that of the acutely honed *E. coli*.

Acknowledgements

This research is funded by an EPRSC studentship.

References

1. Brooks, R.A., Mataric, M.J.: Real robots, real learning problems. In: Connell, J.H., Mahadevan, S. (eds.): Robot Learning. Kluwer (1994) 193-213
2. Prem, E.: Epistemic autonomy in models of living systems. In: Husbands, P., Harvey, I. (eds.): 4th European Conf. Artificial Life. MIT Press, Cambridge, MA (1997) 2-9
3. Dautenhahn, K.: The art of designing socially intelligent agents: science, fiction, and the human in the loop. Applied Art. Int. 12 (1998) 573-617
4. Langton, C.G.: Artificial Life. In: Langton, C.G. (ed.): Artificial Life. SFI Studies in the Sciences of Complexity, Vol. 7. Addison-Wesley, Reading MA (1987) 1-48
5. Ray, T.S.: A network-wide biodiversity reserve for digital organisms. Imagina 96. Institut National De L'audiovisuel, Bry-sur-Marne, France (1996)
6. Pfeifer, R.: Cognition - Perspectives from autonomous agents. Rob. Auton. Sys (1995) 47-70
7. Smithers, T.: On quantitative performance measures of robot behaviour. Rob. Auton. Sys. 15 (1995) 107-134
8. Beer, R.D.: A dynamical systems perspective on agent-environment interaction. Art. Int. 72 (1995) 173-215
9. Chiel, H.J., Beer, R.D.: The brain has a body: adaptive behavior emerges from interactions of nervous system, body and environment. Trends in Neurosciences 20 (1997) 553-556
10. Kushmerick, N.: Software agents and their bodies. Minds & Machines 7 (1997) 227-247
11. Franklin, S.: Autonomous agents as embodied AI. Cybernetics & Sys. 28 (1997) 499-520
12. Etzioni, O.: Intelligence without robots: A reply to Brooks. AI Mag. 14 (1993) 7-13
13. Maturana, H.R., Varela, F.J.: Autopoiesis and Cognition. D. Reidel, Holland (1980)
14. Varela, F.J.: Principles of Biological Autonomy. North Holland, New York (1979)
15. Dautenhahn, K.: Biologically inspired robotic experiments on interaction and dynamic agent-environment couplings. In: Horst-Michael, G. (ed.): Workshop SOAVE'97, Selbstorganization von Adaptivem Verhalten. Llmenau (1997) 14-24
16. Nehaniv, C.L., Rhodes, J.L.: Krohn-Rhodes theory, hierarchies, and evolution. In: Mirkin, B. et al. (eds.): Mathematical Hierarchies and Biology. Amer. Math. Soc. (1997) 29-42
17. Crutchfield, J.P.: The calculi of emergence: Computation, dynamics, and induction. Physica D 75 (1994) 11-54
18. Wolfram, S.: Statistical mechanics of cellular automata. Rev. Mod. Phys. (1983) 601-644
19. Langton, C.: Studying Artificial Life with Cellular Automata. Physica D 22 (1986) 120-149
20. Varela, F.J., et al.: The Embodied Mind. MIT Press, Cambridge, MA (1991)
21. Sims, K.: Interactive evolution of dynamical systems. Varela, F., Bourgine, P. (eds.): 1st European Conf. Artificial Life. MIT Press (1991) 171-178
22. Levin, M.D., Morton-Firth, C.J., et al.: Origins of individual swimming behavior in bacteria. Biophys. J. 74 (1998) 175-181
23. Bray, D., Levin, M.D., Morton-Firth, C.J.: Receptor clustering as a cellular mechanism to control sensitivity. Nature 393 (1998) 85-88
24. Menczer, F. and Belew, R.K.: Adaptive information agents in distributed textual environments. 2nd Int. Conf. on Aut. Agents (1998)
25. Moukas, A.: Amalthaea: Information discovery and filtering using a multiagent evolving ecosystem. Applied Art. Int. 11 (1997) 437-457
26. Quick, T., Dautenhahn, K., Nehaniv, C., and Roberts, G.: The essence of embodiment: A framework for understanding and exploiting structural coupling between system and environment. 3rd Int. Conf. Comp. Anticipatory Syst. Liege, Belgium (1999) To appear

Populating the Virtual Worlds with Interactive Perceptive Virtual Actors

Daniel Thalmann

Computer Graphics Lab
Swiss Federal Institute of Technology
CH 1015 Lausanne, Switzerland
thalmann@lig.di.epfl.ch

Telepresence is the future of multimedia systems and will allow participants to share professional and private experiences, meetings, games, parties. The concepts of Distributed Virtual Environments are a key technology to implement this telepresence. Using Virtual Humans within the shared environment is a essential supporting tool for presence. Real-time realistic 3D avatars will be essential in the future, but we will need interactive perceptive actors to populate the Virtual Worlds. The ultimate objective in creating realistic and believable virtual actors is to build intelligent autonomous virtual humans with adaptation, perception and memory. These actors should be able to act freely and emotionally. Ideally, they should be conscious and unpredictable. But, how far are we from such a ideal situation? Our interactive perceptive actors are able to perceive the virtual world, the people living in this world and in the real world. They may act based on their perception in an autonomous manner. Their intelligence is constrained and limited to the results obtained in the development of new methods of Artificial Intelligence. However, the representation under the form of virtual actors is a way of visually evaluating the progress. In the future, we may expect to meet intelligent actors able to learn or understand a few situations.

1. P. Kalra, N. Magnenat Thalmann, L.Moccozet, G.Sannier, A.Aubel, D.Thalmann. Real-time Animation of Realistic Virtual Humans. *IEEE Computer Graphics and Applications*, 5, 42–55, 1998.
2. L. Emering, R. Boulic, and D. Thalmann. Interacting with Virtual Humans through Body Actions. *IEEE Computer Graphics and Applications*, 18(1):8–11, 1998.
3. S. R. Musse and D. Thalmann. A Model of Human Crowd Behavior. *Computer Animation and Simulation '97, Proc. Eurographics workshop*, Springer Verlag, Wien, 1997.
4. D. Thalmann. Virtual Humans in Virtual Environments: A New View of Multimedia Applications. *Proc. Intern. Workshop on Synthetic - Natural Hybrid Coding and Three Dimensional Imaging*, pages 3–7, 1997.
5. S. R. Musse, C. Babski, T. Capin, and D. Thalmann. Crowd Modelling in Collaborative Virtual Environments. *ACM VRST '98*, 1998.
6. C. Bordeux, R. Boulic, and D.Thalmann. An Efficient and Flexible Perception Pipeline for Autonomous Agents. *Proc. Eurographics '99*, to appear.

Level of Autonomy for Virtual Human Agents

Soraia Raupp Musse, Marcelo Kallmann and Daniel Thalmann

Computer Graphics Lab. Swiss Federal Institute of Technology
EPFL, DI-LIG, CH 1015 Lausanne, Switzerland
{soraia,kallmann,thalmann }@ lig.di.epfl.ch

Abstract. This paper proposes a paradigm for specification of virtual human agents' level of autonomy. The idea we present in this paper aims at optimi s-ing the required complexity of agents in order to perform realistic simulations. The paradigm is based on the distribution of the required virtual human agent "inte lligence" to other simulation entities like groups of agents, and objects.

1 Introduction

Several methods have been introduced to model learning processes, perceptions, actions, behaviours, etc, in order to build more intelligent and autonomous virtual agents. Our goal in this paper is to propose a paradigm to define virtual agents en-dowed with different degrees of behavioural autonomy.

First of all, we present some useful concepts assumed in this work. A *virtual human agent* (here after just referred to as an agent) is a humanoid whose behaviours are inspired by those of humans' [15]. The term *group* will be used to refer to a group of agents, and the term *object* for an interactive object of the environment. Agents, groups, and objects constitute the *entities of the simulation*. A high-level *behavioural autonomy* concerns the ability to simulate complex behaviours. In this paper, we consider that the ability of agents for autonomously acting can be included in the agents (agents-based application), groups (groups-based application) or in the objects (objects-based application). Among others, interactivity, complex behaviours, intelligent abilities and frame rate of execution are directly related to the *level of autonomy* (LOA). Table 1 presents this relation using three kinds of behavioural control: *guided* represents the lower level of autonomy where the behaviours have to be provided by an external process. Yet, in Table 1, *programmed* control implies to use a notation (language) to define possible behaviours. The *autonomous* behaviour concerns the capability of acting independently exhibiting control over their internal state [28].

Table 1. Characteristics of different levels of autonomy (LOA).

BEHAVIOUR CONTROL	GUIDED	PROGRAMMED	AUTONOMOUS
Level of Autonomy	Low	Medium	High
Level of Intelligence	Low	Medium	High
Execution frame-rate	Low	Medium	High
Complexity of behaviours	Low	Variable	High
Level of Interaction	High	Variable	Variable

2 Related Work

Several works have discussed the various manners to simulate and interact with virtual agents. Zeltzer [29] presents a classification of levels of interaction and abstraction required in different applications. Thalmann [4] proposes a new classification of synthetic actors according to the method of controlling motion. Reynolds [22] presented the aggregated motions modelling. In recent work, a crowd model has been introduced using different abstractions of behaviours, like the term *guided crowd* [17]. Considering agent-object interaction tasks, some semantic information has been included within the object description. In particular, the *object specific reasoning* [11] creates a relational table to inform object purpose, and *smart objects* were introduced [10] containing interaction information.

3 LOA Related to Individuals

Several works agree with the concept of autonomous or "intelligent" agent requirements: autonomous behaviour, action, perception, memory, reasoning, learning, self-controlled, etc [15], [18], [22]. Yet, a lot of methods have been developed in order to model autonomous agents: L-systems [18], vision systems [21]; rule-based systems [22]; learning methods [25], etc. Yet, guided or programmed agents can also be useful depending on the application. Table 2 exemplifies the three kinds of agent autonomy using two different agent tasks.

Table 2. LOA present in different agent-oriented tasks.

LOA/TASK	AGENT GOES TO A SPECIFIC LOCATION	AGENT APPLIES A SPECIFIC ACTION
Guided	Agent needs to receive during the simulation a list of collision-free positions	Agent needs to receive information about the action to be applied
Programmed	Agent is programmed to follow a path while avoiding collision with other agents and programmed obstacles	Agent is programmed to manage where and how the action can occur
Autonomous	Agent is able to perceive information in the environment and decide a path to follow to reach the goal, using the environment perception or the memory (past experiences)	Agent can decide about an action to be applied. This action can be programmed, imitated or existent in the memory (past experiences)

4 LOA Related to Groups of Agents

In the case of crowd simulation, usually we intend to have lots of virtual human agents avoiding dealing with individual behaviours. Contrary to the last section, our goal here is to describe methods to provide intelligence focused in a common group entity that controls its individuals. We have called *groups-based application,* the crowd and group applications, where individual complexity is less required. In this case, the intelligence abstraction can be included in the groups providing more autonomy to the groups instead to the individuals.

Considering levels of autonomy (LOA), we have classified the crowd behaviours in three kinds: i) Guided crowds, which behaviours are defined explicitly by the users; ii) Programmed crowds, which behaviours are programmed in a script language; iii) Autonomous crowds, which behaviours are specified using rules or others complex methods. Table 5 exemplifies this classification of crowd autonomy using two different crowd tasks.

Table 3: LOA present in different group-oriented tasks

LOA/TASK	Group goes to a specific location	Group reacts to matched event
Guided	Group needs to receive during the simulation a list of positions "in-betweens" in order to reach the goal	Group needs to receive an information about the matched event and the reaction to be applied
Programmed	Group is programmed to follow a path avoiding collision with other agents and programmed obstacles.	Group can manage events and reactions, which are programmed.
Autonomous	Group is able to perceive information in the environment and decide a path to follow to reach the goal, using the environment perception or the memory (past experiences).	Group can perceive a matched event and decide about the reaction to be applied. This reaction can be also programmed or existent in the group memory.

5 LOA Related to Objects

Whenever the simulation needs to handle complex agent-object interactions, many difficult issues arise. Such difficulties are related to the fact that each object has its own movements, functionality and purposes. One can consider that agents' perceptions can solve some simple tasks, as for instance a single-hand automatic grasping of small objects. But this is no more possible for interactions with objects that have an intricate proper functionality, as the lift example. In fact, each time more information related to the object is given, its level of autonomy (LOA) is increased. Table 6 illustrates how an agent must proceed according to the different LOAs for three different interactive objects of the environment.

Table 4: LOA present in different objects-oriented tasks

LOA/OBJECT	Door	Sign	Lift
Guided	The agent have to move its arm to a attainable and meaning location of the door, and control its movement until open it.	The agent recognises that the sign has an arrow and recognises the showed direction.	The agent recognises where is the call button, how and when the door opens, how and where to enter inside the lift, when and how to go out, etc.
Programmed	The agent has to move its arm to the right place but the door opens by itself.	The agent recognises the sign, but the direction is given with no recognition.	The agent accesses the current lift state and decides only its moves accordingly.
Autonomous	The door takes control of the agent telling exactly the place to put its hand and the complete movement of the door	The sign gives a new direction to go for each agent that passes nearby.	The lift takes control of the movements of the agent and gives him a complete plan, based on primitive actions, to perform the interaction.

6 The Proposed Paradigm

As presented in the last sections we considers that the "intelligence" is not only included in the virtual human agents, but can be also included in groups and objects. Considering the abstraction levels: guided, programmed and autonomous behaviours, we present a schema that includes the entities group and object, as showed in Figure 1. We can so classify a simulation in terms of the autonomy distribution among its entities, i.e., a simulation (S_i), can be translated as a function of three components: agents, groups and objects: $S_i = f(LOA(Agents), LOA(Groups), LOA(Objects))$

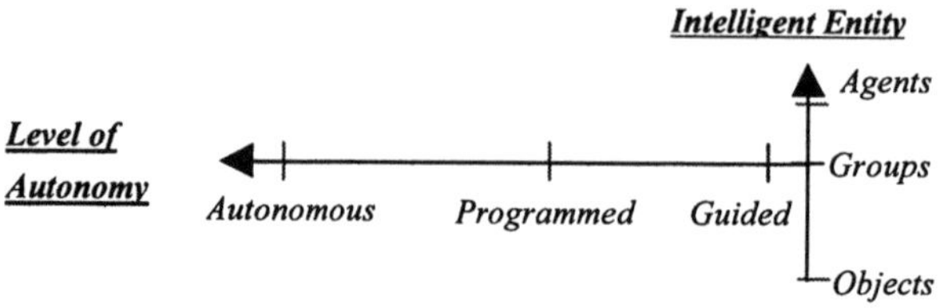

Figure 1: Level of autonomy vs. intelligent entity.

7 A Case Study

The chosen case study was designed to deal with different kinds of control. We consider the environment of a virtual city [6] containing some streets, a supermarket (S), a train station (TS), autonomous objects (direction signs to go to the TS) and other buildings. Let G be a group of virtual agents endowed with different LOAs, which can be guided, programmed or autonomous. The goal of group G is to go from the supermarket to the train station. We dealt with four kinds of simulations having different group controls, interacting or not with objects.

Facts: ~ Goal (G) = Go From S to TS	
Simulation 0 ~ G = Guided Group ~ Autonomous objects exist but G can not recognise them. ~ G does not know its goal ~ Initially, G receives a location to reach ~ G is able to walk to reach this location ~ G is not able to avoid collision with obstacles	Simulation 1 ~ G = Programmed group ~ Autonomous objects exist but G can not recognise them, because G is not programmed for that ~ G knows the programmed goal ~ Initially, G is able to translate a programmed goal (TS) in a path to be applied
Simulation 2 ~ G = Autonomous group. ~ Autonomous objects exist and G can recognise ~ G knows its goal ~ Initially, G is able to recognise the autonomous objects, go to a location near to it and interact with. The autonomous object is able to recognise where G wants to go and to give the correct direction ~G is able to follow object instructions	Simulation 4 ~G = Autonomous group ~ Autonomous objects do not exist ~ G knows its goal ~ G is provided with vision and environment knowledge ~ G can find a path to reach the goal by perceiving the environment (signs) by its own ~ G is able to perceive and avoid collision with obstacles

Figure 2: (left): the starting and goal points; (center) group G going to interact with autonomous object; (right): comparison data between the four simulations.

In Figure 2 (right), some parameters (except execution time) represent subjective data to be measured, then we decide to intuitively quantify it in four levels: 25 (Low), 50 (inferior medium), 75 (superior medium) and 100 (high).

8 Conclusions

We propose in this paper a paradigm to distribute the autonomy among the entities of the simulation. The idea we dealt here concerns the possibility to improve the frame rate of execution as well as to optimise the complexity required, by distributing some knowledge and autonomy to others entities of the simulation: groups and objects. This paradigm has been tested in the context of a Virtual City project [6] because we have to simulate several virtual human agents that can act in differently ways and apply different actions.

Acknowledgements

The research was sponsored by the Swiss National Research Foundation, the Federal Office of Education and Science in the framework of the European project eRENA [8], FUNDEPE, CAPES and CNPq - Coordenação de Aperfeiçoamento de Pessoal de Nível Superior (Brazilian offices of Education and Science).

References

1. Cavazza, M., Earnshaw, R., Magnenat-Thalmann, N. and Thalmann, D. "Motion Control of Virtual Humans". IEEE CG&A, Vol. 18, No. 5, Sept/Oct 1998, pp.24-31.
2. Farenc, N; et al. "One Step Towards Virtual Human Management for Urban Environment Simulation". ECAI'98, Brighton, England. 1998.
3. Kallmann, M. and Thalmann, D. " Modeling Objects for Interaction Tasks", Proc. Eurographics Workshop on Animation and Simulation, 1998.
4. Levison, L. "Connecting Planning and Acting via Object-Specific reasoning", PhD thesis, Dept. of Computer & Information Science, University of Pennsylvania, 1996.
5. Meyer, J.A.; Guillot, A. "From SAB90 to SAB94: Four Years of Animat Research". In: Proceedings of Third International Conference on Simulation of Adaptive Behavior. Brighton, England, 1994.
6. Musse, S.R., Babski, C., Capin, T. and Thalmann, D. "Crowd Modelling in Collaborative Virtual Environments". ACM VRST /98, Taiwan
7. Noser, H. "A Behavioral Animation System based on L-Systems and Synthetic Sensors for Actors". PhD thesis. EPFL, Lausanne, Switzerland, 1996.
8. Renault, O., Magnenat-Thalmann, N. Thalmann, D. "A Vision-based Approach to Behavioral Animation", The Journal of Visualization and Computer Animation, Vol. 1, No. 1, pp. 18 – 21.
9. Reynolds, C. "Flocks, Herds and Schools: A Distributed Behavioral Model". Proc. SIGGRAPH '87, Computer Graphics, v.21, n.4, July, 1987.
10. Sutton, R.S. "Reinforcement Learning Architectures for Animats". In: Proceedings of First International Conference on Simulation of Adaptive Behavior. MTI Press, 1990.
11. Wooldridge, M. and Jennings, N. "Intelligent Agents: Theory and Practice". Knowledge Engineering Review, 10(2), June 1995. Cambridge University Press, 1995.
12. Zeltzer, D. "Task-level Graphical Simulation: Abstraction, Representation and Control". Making them Move: Mechanics, Control and Animation of Articulated Figures. Edited by N. Badler, B. Barsky and D. Zeltzer. Pp 3-33. 1991.

Emotional Disorders in Autonomous Agents ?

Aapo Hyvärinen[1] and Timo Honkela[2]

[1] Helsinki University of Technology
Laboratory of Computer and Information Science
P.O. Box 5400, FIN-02015 HUT, Finland
aapo.hyvarinen@hut.fi

[2] Media Lab, University of Art and Design Helsinki,
Hämeentie 135 C, FIN-00560 Helsinki, Finland
timo.honkela@uiah.fi

Abstract. It has been recently suggested by a number of authors that modelling of emotions and related motivational systems in agents might have great practical value, apart from the interest of providing possible explanations for the emotional mechanisms of human agents. Emotions, or needs, may be used as signalling mechanisms between different subsystems (subagents) inside an agent, as well as between different agents. In this paper, we investigate some problems that may arise with emotional agents. Since needs and emotions are largely global, stable reaction tendencies, they may exhibit rigidities that lead to different forms of maladaptive behavior, i.e. behavior that is not well suited to the present environment of the agent. We investigate emotional learning in agents by an utterly simplified decision-theoretical model. We show that even in this very simple model agents may develop maladaptive patterns of behavior that closely resemble patterns found in emotional disorders in humans. The maladaptive behavior patterns are due to non-optimal values for the two decision parameters, which are functions of the prior beliefs of the agent.

1 Introduction

A central issue in artificial life, as well as in artificial intelligence, is the study of adaptive agents. Adaptive agents learn from their environment, possibly by constructing a model of the world, so as to maximize some desired criteria. The agents' model of the world, and the ensuing behavior tendencies are not rigid, but a product of experience and interaction with the environment. This implies that agents have different models due to differential learning by the agents. Each has its own history which has led to a different model of the agent world.

It has been recently suggested by a number of authors, e.g. [2, 6], that modelling of emotions and related motivational systems in agents might have great practical value, apart from the interest of providing possible explanations for the emotional mechanisms of human agents. For example, emotions, or needs, may be used as signalling mechanisms between different subsystems (subagents) inside an agent, as well as between different agents [5, 1, 4, 6]. A seminal proposal in this line of research was Simon's interruption theory [5], where emotions interrupt ongoing goal processing in order to direct processing resources to more urgent goals.

In this paper, we investigate problems that may arise with emotional agents. Since needs and emotions are largely global, stable reaction tendencies, they may exhibit rigidities that lead to different forms of maladaptive behavior, i.e. behavior that is not well suited to the present environment of the agent. In humans, such maladaptive emotional states, and the ensuing maladaptive behavior patterns, have been a subject of a large body of research in psychology and psychiatry. Modelling emotional learning in agents, we show how agents may develop maladaptive patterns of behavior that closely resemble patterns found in emotional disorders in humans.

We show that maladaptive emotion-based behavior can emerge from a very simple model of the agents' decision-making system. The basic emotional system in our model can be reduced to two parameters that can be learned e.g. by reinforcement learning. Non-optimal values for these parameters may lead to maladaptive behavior patterns that resemble such disorders as depression, anxiety, and mania. A cognitive interpretation of the parameters in terms of (Bayesian) prior beliefs shows how emotional disorders are related to erroneous beliefs on the world. The results are relevant to the design of emotional agents, and may give insight into the processes involved in human maladaptive behavior.

2 Minimal Agent model

To begin with, we describe an very simple agent model that we use to illustrate maladaptive emotion-based behaviors. We accomplish this with an agent that has just one long-term goal, three internal needs (emotions), two types of sensory perceptions, and three action alternatives.

Goal and needs. In our highly simplified agent model, the agent has (or seems to have) just one ultimate long-term *goal*. This is the combination of actions, states or objects that the agent needs to perform or obtain to be useful for its designer (in the case of a robot) or to spread its genes (for biological agents). Typically, the satisfaction of the goal is not a binary truth value, but rather a scalar quantity S that measures how well the agent performed during its lifetime.

Though the agent has only one (ultimate) goal, we consider that its decision-making system consists of several modules that correspond to different *needs*. (We ignore here the modules needed for perception, action etc.) The needs are closely connected to emotions because it may be considered that emotions are expressions of current needs. Thus we model emotions in our system indirectly, via the underlying needs. Biological systems show that it is useful for an agent to construct the decision-making device of components, each corresponding to one type of constraint of the environment, or one type of subgoal. A need corresponds thus in our terminology to one of those subgoals. An emotion can be considered as a signal that expresses the urgence of a given need.

Firstly, to generate behavior that is useful for directly satisfying the (long-term) goal, the agent may be considered to generate a set of internal states that we may call the *goal need*.

If the agent just had its goal need, its behavior would consist of trying to do anything that lets it satisfy its goal need at the present instant. Such a short-sighted behavior would, however, lead to rather small goal satisfactions in a hostile environment. Thus we include in our agent model a second need which consists of trying to avoid any external events that might destroy the agent. We call this need the *safety need.*

Another need that in often in conflict with the goal need is the *energy need.* This means the need to get energy when energy level is low, by means of drinking, eating, charging batteries, etc. For simplicity, we assume in our model that the agent gains energy automatically with time.

Perception of environment. When some need-relevant event occurs, we assume the perceptual system computes two conditional probabilities: First, $P(event|goal)$, the conditional probability that an event that satisfies the goal need would produce such a perception, and second, $P(event|danger)$, the conditional probability that an event that threatens the safety need, i.e. is dangerous, would produce such a perception. Such probabilities are typically produced by most pattern-recognition methods, e.g. by computing distances from a template. It is here assumed for simplicity that all events belong to two uniform categories: those satisfying the goal need and those threatening the safety need.

Actions. On the basis of the perceptual quantities, the agent must decide its course of action among the following alternatives. 1) *Approach*: This means that the event or object is explored further, and the possible increase in goal satisfaction is enjoyed. 2) *Avoidance*: This means that the agent tries to avoid the event or object. 3) *No action.* It is assumed that both actions consume the same amount of energy, whereas the no action alternative does not consume any. Since the energy supplies are constantly replenished, this means that not doing anything really means resting to get energy.

Competition of needs. Choosing among the action alternatives is here done using a simple competition of need signals. (A more sophisticated justification of this procedure is given below.) The need modules have certain weight parameters w_{goal} and w_{danger} that are based on information they have about the context and the possible urgencies of the needs. On the basis of the perceptions and these parameters, the need modules transmit their priorities or need signals in the form of quantities $w_{goal}P(event|goal)$ and $w_{danger}P(event|danger)$. In other words, they multiply the perceptual probabilities by some quantities that express the importance of the need. Similarly, the energy need transmits a signal, which does not depend on perceptions, but possibly on some internal state of the system. We can normalize the signals so that the signal given by the energy need is equal to 1 always. Thus the decision between different action alternatives is made by choosing the maximum among the quantities

$$needsignal(goal) = w_{goal}P(event|goal) \tag{1}$$

$$needsignal(danger) = w_{danger}P(event|danger) \tag{2}$$

$$needsignal(energy) = 1. \tag{3}$$

Utility-theoretical interpretation. The decision rule given above can be interpreted in the framework of utility theory. Due to lack of space, we present here only the results of the analysis, details can be found in [3]. To decide between the three alternatives given above, the agent uses simple Bayesian decision procedures. We assume that the agent has some estimates of the following quantities. First, $\Delta S(goal)$: the increase in (life-time) goal satisfaction due to encounter with a goal satisfying event. Second, $\Delta S(danger)$: the expected decrease in (life-time) goal satisfaction due to destruction caused by encounter with dangerous event. Moreover, the agent has estimated from the environment the following prior probabilities, i.e. probabilities that express the prior belief on the occurance of the two kinds of events that the agent has before observing the present environment : $P(danger)$: the prior probability that a dangerous event occurs and $P(goal)$: the prior probability that a goal satisfying event occurs.

By some algebraic manipulation, we see that the weights w_{goal} and w_{danger} are essentially given by two quantities, which are the *prior* expected utilities $P(goal)\Delta S(goal)$ and $P(danger)\Delta S(danger)$. These quantities can be interpreted as the level of danger or goal satisfaction that the agent expects to receive on the average. They contain the essence of the prior beliefs that the agents holds on the environment, and individual differences can be traced to different estimates of these prior expectations.

3 Maladaptive emotion-based behavior

By maladaptive emotion-based behavior, we mean here configurations where the emotional components of the agent do not allow as large a goal satisfaction as were otherwise possible. Such maladaptive behavior patterns are closely related to emotional disorders, sometimes called neuroses.

In our model, we have two parameters, w_{danger} and w_{goal} (and the corresponding parameters in the Bayesian framework) that govern the behavior of the agent. Suboptimal values of these parameters are the main cause of maladaptive behavior. There might be several *reasons* why the parameters have suboptimal values. First, the environment may change. In particular, the environment where the parameters were first learned may not be actual anymore. In biological agents, this phenomenon seems to be of great importance especially because many parameter values seem to be fixed during childhood learning. Second, the learning method might be faulty. This is usually not so much a problem in biological agents, where evolution has provided quite efficient learning methods, but in artificial agents, the learning algorithms are not always efficient. It is also possible that learning of biological agents might be disturbed by diseases and environmental toxins. Third, there might be discrepancies in the long-term goals. Especially in artificial agents, there may exist several possible definitions of goal satisfactions with correspondingly different optimal values for the parameters. We can identify three classical types of maladaptation, as discussed next

Anxiety. If the coefficient w_{danger} is too high with respect to both w_{goal} and the constant energy need, this leads to a preponderance of avoidance behavior, reducing approach behavior and inactivity. Thus the agent avoids even events whose probability of being dangerous is relatively small. This parallels what is called anxiety (or the generalized anxiety syndrome) in humans.

In a cognitive interpretation, the weight w_{danger} is essentially the prior expected loss of utility due to dangerous events. An elevated value for w_{danger}, and the associated tendency to avoidance, can thus be caused by a strong belief in the occurrence of dangerous events, or on the high value placed on safety. Both of these give rise to a cognition that exaggerates the importance of dangerous events.

Depression. If w_{goal} is too low compared to the constant signal emitted by the energy need, we have a situation characterized by a relative rarity of approach behavior that is not unlike the prevalent behavioral symptom of depression. Especially when combined with a low w_{danger}, this lead ultimately to inactivity and stupor as in severe depression in humans. In the cognitive, utility-theoretic, interpretation, a very low relative value of w_{goal} may be due to a pessimistic stance in which the prior belief on goal satisfaction low, or due to the low value assigned to goal satisfaction.

Mania. The third pathological state is observed when w_{goal} is too high and w_{danger} is too low. In this state, the agent ignores dangers and is excessively active in approach behavior, which resembles the symptoms of the manic states in humans.

4 Conclusion

We modelled the motivational or emotional system of a simplified agent using a system of three needs and three behaviors. It was shown how the behavior of the agent depends basically on two parameters. Suboptimal values for these parameters lead to behavior tendencies that resemble some of the behavior patterns encountered in humans with emotional disorders. The model gives explanations for the etiology of these disorders that are not unlike those given in the psychological literature on humans, and gives a new framework for studying adaptation phenomena related to emotions.

References

1. N. H. Frijda. *The Emotions.* Cambridge University Press, 1986.
2. T. Gomi, J. Vardalas, and K.-I. Ide. Elements of artificial emotion. In *Proc. Robot and Human Communication (RO-MAN'95)*, Tokyo, Japan, 1995.
3. A. Hyvärinen and T. Honkela. Maladaptive emotion-based behaviors in autonomous agents. In *Proc. STeP'98*, Jyväskylä, Finland, 1998.
4. K. Oatley. *Best Laid Schemes: The Psychology of the Emotions.* Cambridge University Press, 1992.
5. H. A. Simon. Motivational and emotional controls of cognition. In *Models of thought*, pages 29–38. Yale University Press, 1979. Reprint, originally appeared in 1967.
6. A. Sloman and M. Croucher. Why robots will have emotions. In *Proc. 7th Int. Joint Conf. on Artificial Intelligence*, Vancouver, Canada, 1981.

Integrating Unsupervised Learning, Motivation and Action Selection in an A-life Agent

Mark Witkowski

Department of Electrical and Electronic Engineering
Imperial College of Science Technology and Medicine
Exhibition Road
London SW7 2BT
United Kingdom
m.witkowski@ic.ac.uk

Abstract.

How can we expect an A-life Agent to learn how to perform tasks when it is not told what those tasks are, and it is not provided any indication or feedback as to its performance? This is at the heart of the unsupervised learning problem. If the Agent were able to learn in this manner, how could specific tasks be communicated to it? This is the Goal setting problem. Having been set a task, how would the Agent go about choosing things to do that will lead it to perform those tasks in an orderly manner? This is at the heart of the action selection problem.

1 Introduction

This paper presents an ingenious approach to solving these three closely related problems, unsupervised learning, goal setting and action selection, for a class of A-life Agents. The method presented, the *Dynamic Expectancy Model* (DEM), builds on the "sensori-motor" ([9]), "intermediate level cognitive" model ([3]), or "cognitive action theory" ([11]) approaches. Each uses a three-part representation for "knowledge" within the Agent – the "Schema". Such schema are formulated as "context – action – outcome" triples. The method presented here overcomes some significant technical problems of these earlier models. In particular we present considerably improved strategies for the creation of new schema-like objects and a robust and flexible mechanism for action selection.

This paper builds on the conjecture that the proper interpretation of this triple is that of an *expectation*, and that the strength of the connection between context and action, and the outcome should depend only on the predictive performance of the unit. A conventional view would hold that the strength of the connection should be related to any goal or task specific "desirability" of the outcome. In the DEM these expectation triples are referred to as *μ-hypotheses*. So called because each is a "micro observation" about the Agent and its world that can be created, verified ("corroborated") and used by the DEM. By adopting the predictive view strength changes can be made internally, and task independently, just by testing whether the predicted event did or did not occur, independently of reward or reliance on an external agent to indicate correctness. Strength changes can be applied immediately the prediction is corroborated, and are always attributed to the specific μ-hypothesis that made the prediction. This leaves the μ-hypothesis uncommitted to any particular goal; learning is not task directed. In this manner the unsupervised learning problem is addressed. Prediction is used in two ways to learn. First to corroborate μ-hypotheses – *tactical learning*, and second, by detecting unexpected events, to trigger the creation of new μ-hypotheses – *strategic learning*. This ability to learn in the absence of motivation is closely related to the *latent learning* phenomena, ([6], [10]).

From time to time the Agent will be called upon to perform certain tasks. By associating a degree of motivation to any particular outcome the Agent is signalled that it should try and achieve that outcome. In this manner, the goal setting problem is addressed.

There may be many possible choices of context and action that will lead to the desired outcome. The Agent may select amongst these primarily on the basis of the corroboration between these components. Where no action may be selected because none of the required contexts are available, then, as context and outcome are uniformly represented, the Agent may chain μ-hypotheses to form a "policy map" of options where μ-hypothesis contexts match the current circumstances. This chaining process is referred to as *spreading activation* ([8], [10]) in the DEM. The DEM provides a uniform measure (the *policy value*) which allows the Agent to select between competing alternatives for selection. In this manner the action selection problem is addressed.

Action selection and learning models based on Reinforcement Learning ([7], [13], [16]) propagate the effects of occasional reward backward to rank sense-act pairings in the form of a "policy map" according to iteratively developed estimates of future reward. Neural networks provide a sensed input to output mapping function which may be refined by a variety of well known learning methods ([5], [14]). Genetic Learning algorithms rank and refine sets of condition-action rules according to some "fitness function" (past payoff in the case of [4]). Classifier Systems ([1], [10]) combine the notion of propagating the effects of occasional reward, using a "bucket-brigade" method, for behavior selection with a genetic algorithm approach to create new classifier selection elements. Alternatively, action selection may be pre-defined, with no little or no learning component ([2], [8] and [15] for a useful review).

The primary purpose of this paper is to describe and define the Dynamic Expectancy Model mechanism, and the next sections provide a description of the individual processes that comprise the DEM. Each process is performed once in an extended "sense, corroborate, evaluate goals, create policy map, act, predict, learn" cycle. All the activities of the Agent are encapsulated into this cycle, which repeats *ad infinitum*. The paper concludes with a simple, but illustrative, example of the DEM in operation, demonstrating both latent learning and its ability to respond rapidly to changing motivations, followed by some discussion.

2 μ-Hypotheses, Signs and Actions - Making Predictions

All working information in the DEM is held in five main "Lists", the Hypothesis List, Sign List, Action List, Goal List and the Prediction List[1]. During each cycle the system tests the sensory apparatus of the Agent and every element of each list, modifying List and List elements (and selects an action to perform) according to the processes described next.

All μ-hypotheses known to the DEM at any time are retained in the *Hypothesis List* (abbreviated to $\mathcal{H}$). The form of the μ-hypothesis (h, $h \in \mathcal{H}$) is:

$$h\colon\ s' \wedge a \rightarrow^{t \pm \tau} s'' \qquad \text{(eqn. 1)}$$

Each μ-hypothesis should be read as an expectation of the form "performing the action a in the context of s' predicts the occurrence of the condition s'' at time t in the future". The time t is bracketed by $\pm\tau$, forming a range a times to generalise the prediction in the temporal domain ($t >> \tau$) and to overcome the effects of sensor sampling aliasing. The overall predictive ability for each μ-hypothesis, the strength of the predictive connection "$\rightarrow$" is recorded in a numeric variable, the *corroboration measure* (C_m).

The context (s') and outcome (s'') terms in eqn. 1 are *Signs* drawn from the *Sign List* (abbreviated S, thus $s' \in S$, $s'' \in S$). Signs both define and detect *situations* that can be recognised by the Model. Signs are derived from, and form the interface to the sensory apparatus available to a physical Agent. Signs are compound items, conjunctions of elemental sensory items ("Tokens"). At each cycle in the execution of the algorithm each Sign is either detected or is absent and so evaluates to *active* or *inactive* for that cycle. All active Signs for a cycle are held on the Active Sign List S^*, a subset of S. The first time a Token is encountered the DEM automatically creates a Sign (containing only that Token) and adds it to the Sign List. New Signs are also appended to a working list S^{new}. This list drives the structural learning process. Tokens, Signs and actions have no *a-priori* meaning to the learning mechanism in the Model, and may be named arbitrarily to suit experimenter or user.

As a special case Signs will equate directly to a "State", a unique situation detected reliably by the Sign. A Sign may also equate to a "partially observed state" (P.O. State), a unique situation unreliably detected by the Sign. Equally, a Sign may just represent a collection of sensory conditions available to the Agent, from which it must create predictions of increasing reliability and repeatability. This notion of "State" is useful when evaluating the model formally, but it is a poor metaphor for a realistic agent problem. The latter case allows Signs to act as abstractions. Adding tokens to a Sign makes it more specific; deleting them makes it more generally

---

[1] *A note on notation.* Each of the lists is denoted by a single, upper case, calligraphic letter ($\mathcal{H}$ (Hypothesis), S (Sign), A (Action), G (Goal) and P (Prediction)). Individual elements are denoted by lower case letters (h, s, a, g and p respectively). Sub-sets of lists, and individual elements that must be identified across steps or cycles are indicated by superscripts (e.g. S^*). The symbol '$\in$' may be read as "member of", '$\cup$' as "union of", '$\cap$' as "intersection of" and '$\wedge$' as "concurrently with". Other symbols are explained as they are encountered.

applicable. In these circumstances we would expect the formation of many "candidate" μ-hypotheses to be created, verified and possibly discarded before a viable Hypothesis List is formed.

Actions, (a; $a \in \mathcal{A}$, the *Action List*) define the activities the Agent may perform. Where Signs defined the interface to the sensory apparatus, actions connect the DEM to the Agent's physical actuators. Selected actions are placed onto the $\mathcal{A}^*$ (active) sub-list. Every action a has associated with it an *action cost*. The action cost measure indicates the relative effort that will be required to complete the action. Action costs may be expressed in any units (such as elapsed time or energy expended) that may be determined and consistently applied across all the elements of $\mathcal{A}$. The action cost measure will be used in the "cost estimation" process for goal directed action selection. The DEM also maintains a memory of recent activations and their associated timings for both Signs (from $\mathcal{S}^*$) and actions (from $\mathcal{A}^*$). Information held on these *activation traces* is used by the structural learning component to construct new μ-hypotheses.

A μ-hypothesis is deemed active (and so placed on $\mathcal{H}^*$) whenever both its context Sign (s') and its action (a) are active simultaneously, ($s' \in \mathcal{S}^* \wedge a \in \mathcal{A}^*$). A new prediction p is created and added to the *Prediction List* $\mathcal{P}$ for every instance of an activated μ-hypothesis. Note in particular that this mechanism is invoked for all μ-hypotheses that meet these criteria and that a prediction records three items. First the identity of the Sign predicted (s^p, s'' from the active μ-hypothesis). Second the time (derived from *t*, eqn. 1, and the current time) that the Sign is predicted to occur and third the identity of the μ-hypothesis that made the prediction (h^p). Elements of the Prediction List are active (and so placed on $\mathcal{P}^*$) when the time element recorded matches the current time within the bounds defined by τ. The presence of active predictions drives the corroboration process.

3 Corroboration - Tactical Learning

For each active prediction, the corroboration measure (C_m) of the μ-hypothesis responsible for the prediction (h^p) is modified according to:

$$C_m = C_m + \alpha(1 - C_m) \qquad \text{(eqn. 2)}$$

where the prediction was successful (that is, when $s^p \in \mathcal{S}^*$, s^p being the Sign recorded at the time of making the prediction), and

$$C_m = C_m - \beta(C_m) \qquad \text{(eqn. 3)}$$

where the prediction was unsuccessful ($s^p \notin \mathcal{S}^*$). Active predictions are discarded from $\mathcal{P}$ once this step is complete.

The positive *reinforcement rate*, α ($0 \leq \alpha \leq 1$), defines the rate at which successful predictions will strengthen C_m. Similarly, the *extinction rate*, β ($0 \leq \beta \leq 1$), defines the rate at which C_m will be weakened by failed predictions. Where no prediction was made the value of C_m remains unchanged. Sequences of successful (or unsuccessful) predictions give rise to the familiar negatively accelerating learning curve, the values being normalized such that C_m rises asymptotically toward 1.0 (or falls toward 0.0).

4 μ-Hypothesis Acquisition - Structural Learning

Prediction, or rather the failure to predict an event, drives the structural learning component of the DEM, which is responsible for forming new μ-hypotheses. The opportunity to create new μ-hypotheses is indicated by appearance for the first time of a previously unknown ("novel") Sign or by the appearance of a known but unpredicted ("unexpected") Sign. New, unknown, Signs trigger the *creation by novelty* method, recall that the first occurrence of a previously unknown Sign is recorded in $\mathcal{S}^{new}$ specifically to invoke this learning method. The appearance of an unpredicted, but previously known, Sign invokes the *creation by unexpected event* method. Unexpected Signs are detected by comparing the active Prediction List to the active Sign List and applying the method to the unpredicted residue ($\mathcal{S}^*$ - ($\mathcal{P}^* \cap \mathcal{S}^*$)).

In either method a new μ-hypothesis is constructed from the novel or unpredicted Sign as 's''', and a Sign (s') and action (a) drawn respectively from the recorded activation trace of values in the Sign and Response Lists. The timing relationship (t and hence τ in eqn. 1) is derived from their relative positions in the respective memory traces. Note that the structural learning mechanism is independent of the source of the Signs and actions it will employ. Riolo ([10]) and Shen ([12]) have described broadly similar strategies for "rule" creation triggered by "surprise" events.

To limit the rate at which new μ-hypotheses are created the user may specify a *learning probability rate*, λ, which determines the probability with which a new μ-hypothesis will be formed given one of these opportunities to do so. The Dynamic Expectancy Model also defines methods for differentiating partially effective μ-hypotheses by making their component Signs more or less specific by adding (or removing) Tokens, and removing ineffective μ-hypotheses. The requirements for such additional processes are considered further in [17].

5 The Action Selection Mechanism

At each execution cycle the Agent must have some action to perform. Normally, the Dynamic Expectancy Model operates in two distinct modes for action selection (1) *Goal directed selection*, (2) *Exploratory selection.* Whenever goal directed action selection is selected, the algorithm attempts to construct a *Dynamic Policy Map* (DPM) from which it may select an action. The construction and use of the DPM is described later.

Where no goal is set the system selects actions in exploratory mode. In the current implementation, exploratory selection is made on a random basis. Regardless of how the action was selected the learning mechanism continually monitors the activities of the Agent, and corroborates existing μ-hypotheses and creates new ones according to the learning strategies described.

Goals - instructions to the system to select actions with respect to some purpose - are held on the *Goal List* ($\mathcal{G}$). Individual goals are drawn from the Sign List. Placing a Sign on the Goal List is a signal to the Agent that it should be motivated to select actions that cause that Sign to become activated (i.e. appear on $\mathcal{S}^*$). Each goal Sign on $\mathcal{G}$ is assigned a *priority*. The goal with the highest priority at any time is referred to as the *top-goal*. Once a goal Sign has been activated (i.e. it appears on $\mathcal{S}^*$), it is deemed *satisfied* and removed automatically from the Goal List. The next highest priority goal becomes top-goal, or the Goal List becomes empty.

Purposive goals are, by definition, largely domain specific - they serve some purpose. The Dynamic Expectancy Model provides two distinct routes to setting goals. (1) Goals may be programmed into or be inherent to the Agent, or (2) they may be imposed externally by directly manipulating $\mathcal{G}$. The former route equates directly to our intuitive notion of primary reinforcer. Some things, such as food for a hungry animal, or water for a thirsty one, inherently motivate because they are "programmed" to do so. In a mobile robot context the detection of a "battery_low" Sign may cause an "on_charge" Sign to be placed on the Goal List, possibly with a priority related to the extent of battery discharge. The latter route provides an experimenter with a method with which to manipulate the goal-driven behaviour of the Agent directly.

6 Building the Dynamic Policy Map

Whenever a top-goal is available, the DEM will attempt to create a Dynamic Policy Map (DPM) to form a sequence of links from every other Sign in $\mathcal{S}$ to the Sign currently set as top-goal. The DPM is conveniently represented as a graph where context Signs associated with individual μ-hypotheses represent the nodes and actions embedded within individual μ-hypotheses the arcs. The DPM is created by a process of *spreading activation* from the top-goal. The method used to construct the DPM is a modified form of the standard breadth-first graph search/construction algorithm. Each arc has associated with it a *cost estimate*, C_e, value. This cost estimate is computed from the given action cost of α and the C_m (eqns. 2 and 3) value defined earlier:

$$C_e \leftarrow \text{action\_cost}(\alpha) \,/\, C_m \qquad \text{(eqn. 4)}$$

Consider a situation where C_m is simply p(number of successful predictions|total predictions made) by a μ-hypothesis - the probability that the μ-hypothesis predicts correctly. The cost estimate value C_e is then reasonably interpreted as the total estimated cost for the average number of attempts that must be made with the given μ-hypothesis to achieve the transition. A similar interpretation may be placed on the case for C_m shown in eqn. 4, with the proviso that the "averages" are now biased towards recent experiences.

Each node (Sign) in the graph will acquire a *valence level*, *v*, indicating the number of arcs, *n*, that must be traversed to reach the top-goal "node". The top-goal has a valence level of zero, the s' Sign of any μ-hypothesis that leads directly to the goal (i.e. where s'' = top-goal) a valence level of 1, and so on. The *policy value*, P_v, of any node s' at level n in the DPM is then expressed as a summation of individual estimated costs $((C_e)^v)$ by:

$$P_v(s') \leftarrow \min \left(\sum_{v=1}^{v=n} (C_e)^v \right) \qquad \text{(eqn. 5)}$$

The policy value for each Sign s' implicated in the DPM is computed by adding the cost estimate for its transition to the minimum cost of the path to its s'' node. If a lower cost path is encountered the spreading activation is re-activated for that node to minimize path costs at higher valence levels.

Construction of the Dynamic Policy Map is complete when there are no further μ-hypotheses that can be implicated, and no further path cost minimization can occur. Following construction of the DPM the Agent has an estimate of the total "cost" to attain the top-goal for every s' implicated in the map. Once DPM construction is complete, the DEM may simply select the action associated with $(\min(P_v(s' \in S^*))$ in the μ-hypothesis containing that s' and route it to the Agent.

If a currently active Sign is included as a node in the DPM (DPM $\cap$ S^*), the action a included in the μ-hypothesis arc associated with the Sign node with the lowest P_v (eqn. 5) is selected. This is the action with the lowest overall estimated cost to achieve the top-goal. Where there is no intersection between the set of active Signs and nodes on the DPM, an exploratory action is selected. These new actions will either (1) reach the goal directly, (2) lead to a situation where a action selection from the DPM may continue, or (3) cause new μ-hypotheses to be created, which in turn expands scope of the DPM. The DPM is recomputed frequently, whenever goals change, new μ-hypotheses are formed or existing ones have undergone sufficient corroboration to indicate that a different solution path may be preferable.

7 An Example

To illustrate the two essential of the properties of the DEM algorithm, unsupervised learning and policy map generation, figure 1 shows a simulated robot learning task for navigation. The robot may recognize some 74 individual locations on a grid within the environment. These equate directly to individual Signs (and, in this instance, to "states"). The robot is supplied with four actions with which to traverse between these locations. These experimental conditions (but not the actual layout) accurately reflect those described by Sutton ([13]). We note that, in simulation, each action takes 2.66 seconds on average. This is used as the action cost. Initially the robot is allowed 2000 exploratory knowledge gathering (randomly selected) actions. This is a period of latent learning, no goal is set nor any other form of reward provided during this period. Nevertheless a Sign List and a corpus of μ-hypotheses is constructed in the Hypothesis List by the novelty and unexpected event methods, and subsequently corroborated by the tactical learning method. Initially both Lists were empty. Random exploration is inefficient in this environment, the robot tending to become "trapped" in "rooms" for extended periods, but this number of actions ensures that every location is visited more than once. Ordered exploration techniques can improve the time to complete the exploration procedure, but otherwise have marginal impact on the underlying learning mechanism.

Immediately following this period of exploration the Sign detecting location "A" is established as top-goal by the experimenter. The DEM computes the Dynamic Policy Map cost estimates, which are visualized in Figure 1a. The column heights corresponding to the total cost estimate from that Sign/location (their position) to the goal. After so much exploration, the task is learnt well. The Dynamic Policy Map corresponds closely to our intuition of a "cost gradient", flowing from "room" edges, through "doors", along the central "corridor" etc. The use of a navigation based example here allows us to think of the Dynamic Policy Map as a "Cognitive Map" ([11]) in a quite literal sense. In other agent based applications there is no clear mapping of cost estimate to place and satisfactory visualization is harder to achieve. Next the Sign detecting location "B" is made top-goal, and the DPM immediately adopts the cost estimate configuration of Figure 1b. Figure 1c shows the DPM cost estimates when the Sign for location "C" is established as top-goal.

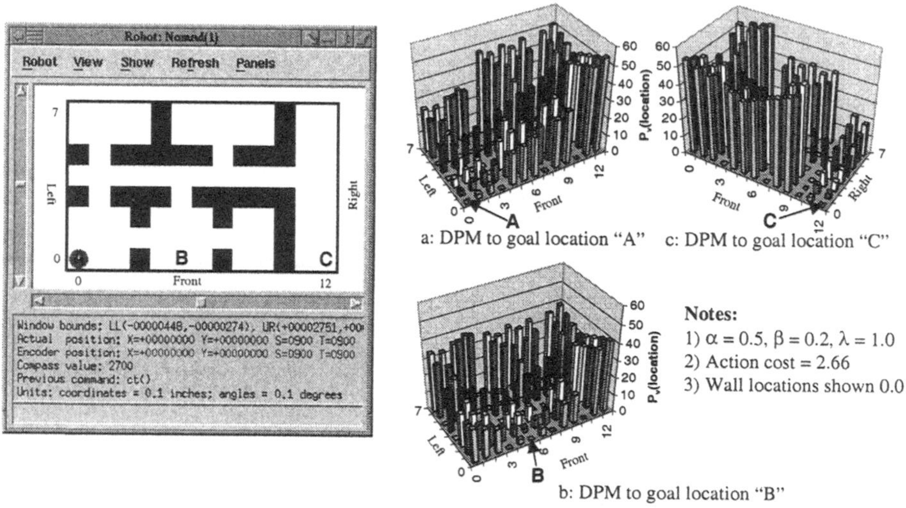

Figure 1: Robot Task and Dynamic Policy Maps for Different Goal Locations

Not shown in these visualizations is the action that is associated with each Sign for each of the three different Dynamic Policy Maps. In each case the action that will be selected with the agent at any particular location represents the first action on the path with the lowest total estimated cost. As the goal, and hence the DPM changes, the action selected in response to any particular Sign stimulus may therefore change dramatically. The spreading activation algorithm is fast, the DPM in these examples being computed in "real-time" (<10mS on a 166MHz Pentium P5 running Linux). [17] describes an extensive series of investigations using the DEM under a variety of learning conditions, imposed uncertainty and varying environments.

8 Discussion

The Dynamic Expectancy Model addresses and unifies three issues central to the construction of a fully autonomous A-life Agent. First, that of true unsupervised learning based on particular properties of the prediction process. Second, the ability to define motivations that drive the overt behavior of the Agent. Third, a mechanism to select actions according to a uniform measure relative to the Agent's current state of knowledge (the Hypothesis List), its current motivations (the top-goal) and the situation as the Agent perceives it (S^*).

The contribution afforded by this work is not so much in the knowledge that all the things it does are possible, but in the careful selection and tight integration of each of the parts to produce an Agent controller capable of robust action selection performance and self-sustaining, self-sufficient learning. The DEM is distinguished from the overwhelming majority of learning reactive systems[2] in its decisive adoption of internal prediction to corroborate schema like connections and away from reliance on external or task specific reward.

The structural learning component of the DEM represents a significant advance over that used by its (arguably) closest precursor system, due to Drescher ([3]). Drescher's computationally intensive "marginal attribution" schema learning process is discarded in favor of the use of the learning by novelty and unexpected event. Marginal attribution requires extended periods of exploration to first construct viable action – outcome pairs, followed by further periods of exploration to establish and subsequently extend the context part of the new schemas. In the DEM μ-hypotheses (the schema like objects) are created as the opportunity arises, but always from a combination of events that can occur in the environment (they did occur – and were stored in the activation traces.)

This method is also clearly distinct, in its directness, from the mutation and crossover approach to structural learning used in genetic algorithms and classifier systems ([1], [4], [5], [10]). Riolo ([10]) describes CFSC2, extending the classifier system model with a three part schema like representation, which adopts a forward chaining approach to detecting possible future rewards rather than the goal directed backward chaining approach used in the DEM.

The Dynamic Policy Map is not a plan; it is a temporary mapping between sensory conditions and actions to take. It does not define a path from start to finish; rather it is a characterization of the Signs known to the system according to an estimated cost from that Sign to the primary source of motivation, the top-goal. In use, it is similar to a reactive look-up table: if this is the current situation, then select this action. In this respect, it is similar to the policy map of, say, Watkins' ([16]) or Sutton's ([13]) *Q*-learning based algorithms. It is profoundly different in that the policy is computed (and re-computed) frequently and quickly, relative to a specific goal, rather than as an iteratively formed static policy estimating future discounted and anonymous rewards.

In comparative tests, [17] using the highly stylized navigation tasks of the type described in section 7, the DEM can show dramatic performance improvements (up to the order of 40:1) over a conventional *Q*-learning algorithm (compared to results presented in [13]). This is due almost entirely to the fact that the connection strength (C_m) can be updated at every step, rather than occasionally at the end of a long

---

[2] A notable exception being Tani's neural network based "prediction learning" [14] method.

sequence of actions. In more realistic environments other factors tend to mask these apparent gains.

The DEM can be seen as an exemplar of Grefenstette's notion of "Anytime Learning" ([4]). New μ-hypotheses can be formed at any point in the Agent's existence, immediately extending its behavioral repertoire. Corroboration is also an ongoing process throughout the Agent's "lifetime".

The Dynamic Expectancy Model defines a style of autonomous Agent controller, it has been implemented and tested in a range of conditions. It learns autonomously and has been found to be flexible and responsive to changing conditions. Work on the Model continues.

References

[1] Booker, L.B., Goldberg, D.E. and Holland, J.H: Classifier Systems and Genetic Algorithms, in: Carbonell, J.G. (Ed.) Machine Learning: Paradigms and Methods, The MIT Press, (1990) 235-282

[2] DeCugis, V. and Ferber, J.: An Extension of Maes' Action Selection Mechanism for Animats, 5th Int. Conf. on Simulation of Adaptive Behavior (1998) 153-158

[3] Drescher, G.L.: Made-up Minds: A Constructivist Approach to Artificial Intelligence, The MIT Press, Cambridge, MA (1991)

[4] Grefenstette, J.J. and Ramsey, C.L.: An Approach to Anytime Learning, 9th Int. Workshop on Machine Learning (1992) 189-195

[5] Grossberg, S.: Birth of a Learning Law, Boston University Technical Report CAS/CNS-TR-97-017 (1997) [http://cns-web.bu.edu/Profiles/Grossberg/Learning.html]

[6] Hergenhain, B.R. and Olson, M.H.: An Introduction to Theories of Learning, Prentice Hall, New Jersey (1993)

[7] Humphrys, M.: Action Selection Methods using Reinforcement Learning, 5th Int. Conf. on Simulation of Adaptive Behavior (1998) 135-144

[8] Maes, P.: A Bottom-up Mechanism for Behavior Selection in an Artificial Creature, 1st Int. Conf. on Simulation of Adaptive Behavior (1991) 238-246.

[9] Mott, D.H.: Sensori-motor Learning in a Mobile Robot, Dept. Comp. Science and Statistics., Queen Mary College, University of London, Ph.D. Thesis (1981)

[10] Riolo, R.L.: Lookahead Planning and Latent Learning in a Classifier System, 1st Int. Conf. on Simulation of Adaptive Behavior (1991) 316-326

[11] Roitblat, H.L.: Cognitive Action Theory as a Control Architecture, 1st Int. Conf. on Simulation of Adaptive Behavior (1991) 444-450

[12] Shen, W-M.: Autonomous Learning from the Environment, Computer Science Press, New York (1994)

[13] Sutton, R.S.: Integrated Architectures for Learning, Planning, and Reacting Based on Approximating Dynamic Programming, Proc. 7th Int. Conf. on Machine Learning, 216-224 (1990)

[14] Tani, J. and Nolfi, S.: Learning to Perceive the World as Articulated: An Approach for Hierarchical Learning in Sensory-Motor Systems, 5th Int. Conf. on Simulation of Adaptive Behavior (1998) 270-279

[15] Tyrrell, T.: Computational Mechanisms for Action Selection, University of Edinburgh, Ph.D. thesis (1993)

[16] Watkins, C.J.C.H.: Learning from Delayed Rewards, King's College, Cambridge University, Ph.D. thesis (1989)

[17] Witkowski, M.: Schemes for Learning and Behaviour: a New Expectancy Model, Dept. of Computer Science, Queen Mary Westfield College (Univ. London), Ph.D. Thesis (1997)

Can Computers Have Sentiments? The Case of Risk Aversion and Utility for Wealth

George G. Szpiro

Israeli Center for Academic Studies
(affiliated with the University of Manchester)
Kiriat Ono, Israel

george@netvision.net.il

Abstract. Genetic algorithms display at least one characteristic that is typical of the economic behavior of human decision-makers. I show that if a choice problem involves uncertainty, genetic algorithms may produce results that are consistent with an aversion to risk.

1. Introduction

Genetic algorithms are generally employed to optimize an objective function in a high-dimensional search space. In this paper, a further dimension is added to the algorithm's chore: uncertainty. As I show, a very important consequence of uncertainty is that the genetic algorithm produces solutions that exhibit risk aversion. In fact, risk-aversion may have evolved in decision making behavior of humans in the manner suggested by the algorithm. (See Szpiro [1997a].) Since risk averse behavior in humans is a consequence of the existence of utility function, I make the bold (and somewhat tongue-in-cheek!) claim that genetic algorithms exhibit utility for wealth.

Applications of genetic algorithms in the field of economics are, for example, Holland and Miller [1991], Arthur [1991], Palmer et al. [1994], Arifovic [1994], Allen and Karjalainen [1993], Vriend [1994], Andreoni and Miller [1995], and Szpiro [1997a and 1997b]. Researchers and practitioners who seek solutions to optimization problems must be aware of the fact that genetic algorithms are not "risk-neutral", but that they find solutions that are consistent with aversion to risk. Apart from shedding more light on the use of genetic algorithms for economic decision problems, the results of this paper can be considered a further example in the burgeoning literature on emerging behavior and artificial life. The application of evolutionary thinking and of genetic algorithms to problems in biology, psychology, linguistics, political science, economics, management science and other disciplines have shown that complex phenomena, which resemble the behavior of human beings or the attributes of societies, can spontaneously evolve in such computer simulations. (See, for example, Langton [1989] or Kauffman [1995].) The emergence of risk aversion and utility functions is an additional instance of such a phenomenon.

The next section gives a short description of genetic algorithms and presents an investment choice model to which the algorithm will be applied. Section 3 presents the evidence on risk aversion that emerges through the genetic algorithm. I also analyze the reasons that underlie this phenomenon and then show that the algorithm's risk-aversion is linked to the selection against bankruptcy risk. In Section 4 I assess the algorithm's degree of risk aversion and determine its "utility for wealth". Section 6 concludes with a short summary.

2. Choice under Uncertainty

I will not discuss genetic algorithms in detail and, instead, refer the reader to Szpiro (1997a). Let me now introduce a simple choice model to which we apply the algorithm. Agents must divide their wealth between cash and stock. Cash returns a yearly, certain interest i, stock carries yearly dividends. In each period an exogenous price is given for the shares, which is normally distributed with mean value P and standard deviation σ. The dividend rate is also a random variable, normally distributed with mean δP and standard deviation $\delta\sigma$. We allow the agents to go into debt and to hold stock short.[1] In this case debtors are *charged* interest, short holders of stock must *pay* the dividends.

In each period all agents have to make a decision, as to how many shares to buy or sell. We assume that the agents' decisions are made according to the formula

$$B_t = (\beta - P_t) \tag{1}$$

where B_t is the number of shares which are purchased, or - if negative - sold, in period t, and where β is a parameter particular to each agent. (The lower the market price, the more stock do the agents buy, and vice versa.)

The agents start with an initial cash position C_0, trade, and get or pay interest and dividends at the end of the period. Without going into details, we state that an agent's wealth at the end of t periods is

$$\begin{aligned} W_t &= C_0(1+i)^t + \sum_{\tau=1}^{t} S_{\tau-1}\delta_\tau(1+i)^{t-\tau+1} - \sum_{\tau=1}^{t}(\beta - P_\tau)P_\tau(1+i)^{t-\tau} + \sum_{\tau=1}^{t}(\beta - P_\tau)P_t \\ &= C_0(1+i)^t + (\beta - P - \tilde{z})\frac{P+\tilde{z}}{i}\left[\left((1+i)^t - 1\right)\left(\delta\frac{1+i}{i} - 1\right) - \delta t(1+i)\right] + (\beta - P - \tilde{z})tP_t \end{aligned} \tag{2}$$

where P_τ is the share price, δ_τ the dividend, and S_τ the accumulated amount of shares in period τ, P_t the share price at the end of the horizon, P the mean market price of the share, and $z^{\sim}$ a normal deviate with standard deviation σ. The agent's expected wealth becomes,

[1] Holding stock "short" means that the agent sells stock he/she does not yet own, in the hope that its price will fall. Then, at a later point in time, the stock is bought at a lower price.

$$E(W_t) = C_0(1+i)^t + (\beta - P - \frac{\sigma^2}{P})\frac{P}{i}\left[\left(\delta\frac{1+i}{i} - 1\right)\left((1+i)^t - (1+ti)\right)\right]. \qquad (3)$$

In order for expected wealth to be positive, β must be greater than $P + \sigma^2/P$ whenever,

$$\delta > \delta^* = \frac{i}{1+i}, \qquad (4)$$

and vice versa for

$$\delta < \delta^* = \frac{i}{1+i}. \qquad (5)$$

In particular, in order to *maximize* expected wealth we expect corner solutions, since the latter is a linear function of β. (See equation 3.) Hence the β-values of "rational" agents should evolve towards β_{max}, whenever equation (4) holds, and towards *zero* when equation (5) holds.

We now describe the subroutines of the algorithm that has been designed to solve the problem in an evolutionary manner:

- *Initialization.* Each agent receives an initial endowment of cash, C_0, (which could be zero), and a β-value, randomly chosen between zero and β_{max}.
- *Trading.* A market price for the stock and a dividend rate are given exogenously. The agents buy or sell stock according to their trading formula (equation 2). They then receive (or pay) the interest on their cash holdings and the dividends on their stock holdings. This is repeated for *t* trading periods.
- *Computing the fitness.* After *t* trading periods, the wealth of the agents is computed, based on the current market price of the stock.
- *Ranking of the agents.* The agents are ranked according to their market wealth.
- *Choice of mates.* The fittest agent has first choice. It chooses a mate from among the remaining agents, the probability of any one being chosen being equal to its wealth. Then the next fittest chooses, etc., until half the agents have formed pairs The remaining half die.
- *Reproduction.* Each pair has four offspring. We chose the following way of passing "genetic information" about β: two offspring receive values of β identical to that of their parents; the β-values of the two other offspring are computed as the arithmetic mean of the β-values of the two parents.[2] (With the first two offspring, fitness is preserved, while provisions for improvement are made with the second pair of offspring.) The combined cash and stock holdings of the parent-pair are divided equally among the four offspring.
- *Mutation.* A small percentage of the agents' β-values are mutated at random. The mutated β's again lie between zero and β_{max}.

---

[2] Genetic algorithms generally use the so-called crossover operator to pass on information to the next generation. This involves translating the parents' β-values into binary notation, interchanging parts of the binary strings, and re-translating. I use the less sophisticated, but simpler method of averaging the parents' β-values. This creates strong pressure for reversion of the β-values to mean, however, and further studies should investigate the effect of binary encoding and bit-string crossover.

The subroutines of the algorithm, except *Initialization*, are re-run for a number of generations.

3. Certainty vs. Uncertainty

The genetic algorithm is now applied to the investment choice problem described in the previous section. The population consists of 120 agents, who are each endowed with an initial value of β, randomly distributed between zero and 100. Initial cash is zero, the interest rate is 15 percent. Hence $\delta^* = {}^{0.15}/_{1.15}$. The algorithm goes through fifty generations of trading, ranking and reproduction, and after the agents of the last generation have been ranked, the top 90 percent are "polled", i.e., the mean of their β-values is computed.[3] In order to eliminate random variations, the algorithm is rerun a few dozen or a few hundred times, and the average β of these runs is computed. The mutation rate in each generation is 2.5%, and initially the agents' time horizon is two trading periods.

Let us first investigate what happens under conditions of certainty. ($P = 50$, and δ = 5, 6, ..., 25%.) Since $\sigma = 0$, the market price and the dividend rate is given with certainty in every period. Figure 1 presents the β-values that evolve in this situation. To the uninitiated it may seem quite surprising at first, that the genetic algorithm produces exactly the results that are predicted by the theory: for $\delta < 13\%$, the average value of β is close to zero, at $\delta = 13\%$, β abruptly jumps to a value close to one, and stays there thereafter.

Fig. 1. Certainty

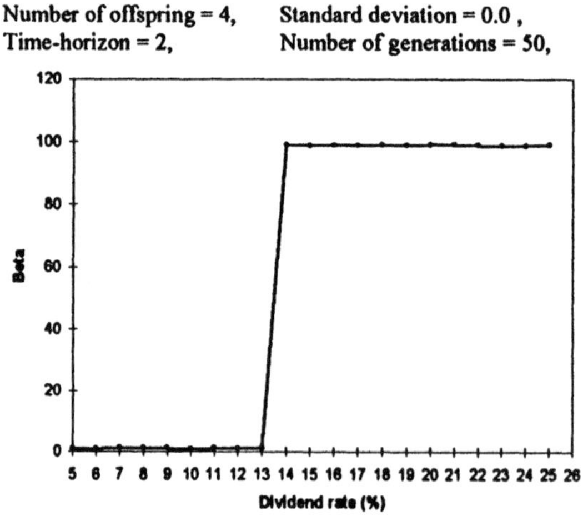

[3] Taking less than 100 percent of the surviving population eliminates some of the worst performers, who would probably have been weeded out in the following generation anyway.

Moreover, the results bred by genetic algorithms are very precise: the program finds the point of indifference between buying and selling stock with an accuracy that is equal to the precision of a pocket calculator. When running the algorithm with increments in the dividend rate of 0.00001% the jump from $\beta \approx 0$ to $\beta \approx 1$ takes place at 13.04348%, which is the precise value of δ^*, to five digits after the decimal point.

But the more interesting phenomenon occurs when uncertainty enters the model. Let the price of the stock and the dividend rate be normally distributed around P and δ, with standard deviations σ and $\delta\sigma$, respectively. (P = 50, and δ = 5, 6, ... 25% .)[4] Figure 2 depicts the results for standard deviations of 1, 3, and 5.

Fig. 2. Uncertainty

Details as in Figure 1, except Standard deviation = 1.0, 3.0, and 5.0

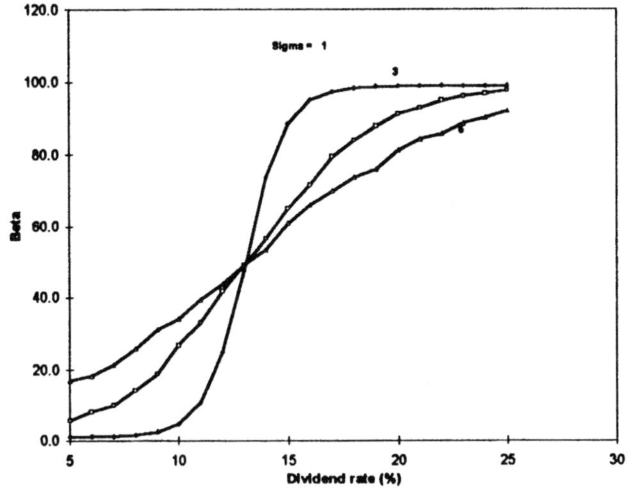

We note that the β-values are not where we expect them to be when compared to Figures 1 and 2 ! For $\delta < 13\%$ the β-values are close to, but distinctly greater than zero, for $\delta > 13\%$ they are close to, but distinctly smaller than unity. The discrepancies are larger the greater the uncertainty. This means that whenever our artificial economy is governed by uncertainty, the agents do not buy or sell shares to the full extent of their possibilities. They have become cautious, that is, risk averse.

How does the phenomenon of risk aversion arise in genetic algorithms that, in principle, do nothing but carry out computations in a "disinterested", that is in a risk-neutral fashion? The answer to this question goes to the heart of evolution, of genetic algorithms, and of the realities of business life. After computing the fitness, that is the wealth of the 120 agents, risk neutral agents, who bought (or sold) the maximum possible amount of stock, sometimes find themselves at the top of the ranking, at

---

[4] The stock markets could be more appropriately modeled with the log-normal distribution. For the sake of simplicity, I use the normal distribution..

other times at the bottom, depending on whether they face a bull market or a bear market.[5] Let us assume a bull market. Obviously risk-neutral agents come out on top. But unless the direction of the market movement persists for an inordinately long time, so that risk neutral agents can accumulate an extreme amount of wealth, these risk neutral agents inevitably fall from the top of the list to the bottom when the market turns against them. In this case they die, and their β-values will no longer be transmitted to subsequent generations. Even though random mutations may produce risk-neutral agents again in a future generation, sooner or later they will disappear too.

Now let us investigate the fate of risk averse agents. Cautious as they are, they buy and sell less stock than their risk-neutral colleagues. In an advantageous market they probably will not make it all the way to the top of the list. On the other hand, they usually do not fall quite to the bottom of the ranking either when the market turns against them. Hence, they and their β-values are the candidates most likely to survive for 50 generations. *A fortiori*, risk loving agents are weeded out even faster than the risk neutral ones.

Hence the emergence of risk aversion is a consequence of the risk of death, the risk of an evolutionary dead-end. Only the agents who are ranked in the top half of the list choose partners and will reproduce, while the lower-ranked ones tend to die out. And with their demise the features that characterized their behavior also disappear. In the framework of economics the risk of death corresponds to the risk of bankruptcy. If a certain business-strategy cannot assert itself vis à vis the competition, if it is unable to cope with adverse market conditions, the corporations that adhere to it generally become insolvent. They eventually go bankrupt, more often than not without having produced any spin-off companies. As a consequence the characteristics that are responsible for the failure are likely to disappear in future generations. With genetic algorithms the situation is similar: risk-neutral or risk-taking behavior implies that agents adhering to it die whenever exogenous circumstances turn against them. Risk aversion is prevalent in genetic algorithms whenever the following two circumstances are present: there must be uncertainty, and the fitness-based selection must contain a risk of death for badly adapted agents. The existence of uncertainty implies that there is no dominant strategy; a choice-strategy that is optimal under certain conditions must lead to a low ranking and to failure when conditions change. And failure implies that the choice-strategy to which unsuccessful agent adhered eventually disappears. Hence the emergence of risk-aversion is robust: whenever these two conditions are satisfied, a genetic algorithm will exhibit risk aversion.

4. What's a Computer's Utility for Wealth ?

In economics one assumes that human decision makers have a utility for wealth *U(W)*. Two conditions on utility functions are usually stipulated:

$$U'(W) > 0 \quad \text{and} \quad U''(W) < 0\,. \tag{6}$$

[5] Brokers call a period in which most prices on the stock exchange rise, a "bull" market. As

The first condition, which expresses the fact that more wealth is preferable to less wealth, can be considered a truism. The second condition is more questionable. Risk aversion implies that the marginal utility for wealth decreases, i.e., that the second condition holds. (For a short explanation see Szpiro 1997a). However, exceptions are conceivable. (See, for example, Szpiro [1992].) In spite of this, it is generally accepted that the vast majority of economic agents are risk averse in most choice situations and, as we showed above, genetic algorithms also subscribe to this behavior. Hence the question may be asked "how risk averse is a genetic algorithm?" As I will show presently, an answer to this question implies an answer to the question "what's a genetic algorithm's utility for wealth?".[6]

At first it may seem quite absurd to attribute human-like sentiments to an inanimate object, like the computer as, for instance, the preference for more money to less, or the aversion to risk. A calculating machine has no use for wealth, nor, by the same token, does it have more utility for large numbers as opposed to small ones. If one nevertheless chooses to preoccupy oneself with such arcane reflections as "what's a computer's utility for wealth?", say in a maximization problem, one would be forgiven for assuming that a computer should exhibit behavior consistent with linear utility for wealth. After all, the machine simply carries out numerical instructions - a linear programming algorithm, for example, to maximize an objective function - so how could it mimic anything but linear utility.

But by now we are no longer surprised to find that computers – when faced with uncertainty – exhibit what can duly be described as risk aversion. Thus computers display what is known in economics as "marginally decreasing utility for wealth". The utility function which seems to govern the computer's behavior can even be estimated, and I will show in this section that under certain conditions, a computer's "utility" for wealth, $U(W)$, is given (up to integration constants) by the function,[7]

$$U(W) = -e^{-0.00032 \cdot W} . \qquad (7)$$

Further numerical experiments show, that the degree of risk aversion varies with the length of the time-horizon, the probability of survival, and possibly with other parameters. (See Szpiro 1997a.) Even the value of β_{max} , which determines the maximum number of shares that can be bought in a single period, has an effect on the degree of risk aversion. This is a bit surprising, and we will investigate the implication a little further. When β_{max} is, say, 100, risk averse agents may, for example, exhibit a β of 95 under certain conditions. When β_{max} is increased to 1000, the surviving agents will have β-values of, say, 950. So, in both cases risk aversion evolves, but its degree is related to the value of β_{max}. The agents obviously discover that more money can be made if more stock can be bought. Caution remains an essential virtue, however: when pitched against risk-neutral individuals, the risk-averse ones always win. How risk averse the agents should be in order to survive,

---

"bear" markets they designate markets with falling prices.

[6] The conclusions of this section are still somewhat speculative, and more extensive studies are warranted to confirm the numerical results.

[7] More precisely: the algorithm gives results which are consistent with a utility function of this form.

depends on what the rest of the population does. Hence, risk aversion is not an absolute, innate phenomenon of genetic algorithms as, for instance, the maxim "more is preferable to less",[8]. it is a relative phenomenon that emerges because of competition with others.

I choose a time-horizon of 15 periods, which seems a reasonable period for medium-range financial decisions, β_{max} is 100, and the probability of survival is fifty percent. The β-value which is preferred by an individual who has utility function $U(W)$ is given implicitly by the solution to the problem of maximizing expected utility of wealth,

$$EU(W_t) = \int U(W_t)\Delta(P_t)dP_t \, . \qquad \textbf{(8)}$$

where W_t is derived from equation (3), P_t is normally distributed around P, with standard deviation σ, and $\Delta(P_t)$ indicates the density of the normal distribution.

250 runs of the genetic algorithm were performed for $t = 15$, $\delta = 12.5\%$, $\sigma = 5$, $P = 50$, and for different initial cash positions ($C_0 = 0, 5000, 50000, 100000$). There is no reason that initial wealth should have any effect on the value of β and, indeed, numerical experiments showed that no significant change is discernible when C_o is varied. Thus, our first finding is that constant absolute risk aversion prevails, and therefore the utility function is (up to multiplicative and additive constants),

$$U(W) = -e^{-kW}, \qquad \textbf{(9)}$$

where k is the degree of absolute risk aversion. (For details, see Pratt [1964] or Arrow [1965].) With W_t from equation (3), this is the function that is inserted into the integrand of equation (8). Since this integral can not be solved in closed form, a software package is used to perform the integration numerically for different values of β. The β-value that gives the maximum expected utility is the one that individuals with utility function (9) prefer.

The mean β-value of the 1000 runs is 14.1, with standard error of 8.8. Actually, since the distribution of the results is asymmetric, and due to some outliers, more than 78% of the β-values lie within one standard error of the estimate, i.e., within the interval [5.3 , 22.9]. Inserting the figures into equation (8) and solving numerically, results in a value $k = 0.00032$. The 78% confidence level puts k in the interval [0.000251, 0.000442]. The measure of absolute risk aversion was defined by Pratt [1964] and Arrow [1965] as

$$r_a(W) = -U''(W)/U'(W) \, . \qquad \textbf{(10)}$$

Most empirical studies on human decision makers have shown that the degree of *relative* risk aversion,

$$r_r(W) = r_a(W)W \, , \qquad \textbf{(11)}$$

[8] "More is preferable to less" is innate to the genetic algorithm insofar as it is incorporated into the fitness function: the wealthiest agents are ranked first.

is constant for wealth, and approximately equal to two, while some studies put this value as high as 10. (See, e.g., Szpiro [1986].) Hence the degree of risk aversion of a genetic algorithm corresponds to the one of a relatively poor person, whose net assets are about $4,500 to $8,000. With wealth of, say, $10,000 the genetic algorithm exhibits a degree of relative risk aversion of 3.2, with the 78% confidence level interval being [2.5 , 14.4]. Hence, a genetic algorithm with the above parameters, exhibits greater risk aversion than does the average human decision maker, but the order of magnitude is at least similar to the higher empirical estimates.

Even though the manner in which the question at the beginning of this section was formulated was somewhat tongue-in-cheek, it is nevertheless imperative to know what degree of risk aversion the genetic algorithm exhibits, in order to assess the solutions that are found to problems containing uncertainty. On the other hand, by varying the parameters of the program one can custom design the genetic algorithm to mimic a certain attitude towards risk. In this manner the algorithm can be endowed with a pre-determined degree of risk aversion. Table 2 exhibits the β-values of some of the possibilities. By varying the number of offspring between six and 120, and simultaneously varying the time-horizon between ten and thirty periods, β-values between 6.7 and 39.0 can be simulated with appropriate combination of parameters. This corresponds to degrees of absolute risk aversion of between about 0.00028 to 0.00145, or – for a person with wealth of $10,000 – to relative risk aversion between 2.8 and 14.5.

5. Conclusions

The paper started by displaying the power of so-called genetic algorithms for solving problems of choice under certainty, and then went on to discuss the algorithms' behavior in situations of uncertainty. Seemingly incorrect answers given by the computer program could be traced back to risk aversion which, it turned out, is inherent to such algorithms as an evolutionary strategy against the risk of bankruptcy .

It may be argued that the attribute of risk aversion has evolved in the human race during the millennia of its history, because *homo oeconomicus* has learnt that with limited time-horizons it does not pay to be risk-neutral or risk-seeking. With genetic algorithms – computer programs that are designed to search for optimal solutions by emulating the Darwinian theories of evolution – the situation is identical. The first condition on the utility of wealth, that more is preferable to less, enters the algorithm through the fitness function (agents are ordered according to their wealth, the richest ones being ranked first). The second condition, which corresponds to risk aversion, is an outcome of evolutionary pressures.

References

Allen, Franklin and Risto Karjalainen [1993]
"Using genetic algorithms to find technical trading rules", Working paper 20-93, Rodney L. White Center for Financial Research, The Wharton School, University of Pennsylvania.

Andreoni, James and John H. Miller [1995]
"Auctions with artificial adaptive agents", **Games and Economic Behavior, 10**, 39-64.

Arifovic, Jasmina [1994]
"Genetic algorithm learning and the cobweb model", **J. of Economic Dynamics and Control, 18**, 3-28.

Arrow, Kenneth J. [1965]
"Aspects of the theory of risk bearing", Yrjo Jahnssonin Saatio, Helsinki.

Arthur, W. Brian [1991]
"Designing economic agents that act like human agents: a behavioral approach to bounded rationality", **American Economic Review: Papers and Proceedings**, 353-360.

Goldberg, David E. [1989]
"Genetic algorithms in search, optimization and machine learning", Addison-Wesley, Reading, MA.

Holland, John. H. [1975]
"Adaptation in natural and artificial systems", University of Michigan Press, Ann Arbor. 2nd edition 1992, MIT Press.

Holland, John. H., and J. H. Miller [1991]
"Artificial adaptive agents in economic theory", **American Economic Review: Papers and Proceedings**, 365-370.

Kauffman, Stuart A. [1995]
"Escaping the red queen effect", **McKinsey Quarterly 1995 Nr 1**, 110-129.

Koza, John R. [1992]
"Genetic programming", MIT Press, Cambridge.

Palmer, R. G., W. Brian Arthur, John H. Holland, Blake LeBaron, and Paul Taylor [1994]
"Artificial economic life: a simple model of a stockmarket", **Physica D, 75**, 264-274.

Pratt, John W. [1964]
"Risk aversion in the small and in the large", **Econometrica, 32**, 122-136.

Langton, Christopher G. [1989]
"Artificial life", Santa Fe Insitute Studies in the Science of Complexity, Christopher Langton (Ed.), Addison-Wesley, Reading, MA.

Szpiro, George G. [1986]
"Measuring risk aversion: an alternative approach", **Review of Economics and Statistics, 68**, 156-159.

Szpiro, George G. [1992]
"Insurance buying gamblers", **Theory and Decision, 32**, 203-207.

Szpiro, George G. [1997a]
"The emergence of risk aversion", **Complexity, 2**, 1997, 31-39.

Szpiro, George G. [1997b]
"A search for hidden relationships: data mining with genetic algorithms", **Computational Economics, 10, 267-277**. 1997.

Self-Replication, Self-Maintenance, and Gene Expression

Towards Robust Bio-inspired Circuits: The Embryonics Approach

Daniel Mange

Logic Systems Laboratory
Swiss Federal Institute of Technology
Lausanne, Switzerland
daniel.mange@epfl.ch

The growth and operation of all living beings are directed by the interpretation, in each of their cells, of a chemical program, the DNA string or genome. This process is the source of inspiration for the Embryonics (embryonic electronics) project, whose final objective is the design of highly robust integrated circuits, endowed with properties usually associated with the living world: self-repair (cicatrization) and self-replication. The Embryonics architecture is based on four hierarchical levels of organization: 1) the basic primitive of our system is the molecule, a multiplexer-based element of a novel programmable circuit; 2) a finite set of molecules makes up a cell, essentially a small processor with an associated memory; 3) a finite set of cells makes up an organism, an application-specific multiprocessor system; 4) the organism can itself replicate, giving rise to a population of identical organisms. Our ongoing research efforts try to meet three challenges: a scientific challenge, that of implementing the original specifications formulated by John von Neumann for the conception of a self-replicating automaton; a technical challenge, that of realizing very robust integrated circuits capable of self-repair and self-replication; and a biological challenge, that of attempting to show that the microscopic architecture of artificial and natural organisms, i.e., their genomes, share common properties.

References

1. D. Mange and M. Tomassini, editors. *Bio-inspired Computing Machines: Towards Novel Computational Architectures.* Presses Polytechniques et Universitaires Romandes, Lausanne, Switzerland, 1998.
2. D. Mange, D. Madon, A. Stauffer, and G. Tempesti. Von Neumann Revisited: A Turing Machine with Self-Repair and Self-Reproduction Properties. *Robotics and Autonomous Systems*, 22(1):35–58, 1997.
3. D. Mange, E. Sanchez, A. Stauffer, G. Tempesti, P. Marchal, and C. Piguet. Embryonics: A New Methodology for Designing Field-Programmable Gate Arrays with Self-Repair and Self-Replicating Properties. *IEEE Transactions on VLSI Systems*, 6(3):387–399, 1998.
4. G. Tempesti, D. Mange, and A. Stauffer. Self-Replicating and Self-Repairing Multicellular Automata. *Artificial Life*, 4(3):259–282, 1998.

5. G. Tempesti, D. Mange, and A. Stauffer. A Robust Multiplexer-Based FPGA Inspired by Biological Systems. *Journal of Systems Architecture: Special Issue on Dependable Parallel Computer Systems*, 43(10), 1997.
6. M. Sipper, D. Mange, and E. Sanchez. Quo Vadis Evolvable Hardware?. *Communications of the ACM*, 42(4):50–56, 1999.

Designing a Simulation Model of a Self-Maintaining Cellular System

K. Ruiz-Mirazo[1], Alvaro Moreno[1], Federico Morán[2], Juli Peretó[3], and J.J. Merelo[4]

[1] Department of Logic and Philosophy of Science, University of the Basque Country, Post Box 1249, 20080 Donostia, Spain
{sfbrumik, ylpmobea}@sf.ehu.es
[2] Department of Biochemistry, University Complutense of Madrid, Spain
fmoran@solea.quim.ucm.es
[3] Department of Biochemistry and Molecular Biology, University of València, Spain
Juli.Pereto@uv.es
[4] Department of Computer Architecture and Technology, University of Granada, Spain
jmerelo@kal-el.ugr.es

Abstract. This paper deals with the problem of finding a suitable framework for designing computer simulations that could help us determine the minimal requirements (both material and organizational) for the origin of the first full-fledged autonomous systems. The design of a particular model that takes into account some fundamental thermodynamic requirements is offered and discussed. Behind this work, there is a belief that Artificial Life models can inform biology on several fundamental questions (such as the origin and definition of life) but only provided that they assume more realistic and grounded premises to lead us to more conclusive results.

1 Introduction

The gap between complex self-organizing phenomena (physico-chemical dissipative structures) and the simplest biological entities we know of today is too big to be bridged without some intermediate stage(s). Therefore, we will focus on a hypothetical stage in which (compartmented) chemical reaction networks self-maintain recursively and operate in the environment but are not yet complex enough to self-reproduce reliably and begin a process of open-ended (Darwinian) evolution. Such a standpoint is supported by the work of different authors who defend that some sort of acellular [21] or cellular [15] metabolism must precede systems that synthesize early polymers exhibiting template activity.

We claim that these chemical reaction networks differed from previous ones in the sense that they could *manage* actively the flows of matter and energy necessary to sustain their own organization. Therefore, they should be regarded as the first *autonomous* systems. However, the problem of the origin of these 'minimal autonomous systems' is much more complex than could have been suspected years ago. In recent articles ([13], [17]) some of us have criticized Artificial Life models of (basic) living organization, arguing that certain material and energetic aspects which are crucial to understand the logic of such organization have been systematically disregarded in the field. Now we try to move forward along the same critical lines offering a concrete proposal: the design of a simulation model of a minimal (proto-biological) autonomous system that takes into account a few central issues pointed out in those previous papers.

2 Autonomy: a Fundamental Concept for Biology

The main theoretical assumption of this paper is that 'proto-metabolic' autonomous systems --even if they disappeared very quickly in a geological time scale-- played a crucial role in the origin of the first living beings. By 'autonomous system' here we mean a system in which a set of processes self-constitutes recursively through functional/adaptive interactions with its environment. The fundamental difference in relation to more simple forms of self-organization is that autonomy involves an active participation of the system in the construction and modification of its own boundary conditions and constraints as a response to external perturbations.

In the context of prebiotic chemical reaction networks the appearance of autonomous entities may be achieved only if some components (or component aggregates) in the network act as rules/constraints on other components and on the transformation processes they all go through. The result of that multiple constraining action constitutes an autonomous entity when a new set of components and aggregates is produced in a recursive way, creating a physical border that establishes the mechanisms through which the system will interact with the environment. Since autonomous systems are able to carry out functional actions --like modifying their own boundary conditions-- in a variable environment, we can say that they are able to self-maintain "adaptively" in it. Self-construction and robust self-maintenance involves achieving some active control over the flows of matter and energy through the system in order to sustain its far-from-equilibrium self-organizing processes.

Therefore, the key to basic autonomy lies on the generation of a suitable set of constraints: those that actually define the new rules of behavior of the system, its boundaries, and the terms in which it will keep the relationship with its environment. Such a set must include global and local constraints whose different actions are well coordinated, ensuring the conditions under which the system is physically viable [20]. This implies, on the one hand, developing mechanisms to solve possible physico-chemical problems (like the osmotic problem) and, on the other hand, articulating the main coupling mechanisms on which the energetic maintenance of the system is based: couplings with the external resources (transduction and transport mechanisms) and couplings among exergonic and endergonic processes within the system (by means of various energy currencies). The fundamental constraint that allows all this to be achieved is a *membrane*.[1]

[1] The relevance of cell individuality in the origin of life was already highlighted by Oparin [16]. Since then, the emergence of membrane-enclosed complex chemical networks as a crucial step in the appearance of life has been defended by numerous authors (see [7] and references therein). In the field of theoretical biology, the autopoietic school ([9][20]) has also claimed that the role of the membrane is central in the characteristic organization of the living.

Two types of basic components are necessary to generate and integrate a membrane: (i) structural components that can self-assemble and define the physical boundary of the system (global constraining action) and (ii) functional components apt to carry out different tasks --catalysis, transport, absorption of energy-- at particular loci (local and specific constraining action). An adequate combination of the constraining action of these two types of components can potentially generate a membrane that is not simply a physical border which marks a spatial distinction/asymmetry between what can happen inside and outside, but also a device which may modify functionally some of the external conditions of the system.

Thus, an autonomous system is able to carry out functional actions on its environment (active transport actions, for instance), whereas the environment only interacts with it in a physico-chemical sense. This is why autonomy is deeply connected to the concept of 'agent' (defined as a system whose influence *on* and *from* the environment is not of the same type). Acting functionally on its environment allows a system to achieve autonomy with respect to some external conditions and preserve its organization even in a situation of temporary absence of material and/or energetic input.

This dimension is crucial to understand the agency of autonomous systems, and is made manifest especially if we look at the problem of self-maintenance from a thermodynamic perspective. Purely relational-constructive approaches (like some which are briefly reviewed in the following section) miss the logic involved in the mechanisms that allow a recursive control of the flows of matter and energy needed to carry out functional autonomous actions. That leads them to hold an internalist view on autonomy, which we think must be thoroughly revised.

3 Critical Review of Artificial Simulations and Realizations of Self-maintaining (Cellular) Systems

It is difficult to find in the literature simulation models that share the approach we take in this paper. Most computational systems do not even consider energetic/thermodynamic requirements, nor spatial ones (although there are some interesting exceptions, as we will point out here). An alternative way to produce artificial proto-organisms is through experiences 'in vitro', which have the advantage of not needing to be explicit about those aspects, since experiments are carried out in physical environments. Even if they lack the plasticity of computational models, and their results are normally obtained under extreme experimental conditions, 'in vitro' approaches seem to have a promising future, as we shall see below.

We cannot begin a survey on computational models of self-maintaining systems without some reference to the (now classical) work on 'autocatalytic sets' [5] and 'algorithmic chemistries' [6]. Despite the big differences between them, both approaches look into the problem of understanding the general rules of chemical reaction combinatorics (in order to generate computational autocatalytic networks or self-maintaining function algebras) and they both get interesting results. However, apart from thermodynamic aspects, these models do not consider fundamental spatial constraints like those related to the presence and action of a membrane.

Computational autopoietic models give a step forward in this direction. These try to implement the powerful idea of autopoiesis in an artificial environment, tackling the problem of how to show through a computer program the complementary relationship that membranes and chemical reaction networks hold in the natural (biological) world. The "qualitative chemistry" they propose [11, 12] takes place in a discrete, two-dimensional space where different types of 'particles' move and interact according to certain rules. A membrane is introduced or formed in that environment and some "self-maintaining" process can begin, provided that an appropriate set of particles is enclosed by it. This process basically consists in "repairing" possible breaks, because, once it gets constituted, the membrane cannot change its shape, nor grow or shrink. Besides, transport is merely passive (it absorbs and emits substrates according to spatial constraints rather than to internal/external concentrations) and, most important, energetic/thermodynamic requirements are completely disregarded.[2]

There is another type of computational model in which the problem of how (proto)cellular structures form is addressed, and in which physical aspects are seriously taken into consideration. These models try to simulate the way lipid aggregates self-assemble, assuming realistic features of the lipid molecules, of their interactions, and of the environment where they coexist (or could have coexisted in the past). Recently, quite interesting results have been obtained through this approach [4, 11]. Nevertheless, the focus of these models is not on the interrelation between a self-assembled membrane and a network of chemical reactions enclosed and integrated by that membrane. They provide solid computational evidence to support the fact that protocell structures (like different types of micelles) are likely to be found in aqueous prebiotic environments, but little more than that in the direction we want to explore. That is to say, they demonstrate the plausibility of structural membranes but do not show their potential to assume active roles in prebiotic scenarios.

Finally, let us include among computational models some recent work that marks a very interesting research direction. This is the design of a model (called Nidus [19]) which aims to investigate the essential components and interactions required to support the origin *and* evolution of living organizations. The author makes a serious attempt to introduce energy in his scheme, but he deals with artificial evolutionary issues on top of basic organizational ones, and that makes the whole thing quite more difficult. Our main criticism is that the model does not take into account properly the question of the physical border/membrane, although its relevance is theoretically acknowledged.

Models of prebiotic systems can also be studied in real environments, as we said above. Realizations 'in vitro' are more reliable and straightforward than computational simulations in the sense that they do not have to make explicit assumptions about material and energetic requirements (further than the actual

[2] This is a criticism that can be extended to the whole theory of autopoiesis. Maturana and Varela may claim that any realization of autopoiesis (as a chemical system) must keep away from equilibrium and, thus, be open to flows of matter and energy, and all the rest of it [9]. However, they assume that this has no influence on the logic of the system; that is, they do not look into its possible theoretical consequences (considering 'organization' fully abstracted from and prior to 'structure'). Hence, their perspective on autonomy becomes strongly internalist, disregarding the interaction with the environment as a crucial factor for creating and maintaining biological organizations.

specifications of the experiment) in order to set up the rules of the test. In this context, the idea would be to generate autonomous chemical reaction networks, but without using techniques/components directly borrowed from biological systems. This is a very hard task and, so far, achievements are only partial (like medical implant chemical devices, which are functional in so far as they get inserted in a fully integrated reaction network). Nevertheless, even though the technology to create 'chemical reaction automata' is just making its first moves, some interesting results have already been achieved.

A particularly relevant result was the production 'in vitro' of micellar systems hosting a reaction which yields a surfactant product that later assembles spontaneously in the boundary [1]. In these experiments, micelles or vesicles --once formed-- have a catalyzing effect on the internal production of the component that makes them up, and, consequently, they induce their own replication. Such systems are considered to be good examples or realizations of (minimal) chemical autopoiesis [8], since they show in a very simple and clear way the complementarity between the physical boundary and the reaction processes that take place within that boundary. However, these vesicles do not perform any kind of specific transport process.

Although aggregates of amphiphilic molecules (micelles, vesicles, liposomes) do not perform any specific transport processes, D. Deamer and his colleagues have recently implemented different systems to overcome the problem of the accessibility of substrates to the encapsulated environment (see [3] and references therein).

However, neither Luisi's nor Deamer's systems would overcome the problem of an osmotic crisis derived from the accumulation of polyanions inside the vesicle once this has initiated a replicative cycle, unless adequate experimental conditions are externally established.

For a good discussion and revision on 'chemical reaction automata' we direct readers to an article by P. Bro [2]. This author defends the relevance of thermodynamic aspects and of the role of "permselective enclosures" to achieve autonomy. Our approach is akin to this, as the reader will be able to see next.

4 Description of the Model and Discussion

This model addresses the problem of how a cellular system may become a self-maintaining autonomous one. The idea is to begin with a set of components "enclosed" by a membrane, establishing the basic reactions and transport processes that these components can undergo. Then, given a variable input of energy, the system must find how to manage it adequately to be able to self-maintain. This means that it has to opt for a particular set of couplings between the different reactions/processes that take place in it; otherwise it will not have any chance to overcome the two basic problems it faces: (i) preserving the membrane (and its dynamic behavior) and (ii) avoiding an "osmotic crisis". Furthermore, an additional problem could be relevant, depending on the type of energetic input provided to the system: the generation of energy reserves.

4.1 Basic Components

We propose eight different components: l, L, e, E, a, A, R, x. The small-letter components l, e, a, are the precursors of the capital letter components L, E, A. The membrane is a chain of L components (with a rather structural role), where E components (with more active functions) can be inserted. R components, energy-reserve compounds, are made of a certain number of A components. There are two possible 'energy currencies' in the system: a "chemical" one, based on the reaction between a and A components, plus an "osmotic" one, based on the concentration gradient of x. This is fully coherent with Skulachev's laws of Bioenergetics [18], which state that the viability of a biological system requires, at least, one currency related to membrane processes and one to internal reactions.

Capital-letter components cannot cross the membrane (L and E can only be added to or subtracted from it), whereas small-letter ones can. However, the x component can only be transported by means of the E components inserted in the membrane (mediated transport, see Table 1 below). Therefore, regarding possible spatial states, small-letter components can be 'in' or 'out'; L and E components can only be found inside the system or as part of the membrane ('mem' state); and A and R components just have one possible spatial state ('in'). Accordingly, the "movement" of such components is restricted to changes in those spatial states. Finally, the system can only receive the input of energy from the environment through some of the E_{mem} components, which become activated into E^{*}_{mem}.

4.2 Fundamental Processes

There are two main types of interaction processes: transformation reactions and transport processes. Although they may become more complicated as a result of later couplings, in principle both reactions and transport processes will involve two different components. In the first case, one component will transform into another; in the second, one will change its spatial state in the presence of a membrane component (L or E, depending on which kind of component is being transported).

Processes will be classified as exergonic ($\Delta G<0$) or endergonic ($\Delta G>0$), according to whether they may occur spontaneously or not. For instance, degradation reactions will generally be exergonic, whereas synthetic ones will need to be coupled to have some chance to occur. Concentration gradients set the spontaneous direction of passive transport processes (those in which l, e, a components get "moved" in or out), but active transport escapes this rule and depends on other factors (like the external input of energy --$\Delta\varepsilon$-- that E_{mem} components may receive). In addition, each fundamental process is assigned a 'probability function'. This determines the rate or frequency with which the process will actually take place.

Certain exergonic processes (when not coupled with endergonic ones) might give away some heat. This heat contributes to a global increase of the temperature of the system. However, heat does not accumulate in the system, since the temperature gradient established between the system and the environment provokes a global loss of heat from the system, and a subsequent decrease of its temperature. This reflects the important idea of having a *flow* of energy through the system (rather than just an input of energy) to keep it running.

Table 1. Fundamental processes

'Synthesis'		'Degradation'	
$l_{in} \text{ -----> } L$	$\Delta G > 0, P_{l\text{-->}L}$	$L \text{ -----> } l_{in} + q$	$\Delta G < 0, P_{L\text{-->}l}$
$e_{in} \text{ -----> } E$	$\Delta G > 0, P_{e\text{-->}E}$	$E \text{ -----> } e_{in} + q$	$\Delta G < 0, P_{E\text{-->}e}$
$a_{in} \text{ -----> } A$	$\Delta G > 0, P_{a\text{-->}A}$	$L_{mem} \text{ -----> } L$	ΔG dep. on $\lvert L \rvert, P_{L}$
$L + L_{mem} \text{ -----> } L_{mem} + L_{mem}$	ΔG dep. on $\lvert L \rvert, P_{-L}$	$E_{mem} \text{ -----> } E + q$	$\Delta G < 0, P_{E}$
$E + L_{mem} \text{ -----> } E_{mem} + L_{mem}$	$\Delta G > 0, P_{-E}$	$E^*_{mem} \text{ -----> } E_{mem} + q$	$\Delta G < 0, P_{E^*\text{-->}E}$
$E_{mem} + \Delta\varepsilon \text{ -----> } E^*_{mem}$	$\Delta G < 0, P_{E\text{-->}E^*}$	$A \text{ -----> } a_{in} + Q$	$\Delta G < 0, P_{A\text{-->}a}$
$nA \text{ -----> } R$	ΔG dep. on $\lvert A \rvert, P_{A\text{-->}A^*}$	$R \text{ -----> } nA$	ΔG dep. on $\lvert A \rvert, P_{A^*\text{-->}A}$

'Non-mediated Transport'		'Mediated Transport'	
$l_{in} + L_{mem} \text{ -----> } l_{out} + L_{mem}$	ΔG dep. on $\Delta \lvert l \rvert_{in\text{-}out}, P_{l,\,io}$	(a) if $\lvert x \rvert_{in} > \lvert x \rvert_{out}$	
$l_{out} + L_{mem} \text{ -----> } l_{in} + L_{mem}$	ΔG dep. on $\Delta \lvert l \rvert_{in\text{-}out}, P_{l,\,oi}$	$E_{mem} + x_{in} \text{ -----> } E_{mem} + x_{out}\ (+ q)$	$\Delta G < 0, P_{x,\,io}$
$e_{in} + L_{mem} \text{ -----> } e_{out} + L_{mem}$	ΔG dep. on $\Delta \lvert e \rvert_{in\text{-}out}, P_{e,\,io}$	$E^*_{mem} + x_{out} \text{ -----> } E_{mem} + x_{in}$	$\Delta G < 0, P_{Ex,\,oi}$
$e_{out} + L_{mem} \text{ -----> } e_{in} + L_{mem}$	ΔG dep. on $\Delta \lvert e \rvert_{in\text{-}out}, P_{e,\,oi}$	(b) if $\lvert x \rvert_{in} < \lvert x \rvert_{out}$	
$a_{in} + L_{mem} \text{ -----> } a_{out} + L_{mem}$	ΔG dep. on $\Delta \lvert a \rvert_{in\text{-}out}, P_{a,\,io}$	$E^*_{mem} + x_{in} \text{ -----> } E_{mem} + x_{out}$	$\Delta G < 0, P_{Ex,\,io}$
$a_{out} + L_{mem} \text{ -----> } a_{in} + L_{mem}$	ΔG dep. on $\Delta \lvert a \rvert_{in\text{-}out}, P_{a,\,oi}$	$E_{mem} + x_{out} \text{ -----> } E_{mem} + x_{in}\ (+ q)$	$\Delta G < 0, P_{x,\,oi}$

4.3 Main Variables, Parameters of Control and Rules of the Model

The main *variables* of the model are the concentrations of all components inside the system, and the size and composition of the membrane (number of L_{mem} and E_{mem} components, from which a "volume" of the system can be derived).

The probability functions associated with each type of process could be taken as external parameters initially, but it is probably better to work with them as variables that depend on the concentrations of the components involved, and on the temperature. In this context, temperature is defined as a global variable related to the loss of heat that can take place in certain uncoupled exergonic processes. This variable would have influence not only on the values of the probability functions of the different processes that are considered in the simulation, but also on a general probability function P_C that can induce changes in the tendency of the system to establish couplings.

Among the variables we also introduce the osmotic coefficient (Θ), which is directly related to the balance of concentrations of all components inside and outside of the system. If Θ crosses a certain threshold value the system will "burst".

The number of *parameters of control* and the type of conservation rules that will be used, of course, depend on how the environment is conceived. In a first approximation, we shall consider the environment as infinite (or very large compared to the system). Hence, external concentrations will be included as parameters of control, and so will be the external "temperature". In addition, the following parameters are considered: the initial concentration of all components inside the membrane, the size and composition of the initial membrane and its minimum/maximum sizes, the free-energy differences of the different processes, the losses of heat associated with various exergonic processes, the initial values of the probability functions, the critical value for the osmotic coefficient, and the number of A components needed to make up one R. Last but not least, a special parameter of control is included: the energy-input function. At first, on simplicity grounds, it can be set to a constant value. However, in order to analyze the behavior of the system under variable external conditions this function would be modulated accordingly.

Taking the environment as an infinite reservoir of basic components, *conservation rules* will only apply to internal processes. This is coherent with the idea of defining the system as thermodynamically open. Therefore, although external concentrations remain constant, within the boundaries of the system the number of components will be limited. A similar criterion is considered for energy balance in the model.

Finally, several *rules for coupling processes* should be introduced. These are meant to restrict the high number of possibilities that the system has to establish couplings. As a starting point, we may assume that: (a) Only two processes can be coupled at the same time, excluding the possibility of one being be the inverse of the other; (b) Only two reactions together or a reaction with transport process can couple; (c) The combination of the two processes must be exergonic (i.e., $\Delta G_{Total} < 0$) and is assigned a new probability function; (d) Once established, the couplings remain fixed.

4.4 Restatement of the Problem and Main Goals of the Model

After presenting the basic features of the model, we can pose the problem it addresses in a more concrete way. The fundamental aim could be stated as follows: given (i) a membrane and a set of components with their initial concentrations plus a set of possible interactions among them with their corresponding probability functions, and (ii) the values of all the other control parameters, try to explore which are the possible sets of couplings that allow the system to reach self-maintenance.

In principle, we expect four different outcomes:

a) Disintegration, either by progressive decay (the membrane shrinking to its minimum size) or by annihilation of any of the system variables.

b) "Burst", if the osmotic coefficient crosses its critical value.

c) A stationary state in which the system self-maintains without growth.

d) Progressive growth until the system reaches its critical maximum size. This would require the introduction of a division process in the model.

Hence, the idea is to create a computational tool through which we can study how real self-maintenance is affected by various aspects of the model. This means analyzing the influence of different factors on the system dynamics, such as: the energy input, the initial and external concentrations of free components, the original size-composition of the membrane, the temperature, the osmotic coefficient, the critical size values, etc.

5 Discussion and prospects

This model framework is meant to be a step forward in our attempt to develop a more realistic and grounded ALife that may contribute to our understanding of some fundamental problems of Biology. Here the issue of self-maintenance in a cellular system has been specifically addressed. Nevertheless, being aware that our model assumes a set of working hypotheses that should be continuously revised, let us discuss briefly in this final section some problematic points.

In the present model, space is highly abstracted. This could be a possible criticism when compared to other options that do take into account spatial constraints and movement, even if it is in an unreal two-dimensional environment. However, the issue is not completely disregarded. Our model takes into account that there must be a significant distinction between an 'inside' and an 'outside' of the system. Furthermore, transport processes are introduced and, in fact, they hold the key to avoid the problem of the osmotic crisis.

Another simplification of the model is that most of what relates to time and process rates is externally controlled. Nevertheless, the problem of synchronizing all the internal physico-chemical processes involved in the metabolic network is also important for autonomy [20]. In real biological systems, for instance, a complex web of enzymatic compounds is responsible for sorting out that problem. However, on simplicity grounds, free catalysts have not been included as fundamental component-constraints of the system.

Of course, more complicated virtual scenarios for subsequent development and improvement of this type of model could be easily imagined. For instance, a possible way to take into account external restrictions on material/energetic resources would be to introduce a situation where a population of different cellular systems --like the ones proposed here-- share a finite environment and struggle to self-maintain in it, competing for the (limited) resources, growing and self-reproducing by simple division, disintegrating and giving out their components and energy to the environment, and so on. New issues that might alter some of the working hypotheses we assumed here will surely arise in that scenario.

Acknowledgments: Authors acknowledge funding from the Research Project Number PB95-0502 of the DGICYT-MEC, from the BIO96-0895 of the CICYT, and from the EX-1998-146 and the HU-1998-142 of the Basque Government. Kepa Ruiz-Mirazo was supported by a Ph. D. Research Grant of the University of the Basque Country (UPV/EHU). Tim Taylor read a previous draft of this paper and made several suggestions to improve the English.

References

1. Bachman, P.A., Luisi, P.L., Lang, J.: Autocatalytic self-replicating micelles as models for prebiotic structures. Nature. **357** (1992) 57-59
2. Bro, P.: Chemical Reaction Automata. Complexity. **2 (3)** (1997) 38-44
3. Deamer, D.W.: Membrane compartments in prebiotic evolution. In: Brack, A. (ed.): The Molecular Origins of Life. Assembling the Pieces of the Puzzle. Cambridge University Press, Cambridge (1998)
4. Edwards, L., Peng, Y., Reggia, J.: Computational Models for the Formation of Protocell Structures. Artificial Life. **4** (1998) 61-67
5. Farmer, J.D., Kauffman, S. & Packard, N.H.: Autocatalytic replication of polymers. Physica **22 D** (1986) 50-67
6. Fontana, W.: Algorithmic chemistry. In: Langton, C. G., Taylor, C., Farmer, J.D., Rasmussen, S. (eds.): Artificial Life II. Addison-Wesley, Redwood City (1992) 159-209
7. Lazcano, A., Fox, G.E., Oró, J.: Life before DNA: The origin and evolution of early Archean cells. In: Mortlock, R.P. (ed.): The Evolution of Metabolic Function. CRC Press, Boca Raton (1992)
8. Luisi, P.L.: Chemical models of minimal autopoietic systems. In: Stein W., Varela, F.J. (eds.): Thinking about biology. Addison-Wesley, New Jersey (1993)
9. Maturana, H. R., Varela, F.J.: Autopoiesis and Cognition. The Realization of the Living. D. Riedel Publishing Company, Dordrecht (1980)
10. Mayer, B., Köhler, G., Rasmussen, S.: Simulation and dynamics of entropy-driven molecular self-assembly processes. Physical Review E, **55(4)** (1997) 4489-4499
11. McMullin, B.: SCL: An Artificial Chemistry in *Swarm*. Santa Fe Institute Working Paper. (1997) Available at: http://www.santafe.edu/sfi/publications/Working-Papers/97-01-002
12. McMullin, B., Varela, F.: Rediscovering Computational Autopoiesis. In: Husbands P., Harvey I. (eds.): Fourth European Conference on Artificial Life. MIT Press, Cambridge, Mass. (1997) 38-47
13. Morán, F., Moreno, A., Montero, F., Minch, E.: Further steps towards a realistic description of the essence of life. In: Langton, C. G., Shimohara, K. (eds.): Artificial Life V. MIT Press, London (1997) 255-263
14. Moreno, A., Ruiz-Mirazo, K.: Metabolism and the problem of its universalization. BioSystems **41 (1)** (1999)
15. Morowitz, H.J., Heinz, B., Deamer, D. W.: The chemical logic of a minimum protocell. Origns Life Evol. Biosph. **18** (1988) 281-287
16. Oparin, A.I.: Proiskhozhdenie zhizny. Izd. Moskovshii Rabochii, Moscow (1924) Reprinted in English in: Deamer, D.W., Fleischker, G.R. (eds.): Origins of Life. The Central Concepts. Jones and Barlett, Boston (1994) 31-71
17. Ruiz-Mirazo, K., Moreno, A., Morán, F.: Merging the energetic and the relational-constructive logic of life. In: Adami, C., Belew, R., Kitano, H., Taylor, C.: Artifical Life VI. MIT Press, Cambridge Mass. (1998) 448-451
18. Skulachev, V.P.: The laws of cell energetics. Eur. J. Biochem. **208**, (1992) 203-209
19. Taylor, T.: Nidus Design Document. Unpublished discussion document, available at: http://www.dai.ed.ac.uk/daidb/people/homes/timt/
20. Varela, F.: Principles of Biological Autonomy. Elsevier, New York
21. Wächtershäuser, W.: Before enzymes and templates: theory of surface metabolism. Microbiol. Rev. **52** (1988) 452-484

Design Patterns for an Object-Oriented Computational Chemistry

Hugues Bersini
IRIDIA – CP 194/6
Université Libre de Bruxelles
50, av. Franklin Roosevelt
1050 Bruxelles – Belgique

Abstract. Chemistry is basically the formation of complexes by combining together basic elements and/or simpler complexes, or by splitting bigger complexes as a result of chemical reactions. The basic elements are glued as a consequence of the affinity that exists between them. This affinity can be explicit, implicit, symmetric or asymmetric. Object-Orientation offers a natural computational framework to capture the very simple and recurrent mechanism which underlies chemistry. The basic elements and the complexes constitute the low-level classes, which can be naturally organized in a hierarchical structure. Every complex is an objet, constructed on the basis of simpler complexes or basic elements, and structured as a special type of ordered computational tree (when affinity is symmetric) and graph (in the asymmetric case). Among the attributes of these objects, one finds their concentration and their reactivity, which are essential values to simulate the dynamics of the whole system. This paper introduces, in a UML type of representation, some basic computational structures and mechanisms which should find a natural place in all simulation interested in complex formation and the time variation of the complexes concentration and reactivity.

1 Introduction

From its origins, Alife has always taken as fundamental the conception of software environments able to model the appearance and the disappearance of complexes constructed from simpler complexes or basic elements. This preoccupation leads to a succession of keystones in the Alife production such as Fontana's alchemy [3], Kauffman's autocatalytic networks [5] or Holland's echo models [4]. These software environments, like the major part of work being done (or which should be done) in the "Alife spirit", could have two major applications. Following an adequate parameterization, they could be proposed to chemical or biological practitioners who could find in these "computational platform" a helpful way to model a particular system. We then can speak of a set of computational design patterns, or software shell, which should easily be adapted to the simulation of a particular chemical or biological environment. On the other hand, these simulations can stand on their own and allow the discovery of generic laws, expressed in mathematics or linguistic terms, characterizing the behaviour of emergent and complex systems. For instance, in Kauffman's simulation of Boolean nets [5], some laws establish the stability or instability of generic networks as a function of the range of their connectivity. Others set the number of attractors as a function of the number of units in the network. In our specific case, these laws could connect, for instance, the number and the maximal size of possible complexes with the number of basic elements, with the number of affinity sites, the number of sources or any time constant for the concentration/reactivity variation, etc.

This work departs from its ancestors by essentially relying on Object-Oriented (OO) computation instead of indifferent or ad-hoc computational tricks. Its originality lies in the way the chemical environment is conceptualized in a network of classes and their specific interactions. From its origins, OO computation has allowed programming to come closer to physical simulation (the first OO language was indeed called "simula") instead of being constrained by the processor set of elementary instructions. There is a today trend which make more and more possible to abstract software engineering from the processor by naturally using high-level natural concept making up the problems as the bricks of the resolution. This goes together with the increased use of visual modeling language such as UML [6]. UML proposes a set of well defined diagrams (transcending any specific OO programming language) to naturally describe and resolve problems with the high level concepts inherent in the formulation of the problem. It is enough to discover the main actors of the problem and how they mutually relate and interact in time to build the algorithmic solution of this problem. It is beyond the scope of this paper to present UML although some of its symbols will be used to describe our chemical software environment. A simple and introductory overview of the UML language can be found in [6]. However, by deliberately restricting our use of UML to the only class diagram, readers familiar enough with OO programming should not have any understanding problem.

The first section of the paper will describe and discuss the basic classes of our chemical software environment, their main attributes and methods, and how they do relate to each other i.e. putting in words what compose the class diagram. The second section will describe in more details what is a complex and how it is structured by means of classical computational constructs such as trees and graphs. The third section will discuss the key notion of the identity of a cell and a complex, and how the latter can easily be duplicated as the result of a chemical reaction. The fourth section will tackle the dynamic aspects: What are the time-changing variables and how do they change. The last section will describe some preliminary experimental simulations, discuss problem posed by visualizing the simulation, and sketch some first observations.

2 The Chemical 00 Class Diagram

Like indicated in figure 1, in the UML class diagram, the 6 basic classes of our software environment are Component, SourceComponent, CellularComponent, Cell, Complex and CellInComplex.

Component: It is the highest super-class and presents as methods and attributes the properties to be inherited by all components of the environment, whether sources, cells or complexes: essentially aspects like their concentration and how this concentration change in time. Concentration is a very essential property of any object of the simulation because, for obvious memory size constraints, an object of the class component is not a single chemical object but rather the set (whose cardinality is the concentration) of all identical chemical objects. The concentration will be modified either by natural decrease or increase, or as a result of chemical reactions and their specific rate.

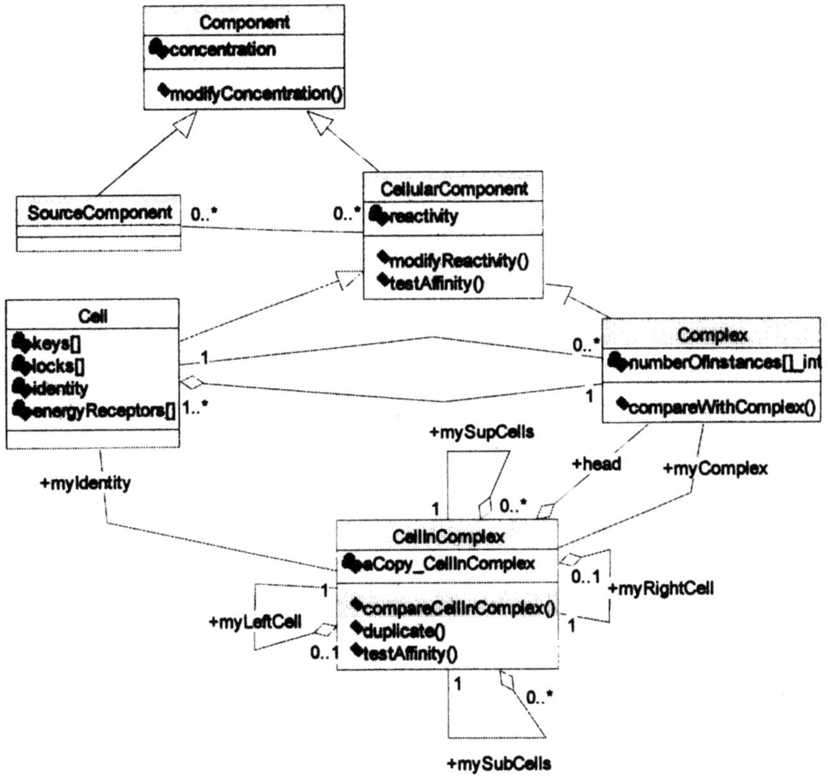

Fig.1. The UML class diagram of our chemical OO environment

SourceComponent: It is the first sub-class of Component. The energy sources are objects of this class. Their role is to supply energy to the cellular components (this justifies the diagram link between SourceComponent and CellularComponent), which by absorbing this energy, can increase their concentration and their reactivity. Each cell presents specific receptors to the sources. The sources can remain at constant concentration or decrease as a function of the cellular component absorption.

CellularComponent: It is the second sub-class of Component and is the super-class of two further sub-classes: Cell and Complex, which are the essential elements to be combined as a product of chemical reaction. Their potentiality to combine is a function of a new attribute called reactivity. For every complex to form or any reaction to occur there is a certain minimum reactivity that cells or complexes must possess. Since the main activity of these cellular components is to connect to each other, one essential method is the testing of their mutual affinity to detect possible connection.

Cell: the first sub-class of CellularComponent describes the basic objects of the whole system: the cells. The fundamental structural attribute of the cell is its *list of affinity sites* which allow cells to combine together in order to form complexes. A lot of variations is possible in the way these affinity sites characterize the cell. Depending on the choice, the affinity between two cells will be *explicit, implicit, symmetric* or *asymmetric*. Suppose the simplest case, explicit and symmetric, where every cell has a set of sites, which simply have to match any of the sites of any other cell. A vector of integer values $S_i[m]$ could characterize cell *i*, such that for cell *i* to connect with cell *j* it would be enough to have for one of the vector element: $S_i[m] = S_j[n]$. The affinity between cell *i* and *j* will then be symmetric i.e. The complex i ←>j is symmetric. Here we will restrict to *yes or no* type of connectivity. In previous works dealing with immune networks, we have allowed smoother type of connectivity [2]. The asymmetric situation occurs as soon as the cell presents two sites of affinity (typical in biology) naturally called the keys $K_i[m]$ and the locks $L_j[n]$, and when connection turns out to be only possible between a key of the first cell and a lock of the second. In such a case the complex i → j will be asymmetric so different from the complex j → i.

The implicit type of affinity occurs when two cells connect yes or no as a result of a function of their respective affinity. For instance, there could be a connection between i and j only if $S_i[m] + S_j[n] = 8$ (we are closer to chemistry now, with the well-known "octet rule", where affinity between basic elements is implicit and symmetric). Irrespectively of its implicitness or explicitness, which finally always lead to a yes or no affinity, the real big difference for the rest of the paper is the symmetric or asymmetric version of the affinity. As shown in the class diagram, the cells are associated with the energetic sources and have an additional attribute (the "1-0..*" association with the complex class) resuming all the complexes in which they appear.

A final attribute of the cell is its *identity*, which simply needs to be an ordered index. It is clear that the way this identity is defined depends on what we take to be unique to any cell (the identity would correspond to the key in the data base language). One natural option would be to relate this identity with the affinity sites, and to uniquely calculate this identity index as a function of the affinity (whether symmetric or not). However this is not an obligation and one could just impose an identity besides the way they mutually connect i.e. two cells with the same affinity sites could still be different (it seems to be the case when one looks at the chemical periodic table where different elements with a same affinity mechanism are grouped in the table).

Complex: the second capital sub-class of CellularComponent describes the second essential family of components: the complexes. These complexes are product of the cells or simpler complexes aggregations. The following section will see how are the complex structured and what in this structure make them unique. As shown in the class diagram, complexes are aggregate of cells. An attribute called *numberOfInstances* is a vector of integer whose elements are the number of times one specific cell appear in the complex. Like all computational objects, complexes need to be "constructed". Here at most four types of constructor methods are necessary to construct complexes: 1 – from two cells (giving rise to the most basic complex) ; 2 – from one cell and one complex (it can still be splitted in two different constructors when the affinity is asymmetric); 3 – finally from two complexes. Complex relates with their most direct partner: the CellInComplex which are responsible for the structural description of the complex. Each complex has one and only one CellInComplex called the “headCell” and which can be seen as its “front door”.

CellInComplex: as soon as a cell gets in a complex, it is transformed into a CellInComplex object. CellInComplex relate to cell since the identity of a such an object is the same as its associated cell. CellInComplex are responsible for coding the tree or the graph structure of the complex, to be described in the next section. It is enough to know that, as classically done for trees and graphs [1], and like shown in the class diagram, CellInComplex has pointer attributes (my subCells, mySupCells, myRightCell, myLeftCell) pointing to other CellInComplex objects. This is the well-known computational trick which recursively allows to treat tree or graph structures in the easiest way. Finally all natural methods associated with complex and allowing to test affinity between two complexes, to compare two complexes and to duplicate one complex have natural counterpart in the CellInComplex class. Thus two CellInComplex objects can test their mutual affinity, be compared and be duplicated.

3 The structure of a Complex

Take the simplest case where two cells “1” and “2” (in the following we will label the cell by just their integer index) present asymmetric affinity (for instance one key of “1” matches one lock of “2”). A new complex appear,“1” is the headCell of this complex (an object of class CellInComplex) and it has one subCell “2” (also an object of class CellInComplex) (fig.2). Suppose then that “1” connects to a second cell “5”, the new complex is depicted fig.3 where now “1” has a vector of two subCells: “2” and “5” (a vector of two CellInComplex). It is important to keep the subCells ordered so that “2” is on the left of “5”. Instead of a vector of subCells, one possible alternative would be to keep only “2” as a subCell, and to make “2” having the rightCell “5”. “1”, “2” and “5” are all CellInComplex with, as subCells and rightCells, pointers to still to come CellInComplex.

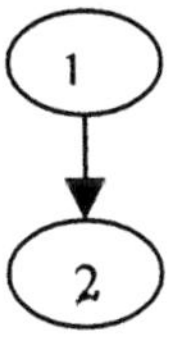

Fig.2. The simplest complex

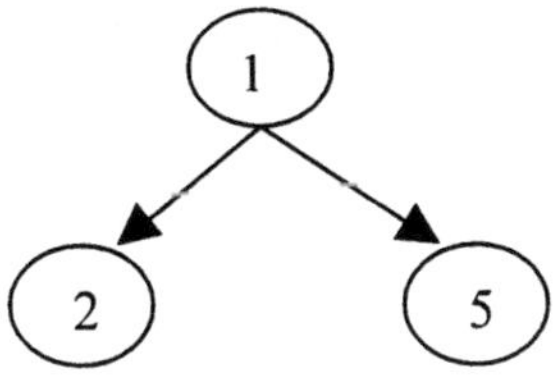

Fig.3. A three cells complex

So far complexes seem to be structured as “Letfmost-Child-Right-Sibling” tree [1]. The reason for maintaining a left-to-right increasing order in the position of the subCells is to ease the detection of equal complexes. Can the structure of the cell be kept as a simple ordered tree? Let’s go one step further and suppose that we now have two complexes like indicated in fig.4, and the cellInComplex “5” of the first complex has affinity with the cellInComplex "6" of the second complex (a key of "5" matches a lock of "6"). In the figure you can see the reaction and its result:

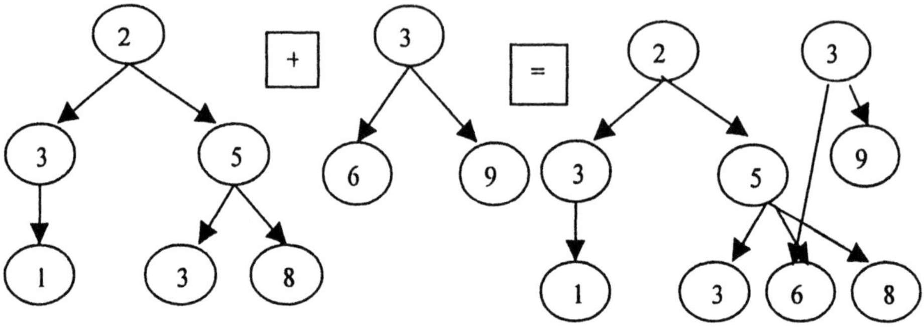

Fig.4. Merging two tree complexes to obtain a new graph complex.

It is clear that any cellInComplex in the structure (except the headCell), and in contrast with classical tree, can be pointed by more than one pointer (here "6"). Also the headCell in the resulting complex "2" has a new rightCell "3". Detailing the resulting complex: "2" has rightCell "3" and has subCells "3" and "5", the first "3" has subCell "1", 5 has subCells "3" and "6" and "8" (the subCells are well ordered), the second "3" has subCell "6" and "9" (here again it is important to check the order of the subCells). For asymmetric complexes, the two rules to be obeyed in order to preserve the identity of the complex are:

1: a head Cells can have rightCells so that a cellInComplex has, in fact, an ordered succession of front doors.
2: the whole vector of subCells must be ordered.

Suppose that a similar reaction occurs again, but now in the symmetric affinity case. The good news in the symmetric case is that it is quite easy to preserve the tree structure (like shown in fig.5). The order must be kept at the level of the subCells. There is only one headCell with no rightCells. Due to the "vertical symmetry" (besides the horizontal symmetry which entails the subCells order), the only novelty is to keep as headCell, the one with the lowest integer index from all the cells on the extreme left of the complex. So the two new rules are:

1. like for the asymmetric case, all the subCells must be ordered
2. the head Cell is the cell with the lowest integer index from all the left most cells.

Both in the symmetric and asymmetric cases, there is no additional difficulty when loops appear in the tree due to transversal connections among the cells.

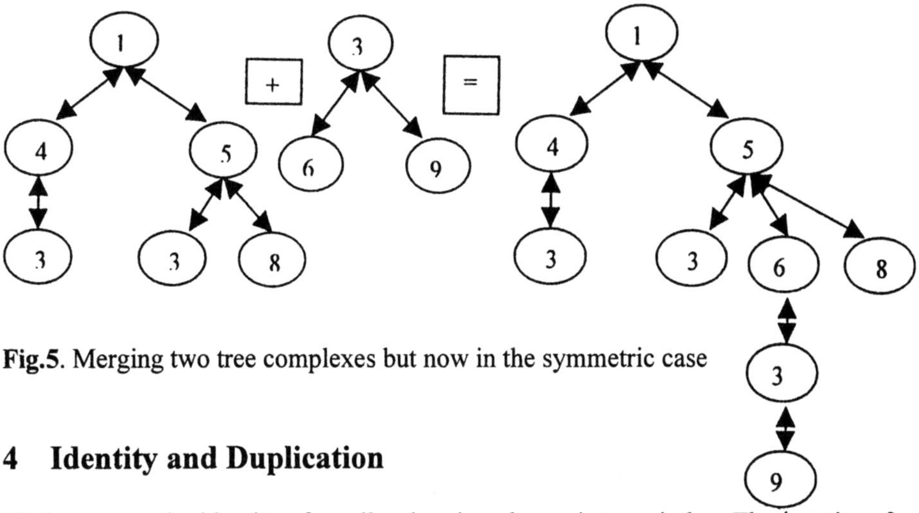

Fig.5. Merging two tree complexes but now in the symmetric case

4 Identity and Duplication

We have seen the identity of a cell to be given by an integer index. The identity of a complex is given by the cells inside of it, the number of instances of each of it and by the way the connectivity is organized in a tree-like or graph-like structure. Once two complexes must be compared to see, for instance, whether a possible new complex, resulting from a reaction, is not already existing in the system, the comparison mechanism is defined in a recursive way, departing from the headCell of each of the complexes. In order to make this recursive testing possible, the tree in the symmetric case and the graph in the asymmetric case must comply with the two architectural rules given in the previous section.

Any time a reaction occurs between two complexes i.e. when a cell of the first connects with a cell of the second, two duplications must take place, roughly: A + B = A + B + AB. AB is constructed by first duplicating A and B and then by connecting these two duplicates, adding a link between the two affine cells. Like the affinity testing and the comparison mechanism, the duplication is done recursively, departing in all cases from the headCell. In the cases of graph, since any cellInComplex could be visited more than one time, they contain a very helpful attribute called "aCopy" which is assigned an exemplar of their last duplication. Thus, whenever any cellInComplex is visited a second time in order to be duplicated, it is enough to return its copy and just stops the duplication propagation. Once a whole complex is duplicated all "aCopy" objects are reset to null.

5 The dynamical aspects

Beyond the metadynamics just discussed, dynamics is a key addition in our software environment. In a lot of Alife simulations, the dynamics (how variables associated with objects change in time) and the metadynamics (how the set of object changes in time by appearing and disappearing) aspects are kept separated. In this work, they can be taken into account in a simultaneous way. However and similarly to all the software mechanisms and constructs proposed here, these dynamical aspects are something to adapt and to parameterize for any particular application. In the following, only some very simple and preliminary mechanisms, set by default, will be proposed. The two main dynamic variables, associated with all cellularComponents, are the concentration and the reactivity.

For sake of clarity, the complex will be indicated by upper case and the free cells (cells not in complex) by lower cases. The reactivity variable of a complex A i.e. R(A,t) can change in time and do play the following role: a reaction between two complexes will be possible only if the lowest reactivity of the two complexes is beyond a certain threshold: $min\ (R(A,t),R(B,\ t)) > threshold$ to obtain a complex AB. The reactivity of any cell is a function of the sources of energy and the energy receptors of the cell. This reactivity changes in time as a function of the sources of energy: R(a,t+dt) = f(R(a,t), sources(t)). The reactivity of a complex is obtained from the reactivity of the cells inside, either by summing or averaging the cell reactivity.

The concentration of a complex A i.e. C(A,t) can change in time. When two complexes could possibly merge, the probability of their merging will be dependent on their respective concentration (to keep simple, here we have considered the product), so that: P(A+B->AB) = k*C(A)*C(B) (P() being the probability that the reaction occurs). The concentration of a free cell changes in time as a function of the sources: C(a,t+dt) = f(C(a,t),sources(t)). Now when a complex is created, the initial concentration of the complex is given by a part of the product of the two concentrations of the components to be merged:
C(AB,t) = 0; C(AB,t+dt) = k*C(A,t)*C(B,t) (typically a first order chemical reaction)
Also the concentration of the two merging complexes decreases consequently:
C(A,t+dt) = C(A,t) – C(AB,t+dt) ; C(B,t+dt) = C(B,t) – C(AB,t+dt)

The additional change in time of the concentration of a complex will be function of the cells inside the complex. One of the obvious benefits provided by the formation of a complex is that all cells inside can take advantage of the others. So a cellInComplex even poorly affected by the energy sources can grow in concentration or reactivity because other cellInComplex are fed by the sources.

6 Visualizing and first results

For various reasons it might be profitable to visualize some simulation run in order for instance to check the correct running or, more profitably, to detect some emergent phenomena like threshold or percolation effects. Fig.6 shows a snapshot of the running with the simple visualizing adopted here. Every cell position is a function of its identity (the cells are the squares with the color indicating their concentration, sources are shown by an additional cross). Complexes are shown by the links existing between cells. The cellInComplex have a white circle inside. The simulation initiates with a certain number of free cells and a certain number of sources. Some precious information such as the current number of complexes, the size of the biggest complex, the average concentration, and their evolution in time can be shown on the right side of the main window.

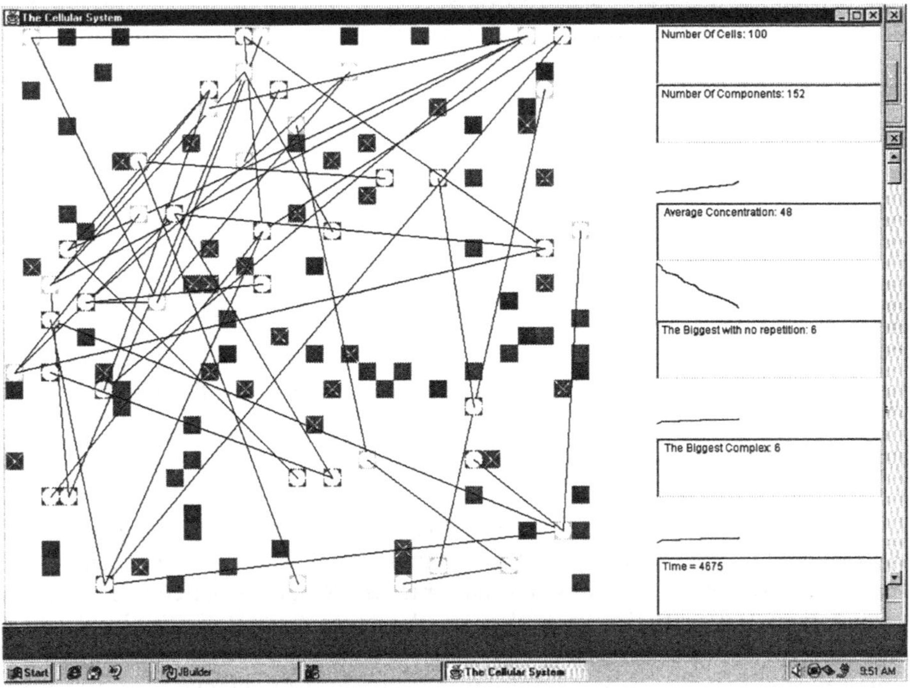

Fig.6. A snapshot of the running software

Two preliminary interesting observations are very reminiscent of qualitative results already discussed in previous Alife works. The figures 7 show the growing in time of the size of the complexes: in both figures, the first plot just indicates the number of distinct cells, the other plot indicates the number of all cells (accounting for cells appearing several times). For certain values of the number of affinity sites and the number of initial cells, like in Kauffman's autocatalytic networks [5], one can attend some kind of phase transition in the number of complexes and the size of the biggest complex (very similar to percolation phenomena) that the simulation produces (fig. 7a). All of a sudden most of the complexes have become cross-connected into one giant structure. However, by playing with the sources concentration and the time constants, this phase transition effect can become a dynamic, instead of a static structural, phenomenon, i.e. occurring only after some time, the time needed to reach a sufficient reactivity or a sufficient concentration.

Certain complexes, once constructed, have the possibility to infinitely grow in size, if not stopped by their concentration and the concentration of the ingredients they need to grow. They are all the complexes that simply loop on themselves. They can very easily grow infinitely by constantly adding replication of themselves. The simplest case is obviously the complex composed of two times the same cell, one as the receptor and the other as the connector, but it is enough to have one cell of a chain connecting back to a previous element of this chain. In fig. 7b only the plot showing the total number of cells explodes in time. Fontana [3] made similar observations in his alchemy simulations, and recovered by suppressing these self-replicating structures as soon as they appear in his simulations.

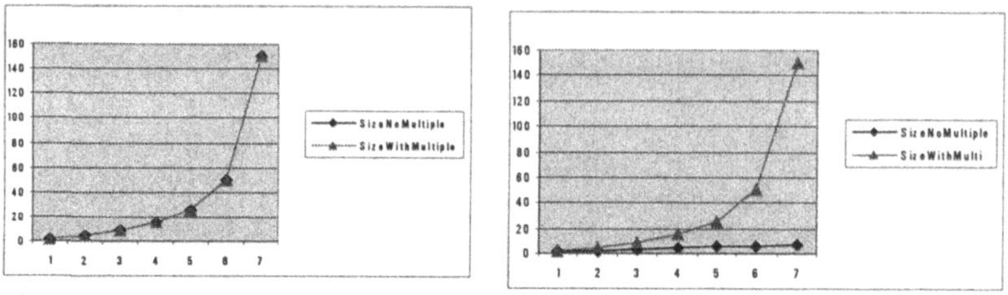

Fig. 7a) and b) The evolution of the size of complexes with and without multiple

7 Conclusions

Several important works have marked Alife by their attempt to computationally replicate the formation of complexes from simpler elements. However because these works are either based on formal but not practical computational constructs [3] or because they lack sufficient precision on the computational mechanisms which underlie them, it seems to remain large methodological gaps to fill with the natural computational friendliness and flexibility allowed by OO computation. The work described here attempts to indicate some very preliminary steps on the road to a more complete, natural and coherent OO chemical environment. These first steps have mainly been restricted to the definition of the basic classes, the way they connect to each other, and the recursive computational structure underlying symmetric or asymmetric complexes. Current and future developments of the software will allow extending the construction of complexes through the cleavage of existing ones. Whereas one obvious source of novelty in chemistry is obtained by the aggregation of complexes, new aggregates, not existing before, can also appear following an internal transitory re-organisation of the complex, and then its splitting.

References

1. Aho V. A., Ullman, J.D. : Foundations of Computer Science – Computer Science Press - W.H. Freeman and Company – New York (1995)
2. Detours, V., Bersini, H., Stewart, J. and Varela, F. : Development of an Idiotypic Network in Shape Space – Journal of Theor. Biol. (170), 401-404 (1994)
3. Fontana, W.: Algorithmic Chemistry. In Artificial Life II: A Proceedings Volume in the SFI Studies in the Sciences of Complexity (C.G. Langton, J.D. Farmer, S. Rasmussen, C. Taylor, eds.), vol. 10. Addison-Wesley, Reading, Mass (1992)
4. Holland, J.H.: Hidden Order – How adaptation builds complexity – Helix Books – Addison Wesley Publishing Company (1995)
5. Kauffman, S.: The Origins of Order: Self-Organisation and Selection in Evolution – Oxford University Press (1993)
6. Eriksson, H-E, Penker, M.: UML Toolkit – John Wiley and Sons (1998)

Model of Self-Replicating Cell Capable of Self-Maintenance

Naoaki Ono and Takashi Ikegami

Institute of Physics, The Graduate School of Arts and Sciences,
University of Tokyo, 3-8-1 Komaba, Meguro-ku,Tokyo 153, Japan
nono@sacral.c.u-tokyo.ac.jp
ikeg@sacral.c.u-tokyo.ac.jp

Abstract. We have constructed a simple model of a proto-cell that simulates stochastic dynamics of abstract chemicals on a two-dimensional lattice. We have assumed that chemicals catalyze their reproduction through interaction with each other, and that repulsion occurs between some chemicals. We have shown that chemicals organize themselves into a cell-like structure that maintains its membranes dynamically. Further, we have obtained cells that can divide themselves automatically into daughter cells.

1 Introduction

The emergence of cells is one of the major transitions in the evolution of life. When primitive self-replicators such as a hypercycle of RNA enzymes evolve into a living cell, they must acquire membranes that will separate them from their noisy environment. It is well known that a hypercycle system can easily be broken down by the occurrence of parasites. Compartmentalization of a hypercycle system is a simplest way to avoid the disaster [2] [14] [15]. At the same time, however, it should be noticed that it is also true that parasites can drive the increase of diversity and complexity of the replicator network [6]. In order to examine the balance between stable reproduction and diversity, we should study a relationship between an internal replication and a cellular structure which enclose it.

Many models of proto-cell structures have been proposed. For example, it is well known that long-chained fatty acids spontaneously form micelles or vesicles when submerged in water. Luisi and his group demonstrated experimentally the self-organization and self-reproduction of liposomes, and showed that such vesicles maintain self-replicating RNA within them [16]. Theoretical models for self-organization and self-reproduction of micelles are studied well [1] [13].

There is an another essential feature of cells, that is, self-maintenance. Living cells metabolize and sustain their membrane by themselves, and the boundary of cells are defined by the membrane. This mutual dependence enables the coevolution of internal chemical networks and the membranes. The coevolution between the two is presumed in the early stages of the cell evolution. With respect to this point, Gánti proposed a model for primitive life termed 'the chemoton', which presents three indispensable functions of the proto-cell: it has a metabolic cycle for assimilation; it maintains its membrane; and it replicate its genetic information [3] [4]. Varela also insisted that the boundary of cells (i.e. the cell membrane) must be organized and maintained by the cell itself [9] [10]. He presented a model on a two-dimensional lattice of an autopoietic cell that can maintain its membrane.

Our purpose in the present study is to demonstrate how such primitive cells can emerge and evolve from a simple set of chemical network. A model of self-maintenance and self-replicating cells in one-dimensional space was proposed by the same author [11]. In the model, we have shown that self-reproduction of the cellular structure emerges spontaneously and there are two distinct processes of replication showing potentially different heredities.

In this paper, we extend our previous model for application to two-dimensional cases and showing that this system has a potential for further evolution.

2 A stochastic particle model

We simulate a discrete space-time dynamics of chemicals in a two-dimensional space, where chemicals catalyze each other's reactions. Each chemical is given as a particle with/without anisotropic shape that moves around on a triangular lattice. Particles demonstrate two basic motions: hopping to neighbouring sites and rotating at one site. In addition to this behavior, a particle can change its chemical qualities. The former is termed a mobile transition, the latter a chemical transition, and both are determined by the potential energy of the particle.

We assume that there is a repulsive force between some chemicals, thus the physical potential of a chemical C at the site x ($E_C(x)$) is computed by summing up the the repulsion potential of all chemicals at the site x and its six neighbouring sites. The mobile transition probability $P_C(x, x_0)$ from site x to x_0 is computed from the difference in the potential magnitudes as

$$P_C(x, x_0) = R_{dif}\, f(E_C(x_0) - E_C(x)),$$

where $E_C(x)$ gives the potential energy of the particle C at the site x. The diffusion parameter R_{dif} is fixed for all particles.

The chemical transition probability $P_{C \to C'}(x)$ from the state C to C' at the site x is given as

$$P_{C \to C'}(x) = R_{C \to C'}(x)\, f(G_{C'} - G_C + E_{C'}(x) - E_C(x)),$$

where G_C represents the chemical potential of C. The reaction parameter $R_{C \to C'}(x)$ is controlled by a catalyst found on the site x. One constraint is given to the form of the function f in order to satisfy the thermal equilibrium condition, as follows:

$$\frac{f(\Delta E)}{f(-\Delta E)} = e^{-\Delta E} \tag{1}$$

We define five different kinds of chemicals, $A; M; W; X$ and Y, in a system. Each particle can belong to any one of these chemicals. W plays the role of abstract 'water', and cannot change into any other chemical. X is the material with high chemical potential, though it is not an autocatalytic chemical. A is a unique autocatalytic chemical in the system. Their reaction processes are,

$$A + A \leftrightarrow AA$$
$$and$$
$$X + AA \leftrightarrow A + AA.$$

These chemical reactions only occur among particles that occupy the same site [1] The forward and backward reactions have an equal reaction parameter, which is given by the following formula:

$$R_{X\to A}(x) = R_{A\to X}(x) = B_{X\leftrightarrow A} + C_A A(x)^2,$$

where $A(x)$ denotes the number of A on the site x. In the above equation, $B_{X\leftrightarrow A}$ and C_A are the base rate and the catalysis coefficient, respectively. As a secondary process, A produces M as a co-product of the total reaction network.

$$X + AA \leftrightarrow M + AA.$$

The reaction parameters are given by

$$R_{X\to M}(x) = R_{M\to X}(x) = B_{X\leftrightarrow M} + C_M A(x)^2.$$

In addition to the above reactions, we introduce the natural decay of chemicals into Y where Y has the lowest chemical potential.

Consider that there is a source of material X in this system so that the reaction parameters between X and Y break the pattern of symmetry, as

$$R_{X\to Y} = B_{X\leftrightarrow Y}$$
$$R_{Y\to X} = B_{X\leftrightarrow Y} + S_x$$

where S_x denotes the strength of the source X.

We assume there is repulsive force between M and other chemicals like oil in water. In the following simulations, we examine three different kinds of potentials on the chemical M. First, M equally repels all the other chemicals around it. Second, M has an anisotropic repulsion regardless of the kinds of chemicals. This feature will be described latter. Third, in addition to the anisotropic repulsion, the repulsion force also depends on the kinds of chemicals.

[1] Note that a number of chemicals can occupy each site. In this study, the average is one hundred.

3 Simulation Results

3.1 Formation of Cells

First, we simulate the case where the repulsion force depends on neither the kinds of chemicals nor the form of M. Starting from the homogeneous initial state which has rich amount of A, a system can maintain replication of A and reproducing co-product M. Chemicals A and M can aggregate to form Turing-like patterns. Figure 1.a shows an example of the pattern generated; the spots of M are formed among W and A.

The second case, where the repulsion force depends on the orientation of M molecule gives different observation. Here M has an anisotropic potential, illustrated in Fig. 1.b. When these Ms are placed on a triangular lattice, the 'head' can be aligned in any of three directions (e.g. $0,\pi/3$ and $2\pi/3$). We assume that it can change its direction stochastically, with the transition probability given by,

$$P_{MM'}(x) = R_{rot}\, f(E_M(x) - E'_M(x)),$$

where M' denotes M which has another orientation.

Figure 1.b shows the repulsion potential generated by M for other chemicals A,W,X and Y. The repulsion force from M becomes strongest when M and other chemicals are on the same site (indicated by black in Fig. 1.b). The repulsion is the second strongest at the front of or behind M M (dark gray sites) and relatively weaker at the other four side-sites (light gray). We also assume there is repulsion between Ms when their directions are different. Thus, the M molecule tends to take the same direction as neighbouring Ms.

When M has this kind of anisotropic repulsive force, the clusters are organized differently than they are in the isotropic cases. The clusters of M become thin films that we simply name 'membrane' (see Fig. 1.c). The difference in the repulsion potential between front- and side-sites of M affects the thickness of membrane. Also, when the repulsion between M with different direction is stronger, the membranes tend to run straighter. These effects allow us to get membranes that have various degrees of flexibility.

When we start from a single 'cell', that is, a spot of A enveloped by membranes as shown Fig. 1.d, this structure can maintain itself stably because A within the cell keep reproducing themselves and sustain the membranes by supplying M; simultaneously, the membranes keep A from diffusing outward. Note that this structure collapses when the membranes are broken (see Fig. 1.e). Chemical A cannot sustain reproduction because they leak away through the defect in the membrane. In the absence of a supply by A, the membranes decay and disappear.

3.2 Cell division

Living cells are not closed systems. They must ingest nutrients and excrete wastes through their membranes. In this section, we study the case where M shows

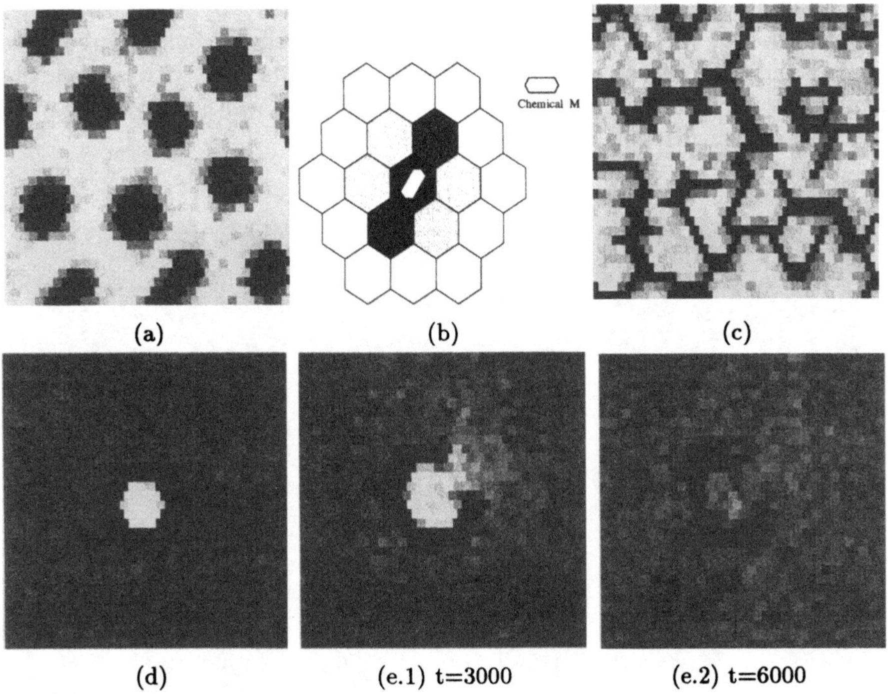

Fig. 1. Cluster formation. Picture (a) presents an example of the patterns formed by molecules with isotropic repulsion. Picture (b) illustrates the anisotropic field of repulsion around M. The depth of gray denotes the intensity of repulsion. Picture (c) shows that M forms thin films and separates the domains of A. The white domain contains rich amount of A. The gray and black domains are dominated by W and M, respectively. In picture (d), a cell structure maintains itself stably. Pictures (e.1) and (e.2) are the snapshots of the collapse of a cell starting from a cell which lacks the upper right membrane.

selective repulsion depending on the kind of molecules. We assume that the repulsion between M and two chemicals X and Y is much weaker than that of the other chemicals.

In this case, X and Y can permeate through the membranes at a rate proportional to the gradient of their density. Because there are more X in the environment than on the inside of the cell, the cells can absorb the external chemical X and grows gradually. When the cell reaches a certain size at which it has outgrown its stability, it begins to generate a new membrane inside. This finally divides the mother cell into daughter cells (see Fig. 2). These new cells repeat the process of growing and dividing. Sometimes a cell fails to sustain its membrane structure and dies, due to a shortage of materials or to interference from other cells.

We can change the flexibility of the membranes by altering the repulsion strength between Ms. This has the result of varying the division dynamics of cells. Examples are presented in Fig. 3. A strong repulsion between M results in the formation of stiff membranes, and the shape of cells becomes more regular (Fig. 3.a and 3.b). On the other hand, Fig. 3.c represents a cell with more flexible membranes, which form in the presence of low repulsion values for M. These cells divide themselves irregularly at some narrow part.

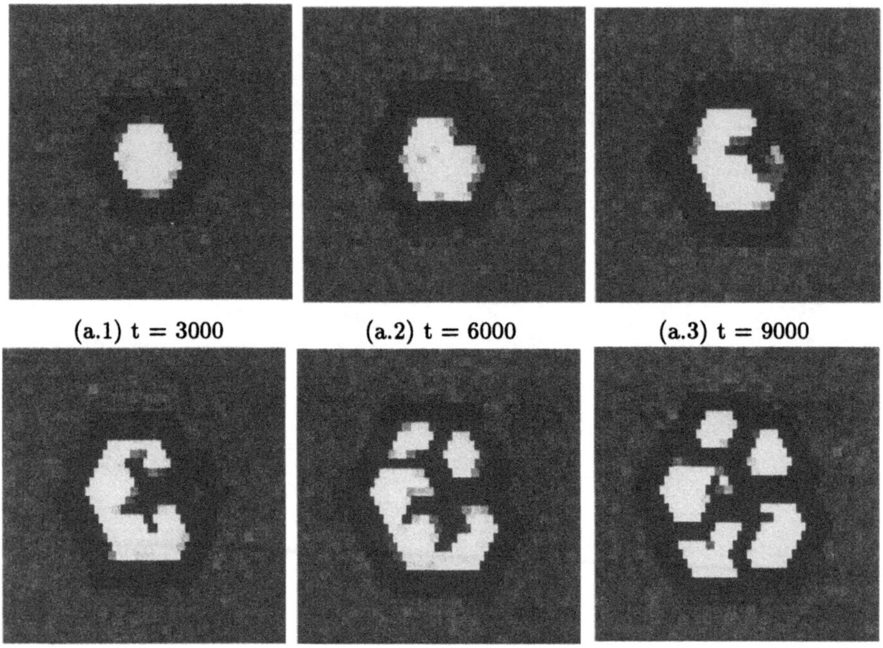

(a.1) t = 3000 (a.2) t = 6000 (a.3) t = 9000

(a.4) t = 12000 (a.5) t = 15000 (a.6) t = 18000

Fig. 2. Snapshots of cell division. The cell grows by ingesting X. Next, the membrane grows inward. Finally, the mother cell divides into five daughter cells.

4 Discussion

We have demonstrated the model of a self-maintaining cell. The cell has an internal autocatalytic cycle of chemicals, which maintains the membrane by itself and the membrane keeps the cell from collapse. We have also shown that the self-maintaining cell can replicate itself spontaneously; a transition is made from molecular reproduction to cellular reproduction.

In real life, the earliest membranes may have been simpler and rougher than the phospholipid membranes. The marigranule [5] represents an example of a primitive cell. It has rough shell and can ingest amino acids from the environment. Though the materials of which marigranules consist are very elementary

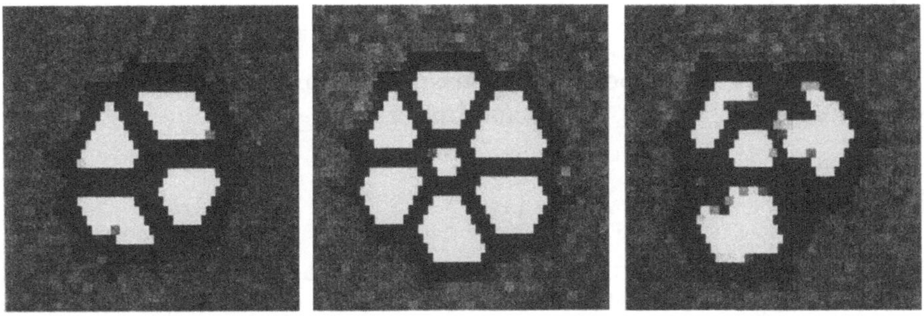

Fig. 3. The variations of the cell. The shapes of cells and their manner of division depend on the values of repulsion between M. Pictures (a) and (b) show cells with stiff membranes. Picture (c) shows cells with flexible membranes.

molecules, there is no linkage between the organization of shells and its internal dynamics. If such structure established a symbiotic relationship between its embedded chemical network and its membrane, the membranes could become the targets of the Darwinian selection and evolve into more complex structures.

In this study, we have shown that the cell divides in a different manner according to variation in the interaction strength between Ms. It is also true that we can change some properties of biological membranes by varying their components. For example, unsaturated fatty acids components soften the membrane and cholesterols do the opposite.

Self-replicating spots in the two-dimensional reaction-diffusion system [12] [8] have been well studied. However, the kinds of replicating pattern cannot be as diverse as the patterns generated by cells with membranes. The cellular membrane can function as a boundary condition to the internal chemical network, and conversely, the internal reactions determine the cell shape. In this sense, the chemical reactions within cell membranes can be richer than those without membranes. The compartmentalization of chemicals will allow cells to be regarded as units of evolution, because it maintains the identity of their contents during reproduction.

Koch previously discussed the division mechanisms of phospholipid vesicles by considering the property of mechanical energy [7]. Our model demonstrates an analogous dynamic division mechanism.

The evolution of selective permeability of the membrane must be considered in future studies. Cell membranes determine how cells communicate with the environment, including other cells. Cells selectively receive stimuli from the environment and from other cells, and respond to these stimuli. Our cell model provides several possible approaches by which to observe the formation and evolution of membrane functions, and of how the interaction between cells is generated by each cell's own internal dynamics.

Acknowledgments

This work is partially supported by Grant-in aid (No. 09640454) from the Ministry of Education, Science, Sports and Culture.

References

[1] P. V. Coveney, A. N. Emerton and B. M. Boghosian. Simulation of self-reproducing micelles using a lattice-gas automaton. *J. Amer. Chem. Soc.*, 118:10719–10724, 1996.

[2] M. Eigen, P. Schuster, W. Gardiner, and R. Winkler-Oswatitsch. The origin of genetic information. *Scientific American*, 244(4):78–94, 1981.

[3] T. Gánti. Organization of chemical reactions into dividing and metabolizing units: the chemotons. *BioSystems*, 7:15–21, 1975.

[4] T. Gánti. Biogenesis itself. *J. theor. Biol.*, 187:583–593, 1997.

[5] H. Yanagawa, Y. Ogawa, K. Kojima and M. Ito. Construction of protocellular structures under simulated primitive earth conditions. *Origins of Life and Evolution of the Biosphere*, 18:179–207, 1988.

[6] T. Ikegami and T. Hashimoto. Replication and diversity in machine-tape coevolutionary systems. *Artificial Life V*, 5:426–433, 1997.

[7] A. L. Koch. Primeval cells: Possible energy-generating and cell-division mechanisms. *J. Mol. Evol.*, 21:270–277, 1985.

[8] K. J. Lee, W. D. McCormick, J. E. Pearson, and H. L. Swinney. Experimental observation of self-replication spots in a reaction-diffusion system. *Nature*, 369:215–218, 1994.

[9] H. R. Maturana and F. J. Varela. *Autopoiesis and Cognition: the Realization of the Living*. D.Reidel Publishing, 1980.

[10] H. R. Maturana and F. J. Varela. *The Tree of Knowledge*. Shambhala Publications, 1987.

[11] N. Ono and T. Ikegami. A model for the origins of cellular replication. in preparation.

[12] J. E. Pearson. Complex patterns in a simple system. *Science*, 261(9):189–192, 1993.

[13] C. Barrett S. Rasmussen, N. Baas and M. Olesen. A note on simulation and dynamical hierarchies. In F. Schweitzer, editor, *Self Organization in Complex Structures - From Individual to Collective Dynamics*, pages 83–90. Gordan & Breach Publiching, 1997.

[14] E. Szathmáry and L. Demeter. Group selection of early replicators and the origin of life. *J. theor. Biol.*, 128:463–486, 1987.

[15] E. Szathmáry and J. M. Smith. From replicators to reproducers: the first major transitions leading to life. *J. theor. Biol.*, 187:555–571, 1997.

[16] P. L. Luisi, T. Oberholzer, R. Wick and Ch. K. Biebricher. Enzymatic rna replicatioin in self-reproducting vesicles an approach to a minimal cell. *Biochem. Biophys. Res. Comm.*, 207(1):250–257, 1995.

Evolution at the Origins of Life: Modelling Simple Persistent Replicating Objects

Oliver Sharpe

School of Cognitive and Computing Sciences
University of Sussex, Brighton BN1 9QH, UK
olivers@cogs.susx.ac.uk

Abstract. This paper looks at concerns for very early evolution near the origins of life. At the very least the objects that are our most distant ancestors must have been objects that could persist for some time, and replicate. The models presented here examine the evolution of such proto-life objects whose only characteristics are how long they exist for and how quickly they can replicate. No explicit selection regime is needed, instead these objects evolve due to their own simple dynamics until a persistence error threshold is reached. The last part of the paper discusses ways in which this error threshold can be first moved and then overcome by the respective evolution of mutation rates and then predation.

1 Introduction

This paper is part of on-going research into how the notion of persistence can be usefully applied in research into dynamical systems, especially those capable of modelling living systems. This research has taken persistence to be a fundamental and important characteristic of objects within the system. For the object to be able to replicate or perform other interesting behaviours, it must at least persist for enough time to perform those behaviours. Hence the ability to persist is seen as a prerequisite for enabling more complex behaviours. If the objects can replicate then there is the possibility that a persistence ratchet [1] will be in place and hence that effective evolution can occur.

The models presented here take a preliminary look at the necessary conditions to enable persistence ratchets at the very origins of life. What they do not address is the question of how replication emerged, but rather they look at the dynamics of evolution the moment after replication has first been made possible. In particular the first model assumes that no predation occurs and that the proto-life objects have no significant influence on their environment. Hence the only characteristics of these simple units of replication are their time for persisting and their replication rate.

Whilst the model presented here is best seen in the context of the nude gene or proto-cell versions of the origin of life, similar concerns about the ability of 'replicating' autocatalytic cycles to persist will also arise (for details of current theories about the origins of life see either [2] [3] or [4]). In any event, if life is seen as requiring a unity, as in the autopoietic view [5], then it is upon the initial persistence ability of these units that this model is based.

2 The model and first results

The world is made out of a number of 'cells'. Each cell contains one object that is described simply by its ability at persisting and replicating. Its ability at persisting is a number that represents how many time steps it is capable of existing for. If this number is ever 0 then the cell is considered to be empty as no object can exist for no time. Replication ability is represented as the number of time steps between replication events. Each object also has an associated age counter and replication counter, counting the time since birth and since the last replication event respectively. These counters are not globally synchronised but are locally measured by the number of update events that each object has had applied to it. Update events are stochastically distributed across the cells. One global time unit is said to have occurred every N update events.

When the age counter reaches the same value as the persistence ability then the object dies, its persistence value (and other variables) is set to zero. When the replication counter reaches the replication ability value (or higher) then replication is attempted. If there are any empty cells then an exact copy of the replicating object is placed in an empty cell, the offspring's counters are set to zero as well as the parent's replication counter. If there are no empty cells then nothing happens. In the first experiments presented here the replication ability is set to attempt a fixed number, X, of offspring per lifetime. So an object with persistence ability Y will always have a replication ability of Y/X.

A mutation operator is applied to each cell when it is updated and can either increase or decrease the persistence value of the object in the cell. In the first experiments the replication ability is also changed accordingly. Negative mutations are set initially at an average of 1 mutation every 10 time units. Positive mutations occur at an average of 1 mutation every 100 time units. A mutation that takes the value of a cell from 0 to 1 has 'created' a replicating object from an empty cell. Each run starts with N empty cells.

Figure 1a), shows how the model behaves when $N = 1000$ and the replication ability is set to attempt 3 offspring per lifetime. The three lines shown on the graph represent the best, average and worst levels of persistence of the objects at a given time during the run. As can be seen, the 'population' of objects soon reaches a persistence error threshold [6] beyond which they are unable to evolve.

A simple explanation of this error threshold can be gained through looking at the necessary requirements to maintain a persistence ratchet [1]. A persistence ratchet is said to be in place when the replication rate is higher than the net negative mutation rate as this guarantees that the level of persistence will at least not decrease and is likely to increase, on average, over time. Once there are no empty cells, then no object can replicate until another object dies and hence the average replication rate falls to the average death rate which, here, is determined only by the average ability at persisting. The mutation rate cannot be altered in this first model, so to maintain the persistence ratchet the replication rate must be kept high by keeping the death rate high which in turn requires the persistence ability to be low. Hence there is a level of persistence above which the persistence ratchet will not be in place because replication cannot occur

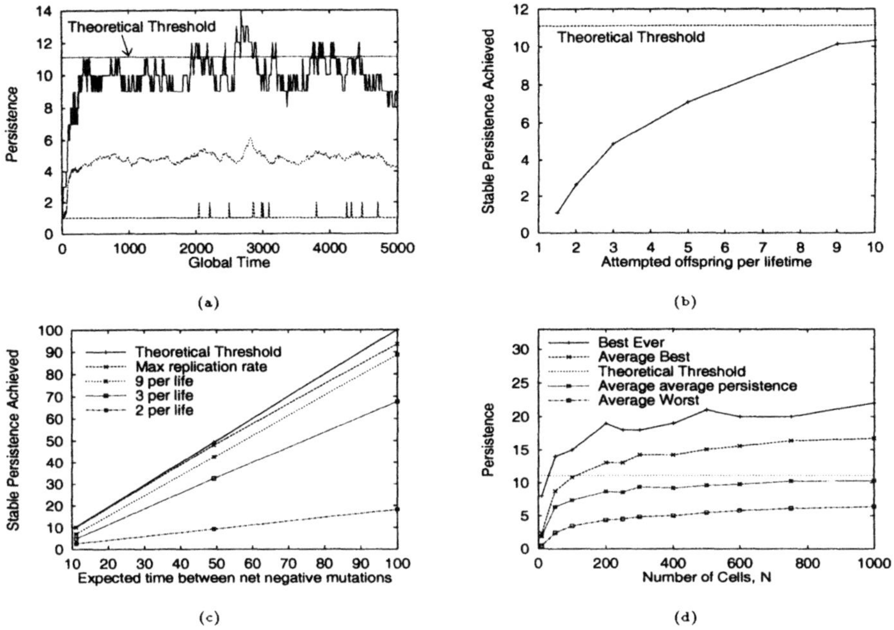

Fig. 1. Getting closer to the persistence error threshold

fast enough to ensure that negative mutations do not accumulate. Hence we can calculate the threshold by seeing that the average persistence ability must be less than the expected time between net negative mutations.

So, with the mutation rates as given above we can calculate the expected average persistence as being less than 11 units of time. Note that this value is independent of the replication ability. Figure 1b) shows the average persistence level reached when the attempted number of offspring per life is changed from 1.5 to 9 and then the final point comes from attempting replication every time step. As the number of offspring per generation is increased, so the average level of persistence reached gets closer to the calculated threshold of 11 units of time.

Figure 1c) shows a graph of average persistence level reached against different net negative mutation rates, with each line representing a different replication rate. What can be seen from both of these graphs is that changing the replication rate enables the expected level of persistence to get closer to, but never more than, the theoretically calculated error threshold. As we want to examine what does affect this error threshold we will, for simplicity, set the replication ability at the maximum possible (one attempt per time step) for the rest of the paper.

Figure 1d) shows the effects of changing the number of cells available, N, on the distribution of persistence values achieved. As N is increased so the average persistence level reached gets closer to, but never exceeds, the calculated error threshold. Interestingly, the maximum persistence level in the population does get higher than the threshold. So one way to achieve a higher level of persistence in some of the population would be to increase the space available.

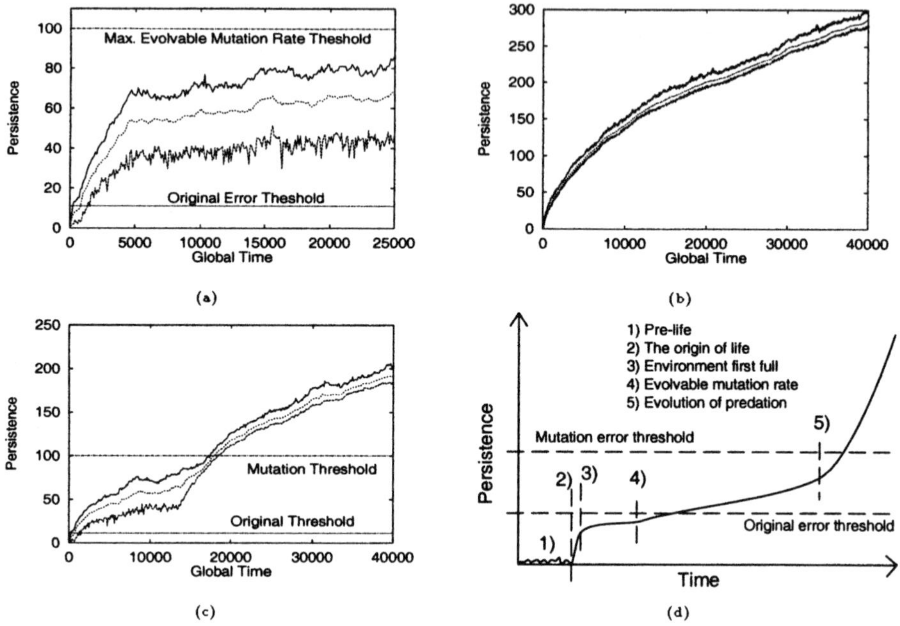

Fig. 2. Moving, then overcoming, the persistence error threshold

3 Adding to the model

As a first attempt to get beyond this error threshold we'll look at making the mutation rate evolvable. In other words objects can evolve to protect themselves against some negative mutations. So each object holds a variable for the average number of time steps between its own negative mutations. The value is copied during replication events, and mutated during update events both positively and negatively at the rate it represents[1]. This value is not allowed to exceed 50 time steps, half the fixed value for positive mutations.

The effect on the dynamics of the system are clearly visible in figure 2a). Through evolution the objects are able to 'push' the error threshold towards a higher limit by making the net negative mutation rate as low as possible, and thereby the objects can reach new heights of persistence. Whilst clearly a useful innovation, we still have a persistence error threshold to get beyond.

Next we look at what happens when objects are able to 'eat', or rather take the cell of, other objects. More specifically each object has an associated **can_eat** variable and the object is allowed to 'eat' and take the cell of any object whose ability at persisting is not larger than this **can_eat** value. The variable's value is copied during reproductive events and is subject to the same mutation rates as the persistence value and hence can evolve. With this added ability to

[1] To do otherwise would allow a net negative mutation rate to feed back on itself. This is a serious issue that must be addressed in more detail in future models.

evolve predation (figure 2b), there is no upper limit to the level of persistence that is attainable within this model. This is because a population of objects with persistence value X, who can eat objects of value $X - 1$ will be able to rapidly replace all negative mutations with 'healthy' objects of value X. So the persistence level can't decrease below X and positive mutations will easily be able to take hold, thus allowing persistence levels to increase forever.

For real organisms, who can eat who is much more complex than as modelled here. Hence predation in real evolution will not necessarily allow ever increasing levels of persistence. However, the point of this model is to show the way in which predation can remove the importance of the persistence error threshold.

Putting them all together we can form one over arching picture of what has been achieved in these models. In this last model we assume that a certain minimum level of persistence is required to enable complex behaviours. When objects can persist for 15 or less time steps, then they can neither evolve their mutation rate nor their predation ability, they are too simple. When they exceed 15 they can evolve their mutation rate. Once over the persistence level of 80 they can finally evolve their ability at predation. Figure 2c) shows the resulting graph of the system's dynamics. So together these models give some indication as to how early evolution was able to push back and then overcome the early persistence error threshold. Once predation has been evolved then the abilities to predate and/or to survive in a world of predators become more important factors in the evolutionary story than simply how long you could persist for on your own. The picture is summarised in figure 2d).

4 Conclusions

The minimum requirement on proto-life objects, that they can persist and replicate, already gives rise to a persistence error threshold that must be overcome. Replication rate and available space can bring a population closer to the theoretical threshold, but neither can change it. Evolvable mutation rates can make the threshold higher, whilst the ability to predate removes the threshold altogether. Therefore, predation must have been an important innovation in early evolution. Future work will analyse these models in more detail.

References

1. Sharpe, O.: Persistence, search and autopoiesis. In: GECCO-99: Proceedings of the Genetic and Evolutionary Computation Conference. (1999)
2. Maynard Smith, J., Szathmary, E.: The Major Transitions in Evolution. W. H. Freeman (1995)
3. Kauffman, S.: Origins of Order. Oxford University Press, New York (1993)
4. Dawkins, R.: The Selfish Gene. Oxford University Press (1989)
5. Maturana, H.R., Varela, F.J.: Autopoiesis and Cognition : the Realization of the Living. Reidel Dordrecht (1980)
6. Eigen, M.: Self-organization of matter and the evolution of biological macromolecules. Naturwissenschaften **58** (1971) 465–523

A Simulation of Life Using a Dynamic Core Memory Partitioned by Membrane Data

Hideaki Suzuki

ATR Human Information Processing Research Laboratories
2-2 Hikaridai Seika-cho Soraku-gun Kyoto 619-0288 Japan
hsuzuki@hip.atr.co.jp

Abstract. An evolutionary core memory system is proposed which simulates reactions between biological molecules in membrane bound compartments in a solvent. In the imitation of biological substances, every word in the core is classified as a Membrane, Nutrition, DNA, or Protein; and the actions of the core proceed by metabolic reactions catalyzed by Proteins. The core is partitioned into small sections by Membranes. A creature in this space is represented by a region delimited by a pair of Membranes, and if it has an appropriate DNA sequence and initial Proteins, it can reproduce itself and dominate the core. In an experiment, a small ancestor creature that the author designed is inoculated and made to evolve in the core. Creatures able to replicate DNA more efficiently than that designed are successfully bred.

1 Introduction

Recently, Suzuki proposed a core memory system named 'Semar' (the sea of matter) which involves correspondences between core words and biological substances [1,2]. He compared an operator (an instruction) to a protein in a cell, and using this analogy, prepared four types of data words on the analogy of biological molecules. They are Membrane (**MEM**), Nutrition (**NUT**), DNA (**DNA**), and Protein (**PRO**). A program (creature) in the core was delimited by the Membrane data, and made to evolve with genetic algorithms (GAs) according to the fitness values calculated from the excreted data. With this system, he succeeded in breeding programs able to solve a problem prepared in the environmental section; and yet this system was unsatisfactory in the following points. First, since GAs were used to evolve the programs, the creatures could not choose their own reproduction units by themselves. The program length (the number of instructions included in a creature) was fixed over a simulation run. Second, the actions of the core were driven not only by Proteins but also by DNA. This was evidently more complicated than a biological system, in which metabolic reactions are catalyzed only by enzymes (proteins).

The aim of this paper is to remedy the above drawbacks of previous works and to present a more life-like system on the Semar core. All creatures are put into one core and instructions necessary for the replication and transcription of DNA are prepared. The actions of the creatures are accomplished only by Proteins and an outer loop using GAs is eliminated.

2 The Model

Semar's distinctive features are summarized as follows.

- The core is addressed with logical addresses which are prepared apart from physical ones, and allows the insertion/deletion of arbitrary numbers of data words.
- All data words are classifed into four groups: Membrane (**MEM**), Nutrition (**NUT**), DNA (**DNA**), and Protein (**PRO**).
- The core is hierarchically partitioned by Membrane data, and almost all actions (computational operations) are confined to a section.
- All actions in the core are activated by Proteins which are created by the transcription of DNA units. The proteins' actions are parallel and independent of each other.

Each word in the core is represented by a 64- to 192-bit string composed of 16-bit data units, called 'header', 'type', 'label', 'value' or 'mcd' (machine code), and 'temps' (templates) in turn. These units are denoted by mnemonics punctuated by colons. The mcd of Membrane data is 'Bgn' or 'End'. The core range delimited by a pair of Membranes (`[MEM:Bgn]` and `[MEM:End]`) constitutes a section.

The actions of a Protein are as follows. First, the Protein searches for ligand Nutrition using the bit-matching between a Nutrition's label and the Protein's temp. Like ligand molecules which regulate allosteric proteins in a living cell, the selected ligand data work as an activator or inactivator of the Protein, and change the Protein's type into 'Actv' or 'Inac'. After this process, if the Protein's type is 'Actv', the Protein is put into action. The protein begins searching for operand data using the bit-matching, and changes, copies, creates, moves, or deletes operands depending upon its function. In one clock (which measures time in the core), the Protein can modify all matched operands in the same section; and yet its influence does not exert beyond Membranes. Although all Proteins in the core are practically put into action one by one, the actions of the Proteins are accomplished logically in parallel using marks attached to the words.

Table 1 shows 20 elementary instructions (machine codes, or genes) that the author has prepared. They are classified into three groups. The first group (`RC_o` and `RC_e`) consists of receptors for regulatory Proteins. `DNA:RC_o` and `DNA:RC_e` are not transcribed and serve only as starting or terminating signals for consecutive DNA units whose types are regulated simultaneously. (In the imitation of biology, the author calls such a set of DNA an 'operon'.) The type of DNA is either 'Utrn' or 'Trns'. DNA can be transcribed to create a Protein only when its type is 'Trns'. The second group (`Prom` and `Repr`) consists of regulatory Instructions. Like regulatory genes in a living cell, when transcribed, `DNA:Prom` and `DNA:Repr` create `PRO:Prom` and `PRO:Repr`, respectively. These Proteins search for the matched `DNA:RC_o` or `DNA:RC_e` (operands) and change the type of the operons. The third group (`creN` to `CHoP`) consists of structural Instructions. After being transcribed, a structural Protein modifies operand data according to the defined function.

Mnemonic	Operand	Function as a Protein
RC_o		(Not transcribed to a Protein.)
RC_e		(Not transcribed to a Protein.)
Prom	DNA:RC_o	Changes the type of operons into 'Trns'.
Repr	DNA:RC_e	Changes the type of operons into 'Utrn'.
creN	MEM:Bgn	Creates NUT after MEM:Bgn.
creS	MEM:Bgn	Creates a new inner section after MEM:Bgn.
prey	MEM:Bgn	Converts (Bgn,–,End,Bgn) to (Bgn,Bgn,–,End).
bred	MEM:End	Moves the operand inner section before MEM:Bgn.
invd	MEM:End	Converts (End,Bgn,–,End) to (Bgn,–,End,End).
dgst	MEM	Deletes Bgn and End delimiting an inner section.
ctbN	NUT	Deletes operand NUTs.
CHiN	NUT	Copies operand NUTs to the inner section.
CHoN	NUT	Copies operand NUTs to the outer section.
trDP	DNA	Transcribes DNA to PRO:Inac after the sequence of DNAs.
ctbD	DNA	Converts operand DNA to NUT with value zero.
CHiD	DNA	Copies operand DNAs to the inner section.
CHoD	DNA	Copies operand DNAs to the outer section.
ctbP	PRO	Converts operand PRO to NUT with value zero.
CHiP	PRO	Copies operand PROs to the inner section.
CHoP	PRO	Copies operand PROs to the outer section.

Table 1. The basic instruction set (gene set)

The simulation of the Semar core is begun by inoculating the core with the ancestor creature designed by the author and a few words of environmental Nutrition. The ancestor can reproduce itself every three clocks. After the start, the ancestor and its offsprings reproduce themselves and the size of the core grows swiftly. The passage of time in a run is measured in terms of the number of clocks, and at each clock, the following three operations are executed sequentially; the operations of Proteins, mutation, and weeding. Mutation is operated in an interval of time during which the size of the core grows twofold. When this happens, a bit is chosen randomly from among all of the core words (which include all significant data from the header to templates) and inverted. This process is repeated an appropriate number of times so that the occurrence rate of bit inversions per bit per clock might be equal to the constant value u. The weeding operation is executed only when the size of the core exceeds the predefined constant CPT_{reap0}. When this happens, the outermost sections (sections that are not included by other sections) are chosen and eliminated from the core until the total size of the core becomes smaller than the constant number CPT_{reap1}.

3 Results of experiments

Figure 1 shows a typical result of the simulation using $u = 3.47\times10^{-5}$, $CPT_{\text{reap0}} = 24576$, and $CPT_{\text{reap1}} = 19661$. According to Fig. 1(a), the average amount of

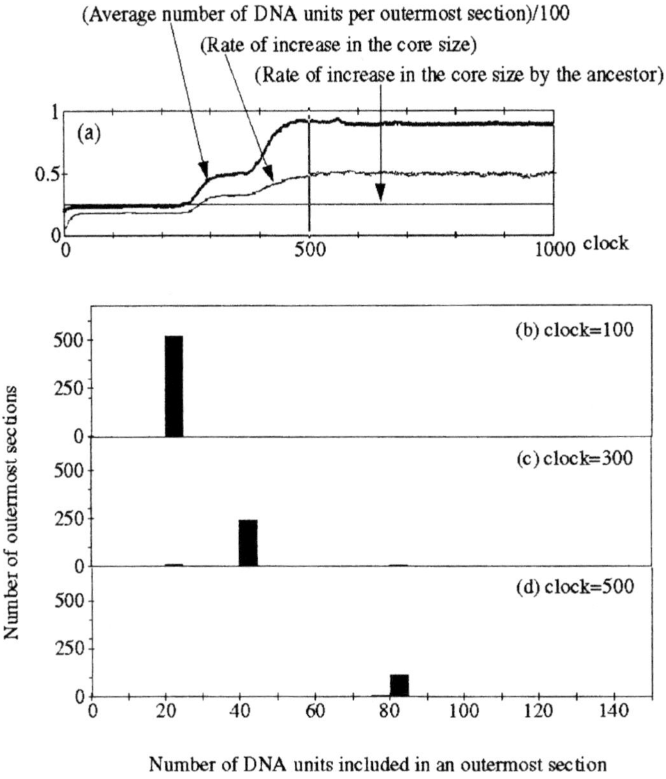

Fig. 1. A typical result of the simulation. (a) The two thick and thin curves are the results of the simulation, and the horizontal straight line is the theoretical value for the ancestor ($\sqrt[3]{2} - 1 = 0.26$). (b) to (d) are histograms showing distributions for the amount of DNA included in an outermost section at each clock.

DNA per outermost section begins from 20 (the amount of DNA included in the ancestor), increases, and reaches about 90 after 500 clocks. The rate of increase in the core size is also increased compared to the value of the ancestor; hence, we can conclude from this graph that the ability of self-reproducing creatures is improved during the run. Figures 1(b)∼(d) suggest the mechanisms for this improvement. After 300 clocks, the original creature (which holds 20 DNA units) and its variants are completely lost from the core, and the core becomes dominated by creatures that hold 40 DNA units (twice as large as 20). A similar improvement happens after about 400 clocks. In the history of evolution, the amount of DNA in a section is doubled or quadrupled.

To see this improvement more closely, several creatures were chosen during the run and bred in a blank core without mutation. The results of these experiments are shown in Table 2. The creatures chosen from the core at clock number 300/500 were able to bear two/four daughter sections every three clocks.

Chosen from the core at the clock number	Number of DNA units included	Initial growth of the section numbers	Rate of replication of DNA units per clock
0 (ancestor)	20	$1, 1, 1, 2, 2, 2, 3, \cdots$	$\sqrt[3]{2} - 1 = 0.26$
300	40	$1, 1, 1, 3, 3, 3, 5, \cdots$	$\sqrt[3]{3} - 1 = 0.44$
500	80	$1, 1, 1, 5, 5, 5, 9, \cdots$	$\sqrt[3]{5} - 1 = 0.71$

Table 2. Typical self-reproducing creatures chosen from the experiment shown in Fig. 1. The initial growth of the section numbers represents the number of outermost sections at the first clock, that at the second clock, and so on.

This multiple reproduction was achieved by the execution of plural numbers of `[PRO:bred]`s held in a section.

The author finishes this section by describing the result of an observation to clarify the mechanisms causing the doubling of DNA sequences. In the experiment of Fig. 1, the first section with a doubled DNA sequence appears as follows. At 74 clocks, the mother section missed the breeding of the daughter and the daughter remained in the mother section after the breeding procedure. At 76 clocks, `PRO:CHiD` replicated DNA into this remaining daughter section and the DNA sequence was doubled. At the next clock, the daughter section was bred. At this example, the doubling of the DNA sequence occurred by an error in breeding the daughter section.

4 Conclusion

A novel ALife system called Semar was proposed and simulated. Using data words designed in the imitation of biological substances, biological reactions were emulated in the core, and the breeding of creatures able to replicate their DNA more efficiently than the ancestors was successfully carried out. The mechanisms causing the evolutionary improvement were also discussed.

Acknowledgements

The author would like to thank Dr. K. Shimohara and Dr. Y. Ichinose of ATR HIP laboratories for their continual encouragement.

References

1. Suzuki, H.: One-dimensional unicellular creatures evolved with genetic algorithms. In: JCIS '98: The Fourth Joint Conference on Information Sciences, Proceedings Vol.II. Association for Intelligent Machinery Inc., USA (1998) 411–414
2. Suzuki, H.: Unicellular core-memory creatures evolved using genetic algorithms. In: InterJournal of Complex Systems, BArticle, Submitted. Available at http://www.interjournal.org/ or http://www.hip.atr.co.jp/ hsuzuki/body/papers/1998.12_InterJournal/index.html

Molecular Evolution in Time-Dependent Environments

Claus O. Wilke, Christopher Ronnewinkel, and Thomas Martinetz

Institut für Neuroinformatik, Ruhr-Universität Bochum
D-44780 Bochum, Germany
email: wilke,ronne,martinetz@neuroinformatik.ruhr-uni-bochum.de,
WWW home page: http://www.neuroinformatik.ruhr-uni-bochum.de

Abstract. The quasispecies theory is studied for dynamic replication landscapes. A meaningful asymptotic quasispecies is defined for periodic time dependencies. The quasispecies' composition is constantly changing over the oscillation period. The error threshold moves towards the position of the time averaged landscape for high oscillation frequencies and follows the landscape closely for low oscillation frequencies.

The quasispecies theory, put forward by Eigen in 1971 [1], and subsequently studied by Eigen, Schuster, McCaskill and coworkers [2, 3, 4], is nowadays one of the classical theories of self-replicating entities. Its prediction of an error threshold, above which the self-replication ceases to produce useful offspring, has important implications for the origin of life. The error threshold effectively limits the amount of information the entities can carry, thus placing an upper bound on the complexity self-reproducing information carriers can achieve without sophisticated error correction mechanisms.

Although completely static environments are unrealistic in any case apart from experiments in perfectly controlled flow reactors, the quasispecies theory has so far been considered mainly in static replication landscapes. Nevertheless, even under fixed environmental conditions can the replication rates of RNA molecules, for example, change because of changing concentrations of template and replica [5]. Jones [6, 7] has studied underlying time-dependencies which are identical for all sequences. Contrasting to that, we want to focus on replication landscapes with individual time-dependency for each sequence. One of the reasons for the neglect of individually changing replication coefficients in earlier work is probably the fact that for arbitrary temporal changes an asymptotic quasispecies cannot be defined. However, a meaningful definition is at hand for time-periodic replication landscapes, as we are going to show below.

We start from the discretized form of Eigen's evolution equation [8], linearized with the appropriate transformation [9, 10]. Due to space limitations, we cannot repeat the arguments leading to that equation here. For details about this calculation, the reader is referred to [4]. We use the same notations as are used there. Additionally, we define the error rate $R = 1 - q$, which gives the probability that a single symbol is copied erroneously. The string length will be denoted by l throughout this paper.

The vector of the unnormalized sequence concentrations $\boldsymbol{y}(t)$ evolves according to

$$\boldsymbol{y}(t+\Delta t) = \Big[\Delta t\, \mathrm{W}(t) + \mathbf{1}\Big]\boldsymbol{y}(t). \tag{1}$$

Here, $\mathrm{W}(t)$ is the replication matrix $\mathrm{W}(t) = \mathrm{QA}(t) - \mathrm{D}(t)$. We assume the matrix $\mathrm{W}(t)$ is periodic with period $T = n\Delta t$, $n \in \mathbb{N}$, with some chosen discretization time step $\Delta t \ll T$. After iteration of Eq. (1), we obtain for $t' = t + \zeta\Delta t$, $\zeta \in \mathbb{N}$

$$\boldsymbol{y}(t') = \mathcal{T}\left\{\prod_{\nu=0}^{\zeta-1}\Big[\Delta t\, \mathrm{W}\left(t+\nu\Delta t\right) + \mathbf{1}\Big]\right\} \boldsymbol{y}(t). \tag{2}$$

where $\mathcal{T}\{.\}$ stresses that the product has to be evaluated in the time order given by the iteration. With the definition of the matrix (or operator)

$$\mathrm{X} := \mathcal{T}\left\{\prod_{\nu=0}^{n-1}\Big[\Delta t\, \mathrm{W}\left(\nu\Delta t\right) + \mathbf{1}\Big]\right\}. \tag{3}$$

which maps $\boldsymbol{y}(0)$ onto $\boldsymbol{y}(T)$, we are now able to write down the solution of the discretized differential equation Eq. (1) for the initial condition $\boldsymbol{y}(0)$ as

$$\boldsymbol{y}(t) = \mathcal{T}\left\{\prod_{\nu=0}^{\zeta-1}\Big[\Delta t\, \mathrm{W}\left(\nu\Delta t\right) + \mathbf{1}\Big]\right\} \mathrm{X}^m \boldsymbol{y}(0). \tag{4}$$

where the time t has been subdivided into $t = mT + \zeta\Delta t$, with $\zeta < n$ and $m, \zeta \in \mathbb{N}$.

If we observe the system in time steps of the period length T, the system appears to evolve in a static replication landscape, which is defined by X. The asymptotic steady state for the oscillation phase $\zeta = 0$ is therefore given by the normalized Perron eigenvector ϕ_0 of X [11]. For $0 < \zeta < n$, the steady states are found by application of $\mathcal{T}\{\dots\}$ from Eq. (4) to ϕ_0 and subsequent normalization.

Let us now study quantitatively the effects a periodic replication landscape has on the prominent quasispecies. As the first step into that direction, we start from the Swetina-Schuster landscape [12] and introduce small oscillations in the master sequence's replication coefficient A_0. For reasons of simplicity, we set all decay constants equal $D_i(t) = D$, because then they drop out of Eq. (1) during the foregoing linearization.

We will write the time-dependent replication coefficient $A_0(t)$ as

$$A_0(t) = A_{0,\mathrm{S}} \exp[\epsilon f(t)]\,, \tag{5}$$

where $A_{0,\mathrm{S}}$ is the replication coefficient in the static landscape, $f(t)$ is a T-periodic function and ϵ is the oscillation amplitude. For $\epsilon = 0$ the corresponding static landscape is reached. The other replication coefficients are equal $A_1 = \dots = A_l = A$ and constant. We will choose A so small that the condition $A \ll$

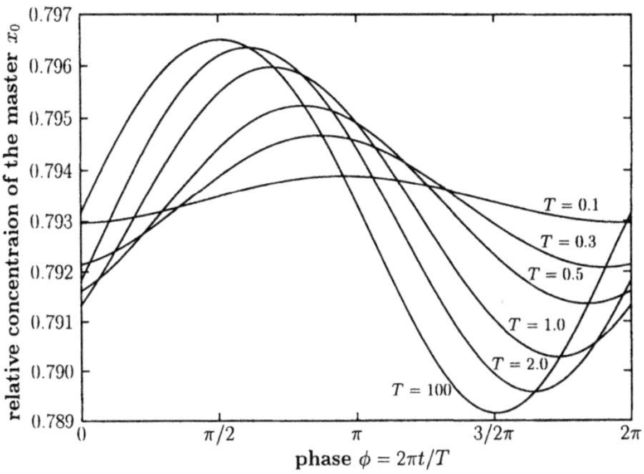

Fig. 1. The steady state oscillations of the master sequence in cyclically changing environments with different oscillation periods T. Parameters used are $l = 2$, $A_{0,S} = e^{2.4}$, $A_1(t) = A_2(t) = 1$, $\epsilon = 0.2$.

$A_0(t)$ is satisfied for all t and $\epsilon \ll 1$. This assures that we see a clear transition from the static case to the dynamic case, and additionally, that the changes in the master sequence's abundance can be directly related to the changes in $A_0(t)$.

One of the simplest forms the function $f(t)$ in Eq. (5) can take on is

$$f(t) = \sin(\omega t) \qquad \text{with} \quad \omega = 2\pi/T. \tag{6}$$

In the following, we will shortly discuss the influences of different frequencies ω and amplitudes ϵ for this time dependency.

As response to the oscillation of the replication coefficients, a modified oscillation is found in the concentration x_0 of the Master sequence (Fig 1). For increasing frequency ω, the amplitude of the x_0 oscillation decreases and a phase shift strengthens. This behaviour is due to the finite time a reaction system as described by Eigen's equation needs to settle into equilibrium. In constant environments, the asymptotic species distribution is approached in exponential time, with the relaxation time scale τ set by the difference between the largest and the second-largest eigenvalue of W. For the oscillating environments the relaxation time needs to be compared to the period T. If $T \gg \tau$, the system is virtually in equilibrium for arbitrary (asymptotic) times t, whereas for $T \approx \tau$ the changes cannot be tracked anymore and phase shift as well as amplitude damping of the response sets in. For $T \ll \tau$ the response amplitude gets fully damped and the system gets identical to one with the time-averaged replication coefficients. Interpreting the A_0 and x_0 time-dependence as input and output signal, the system acts as a low pass analog filter, in analogy to observations made in population genetics models with dynamic fitness landscapes [13, 14, 15, 16]. Moreover, for small ϵ, the filter works linear, which means that a sinusoidal oscillation is found in x_0, whereas this linearity is quickly destroyed for increasing ϵ[1].

[1] Details on this analysis can be found in [17].

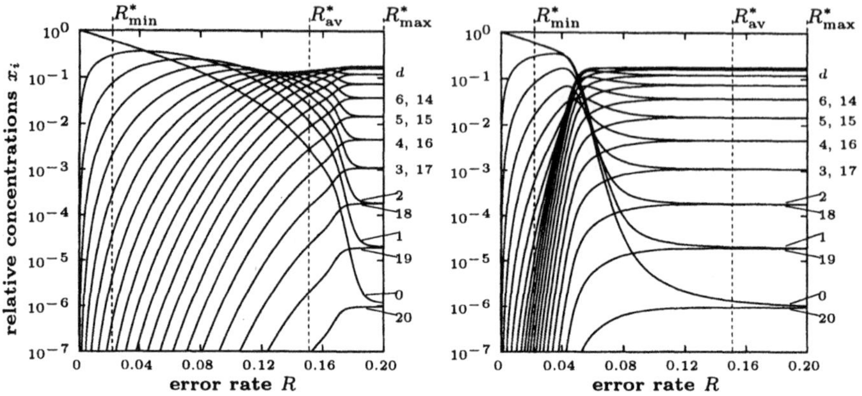

Fig. 2. The quasispecies distribution as a function of the error rate R. $A_{0,S} = e^{2.4}, A = 1, \epsilon = 2, T = 100$. Two different oscillation phases are shown. *left*: $\zeta = n/2$, *right*: $\zeta = 0$.

We will now focus on the influence of a time-dependency as given in Eq. (5) onto the error-threshold. In accordance with the above, we have to distinguish between different dynamic regimes. For $T \ll \tau$, a sharp error-transition occurs at $R^*_{\rm av}$, which denotes the error-threshold of a system with time-averaged $\tilde{A}_0 = \frac{1}{T}\int_0^T A_0(t)\,dt$. Contrasting to that, a moving error-transition can be found for $T \gg \tau$ at approximately $R^*(t)$, which denotes for any given t the error-threshold in a constant landscape with $\tilde{A}_0 = A_0(t)$. $R^*(t)$ lies between $R^*_{\rm max}$ and $R^*_{\rm min}$, which correspond to $\tilde{A}_0 = \max_t A_0(t)$ and $\tilde{A}_0 = \min_t A_0(t)$, respectively. In the intermediate cases $T \approx \tau$, the numerical simulations (see Fig. 2) show that the error-threshold $R^*(t)$ oscillates within a smaller interval than $[R^*_{\rm min}, R^*_{\rm max}]$.

These findings allow to draw a phase diagram as displayed in Fig. 3. For low T, we observe the standard separation into an ordered phase (below the error threshold) and a disordered phase (above the error threshold). With increasing oscillation period T, a third, new phase appears between the two. In this phase, we observe—for a fixed error rate R—an alternation between a fully developed

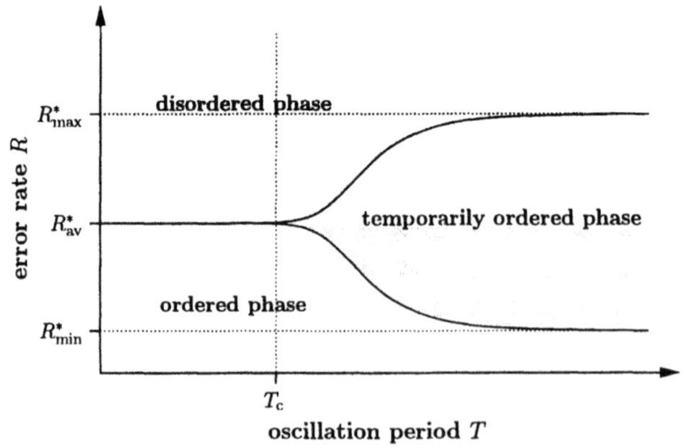

Fig. 3. Schematic phase diagram for a time-dependency like Eq. (5).

quasispecies and a completely disordered system. The population seems to be moving back and forth over the error threshold. Therefore, we call this new phase the temporarily ordered phase. Since a similar phase diagram can be expected for any periodic landscape with finite R^*_{av} and $R^*_{\text{min}} \neq R^*_{\text{max}}$, we believe that such observations could also be made in typical AL simulations such as Tierra or Avida [18] provided with the appropriate replication landscape. The temporarily ordered phase would for a finite population in a rugged landscape have the effect of causing a random drift over the landscape at some times and a localization around a local master sequence at other times.

Upon completing this work, we became aware of Ref [19], in which a different approach towards dynamic replication landscapes is given, using a stochastic time dependency in the landscape. The results presented there cannot directly be related to our findings here, because the equivalent to R^*_{av} vanishes in [19], while the equivalents to R^*_{max} and R^*_{min} take on the same finite value. This leads to a different phase diagram than the one we observe here.

References

[1] M. Eigen, Naturwissenschaften **58**, 465 (1971).

[2] M. Eigen and P. Schuster, *The Hypercycle—A Principle of Natural Self-Organization* (Springer-Verlag, Berlin, 1979).

[3] M. Eigen, J. McCaskill, and P. Schuster, J. Phys. Chem. **92**, 6881 (1988).

[4] M. Eigen, J. McCaskill, and P. Schuster, Adv. Chem. Phys. **75**, 149 (1989).

[5] C. K. Biebricher, M. Eigen, and W. C. Gardiner, Jr., Biochemistry **22**, 2544 (1983).

[6] B. L. Jones, Bull. Math. Biol. **41**, 761 (1979).

[7] B. L. Jones, Bull. Math. Biol. **41**, 849 (1979).

[8] L. Demetrius, P. Schuster, and K. Sigmund, Bull. Math. Biol. **47**, 239 (1985).

[9] C. J. Thompson and J. L. McBride, Math. Biosci. **21**, 127 (1974).

[10] B. L. Jones, R. H. Enns, and S. S. Rangnekar, Bull. Math. Biol. **38**, 15 (1976).

[11] O. Perron, Math. Ann. **64**, 248 (1907).

[12] J. Swetina and P. Schuster, Biophys. Chem. **16**, 329 (1982).

[13] A. Sasaki and Y. Iwasa, Genetics **115**, 377 (1987).

[14] K. Ishii, H. Matsuda, Y. Iwasa, and A. Sasaki, Genetics **121**, 163 (1989).

[15] T. Hirst, in *Fourth European Conference on Artificial Life*, edited by P. Husband and I. Harvey (MIT Press, Cambridge, MA, 1997), pp. 425–431.

[16] A. J. Hirst and J. E. Rowe, J. theor. Biol. (1998), submitted.

[17] C. O. Wilke, Evolutionary Dynamics in Time-Dependent Environments (Ph.D. thesis, Ruhr-Universität Bochum, 1999). http://www.neuroinformatik.ruhr-uni-bochum.de/ini/PEOPLE/wilke/ps/PhD.ps.gz.

[18] C. Adami, *Introduction to Artificial Life* (Telos, Springer-Verlag Publishers, Santa Clara, 1998).

[19] M. Nilsson and N. Snoad, eprint physics/9904023 (1999).

Evolution to Complexity: Replication, Elongation and Assembly in an RNA World

Tomoyuki Yamamoto[A] and Paulien Hogeweg[B]

[A,B] Bioinformatica, Utrecht University, Padualaan 8, 3584 CH Utrecht, The Netherlands
[A] Department of Mathematics, Hokkaido University, Sapporo, 060-0810 Japan
E-mail: ty@math.sci.hokudai.ac.jp

Abstract. A spatially resolved model of RNA world is studied, where primer induced replication, concatenation and random cutting are considered. The increase of diversity of sequences and complexity of shapes are observed; A hierarchically orgnized "replication network" is formed, and evolution to long sequences and assembly of further long sequences are obtained through it. These results suggest a scenario for overcoming the error threshold and for the evolution of enzymatic activity.

1 Introduction

The RNA world is a hypothesis for the origin of life (see [1] for review) and several attempts to simulation it have been made(e.g.[2]). An important problem here is that the error threshold [3] [4] prevents long molecules to increase in the population.

In the present paper, we will investigate this problem by simulation of an RNA-replicator model on a plane. The secondary structure is included. We obtain a replication network with high diversity and an increase of the variety of shapes. Then our model appears to suggest a way to pass the information threshold.

Single-stranded RNA molecules can replicate by themselves (e.g. [5]), but the length is bounded by the error threshold. Then, we assume the concatenation of them is also possible, as is done in DNA [6].

Spatial resolution has several merits for the generation of diversity of sequences. Especially, we believe the spatial clustering supports elongation and repair of molecules via the replication network (see §3.1).

Recently, McCaskill et al. studied DNA/RNA amplification system (e.g. [7]). They have shown a formation of cooperative amplification network and reaction-diffusion like patterns on a plane. While this work is similar to ours, we concentrate on RNA: we include RNA secondary structure and different kinds of phenomena are obtained, like assembly of molecules.

2 Model

The model consists of two parts. One is the plane where RNA polymers exist, the other is the "soup" which supplies oligomers and monomers.

The plane is descretized to a grid (typically 100×100) and each gridcell can be occupied by a molecule, which satisfies the minimum length limit (typically 5). The molecules are assumed to be attached to the plane, but limited diffusion by Margolus' method [8] does occur.

The soup supplies oligomers (typically, length 2 to 5) and monomers as substrates for the replication. The secondary structure of molecules is evaluated by the Vienna RNA package by Hofacker et. al [9].

We apply three reactions, replication, concatination and random cut.

replication We assume the primer is required to start replication and the replication proceeds only from 5' to 3' ends. Dangling 5' end is allowed only for the template (i.e. template do not elongate). Folded, double-stranded molecules cannot be replicated. Point mutations are applied on the part being replicated. The threshold value for the binding energy (typically -2.50 kcal/mol, averaging the number of bonds) is applied. When no matching primer molecule is found in the nearest neighbor, a random oligomer is taken from the soup.

concatenation The molecules can be concatenated when a molecule pairs two molecules in its neighborhood to form a hemiduplex strand, bridging over two strands. The threshold for the binding energy is applied for both matching parts.

random cut The single stranded areas in the molecules can be cut (typically 1.0×10^{-5} per nucleotide). The double-stranded region cannot be cut and a folded molecule is more stable than a single-stranded molecule.

3 Results

Both the population growth and the increase of diversity are frequently obtained. Although sometimes the plane will be ocuppied by trivial sequences (see §3.3), variety of sequences and complexity of secondary structures are obtained around 30% of the simulations (see §3.4).

To illustrate the behavior, let us see the case when the replication fidelity is 1.0. In fig.1a, the sequences of frequent molecular species, whose populations are larger than 5, are shown. In fig.1b, some of the longest sequences are shown. There are about 3000 molecules after 100000 time steps from random initial condition where we start with 100 short molecules. The secondary structure is shown by the bracket representation [10].

3.1 Replication network

In fig.1a, some subsequences are shared by some of the freqent sequences. Such molecules may be orderd, as is in fig.2. In this figure, starting from the longest a)

```
GGAGG ..... 40           CCACCGG ....... 10           GGGCCGGAGG .......... 11
CGGAG ..... 10           CCUCCGG ....... 11           CGGGCCGGAGG ........... 26
GGUGG ..... 7            GCCGGUGG ........ 9          GGCGCCGGUGG ........... 8
CGGUG ..... 8            GCCGGAGG ........ 26         GCGGGCCGGAGG ............ 48
CGGAGG ...... 39         CCGGAGGG ((...)). 6          CGGCGCCGGUGG ............ 9
CGGUGG ...... 14         GGCCCGCG ........ 9          CCACCGGCGCCG ............ 6
CCGGAG ...... 19         GGCGGGCCG ((....)). 12       CCUCCGGCCCGCG ............. 16
CCUCCG ...... 51         CGGCCCGCA ......... 6        CGCGGGCCGGAGG ............. 10
CCACCG ...... 25         GGCCGGAGG ......... 12       CCUCCGGCCCGCA ............. 7
GCCCGC ...... 6          CGGCCCGCG ......... 6        GGCCCGCGGGCCGGAGG
CCGGUG ...... 9          CGCCGGUGG ......... 7        ((((....)))).....  7
CCGGAGG ....... 44       GCGCCGGUGG .......... 13     CCUCCGGCCCGCGGGCCGGAGG
CCGGUGG ....... 31       CCGGCCCGCG .......... 7      (((((((((....))))))))) 15
```

b)

```
GGCCGGAGGCGGGCCUCCGGCGCGCGCCUGCGGGCCGGAGGGG .(((...)))...((((((((.(((......))).)))))))))..
GGCCCGCGGGCCGGAGGGGCGCCGGGGGCGGGCCGGAGGGCGCGCCG .((((.((((((.....))).))).)))).(((((.....)).))).
CGGCGGGCGCCCGCGCGCGGGUGGCGGCCCGCGGGCCGGCGCCGGUGG ..((.((((((.((.((((((((....))))))).)).)))))).))..
AGGCGGCCCUCCGGCCCGCGCCGGGCCUCCGGCCCGCUUGCGGGCUGGGCCU
....(((((...((((((((.((((((...))))))..)))))))))))))).
```

Fig. 1. a: A list of frequent sequences from a snaphot of a simulation. On each line, a sequence, its secondary structure and its population size are shown. b: The longest sequences, from the same snapshot. Some of them have 2 forks.

reverse-complementary sequence pairs (at the top), subsequences whose tail part are lost are ordered on either row, with their reverse-complementary sequences.

In some pairs, the population sizes are asymmetric (e.g. there are 2 CCUC-CGGCCCG and 26 CGGGCCGGAGG). This implies an "elongation mechanis-m". CCUCCGGCCCG is replicated 217 times and made by random cut 108 times. 297 of them are used as primers for replicating longer sequences. Although this mechanism is limited to single-stranded molecules, we did obtain population growth of some long sequences by assembly (see §3.2). Note that the reverse-complimentary sequence, CGGGCCGGAGG cannot elongate, because no dangling 3' end when it matches CCUCCGGCCCGCG.

If we focus on the relationship between a long sequence and its subsequences with loss of tails, above elongation mechamism is regarded as a "repair mechamis-m". While long sequences are more fragile than short sequences, this mechamism supports them and the diversity of the whole system is maintained.

The replication network is an important driving force for the "evolution to complexity". As described above, elongation and repair processes maintains the variety of sequences and long sequences are formed and maintained. While we have not investigated spatial pattern formation, we think spatial clustering supports above elongation and repair. Since the reaction is local, the formation of cluster is trivial. Then, if a molecule elongate, its reverse-complementary sequence is expected to found in the neighborhood and whole cluster can be elongated. Also, the repair can be done in a similar manner.

3.2 Assembly

The long molecules (more than 20) usually have secondary structure and they cannot replicate but are formed via concatenation. However, in fig.1a, some of

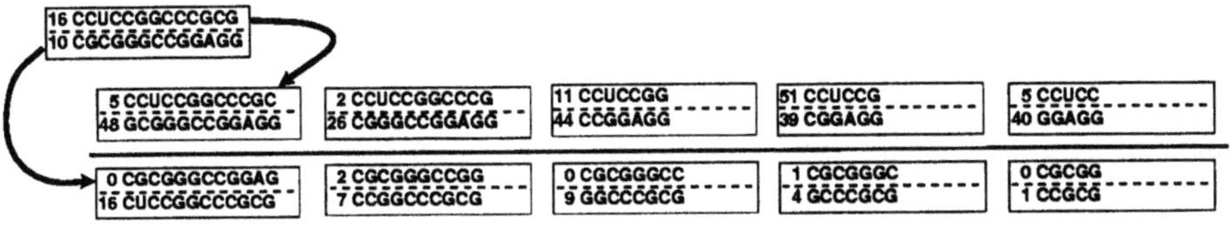

Fig. 2. An example of the replication network. Each box shows reverse-complementary pair. From the longest pair (the box at the top left), their subsequences are shown on the either row (separated by the line) , orderd by the length. The number on the left of the sequence is the population size.

folded sequences are reproduced. Such long sequences are "assembled", by frequent sequences. For example, CCUCCGGCCCGCGGGCCGGAGG is a combination of CCUCCGGCCCGC and GGGCCGGAGG. Further elongation is possible, if there are dangling ends. Thus we may hope a step-by-step extention of sequences.

In fig.1b, some of the longest sequences are shown. Some molecules have multiple forks. About 30% of simulations have such complex secondary structures. The formation of long, multi-fork molecules can lead to the emergence of "ribozymes". Indeed, we have obtained some interesting molecules, which are about 100 nucleotides long, and three or more forks.Also, another merit of folded molecule is stability, since the double-stranded regions do not break by the random cut.

3.3 Parasites

When the replication fidelity is high, parasitic sequences may take over. Trivial sequences, such as all-C or all-G, can easily replicate and concatnate because of high probability for the ligation of a primer. Once some all-C and all-G sequences meet and start replication, their populations grow rapidly. Such trivial sequences do not allow the formation of "ribozymes" with complicated secondary structures.

3.4 The Effects of Mutation

We have calculated the average ratio of parasitic invasion, maximum sequence length and maximum population size. See table.1. Those values are averaged over 100 samples. The classification is evaluated as follows. **A**: more than half out of 10 longest molecules have multiple forks. **B**: secondary structures are less seen, but not invaded by the parasites (i.e. quickly growing all-G/all-C sequences). **C**: invaded by parasites.

On the right side of the table, averaging only **A**-class runs, total population size (pop), maximum length (L_{max}), maximum population size restrincting length ≥ 10 ($P_{max}1$), maximum population size without restriction ($P_{max}2$) are

fidelity	A B C	pop	L_{max}	$P_{max}1$	$P_{max}2$
1.0	20 24 56	2483.6	67.6	20.8	44.3
0.98	32 28 40	2934.5	61.0	13.0	36.6
0.95	34 38 28	2795.1	64.3	4.3	23.1
0.90	35 61 4	2596.3	55.4	2.8	15.3
0.80	27 63 10	2544.3	48.6	2.2	9.9
0.70	19 68 13	2384.5	43.8	2.0	8.8

Table 1. Classification of simulation results and statistics. See text.

shown. The averaged maximum length and population size decrease as the fidelity decreases; this is a trivial result of mutation. Also, the mutation suppresses the parasitic invasion, by giving a secondary structure to mutants of all-C and all-G sequences. The percentage of **A**-class case have an optimum when fidelity is 0.90, due to mutatants of parasites, which have several forks. However, the maximum population is a few at this point. Both rich diversity and population growth of a molecule are obtained around 0.95 fidelity.

4 Summary

The spatial clustering supports the complexity of replication network, where elongation and repair of sequences occur. Within the replication network, frequent sequences increase the population of assembled long sequences where the complexity increases further, beyond the boundary of error threshold. Then the enzymatic activity may expected to arise. We will study the transition to replicase dominated dynamics in the near future.

One of the authors (TY) is supported by the fellowship from Japan Society for the Promotion of Sciences.

References

1. Joyce, G.F., Orgel, L.E.: Prospects for understanding the origin of the RNA world. In Gesteland, R.F., Cech, T.R., Atkins, J.F., eds.: The RNA world, 2nd edition. Cold Spring Harbor Laboratory Press, Cold Spring Harbor, New York (1999) 49–77
2. Anderson, P.W. Proc. Natl. Acad. Sci. USA **80** (1983) 3386–3390
3. Eigen, M., Schuster, P.: The Hypercycle: A Principle of Natural Self-Organization. Springer-Verlag, Berlin (1979)
4. Swetina, J., Schuster, P. Biophysical Chemistry **16** (1982) 187–203
5. Inoue, T., Orgel, L.E. J. Am. Chem. Soc. **103** (1981) 7666–7667
6. James, K.D., Ellington, A.D. Chemistry and Biology **4** (1997) 595–605
7. Breyer, J., Ackermann, J., McCaskill, J. Artificial Life **4** (1998) 25–40
8. Toffoli, T., Margolus, N.: Cellular Automata Machines. MIT Press, Cambridge (1987)
9. Hofacker, I.L., Fontana, W., Stadler, P.F., Bonhoeffer, S., Tacker, M., Schuster, P. Monatshefte f. Chemie **125** (1994) 167–188
10. Konings, D., Hogeweg, P. J. Mol. Biol. **207** (1989) 597–614

Evolution of Multispecificity in an Immune Network

Kouji Harada and Takashi Ikegami

Institute of Physics, The Graduate School of Arts and Sciences,
University of Tokyo,
3-8-1, Komaba, Meguro-ku, Tokyo 153, Japan
harada@sacral.c.u-tokyo.ac.jp
ikeg@sacral.c.u-tokyo.ac.jp

Abstract. Divergence in antigen response of the immune network is discussed, based on shape-space modelling. The present model extends the shape-space model by introducing the evolution of specificity of idiotypes. When the amount of external antigen increases, stability of the immune network changes and the network responds to the antigen. It is shown that specific and non-specific responses emerge as a function of antigen levels. A specific response is observed with a fixed point attractor, and a non-specific response is observed with a long-lived chaotic transient state of the lymphocyte population dynamics. The network topology also changes between these two states. The relevance of such a long-lived transient state is discussed with respect to immune function.

1 Introduction

The 'lock and key' concept has been central to understanding the specificity of biochemical molecular interactions, from enzyme-substrate relationships to antigen-antibody matchings. However, it has gradually been realized that such a 'lock and key' concept is not strictly valid, particularly in immune systems [1]. Antigen-antibody interactions are found to be plastic or ' multispecific' rather than fixed or single-specific. Namely, antibodies inherently have a flexible recognition capacity. Kearney et al.[2] have confirmed experimentally the existence of such ambiguity of recognition in the antibody binding site of immature B cells. It is generally believed that development from ambiguous to specific recognition is caused by somatic hypermutations [3]. We here propose a new dynamics of specificity evolutions based on Jerne's network hypothesis [4]. Our model is characterized by a meta-dynamics of idiotype specificity on shape-space [5]. We show here that specific and non-specific responses to an antigen are governed dynamically by a fixed point attractor and a chaotic long-lived transient state of an immune network, respectively. The relevance of such a long-lived transient state is discussed with respect to immune function.

2 Modeling with a Meta-dynamics of Specificity

We first introduce the standard idiotypic network model. Each idiotype is characterized by a pair of surface sites, called the idiotope and the paratope. If the idiotope site of a lymph cell is bounded by paratopes of other lymph cells, the recognized lymph cells become inactivated, whereas the recognizing cells become activated. Thus the growth dynamics of clone size $x^n_{k,j}$ of an idiotype of paratope k and idiotope j is given as,

$$x^{n+1}_{k,j} = x^n_{k,j} + \sum_p \sum_q b_{q,k} x^n_{p,q} x^n_{k,j} - \alpha \sum_p \sum_q b_{j,p} x^n_{k,j} x^n_{p,q} - d x^n_{k,j} + s, \tag{1}$$

The idiotope-paratope interaction $b_{i,j}$ is assumed to have an exponential form: $\frac{1}{\sigma} e^{\frac{-|i-j|}{\sigma}}$. We characterize the ambiguity of the antigen-antibody by the deviation parameter σ. The proposed meta-dynamics controls this parameter. First, as a simple example, we quantize σ by the power of 2: $\sigma_m = 2^{M-m}$. The maximum specificity is given by $m = M$.

Now each idiotype is characterized by three variables: idiotope k, paratope j, and the specificity m. We thus describe the evolution of specificity as follows:

$$x^{n+1}_{k,j,m} = (1 - \mu') x^n_{k,j,m} + \mu'/2 \sum_{m'=m-1,m+1} x^n_{k,j,m'} + s_{m=1}, \tag{2}$$

where μ' is the mutation rate of specificity. Here the source term $s_{m=1}$ is added for the least specific antibody. This dependency reflects the fact that the premature B-cells are believed to have lower specificities.

By combining these equations, we establish the complete clone growth dynamics with mutations among idiotypes and the evolution of the specificities.

In our model, there are five different types of idiotopes and of paratopes, so that there are 25 different idiotypes, with $M = 5$ different levels of specificity. The rest of the system parameters (i.e. $\mu' = 0.3$, $s = 1.0$, $d = 0.1$, and $\alpha = 2.0$) are selected so that the size of each clone never diverges.

The following results (especially, the natural tolerance at high amount of antigen) are confirmed not to depend on the values of the mutation rate μ' and the source s. The dependency of system size is still unclear.

3 Dynamical Natures of the Network

We pay most attention to how the idiotype network responds to persistent antigenic stimulations. A static antigen with a binding site k is introduced by adding the constant term $+A_k b_{k,j} x_{i,j,m}$ to the above equation. Estimating the mean network specificity Sp_k by averaging the specificity of all idiotypes bearing paratope type k, we study the antigenic effect on the network dynamics.

An antigen of type 4 is used as an example, but the following result does not depend on the selected antigen type. Because we adopt the periodic boundary condition for the shape-space, each idiotype is equivalent within a network.

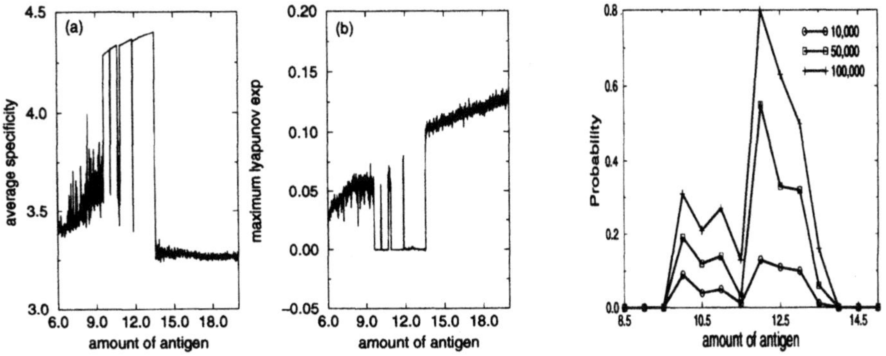

Fig. 1. The average network specificity and the maximum Lyapunov exponents are plotted against the antigen level in Fig. 1(a) and (b) respectively, and in Fig. 1(c), the stability of the type I attractor is plotted against the antigen level. Stability is measured by the proportion of idiotypes in the initial distribution that show a transition from a type I to a type II state before a given time step. The time intervals used are 10,000, 50,000, and 100,000 steps. The number of initial distribution sets is 100.

We show a plot of the averaged specificity ($\bar{Sp_4}$) and the maximum Lyapunov exponent under the antigen stimulations over 10^4 steps (see Fig. 1(a), (b)).

In Fig.1(a), as expected, the network specificity increases when we increase the amount of antigen. At about 9.5 units of antigen, however, the specificity abruptly diverges to a high value. We say that a specific response has occurred at this antigen level. This specific response is observed until the antigen level reaches 13.5 units. Beyond this critical value, the specific response is no longer observed. Inversely, the specificity is sustained at the lower values. This lower sustained response can be compared to natural tolerance to the antigen.

On the other hand, by comparing Fig.1(a) with (b), when the amount of the antigen is set between 9.5 and 13.5 units, we notice that the lower specificity emerges with chaotic dynamics, and the higher specificity emerges with a fixed point dynamics. We shall call the former dynamics a type I attractor and the latter a type II attractor.

However, the type I attractor is not a true attractor. It was found to be a long-lived transient state referred to as a *super-transient state*, which is a common phenomenon in high-dimensional dynamical systems [6]. In Fig.1(c), when the observation period is extended, we observe a transition from type I to type II attractor. There is no inverse-transition from the type II to the type I attractor. The super-transient states are highly dependent on the antigen level. For example, when the antigen level is 11.5 units in Fig.1(c), the transition probability from type I to type II is still less than 12 percent. In such cases, it behaves as an attractor in a practical sense.

From a practical viewpoint, response time is also worth noticing. If we say

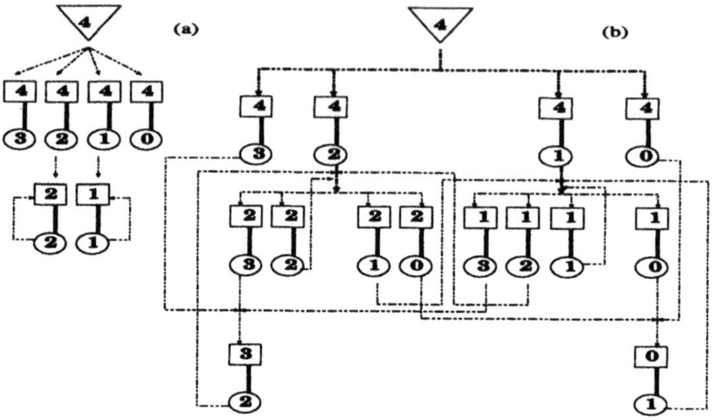

Fig. 2. Topology of an idiotype network of type II and I attractors is shown in Fig.2 (a) and (b), respectively, for an antigen level of 12 units. Only idiotypes that have a population of 1.25 on average of over 10,000 time steps are depicted. Each idiotype in this figure is represented by a pair of symbols, square and circle. The square with the numeral inside denotes the paratope, whereas the circle with the numeral inside denotes the idiotope. The triangle with the number inside represents the injected antigen type. A stimulation wave from idiotope to paratope is shown by a dotted line.

that the relevant time scale for the immune response should be less than 10,000 time steps, in a practical sense there is no specific response even at higher levels of antigen (see Fig.1(c)). Our results suggest that a certain level of antigen causes the super-transient state to suppress fast immune responses under the idiotype network.

Besides the response time, much attention has been paid in the field of theoretical immunology to topological changes of the network [7, 8, 9]. Here we argue that the transition from type I (unspecific) to type II (specific) causes a simultaneous change of network topology. The network topology of each of these two states is shown in Fig.2.

As we see from the figure, a chaotic super-transient state of type I has a more complex network than does type II. Inversely, higher specificity to the dosed antigen is maintained by a simpler network structure. The maintenance of idiotypic diversity can be attributed to chaotic dynamics.

By estimating the amount of specificity of all idiotypes in the type I's distributed state and the type II's localized state respectively, it is found that each idiotype in type I's distributed state has a low specificity on the whole. Namely, each idiotype interacts weakly with many idiotypes in order to have high connectivity. As a result, the stimulation of the network by dosed antigen is distributed over the network, not concentrated only on idiotypes bearing a binding site (paratope with type 4) for the antigen. Thus, the immune response to the antigen has a tendency to be suppressed. This result would support Stewart's extrapolation that 'The higher connectivity among idiotypes, the greater the degree of tolerance' [10].

Recently, a chaotic oscillation was found experimentally in a natural tolerant state. Subsequently, theoretical immunologists have tried to establish 'natural tolerance under chaotic dynamics' against a static antigen [7, 8], though their simulation results show difficulty establishing such a tolerance without assuming a special network topology of an 'odd-loop structure' and so-called 'bell-shaped function' as an activation function. We have shown how such a tolerance can arise naturally under a chaotic dynamics, without these assumptions, by adding an additional flexibility; i.e., meta-mutation dynamics with specificity of idiotype. We have used a simple idiotypic network model, and have not ventured to use the more complex ' bell-shaped function model' because of focusing on capabilities of the meta-dynamics we introduced. Applying the meta-dynamics we introduced here with the bell-shaped function model is left as a future problem.

4 Concluding Remarks

In this paper, we have expanded the possibilities of theoretical immunology by introducing new meta-dynamics. We believe that the immune response should be seen as having a more dynamic nature than allowed by most current models [11], and that the specific antigen-response and the dynamical percolation related to natural tolerance are caused by the meta-dynamics controlling the degree of specificity, as introduced here.

References

1. Ghosh,S., Campbell,A.M.; Immunology today.**7** (1986) 217–222
2. Kearney,J.M., Vakil,M., Nicholson,N.; Evolution and Vertebrate Immunity. G.Kelose and D.Schulze Eds. Texas University Press,Austin. (1987) 175–190
3. Wedemayer,G.J., et al.; Science.**276** (1997) 1665–1669
4. Jerne,N.K.; Ann.Immunol.**125C** (1974) 373–389
5. Segel,L.A., Perelson,A.S.; Theoretical Immunology edited by Perelson,A.S. Addison-Wesley Pub.Company (1987) 321-344
6. Kaneko,K. Phys.Lett.**149 A** (1990) 105–112
7. Bersini,H., Calender,N.; J.theor.Biol.**188** (1997) 187–200
8. Calenbuhr,V., Bersini,H., Stewart,J., Valera,F.J.; J.theor.biol.**177** (1995) 199–213
9. Detoures,V., Bersini,H., Stewart,J., Varela,F.J.; J.theol.Biol.**170** (1994) 401–414
10. Stewart,J. et al.; J.Autoimmune.**2** (1989) 15–23
11. Harada,K., Ikegami,T.; submitted to J.theol.Biol (1998)

Laziness + Sensitivity + Mobility = Structure *

Emergence of Patterns in Lattice Swarms

Andrew Adamatzky, Owen Holland, and Chris Melhuish

Intelligent Autonomous Systems Laboratory, University of the West of England
Frenchay Campus, Bristol BS16 1QY, United Kingdom
{Andrew.Adamatzky, Owen.Holland, Chris.Melhuish}@uwe.ac.uk
http://www.uwe.ac.uk/facults/eng/ias/

Abstract. Natural collective systems, such as social insects, provide us with an existence proof that remarkable feats of construction can be achieved by 'simple' agents. Such feats appear to demand impressive control and coordination which is even more remarkable since the agents are not provided with the overall 'blue-print' for construction. However, as a consequence of the agents carrying out simple rules, an emergent macroscopic structure can develop. In an attempt to understand some of the underlying principles, this paper deals with pattern formations built by a swarm of mobile agents in a lattice. In particular, the morphological classification of the formed structures is provided and classes of simple, complex and non-trivially ordered structures are characterised. For these experiments, all agents start their evolution at the same site in the lattice. This investigation extends the ideas of [1] and [18] by giving agents simple rules, based on neighbourhood characteristics, which govern whether they move or become static. The final outcomes, defined by the immobility of all agents, is studied and the global static structures created are presented and discussed. Since the rules are parameterised, the paper reports on the selection of rule types to generate classes of ordered structures.

1 A problem and the ideology

Structure formation together with self-organization processes have become topics of great importance in many different scientific fields such as chemistry, material science, plasma physics, hydrodynamics and particularly in biological sciences see e.g. [20], [4], [16], [17], [12], [9] and [19]). In all these fields the focus of the investigations was in explaining such phenomena as how a given set of patterns in certain systems could be generated and particularly how the underlying developmental mechanisms functioned.

The recently emerged fields of nanotechnology and nanorobotics (albeit in their infancy)(see e.g. [8], [15] and [14]), collective robotics [18], distributed building and stigmergic behaviour ([10], [5], [21], [7], and [6]), and fabrication of smart

* Some parts of this work were supported by Hewlett Packard Laboratories, Bristol, under the External Research Programme.

matter [13] also employ pattern formation as a key underpinning mechanism with the emphasis on its pragmatic realisation. The problem can be summarised as finding the correspondence between domains of the parametric space which determine the formation of patterns with desirable properties for a given set of discrete entities with a definite set of local parameters

This paper attempts to bring together different strands of previous work such as the distributed spatial sorting of physical objects in robot collectives [18], interval parameterization of the spatio-temporal excitation dynamic in multidimensional excitable media and lattice swarm ([1]); complexity of the automata models of morphogenesis [2]; and designs of swarm based algorithms [3] and employ them in the specific problem of pattern formation in lattice swarms.

The experimental domain is constrained to the following problem:

Consider a set of minimalist agents which can move randomly over a 2D lattice (*mobility*). The agents do not interact with each other but can sense the presence or absence of agents at the nodes immediately around its current position. The agents move according to simple rules and finally stop moving and take up a resultant configuration. The problem is to identify and classify these possible final configurations in terms of order. The rules governing mobility are based on two parameters, firstly, the frequency of checking the immediate neighbourhood nodes for the presence or absence of agents (*degree of activity*) and the number of occupied nodes (*sensitivity*).

The model employed the following features. Firstly, agents only take account of the number of immediately adjacent nodes which are occupied by one or more agents. Secondly, agents can only move one step in a random direction. This is equivalent to a high degree of noise. Thirdly, the size of the final distribution of static agents is polynomially bounded by the number of agents. Fourthly, all trials start with agents injected into one node. This is considered important since the potential implementation of such an approach with real robots may involve small machines being injected at a point source.

2 The model

Let $\mathbf{I}$ be a set of m uniform agents which move at random on the $2D$ lattice $\mathbb{L}$ in discrete time. At every time step automaton i can move or not move to one of the eight nodes of $u(c_i^t)$ neighbouring to its node c_i^t. The agent moves if there is another agent at the same node or if the number of agents in $u(c_i^t)$ lies in some specified interval. At the beginning of the trial all agents are at the same node: c^*: $\forall i \in \mathbb{I} : c_i^0 = c^* \in \mathbb{L}$. The agents do not interact with each other, so every agent simply senses the presence or absence of one or more agents for each of the neighbouring positions and for its current node, i.e. for the nodes c_i^t and $u(c_i^t) = \{v \in \mathbb{L} : |v - c_i^t| = 1\}$.

An agent moves if, after checking its neighbourhood, it finds that its node is occupied by another agent or if the number of the occupied neighbouring nodes belongs to some specified interval $[\theta_1, \theta_2]$, $1 \leq \theta_1 \leq \theta_2 \leq 8$. In some versions of the model, agents do not check their neighbourhood regularly. In this case it will not move even if its own node is occupied by one or more other agents.

When agent i at node c_i^t decides to change its position it walks to a randomly chosen neighbouring node. It does so if: (i) it decides to check its neighbourhood, and one of the following situations takes place: (ii) there is another automaton $j \neq i$ at the same node, $c_i^t = c_j^t$, (iii) the number of nonempty elements of $u(c_i^t)$ lies in the interval $[\theta_1, \theta_2]$.

We represent nodes of $\mathbb{L}$ by their integer coordinates and use additional notations $\alpha = |\{j \in \mathbb{I} : c_i^t = c_j^t\}|$, $\beta = |\{j \in \mathbb{I} : c_j^t \in u(c_i^t)\}|$, $\mathbf{d}$ is the random variable from $\{-1, 0, 1\}^2$, and the function $\nu : \mathbb{Z} \to \{0, 1\}$, such that $\nu(x) = 1$ if $x > 0$ and it equals 0 otherwise. Also we have $\zeta(\alpha, \beta, \theta_1, \theta_2) = 1$ if $\nu(\alpha) \vee \nu(\beta - \theta_1) \wedge \nu(\theta_2 - \beta)$, and it equals 0 otherwise.

The basic equation of motion for agent i therefore is:

$$c_i^{t+1} = c_i^t + \mathbf{d} \cdot \zeta(\alpha, \beta, \theta_1, \theta_2) \tag{1}$$

In this rule all agents act simultaneously. They are active for every step of their evolution. We call the model with agents obeying the equation (1) the S-swarm, or *busy* swarm. In contrast, for some trials an agent will check the number of occupied nodes in its neighbourhood every mth step of the simulation time. These agents are considered to be members of *lazy* swarm (P-swarm). An agent of S-swarm updates its positions as follows:

$$c_i^{t+1} = c_i^t + \mathbf{d} \cdot \zeta(\alpha, \beta, \theta_1, \theta_2) \cdot \nu(1 - |i - (t \mod m + 1)|) \tag{2}$$

3 Convergence

For swarms executing equations (1) and (2) a condition is always reached when there is no further movement of any agent. This is referred to as a stationary global state.

The analyse of the activity patterns of swarms (see, e.g., Figure 1), which is measured in the number of agents which change their positions at current time step, convinces us that lower and upper bounds of the intervals significantly influence the rate of convergence to the stationary global state.

Hereinafter we use the term *pattern* which refers to finite sub-lattice of **L**. Agents in this sublattice are distributed in such a manner that every node of **L** takes the value 1 if there is an agent at the node, and value 0 if there is no agent at a node. The pattern is determined after the swarm converges to the stationary global state.

4 Morphological classification

The set of all activation functions can be subdivided into several classes based on the morphological characteristics of the patterns. The functions are in the same class if, when applied to the agents of the swarm the *same* pattern is formed.

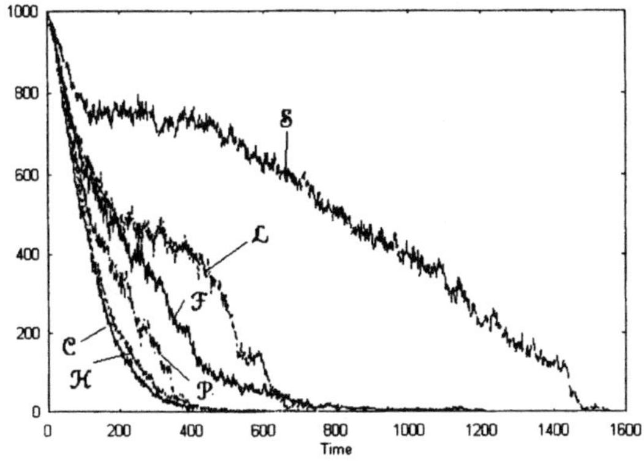

Fig. 1. Examples of activity of agents in the evolutions of swarms from different classes: number of active agents *vs.* time, $m = 1000$.

The *sameness* is determined visually and it is verified with density distributions of agents and an ω-order measure.

The following seven classes (Figure 2) are identified in both P- and S-swarms:

𝒞-class: Condensed patterns. Almost every agent has neighbourhood fully packed with agents. The patterns are solid-like.

𝒮-class: Sparse patterns. Every agent has empty neighbourhood. The patterns are liquid like.

ℋ-class: Halo patterns with condensed core. Agents with fully packed neighbourhood are still in majority but certain amount of the outsiders with less crowded neighbourhood is already visible. The patterns recall us the melting ice; so, we can think on them as two-phase, transitional, structures.

𝒫-class: Porous patterns. They look like the grids.

ℒ-class: Labyrinthine patterns. They are classical (compare with diffusion, magnetic domains, chemicals on the surface of catalyser etc.). Corridors of the labyrinths are usually short. There are many turns and free standing walls.

ℱ-class: Fingerprint like patterns. The patterns are subdivided onto the few large domains with the vertical or horizontal polarization of the walls in labyrinth. Usually there are not so many passes between the domains.

𝒴-class: Porous patterns with large holes inside and some halo outside. They vary in appearances but meristema-like structure is common for all of them.

As one can see from the examples, the classification is not perfect and some ambiguity still remains. Correspondence between the classes and the intervals $[\theta_1, \theta_2]$ of agent activation is shown in Figure 3.

5 Lazy does better: ω-order

Labyrinthine patterns represent the perfect example of the highly ordered structures in the systems under consideration. In the ideal case (Figure 4) the density

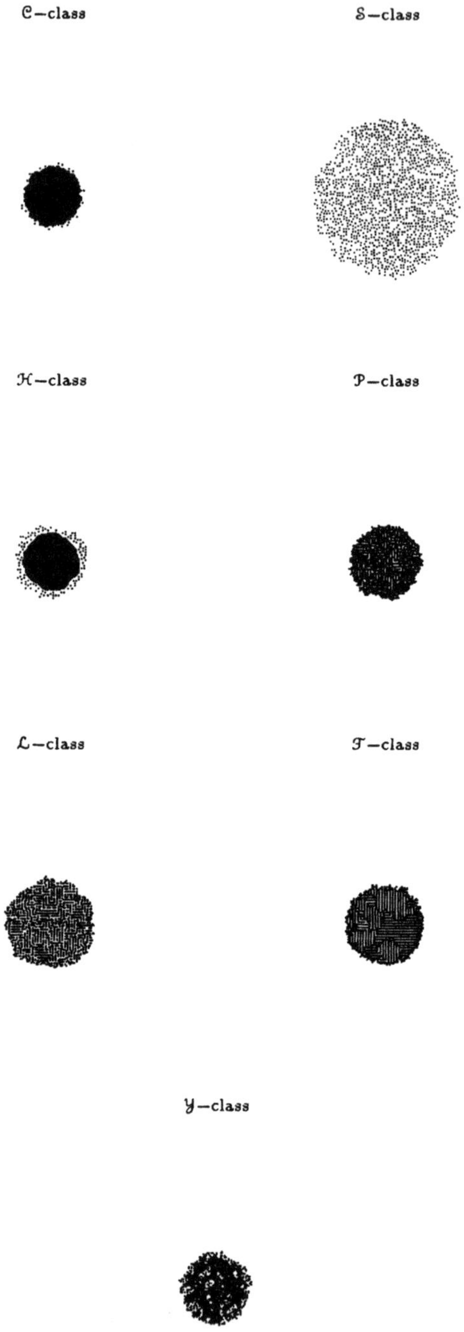

Fig. 2. Representative stationary configurations for the morphological classes of lattice swarm, $m = 1000$.

S-swarm

θ_1 \ θ_2	1	2	3	4	5	6	7	8
1	$\mathcal{C}$	$\mathcal{C}$	$\mathcal{H}$	$\mathcal{S}$	$\mathcal{S}$	$\mathcal{S}$	$\mathcal{S}$	$\mathcal{S}$
2		$\mathcal{C}$	$\mathcal{H}$	$\mathcal{S}$	$\mathcal{S}$	$\mathcal{S}$	$\mathcal{S}$	$\mathcal{S}$
3			$\mathcal{C}$	$\mathcal{L}$	$\mathcal{L}$	$\mathcal{L}$	$\mathcal{L}$	$\mathcal{L}$
4				$\mathcal{H}$	$\mathcal{L}$	$\mathcal{L}$	$\mathcal{L}$	$\mathcal{L}$
5					$\mathcal{L}$	$\mathcal{F}$	$\mathcal{F}$	$\mathcal{F}$
6						$\mathcal{L}$	$\mathcal{F}$	$\mathcal{F}$
7							$\mathcal{P}$	$\mathcal{P}$
8								$\mathcal{P}$

P-swarm

θ_1 \ θ_2	1	2	3	4	5	6	7	8
1	$\mathcal{Y}$	$\mathcal{Y}$	$\mathcal{S}$	$\mathcal{S}$	$\mathcal{S}$	$\mathcal{S}$	$\mathcal{S}$	$\mathcal{S}$
2		$\mathcal{Y}$	$\mathcal{S}$	$\mathcal{S}$	$\mathcal{S}$	$\mathcal{S}$	$\mathcal{S}$	$\mathcal{S}$
3			$\mathcal{L}$	$\mathcal{L}$	$\mathcal{L}$	$\mathcal{L}$	$\mathcal{L}$	$\mathcal{L}$
4				$\mathcal{L}$	$\mathcal{L}$	$\mathcal{L}$	$\mathcal{L}$	$\mathcal{L}$
5					$\mathcal{L}$	$\mathcal{L}$	$\mathcal{L}$	$\mathcal{L}$
6						$\mathcal{L}$	$\mathcal{L}$	$\mathcal{L}$
7							$\mathcal{P}$	$\mathcal{P}$
8								$\mathcal{P}$

Fig. 3. Interval classification of *P*-swarm

distribution shows us that the majority of the agents have two agents in their neighbourhoods. These two agents are in the same column or the same row of $\mathbb{L}$. This gives us an idea to use the following measure of an order: given pattern P generated with activation interval $[\theta_1, \theta_2]$, the so-called wall order is calculated as $\omega = \max\{\omega_H, \omega_V\}$, where ω_V and ω_H denote the number of local configurations as shown below:

```
○ ● ○     ○ ○ ○
○ ● ○     ● ● ●
○ ● ○     ○ ○ ○
```

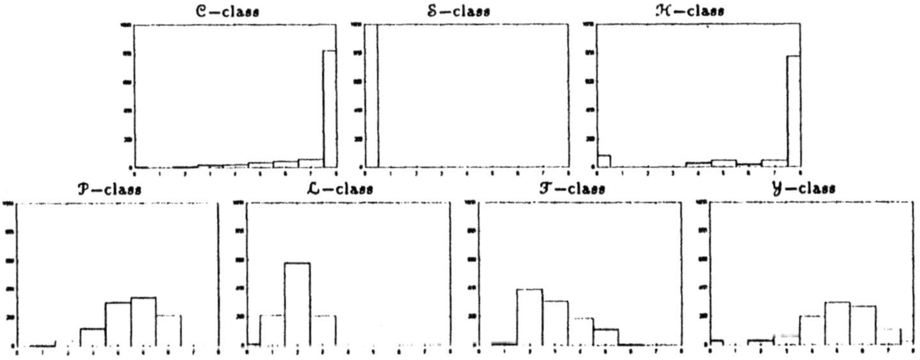

Fig. 4. Density distributions $\mathbf{D} = (D_i)_{0 \le i \le 8}$, $D_i = |\{j \in \mathbb{I} : |\{z \in \mathbb{I} : |s_z^t - s_j^t| = 1\}| = i\}|$, in the stationary patterns for the morphological classes of lattice swarm, $m = 1000$.

For any pattern P formed by m agents $0 \le \omega \le m - 2$. The measure takes its maximum on the line of agents (where agents are arranged along the column or the same row); and it takes its minimum on either densely packed or sparse patterns. Therefore it is concluded that

S-swarms produce more ordered patterns than P-swarms.

It is because ω takes its maximal value on the patterns generated in functions from $\mathfrak{F}$-class. All other classes have sufficiently less values of the ω-order. $\mathfrak{F}$-class was not found amongst the members of P-swarm. Actually the equations (2) and (1) leave some space for the possible criticism. To verify the features of the lazy swarm we show how tuning the laziness of agents will change the order of the structures generated by these agents. Let us look at the following rule:

$$c_i^{t+1} = c_i^t + \mathbf{d} \cdot \zeta(\alpha, \beta, \theta_1, \theta_2) \cdot \rho, \tag{3}$$

where ρ is a random variable that takes a value of 1 with probability $\epsilon \leq p_\rho \leq 1$, ϵ is very small, and it takes value of 0 with probability $1 - p_\rho$. The more p_ρ the more agents decide to check their neighbourhood and, possibly, change their positions at each time step. We can say that agents have random independent delays of switching. There is, therefore, a probabilistic continuum between S-swarm and P-swarm, where lazy swarm is at one end of the continuum and busy swarm is at the other end.

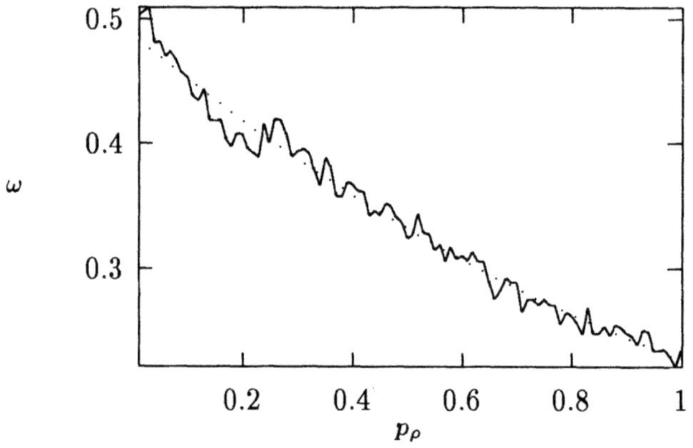

Fig. 5. Order $\frac{\omega()}{m}$ *vs.* degree of activity p_ρ, see rule 3, computed for the swarm of $m = 1000$ agents.

Experiments employing equation (3) (see e.g. Figure 5) show that

Lazy swarms generate more order than busy swarms.

Figure 5 shows that the order of the generated structures is inversely proportional to the degree of agent activity. The remainder of the paper will only deal with S-swarm.

6 Long transients and attractors of order

Size, convergence time and order are the essential parameters, which more or less, successfully characterize spatio-temporal systems. They are attributes of the swarm in general. Sensitivity, in contrast, is an agent based, local, property. How do they relate to each other?

A 8 × 8 node "sensitivity" lattice is indexed by (θ_1, θ_2) pairs. We wish to investigate how changing the $[\theta_1, \theta_2]$ interval will affect the global stationary state.

The convergence time, t_c : for $t \geq t_c$ for any agent i we have $c_i^{t+1} = c_i^t$, is chosen as the temporal attribute. The radius of a pattern of agents, $r = \max_{i \in \mathbb{I}} |c^* - c_i^{t_c}|$, is the size attribute. And, as in the previous section, $\omega(\cdot)$ gives us the measure of order. In computer experiments with the swarms from 1 to 1000 agents we show that $r(m)$ is reasonably well approximated by $\sqrt{\gamma_r m}$, $\gamma_r = \gamma_r(\theta_1, \theta_2)$; $\gamma_t m$, $\gamma_r = \gamma_r(\theta_1, \theta_2)$, is the approximation for $t_c(m)$. So, patterns built by the agents with $[\theta_1, \theta_2]$ intervals of sensitivity are characterised by three dimensionless values: $\gamma_t(\theta_1, \theta_2)$, $\gamma_r(\theta_1, \theta_2)$ and $\frac{\omega(\theta_1, \theta_2)}{m}$.

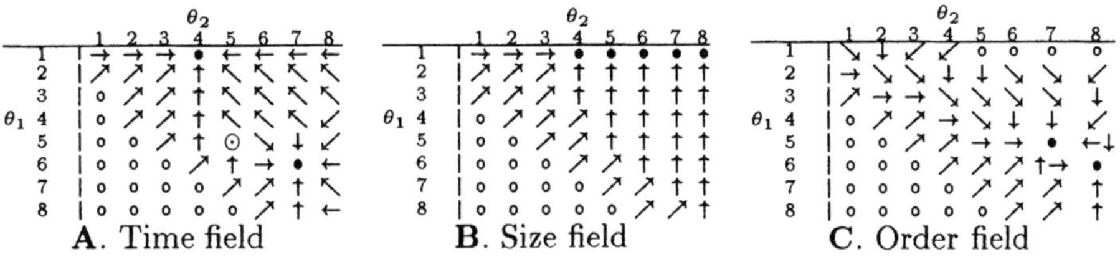

Fig. 6. Time, size and order fields constructed in computer experiments with S-swarm.

Using these values we construct vector-like field where the unit vector at node x points toward the neighbouring node y, which takes the maximal value in the closest neighbourhood of x. From Figure 6 we see that (i) *swarms with wide range and low threshold of sensitivity have longest convergence time*, (ii) *swarms with narrow range and high threshold of sensitivity produce the most ordered patterns.*

In terms of a dynamical system, the vector fields constructed over the sensitivity lattices show the following: (i) both $\mathcal{S}$ and $\mathcal{F}$ classes are the attractors in the time field; Figure 6, A, shows the attractors at nodes [1, 4] and [6, 7], (ii) the $\mathcal{S}$ class is an attractor for a size field ($\theta_1 = 1$ and $4 \leq \theta_2 \leq 8$; Figure 6, B), (iii) the $\mathcal{F}$ class is an attractor in an order field. Figure 6, C, shows attracting nodes [5, 7] and [6, 8].

Assuming that ω-ordered patterns are desirable results of the pattern formation in swarm systems, and calling agents with high threshold of excitation by rough agents, we have the following proposition

Rough and lazy agents are better than sensitive and busy agents.

All our results can be easily applied to $3D$ swarms. We do not have the space to discuss a $3D$ models. However, we have provided one example of the $3D$

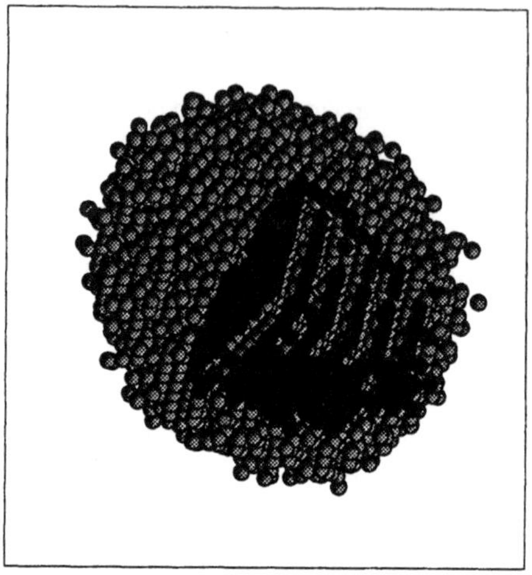

Fig. 7. Stationary pattern of 1000 agents in $3D$ lattice. In the evolution every agent decides to move if number of other agents in its $3 \times 3 \times 3$ node neighbourhood belongs to interval $[20, 25]$. 1/8 of cluster is removed to show the internal structure.

pattern (Figure 7). The interior of the cluster is separated from the surrounding space and is highly ordered inside.

7 Conclusion

A mapping between simple rule sets and emergent global structures have been investigated and presented. The study has attempted to answer some of the questions pertinent to the choice of which parameters of the rules, executed by swarm agents, should be chosen to control the morphology of the emergent global structure. Two basic parameters have been taken into account to parameterise the abilities of swarms to build complex and ordered structures. The first one is the degree of sensitivity measured in the precise amount of occupancy of the neighbourhood of an agent. The second one is the degree of activity that determines how often an agent analyses the state of its vicinity.

In the results of computer experiments we (i) computed the complete mapping between agent parametric space and morphological classes of the formed patterns, (ii) made a correspondence between local agent properties, dynamical features of the swarm systems and morphology, and (iii) show how the ordered non-trivial structures emerge in the families of mobile swarms.

This study highlights the possibility of employing such emergent principles to constructions varying in scale from the very small to the very large.

References

1. Adamatzky A. and Holland O. Edges and computation in excitable media, In: *Proc. ALIFE IV: 6th Int. Conf. on Artificial Life* (MIT Press, 1998) 379-383.
2. Adamatzky A.I. Simulation of inflorescence growth in cellular automata, *Chaos, Solitons and Fractals* **7** (1996) 1065-1094.
3. Adamatzky A. and Holland O. Voronoi-like nondeterministic partition of a lattice by collectives of finite automata, *Mathl. Comput. Modelling* 28 (1998) 73-93.
4. Babloyantz A. *Molecules, Dynamics and Life* (New York: Wiley, 1986).
5. Bonabeau E., Theraulaz G., Deneubourg J.-L., Franks N.R., Rafelsberger O., Joly J.L., Blanco S. A model for the emergence of pillars, walls and royal chambers in termite nests *Philosophical Trans. Royal Soc. London Ser. B - Biol. Sci.* **3**53 (1998) 1561-1576.
6. Bonabeau E. From classical models of morphogenesis to agent-based models of pattern formation, *Artificial Life* **3** (1997) 191-211.
7. Camazine S. Self-organizing pattern formation on the combs of honey bee colonies, *Behav. Ecol. Sociobiol.* **28** (1991) 61-76.
8. Drexler E. *Nanosystems: Molecular Machinery, Manufacturing, and Computation* (Wiley: 1992).
9. Ermentrout B., Campbell J. and Oster G. A model for shell patterns based on neural activity, *Veliger* **28** (1986) 369-388.
10. Franks N.R., Wilby A., Silverman B.W. and Tofts C. Self-organizing nest construction in ants: sophisticated building by blind bulldozing, *Animal Behaviour* **44** (1992) 357-375.
11. Grasse P.-P. La reconstruction du nid et les coordinations interindividuelles chez *Belicositermes natalensis* et *Cubite.* La théorie de le stigmergie: essai d'interpretation du comportement des termites constructeurs *Insectes Sociaux* **4** (1959) 41-84.
12. Swinney H.L. and Krinsky V.I. (Eds.) *Waves and Patterns in Chemical and Biological Media* (North-Holland, 1991).
13. Hogg T. and Huberman B.A. Controlling smart matter *Smart Materials and Structures* **7** (1998) 1-14.
14. Holland O. and Melhuish C. Getting the most from the least: lessons for the nanoscale from minimal mobile agents *Proc. Artificial Life V* (Nara, Japan, 1996) 59-66.
15. Lewis M. A. and Bekey G. A. The behavioural self-organization of nanorobots using local rules *Proc. 1992 IEEE RSJ Intern. Conf. Intelligent Robots and Systems* (Raleigh, NC, 1992) 1333-1338.
16. Markus M., Müller S.C. and Nicolis G. (Eds.) *From Chemical to Biological Organization* (Berlin: Springer, 1988).
17. Meinhardt H. *Models of Biological Pattern Formation* (New York: Academic Press, 1982).
18. Melhuish C.R., Holland O.E. and Hoddel S.E.J. Collective sorting and segregation in robots with minimal sensing *Proc. 5th Intern. Conf. on Simulation of Adaptive Behaviour* (Zurich, 1998) 465-470.
19. Murray J.D. *Mathematical Biology* (Springer, 1989).
20. Nicolis G. and Prigogine I. *Self-Organization in Non-Equilibrium Systems* (New Yorsk: Wiley, 1977).
21. Theraulaz G. and Bonabeau E. Coordination in distributed building *Science* **269** (1995) 686-688.
22. Toffoli T. and Margolus N. Programmable matter *Physica D* 47 (1991) 263-272.

Self-Repairing Multicellular Hardware: A Reliability Analysis

Cesar Ortega[1] and Andy Tyrrell

Department of Electronics
University of York
York, YO10 5DD, UK
{cesar, amt}@ohm.york.ac.uk

Abstract. Artificial Life explores the characteristics of living organisms to understand not only life as it is, but also as it could be. Embryonics is a proposal for a new generation of fault-tolerant Field-Programmable Multicellular Arrays inspired by nature and appropriate for Artificial Life research. Embryonic arrays use hardware redundancy and array reconfiguration mechanisms to achieve fault tolerance. In this paper the k-out-of-m reliability model is used to analyse the reconfiguration strategies used in embryonic arrays. Two schemes are investigated: row- or column-elimination and cell-elimination. The models proposed can also be used to analyse the reliability of systems with spares other than embryonic arrays.

Introduction

Artificial Life pursues a twofold objective: on one side it investigates the properties of natural life by studying emergent behaviours in computer simulations and artificial creatures; and also applies the mechanisms that sustain life to the design of practical solutions in engineering. In this way, Artificial Life studies not only life as it is, but also life as it could be.

The systematic study of artificial cellular systems, like cellular automata, neural networks or processor arrays, has gained momentum during the past few years in Artificial Life studies. The goal is to understand the emergent behaviours observed in natural cellular systems. To borrow the main principles sustaining these mechanisms and applying them to the design of electronic systems could result in a new approach for the design of fault-tolerant systems [1].

In hardware redundancy physical spare components are used to replace the faulty ones. Most hardware redundancy reconfiguration techniques rely on a central processor performing the diagnosis of the cells and executing the algorithms to reconfigure the array in case of failure [2]. An alternative approach inspired by multicellular organisms is to distribute the tasks of diagnosis and reconfiguration among all the cells in the array, eliminating the single point of failure and added complexity introduced by the central processor. A reliability model for such architectures is presented in this paper.

---

[1] Sponsored by Mexico's Government under grants CONACYT-111183 and IIE-9611310226.

Embryonics

Embryonics introduces a new family of fault-tolerant field programmable gate arrays (FPGAs) inspired by nature [3]. Its main ideas come from the mechanisms sustaining the embryonic development of multicellular organisms in nature. When biological multicellular organisms reproduce, the new individual is formed out of a single cell (the fertilised egg). During the days that follow conception, the egg divides itself passing to every offspring a copy of the DNA that corresponds to the individual under development. Cells differentiate according to "instructions" stored in their DNA. Different parts of the DNA are interpreted depending on the position of the cell within the embryo [4]. Before differentiation cells are (theoretically) able to take over any function within the body because each one possess a copy of the DNA.

Correspondingly, every cell in an embryonic array stores not only its own configuration register, but also those of its neighbours. To differentiate every cell selects a configuration register according to its position within the array. Position is determined by a set of co-ordinates that are calculated from the co-ordinates of the nearest neighbours. Every embryonic cell performs self-checking continuously. If a failure is detected, the faulty cell issues a status signal that eliminates some cells according to the reconfiguration mechanism in use, *e.g.* cell elimination. The surviving cells recalculate their co-ordinates and select a new configuration register. By doing so every cell performs a new function and, if the amount of spare cells is enough to replace all the failing cells, the overall functionality of the original array should be preserved [5]. A detailed description of the Embryonics architecture can be found in [6].

The k-out-of-m Reliability Model

In many situations, a system with m units will work properly as long as k out of the m function correctly. For identical units with success probability $p(t)$, the probability of exactly k units working correctly out of m is given by (1) [7].

$$P(k,m,p(t)) = \binom{m}{k} p(t)^k (1-p(t))^{m-k} \tag{1}$$

For the general case, the system remains functional as long as k, k+1... m-1 or m units function correctly. Therefore, the probability of system success is,

$$R_s = \sum_{i=k}^{m} \binom{m}{i} p(t)^i (1-p(t))^{m-i} \tag{2}$$

Failure distribution of electronic systems is generally assumed as,

$$p(t) = e^{-\lambda t} \tag{3}$$

Where λ is a constant known as the failure rate. λ depends on parameters that describe the physical and operating characteristics of a device.

Substituting (3) in (2) yields the reliability of a k-out-of-m electronic system,

$$R_{k-out-of-m}(t) = \sum_{i=k}^{m} \binom{m}{i} e^{-i\lambda t}(1-e^{-\lambda t})^{m-i} \quad (4)$$

Figure 1 shows some graphs of (4) for different values of *k, m* and λ.

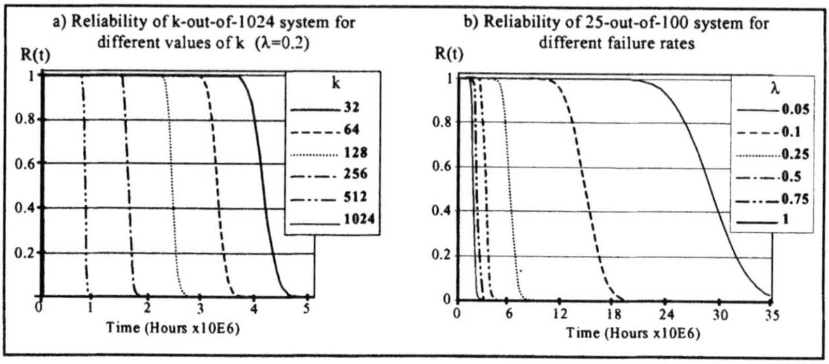

Fig. 1. Reliability of k-out-of-m system for
a) different values of k b) different values of λ

Figure 1(a) shows that in a row with m cells an exponential increment in the number of active cells implies a linear deterioration of reliability. Figure 1(b) shows the high reliability associated to small values of λ, nevertheless decreasing the value of λ requires an improvement in the quality of the system's components and, in the majority of cases, the cost associated with this is too high.

Analysis of Reconfiguration Strategies

In the following analysis an array of size *n*x*m* will be considered in working order only if at least a sub-array of size *r*x*k* is working correctly. In FPGAs *m* and *n* are fixed, whereas *k* and *r* form a logical array mapped into the physical one.

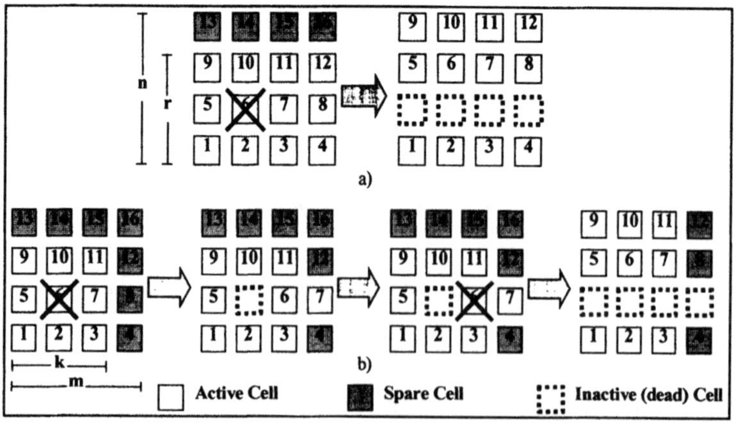

Fig. 2. Fault-tolerance by a) row elimination and b) cell elimination

Figure 2(a) shows that in row elimination the failing of one cell provokes the elimination of the corresponding row, and cells are logically shifted upwards. Cells in a row are connected in series; therefore its reliability is the multiplication of the reliability distributions of all its cells. If all the cells in a row are identical and have an exponential reliability distribution, then row's reliability would be $e^{-m\lambda t}$. The reliability of the whole array $R_{tr}(t)$, would be given by (2),

$$R_{tr}(t) = \sum_{j=r}^{n} \binom{n}{j} e^{-jm\lambda t} \left(1 - e^{-m\lambda t}\right)^{n-j} \tag{5}$$

Even though row elimination logically removes many good cells when a fault occurs, the algorithm to carry it out is very simple and, therefore, fast and easy to implement in hardware. In addition, as the array becomes larger (more than 100 cells), the percentage of cells lost during reconfiguration decreases dramatically. This is because, in square arrays, the size of the array grows quadratically, whereas the number of lost cells grows linearly with respect to the number of cells per side, *n*.

Figure 2(b) shows cell-elimination in an array with one spare cell per row and one spare row. In cell-elimination, spare cells replace faulty cells in two stages. First, spares located in the same row replace faulty cells. When the number of faulty cells in a row surpasses the number of spare cells, then the row is eliminated and rows are logically shifted upwards. Each row of the arrays is itself a k-out-of-m system. The reliability of every cell is given by (3), therefore the reliability for each row $R_{rc}(t)$, would be given by (4). The reliability of the array $R_{tc}(t)$, would be given by,

$$R_{tc}(t) = \sum_{j=r}^{n} \binom{n}{j} R_{rc}(t)^{j} \left(1 - R_{rc}(t)\right)^{n-j} \tag{6}$$

Cell elimination provides a very efficient use of spare cells, but the complexity of cells increases (with the corresponding increment in λ), due to the extra logic needed to re-route data after reconfiguration.

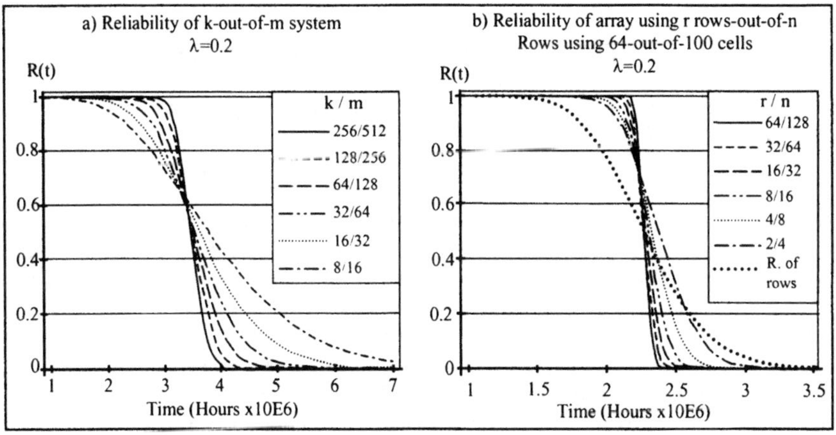

Fig. 3. Reliability distributions for cellular systems with spare units.
a) k-out-of-m system b) r-out-of-n system

For the case when $k=m$ (no spare cells in a row), equations (5) and (6) become equivalent, *i.e.* the row-elimination strategy is a particular case of cell-elimination.

Figures 3(a) and 3(b) show how as the number of cells increases, the reliability curve becomes steeper around the point when reliability equals 0.5. For arrays with hundreds of thousands of cells reliability curves will approximate a rectangle of height 1 and width equal to the time for reliability of 0.5. Figure 3(b) also shows the close relationship between row reliability and array reliability. This indicates that the reliability in rows determines the main features of the array's reliability graph.

Conclusions and future work

Large hardware cellular systems are good candidate platforms to investigate, in real time, the emergent behaviours characteristic of bio-inspired systems. Embryonics offers a good alternative for this kind of research.

The distributed automatic reconfigurability characteristic of embryonic arrays offers considerable advantages over other reconfiguration strategies where, in most cases, a centralised agent, e.g. operating system or central processor, must solve the routing of information problem. For reliability analysis purposes, the effects that this central router imposes to the system must be taken into account.

The main characteristic of embryonic arrays that makes them different to other cellular architectures is the simplicity of individual cells. Simplicity implies small values for λ and consequently, good reliability figures. A design of embryonic cell stressing simplicity can be found in [8].

The reliability models presented in this paper can be adapted to other classes of cellular systems. Further research must be carried out in order to determine to what extent the models proposed hold for any fault-tolerant cellular system with spares.

References

1. Avizienis A., "Toward Systematic Design of Fault-Tolerant Systems", IEEE Computer, April 1997, Computer Society Press, pp. 51-58
2. Negrini R., Sami M. and Stefanelli R., Fault-Tolerance through reconfiguration of VLSI and WSI arrays, MIT Press, Cambridge, Mass., 1989
3. Mange D. and Tomassini M. (Eds.), Bio-Inspired Computing Machines, Presses Polytechniques et Universitaires Romandes, Switzerland, 1998
4. Wolpert L., The Triumph of the Embryo, Oxford University Press, 1991
5. Ortega C. and Tyrrell A., " Biologically Inspired Reconfigurable Hardware for Dependable Applications", IEE Colloquium on Hardware Systems for Dependable Applications, London, November, 1997, Digest No: 97/335
6. Ortega C. and Tyrrell A., "Design of a Basic Cell to Construct Embryonic Arrays", IEE Procs. on Computers and Digital Techniques, Vol.145-3, May, 1998, pp.242-248
7. Misra K., Reliability Analysis and Prediction, Elsevier, 1992
8. Ortega C. and Tyrrell A., "MUXTREE revisited: Embryonics as a Reconfiguration Strategy in Fault-Tolerant Processor Arrays", Lecture Notes in Computer Science 1478, Springer-Verlag, 1998, pp.206-217

Arithmetic Operations on Self-Replicating Cellular Automata

Enrico Petraglio, Jean-Marc Henry, Gianluca Tempesti
Logic Systems Laboratory
Swiss Federal Institute of Technology
Lausanne, Switzerland
enrico.petraglio@epfl.ch

Abstract. In this paper, we present a possible implementation of arithmetic functions (notably, addition and multiplication) using self-replicating cellular automata. The operations are performed by storing a dedicated program (sequence of states) on self-replicating loops, and letting the loops retrieve the operands, exchange data among themselves, and perform the calculations according to a set of rules. To determine the rules required for addition and multiplication, we exploited an existing algorithm for computation in the cellular automata environment and adapted it to exploit the features of self-replicating loops. This approach allowed us to study a variety of issues (synchronization, data exchange, etc.) related to the use of self-replicating machines for complex operations.

1 Introduction

The history of self-replicating cellular automata (CAs) has been marked by two major events. The first is von Neumann's development of his *universal constructor* [1], an automaton capable at the same time of universal construction, that is, of constructing any other automaton given its description (and hence a copy of itself given its own description), and of universal computation, that is, of executing any given application. This automaton, unfortunately handicapped by its great complexity, was the starting point for much of the further research in the field [2,3,4]. The second major event in the study of self-replicating CAs is Christopher Langton's development of the automaton known as *Langton's loop* [5], an automaton where the features of universal construction and universal computation were sacrificed for the sake of simplicity. The result is a small automaton capable exclusively of self-replication, extensively used and ameliorated by Langton's successors [6,7].

The motivations behind the study of self-replication in the environment of cellular automata is not immediately obvious, since this environment presents many features (e.g., the unbalance between the size of the memory required to store the transition rules and the functionality of a single cell) which render it somewhat cumbersome for most practical applications. Nevertheless, CAs do provide a rigid mathematical framework which can be very useful to systematically develop new approaches to the problem of self-replication, approaches which can then be transferred to more "conventional" and "practical" environments. This is, in fact, the motivation of our own research into the field of self-replicating automata. In particular, we have

attempted to re-introduce computation to self-replicating automata in order to develop a mechanism allowing for self-replication in very large scale integrated circuits [8,9,10,11].
In the course of our research, we have developed a set of computationally-useful self-replicating automata, notably by adding a Turing machine to Langton's loop [12], and, more importantly as far as this article is concerned, by developing a "programmable" automaton (Fig. 1), that is, a self-replicating automaton capable of storing and executing a user-specified program [10,13]. The versatility of this automaton, which we used extensively in the development of our hardware systems, was illustrated through a simple self-contained example (that is, a program with no external inputs) which, while useful as a demonstration tool, was nevertheless not very interesting from a computational point of view. In this article, we wish to show that it is indeed possible to perform computationally useful tasks using self-replicating automata by using our programmable automaton to execute some arithmetical operations, notably addition and multiplication, on binary numbers.

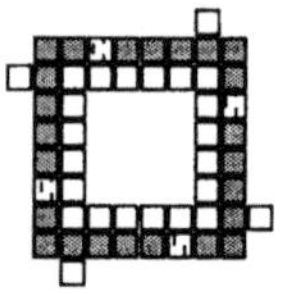

Fig. 1. Our self-replicating automaton

In order to implement these features, we used the *particle model* described by Steiglitz et al. [14], briefly introduced in the next section. We will then describe the implementation of this model in our automaton and its application to the operations of addition and multiplication, as well as to a combination of both. We will conclude with a few observations and remarks.

2 Theoretical Notions

To implement a binary addition function and a binary multiplication function on our automaton, we decided to exploit the model described by Steiglitz et al., since it is a model designed to operate within the CA environment and it can easily be adapted to self-replicating automata.
Obviously, the details of the operation of Steiglitz's algorithm had to be modified to fit the automaton, but we essentially maintained untouched the overall approach to the execution of addition and multiplication.

2.1 Binary Addition

To explain the mechanism used to add binary numbers, we will start with a simple example of a sum of two one-bit numbers. This example is shown in Figure 2.

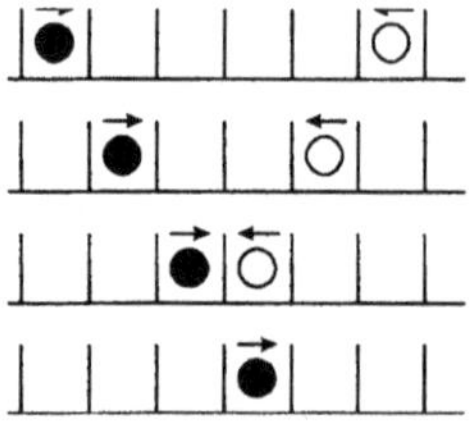

Fig. 2. Sum of two one-bit numbers.

To effect the sum, the two bits are stored in two cells which are moving towards each other. When the two cells collide, the right one is destroyed and the left one is transformed into a new right-moving cell which contains the result of the collision. The carry remains in place in the cell where the collision took place. In our example the left cell represent a logic 1 and the right one a logic 0. On the collision the sum is made and the new right-moving cell represent a logic 1, the result of the computation.
For the sum of binary numbers coded on more than one bit, the left and the right addends are represented by a sequence of cells, each cell representing a bit (one or zero). The two sequences move towards each other. A *processor cell* is placed between the two sequences of cells (Fig. 3). In each number, the least-significant bit lead the sequence, so that when the two numbers collide head-on at the processor cell, this last can add the bits in order of increasing significance.

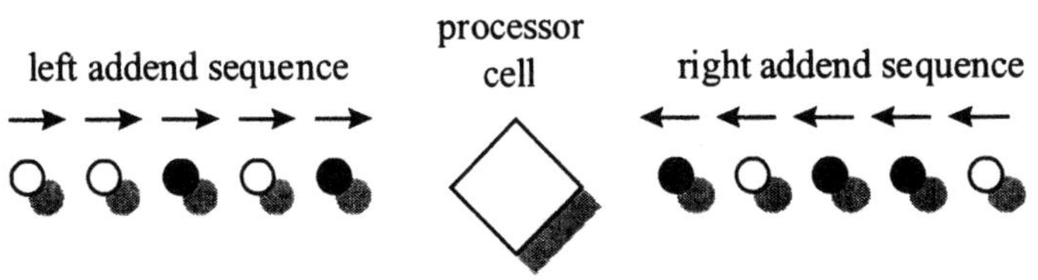

Fig. 3. Two data stream collide on one processor cell

After a collision the two incoming cells are destroyed, the processor cell computes the result and generates a new left-moving cell, which encodes the result of the first addition. After the creation of the "answer" cell, the processor stores the value of carry bit, which it will use to compute the result of the next collision between the bits of the two operands.

2.2 Binary Multiplication

For binary addition, we have seen that a single processor cell was sufficient for the computation. In the case of binary multiplication, we need a stream of processor cells, and more precisely double the number of bits of the multiplicands. Figure 4 shows the starting configuration of the multiplication.

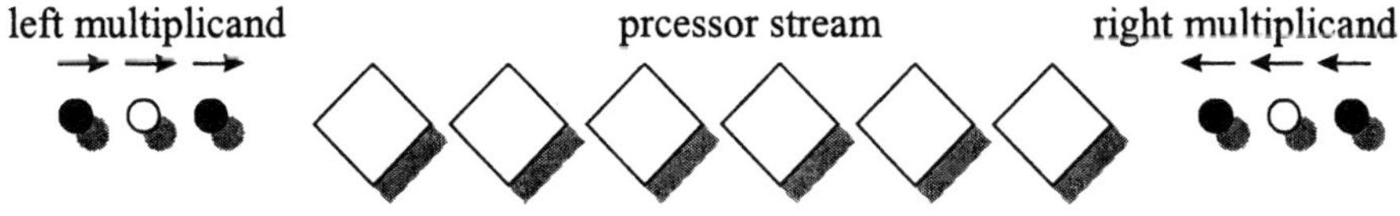

Fig. 4. Two data sequences collide in a processor stream

In this figure the left- and right-moving sequences of cells represent the two multiplicands and the processor stream is placed between the two sequences. To make the multiplication, the two multiplicands travel across all the processor cells. When two cells collide in a processor cell, this last computes the result according to the rules shown in Table 1. The two data cells then continue to travel across the processor stream.

When all the cells have traveled through the entire processor stream, the result of the multiplication is represented by the states of the processor stream's cells. Figure 5 shows an example of the multiplication of two 2-bit numbers.

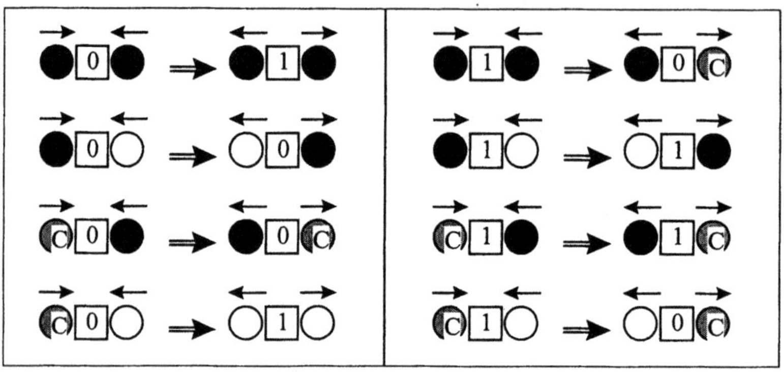

Table 1. Rules for the processor cells

In figure 5, each row represent the state of the multiplication at the time *t*. In this example, the processor cells can have three different states. At the beginning (t=0), the cells' state is empty, while after the first collision the cells' state can be "1" or "0". At the end of the computation all of the processor cells are set to "1" or "0", and we can read the result on the processor stream: 11 x 11 = 1001, that is, in decimal notation, 3 x 3 = = 9.

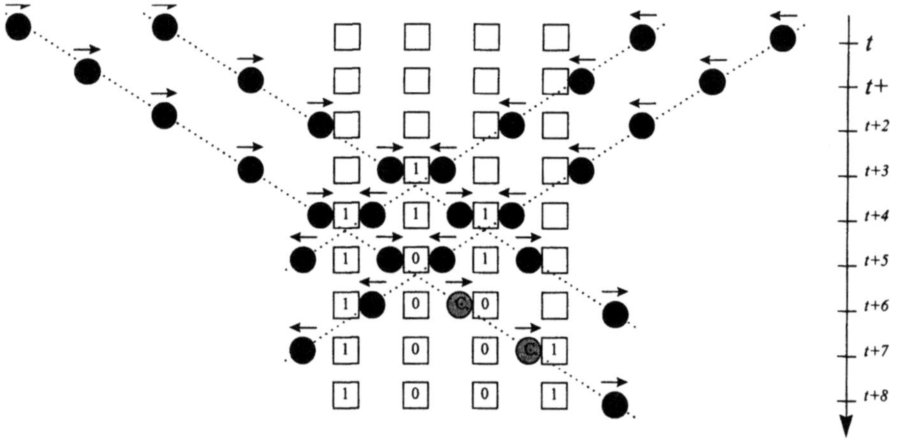

Fig. 5. 3 x 3 binary multiplication

3 Implementation on Self-Replicating Loops

The programmable automaton we mentioned in the introduction (Fig. 1) consists of two concentric square loops: an inert internal loop (the *sheath*) and an active external *program loop*, containing the program to be executed along with the information used to direct the self-replication process. To duplicate itself, the automaton sends out four *constructing arms*, which build four new sheaths in the cardinal directions. When the sheath is complete, the automaton sends out the information contained in

the program loop (the external loop of the original automaton). Finally, the constructing arms retract, completing the self-replication process and letting the four new automata attempt their own self-replication on the four cardinal directions. When an arm finds an obstacle (the border of the cellular array or an existing automaton), it retracts, abandoning the replication attempt. The self-replication will thus end only when all the available space (the cellular array) has been filled.
As we will see in this section, both the operation of our loop and that of Steiglitz's model had to be slightly modified to allow them to be merged, but the modifications were fairly minor and the basic concepts were not in the least altered.

3.1 Addition

To execute this function, one automaton is charged with computing the result of a single collision between two data cells, unlike the original algorithm, in which a single processor cell computed the entire result. The initial configuration of the adder (Fig. 6a) consists of a single loop, containing the program which implements the sum. This first loop is a slightly modified version of the original loop, in that it replicates in one direction only (downwards). As time progresses, a column of loops will be created. The replication process ends when the last automaton finds, in the place where it should replicate, a special cell (Fig. 6b). Upon finding this special cell, the bottom automaton generates a START signal which propagates upwards to the first automaton to tell it to begin the operation.

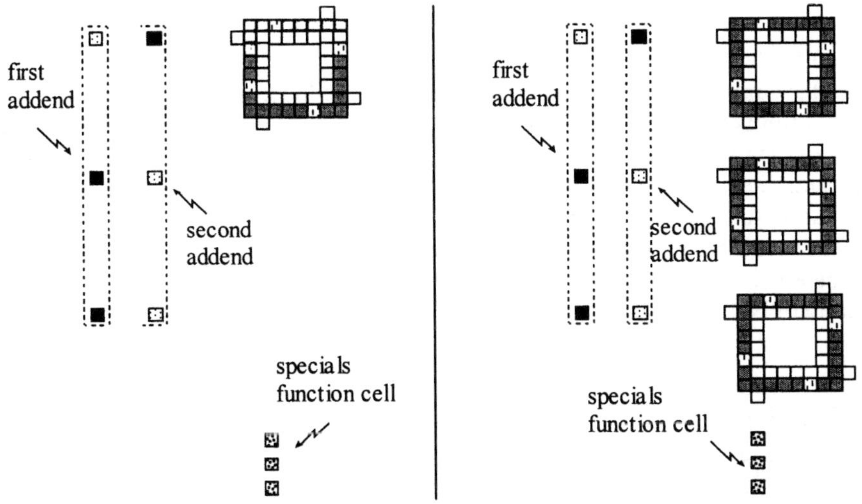

(a) Initial Configuration (b)End of self-replication

Fig. 6. Stream of automata

Once the first automaton has received the START signal, it looks to its left to find the bits it needs to add. It extends its constructing arm (Fig. 7a), retrieves the first bit it finds (least significant bit of the first number) and adds it to the second bit it finds (least significant bit of the second number). The arm then leaves in place the result of the computation and brings the carry bit back to the loop (Fig. 7b), which will

propagate it to the next automaton (Fig. 7c). The process continues until the bottom loop is reached, signaling the end of the sum.
Once the operation is complete, the bottom loop will extend an arm downwards (as if to propagate the carry bit). The arm will meet one of three kinds of cells: a new START cell, which will activate a new sum, an END cell, which halts the operation of the automata, or an ACTIVATE cell, whose functionality will be explained below.

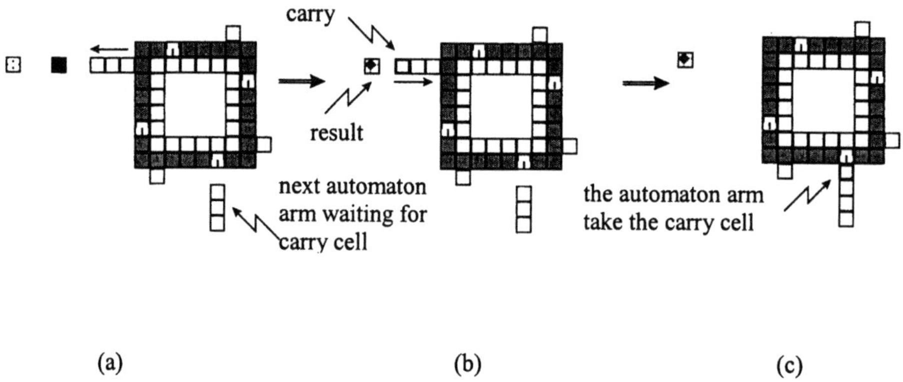

Fig. 7. Computation of a collision

3.2 *Multiplication*

As for the sum, the multiplication starts with a single loop, which replicates towards the right (unlike the sum) to create a stream of 2N automata (where N is the number of bits of the multiplicands).
The multiplication algorithm requires that the first collision between the data cells occur at a specific automaton, notably the Nth automaton from the right. This introduces some synchronization problems which complicate the execution considerably. The first complication is that a sequence of temporization signals, in the form of N-1 *shifting cells*, needs to be added in front of the left operand (Fig. 8).

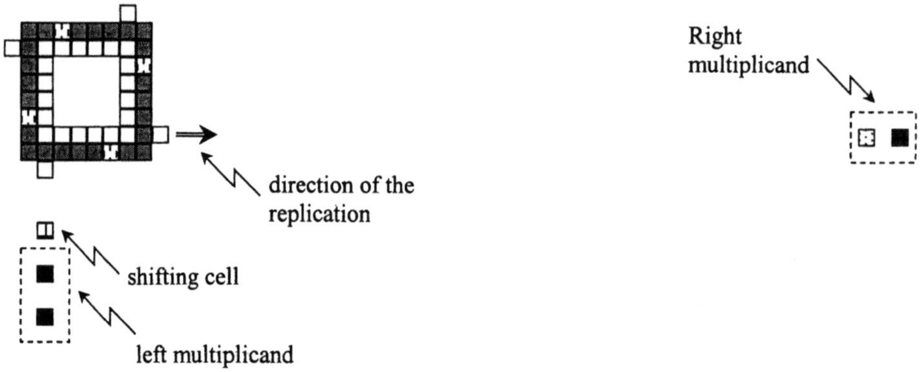

Fig. 8. Starting point of multiplication

The operation begins when the self-replication process has ended (i.e., when the replicating automata have filled all the available space) and the leftmost automaton has received a START signal. At this point, the leftmost automaton (which we will call Loop 1) starts retrieving the data cells of the left operand and propagating them to the right. Throughout the multiplication, Loop 1 will keep retrieving and propagating the data cells at a frequency of one data cell every three time steps (where one *time step* is the time required for an automaton to extend and retract its constructing arm).
The first shifting cell (the first cell of the left operand to be retrieved) propagates then to the right until it reaches the rightmost automaton (which we will call Loop 2N). Upon receiving the shifting cell, Loop 2N retrieves the first cell of the right operand and stores it. Each of the shifting cells traversing the automata will cause the right operand data cells to be shifted from the loop they are on to the loop to its left and a new data cell to be retrieved by Loop 2N.
After N-1 shifting cells have gone through, each of the bits of the right operand (except for the last one) are thus distributed on the N-1 rightmost loops. When the first left operand data cell arrives (behind the shifting cells) on Loop N+1 (the Nth automaton from the right), the first collision occurs (Fig. 9).
The collision process occurs between a data cell A on one loop and a data cell B on the loop to its right, according to the rules shown in Table 1. At the end of the collision process, the result of the collision and data cell B are stored on the left loop, along with a possible carry bit (which will taken into account when computing the next collision), while data cell A has been propagated to the right loop, where it will be used for the next collision. Each left operand data cell will thus collide with each right operand data cell, and the right operand will be shifted by one automaton to the right after being traversed by each left operand data cell. At the end of the multiplication, the right operand, stored on the N rightmost loops of the automaton, will be deleted by a special CLEAR cell, and the result will be stored on each of the loops.

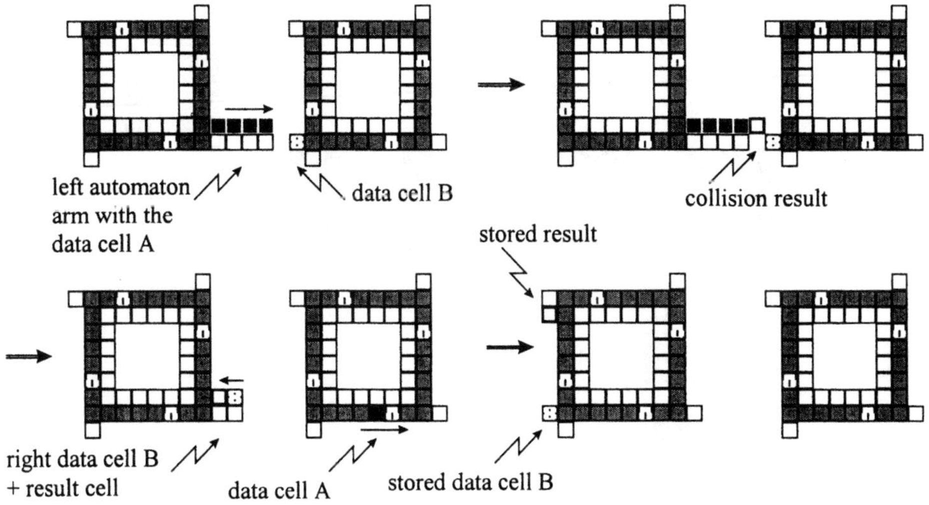

Fig. 9. Collision between two data cells

3.3 Combinations of Multiplication and Addition

In order to render its operation more "useful", our automaton was conceived so as to be able to realize combinations of operations. In particular, it can compute the multiplication of two results of sums. That is, it can compute any function of the form:

$$(A + B + \ldots) * (a + b + \ldots)$$

In order to compute this kind of function, we need a starting configuration similar to Fig. 10, which expands to the machine shown in Fig. 11 after self-replication.

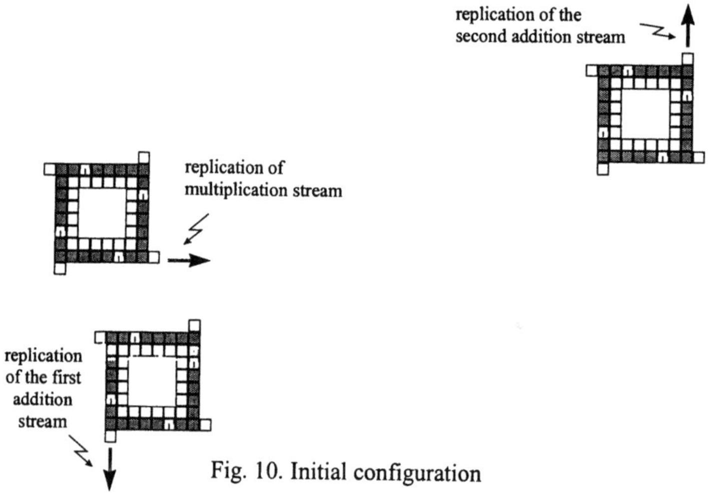

Fig. 10. Initial configuration

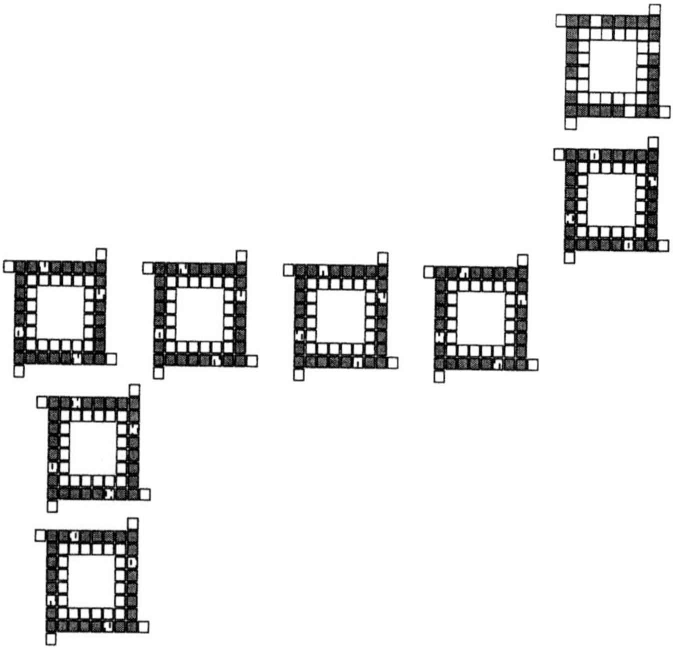

Fig. 11. End of self-reproduction

When the left and the right automata have completed their sums, they leave a special ACTIVATE cell (mentioned above) for the multiplier to retrieve. The latter will interpret this cell as a START signal, and execute the multiplication.

The two operations can thus be chained without difficulty, the only new feature being a *carriage cell* which will "reformat" the data generated by the adders into a form which the multiplier can use as an input (Fig. 12).

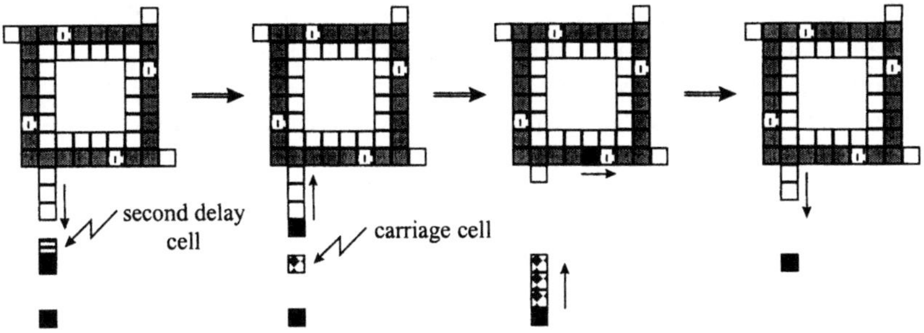

Fig. 12. Operation of the carriage cells.

4 Conclusion

The goal of the work presented in this article was to show that is possible, and indeed not exceedingly difficult, to exploit the capabilities of self-replicating automata (and notably our self-replicating programmable loops) to perform complex mathematical operations. To demonstrate this, we implemented the arithmetic operations of addition and multiplication using the algorithm described by Steiglitz et al. The resulting machines, while relatively complex (the final number of states required for combined sum and multiplication exceeds 30, including the states used only for self-replication), are nevertheless simple enough to be entirely simulated, and the use of the support provided by the programmable loops considerably simplified the finding of the relevant transition rules.

It should be noted that, while the automaton we designed is simple enough for simulation, it is extremely unlikely that such a system would ever be actually used for real-world computing. As we have mentioned, in fact, cellular automata are a useful environment for theoretical research but its real-world applications are few and not usually concerned with complex mathematical operations. Moreover, "pure" cellular automata do not contemplate the existence of external inputs, i.e., of data, such as mathematical operands, which is not present in the cellular space at time 0 (for example, in our system, the operands should clearly be inserted as needed, which would simplify considerably the operation of the automata).

Our aim, however, was not to develop a cellular automaton to be used in real-world applications. As mentioned in the introduction, our goal in studying this kind of structures is to determine what the advantages and constraints are in the use of self-replicating machines for complex operations, so as to be able to transfer these observations to the design of self-replicating integrated circuits. From this perspective, the work we presented is indeed interesting, in that it allows a number of observations:

1 Self-replication can be advantageously exploited to realize application-specific parallel systems by associating a self-replication mechanism and an execution unit.

2 The execution units need not be very powerful, as complex operations can be performed by many small identical units (the fundamental principle of parallelism).
3 Self-replication allows the systems to adapt their architecture to the problem (for example, by producing the correct number of execution units to exactly fit a given problem).
4 The problem of synchronizing the operation of all the units of the system is a major issue, as is the communication between the units.

This kind of information has been, and will be, extremely useful in the development of self-replicating machines and in our attempt to realize von Neumann's dream.

References

[1] J. von Neumann. *The Theory of Self-Reproducing Automata.* A. W. Burks, ed. University of Illinois Press, Urbana, IL, 1966.
[2] E.R. Banks. "Universality in Cellular Automata". In *Proc. IEEE 11th Annual Symposium on Switching and Automata Theory*, Santa Monica, CA, October 1970, pp. 194-215.
[3] E.F. Codd. *Cellular Automata.* Academic Press, New York, 1968.
[4] C. Lee. "Synthesis of a Cellular Computer". In *Applied Automata Theory*, Academic Press, London, 1968, pp 217-234.
[5] C. G. Langton. "Self-Reproduction in Cellular Automata". *Physica 10D*, 1984, pp. 135-144.
[6] J. Byl. "Self-Reproduction in Small Cellular Automata". *Physica 34D*, pp.295-299, 1989.
[7] J.A. Reggia, S.A. Armentrout, H.-H. Chou, Y. Peng. "Simple Systems That Exhibit Self-Directed Replication". *Science*, Vol.259, 26 February 1993, pp. 1282-1287.
[8] D. Mange, M. Tomassini, eds. *Bio-inspired Computing Machines: Towards Novel Computational Architectures.* Presses Polytechniques et Universitaires Romandes, Lausanne, Switzerland, 1998.
[9] D. Mange, M. Goeke, D. Madon, A. Stauffer, G. Tempesti, S. Durand. "Embryonics: A New Family of Coarse-Grained Field-Programmable Gate Array with Self-Repair and Self-Reproducing Properties". In E. Sanchez, M. Tomassini, eds., *Towards Evolvable Hardware*, Lecture Notes in Computer Science, Springer, Berlin, 1996, pp. 197-220.
[10] G. Tempesti. *A Self-Repairing Multiplexer-Based FPGA Inspired by Biological Processes.* Ph.D. Thesis No. 1827, EPFL, Lausanne, 1998.
[11] G. Tempesti, D. Mange, A. Stauffer. "Self-Replicating and Self-Repairing Multicellular Automata". *Artificial Life*, 4(3), 1998, pp. 259-282.
[12] J.-Y. Perrier, M. Sipper, J. Zahnd. "Toward a Viable, Self-Reproducing Universal Computer". *Physica 97D*, pp. 335-352, 1996.
[13] G. Tempesti. "A New Self-Reproducing Cellular Automaton Capable of Construction and Computation". *Proc. 3rd European Conference on Artificial Life*, Lecture Notes in Artificial Intelligence, 929, Springer Verlag, Berlin, 1995, pp. 555-563.
[14] K. Steiglitz, R. K. Squier, M. H. Jakubow. "Programmable Parallel Arithmetic in Cellular Automata using a Particle Model". *Complex Systems* 8, 1994, pp. 311-323.

Dynamics of Gene Expression in an Artificial Genome – Implications for Biological and Artificial Ontogeny

Torsten Reil

School of Cognitive and Computing Sciences, University of Sussex,
Brighton, BN1 9QH, United Kingdom
torstenr@cogs.sussex.ac.uk

Abstract. An artificial genome with biologically plausible properties was developed and the dynamics of gene expression were studied. The model differs from previous approaches, such as Random Boolean Nets [1], in that it is entirely based on template matching in a nucleotide-like sequence. Genes activate or inhibit other genes by binding to their regulatory sequences.
The results of the experiments suggest that many features of real-life development, such as cyclic gene activity, differentiation into multiple cell types, and robusteness may be inherent properties of a template-matching system rather than necessarily designed from scratch by Natural Selection. Moreover, the system may provide a new hypothesis about the role of junk DNA in real genomes. In addition to these biological implications, the approach used here is thought to provide a flexible basis for future simulations of morphogenesis.

1 Introduction

Since the mid 80's, the advances in developmental genetics have been staggering. The number of genes discovered is rapidly growing, and for organisms such as yeast [2] or recently the nematode *C. elegans* [3] the complete genome has been sequenced and the genes mapped. However, despite this wealth of data, a conceptual framework as to how genes interact is lacking [4].

Pioneering work on viewing genomes as complex networks and studying their dynamics was done by Kauffman [1] [5], using Random Boolean Networks as an abstraction. However, his results and the concomitant suggestions – such as viewing cell types as attractors – were only appreciated to a limited extent in the biological community. It can be argued that the reasons for this were a) the abstract nature of Random Boolean Networks, which to many do not exhibit sufficient parallels with real genetic networks, and b) the only limited extendability of the model.

The model presented here therefore tries to provide a biologically more plausible framework for studying gene interactions, and thus to integrate developmental biology and the study of complex systems, which was recently stated as one of the major requirements of for our understanding of ontogeny [6].

2 The Model

The main goal of the model is to study gene activity over time. Genes regulate each other by binding to regulatory sequences.

At the core model is an Artifical Genome consisting of a string of digits, which is randomly created and contains all the information present in the model. Gene are not pre-specified, but identified in the genome after creation. The genome is searched for occurrences of the sequence '0101', and on encountering one, the following *N* digits are defined as a gene. The product of a gene is yielded by a simple transformation of the gene sequence. In the current implementation, each digit is simply incremented by 1, so that the gene sequence '221133' becomes a '332244' protein. Now, for each existing gene product, all matching regulatory sequences in the genome are identified and stored. For example, for the gene above, these would be all occurrences of the sequence '332244', each of which controls the gene immediately following it.

Upon completion of this procedure, each gene will be regulated by 0 to n regulatory elements, or more precisely by the genes coding for the matching gene products. Binding of a gene product to a regulatory sequence can have two effects on the future expression of the regulated gene. The regulatory unit can either act as an *enhancer*, thus activating the gene, or as an *inhibitor*, thus blocking its activation. Which of the two modes is applicable is dependent on the value of the last digit of the gene product. For example, one possible setting is to define all genes ending with '1' as inhibitors. Regulation is not concentration dependent; the presence of single enhancer suffices to activate a gene. However, inhibition has priority over enhancement.

2.1 Biological Plausibility

The concept of a standard promoter ('0101') is used to define genes in the genome string. The equivalent in eukaryotic genomes is the so-called *TATA box* (a succession of 'TA' nucleotide pairs), which is the major component of every gene's promoter (allowing RNA polymerase to bind for transcription). Furthermore, in the model, regulation is achieved by the binding of gene products to matching sequences of the genome. This corresponds to the action of transcription factors in DNA-based genomes. Obviously, in the case of the latter, template matching is more sophisticated since it is determined by the protein structure of the gene product rather than by a simple sequence-sequence correspondence. However, the idealisation in the model retains the fundamental concept or template matching and is therefore sufficient for the purpose.

Genes are under the control of regulatory sequences (also called cis-elements), which in eukaryotic genomes can be either located before (upstream) or after (downstream) the actual gene. Here, regulatory sequences are only located upstream. Again, this is a simplification that does not affect the fundamental principle.

Finally, regulatory sequences can act as enhancers or inhibitors, a concept that is again mirrored in DNA-based genomes. The latter are different however,

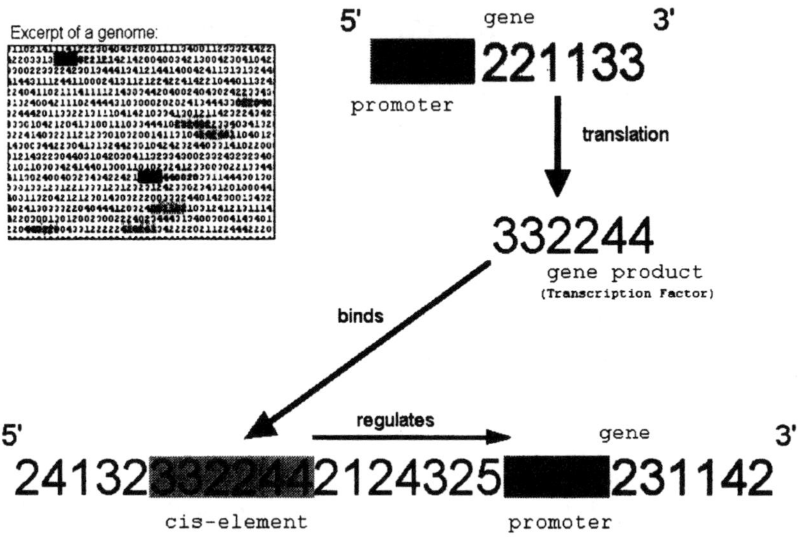

Fig. 1. Gene expression and regulation in the Artificial Genome

in that they allow complex interactions between regulatory elements as well as concentration dependent regulation. For simplicity, this is not the case in the current implementation of the model.

2.2 Studying gene expression

After a genome has been set up according to the rules outlined above, gene expression over time can be studied. To do this, initially all genes but one are deactivated. It is then determined which genes are regulated by the sole active gene, and these are marked as *on* or *off* (according to the mode of regulation, activation or inhibition) and updated accordingly in the next time step. The same procedure is repeated for every active gene of the next time step and analogously throughout a whole run of a specified number of cycles. As a result, a pattern of expression is yielded. This is made visible in an *expression graph* (Fig. 2).

2.3 Parameters

The following parameters can be modified in the current implementation: *genome size*, *gene length*, *base* (i.e. range of digits), *degree of inhibition* [1].

[1] Coded as the fraction of the *base*; i.e. a *base* of 4 and degree of inhibition of 1 means that on average one quarter of genes will be inhibitory

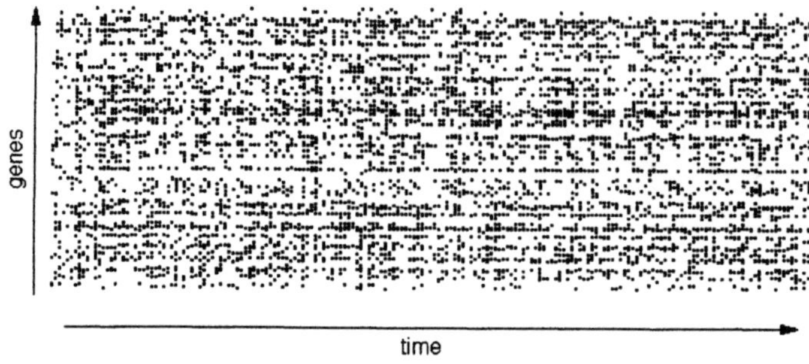

Fig. 2. Gene expression unfolding over time. Each gene is assigned a fixed position on the vertical axis (according to its number)

Furthermore, two options make it possible to influence the system's behaviour manually: *manual toggle* (switching on or off a gene for one time step) and *knock-out* (swtiching off a gene for the duration of an entire run).

3 Behaviour of the Model

The behaviour of the model is highly dependent on the parameter values. Three regimes can be distinguished: *ordered*, *complex*, and *chaotic*. As discussed below, the variable K, the average number of genes that each gene regulates, plays a crucial role in this respect.

3.1 Ordered Gene Expression

Gene expression is called ordered when genes are continuously active or inactive throughout a run (Fig. 3).

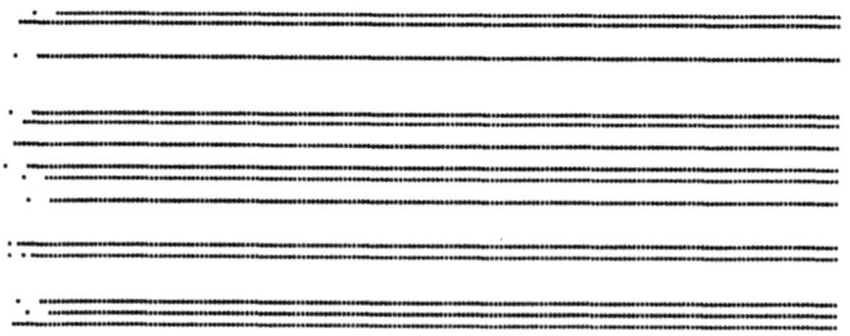

Fig. 3. Ordered gene expression. *size*=10,000; *base*=4; *gene length* =6; *inhibition*=0. *no. genes* = 29; $K = 1.103$

Ordered gene expression is observed if K, the degree of regulation, is low. This state is caused by three parameter settings (or their combinations): a small *genome size*, a high *base*, or a high *gene length*. Each factor exerts its effect by resulting in a low probability that a given gene product finds a matching counterpart.

Naturally, ordered artificial gene expression is of only limited interest due to the lack of resemblance to real-life gene expression, which is considerably more dynamic.

3.2 Chaotic Gene Expression

At the other extreme of possible regimes, gene expression appears to be random and shows no signs of patterns; it is chaotic (Fig. 4).

Fig. 4. Chaotic gene expression. *size*=150,000; *base*=5; *gene length* = 6; *inhibition*=3. *no. genes* = 225; $K = 8.568$

Chaotic behaviour is observed if the number of genes and regulatory sequences is large, thus resulting in a high degree of connectivity (as indicated by a K of 8.568 in Fig. 4).

Analogously to the opposite case (the ordered state), parameter settings with a large genome size, a small base, and short genes cause the system to be in this regime. Like complete order, completely chaotic artificial gene expression is of only little attraction, since for genetic nets to perform a task (e.g. regulate the development of an organism), they must be able to carry information, a feat that is made impossible by complete randomness.

3.3 Complex Gene Expression

If the degree of gene regulation lies in-between the values for ordered or chaotic behaviour, the patterns of gene expression exhibit more comlex dynamics. One striking feature of these complex systems is the tendency to produce cyclic gene activity (Fig. 5). Preliminary results suggest that the period of the cycle (i.e. the number of time steps it encompasses) is positively correlated to the number of genes. For example, while the expression pattern with 56 genes has a period of 10 (Fig. 5), the following one - based on 88 genes - has a period of 30 (Fig. 6).

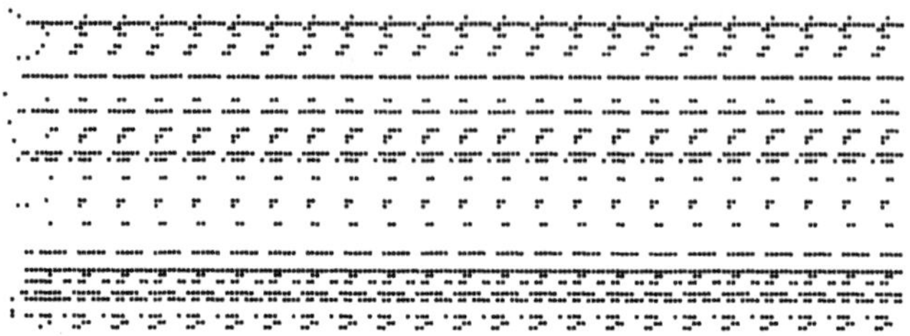

Fig. 5. Cyclic gene expression. *size*=20,000; *base*=4; *gene length* = 6; *inhibition*=1. *no. genes* = 56; $K = 3.285$

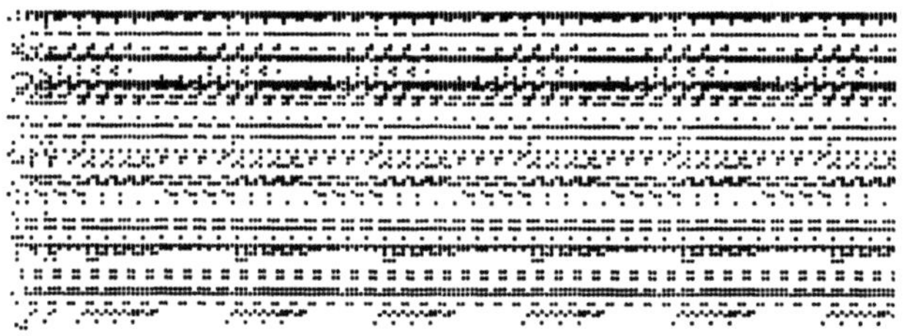

Fig. 6. Cyclic gene expression. *size*=20,000; *base*=4; *gene length* = 6; *inhibition*=1. *no. genes* = 88; $K = 3.303$

For an Artificial Genome in the complex regime, a given cycle of gene expression is converged upon from a number of different start genes (Fig. 7). It can therefore be viewed as a *limit cycle attractor.* Each genome typically exhibits several of such attractors (Fig. 8), the lengths of which were found to be distributed as shown in Figure 9.

The existence of multiple limit cycles in complex Artificial Genomes gives support to the notion developed by Kauffman [1] that cell types may be viewed as attractors of gene expression in biological gene nets. If this is the case, *cell differentiation* can be viewed as a dynamic system moving from one attractor to another. Differentiation of cells in organisms is known to rely to a large extent on external signals (in particular from other cells) that, through chemical cascades, cause the activation or deactivation of genes [7]. It was studied in how far such controlled cell differentiation is possible in the model developed here. To do this, the system was run until it reached a limit cycle attractor, and then manipulated by switching on or off the activation of single genes. It was found that controlled 'differentiation' into a different attractor is indeed possible (Fig. 10).

However, the experiments suggest that only a small proportion of gene activity toggles produced changed expression pattern. In the majority of cases, the system was robust against these disturbances and quickly returned to its original attractors (Fig. 11).

This robustness of gene expression is not just limited to the transient disruptions carried out above, but is also observed if genes are completely knocked out (Fig. 12).

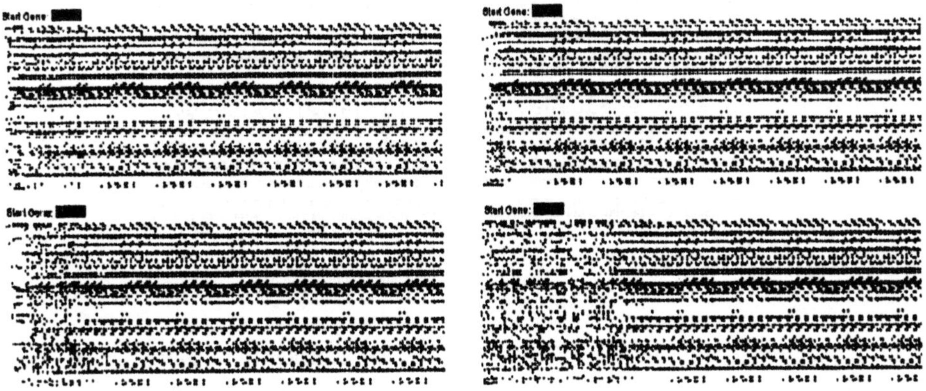

Fig. 7. System converging on the same cycle despite different start genes. *size*=100,000; *base*=5; *gene length* = 6; *inhibition*=2. *no. genes* = 152; $K = 7.374$

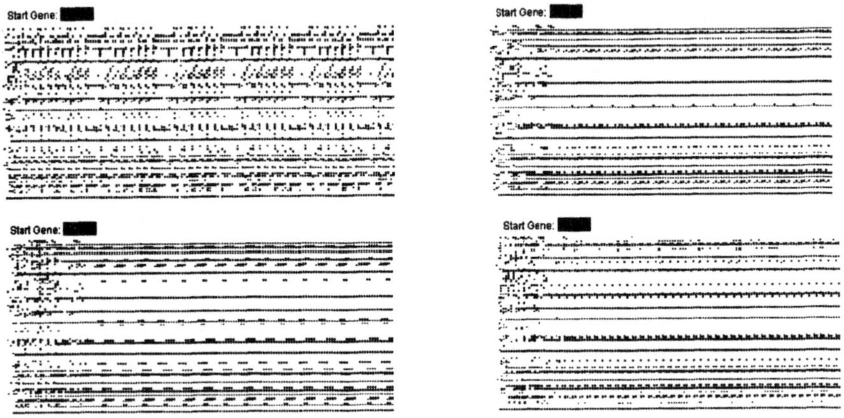

Fig. 8. Multiple attractors of gene expression. *size*=100,000; *base*=5; *gene length* =6; *inhibition*=2. *no. genes* = 175; $K = 4.425$

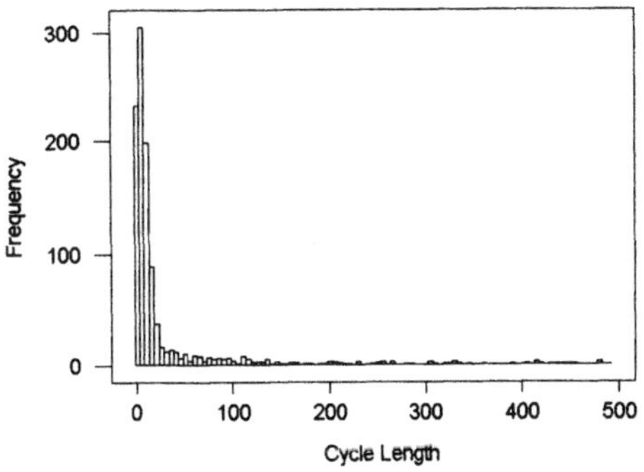

Fig. 9. Distribution of limit cycle lengths of 200 genomes. *size*=100,000; *base*=5; *gene length* = 6; *inhibition* = 2. The average number of attractors per genome is 5.425

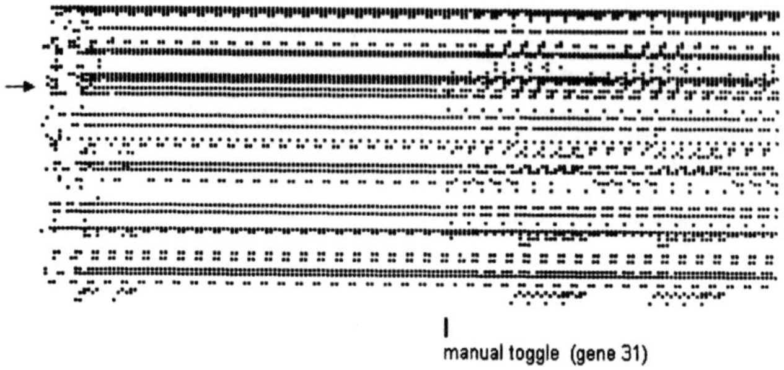

Fig. 10. Differentiation into a different attractor by transiently toggling a gene state. *size*=20,000; *base*=4; *gene length* = 6; *inhibition*=1. *no. genes* = 88; $K = 3.30$

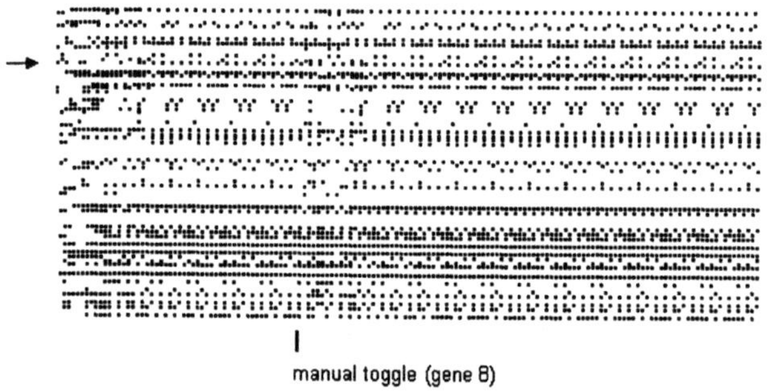

Fig. 11. Gene expression is largely robust agains disturbances. *size*=20,000; *base*=4; *gene length* = 6; *inhibition*=1. *no. genes* = 75; $K = 4.173$

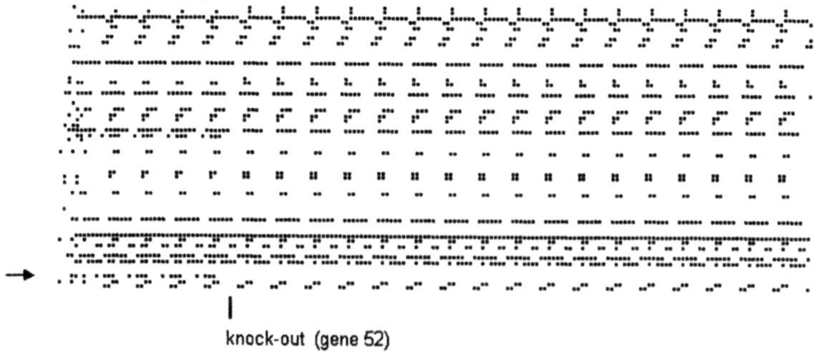

Fig. 12. Robustness against loss of gene function. *size*=20,000; *base*=4; *gene length* = 6; *inhibition*=1. *no. genes* = 56; $K = 3.286$

4 Discussion

The behaviour of random Artificial Genomes mirrors that of real gene expression to a large extent. The model shows that cyclic gene activity is an inherent property of the system and does not have to be especially evolved from scratch by selection.

Furthermore, the experiments suggest that a high degree of robustness is possible without selection. This is manifested in two ways: a) transient disturbances of gene expression (carried out manually in the model) do in the majority of cases not result in altered expression patterns, and b) loss of gene function (achieved by manually knocking out genes) does in most cases not affect the overall pattern of gene expression. While the first case implies that sporadic mistakes in gene expression will not substantially affect the development of biological organisms, the latter case may provide an explanation for the frequency at which, in developmental genetics, mutagenesis expreiments (resulting in the loss of gene function) fail to yield observable deficiencies in development [7]. Traditionally, such robustness is thought to have arisen from evolution [4]; it was shown here that this need not be the case.

A further feature exhibited by the model is the ability to 'differentiate' into different attractors upon receiving external signals. The key to this is the activity of a few crucial genes (*switch genes*), the modulation of which causes the current cycle to become unstable and change into a different expression pattern. In the experiments above, such modulation was achieved by manually switching on or off the respective genes. In biological genetic nets, on the other hand, it seems plausible to view signals external to the cycle as the source for this modulation, such as cell-cell interactions, diffusing morphogens or the unequal distribution of cytoplasmic determinants. Proceeding on the assumption that a cell type represents an attractor, this would open the way for Natural Selection to control cell differentiation by providing such signals, and making the genome responsive to them.

A final interesting observation of these initial experiments is the amount of unused sequence in genomes with complex gene expression. It may be hypothesised that this represents the equivalent to junk DNA in biological systems. The existence of redundant sequence is a result of the central property of the model - template matching. In order for the system to reach the supracritical level [5] above which gene expression is not ordered, the number of genes and regulatory sequences must be sufficiently high. Large genomes increase the probability that gene products (transcription factors) find matching regulatory sequences. A natural by-product of this *modus operandi* is the accumulation of redundant, unused sequence the more complex gene expression becomes. It is conceivable that Natural Selection affects the complexity of gene expression by directly modifying the parameter *genome size*. If this is the case in biological systems, then complexity of gene regulation in organisms should be correlated with the amount of junk DNA. It is certainly too early to make definite judgements as to the validity of this notion, but it seems worth further investigation.

An important next step is a quantitative analysis of the results presented here. This work is currently in progress. Furthermore, the model lends itself to the incorporation of genetic algorithms (GAs). In particular, it will be studied to what extent selection can modify existing gene expression patterns (e.g in terms of cycle lengths or response to external signals). In addition, due to their biological plausibility, it is believed that Artificial Genomes of the type presented here are particularly suited to be embedded into a framework of artificial ontogeny.

5 Conclusion

The approach of studying the dynamics of gene expression with a model based on template matching has proven fruitful. The behaviour of the system mirrors that of biological genetic nets to a considerable extent, including features such as the existence of stable attractors, the ability to differentiate, and a high degree of robustness. However, though these features are inherent properties of the system, they alone do by no means suffice as the source of adaptation. Rather, they should be seen as providing the raw material upon which Natural Selection can act. The opinion is held that extending the model with GAs as well as embedding it into a framework of artificial ontogeny will provide valuable insights into the principles of development.

6 Acknowledgements

Many thanks to Greg Werner, Julian Worby, Stevan Anastasoff and Josh Bongard for helpful suggestions. I am also grateful to the EPSRC for financial support.

References

1. Kauffman, S.A.: Gene regulation networks: A theory for their global structure and behavior. Current Topics in Dev. Biol. **6** (1971): 145.
2. Mewes, H.W., Albermann, K., Bahr, M. Frishman, D., Gleissner, A., Hani, J., Heumann, K., Kleine, K., Maierl, A., Oliver, S.G., Pfeiffer, F., Zollner, A.: Overview of the yeast genome. Nature **387** (1997): 7–8.
3. Wilson, R.K.: How the worm was won - the *C. elegans* genome sequencing project. Trends in Genetics **15** (1999): 51–58.
4. Gilbert, S.: Developmental Biology. Sinauer Associates (1994)
5. Kauffman, S.A.: The large-scale structure and dynamics of gene control circuits: An ensemble approach. J. Theoret. Bio. **44** (1974): 167.
6. Maynard Smith, J.: Shaping Life – Genes, Embryos and Evolution. Weidenfeld and Nicolson (1998).
7. Alberts, B., Bray, D., Lewis, J., Raff, M., Roberts, K., Watson, J.: Molecular Biology of the Cell. Garland Publishing Inc. (1994).

Parameter Optimization in Hierarchical Structures

Shugo Hamahashi[1,2] and Hiroaki Kitano[1,3]

[1] Kitano Symbiotic Systems Project, ERATO, JST
M-31 Suite 6A, 6-31-5 Jingumae, Shibuya-ku, Tokyo 150-0001 Japan
{shugo,kitano}@symbio.jst.go.jp
http://www.symbio.jst.go.jp
[2] Department of Computer Science
Graduate School of Science and Technology, Keio University
3-14-1, Hiyoshi, Kohoku-ku, Yokohama 223-0061 Japan
shugo@shugo.com
[3] Sony Computer Science Laboratories Inc.
3-14-13, Higashi-Gotanda, Shinagawa-ku, Tokyo 141-0022 Japan
kitano@csl.sony.co.jp

Abstract. The genetic-algorithm-based method described in this paper can be used to identify a parameter set whose value defines the gene regulation circuit. To demonstrate the effectiveness of the approach we choose *Drosophila* segmentation processes. In the processes, we search the parameter set of diffusion constant and transcription ratio of each gene. The characteristics of convergence were also investigated in order to find out how to improve the method. The results suggest that (1) when the gene regulatory network is hierarchically structured, genetic algorithm optimize the upstream parameters earlier than that of downstream in the hierarchy structure, (2) some gene network has smooth concave error surface with no local minima, and (3) the method can be used to test appropriateness of the basic model assumed.

1 Introduction

The major questions we wish to address in this paper are the following: (1) given a relatively well understood gene regulatory network, can we identify a set of parameters that can reproduce spatial expression patterns for both wild-type and mutant alleles, (2) what are the characteristics of the gene regulatory network when it is applied to the stochastic parameter optimization method such as GA? Is there any feature that we can take advantage of when we need to work on very large networks? and (3) how does the appropriateness of the basic model framework affect the fitness during the optimization? Would an arbitrary model finds a set of parameters that can fit expression patterns, so that we cannot tell whether the basic model is good or bad? Or is the convergence retarded when an inappropriate model is used, enabling us to, infer quality of the basic model framework from the optimization characteristics?

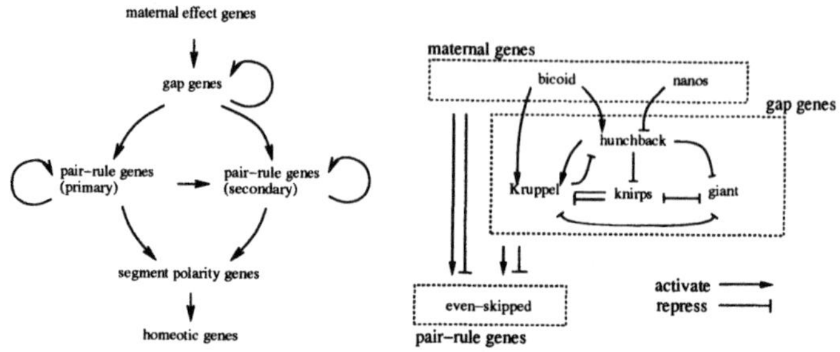

(i) Hierarchy of gene classes (ii) Regulation of genes

Fig. 1. Hierarchical structure of gene regulation

2 Hierarchical regulation of early segmentation genes

Some of the genes involved in the *Drosophila* early segment formation form a hierarchy. A cascade is initiated by the maternal effect genes which regulate the expression of gap genes. The gap genes interact among themselves to control their expression. They also control the patterns of pair-rule gene expression. The pair-rule genes interact among themselves to form the repetitive segmental divisions of the body, and they also control the pattern of segment polarity gene expression. The pair-rule and gap genes also interact to regulate the homeotic genes that determine the structure of each segment. By the end of the cellular blastoderm stage, each segment primordium has gained an individual identity by virtue of its unique constellation of gap, pair-rule, and homeotic gene products. In such manner, the genes mutually interact each other to form the concentration localization in specific areas in the egg. Figure 2 shows the gene expression pattern along the anterior-posterior axis of the egg. The curves of the figure are drew by the computer simulation using a hand tuned parameter set to form the consistent expression pattern with the real expression pattern which is obtained by the biological experiments.

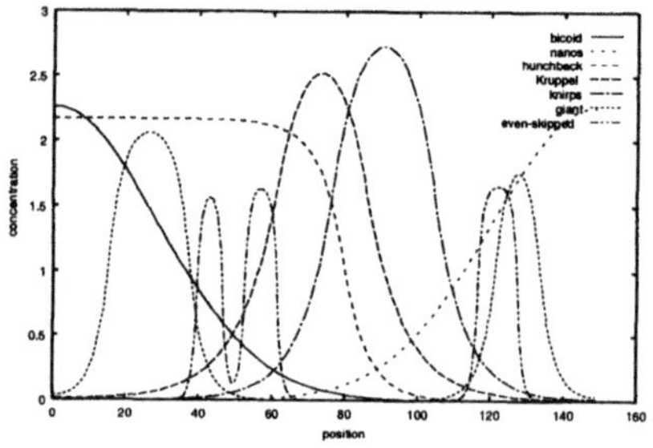

Fig. 2. Expression pattern along the anterior-posterior axis of the *Drosophila* embryo. This graph data is based on real experimentations and a computer simulation.

3 Simulation Model

In this paper we focus on the processes of the cascade of maternal effect genes (*bcd*, *nanos*), gap genes (*hunchback*, *Krüppel*, *knirps*, *giant*), and pair-rule genes (*even-skipped*) — during the early embryogenesis. They are expressed by their own regulation mechanism, such as *hunchback* is activated by *bicoid* but repressed by *nanos*.

In consideration of activation and repression by the activator and repressor, the expression probability t_i is given by the following formula:

$$t_i = \frac{k_a C_{activater_a}}{k_a C_{activater_a} + k_b C_{repressor_b} + \beta_i}, \tag{1}$$

where C is the concentration of activator or repressor surrounding the gene i, k is the binding affinity for the gene i, and β_i is a constant value for the gene i. Expression occurs when t_i is larger than a threshold value is pre-assigned for each gene. As a result of the expression, products are transcribed from genes. The transcription products are contained by the egg frame. As the *Drosophila* egg shape is longest along its anterior-posterior axis and there are no partitions inside it, we approximated it as a cylindrical container.

The free diffusion of the transcription products within the egg is described by the following formula:

$$\frac{\partial U_i}{\partial t} = D_i \frac{\partial^2 U_i}{\partial x^2} + g_i \cdot U_i + T_i(\mathbf{U}), \tag{2}$$

$$\tag{3}$$

where

$$T_i = \begin{cases} \alpha_i \ (t_i > threshold_i) \\ 0 \ \ \text{(otherwise)} \end{cases} . \tag{4}$$

Here U_i is the concentration of the product of gene i, D is its diffusion constant, g_i is its deletion function, and T_i is its transcription function. This function is applied to every gene product contained in the simulator so that the products form the concentration gradient along the x-axis (the anterior-posterior axis).

Genetic algorithm is used to search the optimal parameter set. The answer of the expression pattern is given by the biological information whose pattern is shown in Fig. 2. The fitness is a function of an error value which is figured out by the pattern matching. At each slice along the anterior-posterior axis, the distance between the result value and the answer value is calculated, and the error is the sum of the squares of these distances.

$$(fitness) = \frac{1}{\sum_{genes} \sum_{position} distance^2} \tag{5}$$

The strategy we choose in this paper is tournament selection. The population is 100, tournament size is defined 5, and the tournament selection occurs twice in order to select two agents to mate. After that, the crossover and mutation produce two daughters.

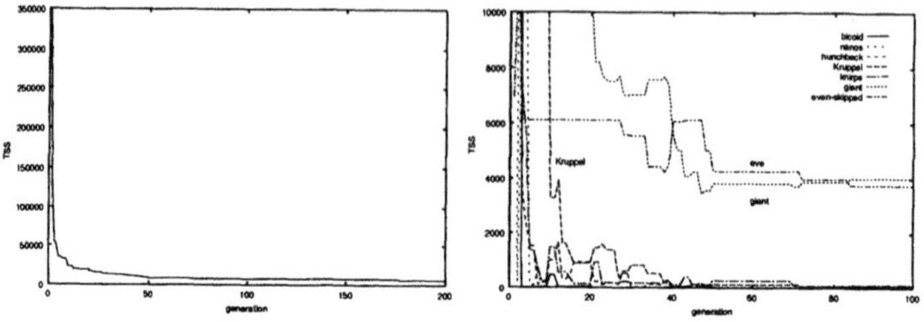

(i) sum of all genes' TSS (ii) TSS of each gene

Fig. 3. TSS transition

4 Result and Discussion

The TSS along the generation is shown in Fig. 3(i). In the early generations, TSS is quickly getting smaller. After the 50th generation, TSS looks almost converged. Figure 3(ii) shows the TSS conversion transition of each gene along the generation. This result shows that the TSS of each gene converges

Figure 4 shows the optimized expression patterns in the generation 10, 20, 30, 40, 50, and 200. According to these results, the parameter set, which is positioned upstream in the regulational hierarchy, is optimized earlier than that of downstream. It is obvious from these graphs that the expression patterns are refined as the number of generations increases.

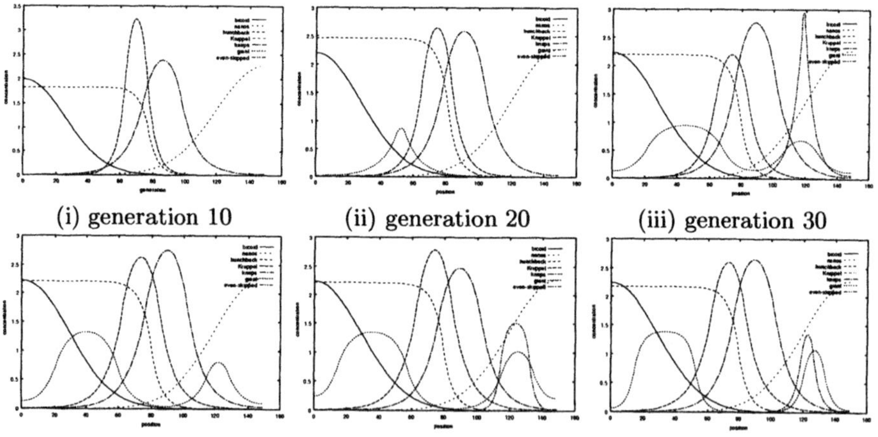

(i) generation 10 (ii) generation 20 (iii) generation 30

(iv) generation 40 (v) generation 50 (vi) generation 200

Fig. 4. Optimized expression patterns

After the TSS is almost converged (generation 200), the parameters of maternal genes and gap genes are properly optimized so that the expression pattern of those genes are successfully reproduced. However, the expression pattern of *even-skipped* is different from the ideal expression pattern.

The simulation results show that only the *even-skipped* expression pattern could not be properly reproduced. We think the regulation of *even-skipped* must be quite complex because it must produce the tree peaks along the anterior-posterior axis and *even-skipped* is controlled in the downstream of regulational hierarchy. In this situation, we need totally different methodology to obtain the optimized parameter set for the downstream parameters. For example, it could work to separate the parameter set relating to their position of the regulational hierarchy and optimize the parameter from upstream to downstream step by step.

Validation of the assumed model: It is suggested that the transition of the TSS can screen out the inappropriate assumptions in the simulation. In this paper, we performed the transcription simulation only by the protein-protein concentration interaction. It is generally used to approximate the transcription in biological simulation. For the *even-skipped* regulation, however, the other micro regulation localizes the *eve* protein in particular positions of the embryo, which is strongly based on the structure of the binding sites in the promoter region. Consequently, the TSS might be able to distinguish the validity of the approximation of the model. At least, it would be appropriate that non-conversion part tells us that the part would be modeled by inappropriate approximation.

Temporal dynamics: Biological experiments commonly took the gene expression data as like qualitative on-off binary data, so normally we can obtain only information about what gene activates or represses what gene. In addition, there is little information about the temporal gene expression transition. In a computer simulation we can observe the temporal dynamics of the transition of the gene expression patterns. To make the value of the parameter set biologically more significant, we need more temporal gene expression transition data with quantitative information.

References

1. Shugo Hamahashi and Hiroaki Kitano: Simulation of *Drosophila* Embryogenesis. In Proceedings of the Sixth International Conference on Artificial Life (1998) 151–160
2. Hiroaki Kitano, Shugo Hamahashi, Jun Kitazawa, Koji Takao, and Shin'ichiro Imai: Virtual biology laboratories. In proceedings of Fourth European Conference on Artificial Life (1997) 274–283
3. Hiroaki Kitano, Shugo Hamahashi and Sean Luke: The Perfect C. elegans Project: The Perfect *C. elegans* Project: An Initial Report. Artificial Life, Vol. 4 (1998) 141–156
4. John Reinitz, Eric Mjolsness, and David H. Sharp: Model for cooperative control of positional information in *Drosophila* by *bicoid* and maternal *hunchback*. J. Exp. Zool., (1995) 47–56

A Model of Axis Determination for the *Drosophila* Wing Disc

Koji Kyoda[1,2] and Hiroaki Kitano[1,3]

[1] Kitano Symbiotic Systems Project, ERATO, JST
6-31-5 Jingumae, M-31 Suite 6A, Shibuya-ku, Tokyo 150-0001 Japan
{kyoda,kitano}@symbio.jst.go.jp
http://www.symbio.jst.go.jp/
[2] Department of Computer Science, Keio University
3-14-1 Hiyoshi, Kohoku-ku, Yokohama 223-0061 Japan
kyoda@kyoda.org
[3] Sony Computer Science Laboratories Inc.
3-14-13 Higashi-Gotanda, Shinagawa-ku, Tokyo 141-0022 Japan
kitano@csl.sony.co.jp

Abstract. In this paper we propose a comprehensive model of axis determination for the *Drosophila* wing disc, and we simulate spatio-temporal gene expression patterning in the *Drosophila* wing disc. The role of interaction among several genes involved has been analyzed systematically in this simulation. The results illustrate that successful reconstruction of the gene expression patterns observed in the actual wing disc requires several hypothetical gene regulations. Above all, we predict the possible existence of a diffusive factor responsible for formation of the D-V boundary in the *Drosophila* wing disc.

1 Introduction

The fruit fly *Drosophila melanogaster* is one of the most popular animal models used in developmental biology, and a substantial amount of knowledge about genetic interactions and cell-cell interactions has been accumulated using this species and various genetic techniques. Scientists working in the field of pattern formation have investigated the *Drosophila* wing formation extensively, and have identified several genes responsible for the formation of the anterior-posterior (A-P), dorsal-ventral (D-V) and proximal-distal (P-D) axes. The mechanism of the D-V boundary formation, however is not well understood and is thus a major research topic related to the *Drosophila* wing disc. Although the development of the *Drosophila* wing disc is so complex that a comprehensive model is required for its systematic understanding, there is very little published literature on the systematic modeling of wing disc.

The goal of this paper is to propose a biologically faithful model of axes determination in *Drosophila* wing disc and to reproduce the gene expression patterns observed in actual wing discs. To take on integrative approach to analyzing the regulation involved in the *Drosophila* wing, we need to model the developmental processes and simulate the spatio-temporal patterning in the development of *Drosophila* wing disc.

2 *Drosophila* Wing Disc

A-P axis determination: Figure 1(a) shows the gene regulations of A-P axis determination. Hh diffuses to the anterior side and activates the *dpp* expression with Ci mediation [1] (in this paper, DNA genes such as *engrailed* are indicated by their short italicized forms (*en*) and the proteins they produce are indicated by capitalized but unitalicized short forms (En)). En represses the *dpp* expression, and *dpp* is expressed at the A-P boundary. Recent evidence has suggested that Dpp acts as a morphogen and determines the A-P positional information.

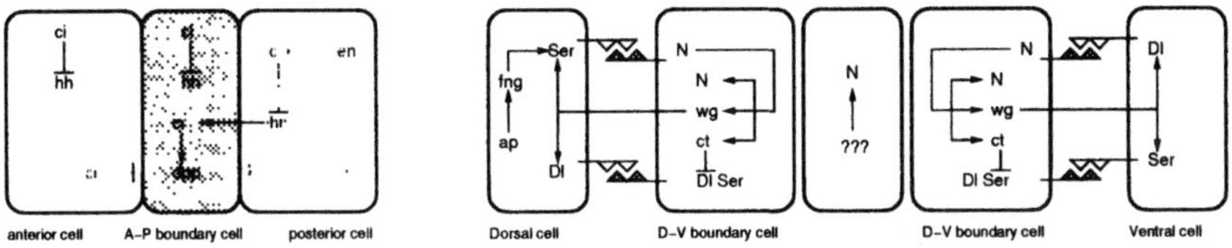

(a) A-P axis determination (b) D-V axis determination

Fig. 1. A-P and D-V axis determination mechanism (a): [Dominguez et al., 96], (b): [de Celis et al., 97]

D-V axis determination: Figure 1(b) shows the gene regulations of D-V axis determination. In the dorsal compartment the sector gene *ap* is expressed continuously and activates the *fng* transcription. Fng encodes a secreted protein, which induces *Ser* expression. Ser is thought to act as a transmembrane protein that interacts with N receptors that are expressed on adjacent cells. N activation at the dorsal-ventral boundary has non-autonomous effects on proliferation and activates the expression of the target genes *N*, *wg*, and *ct* [2]. Diffusing Wg activates the target genes *Ser* and *Dl*, which is thought to act as the N ligand. Since Ct inhibits the expression of *Ser* and *Dl*, or inhibits the function of the Wg, *Ser* and *Dl* cannot be expressed at the D-V boundary [3]. It was observed in a recent study, however that Dl is a potentially a diffusible ligand [4]. These results support the idea that Dl diffuses to the D-V boundary and maintains *N*, *wg*, and *ct* expression, but they cannot explain the initial *N* expression. Both Ser and Dl accumulate in dorsal and ventral cells that flank the D-V boundary. The expression of *vg* is activated by diffusing Wg from the D-V boundary [5].
P-D axis determination: The cells express the *al* gene at the intersection of a stripe of Dpp with a stripe of Wg. The *al* expressing cells are fated to become the most distal cells of the wing. The *dll* expression is activated by Wg, which acts as a long-range morphogen.

3 Modeling Framework

Model Equation: In this simulation, we model and implement transcription/translation, cell-cell interaction, protein diffusion and degradation. We unify these processes with the following equation:

$$\frac{\partial U_i}{\partial t} = D_i \left(\frac{\partial^2}{\partial x^2} + \frac{\partial^2}{\partial y^2} \right) U_i + f(\mathbf{U}) - g \cdot U_i \tag{1}$$

U_i : concentration of protein i	D_i : diffusion constant
t : time	f : protein production function
x : position on x axis	$\mathbf{U}$: concentration vector
y : position on y axis	g : degradation rate ($g = 0.1$)

Transcription/Translation Modeling: Since transcriptional regulation is well known, but translational regulation is not, we simplify the transcription and translation process into one black-box, which means that the amount of transcribed mRNA is the same as the amount of resultant protein. When the rightmost term of equation (2) is greater than a specific threshold value, the transcription function of gene X is determined as follows:

$$f_X(\mathbf{U}) = \frac{\sum_i \alpha_{\mathrm{A}} U_{\mathrm{A}}}{1 + \sum_i \alpha_{\mathrm{A}} U_{\mathrm{A}} + \sum_j \alpha_{\mathrm{R}} U_{\mathrm{R}}} \tag{2}$$

where α_A and α_R are respectively the activation and inhibition rates of gene products i and j, and where U_A and U_R are the concentrations of activator and repressor proteins. In case neither the activator nor the repressor binds to the promoter region of the target gene, we introduce a constant value 1 to adjust the probability of transcription.

Cell-cell Interaction Modeling: The cytoplasmic protein in a cell receives signal from receptor-ligand complexes activated by ligand, and the strength of this signal depends on the concentrations of the receptor and ligand. In this simulation, we determine the activity of cytoplasmic protein as follows:

$$act = \sum_i^{\mathrm{n}} \left[\frac{\alpha_R U_{\mathrm{R}}}{1 + \alpha_R U_{\mathrm{R}}} \cdot \frac{\alpha_L U_{\mathrm{L}}}{1 + \alpha_L U_{\mathrm{L}}} \right]_i \tag{3}$$

where α_R and α_L are the transportation rates of receptor and ligand, where U_{R} and U_{L} are the concentrations of receptor and ligand, and where n is the number of surrounding cells. For example, the concentrations of both N and Dl determine the Su(H) activity, which when high results in the expression of the target genes *wg*, *ct*, and *N*.

4 A Model of Wing Axis Determination

In our model, cells are represented as a square in a two-dimensional lattice, and arranged as a circle. As the gene regulatory relationships, we implement the gene regulation network shown in Figure 1. Some regulatory relationships, however such as Fng activates *N* expression and the dorsally localized protein inhibited *N* expression, are only hypothetical relationships introduced in order to make expression patterns consistent with actual data. In the initial condition, *ci* is already expressed in the whole anterior compartment, *en* is expressed in all of the

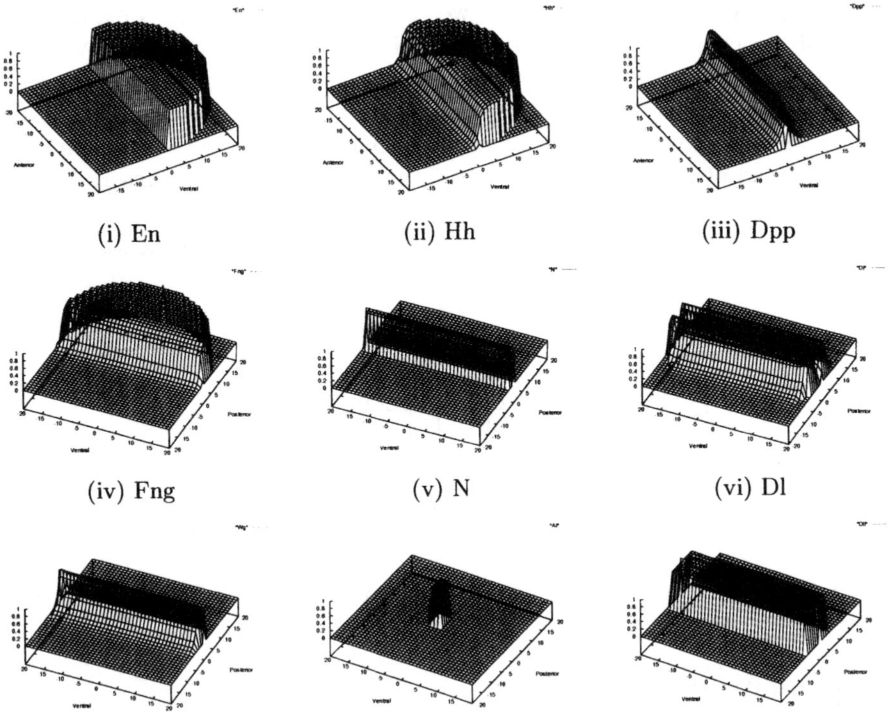

(i) En (ii) Hh (iii) Dpp

(iv) Fng (v) N (vi) Dl

(vii) Wg (viii) Al (ix) Dll

Fig. 2. The protein localization patterns involved in the *Drosophila* wing disc. (i)-(iii): A-P axis determination, (iv)-(vii): D-V axis determination, (viii),(ix): P-D axis determination

posterior compartment, and *ap* is localized in the dorsal compartment. Figure 3 shows simulated protein patterns, reproduced using known gene interaction plus a few hypothetical interactions, for the genes involved in the development development of the third instar *Drosophila* wing disc. The hypothetical interactions introduced are: (1) Fng activates the *N* transcription, and (2) Ap or the Ap downstream gene product inhibits the *N* transcription. These results agree well with experimental biological data. In the actual wing disc, the gene *N* is weakly expressed in the dorsal and ventral compartment, and *Dl* is weakly expressed in the whole ventral compartment. In the actual wing disc, a simulation that implements only known genetic interactions, however cannot reconstruct the expression of *N* and *Dl*. The expression of most genes involved in the axis formation therefore cannot be reproduced, because they are located downstream of N-Dl in the genetic cascade (data not shown, because no gene is expressed at all). It is necessary that other gene regulations be assumed if these patterns are to be reproduced. After a series of simulations, we confirmed that the key to reconstructing the gene expression patterns forming the D-V axis is the initial expression of *N*. For the initial *N* expression, within the limits of the known genes and developmental processes, we assumed Fng activates *N* expression as a diffusive factor, and *N* expression is inhibited by Ap or the Ap downstream gene, which is localized in the dorsal compartment. Using this hypothesis, we can reconstruct the actual gene expression patterns of the *Drosophila* wing disc. Figure 3 shows the comprehensive model of D-V boundary determination for the *Drosophila* wing disc. Although the idea, that Fng activates *N* transcription may surprise knowledgeable readers, the systematic approach leads to the necessity of the diffusive factor or some other mechanisms being responsible for the initial *N* expression.

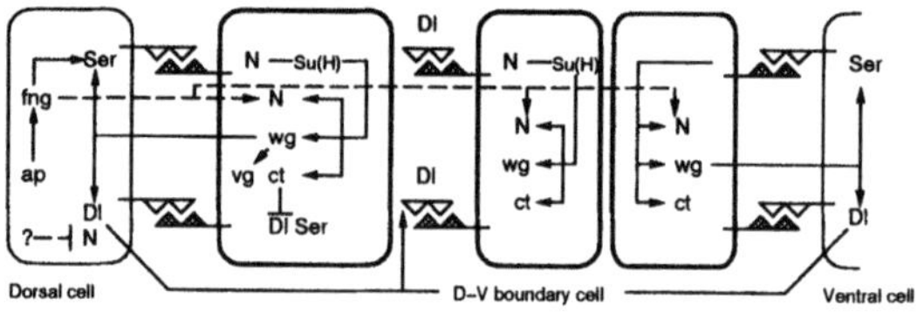

Fig. 3. Comprehensive model of D-V boundary determination. The long-dashed line shows the hypothetical gene regulation

5 Conclusion

In this paper we proposed a comprehensive model of axis determination in *Drosophila* wing discs, and reconstructed in computer simulations the gene expression patterns actually observed in the wing disc. We inferred that Fng protein activates *N* transcription and confirmed by computer simulation that this is a plausible mechanism of the D-V axis determination for the *Drosophila* wing disc. Detailed kinetic analysis and simulation of D-V axis formation led us to a better understanding of developmental processes. We think that our prediction will soon be confirmed by biological experiments.

References

1. Dominguez, M., *et al.*: Sending and Receiving the Hedgehog Signal: Control by the *Drosophila* Gli Protein Cubitus interruptus. Science, **272**, (1996) 1621-1625
2. de Celis, J.F., Bray, S.: Feed-back mechanisms affecting Notch activation at the dorsoventral boundary in the *Drosophila* wing. Development, **124**, (1997) 3241-3251
3. Micchelli, C.A., *et al.*: The function and regulation of *cut* expression on the wing margin of *Drosophila*: Notch, Wingless and a dominant negative role for Delta and Serrate. Development, **124**, (1997) 1485-1495
4. Qi, H., *et al.*: Processing of the Notch Ligand Delta by the Metalloprotease Kuzbanian. Science, **283**, (1999) 91-94
5. Neumann, C.J. and Cohen, S.M.: A hierarchy of cross-regulation involving *Notch*, *wingless*, *vestigial* and *cut* organizes the dorsal/ventral axis of the *Drosophila* wing. Development, **122**, (1996) 3477-3485

Identifying Gene Regulatory Networks from Time Series Expression Data by *in silico* Sampling and Screening

Mineo Morohashi[1] and Hiroaki Kitano[1,2]

[1] Kitano Symbiotic Systems Project, ERATO, JST
6-31-15 Jingumae M-31 Suite 6A, Shibuya-ku, Tokyo 150-0001 Japan
moro@mineo.com
http://www.symbio.jst.go.jp/
[2] Sony Computer Science Laboratories Inc.
3-14-13 Higashi-Gotanda, Shinagawa-ku, Tokyo 141-0022 Japan
kitano@csl.sony.co.jp

Abstract. In order to understand and infer a principle underlying biological phenomena, it is necessary to handle massive data of expression patterns, kinetics, and metabolism, so that plausible regulative mechanisms are revealed. The *in silico* sampling and screening method described in this paper automatically infers possible regulatory network structures using several gene expression profiles. In an experimental evaluation of the feasibility of using this method, each of the possible topologies of three-unit networks were tested exhaustively. After a genetic algorithm was used to identify the parameter set for topology, the plausible topologies were selected by using mutant gene expression data. The experimental results demonstrate that the method can derive a set of possible network structures that includes the correct one.

1 Introduction

Biochemical and genetic approaches have identified the molecular mechanisms of many genetic reactions. Since a series of research projects have been undertaken (e.g., the Human Genome Project), the data of DNA sequencing and expression mapping of genes for each cell in each stage, e.g. of human, will be identified in a few years. In particular, recent technology such as DNA microarrays will reveal the expression pattern of each gene with time series [1–3].

The major question is how we can utilize such data to better understand biological systems. One of the things we would like to do is identifying the gene regulatory network behind specific biological processes. Time series expression data obtained from DNA microarrays is one of the most useful kinds of data that can be use to reconstruct gene regulatory networks. This paper therefore describes an initial attempt to "reverse engineer" a biological system by using the rich data generated from fast-growing high-throughput biology.

Identifying the topology of gene regulatory network is a non-trivial task. There have so far been several attempts to identify or reconstruct genetic networks for specific biological problems [4–7]. Those studies have focused on target-specific networks, and each regulatory relationship is identified with single gene experts, or published literatures on such experiments.

On the other hand, theoretical studies of genetic networks, such as Boolean Network[8, 9] is one of the approaches to identify the network topology. The model simplifies the assumptions about biological systems, and treats gene expression as either completely on or off. It ignores the genes that have different biological regulatory effects at expression levels. Cluster analysis is another approach to let the large-scale data fall into small subgroups [10, 11]. Based on the analysis, using euclidean distance between a pair of points of gene expression pattern, correlated data sets are chosen. Furthermore, there have also been many attempts to model network architecture [12–15]. However, no effective method to reconstruct gene regulation network from time series expression data was proposed so far.

Most previous works regarding genetic regulatory networks have either focused on the modeling of the network or have classified genes over expression pattern. Considering the huge number of components, and functions involved in biological systems, the huge variety of combinatorial regulations possible in the networks is extremely large. Hence, when modeling the networks or analyzing the expression pattern, it is important to carefully take into account redundancy, network combinations, and conditions incompatible with biological phenomena. In addition, since there must be regulation mechanisms and factors not yet identified, the method used to reconstruct the networks should be capable of handling such additional possiblities and constraints.

The work presented in this paper aims at extracting a candidate group of possible topologies from a huge topology space, not uniquely, but as a group. Besides, using the additional data, such as knock-out expression profiles, plausible topologies can be resolved.

2 *in silico* Sampling and Screening

First, it should be noted that what we are trying to identify are genetic regulatory networks, which are networks defining regulatory relationship between genes represented as weight-labeled networks. Although this paper simply considers genes as sole elements, which we refer to as "nodes", in the network for the sake of simplicity, RNA, protein, and other components even tissues can be incorporated when more accurate modeling is desired.

As shown in Fig.1[1], our method consists of two fundamental stages: a sampling stage, and a screening stage. In sampling stage, the possible topologies are chosen, according to the selected target, as virtual initial sampled topologies into a topology pool. If, for instance, the target topology is one in which four nodes

[1] The figure represents a case with up to four genes.

interfere with each other, all possible topologies including up to four units are sampled for the initial topology group. If the scale of the network mechanism is small and tractable, all network topologies can be examined exhaustively. An exhaustive examination, however, is not likely to be feasiable because the mechanism will usually be a scale which takes a considerable amount of time to simulate, thus there are other ways to take instead, such as randomly picking up the network to limit the number of the sampling data. For each possible network, a set of parameters that generates the simulated expression profile which best fits the target data is optimized.

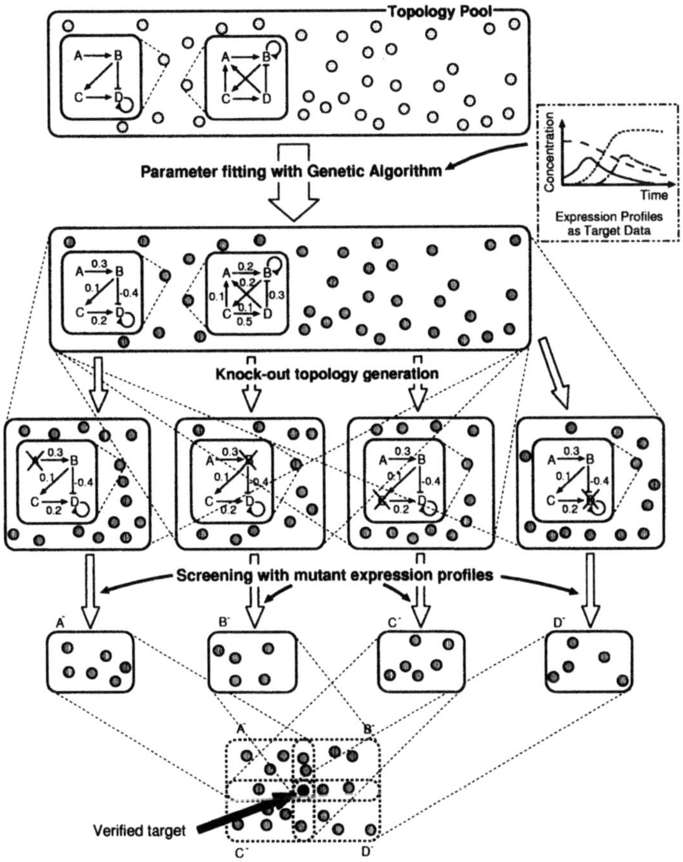

Fig. 1. Schematic representation of our method of *in silico* sampling and screening.

Sampling The aim of the sampling stage is 1) to select, if exhaustive enumeration is not feasible, candidate of network topologies from the entire space of network topologies, and 2) to identify a set of parameters for each network to generate simulated expression profile that best fits the target expression profile. This is done as follows:

1. An expression profile[2] from which plausible topologies are to be identified is selected
2. Possible topologies are chosen (exhaustively, randomly, or using stochastic search) to form a pool of sampled data group D
3. For each network topology in the sample data group D, the set of parameters resulting in the simulated expression profile best fitting the target expression profile is identified. This can be done by using a genetic algorithm (GA) or some other optimization scheme
4. According to the value of TSS(Total Sum Square error) against the target expression profile, the following subgroup D^+ over the fitness threshold T^+ is determined:

$$D^+ = \left\{ D^+ \in D \mid \text{TSS}(D^+) \leq T^+ \right\} \tag{1}$$

Screening Given the set of candidate network topologies and a set of parameters, the aim of the screening stage is to select a set of final candidates using various mutant expression profiles. This is done as follows:

1. Select mutant expression profiles to be used for screening
2. For each mutant expression profiles, the similarities of simulated expression profile generated from each network is measured. In this evaluation, 1) the same weight of the network is used that was identified in step 4 of the sampling stage, and 2) network topology or function in each node is modified to reflect mutant to be used
3. According to the fitness threshold $T^x(x = \mathrm{A}^-, \ldots, \mathrm{N}^-)$ (which can be set independently for each mutant), subgroup $D^x(x = \mathrm{A}^-, \ldots, \mathrm{N}^-)$ is resolved as follows:

$$\begin{aligned} D^{A^-} &= \left\{ D^{A^-} \in D^+ \mid \text{TSS}(D^{A^-}) \leq T^{A^-} \right\} \\ &\vdots \qquad \vdots \qquad \vdots \\ D^{N^-} &= \left\{ D^{N^-} \in D^+ \mid \text{TSS}(D^{N^-}) \leq T^{N^-} \right\} \end{aligned} \tag{2}$$

   where A^-, for example, denotes knock-out mutant of gene A
4. From all subgroups derived in the previous step, a group d of selected network topologies is extracted as below:

$$d = \left\{ d \in (D^{A^-} \cap \cdots \cap D^{N^-}) \mid \text{TSS}(d) \leq T^d \right\} \tag{3}$$

By taking two steps as sampling, and screening, the topologies which have potentials to be plausible ones are derived step by step gradually.

[2] It is a time series data of mRNA, or protein concentration, as shown in Fig.2. All data must be that of wild-type, not supposed to include mutant ones.

3 Experimental Design

3.1 Modeling of Regulative Interaction

Let us assume a network with N nodes. The regulative interactions between the nodes can be represented, with vast simplification, with a fully connected weight matrix W, and an N×N connection matrix C. The value w takes real value between −1.0 and +1.0. Negative value is repressive regulation, while positive value is promoting regulation for target node. In matrix C, an element c_{pq} represents the connection between node p and q with a boolean value (0 means unconnected, and 1 means connected).

The expression state of a genetic network architecture containing N nodes, at certain discrete time t, is represented by a vector $\boldsymbol{x}(t)$ in N-dimensional space. Each element can be described as $x_i(t)(i = 1, \ldots, N)$, and defined by the following rule:

$$x_i(t+1) = F\left[\sum_{j=1}^{N} c_{ji} w_{ji} S_i(t) - h_i\right]$$

$$\text{where} \quad F(u) = \begin{cases} 0 \ (u < 0) \\ u \ (0 \leq u < 1) \\ 1 \ (u \geq 1) \end{cases} \tag{4}$$

where $S_i(t)$ is the concentration of expression product of component i ($i = 1, \ldots, N$), and h_i is threshold for promoting component i to determine the promoting level. As is described above, all regulatory interaction is defined in a simple manner. However, more sophisticated model would be applied in course of the development.

3.2 A Simple Example with Three Interacting Genes

Upon the rules described above, we herein use a small size network, with three nodes as an example. Although the experiment described here is examined under the quite simple condition of three node topology, there is such regulative interaction in biological models, as can be seen in circadian rhythms, which consist of three genes[16]. As for *Drosophila* circadian system, dCLOCK(dCLK)-CYCLE(CYC) heterodimers activate the expression of PERIOD(PER), and TIMELESS(TIM). The heterodimers of PER-TIM, in turn, block the the activation of dCLK-CYC heterodimers. The cycle thus closes as transcription/translation-based negative feedback loop. There are many case in which even simple gene interaction plays a critical role, yet exact regulatory network is not identified.

If each node takes two states, either connect or unconnect, the number of possible topologies for a three-node network is $(2^3)^3 = 512$. Let us consider the topology shown in Fig.2(b). Node A activates B, and also has an autoregulation function. Node B activates C, while node C inhibits A. The initial level expression is set to A=0.5, B=0.0, and C=0.0. This network with this initial value generates the artificial expression pattern is generated as in Fig.2(a).

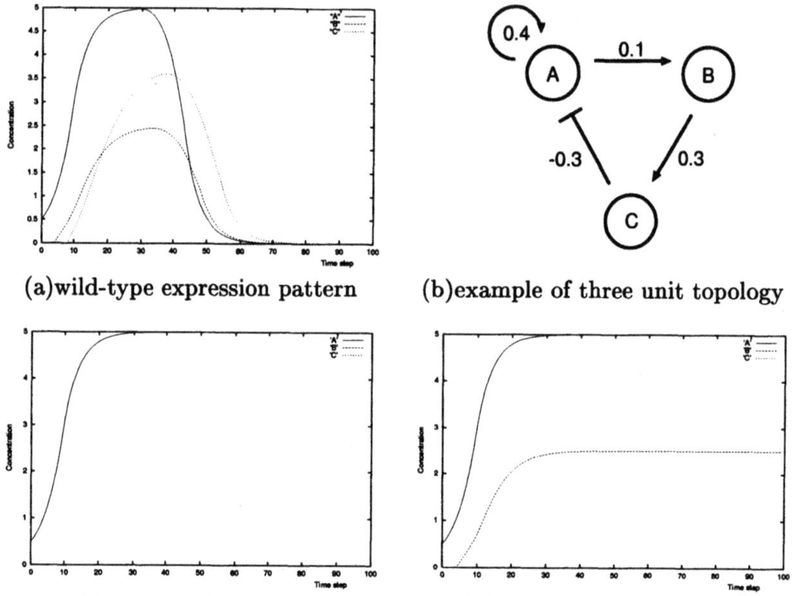

(a)wild-type expression pattern (b)example of three unit topology

(c)B^- expression pattern (d)C^- expression pattern

Fig. 2. Sample of time series data, the topology as target data, and knocked-out gene expression time series data used for screening

We used a GA to fix the weight and threshold parameters. Evaluating strategy of our analysis is under the condition of 2-elite conservation and tournament selection with tournament size 5 to mate. Population size is 300, generation 300, mutation rate 0.01, and one crossover point are applied to the GA.

3.3 *in silico* Sampling

For sampling, the expression patterns of Fig.2(a) were used as target data to resolve the group D into subgroup D^+. We tested this method exhaustively, examining all possible topologies. For each network topologies, a parameter set to best reproduce the target expression profile is searched using GA.

3.4 *in silico* Screening

As for screening, expression patterns for B^- and C^- were prepared as sample patterns in advance. They are shown in Fig.2(c), and Fig.2(d). Expression profiles of B^- and C^- are time series gene expression data when the transcriptions of B and C are disabled. The threshold parameters T^+, T^{B^-}, and T^{C^-} for the screening were respectively set to 1000, 500, and 500.

4 Results and Analysis

As for the results of sampling, the histogram of a topology, which showed minimum TSS(the value was 29.7276) in wild-type, are shown in Fig.3, the topology and its expression pattern are shown in Fig.4(a) and Fig.??(b). Although both Fig.3(a),(b) partially show gentle slopes, the results indicate to converge in fast way.

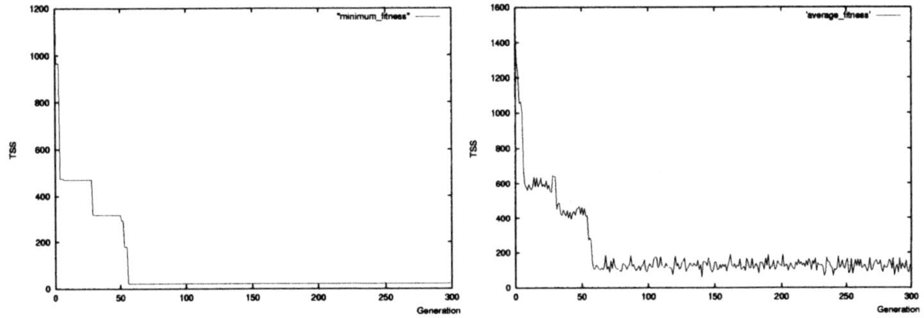

(a)best fitness (b)average fitness

Fig. 3. Convergences of fitness

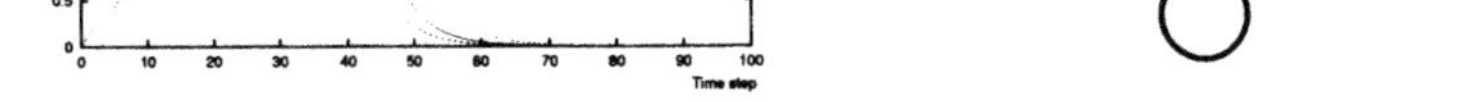

(a)expression patterns (b)network topology

Fig. 4. Expression transition and the regulative description of topology. The topology showed the best fitness among all topologies in comparing with wild-type expression pattern

The histogram of number of topologies against TSS range among whole topologies is shown in Fig.5. The number of topologies are plotted against TSS up to all topologies(=512). The data from the sampling indicate that the distribution of topologies is constant with the value up to 800, above which it increases abruptly. As a result of the sampling, 283 topologies were derived as D^+.

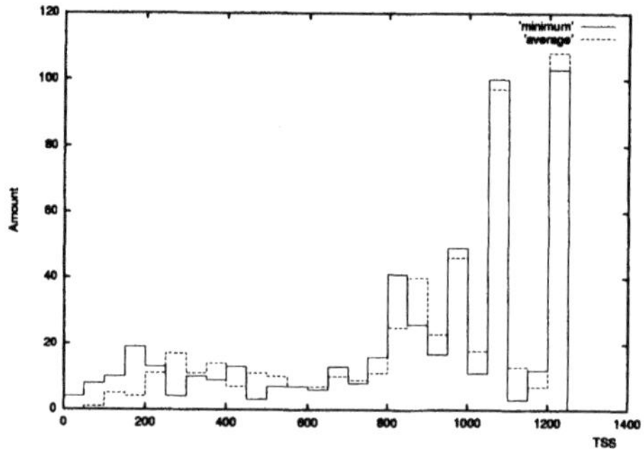

Fig. 5. Histogram of all topologies against TSS range in wild-type. *minimum* is the distribution of the value of best fitness among individuals, whereas *average* takes the average of all individuals for each topology

As the result of the screening process, the histogram of number of topologies for each TSS range for B$^-$ and C$^-$ are described in Fig.6(a), and Fig.??(b), respectively. Subgroups D^{B^-} and D^{C^-} respectively had 37 and 18 topologies. Consequently, 8 topologies were converged into the final subgroup d as shown in Fig.7. The encouraging fact is that the correct topology, shown in Fig.2(b), is included in this subgroup.

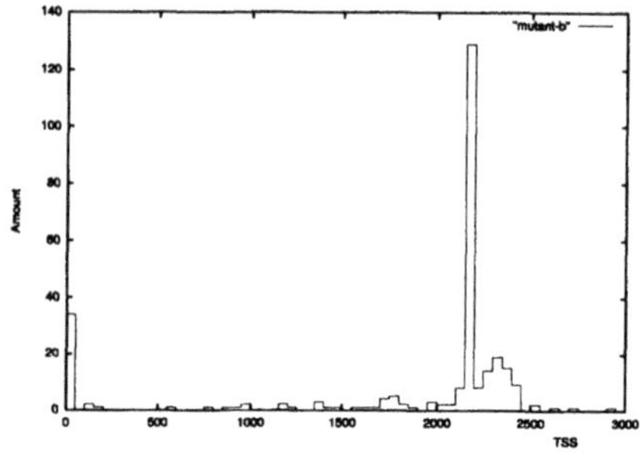

(a)B$^-$ mutant

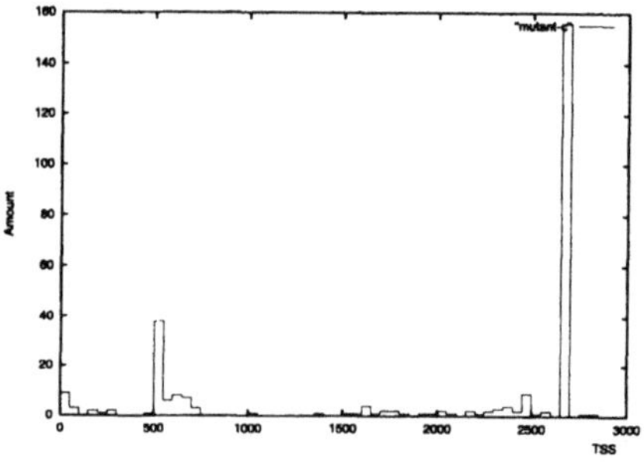

(b)C$^-$ mutant

Fig. 6. Histogram of all topologies against TSS range in B^- and C^-

It is not possible, with only three profile data, to select only one topology from these candidates, because all of the data screened can generate equally good simulated expression profile. The used additional mutant profile data should be able to further reduce the number of final candidates.

At the same time, close example of the candidate topologies revealed that there are common features in many of candidate topologies, such as autoregulative loop of A, inhibition of A from C, and so forth. Extracting basic common features in the final candidates might be feasible way to further reduce the candidate topologies.

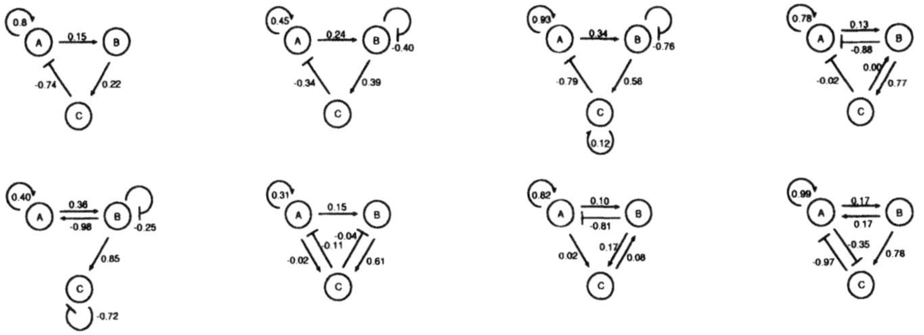

Fig. 7. Eight topologies screened as subgroup d

In order to confirm the stability of the approach, we experimentally examined another expression pattern that had a different type of dynamics. The result was that subgroups D^{B^-} and D^{C^-} respectively had 42 and 56 topologies, and that d had only 20 topologies. In this experiment too, the correct topology was in the subgroup d. Although our model used for regulative interaction is based on linear equation with weight matrix, and is abstract, the results of the experiments exhibited the validity of our method to resolve the large-scale data into a correct topology.

In order to demonstrate the efficiency and scalability of our method, we need to further investigate our method by using more complex, large-scale data. In addition, another criteria is necessary to resolve the subgroup into a unique plausible topology.

5 Concluding Remarks

In this paper, we have described an *in silico* sampling and screening method that can be used to screen out plausible network topologies from large-scale expresion data. The results of an experiment in which plausible three-node genetic network topologies were selected demonstrated that this method can screen out a small set of topologies that includes the correct topology. Embedding some hypothetical factors in the network, we believe that the method can be applied not only to screening plausible topologies, but also to verification of the hypothetical regulation.

References

1. P.O. Brown and D. Botstein. Exploring the New World of the Genome with DNA Microarrays. *Nature Genetics*, 21:33–37, 1999.
2. P.T. Spellman, G. Sherlock, M.Q. Zhang, V.R. Iyer, K. Anders, M. Eisen, P.O. Brown, D. Botstein, and B. Futcher. Comprehensive Identificatino of Cell Cycle-regulated Genes of the Yeast *saccharomyces cerevisiae* by Microarray Hybridization. *Molecular Biology of the Cell*, 9:3273–3297, 1998.
3. J.L. DeRisi, V.R. Lyer, and P.O. Brown. Exploring the Metabolic and Genetic Control of Gene Expression on a Genomic Scale. *Science*, 278:680–686, 1997.
4. M. Morohashi and H. Kitano. A Method to Reconstruct Genetic Networks Applied to the Development of *drosphila*'s Eye. *Proc. of the 6th International Conference on Artificial Life*, pages 72–80, 1998.
5. S. Hamahashi and H. Kitano. Simulation of *drosophila* Embryogenesis. *Proc. of the 6th International Conference on Artificial Life*, pages 151–160, 1998.
6. K. Kyoda and H. Kitano. Simulation of Genetic Interaction for *drosophila* Leg Formation. *Proc. of Pacific Symposium on Biocomputing'99*, pages 77–89, 1999.
7. H.H. McAdams and L. Shapiro. Circuit Simulation of Genetic Networks. *Science*, 269:650–656, 1995.
8. S. Liang, S. Fuhrman, and R. Somogyi. REVEAL, A General Reverse Engineering Algorithm for Inference of Genetic Network Architectures. *Proc. of Pacific Symposium on Biocomputing'99*, pages 18–29, 1999.
9. T. Akutsu, S. Miyano, and S. Kuhara. Identification of Genetic Networks from a Small Number of Gene Expression Patterns under the Boolean Network Model. *Proc. of Pacific Symposium on Biocomputing'99*, pages 17–28, 1999.
10. G.S. Michaels, D.B. Carr, M. Askenazi, S. Fuhrman, X. Wen, and R. Somogyi. Cluster Analysis and Data Visualization of Large-scale Gene Expression Data. *Proc. of Pacific Symposium on Biocomputing'98*, pages 42–53, 1998.
11. P. D'haeseleer, X. Wen, S. Fuhrman, and R. Somogyi. Linear Modeling of mRNA Expression Levels During CNS Development and Injury. *Proc. of Pacific Symposium on Biocomputing*, pages 41–52, 1999.
12. D.C. Weaver, C.T. Workman, and G.D. Stormo. Modeling Regulatory Networks with Weight Matrices. *Proc. of Pacific Symposium on Biocomputing'99*, pages 112–123, 1999.
13. H. Ueda and H. Kitano. A Generalized Reaction-Diffusion Simulator for Studying Molecular Basis of Pattern Formation in Biological Systems. *Proc. of the 6th International Conference on Artificial Life*, pages 462–466, 1998.
14. H.H. McAdams and A. Arkin. Simulation of Prokaryotic Genetic Circuits. *Annu. Rev. Biophys. Biomol. Struct.*, 27:199–224, 1998.
15. R. Somogyi and C.A. Sniegoski. Modeling the Complexity of Genetic Networks: Understanding Multigenic and Pleiotropic Regulation. *Complexity*, 1(6):45–63, 1996.
16. J. Dunlap. An End in the Beginning. *Science*, 280:1548–1549, 1998.

Societies and Collective Behaviour

Simulating the Evolution of Artifacts

Massimiliano Ugolini [1], Domenico Parisi [2]

Department of Communication, University of Siena
15 Via del Giglio, Siena, Italy
ugolini@caio.irmkant.rm.cnr.it

Institute of Psychology, National Research Council, Rome
15 Viale Marx, Rome, Italy
parisi@ip.rm.cnr.it

Abstract. A population of artifacts that increase the energy extracted from the environment by the individuals that use them evolves through selective reproduction of artifacts and the addition of randomly generated new variants. In a series of simulations we explore the consequences of adopting different criteria for selecting artifacts for reproduction and the effects of importing better quality artifacts from a technologically more adavanced population to a less advanced one. The main results are: (a) the best selection strategy is to have individuals with the same energy extraction capacity test all the artifacts; (b) artifacts tend to amplify interindividual differences in extracted energy; (c) better imported artifacts accelerate technological progress only if they are not too few in number; in any case the acceleration effect appears to be temporary.

1. Introduction

Cultural and technological evolution is a crucial component of the adaptive pattern of humans that cannot be ignored if Artificial Life must be able to account for the specific properties of human behavior. While many important formal and mathematical models of cultural transmission and evolution have been published [1], [2], [3], there has been little work on agent-based models and simulations in which one is less interested in the relationships among aggregate variables but expects aggregate phenomena to emerge from the local interactions of many simple "agents". In this paper we present simulations of some limited aspects of the evolution of artifacts using an Artificial Life paradigm. We intend this paradigm to mean not only an agent-based approach to the evolution of culture and technology, but also the study of cultural and technological evolution in populations of biological agents [4]. In our simulations, agents are neural networks living in a physical environment and inheriting not only artifacts but also genotypes. In other words, technological evolution is studied against a backdrop of biological evolution.

A population of individuals lives in an environment that contains food. The reproductive chances of each individual depend on the individual's ability to collect food. As a result of the selective reproduction of the best individuals and the random mutations that occasionally result in better offspring than their parents, the population tends to exhibit a good capacity to procure food. After the population has reached some degree of adaptation to the environment, something new happens. Each individual is provided with a certain number of artifacts (food containers) which are used to store, transport, or cook the food. The use of the artifacts increases the amount of energy obtained from the collected food. The artifacts differ in shape, material, technique of production, etc., and their different properties cause the artifacts to differ in their "added energy value" for the individuals using them. Assuming that each food element yields one unit of energy, the quality of any particular artifact can be described as a number that multiplies the energy contained in the food. The higher the number, the better the artifact. The "population" of artifacts reproduces selectively and with the addition of random mutations in the same way as the population of individuals.

The reproduction of artifacts consists in the fact that new artifacts are produced by each generation of individuals by copying the artifacts of the preceding generation. The reproduction of artifacts is selective and copies of "model" artifacts are not identical to their models but random changes are added during the copying process. This results in the evolution of artifacts. Although the early generations of artifacts are not of very good quality, the selective reproduction of the best artifacts and the addition of new variants cause the progressive improvement of the quality of the artifacts used by the population of individuals.

In this paper we describe a number of simulations that study the evolution of artifacts. We are specifically interested in the mechanisms for selecting the artifacts for reproduction and in the effects of introducing high quality artifacts of a technologically more advanced population into the artifact pool of a less advanced population.

2. Simulations

2.1 Environment and task

The environment is a square grid of 20x20 cells that contains 20 randomly distributed food elements, each providing one unit of energy. An environment is inhabited by a single individual and, since in each generation there are 100 individuals, there are 100 identical environments.

When the individual happens to step on one of the 20 food elements, the food element disappears and the individual's energy level is increased by 1 unit. Then a new food element is introduced in a randomly selected position in the environment so that the environment always contains 20 food elements. The food-seeking behavior of the individual is controlled by a simple feedforward neural network with 1 input unit encoding the position of the nearest food element with respect to the individual, 3 internal units, and 2 output units encoding the movements of the individual in the environment. The simulation begins with a population of 100 individuals each living in its own copy of the environment and with randomly assigned connection weights. The life of all individuals lasts 1,000 input/output cycles. At birth the energy level of all individuals is set to zero. At the end of life the 20 individuals with the highest energy level are selected for reproduction.

Each of these 20 individuals generates one copy of its connection weights which is assigned to the neural network of one offspring, plus one or more additional copies that are assigned to a randomly selected number of additional offpring. Hence, one individual can have 1, 2, or more offspring. However, the number of offspring is subject to the constraint that the next generation must include the same total number of individuals of the preceding generation, i.e., 100 individuals.

While the connection weights of the first offspring are exact copies of the parent's weights, the weights of the additional offspring are modified by adding a number randomly selected in the interval –0.1/+0.1 to each weight's current value. The process is repeated for a certain number of generations until the energy level of the average individual at the end of life is 15 energy units, i.e., 15 food elements are captured by the individual. When the population has reached this level of capacity to procure food (which takes from 10 to 15 generations), the artifacts are introduced to the population. Each simulation is replicated 4 times starting each time from a different 'seed' for generating the initial set of connection weights, the random mutations, and the spatial distribution of food elements.

2.2 Artifacts

An artifact possesses 10 properties (shape, material, technique of production, etc.). Each of these properties can have one of two values: 0 or 1. Hence, each individual artifact is completely described by a sequence of 10 bits. One arbitrarily selected sequence of 10 bits is considered as describing the best possible artifact. If an individual were to use this artifact it would obtain twice the quantity of energy from each food element collected, i.e., 2 energy units instead of 1. An artifact with 5 bits having opposite values with respect to the corresponding 5 bits of the best artifact would allow an individual to obtain 1,5 units of energy from each collected food element. Hence, the quality of each individual artifact can be described as a number ranging from 1 to 2. This number multiplies the quantity of energy provided by the food. An initial population of 100 artifacts is generated by assigning to each artifact randomly selected values to its 10 bits with the constraint that no artifact in this initial population can have an Hamming distance from the best artifact smaller than 6. Hence, no artifact in the initial population of artifacts can have more than 1.4 utility. Five of these 100 artifacts, randomly selected, are assigned as models to be reproduced to each of the 100 individuals composing the first generation following the one in which the population has reached the average level of 15 energy units. While the same artifact is likely to be assigned as a model to be reproduced to different individuals, all five artifacts assigned to the same individual are different.

An individual reproduces an artifact by observing the properties of the model and creating a new artifact with the same properties. To reproduce an artifact the individual has available a separate neural network, distinct from the neural network allowing the individual to perceive and approach the food elements. This network reproduces an artifact by learning to generate an output identical to the input (auto-association task). To add new variants to the population of artifacts some noise is introduced in the copying process by randomly modifying to some extent the perceived properties of the "model" artifact. Therefore, the copies that are generated at the end of learning are similar but not identical to the inherited models. (To determine the quality of an artifact a continuous variant of the Hamming distance is employed.) After learning is completed each of the 100 individuals lives an ordinary life (1,000 input/output cycles) searching for food and using the 5 artifacts it has produced by copying the models. At the end of life the 20 individuals with the highest energy are selected for reproduction using the same procedure as in the preceding generations without artifacts. However, the energy level of an individual is now computed by multiplying the number of food elements collected by the individual by the average utility of the 5 artifacts used by the individual. Therefore, the 20 individuals that are selected for reproduction are identified on the basis of two factors: (a) the individual's ability to collect food, and (b) the quality of the artifacts used by the individual.

The 100 offspring of the reproducing individuals of the first generation using artifacts constitute the second generation using artifacts. What artifacts do these individuals inherit from the preceding generation as models to be reproduced? Each generation uses a total of 500 different artifacts (5 artifacts per individual). These artifacts tend to be all different because of the differences among the individuals that have reproduced the artifacts and because of the random noise introduced into the copying process. How the 100 artifacts that will function as models for reproduction in the next generation are selected from these 500 different artifacts? The criteria for deciding which artifacts are selected for reproduction by the next generation vary in the different simulations we will now describe. The simultaneous evolution of individuals and of their artifacts continues until generation 600.

3. Results

3.1 Perfect knowledge of the quality of artifacts

In this simulation we imagine that the individuals have perfect knowledge of the quality (utility) of each artifact and they decide to reproduce (and use) the best artifacts of the preceding generation. Of the 500 artifacts of each generation the 100 artifacts with the smallest Hamming distance from the prototype are selected for reproduction and each individual of the following generation is assigned 5 of these 100 artifacts as models to be reproduced.

The results of the simulation show that for the 20 best individuals there is a very rapid increase in fitness during the first 25 or so generations and then fitness remains practically invariant for the rest of the simulation, whereas for the total population fitness also increases rapidly in the early generations but then it continues to grown, albeit more slowly, until the end of the simulation. Moreover, if we compare fitness with and without artifacts we find that there is a huge difference between fitness with and without artifacts for the 20 best individuals compared with a much smaller difference between the two fitnesses for the total population.

With respect to the evolutionary improvement in the quality of artifacts the results show that that there is an extremely rapid improvement of the quality of artifacts in the first generations followed by a very stable state.

3.2 The artifacts selected for reproduction are those used by the best individuals of the preceding generation

Perfect knowledge of the quality of artifacts is an ideal case. To know perfectly the quality of an artifact an individual should know what is the absolutely best artifact, i.e., the bit sequence selected by us as definining the best artifact. But we cannot suppose that this knowledge is accessible to our individuals. Therefore, in this and the following simulations we experiment with various ways in which our individuals can judge the quality of the artifacts used by the individuals of the preceding generation in order to decide which artifacts to reproduce. We assume that an individual can infer the quality of the artifacts used by the individuals of the preceding generation by examining the energy level of these individuals. The individuals with more energy can be supposed to have used better artifacts. In the present simulation the artifacts selected as models to be reproduced are those used by the 20 individuals with the highest energy in the preceding generation, and that therefore are the parents of the individuals of the current generation. Each of these 20 individuals has used 5 artifacts during its life and their 20x5=100 artifacts are selected as models for reproduction.

The results show that for both the 20 best individuals and for the entire population (100 individuals) the total energy level (number of food items multiplied by the average quality of the artifacts used by an individual) is lower than in the previous simulation with perfect knowledge. This indicates that the less than perfect knowledge of the quality of artifacts, which in this simulation can only be inferred from the energy level of the individuals using them, makes the selection process for the artifacts less effective, and this is reflected in the lower energy level of the individuals using the artifacts. On the other hand, the ability to collect food does not change in this simulation with respect to the preceding one.

The results concerning the evolution of the quality of artifacts provide a direct confirmation that less than perfect knowledge of the quality of artifacts makes the selection process of the artifacts less effective. At the end of evolution the quality of the average artifact is around 1.65 whereas in the preceding simulation with perfect knowledge the average artifact had a quality greater than 1.85. It is also interesting that the evolutionary improvement of the quality of artifacts is slower in the present simulation than it was in the previous one.

3.3 All the artifacts are tested by all individuals

The method used in the preceding simulation for selecting the artifacts for reproduction has the limitation that the artifacts selected are those used by the 20 individuals with the highest energy. But the individuals with the highest energy are likely to be especially good at procuring food and therefore their high energy level may be due more to their ability to procure food than to the high quality of their artifacts. In order to better evaluate the quality of the artifacts it might be more appropriate to have many individuals, with different levels of food procurement ability, use the same artifact and then judge the quality of the artifact on the basis of the average energy level of the individuals that have used the artifact. This should make it possiblc to separate the role of the quality of the artifact in determining the energy level of an individual from the contribution of the individual's level of ability in procuring food.

The present simulation implements this idea. In each generation we allow all the 100 individuals constituting one generation to live for another 1,000 cycles. These additional cycles have the purpose to evaluate all the 500 artifacts used by the individuals of that generation. Each of the 100 individuals is assigned 20 different artifacts, the 5 artifacts the individual has used during its regular life and 15 additional artifacts randomly selected from the 495 artifacts used by the other individuals. Each of the 20 artifacts is used by the individual for 50 cycles. In this way each of the 500 artifacts is likely to be used by many different individuals (by 4 individuals on average) and these individuals are likely to have varying levels of the ability to procure food. At the end of this evaluation period, the 500 artifacts are ranked on the basis of the average energy level of the individuals that have used them during the evaluation period and the 100 highest ranking artifacts are selected as models to be reproduced by the individuals of the next generation.

The results of this simulation are not very different from those of the preceding simulation. If we added the period of artifact testing in order to obtain better evaluations of the artifacts, the results are disappointing.

3.4 All the artifacts are tested by the 20 best individuals

The method of testing the quality of the artifacts used in the preceding simulation may be too noisy. Since all the 100 individuals of each generation are involved in evaluating the artifacts, even individuals with very low levels of the ability to procure food evaluate the artifacts and therefore good artifacts may not be selected just because they happen to be evaluated by individuals with very low levels of the food collecting ability. We have therefore run another simulation in which the individuals that are asked to evaluate the 500 artifacts during the evaluation period are not all the 100 individuals constituting an entire generation but only the 20 best individuals. In this way we hope to factor out the role of an individual's food getting ability in determining the evaluation of an artifact since the best 20 individuals all tend to have high levels of food collecting ability and, at the same time, we obtain evaluations for all the artifacts used by each generation.

The new system for selecting the artifacts to be reproduced gives better results than the preceding system. With the new system the energy level of both the 20 best individuals and the entire population reaches higher levels and reaches these higher levels more quickly than with the preceding system in which all individuals, good and bad, tested the artifacts. Furthermore, the average quality of the artifacts reaches 1.7 in this simulation compared with slightly above 1.6 in the preceding simulation.

4. Cultural diffusion of artifacts

Imagine two separate populations that are both evolving their artifacts. The two populations are not at the same stage of technological evolution. Population A is technologically more advanced than Population B. In fact, the measured utility of the artifacts of Population A has already converged to a steady state of maximum utility typically observed in this type of simulations whereas the artifacts of Population B still have a long way to go in terms of increasing their utility. At this point for some reason one or more artifacts of Population A are moved to Population B and become members of the pool of artifacts of Population B. What are the consequences of this phenomenon of cultural diffusion?

We assume that the best artifacts of Population A are moved to Population B but we vary the number of artifacts that are moved. In one simulation only the single best artifact used by the individuals of the last generation (generation 600) of Population A is moved to Population B. In other simulations either the third best or the 4 best artifacts of Population A are moved to Population B. When an artifact of Population A is moved to Population B, the artifact simply replaces a randomly selected artifact of Population B. Everything else remains the same. The alien artifacts become members of the pool of artifacts of Population B when Population B is at generation 15 of its technological evolution. The technological evolution of Population B continues until generation 600 is reached.

The results can be summarized as follows:

A. The imported artifacts do not always become permanent members of the technological pool of Population B. If we reconstruct the artifact lineages we find that in some cases the lineages of imported artifacts disappear from the technological pool of the host population.

B. Lineages of imported artifacts are more likely to disappear the smaller the number of originary imported artifacts.

C. If one or more alien lineages become permanent members of the technological pool of Population B, they may cause an acceleration in the evolutionary improvement of the average quality of the artifacts of Population B. Hence, given (2), we can expect to observe a positive effect of the introduction of better external artifacts only if a sufficient number of such artifacts are introduced in the host population.

D. Even when there is a positive effect of the introduction of better alien artifacts, this effect is likely to be progressively cancelled in the further evolution of the host population.

5. Discussion

A population with artifacts is able to extract more energy from the environment than a population without artifacts. Although in the simulations we have described population size is fixed, if population size varies as a function of energy a population with artifacts would outnumber and possibly drive to extinction a population without artifacts. In our simulations the artifacts are of different quality, i.e., they vary in their capacity to increase the amount of energy obtained from the collected food. If there is a tendency for better artifacts to be reproduced and if new variants are periodically added to the pool of artifacts, the average quality of artifacts tends to increase.

5.1 How are artifacts selected for reproduction?

One problem explored in our simulations is how artifacts are selected for reproduction. Since the properties of individual artifacts correlate with artifact quality, if one were to know the properties of the best artifact one would select for reproduction the artifacts possessing those properties. We have shown that if selection of artifacts is based on perfect knowledge of which properties of artifacts correlate with artifact quality, we obtain the best results in terms of evolutionary increase of artifact quality. But how can knowledge about the properties of artifacts correlated with artifact quality be acquired? The performance of an artifact cannot be devined by just examining the artifact's properties. To decide how good is an artifact one has to measure the performance of an individual using the artifact. In our simulations the performance of an individual is measured by the amount of energy the individual is able to obtain from collecting food and using the artifacts. But this poses the problem of separating the role of the individual's food collecting ability from the role of artifact quality in determining the total amount of energy obtained by the individual.

We have explored various strategies that can be used to achieve this separation: (1) the best artifacts are those used by the best individuals, i.e., those that obtain the highest amount of energy during their life; (2) the best artifacts are those that give the best results when they are tested by all individuals, i.e., individuals falling within the entire range of amounts of energy collected during life; (3) the best artifacts are those that give the best results if they are tested by the best individuals.

Selecting for reproduction the artifacts used by the best individuals (first strategy) does not seem to be a good choice because the best individuals may obtain the large amounts of energy that allow them to reproduce not because of the quality of the artifacts they use but because of their high level of ability in collecting food. This result seems to indicate that the strategy called by Boyd and Richerson "indirect bias" in cultural evolution [1] may not be a very efficient one. "Indirect bias" is the strategy of selecting for reproduction those artifacts (behaviors, ideas, etc.) that are used (exhibited, produced, etc.) by the best individuals in the population. Given the impossibility or difficulty of directly assessing the quality of an artifact, one chooses to reproduce the artifacts used by the best individuals with the implicit assumption that the best individuals obtain their high level performance because of the high quality of the artifacts they use. But in fact the best individuals may reach high perfomance levels for other reasons than the quality of artifacts they use.

The second strategy may not be appropriate for the opposite reason. If one has the entire population test all the artifacts, one risks that good artifacts may not be reproduced because they have been tested by individuals with very low level of food collecting ability. An artifact may get bad scores not because it is of inferior quality but because the individuals that test the artifact are not very good at collecting food and therefore they obtain low scores even with good artifacts.

The results of our simulations seem to show that the another strategy may be a better one. This strategy is based on observing the performance of the best individuals of the population while each of them is using a randomly selected sub-set of all the artifacts and on selecting for reproduction the artifacts that on average result in better performances. (This is a form of "direct bias" in Boyd and Richerson's terminology.) This strategy factors out the role of an individual's food getting ability in determining the individual's performance because all the individuals that test the artifacts are very good at getting food and leaves the quality of the artifact as the only factor that determines the average performance obtained by different good individuals using the same artifact. Of course, this strategy has to pay the cost of observing repeated performances of different individuals using the same artifact. This may explain why another strategy may be the preferred one by real individuals, that is, the strategy called by Boyd and Richerson the "frequency bias" strategy. According to this strategy, the artifacts (behavior, ideas, etc.) selected for reproduction are those that appear to be used more frequently in the population (cf. also [5], [6]). However, the scenario used in our simulations did not allow us to test this strategy.

5.2 Artifacts amplify interindividual differences

An interesting result of our simulations is that, apparently, artifacts tend to amplify interindividual differences. When we compare the distance between the evolutionary curves of energy for the 20 best individuals and for the entire population of 100 individuals with and without artifacts, we find that this distance is greater with artifacts than without artifacts. In other words, the difference in total energy between the best individuals and the average individual in the population tends to be greater when one includes the role of artifacts in determining this total energy than if one only considers an individual's food collecting ability. To exemplify, two individuals that are able to collect 100 and 150 food elements, respectively (difference in energy = 50), will have 150 and 225 actual energy (difference of energy = 75) if they both use artifacts of 1.5 quality.

This simple result can have some interest for interpreting the evolution of human societies. Technological evolution can have had a role in increasing social stratification in human societies. In a population with little or no technology the differences in energy among the individuals depend exclusively on the differences in the personal characteristics of the individuals. Since these differences are not in general too great in absolute terms, a population with little or no technology tends to be egalitarian in terms of interindividual differences in available energy. However, artifacts may have the role of amplifying interindividual differences in energy. Hence, the richer and more sophisticated technology associated with the Upper Paleolithic and, especially, with the Neolitic and the development of metallurgy may have had a role in increasing the economic and social stratification of human societies in these periods.

5.3 Diffusion of artifacts

Our simulations show that if the artifacts of a population with more evolved artifacts are brought into the artifact pool of another population with less evolved artifacts, this may result in some sort of boost for the technological evolution of the second population. Both the total energy of the host population and the quality of its artifacts increase more rapidly than if the population continues its evolution with no external influence.

This result must be qualified in two ways, however. First, the more evolved imported artifacts have an influence only if they are not too few in number. The second qualification is that the boost in the technological evolution of the host population may be a temporary one. A population which does not experience the beneficial influence of external artifacts in the long run tends to reach the same level of technological development of another population whose technological pool has been beneficially "infected" by better external artifacts. This result, however, can be restricted to our simple simulation scenario. In more complex, and realistic, scenarios a termporary boost in technological development for a certain class of artifacts can trigger the development of new artifacts in such a way that the population with no experience of better external artifacts may never abolish the distance from a population which has had this experience. In any case, even if the boost in technological development due to the arrival of better external artifacts is temporary, it still can have important long term consequences. In a simulation scenario with variable population size a population experiencing a technological acceleration due to better imported artifacts may have a time window during which it may outnumber and perhaps drive to extinction another less technologically advanced population.

References

1. Boyd, R., Richerson, P.J.: Culture and the Evolutionary Process. Chicago University Press, Chicago (1985)
2. Cavalli-Sforza, L.L. Feldmann, M.W.: Cultural Transmission and Evolution. A Quantitative Approach. Princeton University, Princeton (1981)
3. Durham, W.H.; Co-evolution. Stanford University Press, Stanford (1991)
4. Denaro, D., Parisi, D.: Cultural evolution in a population of neural networks. In: Marinaro, M., Tagliaferri, R. (eds.): Neural nets. Wirn-96. Springer Verlag, Berlin Heidelberg New York (1996)
5. Lane, D.A.: Is what is good for each best for all? Learning from others in the information contagion model. In: Arthur, W.B., Durlauf, S.N., Lane, D.A. (eds) The Economy as an Evolving Complex System. Addison-Wesley, Reading, Mass. (1997)
6. Narduzzo, A., Warglien, M.: Learning from the experience of others: an experiment on information contagion. Ind. & Corp. Change **5** (1996) 113-126

The Evolution of Non-reciprocal Altruism

Martijn Brinkers, Paul den Dulk

University of Amsterdam, Psychonomics department,
Roetersstraat 15, 1018 WB, Amsterdam, Netherlands
{brinkers, dendulk}@psy.uva.nl

Abstract. This paper presents a simulation of the evolution of altruism in a population of agents. The agents can behave altruistic by stepping into water at the cost of drowning, thereby forming a bridge with their body for other agents. We show that altruism evolves only when relatives have a higher probability to benefit from it. By mixing up the population no altruism evolves even though the population goes extinct.

1 Introduction

Altruistic behavior benefits others at the expense of the altruist. How could such behavior evolve, if natural selection selects for the fittest individual? One explanation comes from the field of game theory, where it has been shown that *reciprocal* altruism, which benefits the altruist in the long run, can be evolutionary stable [1]. This paper, however, focuses on *non-reciprocal* altruism[1], which is more problematic from the point of view of evolutionary biology. Some simulations of non-reciprocal altruism have been done [2,3], but with a very different design. The setup of this simulation is particularly relevant for the controversy over the unit of selection, which focuses on the question 'on what unit does selection operate, the gene, the individual, or the group?' In our simulation a population of agents, has to survive and reproduce in a world consisting of islands. In order to survive they have to gather food from the islands, but they can only reach these by using another agent's body as a bridge. But, when this other agent steps into the water to form a bridge, it dies instantaneously.

2 Simulation Setup

The world consisted of 110 by 110 cells placed on a grid (torus). Each cell could either be a water cell or a land cell. Within this grid, 100 islands, each 10 by 10 cells,

[1] The term non-reciprocal altruism is used to distinguish it from reciprocal altruism, while avoiding the term genuine or true altruism, which tends to be associated with intentional altruism, which is a different topic altogether.

were placed. Canals of 1 cell in width separated the islands. Land cells could contain food or an agent, and water cells could contain a dead agent forming a bridge. A living agent could cross the canal by using this bridge.

Agent behavior: The agent's behavior was determined by four parameters: *p-water, p-food, p-land, p-cross*. Each of these parameters corresponded with one of four cells the agent could see in front of it: empty water, food on land, empty land, or a floating agent forming a bridge. The parameters represented the probability of stepping forward when in front of the corresponding cell. A random number, from the interval [0,1] was drawn. When this number was below the corresponding *p* value the agent would step forward, otherwise it would randomly turn left or right on the spot. The agent used no other information apart from the cell in front of it, and it had no internal state. Agents could not step on top of other living agents. When one was in front of it, it always turned randomly left or right on the spot. The agents were synchronously updated.

Food and Energy: After each cycle, hundred cells were randomly chosen anywhere on the map. On these cells a food patch was placed when it was an empty land cell. When stepping on a food patch the energy of the food patch (*EnergyFood*) was added to the agent's energy and the food patch was removed. The agent always lost one unit of energy per iteration. When the agent used up all its energy it died of *starvation.*

Water: When stepping into water the agents immediately drowned. The agents remained 'floating' during a number of iterations as determined by the *BridgeTime* parameter. While floating, other agents could step on top of the agent using it as a bridge.

Reproduction: The agents could multiply through asexual reproduction. The agent gave birth when it had gathered enough energy (*EnergyToGetchild*). When a child was born the parent's energy was reduced by a certain amount of energy (*EnergyCostChild*) and this formed the starting energy of the child. The child inherited all the parent's four *p* values, but for each child a slight mutation was added. For each mutation, a random number was drawn from the interval [-1,1] and this was multiplied by the *Mutation* parameter and was added to the corresponding *p* value. Resulting *p* values outside the [0,1] interval were clipped. This procedure was repeated for all *p* values. A newborn agent was placed on the cell directly behind its parent.

Age: When agents reached the maximum age (*MaxAge)*, expressed in the number of iterations after birth, they died of *old age.*

Initialization: At the start of each simulation, a number of agents (*InitNumberAgents*) and a number of food patches (*InitNumFood*) were placed at random locations, excluding water cells. Depending on the simulation, initial *p* values were randomly chosen or set to a specified value. The initial agent's energy was randomly chosen from the interval [0, *EnergyToGetChild*]. Its initial age was randomly chosen from the interval [0, *MaxAge*].

Simulator: The simulations were performed with a self-build simulator. To rule out any effect of bugs two simulators were independently developed by the two authors, without accessing each other's code, using two different languages.

3 Simulation 1

Two simulations were run, one normal and one control. (The default values for all simulations: *MaxAge* 250, *EnergyToGetChild* 120, *EnergyCostChild* 60, *EnergyFood* 9, *BridgeTime* 30, *Mutation* 0.004, *InitNumFood* 30/Island, *InitNumAgents* 15/Island)

Both simulations were run for 4 million iterations. There is a strong selection for a high *p-food* and high *p-cross* driving them quickly towards 1.0. A high *p-cross* increases the chance a new island is reached once another agent has body-bridged. On average 21% of the islands were uninhabited, and these grew quickly full of food. Reaching such an island would make it possible to give birth to a large number of offspring. The *p-land* parameter reached its stable value at about 0.8. The parameter of our main concern, *p-water*, immediately dropped to a value close to zero. There is a high selection against high *p-water* values because of the lethal consequences of body-bridging. Even though *p-water* was only 0.0080 (all reported values are averaged over the last 100.000 iterations of the simulation), the percentage of agents that died of drowning was high: 20%.

In the control simulation we undermine the effect of kin selection by distributing relatives randomly over the map, so that when an altruist would body-bridge the other agents on that island would have no systematic higher *p-water* than the population average. We mixed up the population with the following procedure: When a parent gave birth, the child was not immediately placed on the map, but only after a second parent gave birth. The first child was then placed next to the second parent, while the second child had to wait for a third parent to give birth before it was placed on the map, etc. In practice the delay of placement was on average 3/4th of a cycle. With this swap method the balance between food and newborns was not affected in any way, and we avoided randomly placing newborns on the map, which would take away the need for island hopping because new agents could then be born onto an uninhabited island. The general pattern of the control simulation was similar to that of the normal simulation, but *p-water* now reached a clearly lower stable value of 0.0046, indicating that at least part of the *p-water* value in the normal simulation is attributable to non-reciprocal altruism.

4 Simulation 2

If kin selection is the driving force for the evolution of altruistic behavior, how crucial is the exact distribution of relatives? In this simulation we varied the probability an agent was swapped (*p-swap*) from 0.0 to 1.0 in steps of 0.1. All conditions were run for 500.000 iterations and were all replicated 10 times. Because *p-water* was our only concern, all following simulations used fixed *p* values except for *p-water; p-food* was fixed at 1.0, *p-land* at 0.8, *p-cross* at 0.8. This was done to decrease the time it took to reach a stable *p-water*. The values were based on an earlier explorative simulation of 1 million iterations. All other parameters were set to the default values, except for the *Mutation* parameter, which was set to 0.0016.

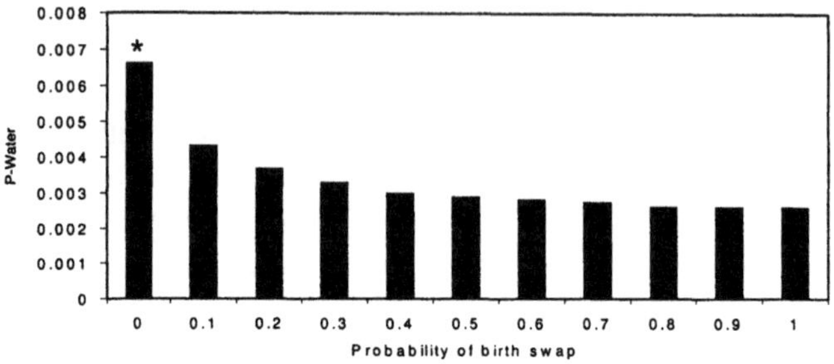

Fig. 1 . The *p-water* values plotted against *p-swap*.

Fig.1 displays the final *p-water* values averaged over the 10 replications for each condition. With the increase of *p-swap, p-water* decreases rapidly, indicating that the amount of altruism is inversely proportional to the amount of kin mix up.

5 Simulation 3

In simulation 2 the percentage of agents drowning in the *p-swap* 1.0 condition was still 9%, while evolution is expected to select against drowning. If altruism indeed depends on the clustering of relatives we should expect no altruism to evolve with the *p-swap* set to 1.0. The drowning rate, we assume, should, thus, be caused by noise added by the mutation parameter. Could we reduce the drowning rate by lowering the mutation parameters? In this simulation the *Mutation* parameter was set to 0.0001 and it was run for 2.500.000 iterations, in both the normal and swap condition.

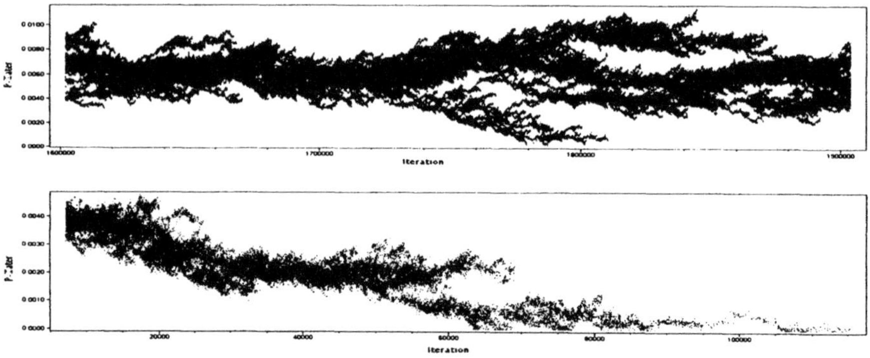

Fig. 2. The *p-water* values plotted for all agents for a number of iterations.

The normal condition showed the same pattern compared to the normal condition in Simulation 1 and 2. The final *p-water* value was 0.0071. To get an impression of how the evolutionary process took place we plotted the *p-water* values for all agents for a number of iterations. Such a plot is a recording of the agent's (artificial) evolutionary history. In the swap condition lowering the mutation parameter did indeed seem to lower the *p-water* value. As can be seen in Fig. 2b, *p-water* reaches values of about 0.0005 on iteration 80.000, much lower than the final value of 0.0046 in Simulation 1b. However, due to the lack of transportation to new islands, the whole population went extinct. This result stresses that the intuition, of altruism evolving because of its value for the whole, is flawed. Without the proper distribution of genes there will be evolution towards egoism, even at the cost of extinction.

6 Discussion

The altruistic behavior in our simulation is not directed explicitly towards kin, and can also benefit unrelated individuals. Unrelated individuals can invade other inhabited islands by using body-bridges, and in simulation 2 also by child swapping. The inhabitants of an island could, therefore, be interpreted as a group according to Williams' [4] definition of a group as 'something other than a family and to be composed of individual that need not be closely related' (p93). However, in the same book Williams refers to kin selection as an alternative to group selection (p97). No such distinction can be made in our simulation. Although our results can be interpreted as a kind of group selection, it is certainly not an alternative process to kin selection. We thus conclude that talk of group selection only makes sense as it is interpreted as another perspective of kin selection (see [5]).

References

1. Axelrod, R., Hamilton, W. D.: The Evolution of Cooperation. Science, Vol. 211. AAAS (1981) 1390-1396
2. Nowak, M. A., May, R. M.: Evolutionary games and spatial chaos. Nature, Vol. 359. (1992) 826-829
3. Parisi, D., Cecconi, F., Cerini, A.: Kin-directed altruism and attachment behaviour in an evolving population of neural networks. In: Gilbert, N. (eds.): Artificial societies. The computer simulation of social life. UCL Press, London (1995) 238-251
4. Williams, G. C.: Adaptation and Natural Selection. Princeton Uni Press, Princeton (1966).
5. Wilson, D. S., Sober, E.: Reintroducing group selection to the human behavioral sciences. Behavioral and Brain Sciences, Vol. 17. Cambridge University Press (1994) 585-654

A Little More than Kind and Less than Kin: The Unwarranted Use of Kin Selection in Spatial Models of Communication

Ezequiel A. Di Paolo

Institute for Autonomous Intelligent Systems, GMD
Schloss Birlinghoven, Sankt Augustin, D-53754, Germany
Email: Ezequiel.Di-Paolo@gmd.de

Abstract. It has become increasingly apparent that spatial and other forms of ecological situatedness can introduce radical differences in the evolutionary outcome of models of conflictive social behavior. Cooperative interactions are often found to have an increased viability in spatially situated models. One possible explanation for this phenomenon makes use of kin-selective arguments according to which high relatedness between neighbors stabilizes cooperation. Unfortunately, in some cases the argument does not go beyond the merely verbal. This paper shows that an explanation in terms of kin selection can easily be tested in a computer simulation and that, in the particular case treated here, the result of such verification is negative thus strengthening previous conclusions regarding the relevance of other factors such as discreteness, stochasticity and ecological organization.

1 Introduction

A possible contribution of Artificial Life to real biology is in the use of relatively novel techniques for addressing already established problems in the biological literature. Such is the case, for instance, of studies that take a problem from evolutionary biology and try to complement what is known about it with extended models mainly in the form of computer simulations (Bullock, 1997; Noble, 1998; Di Paolo 1997 and others).

The type of contribution provided by these models does not differ much from the use of similar techniques by biologists themselves (e.g. Durrett & Levin, 1994; Krakauer & Pagel, 1995). The difference is perhaps one of style and maturity. Theoretical biologists tend to be conservative and address specific issues. In contrast, many Artificial Life models tend to be more exploratory, addressing different issues at the same time and mixing (sometimes, but not always, in a less than clear manner) points regarding evolutionary stability, ecological and physical constraints, cognitive mechanisms, etc.

This wider exploratory attitude is simultaneously a source of excitement and inventiveness as well as a source of poor scientific methodology. With very few exceptions (e.g. Noble & Cliff, 1996) researchers have failed to turn a critical eye on their own work and that of others. This article is intended to be viewed as belonging to this rare class. However, instead of critically addressing other specific models it will concentrate on showing how an assumption has been repeatedly used as an explanation for what goes on in such models without any warrants for its applicability and, worse still, without any confirmation that the conditions were met for its use when such confirmation should be relatively easy to obtain.

It has become apparent that spatial situatedness can enhance the evolutionary viability of cooperative social behaviors under circumstances of conflict of interest (e.g. Krakauer & Pagel, 1995; Nakamaru, Matsuda & Iwasa, 1997; van Baalen, & Rand, 1998). A common situation is that, for certain interaction games, cooperation is unviable in the mixed medium approach but stable if players are spatially distributed and interact locally. There are different possible explanations for this phenomenon, but one of the most favored, at least intuitively, is the use of kin-selective arguments according to which most games are played between highly related individuals (due to the limited dispersal or 'viscosity' of population) and therefore cooperation is understood as the strategy that maximizes inclusive fitness (Hamilton 1964; see also section 3).

Unfortunately, in some cases this explanation has been merely verbal despite the potential for actually confirming its applicability (e.g. Ackley & Littmann 1994, Oliphant 1996 and others). This careless appeal to kin-selective arguments is dangerous. One is tempted to wrongly infer that spatial situatedness always implies the validity of such arguments[1] and, on the flip-side, one is tempted to ignore other possible factors that may play a significant role. It is also bad science since the means exist for verifying whether the chosen explanation is correct or not and there is no justification for not doing so.

This paper will show in some detail that a verification of the applicability of kin-selective arguments can be achieved fairly simply. The study will be based on an evolutionary model presented elsewhere (Di Paolo, 1997, 1999, submitted) but, due to limitations of space, will not go much into the details of that model. Finally, the paper will support the conclusion that cooperative outcomes in spatial games need not be tied to kin-selective arguments and that other ecological factors such as spatial organization, discreteness and stochasticity can be responsible for the selective stability of cooperative behaviors. The paper will, therefore, be less than exciting and perhaps a little on the boring, but necessary, side of scientific research.

2 The game

The model presented in (Di Paolo, 1997; submitted) describes a simple action-response (two-role) game with tuneable degree of conflict of interest, two out-

[1] Quite on the contrary, some models have shown that under certain circumstances the positive effect of higher relatedness between neighbors is actually cancelled by the negative effect of increased local competition (Taylor, 1992a; 1992b; Wilson, Pollock & Dugatkin, 1992). See also (Kelly, 1992; van Baalen & Rand, 1998).

comes (cooperative coordination and non-coordination) and four different strategies. The game is played by a population of individuals distributed in space who interact locally within a fixed neighborhood of their positions. Players subsist on the energy gained by playing the game and reproduce with a rate which is proportional to their capacity for accumulating extra energy. This capacity depends obviously on the strategy they play, on the local pool of strategies of the other players and also on ecological factors such as the density of local players and the density of local resources.

Each time the game is played the energy contained in a local food source is at stake. If the players cooperate they share this energy equally; if they do not, then the first player gets a proportion c which is greater than one half and the second player gets nothing. The parameter c is a measure of the degree of conflict between the first player and the second (low conflict for c near 0.5 and high conflict for c near 1). The outcome of the game depends on the strategies played which remain fixed during each individual's lifetime. Certain combinations of initiating actions (first player) and responses (second player) are cooperative and others are non-cooperative. Players do not choose directly whether to cooperate or not, they just 'choose' (in evolutionary terms) which actions and responses to play.

With a basic repertoire of two possible actions and two possible responses the number of different strategies is four. Of these, two may be called 'self-cooperative' in the sense that two players playing one such strategy will always cooperate independently of the roles they play. The other two are called 'non-self-cooperative' ('non-cooperative' for short). Both simple and extended game-theoretic analyses have been performed for the mixed-medium approach and all of them resulted either in oscillations in the composition of the population or in a combination of non-cooperative strategies as the evolutionarily stable outcome. In both cases the overall proportion of games resulting in coordination was equal to the baseline level that is expected to occur by chance (50%). Never was a single self-cooperative strategy a non-invadable solution.

Similar results were also observed using a partial differential equations approach in which players are spatially distributed and interactions occur within a finite local range (Di Paolo, 1999, submitted). However, in a spatially extended, *individual-based* model of the same game the outcome was different. In this model players are not treated as continuous densities of strategies but are rather implemented as discrete individuals and all interactions are subject to different kinds of noise. Strategies are encoded for each individual in a binary haploid genotype and an evolutionary algorithm is run [2]. It is observed, in contradiction with the game-theoretic results, that for values of c slightly higher than

[2] Other details: 2-D toroidal arena (100×100), square local neighborhood of size 10×10, variable population size, overlapping generations, sexual reproduction, local selection of reproductive mates and local allocation of offspring. Offspring's genotype is built by uniformly recombining the parental genotypes and adding point mutations (probability of no modification after a mutation event: 0.94), see (Di Paolo, 1999; submitted).

0.5 the population tends to be constituted by a single self-cooperative strategy and a small density of a competing non-cooperative one and, as c is increased, the global level of cooperation decreases in a more or less linear fashion until $c \cong 0.65$ where the baseline level is obtained (see Fig. 1).

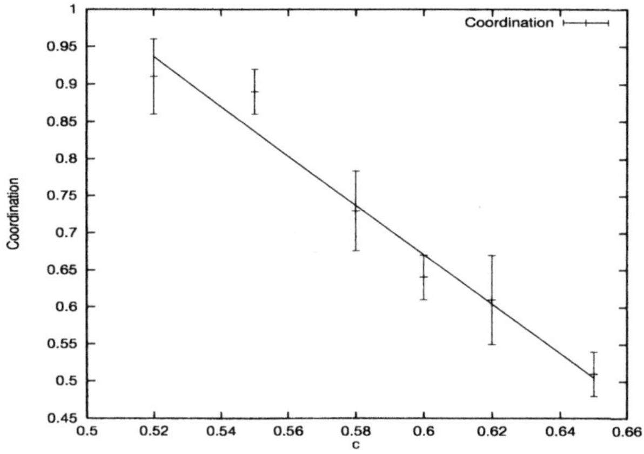

Fig. 1. Average value for the level of cooperative coordination for different values of c. Each point is the average of 5 simulation runs. For each run the level of coordination is averaged in time over the steady state. The line represents a linear regression, (correlation coefficient: -0.982, slope: -3.32, null hypothesis: no variation with c, $P < 0.005$). Error bars indicate standard deviation

In (Di Paolo, 1997), and later in more detail in (Di Paolo, 1999; submitted), an explanation is offered of this and other observations in terms of the effects introduced by the spatial organization of the population into discrete, separate and relatively stable clusters of players. The ecological forces that account for the stability of these clusters are also responsible for breaking many of the inbuilt symmetries between the roles in the game. Thus individuals at the center of a cluster play the second role more often than those at the periphery because of the better opportunities for being selected by other players as partners in the game. Moreover, other geometrical factors enter into play. For instance, the length of genealogies surviving without interruption tends to be longer if they start near the center of the cluster than if they start near the border so that all players tend to reflect the strategy played at the center.

This situation combines with the fact that players are discrete entities and therefore for invasion of a mutant strategy to occur a minimal threshold density must be achieved so that the average extra gain of the invader rises above fluctuations and the new strain is able to assert itself (see for instance de la Torre & Holland, 1990; Tsimring, Levine & Kessler, 1996; Abramson & Zanette, 1998; Goodnight, 1992). The asymmetries introduced by the ecological organization makes the achievement of threshold density easier for self-cooperative strategies if c is near 0.5 and the equilibrium density of non-cooperative strategies increases more or less linearly with c until their capability for invasion is regained. This is how the behavior shown in Fig. 1 is explained. Of course, it is possible to ask whether this complicated picture in which selective arguments are mixed with ecological dynamics is indeed necessary for explaining the results. Perhaps a simpler explanation in the form of a kin-selective argument would be enough and more economical. This is something that has to be tested.

3 Kin Selection

As mentioned above, it has often been found that cooperative interactions under circumstances of conflict can be stable if a spatial dimension is added to the evolutionary scenario even when simpler selective arguments predict the contrary. This has been demonstrated both by mathematical considerations and computer models in the case of the Prisoner's Dilemma (Axelrod, 1984; Nakamaru *et al.*, 1997) and communication games (Ackley & Littman, 1994; Krakauer & Pagel, 1995; Oliphant, 1996). Depending on the particular features of the model there may be more than one possible explanation for this phenomenon. For instance, Axelrod argues that spatial clustering favors reciprocity in the case of the TIT-FOR-TAT strategy (Axelrod, 1984, p. 65 - 69). A different explanation is preferred by Ackley & Littman (1994) and by Oliphant (1996) for their respective models. Such explanation involves the concept of kin selection (Hamilton, 1964). It is argued that, since the spatial regions where mating partners are chosen from and offspring allocated into tend to coincide with areas where game co-participants are selected from, this ensures that players will be highly related so that a cooperative player will tend to increase, on average, the frequency of genes identical to its own in other players, i.e. its inclusive fitness.

If fitness is related to the degree to which an organism can pass copies of its own genes to the next generation, inclusive fitness is a generalization of this concept so that it can also account for the presence of traits that help the transmission of copies of identical genes that happen to be located in other individuals. As the well known example goes, a behavior that is detrimental for an individual but which produces benefits of a same magnitude for two or more siblings will tend, under certain circumstances, to increase the inclusive fitness associated with that behavior and therefore it will be favored by selection. This is so because the probability of finding the same given gene in a sibling is, in haploid and diploid genetic systems, equal to 1/2. Thus, the general condition, known as Hamilton's rule, for a trait or behavior in individual i to be selected is not that it has an individual positive fitness effect W^i_{ind} but that it has a positive inclusive fitness effect W^i_{Inc}:

$$W^i_{Inc} = W^i_{ind} + \sum_j r_{ij} W^j_{ind} > 0 \tag{1}$$

where r_{ij} is the degree of relatedness between i and another individual j (defined carefully below) whose own individual fitness is affected by a quantity W^j_{ind} by the actions of i. The sum is extended over all the other individuals who may be affected by the trait or behavior in question. Even in cases where $W^i_{ind} < 0$, inclusive fitness can still be positive as long as the remaining terms in the sum are large enough.

Here it is clear why the restricted dispersal induced by space can play a relevant role. If interaction and reproduction are local processes, i.e. if co-participants as well as reproductive mates are selected from an individual's vicinity and offspring are allocated into the same vicinity, then the average value for the r_{ij}'s could be expected to be higher than in a mixed population. Therefore, cooperative behaviors at the expense of one individual but which benefit others in the local vicinity could at the same time tend to be beneficial to individuals that are highly related[3]; a situation which may result in $W^i_{Inc} > 0$.

Such is indeed the kind of argument advanced by Ackley & Littman (1994); Oliphant (1996) and others. Unfortunately, none of these authors actually *shows* that this is the case in their models. They just content themselves with finding a good candidate explanation without verifying if condition (1) is fulfilled, despite the fact that such a verification should be easy enough to perform in the computer models involved.

In the model discussed in the previous section, kin selection arguments are not straightforward in the sense that the actions and responses of the players cannot be said to be intrinsically cooperative or non-cooperative; they depend on the context of the strategies used by the rest of the players. However, under the assumption of weak selection pressure it is possible to postulate a situation of quasi-equilibrium in which the context is fixed and then actions and responses could be seen as cooperative in themselves. This is by no means the general situation in this model but this assumption will be maintained in order to see that even if this situation were true, kin selection would not be enough to explain the obtained results.

In order to test the plausibility of kin selection as a valid explanation for the evolution of cooperative coordination, a calculation is performed of the degree of relatedness between individuals and their average partners in the game. Genetic similarity can be caused by descent but also by other factors such as convergence, founder effects, etc. However, relatedness does not intend to measure just genetic similarity, otherwise all sorts of intra-specific conflicts and competition would be inexplicable. To take into account this subtlety relatedness is defined following (Grafen, 1991) as the degree of genetic similarity between two individuals over and above the average similarity within the population in which the individuals interact. In this way, if the difference in genetic constitution between two individuals is zero then their relatedness is equal to one, which means that from the point of view of gene frequencies for an individual to help the other is the same as to help itself. If the difference between the genetic constitution between the players is the same as the difference between one of them and the average genetic constitution in the population, then for that individual related-

[3] However, this need not be the case; see footnote 1.

ness is zero, since in cooperating with the other player, it is not contributing to an increase in the frequency of genes similar to its own.

An estimation of relatedness in the current model is straightforward by taking the perspective of the individual player and keeping track of the average relatedness with the partners it encounters throughout its lifetime. For each game that is played the Hamming distance d_{ij} between the binary genotypes of the participants is calculated as well as the distance between each genotype and the population average genetic constitution[4], d_i^{avg} and d_j^{avg}. The relatedness of individual i to individual j is:

$$r_{ij} = 1 - \frac{d_{ij}}{d_i^{avg}}, \tag{2}$$

Notice that while $d_{ij} = d_{ji}$, in general $r_{ij} \neq r_{ji}$ since for each individual relatedness is defined with respect to its *own* distance to the population average constitution[5]. This quantity is averaged for each individual during its lifetime in order to reflect its mean relatedness to its local partners.

In the present model, if the average genetic constitution is calculated using the *whole* population instead of just the population which is bound to be affected by the actions of a given individual, relatedness will be overestimated. It must be remembered that clusters have a fairly independent evolutionary history. The degree of *extra* similarity that relatedness intends to measure should be in terms of the mean population with which the player and its offspring have a chance to interact. Whereas the average constitution used for the first measure involves the whole population, in a more accurate estimation the average genetic constitution used for the calculation of relatedness is taken as the one 'seen' by each player during its lifetime within its economic neighborhood (Queller, 1994). For each individual, the genetic constitution of all its partners in the game is averaged. The first estimation or upper bound (averaging over the whole population) is kept only for the purpose of comparison since it is assumed that the measure of interest lies much closer to the second estimation.

Figure 2 shows the variations in the two estimations of relatedness for different values of c, each point obtained by averaging the temporal mean after transients of 5 simulation runs with identical parameters. It is observed that

---

[4] The average genetic constitution is simply a string calculated as the sum of all the genotypes in the chosen sub-population and divided by the number of players. Since there are only two possible alleles per locus which are represented by 0's and 1's, the result of this calculation is that every component in the average string gives an indication of the proportion of each allele in that particular locus. In order to calculate the distance from an individual genotype to this real-valued string, a measure is used which reduces to the Hamming distance when applied between two binary strings. This measure involves summing, for each component of the string, the absolute distance between the individual allele and the corresponding population average. For instance, the distance between the string (1,0,1,0) and the string (0.7,0.3,0.9,0.9) is $d = 0.3 + 0.3 + 0.1 + 0.9 = 1.6$.

[5] This estimation coincides with the one proposed in (Queller & Goodnight, 1989) using each allele as a neutral genetic marker.

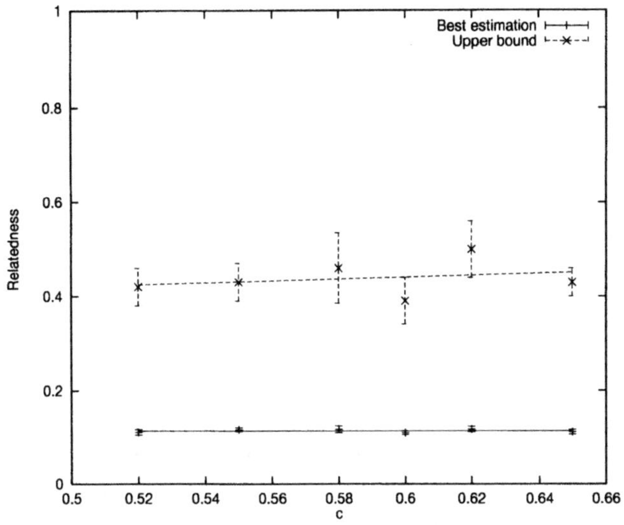

Fig. 2. Average estimations of relatedness vs. c. Lines depict linear regressions. Averages over 5 runs and over the steady state. Error bars indicate standard deviation

neither of the measures varies much with c. This behavior can be understood considering that relatedness is given mainly by the spatial relations between players of different generations which should not be affected too much by c[6].

By calculating the corresponding inclusive fitness per game per unit of energy it is found that for a first player (i) the payoff for no cooperation is $W_{Inc}^{NC,i} = c$, and the payoff for cooperation is $W_{Inc}^{C,i} = (1 + r_{ij})/2$. Using condition (1) and taking the best estimation of relatedness ($r \cong 0.11$) it can be concluded that cooperation should be the favored outcome only if $c < 0.55$, which is not enough to explain the results.

However, if the estimation of relatedness were as high as the upper bound ($r \cong 0.45$) then cooperation would be stable for $c < 0.72$. So it *seems* that if a generous estimation of relatedness could be justified, then kin selection would suffice for explaining the evolution of cooperative coordination in the present model. However, the expected result, if such were the case, would be a *constant* high level of cooperation for any value of c between 0.5 and 0.72, and a step-wise change to no cooperation for $c > 0.72$ which is *not* what is observed. As seen in Fig. 1, the level of cooperative coordination is a linearly *decreasing* function of c, even in ranges where relatedness does not vary. The condition for the stability of cooperation is given by establishing the sign of the inequality between the inclusive fitness for cooperation and for no cooperation. There are two parameters in this condition: relatedness and c. A linear variation of the degree of cooperation with c could only be explained by some spatial variation in relatedness resulting

---

6 The high value of relatedness for the upper bound reflects the fact that intra-cluster similarity is much greater than inter-cluster similarity.

in a linear decrease of the proportion of players for whom cooperation implies the best increase in inclusive fitness. Not only is this variation not observed, but it would also have to be manifested as a decrease in the global average of relatedness. This observation suggests that, whatever the mechanism responsible for the stabilization of cooperation, it must not work equally well for all values of c below 0.72 as kin selection would if relatedness were to be estimated by its upper bound.

Consequently, even after many concessions, the conditions for explaining the results in terms of kin selection are not met.

4 Conclusions

While kin selection remains an important concept for studying the evolution of social behavior, its use is far from being universal or automatic. The theory provides very simple rules for verifying its conditions of applicability and those rules should be used. This paper demonstrates how to do this for an individual-based model instantiated in a computer simulation. Additionally, it shows that the conditions for kin selection fail to be met in this model thus strengthening an alternative explanation in terms of ecological and selective factors.

Independently of this last result, the main point of the paper should be taken as a methodological one. The behavior of computer simulations can be quite hard to understand in itself. There is no reason why anything should be obvious in view of the complexity of contemporary models. Fortunately, the strength of simulations is that any hypothesis regarding what is going on in them can be put to the test with relative ease. To fail to do this is to fail to take advantage of one of the crucial aspects of simulations. In science, a hypothesis or an explanation should be tested against all the available evidence and there is no reason to expect a lesser standard from Artificial Life.

This work was done before finishing my PhD at the School of Cognitive and Computing Sciences, University of Sussex. Many thanks to Phil Husbands, Inman Harvey and John Stewart for their comments. Thanks also to Seth Bullock, Jason Noble and Matt Quinn for lively discussions.

References

Abramson, G., & Zanette, D. H.: Statistics of extinction and survival in Lotka-Volterra systems. *Phys. Rev. E* **57** (1998) 4572

Ackley, D. H., & Littman, M. L.: Altruism in the evolution of communication. In Brooks, R. A., & Maes, P. (Eds.), *Artificial Life IV*, MIT Press, Cambridge, Mass. (1994)

Axelrod, R.: *The evolution of cooperation.* Basic Books, New York (1984)

Bullock, S.: An exploration of signalling behaviour by both analytic and simulation means for both discrete and continuous models. In Husbands, P., & Harvey, I. (Eds.), *Proceedings of the Fourth European Conference on Artificial Life*, MIT Press/Bradford Books, Cambridge, Mass. (1997) 454–463 de la Torre, J. C., & Holland, J. J.: RNA virus quasispecies populations can suppress vastly superior mutant progeny. *J. Virol.* **64** (1990) 6278–6281

Di Paolo, E. A. An investigation into the evolution of communication. *Adaptive Behavior* **6:2** (1997) 285–324

Di Paolo, E. A.: *On the evolutionary and behavioral dynamics of social coordination: Models and theoretical aspects.* Ph.D. thesis, School of Cognitive and Computing Sciences, University of Sussex (1999)

Di Paolo, E. A.: Ecological symmetry breaking can stabilize the evolution of altruism in an action-response game. (Submitted)

Durrett, R., & Levin, S.: The importance of being discrete (and spatial). *Theor. Pop. Biol.* **46** (1994) 363–394

Goodnight, K. F.: The effect of stochastic variation on kin selection in a budding-viscous population. *Am. Nat.* **140** (1992) 1028–1040

Grafen, A.: Modelling in behavioural ecology. In Krebs, J. R., & Davies, N. B. (Eds.), *Behavioural ecology: An evolutionary approach*, (Third edition) Blackwell Scientific Publications, Oxford (1991) 5–31

Hamilton, W. D.: The genetical evolution of social behaviour, I and II. *J. theor. Biol.* **7** (1964) 1–16; 17–32

Kelly, J. K.: Kin selection in density regulated populations. *J. theor. Biol.* **157** (1992) 447–461

Krakauer, D. C., & Pagel, M.: Spatial structure and the evolution of honest cost-free signalling. *Proc. R. Soc. Lond. B* **260** (1995) 365–372

Nakamaru, M., Matsuda, H., & Iwasa, Y.: The evolution of cooperation in a lattice-structured population. *J. theor. Biol.* **184** (1997) 65–81

Noble, J.: Evolved signals: Expensive hype vs. conspiratorial whispers. In Adami, C., Belew, R. K., Kitano, H., & Taylor, C. E.(Eds.), *Proceedings of the Sixth International Conference on Artificial Life (ALife VI)* MIT Press/Bradford Books, Cambridge, Mass. (1998)

Noble, J., & Cliff, D.: On simulating the evolution of communication. In Maes, P., Mataric, M., Meyer, J.-A., Pollack, J., & Wilson, S.(Eds.), *From Animals to Animats 4: Proceedings of the Fourth International Conference on Simulation of Adaptive Behavior* MIT Press/Bradford Books, Cambridge, Mass. (1996)

Oliphant, M.: The dilemma of Saussurean communication. *BioSystems* **37** (1996) 31–38

Queller, D. C.: Genetic relatedness in viscous populations. *Evol. Ecol.* **8** (1994) 70–73

Queller, D. C., & Goodnight, K. F.: Estimation of genetic relatedness using genetic markers. *Evolution* **43** (1989) 258–275

Taylor, P. D.: Altruism in viscous populations - an inclusive fitness approach. *Evol. Ecol.* **6** (1992a) 352–356

Taylor, P. D.: Inclusive fitness in a homogeneous environment. *Proc. R. Soc. Lond. B* **249** (1992b) 299–302

Tsimring, L., Levine, H., & Kessler, D. A.: RNA virus evolution via a fitness-space model. *Phys. Rev. Lett* **76** (1996) 4440–4444

van Baalen, M., & Rand, D. A.: The unit of selection in viscous populations and the evolution of altruism. *J. theor. Biol.* **193** (1998) 631–648

Wilson, D., Pollock, G. B., & Dugatkin, L.: Can altruism evolve in purely viscous populations? *Evol. Ecol.* **6** (1992) 331–341

An Evolutionary Simulation Model of Social Learning about Food by Norway Rats

Jason Noble, Elio Tuci, and Peter M. Todd

Center for Adaptive Behavior and Cognition
Max Planck Institute for Human Development
Lentzeallee 94, D-14195 Berlin, Germany
noble, tuci, ptodd@mpib-berlin.mpg.de

Abstract. We investigate simple mechanisms for social learning in an evolutionary simulation of food-preference copying in Norway rats. These animals learn preferences by interacting with conspecifics, but, unexpectedly, they fail to learn aversions after interacting with a poisoned demonstrator. They also follow each other to food sites. Simulation results show that failure to discriminate between sick and healthy demonstrators may be due to details of food toxicity in foraging environments. A seemingly complex instance of social information transmission is explained through the action of simple behaviours in an appropriately structured environment.

1 Introduction

Animal behaviour can be seen as the problem of what to do next. Natural selection has shaped the behavioural strategies of the animals we see today, but, clearly, it has arrived at different solutions in different species. For an animal facing a particular environmental challenge, three broad sources of strategy can be distinguished: instinct, learning, and social learning. For example, suppose that a foraging animal has to decide whether or not to eat a piece of toxic, unripe fruit that it has found. The decision might be made instinctively: the animal has an inherited tendency to avoid fruit of that colour and texture. Alternatively, the animal might have learned through bitter experience that such fruit is unpalatable. Finally, it could have learned socially: perhaps it has observed conspecifics rejecting this kind of fruit, or has seen conspecifics become ill after eating it. In this paper we focus on the third strategy source, social learning, to explore the way in which simple specific mechanisms of social information gathering can interact with structured environments to yield unexpected behavioural implications.

In recent years there has been some progress towards understanding the adaptive function of social learning. Models of cultural [1] and "highly horizontal" transmission [2] help to delineate the conditions under which it will be advantageous for individuals to learn from others rather than finding things out for themselves. But we are also interested in questions of mechanism: *how* exactly does an animal gain information from the behaviour of its conspecifics? A variety of candidate mechanisms have been proposed [3], and complex possibilities such as intentional imitation have received a great deal of attention. Nevertheless, seemingly complex behaviour may be produced by simpler mechanisms. For example, Galef [3] discusses "stimulus enhancement" in which a tendency on the part of naive animals to approach conspecifics leads to their being more likely to encounter one set of stimuli rather than another, and thus shapes their (individual) learning experience. The power of such simpler evolutionary solutions has probably been underestimated. Several phenomena that were once seen as involving imitation, such as the opening of milk bottles by birds and the washing of food by monkeys, have since been questioned [3], and parsimonious explanations have been offered in terms of processes like stimulus enhancement. However, it is difficult to design experiments that conclusively expose the mechanism at work in particular cases.

We believe that the individual-based simulations characteristic of artificial life can be useful tools in the investigation of social learning, much as they have been for studying the evolution of individual learning. It has long been recognized within artificial life that complex global phenomena can arise from simple local rules, and this is precisely what some researchers suspect is happening in animal social learning: individuals follow a simple rule (e.g., "stay close to your mother") and, in combination with associative learning, the overall pattern of behaviour that arises makes human observers suspect imitation. Although work in artificial life has certainly considered social dynamics in contexts such as foraging, communication, and flocking or schooling, there has been relatively little work on the specific topic of social learning. The model most relevant to our own work is by Toquenaga et al. [4], who constructed a simulation of foraging behaviour in egrets. The authors demonstrate that stimulus enhancement is important in the evolution of flock foraging and colonial roosting, and is more likely to evolve when resources are patchy rather than evenly distributed.

We look at social learning in Norway rats (***Rattus norvegicus***) – an opportunistic, central place foraging species – to see how their specific and rather surprising social learning mechanisms may have evolved in response to environmental features. These rats employ at least two simple mechanisms that allow them to share information about food [5, 6]. Firstly, they have a robust tendency to copy the feeding preferences of their conspecifics. A rat will develop a marked preference for a novel food that it smells on the breath of another, and the effect is strong enough to make a rat choose the novel food type over its normal diet, despite the fact that rats usually avoid new foods. The key stimulus is the detection of the novel food odor in combination with carbon disulfide, a component of rat breath. Rats will not, for instance, develop a preference for foods that an experimenter has wiped onto the fur of another rat. Secondly, rats will spontaneously follow conspecifics on foraging trips out of the nest; this habit is especially pronounced in younger animals. Such behaviour clearly suggests that stimulus enhancement may be occurring. One of the ways a rat could come to exploit a new food source would be simply by following another and learning from the experience.

Galef et al. [7] uncovered an apparent paradox in rats' social learning. They assumed that if rats could acquire food preferences through social interaction, they would probably also be able to learn an aversion to a novel, toxic food by smelling it on the breath of a demonstrator rat and simultaneously noting that the demonstrator was ill. Experiments revealed, however, that this is not the case. Rats are not sensitive to a demonstrator's state of health, and in fact only ever develop a *preference* for the novel food.

This surprising finding was the starting point for our own investigations. The value of distributed intelligence through copying others' food choices seems clear in an opportunistic forager that must deal with new and potentially toxic foods, especially when seen alongside the fact that rats will normally avoid novel foods. But why don't they discriminate between sick and healthy demonstrators? It is not because they can't – rats perform well on a wide variety of discrimination tasks, and are capable of identifying sick conspecifics using odor and behavioural cues [8]. Curiously, other species such as blackbirds [9] and chickens [10] *do* manage to learn both preferences and aversions through observation.

We suspected that the answer might depend on characteristics of the rats' foraging environment; specifically on the probability that eating a toxic food would result in the death of the animal. To test this suspicion we constructed an evolutionary simulation, within which we systematically varied the lethality of toxic foods in the environment, and observed the effect on the evolution of a gene for discriminating between sick and healthy demonstrators. We were also interested in following behaviour, and extended the initial model to include this possibility. In particular, we wanted to demonstrate that these two simple mechanisms – copying and following – could together account for apparently complex social learning. We were also interested in determining whether there was any interaction between the two: are the benefits of copying food preferences and those of following others independent, or do they interfere with each other in some way?

2 The Evolution of Preference Copying

2.1 Modelling Learning Rats

In our simulation an initial population of 100 rats foraged from five centrally located nests. Rats foraged by night, with each night divided into five foraging periods. During each period, a rat could visit one of 25 foraging sites: if it found food at that site, it had to decide whether or not to eat it. If a rat chose to eat, and if there was sufficient food, it would fill its stomach and return to the nest – the night's foraging was over. However, rats that rejected the food they had found, and rats that ate only a partial meal due to competition, would continue to forage until they had eaten their fill or until the night was over. If a rat consumed nutritious food it gained energy, but 10% of the food types in the environment were toxic. If a rat consumed a toxic food there was a parameter governing the probability that it would die at once; otherwise it would lose some energy and would show signs of poisoning when it returned to the nest. Rats that consumed no food at all would eventually die of starvation.

The rats were given a simple memory: any food they encountered would either be novel, familiar or aversive. For newborns, all foods were novel. The first variable under evolutionary control was the probability E that a rat would eat a novel food. After eating a new food, the rat would remember it as either familiar or aversive depending on whether or not it was toxic. Rats would always eat familiar foods, and always reject aversive foods. Rats were also assumed to be capable of remembering where they had foraged last night, and a binary gene P controlled whether or not they would persist and return to their last successful feeding site, if any, during the first period of the next night.

Each individual had two more genes controlling its preference-copying behaviour: a binary gene S indicating whether the rat would learn about new foods by smelling the breath of other rats, and a binary gene D that controlled whether the artificial rat (unlike real rats) would discriminate with regard to the state of health of another rat whose breath it smelled, avoiding foods eaten by sick rats and preferring foods on the breath of healthy rats. Upon returning to the nest each morning, rats could potentially smell the breath of a randomly chosen nestmate as a way of gaining information about new foods in the environment. If a rat had only the smell-based learning ability, S, it would simply smell the breath of a nestmate and become familiar with the type of food the nestmate had eaten that day, if any. But the rat might thereby develop familiarity with and hence preference for a food that was in fact toxic: some rats died immediately after eating a poisonous food, but others made it back to the nest and were ill. Only if the learning rat also had the ability to discriminate, D, would it develop preference or aversion for the new food type depending on whether the nestmate was showing signs of poisoning. (A low level of error could also be associated with this discrimination ability.)

When a rat had accumulated a certain amount of energy, it would undergo simplified asexual reproduction. The carrying capacity of the environment was fixed at 100. If a newborn individual could not take the place of a rat that had recently died of poisoning or starvation, the current oldest rat would be selected for death by old age in order to make room. Newborn rats inherited (with a small chance of mutation) the four-element behavioural strategy of their parent described above, plus a level of starting energy.

Parcels of food appeared in random sites in the environment at a constant rate. In order to ensure that the rats always had to deal with novelty, a total of 100 food types were used, but with a "window" of 10 food types that could appear at any one time. Every 16 days the window would advance, so that one old food type stopped appearing, and a novel one entered the scene. The lifespan of an individual rat (which was an emergent property of the simulation) never grew long enough for it to experience all 100 food types.

Whilst we have tried to make this model reflect part of the lives of real rats, the parameters used in the simulation[1] are not as closely matched to real data as we would like. There is a great deal of information available on the behaviour of rats in the laboratory, but data on the ecology of wild rats is not extensive [5]. We have therefore pitched the simulation at a relatively abstract level, and in so doing we hope to have captured some aspects of the selection pressures impinging on social foragers in general.

2.2 Results

In accordance with our hypothesis, we varied the probability of death due to eating a toxic food. For each level of lethality investigated, ten evolutionary runs, each 200,000 days in length, were performed. The statistics reported below describe the state of populations at the end of these runs.

When the lethality value was zero, i.e., in a benign environment, there was no selection pressure on the rats to smell each other or to discriminate: Fig. 1(a) shows that gene frequencies for S and D remained close to 50%, the value expected by chance. Nor were the rats particularly persistent, as we will see later in Fig. 3(b). At the same time, the mean probability for eating novel foods (E) was high at 89.8% (Fig. 1(b)). The mean lifetime of a rat was 268 days, during which time approximately 17 new foods would have made an appearance. The rats were, on average, familiar with 13.3 food types, and had aversions to 2.1 foods: these frequencies are not too far from the 10% base rate of toxic foods in the environment. Thus, when poison results only in a stomach ache and an aversion to trying that food again in future, simulated rats are open to trying new foods, and pay no special attention to the eating habits of others.

It is clear from Fig. 1(a) that as the lethality level increases there is increasing selection pressure for learning from others. Rats become more likely to smell the breath of conspecifics, and to discriminate between the sick and the healthy when doing so. At the same time, they are still willing to try new foods that they come across – Fig. 1(b). However, when the lethality level reaches approximately 0.3, there is something like a phase transition in the results. The probability of eating novel foods drops dramatically; the overall mean probability for eating a novel food, across all runs with lethality greater than 0.3, was less than 1%. As lethality approaches 1.0, there is a uniform strong selection pressure on the gene for smelling the breath of others. Importantly, though, the selection pressure on discriminating decreases: as poison became more and more dangerous, it was no longer important to pay attention to the health of a demonstrator – we will argue in section 4 that this result explains the failure of real rats to discriminate. Aggregating the results for lethality levels greater than 0.3, the rats were familiar

[1] Rat stomach capacity = 10 food units; cost of living = 1 unit per day; default error level in discrimination = 0.01; mutation rate = 0.05, standard deviation for mutation of real-numbered genes = 0.1; food parcel mean size = 100 units, standard deviation = 40, food parcel input rate = 5 per day; reproduction level = 1000 units, cost of offspring = 500 units, offspring starting energy = 400 units.

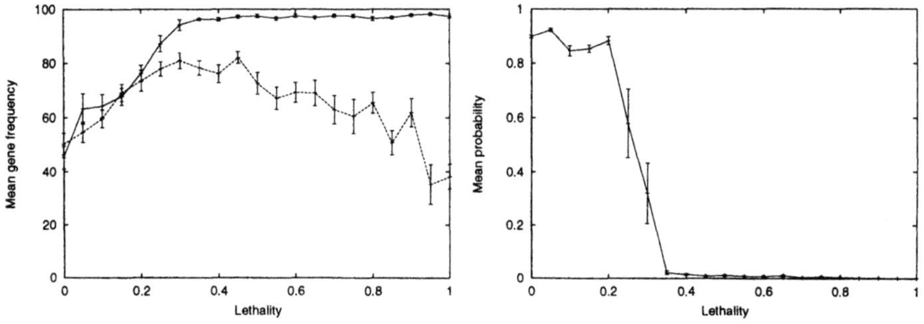

(a) (b)

Fig. 1. (a) Mean frequency of the smelling and discrimination genes S and D by lethality level – solid and dashed lines respectively. In this and subsequent figures, error bars show the standard error across ten simulation runs. (b) Mean value of the real-valued gene E, governing the probability of eating novel food, by lethality level.

with about the same number of foods (10.0), but they developed almost no aversions (0.2). There were other effects: because the rats were more cautious in their treatment of novel foods, they generally ate less, which led to increases in the inter-birth interval and the average lifespan.

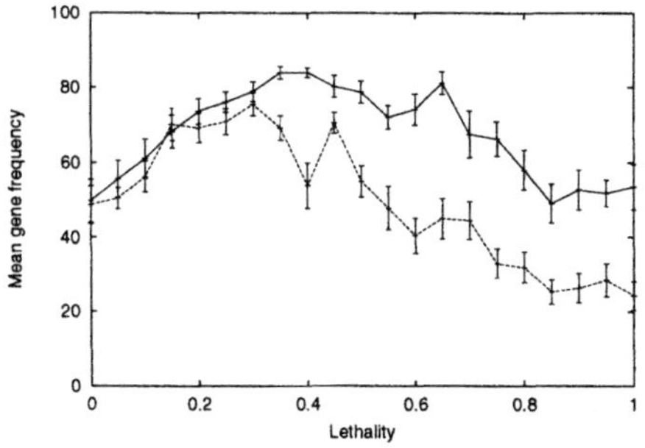

Fig. 2. Mean frequency of the discrimination gene D given zero (solid line) and 5% (dashed line) error rates in discriminative ability.

The finding that selection for discrimination falls off with increasing lethality could have something to do with the error rate of the rats' discriminative ability, normally set to 1%. We therefore considered the evolution of the gene D under zero-error and 5% error conditions. Fig. 2 shows that a higher error rate indeed leads to selection *against* discriminating. But even with zero error, the frequency of the gene D is close to the expected value for random drift (i.e., 50%), given high levels of lethality. Even if one's discrimination ability were perfect, it appears that it would not be a selective advantage in an environment where eating a poisonous food would kill you four times out of five.

3 Adding Following to the Model

3.1 Changes in the Model

In order to make following behaviour possible, the rats were given a fifth gene, F, that determined the probability that they would follow a random conspecific when leaving the nest for the first foraging period. Following was only ever selected if the rat was not being persistent. In other words, the decision tree for the rats was as follows.

1. Check persistence gene, P; if set, was I successful last night? If so, return.
2. Check percentage chance to follow, F; should I follow someone else?
3. If neither of the above selected, choose a random site.

Rats that followed simply went to the same site as a randomly chosen conspecific who was not also following, but there was no way for a rat to tell whether it was following a persistent conspecific or one that had merely selected a site at random. (If *all* the rats chose to follow, they were all sent to random sites.) The extended model was in other respects the same as the basic model.

3.2 Results

The evolution of the genes for smelling, discrimination and eating novel foods were not much affected by the possibility that rats could follow each other to food sites; the results were qualitatively similar to those shown in Fig. 1. The only difference was that the transition between liberal and conservative attitudes to new foods occurred at a lower level of lethality, approximately 0.15.

The results for the evolution of the following gene are shown in Fig. 3(a). We were somewhat surprised to find that following other rats was not strongly selected for (solid line). There is a modest positive relationship between the mean value of F and lethality, but for the more benign environments the data could just as well be the result of a random walk. To find circumstances that could select for following, we ran a variant of the simulation in which food was less uniformly distributed. Following others to food is a kind of stimulus enhancement, which is more likely given patchy food distributions [4]. Instead of five food parcels with a mean size of 100 units arriving per day, one food parcel with a mean size of 500 units was supplied; Fig. 3(a) shows that under these conditions following behaviour was indeed more strongly selected for (dashed line).

Another way of looking at following behaviour is to consider its effects on the gene for persistence. Fig. 3(b) shows the mean frequencies for the gene P for the initial model and for the model with following added. In the initial case (solid line), we can see that persistence becomes more important with increasing levels of lethality. In a dangerous world where you don't want to try anything new, it pays to go back to the site of yesterday's successful feeding. However, when following behaviour is possible (dashed line), increasing levels of lethality no longer produce selection pressure in favour of persistence.

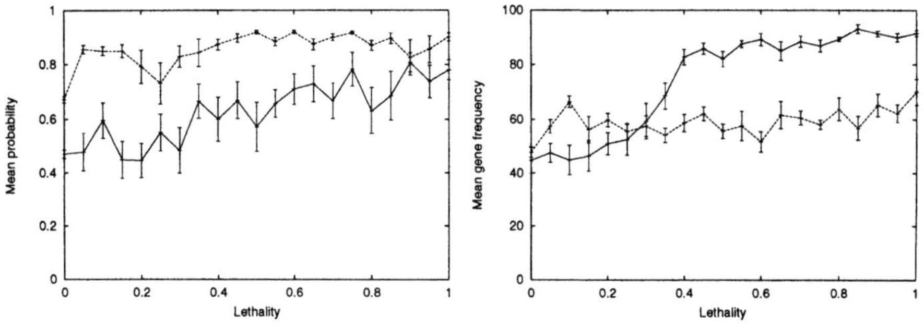

(a) (b)

Fig. 3. (a) Mean value of the real-valued gene F, governing the probability of following other rats to a foraging site. The solid line shows the standard condition, while the dashed line shows the effect of a less uniform food distribution (see text). (b) Comparison of the frequencies for the persistence gene P with (dashed line) and without (solid line) the possibility of following behaviour.

4 Implications for Real Rats

The failure of real Norway rats to discriminate between sick and healthy demonstrators may be due to the details of food toxicity in their foraging environments. In the initial model, when the lethality value crosses the threshold of about 0.3 the whole pattern of behaviour changes. The simulated rats become extremely wary of new foods, but attend closely to what their conspecifics are eating. However, because they "know" that their nestmates are just as conservative as they are about trying new and potentially dangerous foods, the need to discriminate between sick and healthy demonstrators is reduced. Observing a poisoned demonstrator would doubtless provide useful information, but it becomes such a rare event that there is little selection pressure for paying attention to it.

The possibility of error in discrimination just makes things worse. In the terminology of signal detection theory, a hit would be correctly identifying a food as poisonous after observing a sick demonstrator who has eaten that food. A miss would be failing to identify a food as poisonous under these circumstances. Misses would certainly be costly, but false alarms may at times be even more so: given that one has a low likelihood of eating new foods, and that there is a limited number of foods present in the environment, believing falsely that a palatable food is poisonous could deprive a rat of a much-needed food source. As the error rate for discrimination increases, there must come a point when it is better to simply accept all foods detected on the breath of others, thus risking occasional poisoning but ensuring that no palatable food source ever goes unexploited. It is not clear what the error rates for discrimination would be in real *Rattus norvegicus*, but if we recognize that animals can be ill for reasons other than food poisoning, and that they may conceal illness when it exists, the levels of 1% and 5% seem conservative.

Furthermore, discrimination in the real world would not come for free. The animal would have to pay the time and energy costs of developing a sensory system and decision mechanism that allowed it to detect sickness in others and then act accordingly; the blind acceptance of any food odor that is smelled in conjunction with carbon disulfide is clearly a simpler way to do things. In the simulation we did not attempt to model this fact, i.e., we did not include any direct costs on the ability to discriminate. The existence of such direct costs for real rats would make discrimination even less likely to evolve.

There is a clear empirical prediction arising from the initial model. Firstly, if we consider foraging populations with very low levels of danger associated with eating toxic food, animals in these populations should be content to eat novel food and ignore the experiences of others. Next, as lethality increases, individuals will become more likely to pay attention to the eating habits of others, and to discriminate as to the state of health of demonstrators. Finally, above a certain threshold of lethality we expect to find animals with a very low likelihood of eating novel foods, a great interest in what others are eating, and no strong tendency to discriminate between sick and healthy demonstrators. Norway rats appear to fit this last profile. Blackbirds [9] and chickens [10] may fit the second.

The results for the extended simulation with the possibility of following have less clear implications. In increasingly lethal environments, following others to food sources is moderately favoured. When the distribution of food is patchy, there is reasonably strong selection pressure for following behaviour; this is in accordance with an earlier finding that patchy food distributions will promote stimulus enhancement behaviours [4]. But the interaction between following and persistence shown in Fig. 3(b) demands further analysis. The problem is that mean values for the probability of following in the non-patchy environment are close to 50%; this is exactly what would be expected if there was no selection pressure, and suggests that following is not particularly adaptive. Nevertheless, the observed means imply that about half the time rats will be following a conspecific rather than choosing a random food site, assuming they are not being persistent and returning to the scene of yesterday's success. Comparing the initial model to the model with following added, we find that the gene for persistence is selected for in the former case, but not in the latter. The apparently chance levels of following behaviour are affecting the evolution of persistence; one could even argue that following behaviour *takes the place* of persistence. It is therefore not so clear that following behaviour is adaptively neutral.

Presumably there is a trade-off between following others to food but having to share it when you get there, and thus taking a chance that you will get less than a full meal, versus choosing a food site randomly and having it to yourself. (However, there were always more rats than food sites; on average a rat would be sharing with four others in any case.) This may also have something to do with the results for persistence: if some fraction of the population is going to follow you to your current favourite food site, the benefits of persistence may be swamped by the costs of attracting a large and hungry crowd. Clarification of these issues must wait for future work.

5 Conclusions

We see foraging behaviour in rats as a paradigm case in which a seemingly complex instance of social (or even cultural) information transmission can be explained through the action of simple behaviours in an appropriately structured environment. In fact, the behaviour underlying rat food preference copying is even simpler than was originally expected by those looking for simple mechanisms. That is, rats pay no attention to whether the individual they are learning a preference from is suffering from food poisoning. Our model has added to this picture by showing that this strategy may well have evolved just because eating poison lessens the likelihood that a rat will survive to influence any of its conspecifics. Similarly, when following behaviour is possible and occurs at arbitrary levels in the population, rats do not even need to remember the site where they found food the night before.

Acknowledgements

Thanks to Seth Bullock and Jennifer Davis for discussions. E.T. was supported by studentship 204 4801 from the Consiglio Nazionale delle Ricerche (Italy).

References

1. Boyd, R., Richerson, P.J.: Culture and the Evolutionary Process. Univ. Chicago Press (1985)
2. Laland, K.N., Richerson, P.J., Boyd, R.: Developing a theory of animal social learning. [11] 129–154
3. Galef, Jr., B.G.: Imitation in animals: History, definition, and interpretation of data from the psychological laboratory. In Zentall, T.R., Galef, Jr., B.G., eds.: Social Learning: Psychological and Biological Perspectives. Lawrence Erlbaum Assoc. (1988) 3–28
4. Toquenaga, Y., Kajitani, I., Hoshino, T.: Egrets of a feather flock together. Artificial Life **1** (1994) 391–411
5. Lore, R., Flannelly, K.: Rat societies. Sci. Am. **236** (1977) 106–116
6. Galef, Jr., B.G.: Social enhancement of food preferences in Norway rats: A brief review. [11] 49–64
7. Galef, Jr., B.G., Wigmore, S.W., Kennett, D.J.: A failure to find socially mediated taste aversion learning in Norway rats (*R. norvegicus*). J. Comparative Psych. **97** (1983) 358–363
8. Lavin, M.J., Freise, B., Coombes, S.: Transferred flavor aversions in adult rats. Behavioral and Neural Biology **28** (1980) 25–33
9. Mason, J.R., Arzt, A.H., Reidinger, R.F.: Comparative assessment of food preferences and aversions acquired by blackbirds via observational learning. Auk **101** (1984) 796–803
10. Johnston, A.N.B., Burne, T.H.J., Rose, S.P.R.: Observation learning in day-old chicks using a one-trial passive avoidance learning paradigm. Anim. Behav. **56** (1998) 1347–1353
11. Heyes, C.M., Galef, Jr., B.G., eds.: Social Learning in Animals: The Roots of Culture. Academic Press, San Diego, CA (1996)

Effects of Cohesiveness on Inter-sexual Dominance Relationships and Spatial Structure among Group-Living Virtual Entities

Charlotte K. Hemelrijk

AI Lab, Department of Computer Science,
University of Zürich, Winterthurerstr 190, CH-8057
Zürich, Switzerland, Fax 0041-1-363 00 35, Email:
hemelrij@ifi.unizh.ch.

Abstract. Since male primates are bigger and stronger than females, they are by default considered dominant. When in a cohesively grouping ape (but not in its loosely grouping relative), females often appear dominant to males, the static image of female weakness is maintained and female dominance is attributed to high, species-specific co-operation among several females against single males.

In this paper, an individual-oriented model is used to produce a parsimonious alternative: female dominance over males may directly vary with group-cohesiveness without species-specific differences in co-operative tendencies among females. The model consists of a homogeneous world in which entities roam. Entities are so constructed as to have merely a tendency to group and perform dominance interactions. 'Male' entities (StrongTypes) are characterised by a higher initial dominance value and intensity of attack than 'female' entities (called WeakTypes). Dominance values change and evolve due to the self-reinforcing effects of winning and losing contests. In the model, more rank-overlap between both types arises from a stronger feedback between dominance and spatial structure in cohesive than in looser groupings. Biological implications of these phenomena and testable hypotheses for real animals are discussed.

1 Introduction

High dominance rank is supposed to be associated with many benefits and accordingly it is thought to be of central importance in primate social behavior (1). Although many anthropoid primate species live in permanently bi-sexual groups, hardly any data on male-female dominance relations exist (2). Instead, dominance relations are usually studied within each sex separately, and females, being often smaller and of inferior fighting capacity, are considered to be subordinate to males by default. Although case histories of females being dominant to all or some males in some groups of certain species are known (3), they are usually disregarded as abnormal. For those few species in which it is consistently reported that adolescent or adult males have difficulties in dominating females, an additional assumption is made that females of these species have a stronger than usual tendency to form large coalitions against single males (4,5). In this way, the generally accepted image of weak individual females is maintained because female dominance is attributed to collective force of larger numbers.

Although it is common to attribute social characteristics to internal qualities, and the observation of different social characteristics at a group level, automatically leads to the search for a different individual behaviour, an alternative, more integrative approach is possible which we will follow here. In this approach, social characteristics are studied and explained within the context of typicalities of the species. For instance, Deneubourg and co-authors (6) described a very different swarming pattern for each of three ant species. Instead of attributing this to internal differences between the species, they showed, by means of a simulation, that under the different feeding conditions that are characteristic of the three species, one and the same set of rules leads to the three characteristic swarming patterns. Thus, the swarming patterns result from the interactions between the entities and their environment.

The same, contextual approach will be followed in this paper as regards the difference in social behaviour in two related chimpanzee species, the common and pygmy chimpanzee. Female dominance over males manifestly occurs in the pygmy chimpanzee, male dominance is current in the common chimpanzee (7). Following the traditional scenario, female dominance in the pygmys is attributed to the formation of coalitions against males but such coalitions are described also for common chimpanzees, for instance by de Waal (8). Since both species differ significantly in cohesiveness of grouping (the pygmy chimpanzees group more cohesively (7)), we will focus here on explaining the described contrast in female dominance as a side-effect of the difference in cohesiveness per sé, without assuming a higher tendency among females to form coalitions against males.

In our approach we discard the fixed image of dominance, because cherishing it implies the assumption that there is a strong heritable component in dominance rank. This view is supported by some (9), but it is contradicted by the failure to replicate former dominance relations in experiments in which individuals are reshuffled between groups (10).

Instead of being strongly inherited, others, such as Rowell (11) and Chase (12) consider dominance to be due to chance and the self-reinforcing effects of winning and losing. Whereas the winner of the first encounter is decided purely by chance, after winning or losing once, the effects of winning and losing are self-reinforcing. For instance, Chase has shown by means of experiments that after losing, a monkey is more likely to lose again even when it encounters a much smaller opponent whom it would have defeated easily under other circumstances. This points to a strong psycho-physiological, experience-based component in dominance rank.

Furthermore, how dominance may be subject to self-organisation, because characteristics of dominance hierarchy and spatial structure (specifically the central location of dominants) are mutually reinforcing has been shown by means of an individual-oriented model (13,14).

Taking the last-mentioned approach, we are now going to study the effects of cohesiveness on rank-overlap between both sexes by means of an individual-oriented model. Since primate species are described to differ in their intensity of attack (15), we study virtual entities of different species-specific intensities of attack. The present model deals with dominance interactions among grouping entities and was originally inspired by a model designed by Hogeweg (13). It consists of a homogeneous world with entities that are merely aggregating and that, upon meeting each other, perform dominance interactions, if risks are low (14,16-18). The superior fighting capacity of males compared to females is represented by a higher intensity of aggression (following findings on primates by Bernstein 19) and by a higher initial rank value in so-called 'StrongType' entities, higher than in 'female' inspired entities called 'WeakTypes'. Within each type of entity, individuals are completely identical at the beginning of each run. The outcome of the model will be used to produce testable hypotheses for real primates.

2 Methods

In this section, a description of the model and behavioural measures is given.

2.1 The Model

The model is individual-oriented and event-driven (see 20). It is written in object-Pascal, Borland Pascal 7.0 and consists of a 'world' (toroid) with its interacting agents, its visualisation and special entities that collect and analyse data on what happens in the 'world' (cf. the 'recorders' and 'reporters' of Hogeweg, 1988). The 'world' consists of a space of 200 by 200 units. At the start of each run entities occupy random locations within a predefined subspace of 30 by 30 units. Since the space of the world is continuous, agents are able to move in any direction. They have an angle of vision of 120 degrees and their maximum perception distance (MaxView) is 50 units. Activities of agents are regulated by a timing regime, as follows. Each entity draws a random waiting time from a uniform distribution. The entity with the shortest waiting time is activated first. The lapse of waiting time is usually the same for all entities, but if a dominance interaction occurs within NearView of an agent, its waiting time is reduced, thus increasing the probability that the agent will be activated. Agents group and perform dominance interactions according to a set of rules described below (figure 1).

2.1.1 Grouping rules

Usually, two opposing forces affecting group structure are postulated: on the one hand animals are attracted to one another because participation in a group provides safety; on the other, aggregation implies competition for resources, and this drives individuals apart (e.g., 21).

The forces leading to aggregation and spacing are realised in the model by a set of rules that are graphically displayed in Figure 1 (see 14).

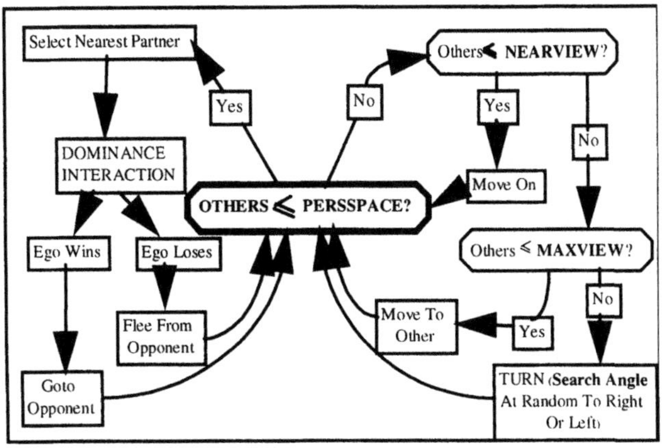

Fig. 1. Flow chart for the behavioural rules of the entities

2.1.2. Dominance interactions

Dominance interactions are competitive interactions over resources that are not specified in this model, but are presumed to include, food, mates and spatial location. Competitive interactions are only initiated if the perceived risks of defeat are low (i.e. so-called risk sensitive system (14). Interactions between agents are modelled after Hogeweg (22) and Hemelrijk (14), as follows:

- Each entity has a variable DOM (representing the capacity to win a hierarchical interaction).
- After meeting one another in their PerSpace, entities 'decide' whether or not to attack following the Risk-Sensitive system. Here, the probability to attack decreases according to the potential risk of defeat as follows. Upon meeting another agent and observing its DOM-value, an entity may predict it will win or lose on the basis of a 'mental' battle, which follows the rules of a dominance interaction as described below. If ego loses the mental interaction, it will refrain from action (thus displaying 'non-aggressive' proximity). If it wins the mental battle, it will start a 'real' dominance interaction.
- If an actual dominance interaction takes place, then entities display and observe each other's DOM. Subsequent winning and losing is determined by chance and values of DOM as follows :

$$w_i = \begin{cases} 1 & \dfrac{DOM_i}{DOM_i + DOM_j} > RND(0,1) \\ 0 & \text{else} \end{cases} \qquad (1)$$

Here w_i is the outcome of a dominance interaction initiated by agent i (1=winning, 0=losing). In other words, if the relative dominance value of the interacting agents is larger than a random number (drawn from a uniform distribution), then agent i wins, else it loses. Thus, the probability of winning is larger for whoever is higher in rank, and this is proportional to the relative DOM-value with its partner.

- Updating of the dominance values is done by increasing the dominance value of the winner and decreasing that of the loser:

$$DOM_i := DOM_i + \left(w_i - \frac{DOM_i}{DOM_i + DOM_j} \right) * STEPDOM$$
$$DOM_j := DOM_j - \left(w_i - \frac{DOM_i}{DOM_i + DOM_j} \right) * STEPDOM \qquad (2)$$

  The consequence of this system is that it functions as a damped positive feedback: a victory of the higher ranking agent reinforces their relative DOM-values only slightly, whereas success of the lower ranking agent gives rise to a relatively large change in DOM. (To keep DOM-values positive, their minimum value is, arbitrarily, put at 0.01.) The change in Dom-values is multiplied by STEPDOM, i.e. as a scaling factor that varies between 0 and 1 and represents intensity of aggression. High values imply a large change in DOM-value when updating it, and thus indicate that single interactions may strongly influence the future outcome of conflicts. Conversely, low STEPDOM-values represent low impact.
- Winning includes chasing the opponent over one unit distance and then turning randomly 45 degrees to right or left in order to reduce the chance of repeated interactions between the same opponents. The loser responds by fleeing under a small random angle over a predefined FleeingDistance.

  From now on, the initiation of a dominance interaction is for short referred to as 'attack'.

2.2 Experimental set-up and Data collection

Here, the same parameter setting (n=8, persSpace=2, nearView=24, FleeingDistance=2 units) is used as in a former study (23).

The present study is confined to a population size of ten entities consisting of two types that differ in fighting capacity. Reflecting the physiologically superior fighting abilities of males (e.g. muscle structure) compared to females, StrongTypes start with a higher winning tendency than WeakTypes (i.e. of 20 versus 10) and display a higher intensity of aggression. I have experimented with three different StepDom values (i.e. 1.0, 0.5 and 0.25 for StrongTypes). StepDom values for WeakTypes were always 80% of those of StrongTypes (respectively 0.8, 0.4, 0.20).

Varying the SearchAngle (i.e. 45, 90 and 180 degrees) produces three degrees of cohesiveness of grouping. For each combination of Stepdom and Searchangle, 5 runs are conducted, resulting in a total of 5 * 3 * 3 = 45 runs.

During a run, every change in spatial position and in heading direction of each entity is recorded. After every time step (consisting of 160 activation), the distance between agents is calculated. Dominance interactions are continuously monitored by recording: 1) the identity of the attacker and its opponent; 2) the winner/loser; 3) the updated DOM-values of the entities.

2.3 Measurements

At intervals of two time steps (320 activation), the degree of rank differentiation and the overlap between the dominance hierarchies of StrongTypes and WeakTypes are measured as follows.

Rank differentiation is measured by the coefficient of variation (standard deviation divided by the mean) of Dom-values (24). For each run the average value is calculated. Higher values indicate larger rank distances among entities.

At the start of each run, all StrongTypes are dominant over each WeakTypes, but during run-time some WeakTypes may become dominant over (some or all) StrongTypes. The degree of dominance of WeakTypes over StrongTypes is estimated by the Mann Whitney U- statistic (25). At the beginning of the run U-values are zero. Later on they may become positive.

Bidirectionality of attack is calculated as a τ_{Kr}-correlation between an actor and receiver matrix of attack (26). This statistic measures the correlation between the corresponding rows of two social interaction matrices. The method reckons with the statistical dependency due to recurrent observations of the same individual. The measure of Unidirectionality of attack corresponds to the sign-changed τ_{Kr}-value.

The clustering together of entities of the same type is measured as a τ_{Kr}-correlation between a matrix of mean distance among entities and a 'hypothesis'-matrix. The 'hypothesis'-matrix reflects Type-segregation because cells belonging to entities of the same type are filled with the number 1 and cells of different types are filled with zeros. Segregation is thus reflected by a positive correlation.

The degree with which dominants occupy the centre is measured by the spatial directions of others around ego. Using circular statistics (27) the centrality of each individual is calculated for each scan by drawing a unit circle around it and projecting the direction of other group members (as seen by ego) as points on the circumference of this circle. Connecting these points with the origin produces vectors. The length of the mean vector represents the degree in which the position of group members relative to ego is clumped; longer mean vectors reflect more clustering in one direction and indicate lower centrality (i.e. lower 'encirclement'). Thus, greater centrality of higher ranking entities is reflected in a stronger positive correlation between rank and encirclement.

To exclude a possible bias brought about by transient values, the correlations for centrality of dominants, for unidirectionality and between social behaviour and rank of the partner, are calculated on data collected after time-step 200.

3. Results

3.1 Cohesiveness of groups.

Cohesiveness of grouping decreases at lower values of SearchAngle, because entities return to others later. This is in line with former findings based on a discrete, lattice-based world (16,17). Different degrees of cohesiveness will be referred to as high (SearchAngle=180°), medium (SearchAngle=90°) and low cohesiveness or loose groupings (SearchAngle=45°). Looser grouping diminishes the frequency of interaction. StepDom affects mean distance much less than SearchAngle and only affects distance for the most cohesive grouping (i.e. via route 1, described below).

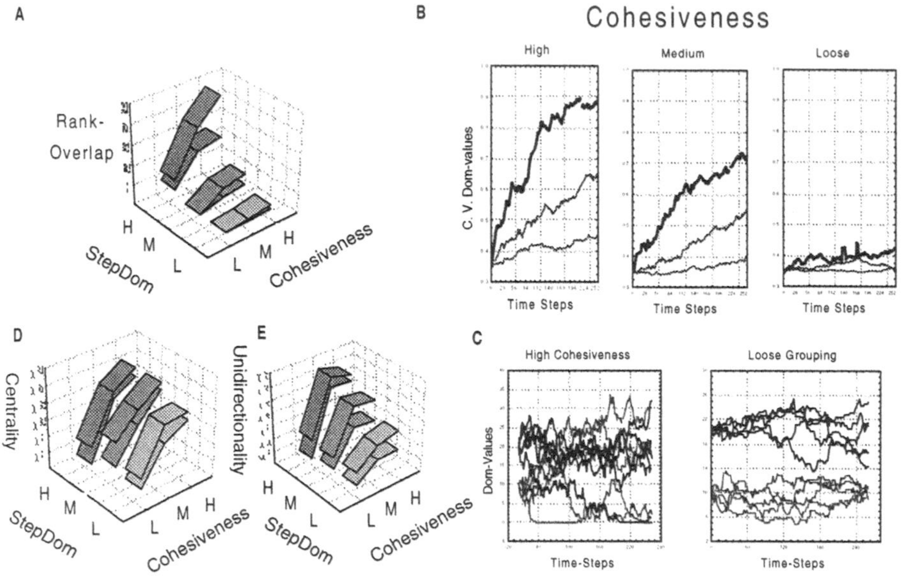

Fig. 2. Means and standard errors are calculated over 5 runs. H = high, M = medium, L = low. **A.** Standard error above and below the mean U-statistic (measuring dominance of Weak- over StrongType) for different degrees of cohesiveness and StepDom-values. **B.** Mean Rank-differentiation measured by the coefficient of variation of Dom-values for different conditions. Lines of increasing boldness correspond to increasing StepDom-values. **C.** Differentiation of Dom-values for Strong and WeakType entities with StepDom = 1. Typical case as observed in one run of highly cohesive and loosely grouping VirtualSpecies. Grey: WeakTypes, black: StrongTypes. Bottom left: Standard error above and below the mean of **D.** centrality measure between Dom-value and Encirclement and **E.** the measure for unidirectionality.

3.2 Rank-overlap and other social-spatial consequences.

In line with expectancies, WeakTypes dominate more over StrongTypes at higher levels of cohesiveness (Figure 2). This holds at all intensities of attack. Furthermore, the degree of rank-overlap between both Types also increases with higher intensity of attack (Figure 2). This can be explained, because both high intensity and cohesiveness cause larger rank-differentiation (Figure 2B). Thus, ranks among both Types, Strong and Weak, diverge more, so that their hierarchies of both types overlap faster (Figure 2C).

The stronger hierarchical differentiation in medium and highly cohesive groups arises from a complex interplay with spatial structure. This interplay can be described as if following three sub-routes (Figure 3).

First, while the hierarchy differentiates in the course of time the entities space out. This is due to the stronger role diversification that accompanies rank differentiation:

some entities become permanent losers and by fleeing again and again from others they move further and further away from their group members. In this way the group spaces out. Spacing out causes the frequency of aggression to decline. This naturally implies a smaller number of opportunities for rank-reversal, and thus a more stable hierarchy. The stability of the hierarchy, in turn, maintains its differentiation. A more differentiated hierarchy is more stable (Figure 3, route I).

Second, differentiation of the hierarchy also increases unidirectionality of attack and this enhances its stability, which in turn supports its differentiation (Figure 3, route II).

Third, strong hierarchical differentiation and stability cause, and are maintained by, a spatial structure with dominants in the centre and subordinates at the periphery. Spatial centrality of dominants emerges because entities of similar rank have equal chances of being defeated by or defeating each other. Since they are also treated in a similar way by other group-members, they will remain on approximately the same location. Lower-ranking entities are chased away by more group-members and therefore end up at the periphery. In turn, spatial structure also maintains the stability and differentiation of the hierarchy, because entities mainly interact with partners that are close in rank. It follows that if a rank-reversal occurs, it will not be a dramatic one (Figure 3, route III).

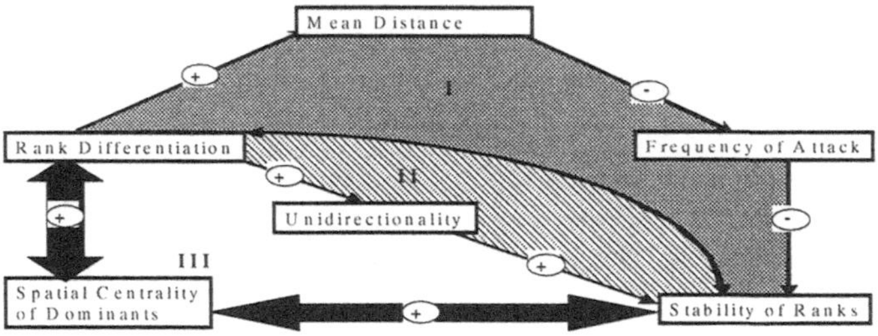

Fig. 3. Summary of interrelations of different variables. Numbers indicate processes leading to increased rank differentiation described in the text.

Although these three routes are interconnected, significant correlations for these effects are mainly found at medium and high values for cohesiveness and intensity of attack (see stronger average spatial centrality (Figure 2D) and average unidirectionality (Figure 2E) found at these conditions).

Cohesive groups differ from loose groups in two respects, the average distance and total frequency of attack. To keep the total frequency of attack under control we run a loose group proportionally longer than a highly cohesive group (i.e. ten times). The hierarchical differentiation, spatial centrality and unidirectionality of attack appear weaker, after the same frequency of interactions, than in a strongly cohesive group (Figure 4) and the three routes of social-spatial structuring (based on correlations over periods lengthened proportionally to contain the same mean frequency of attack as in cohesive groups) are non significant. Yet, unexpectedly, dominance of WeakTypes over Strong ones is similar after the same frequency of attack in highly cohesive groups! This may be due to lower sexual segregation than in cohesive groups (Figure

4), which may lead to relatively higher frequencies of interactions between the sexs. As a consequence, StrongTypes may be 'pulling up' WeakTypes and WeakTypes 'pulling down' Dom-values of StrongTypes, dissolving the rank-differences between them. These details will be tested in a subsequent study.

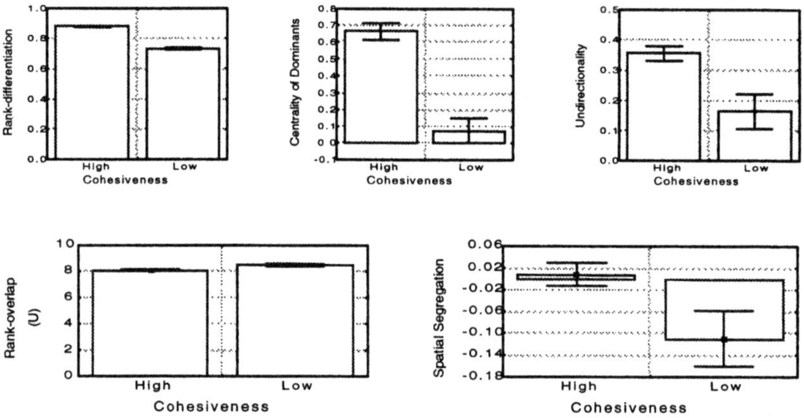

Fig. 4. Comparison of social spatial measures (mean plus/minus S. E.) between highly cohesive and loose groups controlled for the total frequency of attack.

4. Discussion

4.1 The model

Effects of cohesiveness are similar to those of intensity of aggression (for group size 8, see 14,18): both result in a steeper hierarchy and consequently more rank-overlap develops between types.

In more cohesive groups the initial two rank classes to which StrongTypes and WeakTypes belong, clearly dissolve faster due to more frequent encounters. This effect was also described by Hogeweg and co-authors (22,28) in a model on 'BumbleBees'. The Bumble colony consisted of a Queen with a high initial rank and many Workers with a low initial Dom-value. Dissolution of the rank differences was correlated with speed of nest growth. In a slow-growing colony, workers had much time to interact with the Queen and consequently, relatively many workers transfer to the 'elite' rank category of the Queen. In a fast-growing colony, workers are more occupied with rearing young and so are left with fewer opportunities for interaction. Due to the rareness of interactions, the hierarchy among workers remains weak and labile, so that only few Workers transfer to the 'elite' category. Thus, ranks of Workers and the Queen are kept apart more distinctly.

4.2 Biological Implications

Increased cohesiveness not only obscures rank-categories belonging to different types, but also enhances spatial centrality of dominants. In the model, such spatial structure emerges although entities do not prefer a central, or some other, spatial location. Therefore, the model presents us with a parsimonious alternative to the selfish herd theory (29) in which it is assumed that individuals evolved a 'centripetal instinct', because it is safer to be surrounded by others as a protection against possibly approaching predators.

In confirmation of the model, a preference for spatial centrality has so far not been discovered in real animals, not even in the very elegant fish experiments carried out by Krause (30). What is clear, however, is that upon perceiving signals of a predator, fish tighten their shoals. In a shoal consisting of small and large fishes, increased cohesiveness leads to size assortment, with large fishes in the centre and small ones at the periphery (31). Again this is in line with our model: Because large fishes were dominant over small ones, size-assortative shoaling is exactly what is expected to occur, and there is no need for an assumption of a certain preference for any location.

This model represents only the barest essentials of group-living and dominance interactions. It bears no resemblance to highly intelligent and complex apes. Yet, the essential characteristics of this model may hold for any group-living species that performs dominance interactions. Supposedly, it is a general rule that cohesiveness entails more rank-overlap among types. Thus, these results confirm the hypothesis that initially inspired this study: the fact that female dominance is relatively stronger in pygmy chimpanzees than in common ones may, at least partly, be a due to their stronger cohesiveness of grouping. Thus it is unnecessary to invoke additional differences among species in co-operative tendencies among females against males. However, the model does not preclude that coalitions among females contribute to female dominance. However, even if future study confirms a relatively higher frequency of female coalitions among pygmy chimpanzees than common ones, even then it is not clear whether these coalitions are caused by or the result of female dominance: after all, higher female rank, arising from stronger cohesiveness, may allow females to co-operate more intensely against males than in those cases in which females rank lower.

Cohesiveness of groupings varies not only between species (e.g. see, 32), but also within species according to environmental conditions. Chimpanzees, for instance, group more cohesively in less seasonal areas (33). It would therefore be interesting to compare the degree of rank-overlap between both sexes for different environments.

Thus, these individual-oriented models produce new hypotheses for studying social-ecological processes and their consequences and direct our research into a more situated context-based, realistic approach.

Acknowledgements

I am grateful to Rolf Pfeifer for continuous support, to René te Boekhorst for comments on the figures and to Jaap Hemelijk for correcting the English. This work is supported by the Swiss National Science Foundation by a grant from the Kommission zur Förderung des akademischen Nachwuchses der Universität Zürich.

References

1. C. Drews, Behaviour 125 283-313 (1993).
2. P.M. Kappeler, Female dominance in primates and other mammals, Vol. 10, P.P.G. Bateson and e. al. (Eds), 143-158, Plenum Press, New York (1993).
3. B.B. Smuts, Gender, aggression and influence, B.B. Smuts, D.L. Cheney, R.M. Seyfarth, R.W. Wrangham and T.T. Struhsaker (Eds), 400-412, Chicago University Press, Chicago (1987).
4. B. Thierry, Journal of theoretical Biology 145 511-521 (1990).
5. T. Parish, Human Nature 7 61-96 (1996).
6. J.L. Deneubourg, S. Goss, N. Franks and J.M. Pasteels, Journal of Insect Behaviour 2 719-725 (1989).
7. C.B. Stanford, Current Anthropology 39 399-420 (1998).
8. F.B.M. de Waal, Chimpanzee Politics: sex and power among apes, Harper and Row, New York (1982).
9. L. Ellis (Ed.), Reproductive and interpersonal aspects of dominance and status (1 Ed), Vol. 2, Greenwood publishing group, Westport (1994).
10. M.T. McGuire, G.L. Brammer and M.J. Raleigh, Hormones and behavior 20 106-117 (1986).
11. T.E. Rowell, Behav. Biol. 11 131-154 (1974).
12. I.D. Chase, C. Bartelomeo and L.A. Dugatkin, Anim. Behav. 48 393-400 (1994).
13. P. Hogeweg, MIRROR beyond MIRROR, Puddles of LIFE, C. Langton (Ed), 297-316, Adisson-Wesley Publishing Company, Redwood City, California (1988).
14. C.K. Hemelrijk, Proceedings of the Royal Society London B: Biological Sciences. 266 361-369 (1999).
15. B. Thierry, C. R. Acad. Sci. Paris 310 35-40 (1990).
16. C.K. Hemelrijk, Dominance interactions, spatial dynamics and emergent reciprocity in a virtual world, Vol. 4, P. Maes, M. J. Mataric, J-A Meyer, J Pollack and S.W. Wilson (Eds), 545-552, The MIT Press, Cambridge, MA (1996).
17. C.K. Hemelrijk, Cooperation without genes, games or cognition., P. Husbands and I. Harvey (Eds), MIT-Press, Cambridge MA (1997).
18. C.K. Hemelrijk, Adaptive Behavior (in press).
19. I.S. Bernstein and C.L. Ehardt, International journal of Primatology 6 209-226 (1985).
20. O.P. Judson, Trends. Ecol. Evol. 9 9-14 (1994).
21. C.P.v. Schaik, Behaviour 87 120-144 (1983).
22. P. Hogeweg and B. Hesper, Behavioral Ecology and Sociobiology 12 271-283 (1983).
23. C.K. Hemelrijk, Risk sensitive and ambiguity reducing dominance interactions in a virtual laboratory, Vol. 5, R. Pfeifer, B. Blumberg, J-A Meyer and S.W. Wilson (Eds), 255-262, MIT-Press, Cambridge MA (1998).
24. R.R. Sokal and F.J. Rohlf, Biometry: the principles and practice of statistics in biological research. (2 Ed), W.H. Freeman and company, San Francisco (1981).
25. S. Siegel and N.J. Castellan, Nonparametric statistics for the behavioral sciences (second Ed), McGraww-Hill internatioonal editions, New York (1988).
26. C.K. Hemelrijk, Animal Behaviour 39 1013-1029 (1990).
27. K.V. Mardia, Statistics of directional data, Academic Press, London and New York (1972).
28. P. Hogeweg and B. Hesper, J. theor. Biol. 113 311-330 (1985).
29. W.D. Hamilton, Journal of theoretical Biology 31 295-311 (1971).
30. J. Krause, Animal Behaviour 45 1019-1024 (1993).
31. J. Krause, Biol. Rev. 69 187-206 (1994).
32. C.P. van Schaik and J.A.R.A.M.v. van Hooff, Behaviour 85 91-117 (1983).
33. C. Boesch, Social grouping in Tai chimpanzees, W.C. McGrew, L.F. Marchant and T. Nishida (Eds), 101-113, Cambridge University Press, Cambridge (1996).

Simplicity Rules the Roost: Exploring Birdbrain Parental Investment Heuristics

Seth Bullock, Jennifer Nerissa Davis and Peter M. Todd

Center for Adaptive Behavior and Cognition
Max Planck Institute for Human Development
Lentzeallee 94, D-14195 Berlin, Germany

Abstract. Parents raising multiple offspring must decide how to divide resources between them. Much empirical data on the parenting behaviour of particular species has been collected. Birds, in particular, have been shown to follow a number of provisioning rules. However, the adaptive significance of this variation in decision-making strategies has been largely unexplored. Here we present a simulation model of the western bluebird, *Sialia mexicana*, with which we explore the utility of various simple feeding heuristics. The simulated parents face the task of simultaneously raising several offspring who are of differing ages and thus have differing resource needs. We show that the success of simple rules of thumb varies with environmental parameters in a manner which (i) predicts experimental results in the biology literature, and (ii) can be explained using a notion of parental egalitarianism.

A repeated finding within the field of artificial life has been that complex adaptive behaviour may arise from the interaction between simple systems and their environment. This perspective contrasts starkly with received wisdom within the decision-making literature where simplicity is often automatically equated with stupidity. The reason for this prejudice is easy to see. The normative theories which currently dominate research into decision making (probabilistic reasoning, Bayesian reasoning, various propositional logics, etc.) often involve the complex integration of many pieces of information and the construction of elaborate models.

This complexity is geared towards achieving decision-making's gold standard: rational coherence. The proponents of these complex normative models claim that if one follows their prescriptions, and the assumptions that these models require in order to work properly are met, one is sure never to behave intransitively, never to hold inconsistent beliefs, and never to fall foul of reasoning fallacies or other delusions. In contrast, through his simplicity, the naive roulette-wheel gambler who bets his last dime on red thirteen because it has not come up for the entire time he has been sitting at the table, or simply because it is his lucky number, is revealing himself to be irrational and thus exploitable.

The appeal of rationality is very strong, and the failure of human and animal decision-making to live up to its high standards has been the cause of some consternation amongst psychologists and economists [1]. However, two recent research movements have challenged the rationality-based hegemony and suggested that successful, adaptive behaviour in the real world may not require the complex machinery advocated by the orthodoxy, but may rely on simple heuristics which although limited in their ability, are suited to the environments in which they must operate.

The first of these research programmes has become associated with the behaviour-based robotics field as it has tried to extricate itself from good old-fashioned artificial intelligence (GOFAI) [2]. The second is exemplified by decision-making researchers who are beginning to question the benefit of rational coherence if it can only be obtained at the expense of tractability and psychological reality [3].

The animal behaviour literature has proved influential in both of these cases. Roboticists have discovered that organisms have achieved robust, simple solutions to many of the problems they face as engineers, and that these natural solutions reveal that the elaborate planning and modelling which GOFAI assumes goes on inside the heads of even the simplest intelligent agents is to a surprising extent unnecessary [4]. Similarly, decision-making theorists have begun to appreciate that ethologists have had considerable success explaining animal behaviour from a perspective which appreciates that evolution will favour accurate and reliable reasoning, but accepts that this reasoning must be carried out in the real world by real mechanisms [5].

Here we approach an ecologically relevant decision-making task which is typically regarded as intractable from the perspective of normative frameworks and demonstrate that simple ecologically sound rules of thumb can achieve high levels of performance and that their performance predicts empirical data collected by animal behaviour researchers.

1 Parental Investment

A decision-making problem which besets all animals which care for multiple simultaneous young is how to divide resources amongst these offspring. A classical approach to this problem would be to treat it as a game against nature in which the parent must play a strategy which maximises return on investment over the period of care. However, since the moves in such a game are many and the options available at each move are plural, the game tree which must be analysed in order that such an approach lead to a solution is prohibitively large.

For example, Becker [6] provides an economic analysis of how rational human parents should distribute investment among their children, assuming that parents are trying to maximise aggregate child quality as defined by the sum of all the children's wealth as adults. This quality is a function of the resources invested in the child, the child's own skill and abilities, and any extra income he or she might earn through sheer luck. Becker assumes that there are diminishing returns on parental investment, and shows that as long as these diminishing returns are the same for all children, parents should distribute investment such that each child achieves the same degree of wealth. However, if some children are capable of accumulating more wealth per unit of parental investment than others, then parents should of course favour them.

While these conclusions may sound reasonable in general, they are of limited use in making predictions about actual parental behaviour in specific situations. This is because they assume that parents have some means of calculating the effects of each unit of investment on the future payoff they expect to gain from a child. In practice, however, this calculation can require involved manipulations of information that is itself difficult to obtain. Children do not come equipped with investment meters for their parents' convenience.

In response to problems of this kind, biological models of parental investment have only dealt either with single offspring at a time, multiple simultaneous offspring assumed to be identical, or have been limited to cases where parents are assumed to base individual investment decisions on fully informative offspring solicitation signals [7–9]. In addition, previous models have treated parental investment as a series of events that have independent consequences for offspring fitness [10].

These simplified models of the parental investment problem have identified two simple rules which animals can employ in order to successfully raise as many offspring to reproductive age as possible. When each offspring requires the same amount of investment in order to reach a given level of fitness, then one parental solution would be to treat each on the basis of its need. In birds, for example, chicks often beg for food when hungry. If intensity of begging is an honest and accurate signal of need, parents would then be expected to feed their chicks according to this intensity in order to achieve investment equality. This is clearly a very simple decision rule.

Alternatively, in the case that offspring are not born simultaneously (chicks, for example, often hatch at different times), behavioural ecologists have identified a second simple but adaptive parental decision rule: satisfy the oldest offspring first. Since one major predictor of the probability of survival to adulthood is the current age of offspring (the older they are, the closer they are to independence and reproduction, and the more likely they are to make it all the way there), parents are expected to benefit from preferentially investing in their older offspring [7].

However, birds display a variety of parental feeding patterns. Coots preferentially feed the smallest chicks[11]. Pigeons preferentially feed the hungriest [12]. Common swifts preferentially feed their largest/oldest[1] chicks [13]. Fieldfares appear to feed randomly [14]. Despite the amount of published data on this topic, there has been no proposal for why such a variety of strategies should exist.

2 Simulating Parental Investment

Here we model the problem of parental investment as a series of interdependent food-allocation decisions, in which investment is meted out in many small indi-

[1] Feeding the largest of a brood of asynchronously hatched chicks is equivalent to feeding the oldest since it ensures that the oldest chick will remain the largest.

visible doses. Since offspring are of different ages they will have different resource demands. A successful parental investment strategy must balance these demands over the care-giving period in an effective manner despite the stochasticity of the foraging environment. The model we present is an iterative computer simulation that mimics the feeding, digestion, metabolism, and growth of asynchronously hatched western bluebird (*Sialia mexicana*) chicks from hatching until fledging. This model allows us to explore the effects of important aspects of parental decision making which are difficult to capture with more formal approaches. More specifically, it allows us to explore the complex relationship between a proximal decision-rule and the ultimate effects of this rule.

We compared the performance of the various parental strategies reported in the parental investment literature and some which were not (i.e., preferentially feeding the youngest or feeding in a fixed random order), in terms of number and weight of chicks fledged, across a range of different artificial environments. In doing so we assume that begging is an honest signal of need [9] and hence that preferentially feeding the chick begging with the greatest intensity is equivalent to preferentially feeding the hungriest chick.[2]

We chose to model western bluebirds because there is information available about both metabolic rates and growth rates for this species. Equations for metabolic rate, digestion rate, and stomach size of our simulated chicks were derived from published field data on growth and metabolic rates of bluebird chicks across the nestling period [15]. Values for the calorie content of the insects that parents feed to their chicks, and the proportion of metabolisable energy these items contain were taken from Dykstra & Karasov [16].

To mimic environmental differences in food availability, we varied the frequency with which food was found and returned to the nest, using a parameter p governing the probability of repeatedly finding food and q governing the probability of repeatedly failing to find food. The decision rules we tested determined which chick was fed on any given occasion, and it is the sum of these feeding decisions over the course of the nestling period that determines overall parental success in terms of the total size of all fledged chicks.

A total of four simulated chicks hatched at one day intervals, were fed for 20 days, then fledged. Each 24-hour day was divided into 10-minute intervals of simulated time, during which: 1. Any egg due to hatch, hatched. 2. A bug (food) was found by the foraging parent with probability determined by environmental parameters p and q. 3. If a bug was found, the parent's decision strategy was used to choose a chick to feed. 4. If the chosen chick had enough space in its stomach for the bug, it was fed, otherwise the next most preferred chick was chosen. 5. If a chick was fed, the food was added to the chick's stomach. 6. Each chick with food in its stomach gained calories by digesting 10 minutes' worth of that food. 7. Every chick burned 10 minutes' worth of calories in accordance

[2] In reality, rather than always honestly revealing their true level of need, chick begging strategies will have coevolved with parental provisioning strategies. Although modelling this coevolutionary process is beyond the scope of the present paper, it will be pursued in further work.

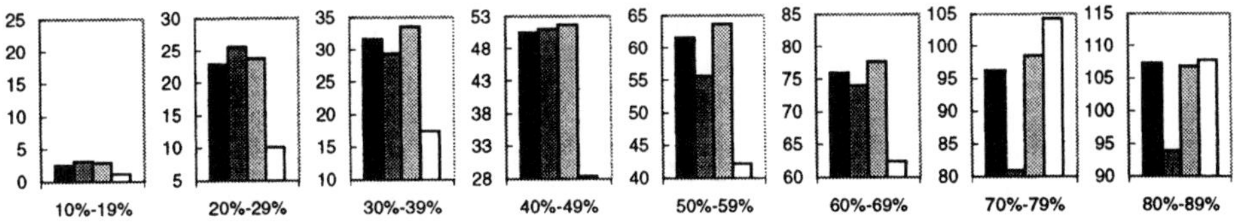

Fig. 1. The mean nest weight in grams achieved by each of four simple strategies (Random, Largest, Smallest, Hungriest, respectively) across eight classes of environment. Each strategy was simulated 500 times in each of 100 environments $\{p, q\}$ with parameters drawn from the set $\{.0, .1, ..., .9\}$. These environments are grouped by food availability, defined as the mean percentage of successful foraging attempts. Small differences in strategy successes can be very important ones. A half gram difference in the weight of a fledged chick is a significant one. The results depicted remain qualitatively the same for clutches of 3 or 5 chicks.

with its metabolic rate, gained or lost weight according to the net calorie change, and if it grew, its stomach capacity increased to accord with its new size. 8. If a chick's weight dropped below a certain age-specific limit it died. Foraging and feeding took place for 14 hours out of every day. As with real bluebirds, during the night no food was gathered or distributed, but, chicks continued to undergo steps 6 through 8.

3 Results

Not surprisingly, the number of chicks that successfully fledged increased with increasing food availability. More interesting was the finding that the mean amount of food available to parents had a strong effect on the success of the different feeding rules (Fig. 1). Each of the four decision rules identified in the parental investment literature on birds was successful for some range of environmental "richness". Furthermore, the manner in which the success of these strategies varied with the abundance of food in the foraging environment is explicable from a perspective of "parental egalitarianism".

For relatively poor environments in which food was found 30% of the time or less, preferentially feeding the largest/oldest chick was the most successful decision rule. This rule, which picks out a single chick and targets it for preferential investment, does particularly well in environments where only a single chick can be raised. In this sense it is non-egalitarian, rejecting impartiality in favour of inequitable prejudice. In harsh environments, this bias is adaptive and successful.

For environments in which between 30% and 70% of parental foraging trips were successful, preferentially feeding the smallest chicks outperformed all other strategies. This strategy is more egalitarian than feeding the largest/oldest because the smallest chick is not fixed, but may change over time dependent on who gets fed. If whichever chick is currently smallest is preferentially fed, no smallest chick is likely to remain the smallest for long. In this way parents may achieve a limited form of equality amongst the chicks without any complicated calculation. Parents using this strategy are letting their environment (in this case the metabolism of their young) do some of the decision-making for them.

For richer environments, in which food is found on between 70% and 90% of foraging trips, the most successful parents fed on the basis of short term need, i.e., they fed the hungriest chicks, as revealed by intensity of begging. Preferentially feeding the hungriest chicks is an even more egalitarian strategy than favouring the smallest. Since the degree of hunger changes rapidly within a nest, if parents have access to some accurate indicator of this information, and the environment in which they forage is rich enough to provide for an entire brood, it makes sense to use this information in order to evenly balance the distribution of resources within the nest.

Finally, when food was even more abundant, all decision rules performed equally well. In such cases we would expect random feeding to evolve since this may make less demands on parental decision-making mechanisms.

Preferentially feeding the youngest chicks and feeding chicks in a fixed random order were never the most successful strategies. Thus our simulation agrees with empirical findings insofar as only the empirically observed strategies are the most successful in any simulated environment. But to what extent do our results concerning the manner in which the success of these strategies varies with environmental richness also agree with experimental and observational data?

The few published reports of species switching provisioning rules as a result of changing environmental conditions are surprisingly consistent with the findings of the simulation model. For instance, pied flycatcher females preferentially feed the smallest chicks under normal food conditions, but when food availability is experimentally reduced they preferentially feed the largest [17]. When food is plentiful, sparrow hawk mothers allocate food resources equally among chicks, when it is scarce they switch to a strategy that favours the largest [18].

4 Combining Cues

All of the simple strategies explored above make decisions based only on a single cue (weight, hunger, or age). As field studies suggest that parents sometimes combine cues [19], we also tested two sets of multi-cue feeding rules. Both sets fed chicks in an order determined by a criterion constructed as a linear weighted sum of the three chick cues (weight, hunger, age). The first set of *rank-based* multi-cue strategies calculated this criterion on the basis of a chick's rank-order within the nest. For every feeding decision made by a parent bird, all chicks were ranked according to each cue, and a chick's ranks were summed after being weighted according to the parental strategy. The parent then preferentially fed chicks scoring highest on this combined cue criterion.

The second set of *real-valued* multi-cue strategies weighted and summed the real cue values (e.g., weight in grams) rather than merely the chicks' ranks. We assessed both sets of strategies bearing in mind that the real values employed by

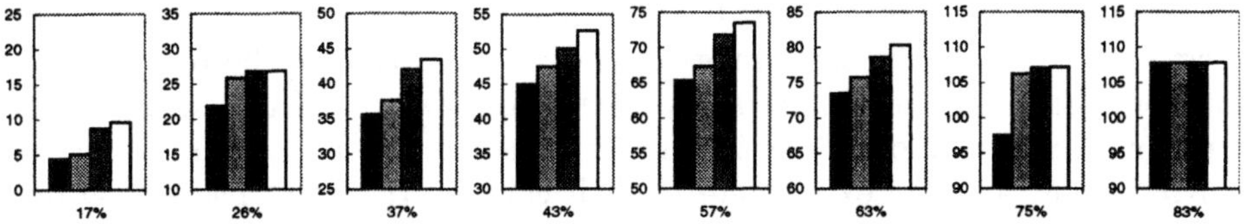

Fig. 2. The mean nest weight (in grams) achieved by the best of the 2197 possible mixed-cue strategies with cue weights drawn from the set $\{-6, -5, ...5, 6\}$. Each strategy was simulated 100 times in each of eight selected environments differing in mean food availability, but not variance. The performance of the best simple decision rule and of random feeding are shown for comparison. Columns show Random, Best Simple, Best Linear Weighted Ranks, and Best Linear Weighted Reals, respectively.

the second set of strategies might not be realistically assessable by parent birds, whereas ascertaining rank-orders is more likely to be within their ability.

Not surprisingly, for each of the environments tested, there was some subset of both the rank-based and real-valued mixed-cue rules that outperformed the single-cue rules (Fig. 2). Whilst the best real-valued strategies also outperformed the most successful of their rank-based cousins, they did not differ qualitatively in that they employed roughly the same weights. Furthermore, the degree of egalitarianism exhibited by the most successful mixed-cue rules increased with environmental quality just as it had done for the single-cue rules.

The strategies employed by the best multi-cue decision-makers can, for the most part, be regarded as refinements of those employed by the best one-cue decision-makers. In poor environments strategies weighted Age positively. In rich environments Hunger was weighted positively. Intermediate environments favoured strategies which either favoured small chicks, or achieved an intermediate degree of egalitarianism through balancing opposing cues (e.g., feeding young *and* heavy chicks). The ability of multi-cue strategies to balance opposing cues in this way can be used to explain the increased performance of these strategies.

5 Optimal Investment?

So far we have used the performance of random feeding behaviour as a lower bound against which to compare the utility of the parental investment strategies we have assessed. However, without knowing the maximum possible success that is attainable in each environment, it is difficult to assess how well these decision rules perform in an absolute sense. As was noted during the discussion of parental investment modelling, in practice finding this upper bound on performance is intractable. We have attempted to approximate the results of optimal investment by assessing feeding rules that integrate as much relevant information into their decisions as is computationally tractable, whether or not such strategies would be realistically achievable by parent birds.

We implemented three short-term optimisation strategies and assessed their performance across the same range of environments used in the previous simulations. Each optimisation strategy relied on knowledge of the equations underlying chick metabolism and growth, the character of the stochasticity in the environment, and the exact state of each chick in the nest. These strategies then utilised this knowledge to make reasonable guesses about the future and thereby construct a limited portion of the full decision tree which characterised the investment problem that they faced.

Under the first maximising strategy (Next Bug), the current bug is offered to the chick whose eating of it would maximise the total weight of all chicks in the nest at the time the next bug is expected to be found. The second strategy (Two Bugs) is identical to the first except that it maximises nest weight at the time the second subsequent bug is expected to be found. The third strategy (1-10 Bugs) maximizes the short-term expected value of nest weight – that is, the sum, over the next 10 time-steps, of the probability of finding the next food item multiplied by predicted total nest weight at that time-step. This third strategy thus copes with variance in the interval between finding bugs. Surprisingly, not only did all three strategies perform worse than the multi-cue and single-cue rules, but by and large they performed worse than feeding chicks at random. In addition, the most sophisticated optimization strategy out-performed its less complex relatives in only one environment.

These three strategies are far more complex than the successful simple decision rules. They require knowledge that actual parents are unlikely to possess and could not directly assess. They integrate this information in an attempt to determine the best possible decision to make, and yet, despite all that, they make terrible decisions. Why are the simple strategies so much more successful than their complicated, computationally expensive competitors?

A general answer to this kind of problem is that the nature of the provisioning problem faced by many parents involves long-term dependencies which ensure that actions which are successful in the short term may have catastrophic impact in the longer run. In the current context, one form of long-term dependency is the difference between day and night. Even the most far-sighted of the short-term optimisation rules that we tested could only base its behaviour on what it expected to happen in the next 100 minutes. For most of every day such a strategist plans and acts in blissful ignorance of the coming dusk. Whilst the short-term future anticipated by such a strategist may look bright, the fact that no foraging can be undertaken during the 10 hours of darkness ensures that any optimism may be misguided.

The hypothesis that the poor performance of these short-term optimisers is due entirely to the unforeseen effect of night falling was tested by shortening the simulation's hours of darkness from 10 to zero. As predicted, in this "land of the midnight sun" the performance of the optimising strategies increased, but only to around that of feeding in random order (Fig. 3). In no environment did an optimising strategy outperform the best simple rule. Clearly the parental

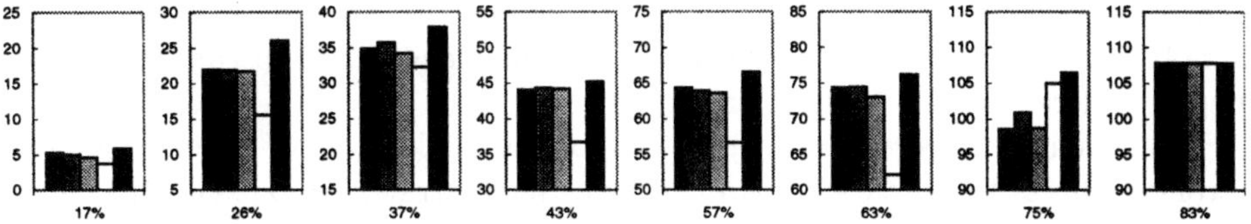

Fig. 3. Columns show the mean nest weight achieved by Random, Next Bug, Two Bugs, 1-10 Bugs, and Best Simple, respectively over 500 runs in eight environments in which each day comprised 14 hours of daylight and an instantaneous period of night.

investment problem features residual long-term dependencies which continue to defeat short-term optimisation.

As these results suggest, the problem is wider than dealing with the cycle of night and day. Many of the decision-making problems faced by humans and other animals exhibit similar kinds of long-term dependencies which invalidate short-term optimisation approaches. The stock market, for example, has proven to be extremely resistant to short-term optimisation approaches largely because of the long-term dependencies and stochasticity which characterise it. However, this does not stop simple rules of thumb such as "only invest in well-known companies" from making money [3].

The failure of short-term optimisation in these contexts might therefore be considered to be symptomatic of a general failure of normative models to apply unproblematically to realistic decision-making problems [20]. Whilst simplified models of these problems may be tractable, adding even the degree of realism modelled here renders them inapplicable in their full-blown form, and hard to approximate using attenuated short-term versions. In this sense, the success of the simple rules which were explored in this study cannot be accounted for by claiming that they approximate some optimal solution to the problem they face, since no such optimal solution, nor any successful approximation to it, can be presented.

6 Conclusion

Unlike Becker's unboundedly rational parents, discussed earlier, our results suggest that parents do not have to carry cumbersome investment equations in their heads. The unsophisticated feeding decision rules we present here were successful despite their irrational simplicity. Combined with an understanding of the role of parental egalitarianism, these rules allow parental investment decisions to be robust with respect to changes in environmental quality. Indeed, within our simulation, it is only through the use of a small set of environmentally contingent decision rules that parents can be most successful.

Although empirical studies demonstrate a diversity of parental feeding rules, to date there has been very little information on the relation between these rules and the richness of the environments in which they operate. Our findings suggest that the notion of parental egalitarianism coupled with an understanding of the simple rules which parents may use to vary this egalitarianism can be used to better understand this body of empirical data.

The results of this research suggest not only that the state of the environment strongly affects the success of parental investment strategies, but also that even under these complex conditions, strategies do not have to be complex to be successful.

Acknowledgements

This paper benefited from the comments of Jason Noble, Henrietta Wilson, two anonymous referees, and the programming skills of Martin Dieringer.

References

1. Kahneman, D., Slovic, P., Tversky, A.: Judgement under uncertainty: Heuristics and biases. CUP, New York (1982)
2. Brooks, R.A.: New approaches to robotics. Science **253** (1991) 1227–1232
3. Gigerenzer, G., Todd, P.M., the ABC Group: Simple heuristics that make us smart. OUP, New York (1999)
4. Webb, B.: A robot cricket. Science **275** (1996) 62–67
5. Green, R.F.: Stopping rules for optimal foragers. Am. Nat. **123** (1984) 30–43
6. Becker, G.S.: A treatise on the family. HUP, Cambridge, MA (1991)
7. Lack, D.: The natural regulation of animal numbers. OUP, Oxford (1954)
8. Parker, G.A.: Models of parent-offspring conflict. V. Effects of the behaviour of the two parents. Anim. Behav. **33** (1985) 519–533
9. Godfray, H.C.J.: Signaling of need by offspring to their parents. Nature **352** (1991) 328–330
10. Winkler, D.W.: A general model for parental care. Am. Nat. **130** (1987) 526–534
11. Horsfall, J.A.: Brood reduction and brood division in coots. Anim. Behav. **32** (1984) 216–225
12. Mondloch, C.J.: Chick hunger and begging affect parental allocation of feeding in pigeons. Anim. Behav. **49** (1995) 601–613
13. Martins, T.L.F., Wright, J.: Brood reduction in response to manipulated brood sizes in the common swift. Behav. Ecol. and Soc. **32** (1993) 61–70
14. Ryden, O., Bengtsson, H.: Differential begging and locomotory behavior by early and late hatched nestlings affecting the distribution of food in asynchronously hatched broods of altricial birds. Zeitschrift für Tierpsychologie **53** (1980) 291–303
15. Mock, P.J., Khubesrian, M., Larcheveque, D.M.: Energetics of growth and maturation in sympatric passerines that fledge at different ages. Auk **108** (1991) 34–41
16. Dykstra, C.R., Karasov, W.H.: Daily energy expenditure by nestling house wrens. Condor **95** (1993) 1028–1030
17. Gottlander, K.: Parental feeding behavior and sibling competition in the pied flycatcher, *Ficedula hypoleuca*. Ornis Scandanavica **18** (1997) 269–276
18. Newton, I.: Feeding and development of the sparrowhawk, *Accipiter nisus*, nestlings. J. Zool. (London) **184** (1978) 465–487
19. Kölliker, M., Richner, H., Werner, I., Heeb, P.: Begging signals and biparental care: nestling choice between parental feeding locations. Anim. Behav. **53** (1998) 215–222
20. Simon, H.A.: Invariants of human behavior. Ann. Rev. Psych. **41** (1990) 1–19

Imitation and Cooperation in Coupled Dynamical Recognizers

Takashi Ikegami 1)
and
Makoto Taiji 2)

1) Institute of Physics, The Graduate School of Arts and Sciences,
University of Tokyo, 3-8-1 Komaba, Meguro-ku,Tokyo 153, Japan
ikeg@sacral.c.u-tokyo.ac.jp
2) Institute of Statistical Mathematics,
4-6-7 Minami-Azabu, Minato-ku, Tokyo 106
taiji@ism.ac.jp

Abstract. A coupled dynamical recognizer is proposed as a model for simulating intelligent game players, who can imitate the other player's behavior. A kind of recurrent neural network called a dynamical recognizer is used as an internal model of the other player to imitate the behavior. The Rashevskyan game is examined, where each player moves along a separate spatial axis to take an advantageous position over the other player. Though the players are egocentric in principle, it is shown that some altruistic behavior will be performed as a dynamical attractor phase. The altruistic behavior is no longer attainable by continually modeling the opponent player merely as a Tit for Tat player. Rather, players have to dynamically change their model of imitation to achieve mutual co-operation, otherwise they go to a static non-cooperative Nash solution. Enhancement of a minute difference in players' action patterns, called the pragmatic paradox, is the key issue throughout this paper.

1 Introduction

The play required by classical game theory generally requires a level of rationality much higher than natural rationality. For example, Anderlin (1990) showed that a Nash equilibrium is not guaranteed even with universal Turing machine players. To overcome such over-rationality, a kind of bounded rationality has been proposed and studied by Rubinstein (1986) and others (see e.g. Binmore 1987) using finite automata models. More recently, epistemic logic has been developed to study what beliefs and inference abilities are required for rational players (Bacharach 1997).

The above studies are more concerned with playing matrix games using static solutions. On the other hand, another type of a game theory is concerned more with the dynamic aspects of game interactions. It was first introduced by Rashevsky (1947) and Rapoport (1947), and recently it has been further developed by Rössler (1994), Akiyama (1998) and the authors of the present paper (1998, 1999). Those approaches can be called the Rashevskian or *dynamical systems* game. In the following, this new approach to game theory is introduced, and a new understanding of dynamic co-operative behavior is proposed. This approach is developed into a possible theory of developmental psychology and language. The paper concludes with a discussion of this on-going work.

2 Coupled Dynamical Recognizers

A kind of neural network with recurrent interactions has been used to imitate human language performance (Elman 1990), to mimic finite automaton behavior (Pollack 1991), and to manipulate robot navigation (Tani 1995). Following Pollack, we call the network a *dynamical recognizer.*

Recently, we proposed a coupled dynamical recognizer to simulate two agents playing the prisoner's dilemma game (Taiji and Ikegami 1999; Ikegami and Taiji 1998). Each agent generates an internal model of the other agent that best imitates the other's past behavior patterns. The model is expressed by dynamical recognizers. Based on the best internal model, players chose their next action by anticipating their future behavior and optimizing it.

However, the prisoner's dilemma (PD) game is a simple matrix game without any external environment. A new problem arises when a certain environmental factor enter the agents' optimality functional (i.e. an individual's satisfaction value) in addition to the inputs from the other agent. This situation was first formalized independently by Rashevsky (1947) and Rapoport (1947). In this paper, we apply our coupled dynamical recognizer model to this problem. We now assume that each agent can move along one spatial axis in discrete time steps, so that the agents now have to consider both players' action patterns and spatial information. Though those agents are egocentric players in principle, we will show that some altruistic behavior will be performed only as a result phenomenon.

3 Descriptions of Games and Dynamics Equations

3.1 Dubey's spatial game

A game we use was originally introduced by Dubey (1986). First, consider a game with two players, where each player has an optimal functional consisting of two parts; one given by the function of its own position x and one given by the function of other player's position y. Given a home position at (a, b) for a player 1, the optimal functional (U_1) of this player is given by,

$$U_1(x,y) = -[(x-a)^2 + (y-b)^2]. \tag{1}$$

The egocentric player 1 is most rewarded when $x = a$ and $y = b$. Similarly, given other player's optimal functional as $U_2(x,y) = -[(x-c)^2 + (y-d)^2]$, the egocentric player 2 wants to have the player 1 at $x = c$ and $y = d$. If $a \neq c$ and $c \neq d$, classical game theory indicates that there is a unique Nash equilibria pair of this game, i.e. $(x = a, y = c)$, assuming that the strategy-set of each player is a point on each axis [1]

This game is converted into the Rashevskyan game if 1)each player is assumed to announce the numerical value of his or her position at that time step, 2) each player can memorize the other player's action sequences, and 3) each player can compute the optimal functional at that position and decide to step forward to or backward from its home position at the next time step. We thus assume a minimal kind of intelligence for our players; that is, memory and inference capability.

3.2 Dynamical recognizer as a model of the opponent

Two "intelligent" players play the Dubey's spatial game by imitating the other's behavior using a dynamical recognizer. The dynamical recognizer we use here is a simple two-layered, 12 connections network which is decomposed into context and function networks. As Pollack (1991) first explicitly showed, a dynamical recognizer can imitate behaviors of some finite automaton. To see how well a dynamical recognizer learned the given automaton, we examine geometrical patterns in the "context space", i.e., a plot of output neuron states against all possible input states. If a dynamical recognizer can successfully imitate the given finite automaton, the context space plotting shows finite islands of clusters. A clear correspondence is observed between each cluster and a node of the finite automaton. When the dynamical recognizer fails to imitate or the opponent is not a finite automaton, the context space plotting shows a stretched and folded, fractal-like structure.

We use this dynamical recognizer to mimic the opponent's behavior. In each game round, both players 1) generate a model of the opponent that best mimics the opponent's past moves, 2) anticipate the opponent's future moves and 3) chose as their next action that which is expected to induce the best future outcomes. These three steps are repeatedly applied in each game round. In the following, we explain those three steps in detail.

1) Past behavior of the opponent will be mimicked by the prescribed dynamical recognizers. We here use only two input neurons (y_0 and y_1) and three output neurons (z_0,z_1 and z_2). One of the input neurons y_1 is fixed in its state as a biasing network, [2] while two output neurons called recurrent outputs (z_1 and z_2) are used recurrently to determine the function network. The dynamics

[1] As long as one player sticks to a point, there is no reason also for the other player to deviate from the point. It is then called a Nsh equilibruim point. Also if it is impossible for both players to take simultaneously advantage from a deviation from the point, it is called a Pareto efficient point. In this game, such Pareto efficient point forms a line segment joining (a,b) and (c,d).(Dubey 1986)

[2] When a player is on the left hand side of his or her home position, we set $y_1 = 0.3$. Otherwise we set $y_1 = 0.7$. This value is fixed through the simulation.

of context and function network is expressed by the following equations at the time step n.

$$z_i(n) = g(\sum_{j=0}^{M} w_{ij} y_j(n)), \tag{2}$$

$$w_{ij} = \sum_{k=1}^{N} u_{ijk} z_k(n-1). \tag{3}$$

where w_{ij} denotes the weight of the function network determined by the context network with its backbone context weights u_{ijk}. In the equations above, nonlinearity exists only in the sigmoid function $g(x) = (e^{-\beta x} + 1)^{-1}$. We can control the nonlinearity by changing the parameter β (through this paper, we set $\beta = 6.0$).

For the purposes of the present study, only one input and one output neuron are necessary since this game has only two actions: stepping forward and backward relative to the home position. One input neuron $y_0(n)$ is used for specifying the imitator's action and one output neuron $z_0(n)$ is used for specifying the associated action of the imitated player. The output is rounded off to 0 (forward) and 1 (backward).

The degree to which the obtained dynamical recognizer can imitate the real opponent is measured by the weighted sum over past behavior patterns. In practice, the error $E(n)$ after the n-th game is computed by

$$E(n) = \sum_{k=1}^{n} \lambda^{n-k} (z_0(k) - d(k))^2, \tag{4}$$

where $d(k)$ is the actual opponent's action in the k-th game, $z_0(k)$ is the action predicted by the network, and λ is a parameter which controls forgetfulness. For most simulations, $\lambda = 0.95$ was used.

In the present paper, we quantify the strength of the connections u_{ijk} as either -1 or 1. Therefore we can exhaustively search all structures to arrive at those that best mimic the opponent's behavior. Since the context networks do not have continuous values, the supposed language class is severely limited. However, we find that not only finite automata but also many other non-finite automata are expressed by those networks. In each game round, we carry out exhaustive search over all possible structures to update the model of the opponent.

2) Using the model of the opponent, each player anticipates any future actions. In practice, by giving all possible combinations of actions against the model over the next T rounds (i.e. $\hat{x}(t+1), \hat{x}(t+2) \cdots \hat{x}(t+T)$), we compute the expected reward averaged over trajectories from the current position $(x(t), y(t))$.

$$< U_1(\hat{x}, \hat{y}) >= \sum_{n=1}^{T} U_1(\hat{x}(t+n), \hat{y}(t+n)) \omega^{T-n}, \tag{5}$$

where the possible combinations are denoted by $\hat{x}$ and the expected opponent's actions associated with $\hat{x}$ are denoted by $\hat{y}$. The future weight ω is set at 0.7 here. The outputs $\hat{y}(t)$ are computed for each time step n by the model obtained in the time step t. In other word, drawing the future spatial trajectories from the current position, players anticipate the possible reward associated with each trajectory up to T time steps ahead. The time step T is set at 10 here.

3) The players choose the action combinations that is expected to bring the highest functional value to itself, with the future weight $\omega = 0.7$ and take the first element of that action pattern (i.e. $\pm\Delta$) as the next action.

We have to introduce one important notion here, called model undecidability. In general if there are two or more networks that are equally likely to mimic the opponent's behavior within the accuracy ϵ from the best imitating model, we define them as equivalent models. The equivalent models have different net structures from each other, and we use $\epsilon = 0.006$ for the present simulation. Yet this may not cause any problem if all of the equivalent models propose to a player the same decision for the next action. If they happen to propose different ones, however, the game becomes undecidable. Instead of introducing a criterion of model selection, we branch a world line. That is, we follow the every possible trajectory of the game's dynamics when such undecidability occurs. If the dynamics meets many such undecidabilities, many trajectories will branch out. Those trajectories are visualized as the spatial trails on a two-dimensional plane, and this will be made clear in the next section.

4 Observation and Simulation

We assume that each player has his or her own home position at (0.1,0.9) for player 1 and (0.9,0.1) for player 2. Starting from points close to the Nash solution (0.1,0.1), player 1 moves along the x axis and player 2 moves along the y axis with a unit step size $\Delta = 0.001$. A combined point (x,y) will be able to move on the two-dimensional plane with the minimal scale Δ. An initial set of action sequences are given to players either randomly or in specific patterns. For each game round, having one new bit of information concerning their opponents' behavioral strategies, players can update their models. Model undecidability is allowed only for the first 100 time steps. Any branching trajectories from the first 100 time steps are investigated. Successive branching trajectories from the already branched trajectories are all taken into account.

4.1 A bifurcation of trajectories

First, we show (Fig. 1) an ensemble of all branching trajectories from the same initial condition on the two-dimensional plane (x,y). Staring from the Nash equilibrium point, most trajectories will soon come back to the Nash point if we do not allow any model undecidabilities. Players will remain around the Nash point. With a finite amount model undecidability ϵ, the situation changes drastically. Some trajectories tend to deviate from the Nash point to make loops as in Fig.1,

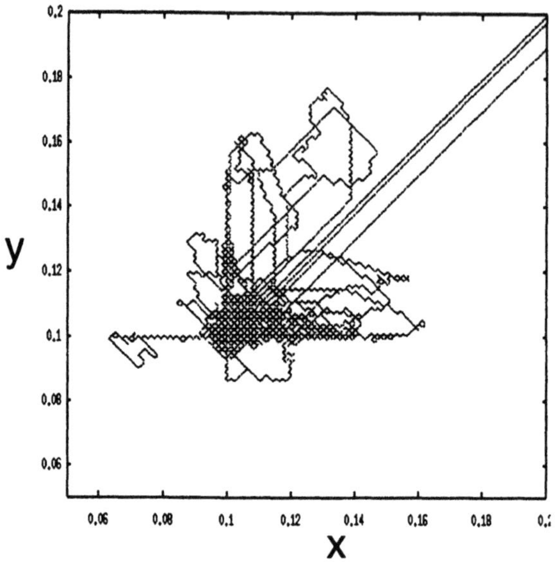

Fig. 1. Trajectories of the game dynamics are displayed as spatial trails on the two-dimensional plane. All trajectories from the same initial states are overlaid on the same plane. Player 1 and 2 move along the x and y axes, respectively. The trajectories which go to the Pareto efficient points are the straight lines in this plane.

most of which finally settle down to the Nash point. Yet some trajectories emerge, which leave from the vicinity of the Nash point to reach the Pareto efficient line (i.e. $x + y = 1$). In this example, the time averages of the players' positions come close to the fair Pareto efficient solution (0.5,0.5). Fig.2 shows the two attractors of this system with respect to the time evolution of player 1's position.

It is interesting to note that a precursory phenomenon is observed for the trajectories. At certain periods of time, one player behaves as 0101010 against other player's 1010101. That is, both players move back and forth every time step alternatively in a succession of rounds. Then, both players simultaneously leave from their home positions and start to step forward. When they have crossed the Pareto efficient line, they start to go back to their home positions in a zigzag manner (i.e. 2 steps back, 1 step forward and so on). After a certain period of time, they again go back to the Pareto efficient line. All this behavior becomes an attractor of some branched trajectories.

At times, players show similar oscillating behaviour around the diagonal line perpendicular to the Pareto efficient line (i.e., the $y = x$ line). However, in this case, players always deviate from the cycle to fall back to the Nash equilibrium point.

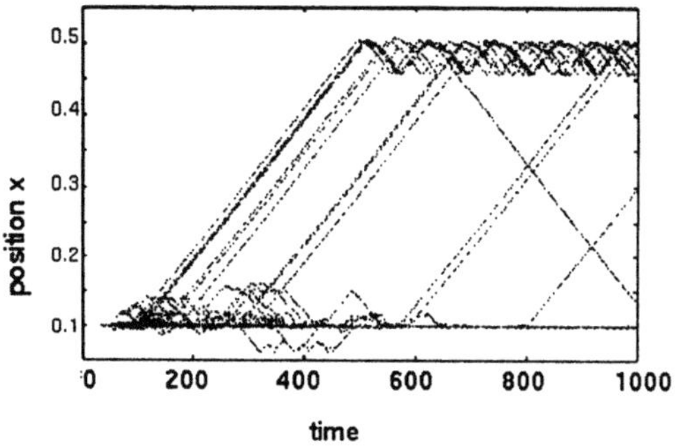

Fig. 2. Overlaid plots of player 1's position as a function of time. It is apparent that there are two attractors: one is a periodic attractor around x=0.5 (Pareto point) and the other one is a fix point attractor around x=0.1 (Nash point). When they fail to generate particular models, players cannot stay in the Pareto point and return to the Nash point.

4.2 Dynamics in the model space

The situation is now analyzed by the model of the players' internal dynamics. The periodic cycle underlying the co-operative motion is decomposed into three phases (Fig.3): climbing up the $x = y$ line, climbing down the $x = y$ line, and a phase of switching from the first phase to the second and back.

After a series of rounds of alternating between 1(backward) and 0(forward) actions, each player comes to 'believe' that the other is a 'Tit for Tat' player who copies its opponent's previous action. They then approach the Pareto efficient line by climbing up the $x = y$ line. During this phase, players come to obtain the same model against each other as they go through the common experience (i.e., repeat the same action sequence and move along the same line). In maintaining this image, however, both players cannot stay in the Pareto efficient line, but have to step back toward their home positions. A simultaneous step back motion is not included in a Tit for Tat algorithm. Therefore, they have to change their images. Complex images are then generated during the switching phase, wherein players, only on average, gradually climb down the x = y line.

Here the complexity of the images is measured by the number of clusters generated in the context space plots. The second phase then emerges where players have "negative Tit for 2 Tats -like" images. Due to these images, at least two successive forward actions are needed to return to mutual co-operation. Then at a certain period of time, they step forward to the Pareto efficient line by recovering the "Tit for Tat" image. But when players fail to perform such processes, however, they cannot come back to the Pareto efficient behavior. It is important to have an image such as the negative Tit for 2 Tats in order to anticipate the future, which retains the hope of recovering mutual cooperation by means of particular action sequences. The context space plots of those images are shown in Fig.4.

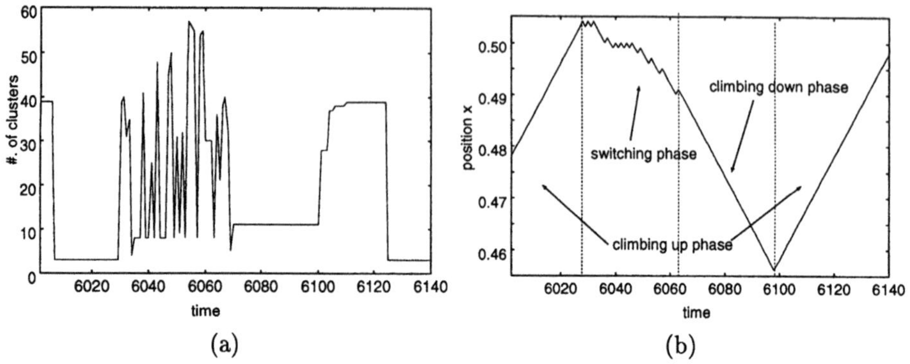

Fig. 3. Time evolution of player 1's model of player 2 at the Pareto efficient attractor. (a) The number of clusters generated in the context space, which is computed by the box counting method with the minimum size 0.1. (b) Corresponding positions of player 1. When players start to step towards their home position, their internal models become diversified. Three phases described in the text are marked in figure (b). This is a scale-up picture of Fig.2 between x=0.45 and 0.5.

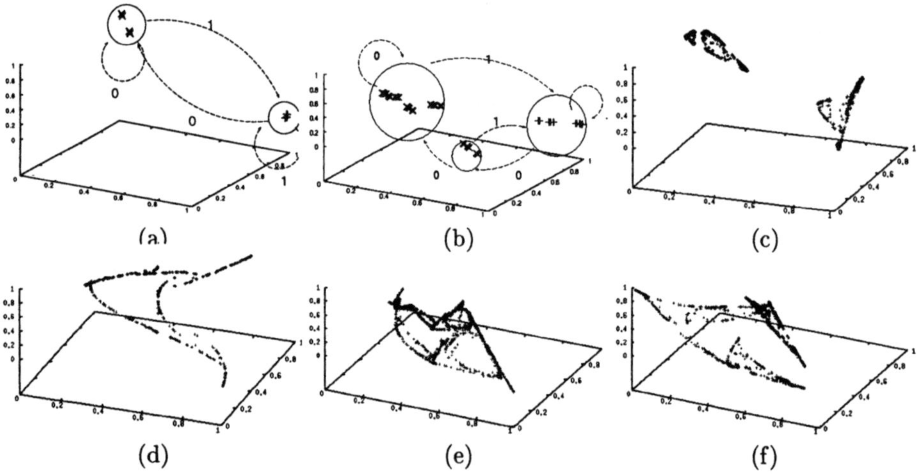

Fig. 4. Context Space plots generated at the Pareto efficient attractor. They correspond to the states in Fig.3 at time step = (a) 6012, (b) 6082, (c) 6117, (d) 6038, (e) 6043 and (f) 6058. The patterns of (a), (b) and (c) are relatively long-lived finite automaton-like images. In particular, the pattern (a) corresponds to Tit for Tat and (b) to negative Tit for 2 Tats in terms of the IPD game. The corresponding nodes of automaton are marked with circles. On the other hand, the patterns of (d), (e) and (f) are described as complex images in the text. They are images with more than 40 distinct islands of clusters.

It is worth comparing the above scheme with the iterated Prisoner's Dilemma (IPD) game situations (Axelrod 1984; Ikegami and Taiji 1998). In the IPD game, playing C immediately implies co-operation. Therefore, the 'Tit for Tat' image becomes the fixed point of this coupled recognizer system with model undecidability . In the present case that is not true. You have to deceive others by stepping back to your home position, which will paradoxically sustain mutual co-operation. To do so, players have to temporarily hold complex, non-automaton images, while changing between two distinct finite automaton images, Tit for Tat and negative Tit for 2 Tats. This model switching behavior is a temporarily periodic dynamics; however, the underlying model images include the 'strange attractor' of the corresponding iterated functional system, i.e. a dynamical recognizer.

Similar findings have been reported in the navigation learning of mobile robot experiments conducted by Tani (1998). Using a recurrent network, a mobile robot incrementally learns that a given environment consists of walls, corners, and obstacles. The image the robot learns of the environment is characterized by the context space plot, as is the case here. The experiment demonstrated that image switching occurs between a simple finite automaton and a chaotic attractor. The switching phenomenon is due to the strong cross-coupling between a real world mobile action and an organizing recurrent network. The model switching in our case is due to the strong cross-coupling between two sometimes competing inputs: the other players' action sequences and the spatial context.

5 Conclusions and Discussion

So far, we have demonstrated a new type of co-operation arising in the Rashevskyan game. A coupled dynamical recognizer has made it possible to simulate dynamic co-operation among players who anticipate their future actions by making internal models of other players. The dynamic co-operation is sustained basically by switching between two finite automaton models and other complex ones. Inability to have a static and unique model of the other player is due to the context dependency of the internal models. It may be a counter-intuitive fact that players cannot make a spatially-*independent* model of the other player.

How players can generate a context-free model may depend on the complexity of the game as well as on the network size used to generate the internal model. Anti-phase synchronization in action sequences between players emerges as a precursory phenomenon of this mutual co-operation phase, but without any intentions.

Model undecidability is a key concept here. A situation occurs when players cannot decide in which direction to move. Classical game theory has made efforts to remove this undecidability. Here, due to the undecidability, some branched trajectories are attracted to the Pareto efficient behavior. At the same time, the model undecidability reveals the pragmatic paradox (Bateson 1972; Rössler 1994) embedded in the *intelligent* players. That is, only minute differences in past action sequences or in spatial context can be interpreted as critical signals used to predict and plan future outcomes.

From our series of works of coupled dynamical recognizers, a connection can be found between a kind of games played and the images of other player generated in other players' mind. We believe that issues such as the cognitive development of early infants (e.g. how to get to know other people's mind) can be also elucidated using a minimal simulation model employing a coupled dynamical recognizer of the type described in this paper.

Acknowledgments: One of the authors (T.I.) would like to thank Dr. Luc Steels and his colleagues, for their courtesy while staying in Sony CSL-Paris. This work is partially supported by Grant-in aid (No. 09640454) from the Ministry of Education, Science, Sports and Culture.

References

1. Akiyama, E., Ph.D thesis *Dynamical Systems Game and Dynamics of games* (The graduate School of Arts and Sciences, University of Tokyo, 1998).
2. Anderelini, L., Some Notes on Church's Thesis and the Theory of the Games. *Theory and Decision* 29 (1990) pp.19-52.
3. Axelrod, R., *The Evolution of Cooperation*, (Basic Books, New York) 1984.
4. Bacharach, M., The Epistemic Structure of a Theory of a Game. in *Epistemic Logic and the Theory of Games and Decisions* (eds. M.O.L. Bacharach et al., Kluwer Acad. publ. 1997) pp. 303–344.
5. Bateson, G., *Steps to an Ecology of Mind* (Ballentine, New York 1972).
6. Binmore,K., Modeling Rational Players I. *Economics and Philosophy* 3 (1987) pp.179-214.
7. Dubey, P., Inefficiency of Nash Equilibria. *Math. Oper. Res.* 11 (1986) pp. 1–8.
8. Elman, J., Language as a Dynamical System in *Mind as Motion* (eds. R.E.Port a nd T.van Gelder, MITpress 1995) pp.195–226.
9. Ikegami, T., Taiji, M. Structures of Possible Worlds in a Game of Players with Internal Models. *Acta Poly. Scan.* Ma. 91. 283–292. 1998.
10. Pollack, J. B., The induction of dynamical recognizers. *Machine Learning*, 7:227–252, 1991.
11. Rashevsky,N., A problem in mathematical biophysics of interactions of two or more players which may be of interest in mathematical sociology. *Bull. Math. Biophys.* 9 (1947) pp. 9–15.
12. Rapoport, A., Mathematical theory of motivation interaction of two individuals. *Bull. Math. Biophy.* 9 (1947) pp.17–27.
13. Rössler,O.E., Fraiberg-Lenneberg Speech, *Chaos, Soliton and Fractals* 4 (1994) pp.1 25–131.
14. Rubinstein, A., Finite Automata play the repeated prisoner's dilemma, *J. Econ. Theory* 36 (1986) pp.83–96.
15. Taiji, M. and Ikegami, T., Dynamics of Internal Models in Game Players. (*Physica D* in press).
16. Tani, J. and Fukumura, N., Self-organizing Internal Representation in Learning of Navigation: A Physical Experiment by the Mobile Robot YAMABICO, *Neural Networks*, Vol. 10, No. 1, pp. 153–159. 1997.
17. Tani, J., An Interpretation of the 'Self' from the Dynamical Systems Perspective: A Constructivist Approach. *Journal of Consciousness Studies*, Vol.5 No.5–6, 1998.

Emergence of Adaptive Society with Competitive Selfish Agents

Takashi Ishida, Hiroshi Yokoi and Yukinori Kakazu

Complex Systems Engineering, Hokkaido University,
N13W8, Kita-ku, Sapporo 060-8628, Japan.
{ishida, yokoi, kakazu}@complex.eng.hokudai.ac.jp

Abstract. In multi-agent systems, agents receive incomplete local information, and they must achieve global tasks. We consider this ability to determine the appropriate action to correspond exactly to "norms of behavior" in the human sense. In this study, we constructed a competitive social system consisting of selfish autonomous agents. Each agent had an independent evaluation table for their actions. The Agents' strategies were adapted on the basis of their individual evaluations, and agents' norms of behavior satisfy. In such systems, only the agents that adapt not only their strategy but also their norms of behavior to the environment were able to survive.

1 Introduction

This paper reports on our attempt to construct a model of a multi-agent system in which the agents interact on a basis of their selfish decision-makings. The main feature of the suggested model is that each agent can generate its own "norms of behavior". The objective of our study was to clarify the behavior of a society constructed from agents that have the ability to change their "norms of behavior" adaptively, and to investigate the process of problem-solving.

In this study, we used the Iterated Prisoner's Dilemma (IPD) [4], a typical model of game theory, as the environment in which the agents would interact. Using simple alternative action selection mechanisms, the agents recognize both the conditions they should obey and the strategies of the other agents as their environment. Then a system in constructed in which the agents have proper norms of behavior. Computer simulations show that self-preservation and cooperative behavior are acquired adaptively. We consider that introduction of the independent norms of behavior (as an internal payoff matrix) to each agent leads to an ordered society from a social dilemma situation.

In the original IPD games, all agents who interact with each other have the same norms of behavior. In contrast, in the suggested model, the agents have free norms of behavior. Because each agent processes the freedom to react to a single phenomenon with various norms of behavior, we expected various functions to emerge. Hyper-game analysis [5] is an example of this concept. The suggested model, however, is fundamentally different from that because it includes the adaptive searching by each agent for norms of behavior.

2 Social Dilemma on Multi-agent Model

Each of the N -agents must decide whether to move a piece of the cargo at each time step. When an agent carries a piece of cargo, it expends energy (E). After that, all of the agents receive a reward (Ri) according to the number of pieces of cargo (i) that they have moved. Note that not only agents who worked but also agents who rested receive the same energy from the environment. Rewards are paid as energy. (Fig.1, Left)

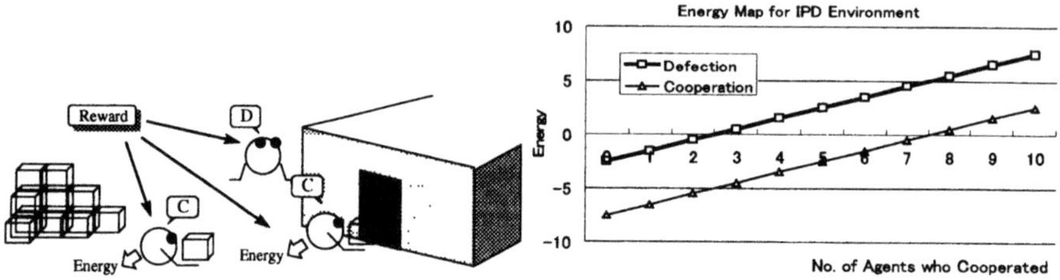

Fig. 1. Prisoner's Dilemma with Multi-Agent Model and Payoff Graph of N-Person IPD. This chart indicates that choosing "D" offers a better payoff than does choosing "C" even if the number of cooperating agents is decreased by the defection of the agent. This rationality corresponds to condition (1) of the Iterated Prisoner's Dilemma. However, if all of the agents choose "D" based on this rationality, the payoff received by each agent is much smaller than that received if all of the agents cooperate. This dilemma corresponds to condition (2) above.

Then, by representing the moving of a piece of cargo by an agent as "C" (Cooperation), and the failure to move a piece of cargo as "D" (Defection), the situation of this multi-agent system can be considered to correspond to the "Iterated Prisoner's Dilemma". That is to say, the payoff table of the system includes the following dilemma.

(1): Whatever actions the opponents choose, the choice of "D" (Defection) always produces good reward. ($R(i-1) > Ri - E$)

(2): the agents receive a better reward if they all choose "C" (Cooperation) rather than "D". ($RN - E > 0$)

The situation of the above-mentioned "Society Dilemma" [3], is extended from a one-on-one payoff table by using the payoff chart (Fig.1, Right) in which the number of cooperating agents is shown as the horizontal axis [2]. Two lines in the chart represent the payoff received by the defecting agents (thick line) and the payoff received by the cooperating agents (thin line).

3 Structure of Competitive Society and Agents

In the computational simulations, when i agents chose "C" at any step, the reward (energy) received by the cooperator is given as $C_i = Ri - E - E_0$, and $D_i = Ri - E_0$ represents the reward received the agents that chose "D". Where, E_0 is the energy required for existence. In this computational simulation, in order to maintain the population balance of the society, It is determined as $E_0 = (RN - E)/2$

Thus, agents compete for existence according to the following energy rules. Initially, agent a is given a certain amount of energy (Eng_a) for existence. The agents then indicate their intention to carry cargo according to their own strategy engines (see below).

The environment (energy rules) provides energy as a reward for the agents based on their decisions $act_a = \{C, D\}$.

$$Eng_a(t+1) = \begin{cases} Eng_a(t) + C_i & \text{if } act_a(t) = C \\ Eng_a(t) + D_i & \text{if } act_a(t) = D \end{cases} \qquad (1)$$

If not enough agents cooperate, the amount of reward energy will be less than that of expended energy. When an agent has expended all of its energy, it is culled from the society system, and a new agent with randomly determined parameters is introduced into the system.

In this model, at the time step t, agent a has an independent payoff table T_a, and the agent generates its action strategies $st_{aj}(t)$ according only to that table.

Since each agent has a different action standard, in this model, we call each agent's payoff table its "norms of behavior" in a human sense. As shown in Fig.2, the payoff table evaluates the action of agent in totally order according to the number of agents in the whole system that chose the action "C" at the same time.

Action	Number of Cooperator 0~5	Number of Cooperator 6~10
C	P^a_{CD}	P^a_{CC}
D	P^a_{DD}	P^a_{DC}

Fig. 2. Norms of Behavior Matrix (Payoff Table T_a) for IPD. P^a_{CC}, P^a_{CD}, P^a_{DD} and P^a_{DC} represent the evaluated values of actions of an agent a.

The action strategies st_{aj} that each agent stores independently are corded to strings based on the history of past actions of all the agents [6]. Those strategies are polished up by genetic operations (GA). Values in matrix T_a are used as fitness of each strategy at every interaction. Fundamental behaviors of simple games between the agents with same payoff table (IPD norms) are researched in [1].

4 Computational Experiments

The above-mentioned parameters R=1, N=10 and E=5 were used for 1000 generations of computational simulation of the competitive social system. The number of action strategies (n) in each agent's gene pool is 20. The initial amount of energy

($Eng_a(0)$) for each newly generated agent is 1000. The number of times that interactions are repeated in each generation is 50.

The results of experiments performed using the above parameters are shown in Fig. 3(Left).

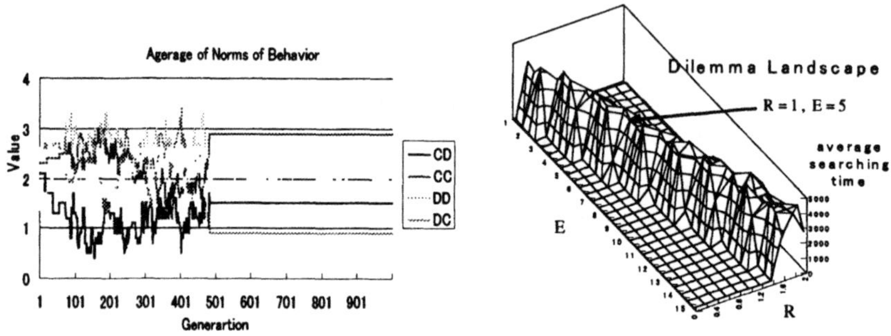

Fig. 3. Average Values of Agents' Norms of Behavior for each Generation (Left) and Landscape of "dilemma" corresponds to the parametric space of multi-agent system. (Right)

The averages of the values in each payoff table of all agents are plotted for every generation in Fig. 3 (Left). Until the 500th generation, the system was not stable because the agents couldn't recognize or adapt to the environment (energy rules), so those agents were continually being replaced by new agents. After that, agents that could adapt to environment began to be accumulated within the system. From that point, no agents became extinct, and system was fixed. At that time, the highest acquired value on the norm of behavior table was P^a_{CC}; i.e., the agents' norms of behavior showed a preference for cooperating with each other. However, when each agent's norms table was investigated, it was found that though eight agents thought highly of cooperative action ("C") and worked hard, two agents received higher reward energy by defecting ("D"). That is to say, it is a society in which a group of sucker (cooperative) agents is dominated by a much smaller group of extortionate (defected) agents.

4.1 Evaluation and Analysis of the problem Field

In the above experiment, although a society in which the agents could achieve their objective was constituted by about the 500th generation, it is thought that the difficulty in determining what is the appropriate society depends on the structure or parameters of the environment. Fig. 3 (Right) shows the landscape of the average of search time in 20 trials for each value of R and E. That is, the average number of generations by which the system was fixed and a one agent society was acquired after the selection of agents with independent norms of behavior. However, the maximum searching time is set as the 5000th generation.

Fig. 3 (Right) shows that the searching time landscape produced a ridge in the parameter space of R and E. The location of this ridge corresponds to the conditions of the dilemma, and the height represents the difficulty of the search for the solution to the dilemma.

On the other hand, in the two domains of the parameters that are separated by the dilemma ridge, the searching time needed to constitute the appropriate society was extremely short; i.e., there is a broad range of stable solutions corresponding to the combination of norms of behavior of agents, so search for society was easy. In the area in front of the ridge (the domain of a small R value and large a E value), the acquired society was lazy ("defecting" was the dominant strategy) because the environment with such parameters was advantageous to defecting. On the other hand, in area behind the ridge (the domain of a large R value and a small E value), the acquired society was composed of mainly hard-working agents.

These results clarify the relation map between IPD and a multi-agent system in which the aim of the agents is to achieve a set task and for self-preservation. At the same time, these results also demonstrate the complexity of that relationship.

5 Conclusion

We attempted to construct and investigate a competitive society model consisting of agents with self-organized norms of behavior, against the complex requirement from environment. In this model, agents competed on the basis of game interaction according to only their own norms of behavior.

The environment for this system consisted of rules of the acquisition of rewards. In this study, the Iterated Prisoner's Dilemma model was introduced as the environment rules. In such an environment, the agent group that can recognize the IPD rules and cooperate with each other is self-organized in the computer simulations. At the same time, the agents form themselves into two groups of norms of behavior that show a kind of dominant relationship just as do real societies.

We consider that the setting of independent norms of behavior enables the emergence of a society among the agents. Further, such a competitive social system model also implies that multi-agent systems can harmonize themselves to their environment autonomously through the interactions of the agents.

References

1. Bankes, Steve. : Exploring the Foundations of Artificial Societies: Experiments in Evolving Solutions to Iterated N-player Prisoner's Dilemma, Artificial LIFE IV pp337-342. (1995)
2. Hamburger, Henry. "N-person prisoner's dilemma" Journal of Mathematical Sociology, vol.3, 1973.
3. Garrett Hardin. 1968. The tragedy of the commons, Science, vol. 162.
4. Ikegami, Takashi. : From genetic evolution to emergence of game strategies, Phisica D 75 310-327. (1994)
5. Kijima, Kyoichi. 1996. Intelligent Poly-agent Learning Model and its Application.: Information and Systems Engineering, 2, 47-61,.
6. Lindgren, K. Evolutionary Phenomena in Simple Dynamics, In C.G. Langton, J. D. Farmer, S. Rasmussen, and C. Taylor. editors, Artificial LIFE II: Proceedings Volume of Santa Fe Conference, volume XI. Addison Wesley: series of the Santa Fe Institute Studies in the Sciences of Complexities, Redwood City, CA, 1992.pp295-312.

Dynamical Systems Game

(The Lumberjacks' Dilemma Model)

Eizo Akiyama and Kunihiko Kaneko

C/O K.Kaneko, Graduate School of Arts and Sciences, The University of Tokyo, Komaba 3-8-1, Meguro-ku, Tokyo 153-8902, Japan,

Abstract. With regard to the dynamics of games, a framework what we call "Dynamical Systems Game" model is presented, where the game itself can be affected and changed by the players' behaviors or states. Relation between the game dynamics and the evolution of strategies is discussed by applying this model. Computer experiments are carried out for a simple model, to show the evolution of dynamical systems with effective use of resources.

1 Main Issues about Game Dynamics

In this paper, we would like to introduce a framework model to deal with the issues of "game dynamics." In the real world, when we decide to select an action and carry it out, our behavior sometimes changes our own game environment. A change in the environment may also have an effect on a player's decision-making process. Further, the utility of the same behavior sometimes varies according to an individual's (or others') current circumstances. In game theory[3], such situations are sometimes represented by one (large) game: that is, from the past into the future, all possible actions of all players at all points of time are taken into account. Thus all possible bifurcation patterns of the game are derived, with this situation as a whole depicted as one huge game-tree. In this way, we can project the course of time into a static game and analyze its solution in the form of a game-tree or a game matrix. "Strategy" here means the action plan for all points in time, and the analysis of the rational solution for a game is possible only when we know all the possible actions from the past to the future. It seems, however, that we ourselves do not always make such decisions even if they are theoretically possible. Here we would like to present another framework model whose formulation can *naturally* deal with the above issues. This is what we call the "Dynamical Systems (DS) Game model," where the game itself is described as a dynamical system[1]. Dtailed description of the DS Game is given in [1].

[1] This type of problems was first considered by Rashevsky[5] and Rapopport[4]. Furthermore, Rössler introduced the abstract model of multiply linked coupled autonomous optimizers[6], as a model for complex biological systems, in particular for brain.

2 Lumberjacks' Dilemma as a DS Game

Modeling As an application of this DS Game framework, we present in this paper what we call the "Lumberjacks' Dilemma (LD) Game." Let us consider the following situation:

There is a wooded hill where several lumberjacks live. The lumberjacks fell the trees for their living. They can maximize their collective profit if they cooperate in waiting until the trees have fully grown before felling them, and sharing the profits. However, any lumberjack who fells a tree earlier will take the entire profit on that tree. Thus each lumberjack can maximize his personal profit by cutting trees earlier. If all the lumberjacks do this, however, the hill will go bald and there will eventually be no profit. This situation inevitably brings about a dilemma.

This LD Game can be categorized into the *social dilemma* that deals with the problem of forming and maintaining cooperation in a society, which is represented by the classical story "the tragedy of the commons" presented by Garret Hardin in 1968[2]. Its structure is logically similar to that of the Prisoners' Dilemma Model if considered at the level of a static game. In other words, it can be represented in the form of an n-person version Prisoners' Dilemma if we project it onto static games. Here we note several important differences. Dynamics of the size of trees should be expressed explicitly in this LD Game. The yield of a tree, and thus the lumberjacks' profit, differs by the timing when the tree is felled. The profits have a continuous distribution, because the yield of a tree can have a continuous value. A lumberjack's decision today can affect the future game environment through the growth of a tree.

By formulating the above game concretely, though the details of modeling is ommited is here, we have analyzed the game dynamics of LD Games and investigated the nature of evolutionary phenomena among players in LD Games based on computer simulations. From among the LD Games under various kinds of conditions (e.g. the number of players, the number of trees, the way of implementation of strategies, etc.), we present in this paper the LD Games of 1-person and 1-tree as the introdion of DS Games, focusing on the relation between game-dyanmics and the player's decision where any confliction among players' decisions does not exist.

Experiments on 1-Person LD Games Let us consider a simplified model of a 1-person 1-tree LD Game. Here we make two simplifications. First, the player never refers to his or her state, that is, the player makes his or her decision only by referring to the size of the tree. Second, the player cuts the tree if the size of the tree exceeds a certain value, called the ***decision value***, x_d. We denote the size of the tree by x and the states of the player by y. Note that x and y are multidimentional vectors in case of n-person m-tree LD Games. The value of x_d uniquely decides the time series of the phase (x, y). The attractor of the time series can be a fixed point, periodic, quasiperiodic, or chaotic motion, which depends on the dynamical law given to the system.

As in the 'bifurcation diagram,' we have plotted in Figure 1-(a) the set of values that x takes in the attractor with the change of x_d. The figure gives a

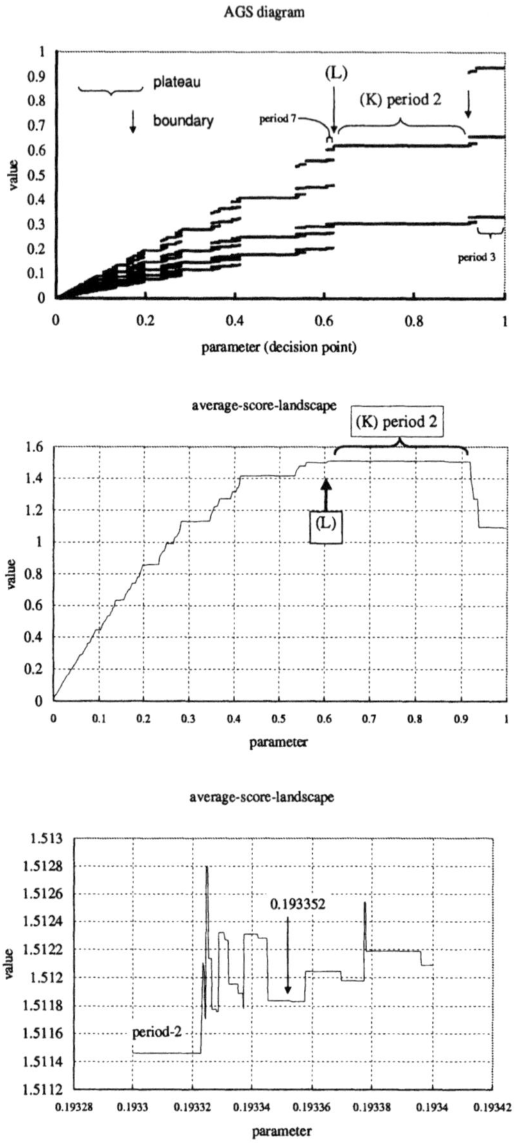

(a)

(b)

(c)

Fig. 1. 1-person 1-tree mountainside-type LD Game — (a) AGS Diagram : Change of the attractor with the strategy is plotted. A set of values of x at the attractor (all the values that x takes between the 200th and 400th rounds), is plotted with the decision value, x_d, given in the horizontal axis. For example, the two parallel straight segments around $x_d = 0.8$ show that the dynamics of x is attracted to the period-2 cycle between the values around 0.3 and 0.6. (b) average score landscape: The average score from 0th round to 400th is plotted with the change of the decision value x_d. (c) complicated structure in the average score landscape of a evolutionary simulation: The average score landscape based on the decision-making function of the player who actually appeared in the 1-person 1-tree simulation (the 5848th generation) is shown.

diagram how the attractor of game dynamics changes with a parameter in the decision-making function. Let us call such a figure the *AGS Diagram* — the transition of the **A**ttractor of the **G**ame dynamics versus the change of the **S**trategy). With the AGS Diagram, one can study how the nature of game dynamics shifts among various states (fixed point/periodic/chaotic game-dynamics, or productive/unproductive game-dynamics, etc.) with the change in decision-making. The following two characteristics in Figure 1-(a) are noted — (1) For each decision value, its corresponding attractor is always a periodic cycle. (2) There are infinite numbers of 'plateaus,' in which the attractors are completely the same over some range of decision values (For examples of such plateaus, see the period-2 and 3 plateaus in Figure 1-(a)).

Corresponding to Figure 1-(a), the average score that the player gets during the attractor part of the game dynamics is plotted in Figure 1-(b), with the change of the decision value x_d. We will call this diagram *average score landscape.* The optimal decision value seems to exist in the period-2 plateau around 0.8 as far as we can see in the scale of this figure. Consequently, the best strategy for the player seems to construct this period-2 dynamics, which was indeed observed in the early stage of the LD Game simulation shown later. However, by examining the close-up of the left-hand side of this period-2 plateau of Figure 1-(a), (indicated by the arrow with (L)), infinite numbers of plateaus are found to accumulate there and they form a fractal structure (so-called *devil's staircase*), where more profitable dynamics than period-2 exist, among which the period-11 is the optimal (productive) for the player.

As for the 1-person "line-type" LD Game, the AGS Diagram shows that the dynamics is attracted to a quasiperiodic motion if $x_d \leq 2/3$, otherwise to a periodic motion. The average score landscape shows that the landscape is not stepwise but a monotonically increasing straight line for $x_d \leq 2/3$, and that the optimal decision value lies in $x_d \geq 0.75$ for this 1-person line-type LD Game.

Simulations on Evolutionary 1-Person LD Games An LD Game in general has several trees and several lumberjacks. Here, let us consider an evolutionary LD Game with a single lumberjack as the simplest example. Note that there exists a dilemma-like situation even in 1-person LD Games (although the meaning of 'dilemma' is different from that of 2-person games). By cutting the tree one always gets a higher score at the next step, which is not usually good for a long-term fitness. Here we need a strategy that takes into consideration the dynamics of the game.

In Figure 2-(a)(b) (each of which is called "fitness chart"), the fitness that the fittest species of each generation ("generationarily fittest species") gets, which we call the "fittest value," is plotted with generation. In a fitness chart, the horizontal axis shows the generation, while the vertical axis shows the fittest value of each generation. Figure 2-(a)(b) show that the fittest value increases monotonically and stepwise with generation. This 'monotonic increase' is rather natural, since in a 1-person game, the fitness is determined solely by its strategy.

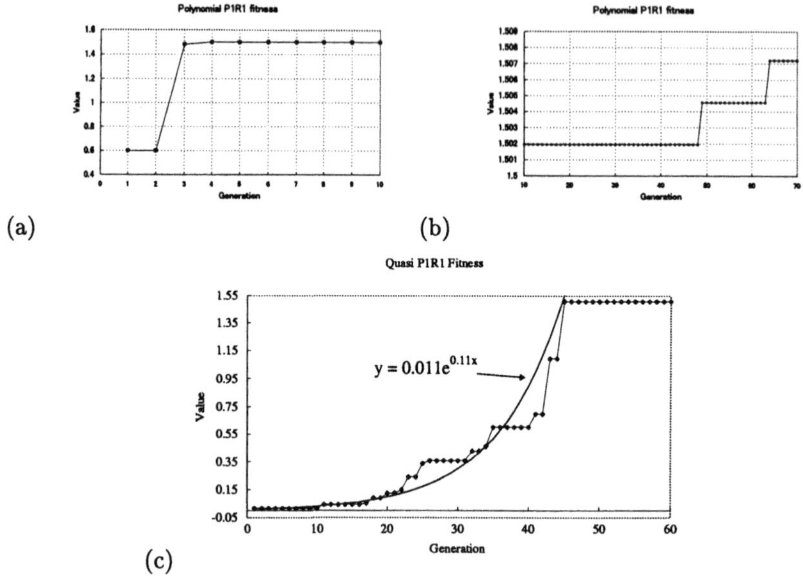

Fig. 2. "Fitness chart" in a 1-person LD Game: The fitness chart of a mountainside-type LD Game is shown in (a) up to the 10th generation, in (b) for 10-70 generation. In all the figures, the horizontal axis shows the generation. The vertical axis shows the "fittest value" of each generation, which is the fitness value of the fittest species of each generation. (c) The fitness chart of the 1-person line-type LD Game from the 1st generation to 70th is plotted with a fitting curve by exponential function.

The dynamics of the player's action and of the states of both the player and the tree observed in this mountain-side type LD Game is attracted to periodic cycles. For example, the period-2 dynamics is shared by the generationarily fittest species over long generations, where the evolution of dynamics is seen only in the transient before the dynamics falls onto the cycle. It lasts from the 49th to the 3605th generation when the management of the period-7 dynamics is achieved. In the later evolution, a new dominant species of period-23 appears at the 5848th generation and that of period-11 (the optimal attractor of game dynamics for the player) appears at the 8984th generation. The time required for the evolution of the periodic pattern gets longer at later stages. This is because the evolution of the species is likely to be trapped by meta-stable states, as is seen by the average score landscape given in Figure 1-(c) that is drawn based on the decision-making function of a player who appeared in the simulation of 1-person 1-tree LD Game.

As for the simulation about the 1-person *line-type* LD Game, the optimal game dynamics is easily realized as early as 45th generation. Fitness chart Figure 2-(c) shows that the fittest value increases gradually but not step by step with generation. Such smooth and quick evolution obviously depends on the structure in the AGS Diagram and consequently on the average score landscape.

3 Discussion

Let us discuss general advantage of the DS Game model over other models. DS Game modeling is suitable to study the evolution/learning of the decision-making subjects living in the world that can be described by the dynamical system. Models of game theory certainly have strength in dealing with the issues of decision-making subjects who interact with each other, but they are not congenial to dynamics by nature, and so, it cannot touch upon the issues that can be studied only at the levels of dynamics. For example, the devil's staircase in the AGS Diagram leads to stepwise innovation in the mountainside-type LD Game, while the evolution is gradual in a simple dynamic structure in the line-type LD Games. In a 1-person games, the game of the line-type version has the same structure as the mountainside-type one from the static point of view, in that a player who cuts the tree as early as possible can take the short-range profit, but loses the long-range gains. However, the fundamental difference appears at the level of game dynamics and strategies' evolution (At the level of plural-person games, the present modification keeps the common social dilemma that exists in the previous verion). Thus, several different DS Games can be categorized into the same static game if modeled by the traditional game theory, but there is also a possibility that these games have completely different nature at the level of the dynamical structure, especially when evolution or learning is involved.

Acknowledgments
The authors would like to thank T.Ikegami and S.Sasa for useful discussions. This work is partially supported by Grant-in-Aids for Scientific Research from the Ministry of Education, Science and Culture of Japan. The first Author is supported by research fellowship from Japan Society for Promotion of Science.

References

1. Akiyama E.: Dynamical Systems Game and Dynamics of Games. submitted to Physica D.
2. Hardin, G.: The tragedy of the commons. Science (Washington, D. C.) **162** (1968) 1243–1248
3. Neumann, J. von and Morgenstern, O.: Theory of Games and Economic Behavior. Princeton University Press. Princeton. (1944)
4. Rapoport, A.: Mathematical theory of motivation interaction of two individuals. Bull. Math. Biophys. **9** (1947) 17–27
5. Rashevsky, N.: A problem in mathematical biophysics of interactions of two or more individuals which may be of interest in mathematical sociology. Bull. Math. Biophys. **9** (1947) 9–15
6. Rössler, O. E.: Fraiberg-Lenneberg Speech. Chaos, Solitons & Fractals. **4-1** (1994) 125–131 Elsevier Science Ltd.

High Speed Hardware Computation of Co-evolution Models

Yoshiki Yamaguchi, Tsutomu Maruyama and Tsutomu Hoshino

Institute of Engineering Mechanics and Systems, University of Tsukuba
1-1-1 Ten-ou-dai Tsukuba Ibaraki 305 JAPAN
yoshiki@darwin.esys.tsukuba.ac.jp

Abstract. Field Programmable Gate Arrays (FPGAs) can provide the most suitable circuits for given problems by reconfiguring its circuits. In this paper, we show that a FPGA chip can achieve about 120 times of speedup compared with a workstation (Ultra-Sparc 200 MHz) in the computation of a co-evolution of strategies and scores in Iterated Prisoner's Dilemma game. This speedup makes it possible to challenge more complex problems beyond the limitation by software.

1 Introduction

Field programmable gate arrays (FPGAs) can be reconfigured for each problem by down-loading configuration data from memory (ROM) or computers. Thus, systems with FPGAs can provide the most suitable circuits for given problems. In this paper, we show that a FPGA chip can achieve about 120 times of speedup compared with a workstation (Ultra-Sparc 200 MHz) in the computation of a co-evolution of strategies and scores in Iterated Prisoner's Dilemma (IPD) game. By processing IPD games in parallel with pipelined circuits implemented in a FPGA chip, the computation time of the model for a initial state can be reduced from several days to about an hour. This drastic speedup makes it possible to study behaviors that appear over a large number of generations with a large number of agents [6].

The co-evolution model consists of two layers. In the first layer, scores for IPD games are evolved using a genetic algorithm. Scores vary within the range of Prisoner's Dilemma game, and scores that attract more agents in the second layer gradually increase. In the second layer, agents play IPD games with all other agents following the scores that they believe and evolve using Lindgren's model[1, 2]. With this model, we can observe how cooperative behaviors and scores emerge, which may lead to the understanding of emergence of social morals.

2 Hardware Computation

In evolutionary computations, same sequences of operations are repeatedly applied to a large number of individuals (agents). With dedicated hardwares, the sequences can be applied to all individuals in parallel, and each computation of the sequences can be pipelined. The expected speedup by hardware is N (number of agents) × P (speedup by pipeline processing, which is almost equal with the depth of the pipeline when the data sequences are long enough). However, dedicated hardware systems with ASICs can not deal with many problems, because details of the sequences depend on given problems and vary considerably.

Field programmable gate arrays can be reconfigured for each problem. Thus, systems with FPGAs can provide the most suitable circuits for given problems. The speed and size of FPGAs are drastically improved recently, and several systems show very good performance in evolutionary computations[3, 4, 5]. In evolutionary computations, the size of FPGAs is especially important, because, as described above, the speedup by hardware is proportional to the number of agents processed in parallel. The size of LSIs (size of FPGAs, off course) is steadily being improved compared with the improvement of the operation speed. Therefore, the speedup by hardware computation will increase, as the LSI technologies progress.

3 Co-evolution of Strategies and Scores

3.1 Co-evolution model

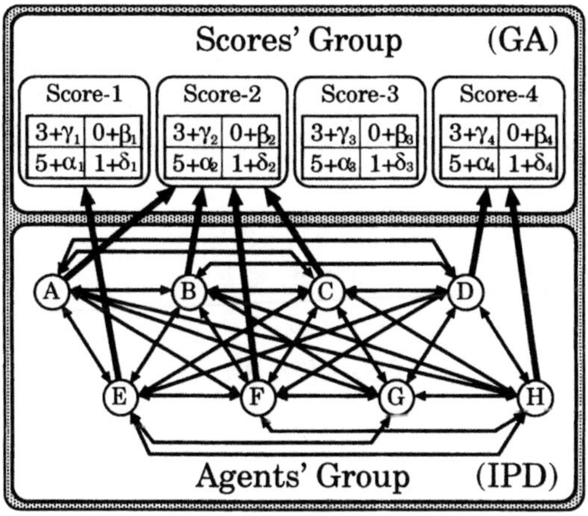

Fig. 1. Co-evolution of Strategies and Scores

In this section, we describe the details of a co-evolution model of strategies and scores. Figure 1 shows the overview of the model. The scores in the payoff matrix ($3+\gamma_i$, $0+\beta_i$, $5+\alpha_i$, $1+\delta_i$) for IPD games are evolved using a simple genetic algorithm. Scores vary within the range of Prisoner's Dilemma game, and a score that attracted more agents in the second layer gradually increases. In figure 1, agents A, B, C and F follow Score-2, and no agent follows Score-3 (each agent chooses one of scores or not to follow any score at random). The fitness of the scores is calculated using the total points gained by IPD agents that follow the scores. Therefore, Score-2 will increase in the next generation, and Score-3 may be deleted (the agents that have followed the Score-3 choose new scores at random).

IPD agents get their points by playing IPD games with all other agents using the scores they follow, and evolve using Lindgren's model[1, 2]. When an agent does not follow any scores, default points are used.

With this model, we aim to get rid of subjective interpretation of behaviors of agents, and we can observe how cooperative behaviors and scores emerge.

3.2 Scores

Table 1 shows a payoff matrix. α_i, β_i, γ_i and δ_i are parameters, and by changing the values of these parameters, categories of games becomes one of the follows.

$$T > P > R > S \qquad \textit{Deadlock} \tag{1}$$
$$T > R > P > S \qquad \textit{Prisoner's Dilemma} \tag{2}$$
$$T > R > S > P \qquad \textit{Chicken} \tag{3}$$
$$R > T > P > S \qquad \textit{Stag Hunt} \tag{4}$$

In our evaluation, we fixed the value of γ_i and δ_i to zero, and α_i and β_i vary within the range of (5) and (6), in order to satisfy the requirement by Prisoner's Dilemma game ((2) and (7)). An agent gets more payoff by defecting another agent when α is larger, and has a more bitter experience by being defected when β is smaller (close to -1).

$$0 \leq \alpha_i < 1 \tag{5}$$
$$-1 < \beta_i \leq 0 \tag{6}$$
$$S + T < 2R \tag{7}$$

Table 1. Payoff Matrix

		Agent-2	
		Cooperate	Defect
Agent-1	Cooperate	Reward $3+\gamma_i$	Sucker $0+\beta_i$
	Defect	Temptation $5+\alpha_i$	Punishment $1+\delta_i$

3.3 Strategies

Strategies (agents) evolve using Lindgren's model. In Lindgren's model, each agent decides next move according to five previous moves. Thus, each agent has 32 bits table (at maximum), and the values and the length of the table are mutated. Noise is assumed in the communication between the two agents. Therefore, a move by an agent are not always reported to another agent correctly.

In our model, agents get their payoff using the payoff matrix (scores) that they follow. In initial state, no agents believe scores, and default payoff matrix (α and β are zero) is used. By mutation, agents begin to follow one of the scores, and stop following the scores.

3.4 An Example of Co-evolution of Strategies and Scores

Figure 2 shows the average points obtained by agents in this model. From generation 0 to 20,000, various strategies appear and vanish. In generation 20,000, defective strategies begin to dominate the world, and are replaced by cooperative strategies around generation 27,500. These drastic changes are caused by emergence of defective and cooperative strategies by mutation.

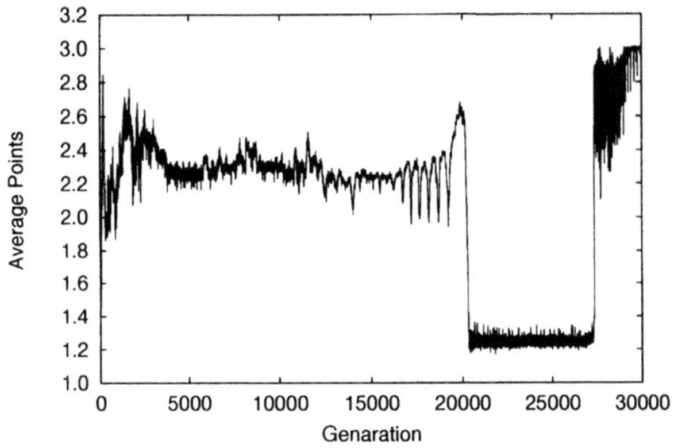

Fig. 2. Average Points of Agents

Figure 3 shows the distribution of the distances of the scores (20 scores are used in this evaluation). The maximum distance of two scores is $\sqrt{2}$, because α and β vary in the range (5) and (6) respectively. In figure 3, the distribution of the initial phase (from generation 0 to 2500) is very large. In this phase, many kinds of strategies are created by mutation and can survive, because the strategies are not evolved enough. However, the distribution becomes smaller as strategies are evolved, and scores with larger α (close to 1.0) and larger β (close to 0) become dominant (around generation 2,500). This pair of α and β gives an agent better point when the agent defects another agent, and gives smaller damage when the agent is defected. The distribution begins to increase around generation 24,000. As shown in figure 2, the world around this generation is dominated by defective strategies. The point obtained by two defective agents is 1 as shown in table 1, and independent with α and β. Therefore, the mutation in α and β is neutral. During this phase, the distribution gradually increases. Around generation 27,500, the defective strategies are replaced by cooperative strategies suddenly. The distribution continues to increase in spite of the change because the point obtained by two cooperative agents is also independent with α and β. Around generation 30,000, some strategies that sometimes defect another agent begin to appear, and the distribution begins to decrease.

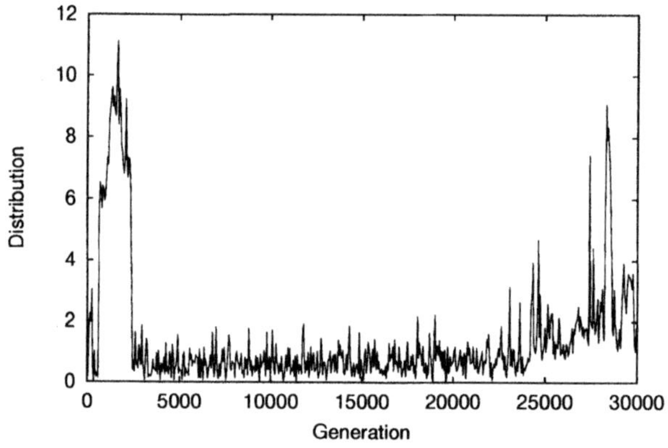

Fig. 3. Distribution of Scores

Figure 4 and 5 show the positions of scores of generation 20,326 and 28,343 on α and β plain. In figure 4, all scores gather around a point (larger α and larger β as described above), while scores are dispersed in figure 5 because of the neutrality of mutation in α and β in the cooperative world.

4 Hardware Implementation

4.1 Parallel Processing of IPD games

The size of a FPGA chip is not large enough to implement the computation of the whole model. Therefore, we have to decide which parts should be computed by hardware. The most time exhaustive part of the model is the computation of IPD games. The order of the computation is $N_{agents} \times N_{agents}$ (N_{agents} is

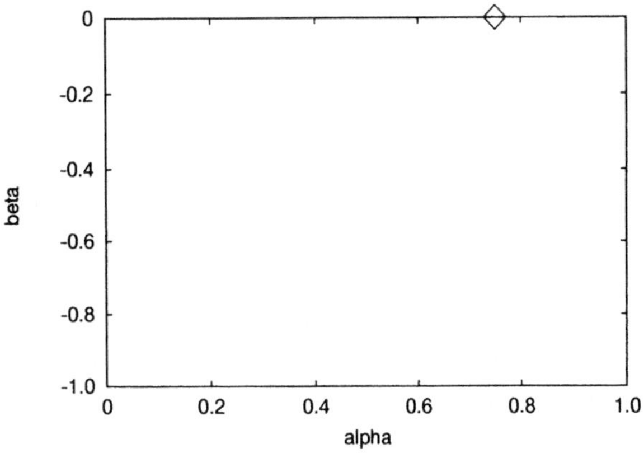

Fig. 4. Location of Scores (Generation = 20326)

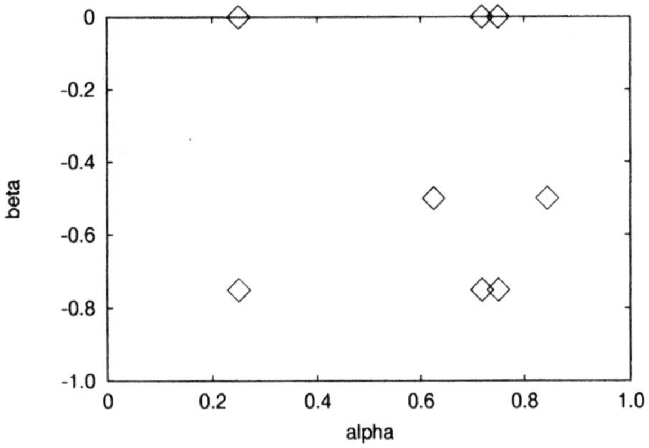

Fig. 5. Location of Scores (Generation = 28343)

the number of agents) because all agents play with all other agents. The orders of computations of other parts are at most N_{agents} or N_{scores} (N_{scores} is the number of genes used in the genetic algorithm). N_{agents} and N_{scores} are 1024 (at maximum) and 20 in the current evaluation, respectively. Therefore, all hardware should be used for the computation of the IPD games, and other parts can be processed by software without decreasing the performance.

In the current hardware implementation, 24 games (48 agents) can be processed in parallel with a FPGA chip (ALTERA EPF 10K100). In order to process more agents, we need external memories to store whole agents, and have to exchange when all the games between 48 agents finish. The time for exchanging agents is also negligible compared with computation time of IPD games.

4.2 Hardware for a IPD game

Figure 6 shows the hardware for a IPD game. In figure 6, next moves of two agents are computed in one clock cycle, and the payoff of both agents are computed and accumulated in the next two clock cycle. These two phases are pipelined into three stages (adders are pipelined into two stages in order to achieve high operation speed), and we can finish N times repetition in $N+2$ clock cycles as shown in Figure 7. In the hardware, the value of α and β are represented as fixed point variables in order to simplify the adders.

Two noises are also computed in parallel using pseudo random number generators. In order to realize high quality (long frequency) of pseudo random numbers without slowing down operation speed, adders in the pseudo random number generators are also pipelined.

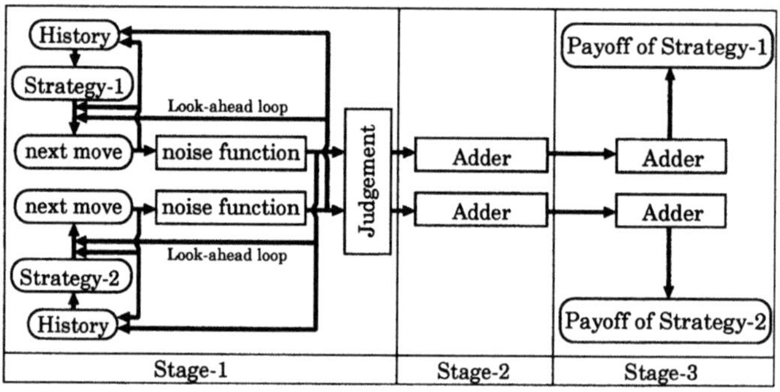

Fig. 6. Block Diagram for a IPD game

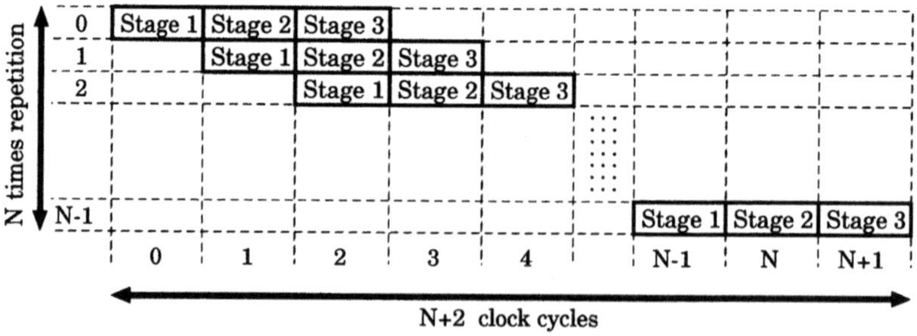

Fig. 7. Pipelining Processing of IPD games

4.3 Details of the Circuits Deciding Next Move

Figure 8 show details of the circuits for deciding next move. In figure 8, the strategy of an agent is implemented on registers, and one of the bits is selected according to the history. The history is also implemented on registers (shift registers), and updated every clock cycle. Noise is implemented using a pseudo random number generator and a Ex-OR gate.

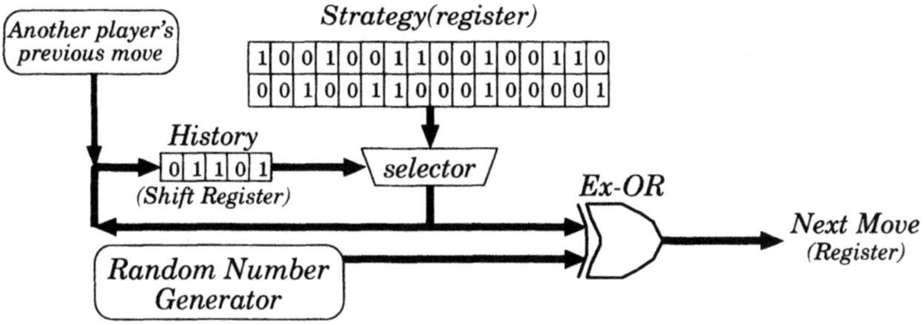

Fig. 8. Details of the Circuits Deciding Next Move

4.4 Scheduling of the IPD games

Figure 9 shows how all combinations of games among the agents are realized in the hardware. In the hardware, it is very important to reduce connections among the circuits in order to implement more units in a single chip. In figure 9, boxes are the circuits for IPD games, and strings in the boxes mean initial positions of agents. All combination of the games are realized by only connecting each units serially and shifting the agents among them. The initial positions of the agents are decided by computers in advance, and given to the hardware.

4.5 Results

Table 2 shows the computation time of all combination of games for 48 agents. Speedup by a FPGA is about 120 times compared with Ultra-Sparc 200 MHz. This high speedup is achieved by parallel processing of IPD games, and pipeline processing of a IPD game.

5 Conclusion

We show that a FPGA chip can achieve more than a hundreds times of speedup compared with a workstation (Ultra-Sparc 200MHz) in the computation of the

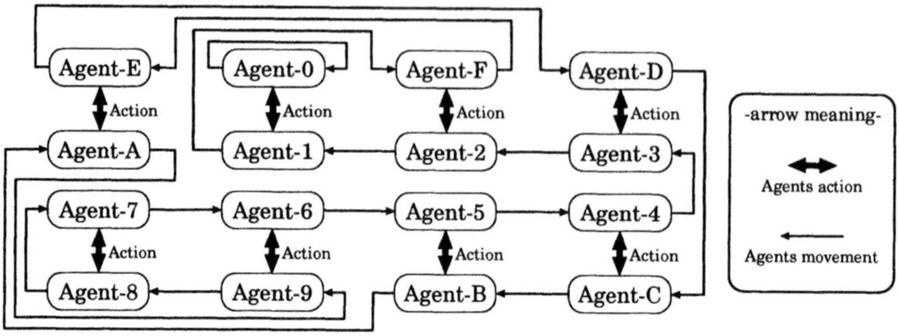

Fig. 9. Scheduling of the IPD games

Table 2. Simulation Time

	time(sec)	speedup
Sun Ultra-Sparc (200 MHz)	119.02	1.00
FPGA (33 MHz)	0.98	121.45

co-evolution model. This drastic speedup make it possible to reduce the computation time of one trial from several days to about an hour. The hardware algorithm is scalable, and performance can be improved in proportional to the amount of hardwares. The size of the latest FPGAs is more than twice larger than the FPGA we used in this paper. With a system consisting of several these FPGAs, it is possible to achieve more than a thousand times of speedup. We believe that this high speed computation of co-evolution of strategies and scores will help to understand the emergence of social morals in the real world.

References

1. K. Lindgren, "Evolutionary Phenomena in Simple Dynamics", Artificial Life II, pp.295-312, 1991.
2. K. Lindgren and M. G. Nordah, "Cooperation and Community Structure in Artificial Ecosystems" Artificial Life 1, pp.15-37, 1994.
3. P. Graham and B. Nelson, "Genetic Algorithm In Software and In Hardware - A Performance Analysis of Workstation and Custom Computing Machine Implementations", FPGAs for Custom Computing Machines, 1996 pp.216-225.
4. Stephen D.Scott, Ashok Samal and Shared Seth, HGA: "A Hardware-Based Genetic Algorithm" ACM/SIGDA 1995, pp.53-59
5. T.Maruyama, T.Funatsu and T.Hoshino, "A Field Programmable Gate-Array System for Evolutionary Computation" FPL'98 pp.356-365
6. David B.Fogel, "Evolving Behaviors in the Iterated Prisoner's Dilemma", Evolutionary Computation Vol.1:1, pp.77-97, 1993.

A Probabilistic Model for Understanding and Comparing Collective Aggregation Mechanisms

Alcherio Martinoli[1], Auke Jan Ijspeert[2,1], and Luca Maria Gambardella[2]

[1] LAMI-EPFL, IN-F Ecublens, CH-1015 Lausanne, Switzerland
[2] IDSIA, C.so Elvezia 36, CH-6900 Lugano, Switzerland
Alcherio.Martinoli@epfl.ch, ijspeert@lamisun1.epfl.ch, luca@idsia.ch
WWW page: http://diwww.epfl.ch/lami/collective/collective.html

Abstract. This paper presents a flexible probabilistic model for the description of aggregation processes in autonomous collective robotics. Two different experiments are considered: one carried out by the authors, the other by Beckers et al. [1] with teams of reactive autonomous robots which differ from a morphological as well as from a control point of view. Rather than simulating robots moving within an environment, the probabilistic model represents the clustering activity as a sequence of probabilistic events during which cluster sizes can be modified depending on simple geometrical considerations and robot control parameters. It is shown that, for both considered robotic platforms, the evolution of the cluster sizes is perfectly described, both qualitatively and quantitatively, by the probabilistic model. By comparing the results at the model level, a better understanding is gained of the influence of the interaction geometry and of the robot control parameters on the collective aggregation dynamics.

1 Introduction

Biologically inspired collective robotics favours distributed solutions, i.e. solutions where coordination is *not* taken over by a special unit using private information sources, or concentrating and redistributing most of the information gathered by the individual robots. Inspired by the so-called collective intelligence demonstrated by social insects [2], bio-inspired collective robotics studies robot-robot and robot-environment interactions leading to robust, goal-oriented, and perhaps emergent group behaviours.

Often, fully distributed control is combined with minimal robotic skills: robots are not able to communicate to each other, to plan their activity or to adapt their behaviour continuously. With such simple controllers, a gathering task becomes essentially a geometrical problem with a probabilistic nature which is well adapted to be described by simple probabilistic models.

The motivations for such a modelling are two-fold. *Firstly*, because of its minimalist essence, it enables the investigation and the determination of which characteristics of the experiment are most influential on the clustering process.

Secondly, working with probabilistic simulations means time saving. In particular, it would be interesting to dispose of a tool which allows the evaluation of critical characteristics of an experiment before the robot's final design is accomplished or before a much more complicated sensor-based simulator is developed.

Collective distributed clustering of spread objects is inspired by studies of aggregation processes with social insects, such as ants for instance [3]. In [1,5, 7], similar experiments were carried out with real robots and reactive control architectures. In the former two papers, a precise statistical analysis was carried out but neither a modelling of the experiment nor a comparison with simulation results were presented. In [7] we presented a probabilistic model and we compared its predictions with data delivered by Webots, a 3D simulator of Khepera robots [8].

This paper firstly presents new results with real robots over long experiments which match perfectly with the predictions reported in [7] and, secondly, it shows how the probabilistic model can describe an experiment with a completely different robotic platform such as the one used in Beckers et al.'s experiments [1].

2 Materials and Methods

This section presents the two experiments at the experimental and at the modelling level. We focus on the experiment with the Khepera robots and we use the experimental data of Beckers et al. to demonstrate that our model is valid for a completely different set-up. The relevant differences between both set-ups are shown in Tab. 1.

a) b)

Fig. 1. a) Khepera equiped with gripper. b) Robots used by Beckers et al. (photo courtesy of Prof. J.-L. Deneubourg)

2.1 Experimental Set-Up

Experiments with Khepera. All Khepera robots [9] used in the experiments are extended with a gripper module (see Fig. 1a) and equipped with an IR

a) b)

Fig. 2. a) Seed scattering at beginning of the experiment and b) at the end of one of the longest experiments (after about 2:05 h)

a) b)

Fig. 3. a) Example of aggregation process with 3 robots in Webots. Seed scattering at beginning of the experiment and b) after about 4:15 h of simulated time

enhanced reflecting band (not shown in Fig. 1a). This band avoids that moving robots can be recognized as objects to grasp and increases the robot-robot avoidance distance (see Tab. 1). The same robot configuration can be obtained using the Webots simulator [8]. Webots is based on the "as realistic as possible" reproduction of the sensor capabilities as well as of the robot-robot and robot-environment interaction kinematics. As term of comparison, the mean acceleration ratio for this experiment between Webots and real time is about 15 on a workstation Ultra Sun 1 with five robots.

We can summarize the robot's behaviour with the following simple rules. The robot moves on the arena in a straight line looking for seeds. When its sensors are activated by an object, the robot starts a discriminating procedure. Two

Table 1. Comparison between Khepera's and Beckers et al.'s set-ups. All geometrical dimensions are given in [*cm*]. Beckers et al.'s robots are equipped with a 17 cm wide C-shaped gripper (*), see Fig. 1b

Set-up	Arena	Seeds			Robots		Detection range			Initial scattering
		∅	h	nb	dim	nb	seeds	walls	robots	
Khepera	80x80	1.7	2.5	20	5.5∅	1–10	1.7	1.7	6.5	arbitrarily irregular
Beckers et al.	250x250	4	2.5	81	17x21	1–5	8.5*	20	20	regular grid

cases can occur: if the robot is in front of a large obstacle (a wall, another robot or an array of seeds), the object is considered as an obstacle and the robot avoids it. In the second case, the small obstacle is considered as a seed. If the robot is not carrying a seed, it grasps the seed with the gripper; if the robot is already carrying a seed, it drops the seed it is carrying close to the one it has found; then, in both cases, it resumes looking for seeds. Note that, because only the two extreme seeds of a cluster can be identified as seeds (in opposition to obstacles) by the robots, clusters are build in lines (see Fig. 2 as an example).

Beckers et al.'s Experiments. In [1], each robot is equiped with IR sensors for obstacle avoidance and a C-shaped gripper. The gripper is provided with a micro-switch which is activated when a certain number of pucks are pushed.

We can summarize the robot's behaviour with the following simple rules. The robot moves on the arena in a straight line looking for pucks and avoiding obstacles (wall or teammates). If the pucks pushed by the gripper are three or more, the micro-switch triggers a puck-dropping behavior: the robot leaves the pucks on place and resumes the searching behaviour. Notice that, because of the sensorial orthogonality in the distinction between pucks and obstacles, cluster are recognized and accessible from any angle. The resulting cluster form is therefore approximatively circular.

2.2 Probabilistic Model

The central idea of the probabilistic model is that instead of simulating robots moving within an environment, their activity is represented as a sequence of probabilistic events. Robots could therefore be seen as dice being thrown into the arena at each iteration, with their random location as well as their current state (i.e. carrying zero, one or two objects), determining their next state and the next state of the environment (i.e. the state of the clusters). The model takes into account physical sizes (arena, robots, and objects), the geometry of robot-robot and robot-environment interactions, the time needed to manage them,and the sensor range for detecting objects, walls or teammates. An interesting feature of this method is that the probabilistic model is easy to implement because it is closely related to the flowchart of the robot controllers.

In the considered aggregation process, the transition probabilities among the different states of each robot controller are conditioned by three (four for Khepera experiments) stochastic processes. The *first* stochastic process assigns a random position to the robot. If this position is inside the detection area of a cluster, the *second* random process is started. According to the robot state, the size of the found cluster is incremented or decremented by one or two objects if the number delivered by the second random process is within the incrementing or decrementing region (calculated with the values of Fig. 4c and Fig. 5c). Notice that Khepera and Beckers et al.'s robots differ not only from a morphological but also from a sensorial and therefore behavioural point of view. On one hand, when Khepera notice an object, it always rotates on place till the object is in

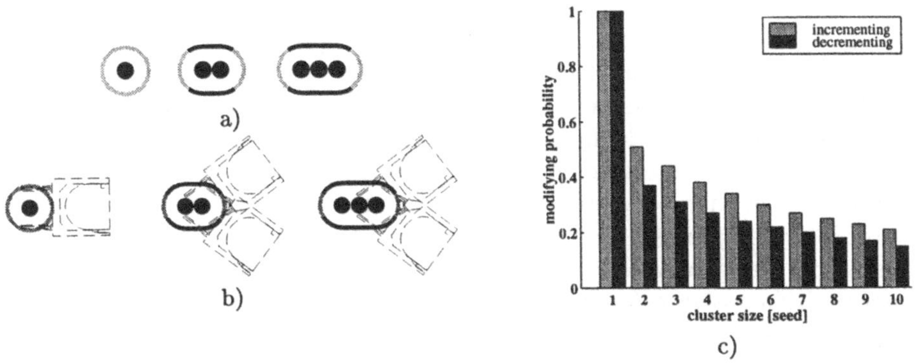

Fig. 4. Khepera (scale 1:10). a) Geometrical representation of the cluster incrementing probability. The ratio between the identification perimeter (*light grey arc*) and the total detection perimeter of the cluster represents the probability to increment the cluster size by one seed. b) Geometrical representation of the cluster decrementing probability. The robot, in order to decrement the size of the cluster by 1 seed, first has to detect the seed on the cluster tip as in Fig. 4a and then grasp it (the angle from which a seed can successfully be grasped from a cluster is slightly smaller than its detection angle, see dark grey arc in Fig. 4b). c) The above mentioned geometrical considerations are translated in probabilities of incrementing or decrementing the cluster size given the state of the robot (carrying or not a seed), once the cluster has been found

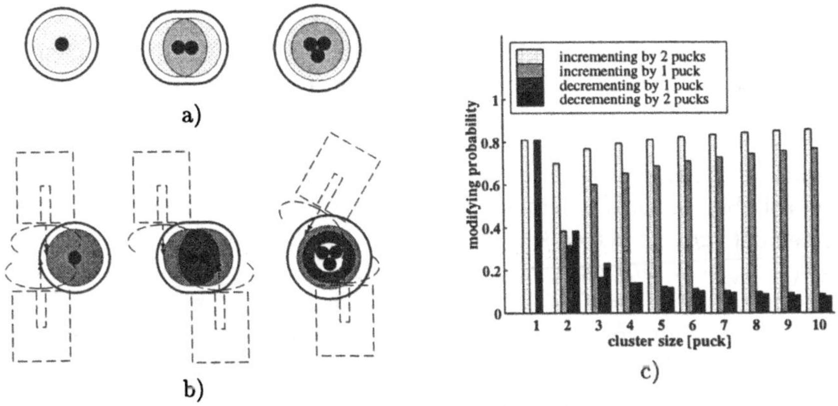

Fig. 5. Beckers et al. (scale 1:20). a) Geometrical representation of the cluster incrementing probability. The ratio between the grey areas and the whole detection area of the cluster represents the probability to increment the cluster size by two pucks (*very light grey*) or one puck (*light grey*). b) Geometrical representation of the cluster decrementing probability. The ratio between the grey areas and the whole detection area of the cluster represents the probability to decrement the cluster size by one puck (*dark grey*) or two pucks (*very dark grey*). c) The above mentioned geometrical considerations are translated in probabilities of incrementing or decrementing the cluster size given the state of the robot (pushing one puck, pushing two pucks or unloaded), once the cluster has been found

front of it. It is able to accomplish this behavior because it is equiped with a belt of IR sensors. As a consequence, the modifying probabilities are calculated from perimeters of clusters. On the other hand, this behavior is not accomplished by Beckers et al.'s robots because the gripper is only able to measure the pressure exerted by the pucks but not the contact point. As a consequence, these robots access the clusters from any point and any angle and we have to consider surfaces rather than perimeters for calculating the modifying probabilities. The *third* stochastic process, the interference with other teammates, can can occur during the search as well as during object pick up or drop activity. If the random position assigned to the robot is inside the detection area of one of the teammates, the robot's activity is frozen for a given number of iterations corresponding to the real time lapse needed for avoiding the teammate. The *fourth* stochastic process, implemented only in the Khepera experiments' modelling, takes into account the object distinguishing efficiency (0.89) of the implemented controller. Each random process is carried out for each robot independently before the next iteration of the program is started.

In order to convert the number of iterations into time, we consider the time needed by a robot to sweep the detection area of a single seed (equivalent to an iteration) and calculate a fixed conversion factor (F) as follows (see [7] for further details). $A_{\mathrm{detobject}}$ = detection surface of a single object, W_{robot} = robot width, $\bar{v}_{\mathrm{robot}}$ = mean forward velocity of the robots.

$$F = \frac{A_{\mathrm{detobject}}}{\bar{v}_{\mathrm{robot}} W_{\mathrm{robot}}} [\mathrm{s/iterations}] \tag{1}$$

As term of reference, the mean acceleration ratio between the probabilistic simulation and real time is about 4000 on a Ultra Sun 1 with five robots.

Probabilistic Modelling of Experiments with Khepera. The cluster modifying probabilities for Khepera are depicted in Fig. 4. Every robot can increment or decrement the size of a cluster by one seed at a time. With the used numerical values ($A_{\mathrm{detseed}} = 20.4\ cm^2$, $\bar{v}_{\mathrm{robot}} = 8.0$ cm/s, $W_{\mathrm{robot}} = 5.5$ cm) the resulting conversion factor is F = 0.46 [s/iterations]). The conversion factor is also used for taking in account the duration of the actions of the robots. For instance, with the implemented discriminating behavior, it takes 3 s for avoiding obstacles and 10 s for modifying the size of a cluster. The algorithm translates these time lapses into number of iterations during which the searching behavior is frozen.

Probabilistic Modelling of Beckers et al.'s Experiments. The cluster modifying probabilities for Beckers et al.'s robots are depicted in Fig. 5. Every robot can increment or decrement the size of a cluster by one or two pucks at a time. With the used numerical values ($A_{\mathrm{detpucks}} = 12.6\ cm^2$, $\bar{v}_{\mathrm{robot}} = 27$ cm/s, $W_{\mathrm{robot}} = 17$ cm) the resulting conversion factor is F = 0.027 [s/iterations]). The robot takes about 0.5 s to avoid an obstacle and about 1.6 s to modify the size of a cluster.

3 Results

We have carried out several sets of experiments with different number of robots and durations (see Table 2). The plots show the mean over all the runs and error bars represent the standard deviation around the mean.

3.1 Experiments with Khepera

All experiments are carried in the three different implementations (real robots, Webots simulator and probabilistic model) and different group sizes, except for the longest experiments which have been realized only with a group of three real robots. It is worth to notice that the about 18 hours needed for all the runs of the long experiments with real robots have been performed without any recharging break thanks a special tool [6, 7] which allows the robots to be supplied from the floor (see Fig. 2).

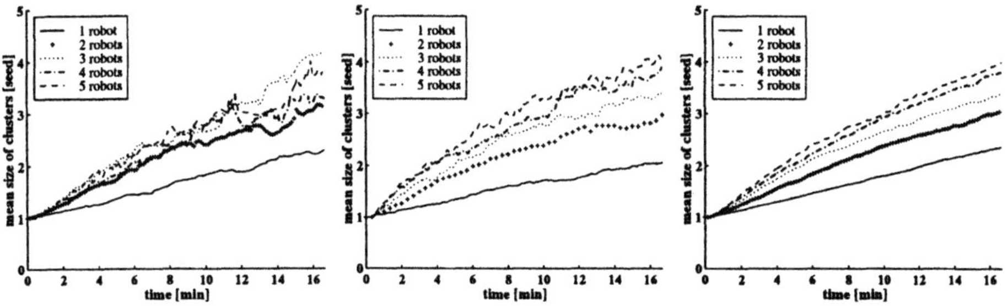

a) b) c)

Fig. 6. Aggregation evolution with increasing number of robots (one to five). a) Results of the experiments with real robots. b) Results of the Webots simulator. c) Results of the probabilistic modelling

In order to check whether or not there is a significant difference between data collected from the simulations and the real experiment, we performed a Mann-Whitney test [4] on the distributions of mean cluster size at the end of the shorter

Table 2. Characteristics of the experiments carried out. *mcs* = mean size of the clusters, *bcs* = size of the biggest cluster, *nc* = number of clusters

Set-up	Robots	Duration	Nb of repetitions			Measurement	Figures
			real	Webots	prob. model		
Khepera	1–5	16 min	5	10	100	msc	6
Khepera	3	600 min	5	10	100	msc, bcs, nc	7
Khepera	1–10	till single cluster	5	10	100	nc	8a
Beckers et al.	1–5	till single cluster	3	–	100	nc	8b

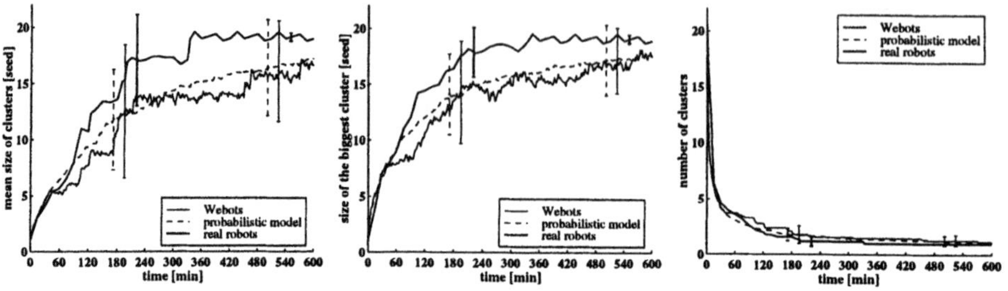

a) b) c)

Fig. 7. Evolution of aggregation process during ten hours. The results of real robots, Webots simulator and probabilistic modelling are overlapped in the same plot. Notice that the resulting plots are the average of a different number of runs with different sample times (15 minutes for real robots, 2 min for Webots and probabilistic modelling). In order to obtain a plot with real robots extended over a pre-established, wide time window, we recorded the aggregation noise during two hours after a single cluster arose. This allowed us to stop a given run once all seeds were gathered in a single cluster and to extend the run data with the recorded noise

experiments and on the time needed to gather all seeds in the longer experiments. With the help of this non-parametric test, we compared the distributions of pairs of data sets. The results show that there is no statistically significant difference ($p<0.05$) over 27 data sets except in three cases (Fig. 6, one robot, Webots vs. prob. model and three robots, real robots vs. prob. model; Fig. 8a, nine robots, Webots vs. prob. model).

Fig. 6 shows the initial clustering evolution for a group of one to five robots. Although the results of both simulations, Webots and probabilistic model, are slightly smoother than those of the real ones (they are namely the average of a larger number of experimental replications), the three plots present a good agreement.

Fig. 7 illustrates longer experiments with a group of three robots. Once again, both the sensor-based simulation and the probabilistic model are in good agreement. Real robots data show a slightly faster evolution of the aggregation process but this is mainly due to the reduced number of runs in comparison to simulation. The mean real robot data are in any case within the mean value extended with the standard deviation of simulation data.

Fig. 8a shows a comparison of Webots simulator and probabilistic model based on mean and variance of the time needed by the robots to gather all the seeds in a single cluster.

3.2 Beckers et al.'s Experiments

The comparison between the results of our probabilistic model and the ones reported in [1] is shown in Fig. 8b. There is a good agreement between both results based on mean and variance of the time needed by the robots to gather

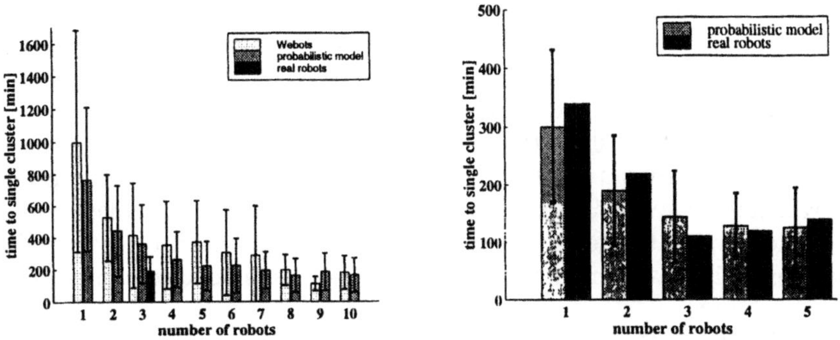

a) b)

Fig. 8. Comparison between the results of simulations and real robots on the time needed to gather all the objects in a single cluster. a) Khepera experiments. b) Beckers et al.'s experiments

all the pucks in a single cluster. However, we were not able to perform any statistical test because single run data were no more available.

4 Discussion and Conclusion

This paper presented a simple and flexible probabilistic model which has been applied to two completely different robotics platforms. The good agreement of the clustering dynamics described by the probabilistic model with data collected with real robots or more sophisticated simulators such as Webots shows that the proposed minimalist model incorporates the essential characteristics of the clustering problem. These characteristics have been identified to be probabilities of modifying the size of clusters and probabilities of having interferences with other robots. These probabilities are essentially based on geometrical considerations and can be derived from the sensory capacity of single robots. Once these probabilities are established and interaction time lapses measured, the probabilistic model has the interesting feature of being a prediction tool of the same quality as a detailed sensor-based simulation, while being significantly simpler and faster.

Another interesting feature of the model is that the identification of the primary characteristics of this particular clustering problem is a step forward towards the understanding of collective mechanisms underlying clustering in general. For instance with the Khepera set-up, the model has predicted, before the long experiments were carried out, that it was possible to gather all the seeds in a single cluster, if enough time is available. In fact, since clusters of isolated seeds are in an irreversible way eliminated during the aggregation process and since aggregation is enhanced by a positive building gradient (the incrementing probability is consistently greater than the decrementing probability, see Fig. 4), these predictions have been confirmed experimentally. This resulting building gradient is even more enhanced in the experiments of Beckers et al.: the greater the cluster size, the more stable the cluster is. This mechanism, simply depending on the robot-environment interaction, speeds up the team performances (Beckers et al.'s robots are able to cluster 4 times more pucks in half the time needed by a Khepera team of the same size).

Finally, the results show that in this kind of experiments, where the coordination between robots is essentially probabilistic, the data obtained present a high variance. One possible solution to reduce variance and increase coordination while keeping the team control fully distributed, would be to introduce a form of explicit local communication (signalling or symbolic communication).

Acknowledgements
We would like to thank Prof. J.-L. Deneubourg for photos and further data on the experiments presented in [1] and M. Yamamoto for the improvement of the distinguishing algorithm of Khepera. A. Martinoli, A. J. Ijspeert and L. M. Gambardella have been partially supported by the Swiss National Research Foundation, project "A Methodology for Collective Robotics Design".

References

1. R. Beckers, O.E. Holland, and J.L. Deneubourg. From local actions to global tasks: Stigmergy and collective robotics. In R. Brooks and P. Maes, editors, *Proc. of the Fourth Workshop on Artificial Life*, pages 181–189, Boston, MA, 1994. MIT Press.
2. E. G. Bonabeau and G. Theraulaz. *Intelligence Collective.* Hermès, Paris, France, 1994.
3. J. C. Deneubourg, S. Goss, N. Franks, A. Sendova, A. Franks, C. Detrain, and L. Chretien. The dynamics of collective sorting: Robot-like ant and ant-like robot. In J. A. Mayer and S. W. Wilson, editors, *Simulation of Adaptive Behavior: From Animals to Animats*, pages 356–365. MIT Press, 1991.
4. W. L. Hays. *Statistics.* Harcourt Brace College, Orlando, US, 1991.
5. M. Maris and R. te Boekhorst. Exploiting physical constraints: Heap formation through behavioral error in a group of robots. In *Proceedings of IEEE/RSJ International Conference on Intelligent Robots and Systems IROS-96*, volume 3, pages 1655–1660, Osaka, Japan, November 1996.
6. A. Martinoli, E. Franzi, and O. Matthey. Towards a reliable set-up for bio-inspired collective experiments with real robots. In A. Casals and A. T. de Almeida, editors, *Proc. of the Fifth International Symposium on Experimental Robotics ISER-97*, pages 597–608, Barcelona, Spain, June 1997. Springer Verlag.
7. A. Martinoli, A. J. Ijspeert, and F. Mondada. Understanding collective aggreagtion mechanisms: From probabilistic modelling to experiments with real robots. *Robotic and Autonomous Systems.* To appear. Preprint available at `http://diwww.epfl.ch/lami/team/alcherio/am_pub.html`.
8. O. Michel. Webots: Symbiosis between virtual and real mobile robots. In *Proceedings of the First International Conference on Virtual Worlds, VW'98*, pages 254–263, Paris, France, July 1998. Springer Verlag.
9. F. Mondada, E. Franzi, and P. Ienne. Mobile robot miniaturization: A tool for investigation in control algorithms. In T. Yoshikawa and F. Miyazaki, editors, *Proc. of the Third International Symposium on Experimental Robotics ISER-93*, pages 501–513, Kyoto, Japan, 1993. Springer Verlag.

Exploiting Domain Physics: Using Stigmergy to Control Cluster Building with Real Robots

Chris Melhuish

Intelligent Autonomous Systems Engineering Laboratory
Faculty of Engineering
University of the West of England
Frenchay
Bristol BS16 1QY
chris.melhuish@uwe.ac.uk

Abstract. This paper explores the application of stigmergy as a mechanism for the control and coordination of cluster building with 10 simple, autonomous, mobile real robots. Social insects collectively carry out extraordinary tasks using simple reactive rule sets with little or no memory and without recourse to internal world models over which they reason. Control and coordination of behaviour can be mediated externally through the environment. In particular, building tasks may employ the configuration of each construction phase to cue the next phase. This paper shows how, using physical robots, the deterministic clustering algorithm of Beckers *et al* [1994] can be modified to create a cluster of objects against the arena wall. The modification takes the form of either applying probabilistic rule selection or by altering the robot sensor morphology. These two approaches illustrate that equivalent qualitative emergent consequences can be generated algorithmically or by exploiting the domain physics.

1 Introduction

Social insects, such as ants, provide an existence proof of minimalist collective control mechanisms. They exhibit characteristics such as, decentralized control, self-organization, redundancy and indirect communication through the environment (stigmergy). This paper develops the idea that these minimalist collective characteristics could be attractive to engineers in the design and implementation of multiple robots. For example, such mechanisms may prove to be usable by designers of milli and micro-scaled robots. At such scales there may be considerable limitations on communication, sensing, mobility, computatation as well as power. It makes sense therefore to consider if there are exploitable characteristics arising from the tight coupling between the environment and the action repertoire of embodied agents. In particular this study explores collective building activity of many simple robots where the global task achieving behaviour is not explicitly represented, that is, no robot has an internal representation of the task. The task is achieved as the emergent consequence of the robots executing simple reactive rules triggered by local conditions. In this context the paper focuses on how stigmergy can be employed to control and coordinate the global building behaviour without direct communication.

The paper is organised as follows. Section 1 reviews the employment of stigmergy in some social insects and how it can be used for real robots. In section 2, the robots and the environment used in this work are described. Section 3 discusses the origins and canonical form of a puck clustering algorithm in real robots. Sections 4 and 5 describe the experiments on non-seeded clustering and present the results, which are discussed in Section 6.

1.1 Stigmergy

The concept of stigmergy was first introduced by Grasse [1959]. He described the indirect communication taking place among individual termites through dynamically evolving features of a structure. His work focused on the mechanism for nest construction where communication and subsequent decentralized control was mediated through the environment rather than through a process of direct communication. Bonabeau *et al* [1994] refer to stigmergy employing stimulating configurations which trigger the building action of a termite worker, which transforms the current configuration of the structure into the next. The new configuration may, in turn, trigger another action of the same or another termite to modify the structure thus creating the next configuration. Thus modification to the environment provides the cue for subsequent changes in behaviour and therefore communication is effected through the environment. Holland has suggested an interesting further classification of stigmergy into active and passive categories [Holland 1996]. However, the term 'sematectonic' has been employed by Wilson [1975] to imply cue based stigmergy in the domain of construction. Bonabeau *et al* [1994] also use the term sematectonic in the description of the use of cue based stigmergy in the construction of comb lattices in social wasps. They use simulation to show how the state of the simulated nest can be used to trigger the next action. In this way no a-priori blue-print is required since the next stage of construction is driven by the current state of the construction. However, some researchers, such as Bonabeau *et al* [1994], do point out that there still are problems in understanding how action triggering stimuli are organized in space and time to ensure that global building behaviour is coherent. Downing and Jeanne [1988] noted that some (non-linear) building activites of wasps can be carried out at the same time but not necessarily performed in a set sequence. Such non-linear construction is more likely to involve a greater array of cues for an animal to filter simply because there are more potential construction sites. These multiple cues, which may change with time, may include re-evaluating particular characteristics of earlier construction such as petiole length in a wasp nest, environmental cues such as temperature and humidity, and cues from developing young such as the growth of larvae.

1.2 Employing Stigmergy

Stigmergy has been observed to effect indirect communication in experiments employing the picking up or placing down of objects. The removal or deposition of such objects can alter the consequent behaviour of other (or the same) animal or robot. In this way a form of global task control can be realised. In the examples below a stigmergic mechanism is realized by employing the removal or deposition of objects which consequently 'controls' the global task achieving behaviours.

Franks *et al* [1992], Franks & Deneubourg [1997] looked at the construction of peripheral walls accomplished by worker ants in Leptothorax colonies that nest in narrow slits in rocks. They showed in a computer simulation model how ants could use a simple self-organizing procedure based largely on bulldozing. Each builder, by pushing its building block into others, adds its work to existing structures. Bulldozing workers do not communicate directly but can communicate efficiently via the products of their successful labours. Beckers *et al* have incorporated these ideas into a group of puck pushing robots in which each robot pushes pucks into others thus adding its work to existing structures.

Deneubourg *et al* [1991] presented in their paper on 'The dynamics of collective sorting: ant-like robots and robot-like ants' a simulation showing that simple agents, specified in terms which could equally well apply to ants or robots, could achieve two generic tasks of fundamental importance: the clustering of scattered objects of a single type, and the grouping and sorting of objects of two different types. For sorting, the agents needed to be able to sense the local densities of the different types of brood items. This was achieved by using a short-term memory augmented by recognition of the type of any brood item an ant was carrying. Clustering was the result of the mechanism operating on only a single type of item. In this way indirect communication through the environment was achieved and behaviour was 'controlled' by the consequences of cumulative deposition and removal of objects, that is, a stigmergic mechanism was employed.

Altenburg & Pavicic [1993] demonstrated puck clearance in an arena with 6 lego robots but puck clustering was first achieved by Beckers *et al* [1994], who showed that puck clustering could be achieved with an extremely simple mechanism They used physical robots which were unable to detect whether or not they were moving any objects, which had no memory, and which could sense the local density of objects only as being below or above a fixed threshold. The clusters formed were always away from the arena perimeter are referred to as np-clusters (i.e. non-peripheral). Melhuish *et al* [1998] reported a further robot study which demonstrated that this technique could also be extended to the sorting of two objects (frisbees) with a more simple mechanism than proposed by the model of Deneubourg *et al.* This paper explores how indirect communication through the environment in the guise of stigmergy can be further exploited in tasks involving the clustering of objects. In this way, it is argued here that, 'artificial' self-organization emerges at a meta-level as a consequence of simple rule sets acting on the environment and the indirect communication (stigmergy), which results from those actions which change the environment and subsequently effect the actions or behaviours of the agents. The resulting global self-organized structure is therefore a product of the tightly coupled interactions between the agents and their environment and is therefore sensitive to the nature of the domain physics such as friction, compliance and texture.

This paper shows how, using physical robots, the deterministic clustering algorithm of Beckers *et al* [1994] can be modified to create a cluster against the arena wall. The modification takes the form of either applying probabilistic rule selection or by altering the robot sensor morphology. These two approaches illustrate that equivalent qualitative emergent consequences can be generated algorithmically or by exploiting the domain physics. Both methods emphasise the importance of the tight coupling between local minimalist actions executed by the robots and the environment.

2 Materials and Methods

The robots and experimental environment used in this study were designed to investigate a range of social insect behaviour, with particular emphasis being placed on building tasks. The design of the robot system was inspired by the work of Deneubourg *et al* [1990] on corpse clustering in the ant Pheidole pallidula and Franks and his collaborators on ant species Leptothorax tubero-interruptus, Leptothorax unifasciatus, and related species (Franks and Sendova-Franks [1992], Franks *et al* [1992]). These ants live in small colonies (typically with a few hundred members) in cracks in rock. This constrained environment means that they are behaviourally adapted to life in two dimensions. It is therefore possible to study the behaviour of a colony by providing it with a particular two dimensional habitat - the space between two glass slides - which allows unrivalled opportunities for observation and recording.

Leptothorax is also peculiarly suitable as a model for robotic investigations of some collective behaviours, because current robots, which are wheeled, operate well in two dimensions but are extremely limited in their abilities to operate in the third dimension. Leptothorax building behaviours tend to involve the movement of single lumps of material (carborundum grit of regular dimensions is usually provided by the experimenters) which are placed next to other lumps rather than piled on top of them. A similar level of functionality can be achieved in robots by simply pushing and pulling building blocks (referred to as pucks but actually modified frisbees) around the floor, rather than lifting them and piling them up.

The robots known as U-bots, were designed and built in our laboratory to provide a flexible and capable platform for a range of collective robot experiments in a 10m diameter arena. The design and operation of the robots are described in detail in Melhuish *et al* [1998].

3 The Canonical Rule Set

Creating np-clusters of small pucks was achieved by Beckers *et al* [1994] with up to six robots and was repeated by Melhuish *et al* [1998] using 10 robots with larger pucks and non-deformable boundary walls. The rule set for np-clustering of Melhuish *et al* is taken as the canonical form as the basis for further modifications in the experiments described below. The canonical form may defined algorithmically as:

Rule 1:
if (gripper pressed & *Object* ahead) **then**
make random turn away from object

Rule 2:
if (gripper pressed & no *Object* ahead) **then**
reverse small distance i.e. drop the puck.
make random turn left or right

Rule 3:
go forward

Fig. 1. The Canonical Rule Set

4. Experiment 1: Unseeded Wall Clustering - Applying a Probabilistic Strategy

The rules to create the single 'central' cluster described above were applied deterministically. It was reasoned that if the likelihood that pucks could be left at the wall could be increased in some manner then this could enhance the possibility of cluster formation at the wall. If the probability of dropping a puck against the wall is too high then it would be expected that all pucks would be deposited at the periphery but not necessarily in a single cluster. In contrast if the probability is too low then a central cluster is likely to form (a probability of 0.0 is the equivalent of the canonical 3 rule clustering algorithm as shown in figure 1). It was reasoned that a 'sweet' spot in the probability continuum existed where sufficient pucks could be dropped which would have the effect of acting as a seed. At such a probability there would be a dynamic equilibrium balancing the tendency to drop pucks at the wall and capturing them such that they are recycled. At this 'sweet' spot a puck would be 'allowed' to remain against the wall for a sufficient time such that other pucks could be deposited against it.

To effect the probabilistic approach to depositing a puck the canonical 3 rule clustering rule set was modified so that rule 1 included a stochastic element as shown in Figure 2. The modification of Rule 1 provided the opportunity for one of two behaviours to be executed. When a robot collided with a wall (or another robot) the scoop is depressed and the robot registers the presence of an object ahead. With a probability of p the 'normal' Rule 1 behaviour of making a random turn away from the object is carried out. Conversely with a probability of $1-p$ the puck is dropped since the robot will reverse a small distance and then make a random turn.

Rule 1:
if (gripper pressed & Object ahead) **then**
with a probability ***p***
make random turn away from object
else (with probability ***1-p***)
reverse small distance i.e. drop the puck.
make random turn left or right

Rule 2:
if (gripper pressed & no Object ahead) **then**
reverse small distance i.e. drop the puck.
make random turn left or right

Rule 3:
go forward

Fig.2. 3 Rule 'Stochastic' Clustering Algorithm

4.1 Materials and Methods

44 pucks were set out in a regular pattern. The overhead camera captured a frame every 5 minutes. The end condition was considered to be when a configuration of 40 pucks (over 90%) were within a puck radius of each other. The following probability p values of the puck being retained were trialled 0.0, 0.5, 0.8, 0.88, 0.9, 0.95 and 1.0. The trials with real robots proved to be extremely time consuming and thus, initially, only one trial was conducted for each setting of p using 10 robots. More trials were then run at the 'sweet' spot. In some cases the trials were terminated when two large clusters had formed. From previous observations it was noted that such a configuration could remain for a very long period before one single cluster would eventually form. Such a configuration was therefore considered to indicate that a single cluster would be formed.

4.2 Results

Table 1 sets out the results for the experiment. As expected probabilities of 0 and 1 produce either pucks around the periphery (Figure 3h) or a central cluster (Figure 4a). With p = 0.5 pucks are still strewn around the periphery (Figure 3g). However at p = 0.8 and 0.85 a major cluster was formed at the edge but with some 15 singletons strewn around the periphery (Figure 3f and e). The size of the major cluster against the wall reached the acceptable final configuration of 40 pucks with p =0.88 (Figure 3d). Probabilities of p = 0.9 and 0.95 produced two main central clusters after 5 hours and 2.5 hours respectively. From earlier experience and observation it was concluded that these would eventually form one large central cluster and the runs were halted (Figure 3c and b respectively).

Thus the experimental evidence suggests that the 'sweet' spot lay very near p = 0.88. 5 more trials were therefore conducted with p = 0.88. Of the extra trials 3 formed a single cluster at the edge (9hrs 25m, 10hrs 35m and 13hrs 10m) and two formed central clusters; one with 40 pucks (6hrs 0m) and the other with some 9 pucks strewn around the periphery (7hrs 35m). Thus of the total 6 trials 4 formed acceptable edge clusters, one formed an acceptable central cluster and one had nearly formed an acceptable centre cluster.

4.3 Conclusions

The results demonstrate that the stigmergic mechanism employed in the Beckers *et al* clustering work can be modified to control whether a single cluster is formed in the arena or at its edge. Low *p* produces no clustering but pucks strewn at the peripheral wall. High *p* produces central clusters and with an intermediate *p* a single cluster at the wall can be produced. This results are qualified in that not all trials were successful (4 out of 6) but the choice of *p* may need to be refined further and/or the very nature of the task, with its concomitant noise and randomness, cannot guarantee 100% success.

Retain Prob *p*	RESULTS
1.0	leads to a central cluster after 6 hours 35 minutes
0.95	Stopped when 2 main central clusters formed. Stopped after 2.5hours .
0.9	Stopped when 2 main central clusters formed. Stopped after 5hours
0.88	1 cluster formed at edge at 9hrs 5m continued to be stable up to 11hrs 20min.
0.85	1 major cluster formed at edge and approx. 15 single pucks around the periphery. stopped after 11hours
0.8	1 major cluster formed at edge and approx. 15 single pucks around the periphery. Stopped after 11hrs
0.5	All pucks taken to periphery after 40minutes but no single cluster formed. Stopped at 11 hrs
0.0	All pucks taken to periphery after 3hr15m but no single cluster formed. Stopped at 3hrs.

Table 1. The Probabilistic Continuum for Unseeded Clustering

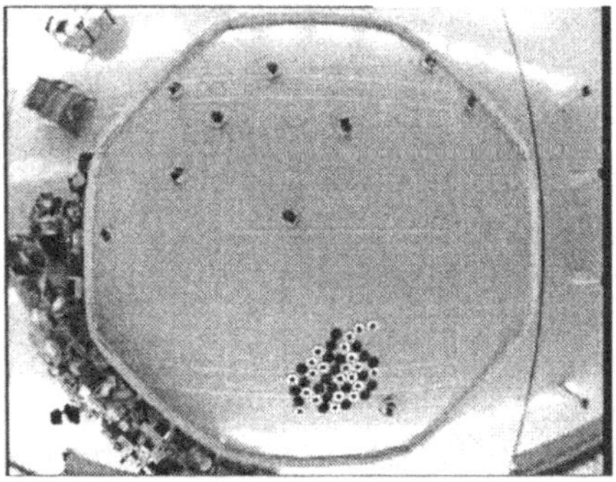

(a) p =1.0

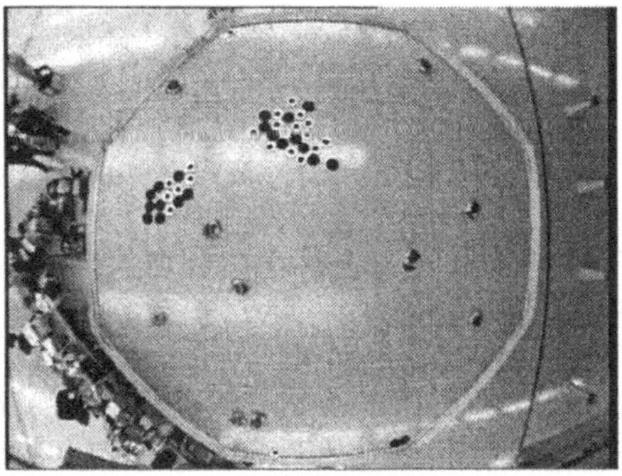

(b) p =0.95

(figure 3 continued below)

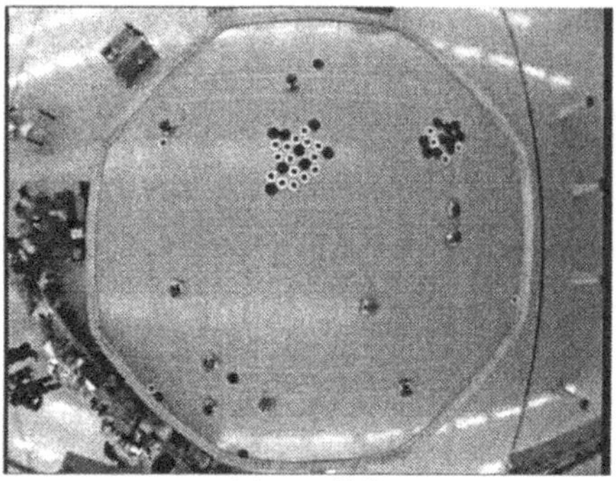

(c) p = 0.9

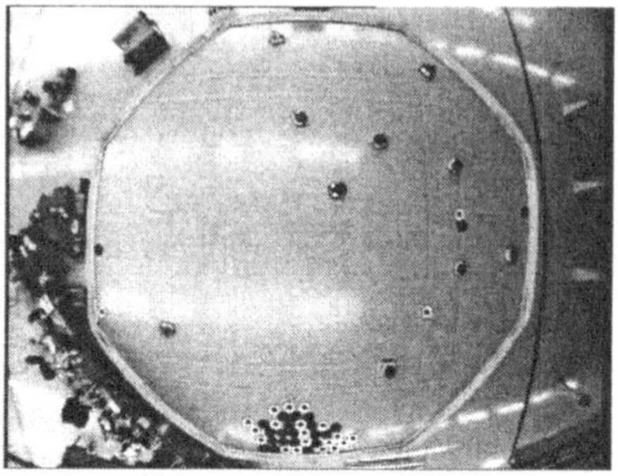

(d) p = 0.88

(e) p =0.85

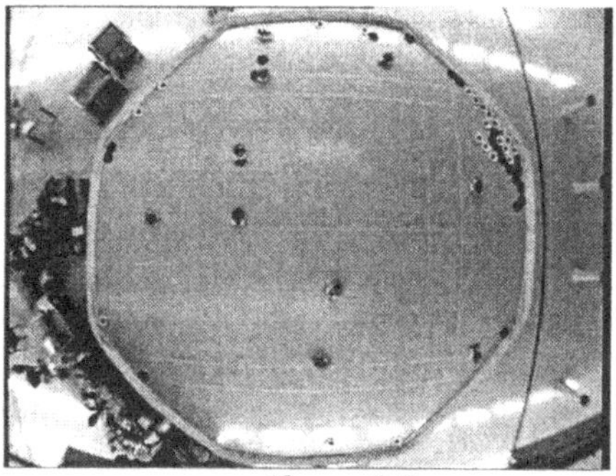

(f) p = 0.8

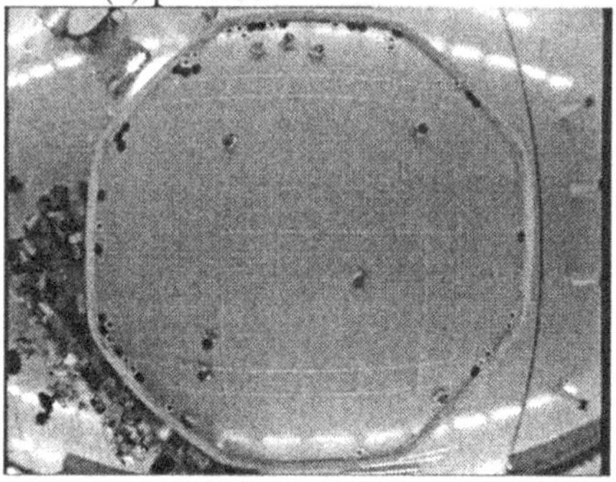

(g) p =0.5

(h) p = 0.0

Fig. 3. Final Configurations for the Probabilistic Continuum

5. Experiment 2: Modifying the Robot Morphology - Computational Equivalence

In the case of applying a probabilistic element to the deterministic canonical 3 rule clustering algorithm a probability of 0.88 puck retention at a wall showed some success in forming a cluster against the wall. It was reasoned that such a probability of puck retention could also be generated by employing the deterministic rule set with the proviso that the efficiency of the obstacle detection sensors would be decreased. Since robots execute random turns after collisions and then continue on a straight path across the arena, the angle at which a robot collides with the wall may also be considered essentially as random. If the effective acceptance angle of the forward sensing infra-red obstacle detectors could be appropriately reduced then it was further reasoned that if the robot with a puck approached the wall at an appropriate angle, then the wall would be in the sensor 'blind-spot' region. In such a case the robot's scoop would be depressed and since no obstacle would be detected Rule 2 would be executed causing the puck to be deposited against the wall. A compromise has to be reached between 'tendencies' to drop the puck and pick up a puck at the wall. If no pucks are dropped (as in the canonical cluster algorithm in Figure 1) then a single cluster will form away from the wall. If a puck is dropped whenever a wall is sensed then all the pucks will eventually be moved to the wall or against pucks which themselves are against the wall. It is therefore expected that when an appropriate compromise between the two behaviours is realised then one cluster will eventually become prominent and will (with a high probability) become the stable 'master' cluster. The experiment described below tests this hypothesis.

The balance of the behaviours is achieved by reducing the range and angle of acceptance of the collision avoidance sensors. In order to reduce the acceptance angle of detection the left and right eyes were disconnected and only the central eye was used. However, to produce acceptable results the range of the eye was further reduced by attaching a single layer of translucent tape over the sensor. Thus in contrast with the 'standard' puck clustering set-up a robot carrying a puck and hitting the wall cannot see the wall from all angles. This increases the probability of dropping a puck at the wall. With only the central sensor functioning the 'normal' retain behaviour is restricted to the central region. For a robot colliding with the wall with a trajectory which enters either of the two blind regions the puck would be dropped.

The retain/pickup mean angle for the 10 robots used with a puck in scoop was 100.3^0 with a s.d. 7.4^0. This compares with the unmodified detection angle of 180^0. That is, with the three sensor arrangement no matter what the angle the robot collides with the wall, which triggers the scoop, the wall is detected.

5.1 Materials and Methods

44 pucks were set out in a regular pattern. The overhead camera grabbed a frame every 5 minutes. The end condition was considered to be when a configuration of 40 pucks (over 90%) were within a puck radius of each other. Three trials were conducted with 10 robots.

5.2 Results and Conclusion

In all three trials the successful end condition of 40 pucks clustered against the wall in a single cluster was achieved. The time to completion for each trial were 10hrs 15m, 13hrs 30m and 14hrs 25m. Figure 4 shows the successful wall clustering after the first trial.

It has been demonstrated that wall clustering can be achieved by altering the robot morphology by modifying the characteristics of the obstacle sensors. By employing the same deterministic 3 rule algorithm which creates a central cluster and, very importantly, reducing the forward obstacle detection capability, wall clustering can also be induced. The subtle alteration of the sensors causes the consequences of the deterministic rule set to have the same gross behaviour near the wall as the probabilistic mechanism described earlier. That is, a robot with a puck colliding with the wall is more likely to drop the puck. This increases the likelihood of other pucks being deposited against it which encourages puck cluster development at the wall. Such an 'equivalence' is interesting in that it demonstrates that a deterministic rule set can generate different outcomes on the macro scale when the 'domain physics' in the guise of robot morphology is altered. Thus, it may be possible to create particular behavioural repertoires by altering the general robot morphology. Conversely, it is interesting to consider that a deterministic rule set operating in different environments (for example, robots bulldozing wet sand or dry sand) could generate tangibly different outcomes. Arguably, as in this case, this could lead to a more minimalist solution.

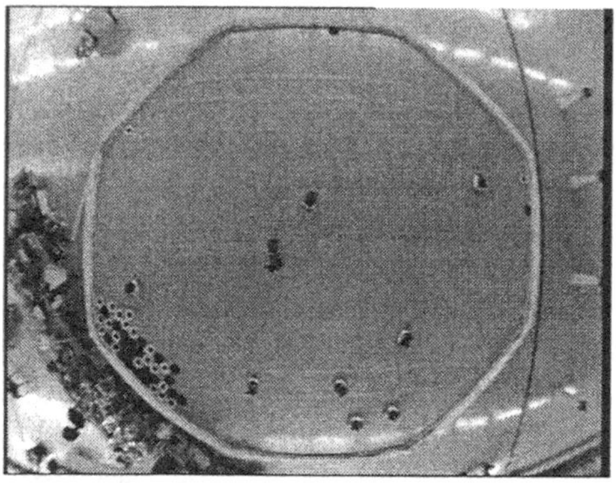

Fig. 4. Wall Clustering with the Canonical Deterministic Algorithm

6 Discussion

It has been argued here that the results of the experiments indicate that, for certain building tasks, control and coordination of a collective of minimalist robots can be realised by employing stigmergic mechanisms. It has been shown that changes to the environment (as in the case of seeding with different types of seeds) or robot morphology (as in the example of restricted acceptance angle for eyes) result in markedly different outcomes. Such observations emphasise the tight coupling and dependence of the stigmergic mechanism between the robots' action rule set and the environment. Such a tight coupling has been exploited by social insects which demonstrates that stigmergic 'solutions' can be evolved by natural systems. A significant challenge for a designer of a system employing a minimalist action reperotire and stigmergic mechanisms is that variations in the domain physics could be considered as problematical or potentially advantageous and thus exploitable.

7 References

Altenburg K. & Pavicic M. (1993). "COCOBOTS: A Collective and Cooperative Robot System" NDSU Technical Report: NDSU-TR-CSOR-9401.

Beckers R., Holland O. & Deneubourg J-L. (1994). "From Local Actions to Global Tasks: Stigmergy & Collective Robots." Proceedings of the 4th International Conference on Artificial Life.

Bonabeau E. (1997). "From Classical Models of Morphogenesis to Agent-Based Models of Patttern Formation." Artificial Life 3: 191-211.

Deneubourg J.-L., Goss S., Franks N., Sendova-Franks A., Detrain C. & Chretien L. (1990). "The Dynamics of Collective Sorting Robot-Like Ants and Ant-Like Robots." 1st International Conference on the Simulation of Adaptive Behavior. . MIT Press.

Deneubourg J.L., Theraulaz G. & Beckers R. (1992). "Swarm-Made Architectures." Proceedings of the 1st European Conference on Artificial Life.

Franks N., Wilby A., Silverman B. & Tofts C. (1992). "Self-organizing Nest Construction in Ants: Sophisticated Building by Blind Bulldozing." Animal Behaviour, vol. 44:357-375.

Franks N. & Deneubourg J.-L. (1997). "Self-organizing Nest Construction in Ants: Individual Behaviour and the Nest's Dynamics." Animal Behaviour 54.

Franks N. & Sendova-Franks A. (1992). "Brood Sorting by ants: distributing the workload over the work-surface." Behav. Ecol. Sociobiol. 30:109-123.

Grasse P-P. (1959). "La Reconstruction du Nid et les Coordinations Interindividuelles Chez Bellicositermes Natalensis et Cubitermes SP. La Theorie de la Stigmergie: Essai D'interpretation du Comportement des Termites Constructeurs." Insectes Sociaux, Tome V1, No1.

Holland O. (1996). "The Varieties of Stigmergy: Indirect Interaction as a Control Strategy for Collective Robotics" IASeL Technical Report 101, UWE

Melhuish C., Holland O. and Hoddell S. (1998) "Collective Sorting and Segregation in robots with Minimal Sensing" 5th International Conference on the Simulation of Adaptive Behavior. Zurich. From Animals to Animats. MIT Press.

Wilson E.O. (1975). "Sociobiology - A New Synthesis." Belknap Press, Harvard University Press.

Adaptive Exploration of a Dynamic Environment by a Group of Communicating Robots

Aude Billard[2,1], Auke Jan Ijspeert[2,1], Alcherio Martinoli[1]

[1] LAMI, Swiss Institute of Technology (EPFL)
IN-F Ecublens, 1015 Lausanne, Switzerland / Aude.Billard@epfl.ch
`http://diwww.epfl.ch/w3lami/team/billard/`
[2] - IDSIA, Corso Elvezia 36, CH-9600 Lugano

Abstract. Is it more efficient to use one or several robots? Will the performance of a group of robots working in a collaborative task be enhanced if the robots can communicate with one another? What learning abilities should the robot(s) be provided with for adapting to a continuously changing environment? We address these three issues in a specific task, namely learning the topography of an environment whose features change frequently. We propose a learning algorithm based on an associative memory which allows a group of robots to keep an up-to-date account of the environmental state when this changes regularly. A probabilistic model is developed which gives an abstract representation of the system. It is used to determine boundaries for the system's variables (the number of robots, the frequency of environmental changes, and the environment's configuration) within which the learning is successful. The predictions of the probabilistic model are confirmed by simulations run in Webots, a 3-D simulator of Khepera robots.

1 Introduction

Numerous works on autonomous robot systems investigate the questions of 1) whether it is more efficient to distribute the area of expertise needed for performing a complicated task between several robots rather than designing a unique expert robot [7]; 2) whether the use of explicit communication could improve the performance of a group of robots in a collaborative task ([2, 12, 14]); 3) what learning abilities should the robot(s) be provided with for adapting to a continuously changing environment [10].

We address these three issues in a specific task, namely learning the topography of an environment whose features, the locations of objects, change frequently. A group of *worker robots* search constantly the environment. The robots are provided with an associative memory which allows them to store the locations of the objects as they detect them. They transmit to each other the coordinates of the objects' locations by locally broadcasting the location when finding an object or when meeting another robot. The information gathered by each robot is also transmitted to a static *database robot* which each robot visits regularly. The database robot keeps an up-to-date account of the global state of the environment of which each robot has only a partial knowledge. We study the system's performance in a dynamic environment, in which the locations of the objects change with a constant frequency. We investigate the influence of the variables of the system, namely the number of worker robots, the frequency of environmental changes and the environment's configuration, on the data collecting performance of the group.

A number of works have investigated multi-robot systems for the mapping of a static environment, e.g. [1, 5, 13]. Our work brings new contributions to this research area. First, we give an abstract representation of the problem by modelling the system as a set of probabilistic equations. In this respect, our work follows current line of research which develops probabilistic models of multi-agents systems, e.g. for representing biological systems (ants' society [6]) or engineering systems ([8, 14]). Second, we validate the model by comparing its predictions to results of simulations (this paper) and physical experiments ([4]). Thus, our study complements previous work in the domain by providing both a theoretical and practical view of the problem. Finally, we study mapping of a *dynamic* environment, that is an environment whose landmarks change location frequently, and propose a distributed control architecture which allows a multi-robot system to dynamically update its map following the observed changes.

2 The simulations

2.1 The experimental set-up

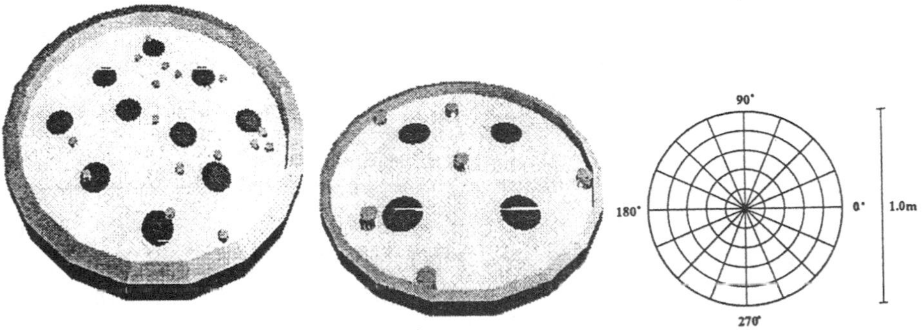

Fig. 1. Arenas of 2 meters (left) and 1 meter (middle) of diameter with respectively 5 and 15 *worker* robots. The database robot stands in the centre of each arena. There are 10 and 4 sources in the big (left) and small (right) arenas respectively which are represented as patches of 0.1m and 0.07m diameter lying on the floor. **Right** Division of the small arena into 80 zones.

Simulations were carried out in Webots [9], a 3-D simulator of the Khepera [11] robots. Simulations use two circular arenas of 1 meter and 2 meters diameter respectively as shown in figure 1. The Khepera robot is round with a diameter of

5.5cm. Thus, we study the exploration strategy in environments which are 1600 and 400 times the robot's size.

The simulator gives a relatively faithful representation of the Khepera robots, by incorporating imprecise movements of the robots' wheels (slipping) and introducing noise in the robots' sensors measurements. Each robot is provided with 9 infra-red (IR) sensors (used to detect other robots and the arena walls, the 9th IR is activated only by the walls and allows to distinguish between robots and walls), a detector of ground colour (used to distinguish between zones with/without sources), a radio transceiver, a compass with 5^0 degrees precision and one odometry counter on each wheel.[1] Compass and odometry sensors are used by the robots to determine their location relative to the centre of the arena. The robots reset their position to the correct one each time they meet the database robot or hit a wall in the arena. The odometry errors are therefore contained within a range of up to 10 percent error. The sources' locations, given as an angle and a distance relative to the centre, are determined following a scaling of the arena into $5 \cdot 16 = 80$ (small arena) and $10 * 16 = 160$ (big arena) zones, see schema of figure 1. Thus, the sources' locations are known within a precision 22.5 degrees (for the angle) and 10 cm (for the distance).

2.2 The robots' controllers

All worker robots have the same controller which is composed of five modules: 1) an *obstacle avoidance module* which consists of a one-layer real value feed forward neural network with eight input units (one for each infra-red sensor measurement) and two output units for the two motors (speed control); 2) a *memory-based exploration module* which determines the robot's direction of travel when crossing the border between two zones of the arena (following the division represented in figure 1); each robot keeps track of the number of times it has crossed each zone; when it estimates that it has reached the border between two zones, the robot turns toward the zone it has less visited so far. 3) a *communication module* which consists of two rules; the robot emits the location of one source (chosen randomly over all locations it knows) when it meets another robot (using pt-to-pt protocol, with acknowledgement from the receiver robot); the robot broadcasts locally (within a limited range) the location of a source when it discovers one; 4) an *odometry module* which calculates the robot's position relative to the database (centre of the arena) given the measurements of the wheels' counters and of the compass; 5) a *learning module* which consists of a bidirectional associative memory; the robots keeps track of the sources' locations by correlating the two outputs of the odometry module which are the angle θ and the distance ρ, the polar coordinates of the robot relative to the centre of the arena. Each connection of the module between an angle and a distance measurement is bidirectional and is associated with two parameters, a weight $w_{ij} = w_{ji}$ and a time parameter $\tau_{ij} = \tau_{ji}$. The associative module takes binary inputs (1/0); each input corresponds to a measure of angle and distance

[1] All sensors used in the simulations exist for the real Khepera robots.

following the arena's scaling (see figure 1). The weights w and time parameters τ are two matrices of 10 by 16 units (for simulations in the big arena) and of 5 by 16 units (for simulations in the small arena). The experiments starts with all weights w and time parameter set to zero. The learning algorithm is a system of three rules:

1. **Learning by seeing:**
   If the robot detects an object, then
   $$w_{\theta,\rho} = 99 \text{ and } \tau_{\theta,\rho} = t$$
   where t is the time measured by the clock of the robot[2].
2. **Forgetting:**
   If the robot crosses a location given by θ, ρ such that $w_{\theta,\rho} > 0$ but does not detect a object, then
   $$w_{\theta,\rho} = 1 \text{ and } \tau_{\theta,\rho} = t$$
3. **Learning by hearing:**
   If the robot hears the location of an object as told by another robot, then:
   $$\text{If } \tau'_{\theta',\rho'} > \tau_{\theta',\rho'} \text{ then } w_{\theta',\rho'} = \frac{1}{2}\cdot(w_{\theta',\rho'} + w'_{\theta',\rho'}) \text{ and } \tau_{\theta',\rho'} = \tau'_{\theta',\rho'}$$
   $\theta', \rho', w'_{\theta',\rho'}, \tau'_{\theta',\rho'}$ are the distance, angle, weight and time parameter transmitted by the emitter robot.

The learning for one robot is evaluated by counting the number of correctly correlated pairs of coordinates. Learning is successful when this number is equal to the number of different locations. A pair of coordinates $\{\theta, \rho\}$ is considered as correctly correlated when the weight $w_{\theta,\rho}$ is greater than a threshold H. H is calculated at each time step as a function of the current value of all the weights w: $H = \frac{\frac{M(w)-m(w)}{2}+M_A(w)}{2}$, where $M(w) = \max_{w>0}(w)$ and $m(w) = \min_{w>0}(w)$ are the maximum and minimum values of weights for all $w > 0$. $M_A(w) = \text{mean}_{w>0}(w)$ is the arithmetic mean calculated over all $w > 0$.

H estimates the threshold between the important weights (close to 99) which correspond to correct correlations and the small weights (close to 1) which are noisy or discarded correlations. The calculation is based on the hypothesis that the distribution of the correct and incorrect correlations are uniform, that is equiprobable on all pairs. This is to some extend correct as there is no a-priori bias on the number of times the robots can observe each object. Note that there is no total forgetting of a source's location, that is, the weight associated to the corresponding pair of coordinates never returns to zero. The minimal value for a weight which has been updated once is 1. This value can never be greater than the threshold; therefore, a location associated with a weight equal to 1 is always discarded as no longer valid. Keeping track of all sources' locations which have once been discovered results in the robots constantly checking the validity of this location (following rule 2). Forgetting and relearning of locations

[2] The clock is incremented at each processing cycle and is set at zero when the experiment starts.

is thus made faster. The learning algorithm described above is a simplification of the DRAMA connectionist architecture [3]. The reader can also refer to [4] for further explanations.

The database robot's controller consists of the same learning module as the worker robot. When worker and database robots meet, the worker robot transmits to the database robot its matrices of weights and time parameters (all w and τ, while in a two worker robots communication, only one pair w, τ corresponding to one location is transmitted). Following rule 3, the database robot calculates the mean value between its current set of weights (collected from another robot) and those newly transmitted, iff the new information is more recent than its current one. The database robot transmits then back to the worker robot the mean matrix of weights and time parameters. After a meeting with the database, a worker robot has therefore the same global knowledge of the environment as the database. This speeds up the forgetting process as the robot can then verify more locations (all the locations which have been recorded by the group) than only those it had stored itself.

3 The probabilistic model

The aim is to define an equation which will allow us to determine T the minimal time for the database robot to learn the locations of N_s objects, given that there are N_r worker robots, that the arena has size A and that an object covers a surface S_s.

We define the building blocks or fundamental probabilities of the model by considering the geometrical configurations of the system. We define the probability of meeting the database robot (P_{db}) as the ratio of the surface of detection of the database robot by another robot S_d over the arena's surface A, i.e. $P_{db} = S_d/A$. Similarly, the probabilities of meeting another robot ($P_r = S_r/A$), of passing across a source ($P_s = S_s/A$) or of being in the range of communication of another worker robot ($P_c = S_c/A$) are the ratios of the surfaces of each of these objects over the arena's surface ($A = \pi \cdot (r^2)$, $r = 0.5[m]$ or $r = 1[m]$ for small/big arenas), $S_d = \pi \cdot 0.1^2[m^2]$ (small arena), $S_d = \cdot 0.15^2[m^2]$ (big arena) $S_r = \cdot 0.1^2[m^2]$, $S_s = 0.0038[m^2]$ (small arena), $S_s = 0.0078[m^2]$ (big arena), $S_c = \cdot 0.3^2[m^2]$.

Let $P_{\text{success}}(N_r, N_s, T)$ be the probability that the event "the database robot has recorded N_s locations" has occurred after a time T. This event is true if each of the N_s locations have been seen by at least a robot and been transmitted to the database robot at least once in a time T, i.e.:

$$P_{success}(N_r, N_s, T) = 1 - (1 - P_{L-success}(N_s, T))^{N_r}$$

$P_{\text{L-success}}(N_s, T)$ is the probability that a first event "all N_s locations have been transmitted" has occurred within a period $T - t_1$, given that a second event "all N_s locations have been learned" has happened in a period t_1. This conditional probability can be expressed as follows:

$$P_{\text{L-success}}(N_s, T) = P(seeDBinT - t_1 | learnobjectint_1)$$

The two events are independent, thus the probability of their co-occurence for a given pair $\{t_1, T - t_1\}$ is the product of each event's probability. The total probability is the normalised sum over all possible pairs $\{t_1, T - t_1\}$ (time is discretised) of this product:

$$P_{\text{L-success}}(N_s, T) = \frac{\sum_{t_1=1}^{T} P_{\text{learn-object}}(N_s, t_1) \cdot P_{\text{see-database}}(T - t_1)}{\sum_{t_1=1}^{T} P_{\text{learn-object}}(N_s, t_1)} \quad (1)$$

The probability of meeting the database robot, $P_{\text{see-database}}(T - t_1)$ in equation 1, is the probability of crossing the surface S_d within a period $T - t_1$: $P_{\text{see-database}}(T - t_1) = 1 - (1 - P_{db})^{T-t_1}$. $P_{\text{learn-object}}(N_s, t_1)$ is the probability that the event "a robot has learned N_s object locations in a time t_1" is true. A robot learns about an object's locations if the robot either sees the object P_S or hears its location from another robot P_H: $P_{\text{learn-object}}(N_s, t_1) = (1 - (P_{\text{not-learn-object}})^{t_1})^{N_s}$; $P_{\text{not-learn-object}} = (1 - P_S) \cdot (1 - P_H)^{N_r - 1}$.

The probability of hearing an object's location from another robot's broadcast is the probability that the three following events are true: 1) the listener robot is within an area S_c around the emitting robot (S_c is the surface within which the communication is audible) and 2) the emitting robot broadcasts the particular location, 3) no other robot out of the $N_s - 1$ (excluding the emitting robot, including the listener robot) is simultaneously emitting in that same area (P_{Interf} is the probability of this event). It follows:

$$P_h = P_{\text{Hear}} \cdot P_{\text{Interf}};\ P_{\text{Interf}} = (1 - P_{\text{Hear}})^{N_s \cdot N_r - 1}\ P_{\text{Hear}} = (\tfrac{S_c}{A}) \cdot (1 - P_{\text{Not-emit}}).$$

Event 2 is true if the emitter robot either sees the object (it then broadcasts the location) or if the emitter robot meets another robot which transmits it that particular location. In the later case, the emitting robot chooses 1 location among the $2 * N_s$ it knows, which comprised the N_s correct and no longer valid locations. The later event can occur only if the robot has seen that object within a time $t_o < t_1$ before meeting another robot. It follows:

$$P_{\text{Not-emit}} = (1 - P_s)^{t_1} \cdot (1 - \frac{1}{2 \cdot N_s} \cdot \frac{\sum_{t_o=1}^{t_1} (1 - (1 - P_s).^{t_o}) \cdot (1 - (1 - P_r)^{t_1 - t_o})}{\sum_{t_o=1}^{t_1} 1 - (1 - P_s)(t_1 - t_o)})$$

In the above equations, the unity of surface is the meter and the unity of time corresponds to the time needed to cover the surface S_s (which is the minimal surface considered in the equations). In order to convert the value of time in seconds, T has to be multiplied by $\frac{S_s}{V_r * D_R}$, where $V_r = 0.16[m/s]$ is the maximal speed of the robots and $D_r = 0.055[m]$ is the diameter of the robot.

4 Results

4.1 Probabilistic model versus simulations

A first set of Webots simulations were carried out in a static environment (i.e. the locations of the objects did not change). 100 and 150 runs were A run simulated

1000 seconds. 10 different runs were done using a different random seed for a given number of robots. The number of worker robots was varied from 1 to 10 and from 1 to 15 in the small and big arenas respectively. We measured the mean time delay after which the database robot knew all 4 (small arena) and 10 (big arena) objects' locations. In figure 2, we compare the prediction of the probabilistic model and the results of the simulations. As one would expect, the more robots the faster the learning. However, the relation between these two variables is not linear and the increase of time efficiency saturates for important numbers of robots. Thus, if one would consider implementing the system in a real robotic set-up based on these results, one would determine the optimal number of robots by comparing the gain in time efficiency to the cost increase when augmenting the number of robots.

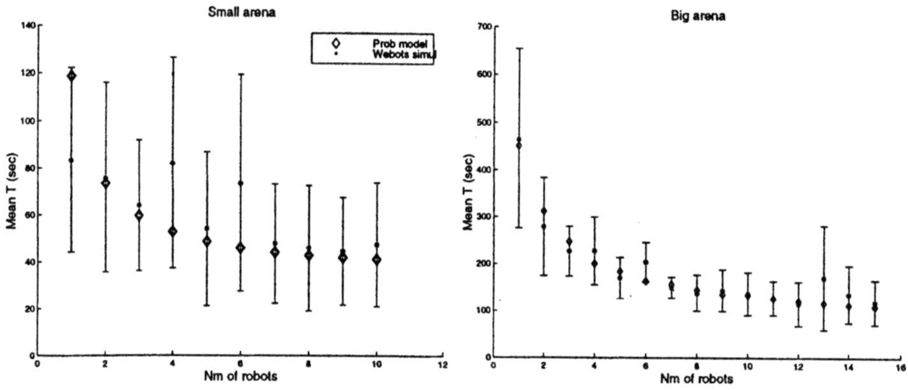

Fig. 2. The Y-axis represents the mean (over 10 runs) time delay T and the X-axis is the number of robots. Each figure compares the prediction of the probabilistic model and of the Webots simulations in the small arena (left) and in the big arena (right) ('$*$' point with error bars).

For both small and big arenas, the results of the probabilistic model and of the simulations are qualitatively and quantitatively similar. This means that, although the probabilistic model is a crude representation of the system, it approximates well the correlations between the main system's variables. Two aspects of the simulations are, however, not represented by the model: 1) the probabilistic model assumes a uniform coverage of the space, where all points of the space are visited with the same frequency; this does not take into account the boundary effects due to the walls which make the centre of the arena (i.e. the database robot's location) a point more often visited than the exterior of the arena (this effect is augmented by the exploration strategy which gives the preference to riding towards the database robot when crossing the border of the 16 central zones); in order to represent this effect in the model, we increased S_{db} compared to its real geometrical value so that we obtained the same probability of meeting the database robot ($P_{db} = S_{db}/A$) as that measured in the simulations. 2) The probabilistic model assumes that learning of a source' location is perfect, i.e. when a robot crosses the source, it learns its location; this does not take in account the imprecise determination of the location due to odometry error[3].

The probabilistic model is, therefore, a good first approximation of the studied system. It allows to determine the optimal efficiency of the system in the ideal case; However, more realistic simulations, such as done in Webots, should be carried out, in order to evaluate the importance of the above mentioned aspects which not represented by the model. The probabilistic model is general to the extend that it makes no requirement on the type of robots and environment used. Thus, it could potentially be applied to other experimental set-ups for the same task. It is a *parameter free* model. In order to apply it to another set-up, one should simply set the parameters defining the robots (speed of movements, size) and the environment (dimension and number of sources) to the experimental ones.

4.2 Learning in a dynamic environment

3 times 15 runs were carried out in a dynamic environment, in which the object changed locations with a constant *update rate* R varied from 1 to 15. Each run lasted $5000 \cdot R$ seconds, during which all the objects' locations changed every $8 \cdot R$ (small arena) and every $28 \cdot R$ (big arena) seconds (that is several hundred changes per runs). We ran simulations with groups of 1, 3 and 5 robots in the small arena and 5, 10 and 15 robots in the big arena. Figures 3 top left and 3 top right show the mean number of correctly and incorrectly learned locations over the whole run for the small and big arenas respectively. The results for each three configuration of robots are superimposed. For R less than 5, the database knows on average about 50% of the correct locations, while still taking for correct almost 50% of the locations which have been updated. As long as R is less than 5, the environment changes faster than the minimal time delay T required for the robots to learn all the locations. The minimal T was measured in the simulations of section 4.1 (see figure 2) as a minimum of 40 and 140 seconds for small and big arenas respectively (the measures were consistent with the probabilistic predictions). For R bigger than 5, the proportion of correctly learnt location increases steadily while the proportion of incorrectly learnt locations decrease by the same proportion. There is almost no difference between the three different robots' configurations in each case. Although the minimal time delay for complete learning decreases with the number of robots (see figure 2), the gain is small (about 10%). In addition, because of the important variance of the measured time values (see the error bars in figure 2), the learning performance of the database robot appears on average similar for the three different robots' configurations. Thus, there is almost no benefit in using 15 rather than 5 robots in the big arena and 5 robots rather than 1 in the small arena.

[3] Thanks to the resetting strategy, the odometry errors are in fact negligible compared to the environmental discretisation, see section 2.1.

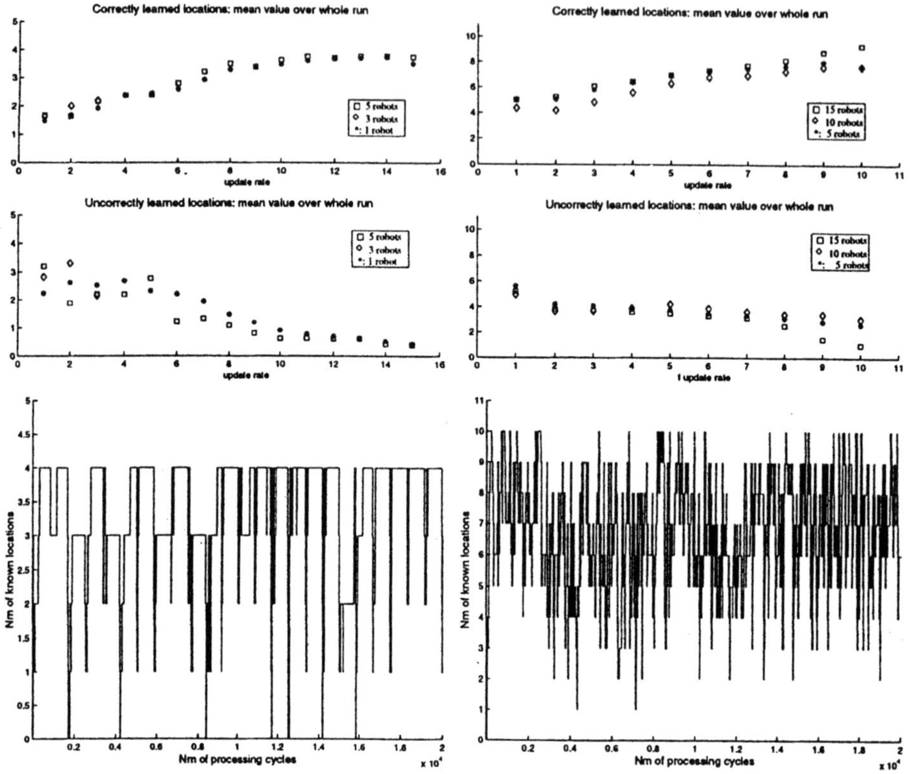

Fig. 3. Top: Mean number of correctly and incorrectly learned locations over the whole run for small (left) and big arenas (right); superposition of the results for three robots' configurations (left: 1,3,5 robots; right: 5,10,15 robots). Bottom State of the database's knowledge (number of known locations) along a run. 5 robots configuration in small arena (left) and 10 robots configuration in big arena (right).

Figures 3 bottom shows the progression of the learning of the database robot along a run (results of the simulation in small (left) and big (right) arenas with 5 and 10 robots respectively for $R = 5$). The curve varies from zero (no locations known) to the maximum (4 and 10 correctly known locations in small and big arenas). In the simulation, the objects are not displaced simultaneously. The rate at which every object is displaced is constant (it is R) but the moment at which the first displacement occurs is different for each object. This explains the fact that the learning curve does not always decrease until zero (the new locations being discovered and transmitted before all locations have been changed).

5 Conclusion

This paper presented a multi-robot system, composed of a group of mobile worker robots and one static database robot. A learning algorithm was proposed, composed of learning and forgetting processes, which allows a group of robots to keep an up-to-date account of the environmental state when this changes regularly. The correlations between the variables of the system, the number of robots, the frequency of environmental change and the environments' configuration, were modelled by probabilistic equations. Simulations were carried out in a 3-D simulator of Khepera robots, which confirmed the predictions of the probabilistic model. These experiments demonstrated that the proposed multiple robots system, which is based on an associative memory learning algorithm, is successful at learning the topography of an environment which changes with a constant frequency.

Acknowledgements Many thanks to Luca Gambardella and the anonymous reviewers for useful comments on these experiments. Lots of thanks to the LAMI, EPFL which provided the facilities and technical support for these experiments. This research was supported by a grant of the Swiss National Research Foundation, project "A methodology for collective robotics design".

References

1. Amat, J., López de Màntaras, R., Sierra, C. "Cooperative Autonomous Low-Cost Robots for Exploring Unknown Environments", In O. Khatib and J.K. Salisbury, editors, Proceeding of the Fourth Symposium on Experimental Robotics ISER-95, Stanford, US, June 30- July 2, 1995, Lecture Notes in Control and Information Sciences, Springer Verlag, pp. 41-49.
2. T. Balch & R. C. Arkin, (1994), 'Communication in reactive multiagent robotic systems', *Autonomous robots Journal*, 1, pp. 27-52.
3. A. Billard & G. Hayes, 1999, "DRAMA, a connectionist architecture for control and learning in autonomous robots", *Adaptive Behaviour Journal*, vol. 7:1.
4. A.Billard, A-J. Ijspeert, A. Martinoli. "A multi-robot system for adaptive exploration of a fast changing environment: probabilistic modelling and experimental study". Submitted to *Connection Science*, special issue on *Adaptive Robots*. March 99.
5. W. W. Cohen, (1996), 'Adaptive mapping and navigation by teams of simple robots', *Robotics and autonomous systems*, 18, pp. 411-434.
6. E. Bonabeau, G. Theraulaz & J-L Deneubourg, (1998), 'Fixed response thresholds and the regulation of division of labor in insect societies', *Bulletin of Mathematical Biology*, **60**, pp. 753-807.
7. M. J Mataric, (1998), 'Using Communication to Reduce Locality in Distributed Multi-Agent Learning', *Journal of Experimental and Theoretical Artificial Intelligence*, special issue on Learning in DAI Systems, Gerhard Weiss, ed., **10(3)**, Jul-Sep, pp. 357-369.
8. A. Martinoli, A. Ijspeert, F. Mondada, (1999), 'Understanding collective aggregation mechanisms: from probabilistic modelling to experiments with real robots', *Robotics and Autonomous Systems*, Elsevier. To appear. Preprint available at `http://diwww.epfl.ch/lami/team/alcherio/am_pub.html`
9. O. Michel, (1998), 'Webots: a Powerful Realistic Mobile Robots Simulator', Proceeding of the Second International Workshop on RoboCup, LNAI, Springer-Verlag.
10. A. Murciano & J. del R. Millan, (1997), 'Learning signalling behaviour and specialisation in Cooperative agents', *Adaptive Behaviour Journal*, **5**:1, pp. 5-28.
11. F. Mondada, E. Franzi & P. Ienne, (1993), 'Mobile Robot Miniaturisation: a Tool for Investigation in Control Algorithms', Proceedings of ISER'93, Kyoto, Japan, October 1993, pp. 501-513.
12. L. E. Parker, ALLIANCE: An Architecture for Fault Tolerant Multi-Robot Cooperation, IEEE Transactions on Robotics and Automation, 14 (2), 1998.
13. Rekleitis, I. M., Dudek, G., Milios E. E. "Multi-Robot Exploration of an Unknown Environment, Efficiently Reducing the Odometry Error", Int. Joint Conf. on Artificial Intelligence, August, 1997, Nagoya, Japan, Morgan Kaufmann, pp. 1340-1345.
14. E. Yoshida, T. Arai, M. Yamamoto and J. Ota, (1998), 'Local communication of multiple robots: Design of optimal communication area for cooperative tasks', *Journal of Robotics Systems* 15(7), pp. 407-419.

Task Fulfilment and Temporal Patterns of Activity in Artificial Ant Colonies

Jordi Delgado[1,2,3] and Ricard V. Solé[2,3]

[1] Dept. Llenguatges i Sistemes Informàtics, Universitat Politècnica de Catalunya, Campus Nord, Mòdul C6, 08034 Barcelona (Spain)
jdelgado@lsi.upc.es,
[2] Complex Systems Research Group,
Department of Physics, FEN, Universitat Politècnica de Catalunya,
Campus Nord, Mòdul B4, 08034 Barcelona (Spain)
ricard@complex.upc.es,
[3] Santa Fe Institute, 1399 Hyde Park Road, Santa Fe, NM87501, U.S.A.

Abstract. Since the discovery of self-synchronization of activity in ant colonies some authors have argued that this phenomenon lacks any adaptive significance, while others have suggested that there may be some functional behaviors related to these rhythmical patterns of activity. In this paper, we introduce a mobile automata model of task fulfilment in ant colonies with self-synchronized activity, and test if these synchronized patterns provide any advantage when compared with non-synchronized activity patterns.

1 Introduction

Oscillatory patterns of activity in ant colonies were discovered independently by N. Franks in *Leptothorax acervorum* colonies [15] and B.J. Cole in *L. allardycei* colonies [7] at the end of the eighties. Since then, self-synchronization has also been found in other species (*L. longispinosus*, *L. ambiguus*, *L. curvispinosus*, *L. allardycei* and *L. muscorum* [22, chap. 2], also in *Pseudomyrmex elongatus*, *P. pallidus*, *Tapinoma littorale*, *Zacryptocerus varians* and *Crematogaster ashmeadi* [9]), so it seems to be a very general pattern of temporal behavior. Furthermore, there is strong experimental evidence that these patterns are a collective property, since the individual activation dynamics was found to be chaotic by Cole [8].

Cole [7] discussed the adaptive significance of these short-term activity cycles arguing that it is unlikely that these cycles contribute to the efficiency of the colony. They are "(...) the inevitable outcome of interactions within social groups"[7, pag. 257]. However, further research has been done in order to clarify the relation between functional behavior and oscillatory patterns and at least two colony activities have been suggested to be enhanced by self-synchronized behavior: brood tending in *L. acervorum* [19] and task allocation [23]. It is the latter [23] that inspired the present work. G.E. Robinson suggested that self-synchronized behavior provides a mechanism for information propagation:

> "Sampling behavior that involves social interactions may be facilitated by synchronous bursts of worker activity, which have been observed in ant colonies (...). The decision of which task to perform would be based on the integration of acquired information, coupled with behavioral biases associated with worker, caste, physiological status and prior experience." [23, pag. 652]

So, according to [23], self-synchronization facilitates the sampling of any information an individual may need from other individuals. Let us try to clarify this point. Assuming that ants cannot be active all the time (which is what is observed in nature, see above), why self-synchronized behavior would be a better (simple) strategy than, say, random (in the sense of "non-synchronized") activity patterns? We will need further assumptions to answer this question: First, the obvious one of locality (an individual is able to get only local information) and second, the quite reasonable (and biologically plausible [15, 7, 8]) assumption that the unique interaction allowed to an *inactive* individual is to be "awaked" by other(s) individual(s); an inactive individual does not carry any information, namely, it is equivalent to a "slept" individual. Now, in this context, it would be clear why we should obtain an increase of efficiency with synchronized patterns of activity: It would maximize the number of simultaneously *active* neighbours of an *active* individual (let us remind that we are assuming that individuals cannot be active all the time). This reasoning, however clear, must be validated with both theoretical models and experiments, since only by doing so can we explore which assumptions are necessary, and then later go out and see what is the real state of things in real ants. The study of a theoretical model is the main point of this paper.

Designers of artificial societies should be also interested in this subject-matter, since if self-synchronization is able to provide a better functioning of collective systems, it would be highly desirable to implement this mechanism in either collective algorithms or collective robotics [3, chaps. 5–7]. A further advantage is that it is quite easy to implement using only simple local interactions, as the work with Fluid Neural Networks suggests [25, 26] (see below).

In section 2 we introduce our model and our first results are detailed in section 3. In section 4 we discuss our results and suggest a simple experimental setting to test them.

2 The Model: Fluid Neural Networks and Finite Threshold Models

In order to start our work, we need to couple two different phenomena: Self-synchronized activity and information spreading. There are several mathematical models to deal with the former [2, 25, 26, 28, 18], among which, for simplicity, we have chosen the Fluid Neural Network (FNN) [10–12, 22, 24–26]. Furthermore, the FNN is the model best suited to our purposes, since it allows one to introduce additional behaviors upon the basic activation behavior of individual automata. The latter, information spreading, is a relatively loose notion that we will make more specific through the notion of task spreading. Broadly, a mechanism based on the Finite Threshold Model of division of labour [4] is built upon each individual in a FNN. Besides, individuals will be able to communicate a certain fraction of task to their neighbours, spreading it through the system (see details in section 2.3). It is similar to the spread of a large protein source within the nest or to the spread of liquid food via trophalaxis [21].

2.1 Fluid Neural Networks

In FNN the standard approach of neural networks is used [1, 17], but a new set of rules defining local movement and individual activation are also introduced. A set of N automata or "neuron-ants" is used. The state of each automaton (say the i-th one) is described through a continuous state variable $S_i(t) \in \mathbf{R}$, at each time step $t \in \mathbf{N}$. Each element can move on a $L \times L$ two-dimensional lattice with periodic boundary conditions.

If $S_j(t)$ is a given automaton (the spatial dependence is omitted for simplicity), the new states are updated following:

$$S_i(t+1) = \tanh\left[g h_i(t)\right] \tag{1}$$

where g is a gain parameter and $h_i(t)$ can be defined in diverse ways in order to get the desired behaviour. We will obtain oscillations in activity (see below) if the term $h_i(t)$ includes interactions with the eight nearest neighbours:

$$h_i(t) = S_i(t) + \sum_{i \neq j \in B(i)} S_j(t) \tag{2}$$

where $B(i)$ are the nearest automata. To get non-synchronized individuals we do not need interaction with any other automata:

$$h_i(t) = S_i(t) \tag{3}$$

We have seen above that one of the properties observed in isolated ants was spontaneous activation [8]. In FNNs this has been included in the following way: each automaton can be either ***active*** or ***inactive*** and, if active, it moves randomly to one of the eight nearest cells (if no space is available, no movement takes place). In our model a given automaton will be active if $S_i(t) > \theta_{act}$ and inactive otherwise. Once an automaton becomes inactive, it can return to the active state (with an ***spontaneous activity level*** S_a) with some probability p_a.

The collective behaviour we measure in FNNs is the mean activity of the system. We define an activity for each individual $S_i(t)$, $a_i^t = \Theta\left[S_i(t) - \theta_{act}\right]$, so the mean activity at time t will be

$$\rho_t^+ = \frac{1}{N}\sum_{j=1}^{N} a_j^t = \frac{1}{N}\sum_{j=1}^{N} \Theta\left[S_j(t) - \theta_{act}\right] \tag{4}$$

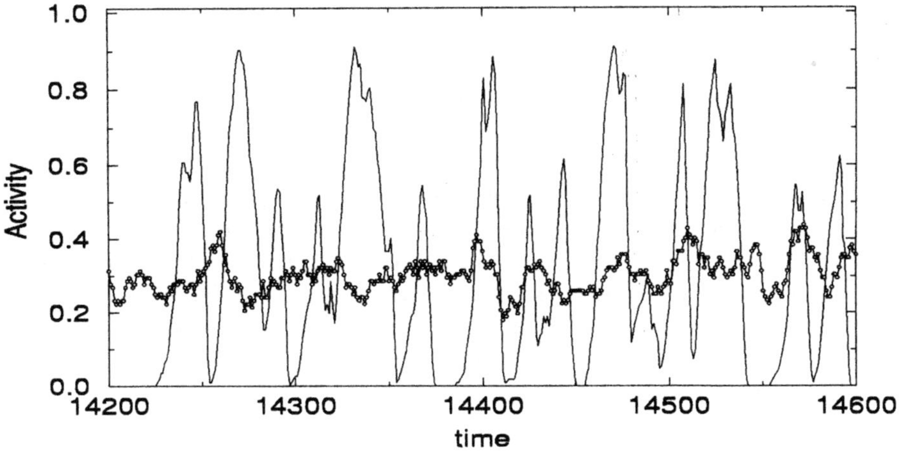

Fig. 1. Evolution in time of ρ_t^+. We get bursts of synchronized behaviour with $S_a = 0.01$, $p_a = 0.001$ and non-synchronized behaviour with $S_a = 0.1$, $p_a = 0.03$ (circles). In both cases the average activity per individual and time step is $\simeq 0.3$, that is, an individual is active approximately the 30% of the time.

where $\rho_t^+ \in [0,1]$ and $\Theta[x]$ is such that $\Theta[x] = 1$ if $x \geq 0$ and $\Theta[x] = 0$ otherwise. We define also the total density of automata as $\rho = N/L^2$.

In this paper we have chosen the FNN parameters N, L, g, θ_{act}, S_a and p_a to get the desired behavior, that is, self-synchronized activity and non-synchronized activity (see figure 1), though in both cases we have imposed the following constraints[1]:

- A number of individuals similar to that observed in colonies with synchronized activity [22].
- The activity level per individual must be around $\simeq 0.3$, that is, each individual is active approximately the 30% of the time, on average, as observed in species with synchronized activity [20, 6, 14].
- The density ρ of the system should be around $\simeq 0.2$, as was observed experimentally in [16], see also [24].

The sort of interactions defined in FNNs (activation among individuals) are currently being the subject-matter of experimental research by B.J. Cole and collaborators [9]. An analysis of FNNs is performed in [10–12].

[1] To be more specific, the parameters we use are: $N = 120$, $L = 25$, $g = 0.1$ and $\theta_{act} = 10^{-16}$. We get synchronized behavior with $S_a = 0.01$, $p_a = 0.001$ and h_i defined as in eq. 2, and random behavior with $S_a = 0.1$, $p_a = 0.03$ and h_i following eq. 3. See [24, 11] for a study of the FNN parameter space

2.2 Finite Threshold Models

Task allocation in ant societies is one of the more astonishing aspects in the life of a colony. How do individuals "know" what to do all the time? Genetic determination would be an explanation, at least in polymorphic species, but only a 17% of the known living ant genera contain species with some degree of polymorphism [21], so most of ant species are monomorphic. It is obvious that "cognitive" capabilities of individuals are not sophisticated enough to decide what task to do next in order to fulfil *global* colony needs. Thus, this question remains a matter of controversy, though there are some hypothesis at hand (see [23, 27] for reviews of division of labour in insect societies). One of them, the Finite Threshold Model (FTM) [4] is the one we have chosen to model task fulfilment, since it is simple enough to couple with FNNs.

The basic assumptions are that some specific stimulus is associated with each task and that each individual has fixed response thresholds to the various stimuli, so that the lower the threshold the more likely the individual will engage in the task, given exposure. There is an experimental basis that justifies this approach, for example it has been proved the existence of response thresholds in honey bees (see [23], and [4] for a more detailed discussion on the experimental basis of the FTM).

Here we will work only with one "abstract", spatially distributed task. Assume that an active, though not working, individual S_i perceives, in some way to be specified below, a quantity s of stimulus. It will engage in the task with probability [4]

$$P\left(S_i^{\text{Not working}} \to S_i^{\text{Working}}\right) = \frac{s^2}{s^2 + \theta^2} \tag{5}$$

where θ is the individual threshold associated to the task. In fact, we will use the stimulus also as a representation of the "amount of task" to be done, so that once an individual is engaged in a determined task, it has some probability $p(s)$ per unit time of completing the task. Throughout this paper we will use the nonlinear response function

$$p(s) = \frac{1}{(1 + s^2)}$$

so that the greater the task, the less likely the individual will complete the task.

2.3 Coupling FNNs and FTMs

Let us assume a $L \times L$ lattice where N individuals are spread out. Each individual will be characterized by a triple $(S_i(t), X_i(t), \theta_i)$ where $S_i(t)$ is the FNN–state of the individual i, $X_i(t)$ is a two-valued variable signaling whether the individual is working ($X_i(t) = 1$) or not ($X_i(t) = 0$) and θ_i is the FTM threshold. Also, a working ($X_i(t) = 1$) and active ($S_i(t) > \theta_{act}$) individual may be doing a certain amount of task $c_i(t)$. Each lattice position will be either void or will contain one individual. Besides, it may also contain a certain amount of stimulus.

Initially our system will be composed of N non-working individuals, with a random initial FNN–state. θ_i will be initially distributed uniformly between θ_{min} and θ_{max} among individuals. A randomly chosen position of the lattice (the "task origin"), will contain a certain amount of total task C_{in} to be performed by the individuals, while lattice coordinates other than the task origin will be initially empty of stimulus.

$S_i(t)$ will evolve in time *exactly* as a FNN (section 2.1), so what remains to be defined is the evolution in time of $X_i(t)$ and eventually $c_i(t)$, that is, the task fulfilment process. This has the form of some rules more or less plausible from a biological point of view (see the initial paragraph of section 2):

1. *Effective realization of the task*: An individual i at time t may be active ($S_i(t) > \theta_{act}$) and working ($X_i(t) = 1$), with a certain quantity of task $c_i(t)$ to be done. At time $t+1$, this individual may accomplish the task with probability $p(c_i(t))$, in which case the total task remaining in the system will be decreased by an amount $c_i(t)$. If the (active and working) individual i does not get the task done at time $t+1$, it may become inactive ($S_i(t+1) \leq \theta_{act}$), in which case the amount of task $c_i(t)$ will be stored in the lattice coordinates of the individual i. Inactive individuals do nothing, though they may be activated by the FNN dynamics of the system. Next rule deals with the remaining case, that of active and non-working individuals.

2. *Propagation of the stimulus*: An active and non-working individual may be stimulated by all its active and working nearest neighbours and by the amount of task stored at its lattice position. Each active and working individual, say the j-th, will be able to provide a quantity of stimulus that will depend on the number of active and non-working neighbours, say n_j. A quantity $\sigma_j = \beta c_j(t)/n_j$ will be the stimulus provided by j to each of its n_j non-working active neighbours. Thus, an active and non-working individual, say the k-th, will receive a quantity of stimulus $\sigma_k = \sum_{i(k)} \sigma_i + \alpha\Sigma_{x,y}$, where $i(k)$ ranges over the active and working neighbours of the k-th individual and $\Sigma_{x,y}$ is the amount of task at the lattice coordinates x, y of k. If this individual becomes a worker, with a probability given by equation 5, the quantities σ_i are substracted from $c_i(t)$ (for all i active and working neighbour of k) and $\Sigma_{x,y}$ will become $(1-\alpha)\Sigma_{x,y}$. The parameters α and β allow one to control the degree of stimuli received from the task stored in the lattice and the spreading of stimuli from individuals to individuals, respectively.

3 Comparing Self-synchronized Activation with Non-synchronized Activation

The model detailed in section 2 has been used to test whether a system with self-synchronized activity patterns is able to perform an "abstract" task better than a system with random activity patterns, both subject to the constraints mentioned in section 2.1. What does "better" mean in this context? It will be equivalent to "faster". Thus, we want to measure how fast a system with certain patterns of activation is able to perform a certain fraction of the initial task C_{in}.

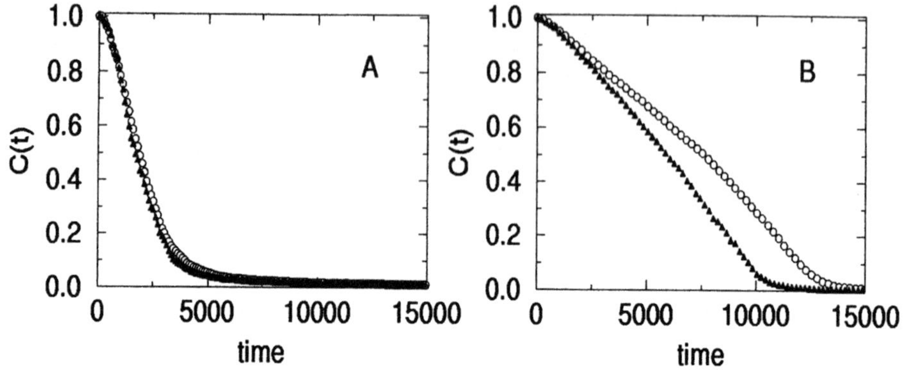

Fig. 2. Single run simulations to compare $C^{SS}(t)$ and $C^{NS}(t)$. (A) $C^{SS}(t)$ (triangles) and $C^{NS}(t)$ (circles) with $C_{in} = 10000$. (B) $C^{SS}(t)$ (triangles) and $C^{NS}(t)$ (circles) with $C_{in} = 50000$. In both cases $\alpha = 0.25$ and $\beta = 0.85$.

We will measure how the total task that remains to be done in the system

$$C(t) = \frac{1}{C_{in}} \left(\sum_{x,y=1}^{L} \Sigma_{x,y}(t) + \sum_{i=1}^{N} c_i(t) X_i(t) \right) \tag{6}$$

evolves in time with two different patterns of activity: self-synchronized (SS) and non-synchronized (NS). This has been done for a system where each individual has a θ_i chosen at random with uniform distribution in the interval $(1, 10)$, reflecting a plausible lack of threshold uniformity in real ant colonies [4]. Figure 2 shows that there is no difference in behaviour with a small C_{in}, though a larger initial task makes the SS behaviour much more efficient than NS, that is, the SS system gets the task done in less time than the NS system. More systematic calculations have been done computing the difference between t^*_{NS} and t^*_{SS}, where t^*_{SS} is the first time step such that the system with self-synchronized behaviour verifies $C(t^*_{SS}) < 0.1$ (some lower bound is needed, since the amount of task remaining decreases asymptotically to zero). We define t^*_{NS} in the same way. As we can see in figure 3 the difference

$$E(C_{in}) = t^*_{NS} - t^*_{SS}$$

averaged over M measures, grows with C_{in}, making clear that the more "quantity" of task needs to be done, the more efficient is the SS behaviour.

Robinson (see section 1 and [23]) gave us some clues to ascertain the causes of the superior efficiency of SS behaviour. The key idea is that the spreading of the task is enhanced by the greater number of *active* neighbors one individual has, on average, in the SS case. This phenomenon allows the individuals to distribute faster the task, "breaking" it up into smaller pieces, so that it is much more likely that an individual completes its task.

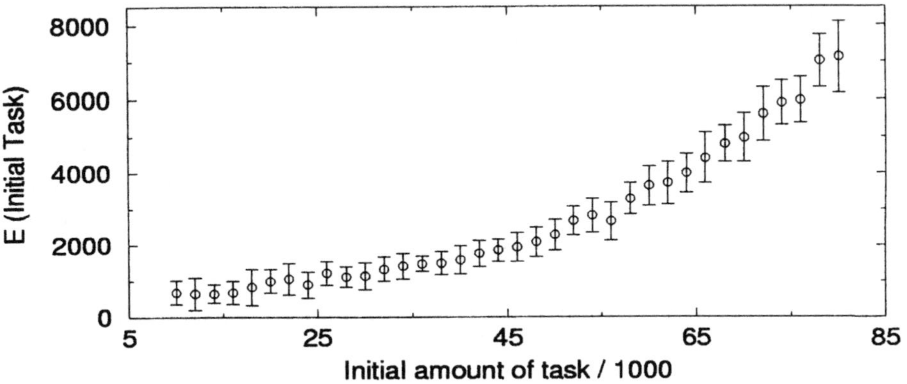

Fig. 3. $E(C_{in})$ (see text) averaged over $M = 25$ samples for each C_{in}, with parameters $\alpha = 0.25$ and $\beta = 0.85$.

4 Discussion

In this paper we have introduced a framework with which to study the relation between patterns of activity in social insects and the ability to fulfil some task. We have seen that self-synchronization enhances the efficiency with which the system performs some sort of "abstract" task. Our first numerical results with the coupling between the FNN and the FTM point out that observations in real colonies, concerning the functional side of synchronized patterns, may reflect a more general relation between task fulfilment and self-synchronization, since, as Robinson [23] tells us in the quote above mentioned, to ascertain the mechanisms of the spreading out of tasks and/or information inside the colony is important to account for the task allocation abilities of social insects.

The phenomenon is not difficult to understand. The key point is the idea that "if it is not active, it does not work" together with the local transmission of information (stimulus, task, etc...) from individuals to individuals. If not active permanently, the only way to ensure that an *active* individual will have as many *active* neighbours as possible that can be stimulated, and consequently to get the task as scattered as possible, is having synchronized activity. Besides, individuals are more likely to start tasks immediately on being activated, since the activating neighbours may be work-carrying. We are currently working on the precise relationship between activity patterns and task spreading, using the model introduced in this paper to study the influence of β in the efficiency of task fulfilment (our first results were recently introduced in [13]).

Related work was introduced in [5] where it was suggested that synchronization may enhance foraging efficiency. Finally, some experiments [16] and previous work with FNNs [24] clearly suggest a dependence of self-synchronization on colony density. Thus, it would be possible to do experimental work with real colonies by modifying artificially their density (as in [16]) inducing non-synchronized behavior. This setting would allow us to compare task fulfiling in colonies with different patterns of temporal activity.

Acknowledgments
This paper was completed while the authors were visiting the Santa Fe Institute, whose hospitality and support we gratefully acknowledge. The content of the paper has benefited from discussions with Brian Goodwin, Jordi Bascompte, Bartolo Luque and Susanna C. Manrubia. Also, discussions with Guy Theraulaz and Eric Bonabeau have been quite illuminating. This work has been supported by DGYCIT grant No. PB94-1195, by UPC grant PR98-11 and by the *Generalitat de Catalunya* grant FI93/3008.

References

1. Amit, D.: *Modeling Brain Function*. Cambridge: Cambridge University Press (1989).
2. Blanchard, G.B., Franks, N.R.: A new mechanism of synchronous activity. *Preprint* (1998).
3. Bonabeau, E., Theraulaz, G. (eds.): *Intelligence Collective*. Hermes: Paris (1994)
4. Bonabeau, E., Theraulaz, G. & Deneubourg, J.-L.: Quantitative study of the fixed threshold model for the regulation of division of labour in insect societies. *Proc. R. Soc. Lond. B* **263** (1996) 1565–1569.
5. Bonabeau, E., Theraulaz, G. & Deneubourg, J.-L.: The synchronization of recruitment-based activities of ants. *BioSystems* **45** (1998) 195–211.
6. Cole, B.J.: The social behavior of *Leptothorax allardycei* (*Hymenoptera, Formicidae*) - time budgets and the evolution of worker reproduction. *Behav. Ecol. Sociobiol.* **18** (1986) 165–173.
7. Cole, B.J.: Short-term activity cycles in ants: generation of periodicity by worker interaction. *Am. Nat.* **137** (1991) 244–259.
8. Cole, B.J.: Is animal behaviour chaotic? Evidence from the activity of ants. *Proc. R. Soc. Lond. B* **244** (1991) 253–259.
9. Cole, B.J., Cheshire, D.: Mobile cellular automata models of ant behavior: movement activity of *Leptothorax Allardycei. Am. Nat.* **148** (1996) 1–15.
10. Delgado, J.: *On Collective Computation*. Ph.D. Thesis, Dept. Llenguatges i Sistemes Informàtics, Universitat Politècnica de Catalunya, Barcelona, Spain (1997).
11. Delgado, J., Solé, R.V.: Noise Induced Transitions in Fluid Neural Networks. *Phys. Lett. A* **229** (1997) 183–189.
12. Delgado, J., Solé, R.V.: Mean field theory of Fluid Neural Networks. *Phys. Rev. E* **57** (1998) 2204–2211.
13. Delgado, J., Solé, R.V.: Self-synchronization and task fulfilment in social insects. *Santa Fe Institute Working Paper 98-08-069* (1998), Santa Fe Institute (Santa Fe, U.S.A.).
14. Franks, N.R., Bryant, S.: Rhythmical patterns of activity within the nest of ants. In *Chemistry and Biology of Social Insects* (Eder, J. & Rembold, H. eds.) pp. 122-123. Munich: Peperny (1987).
15. Franks, N.R., Bryant, S., Griffiths, R., Hemerik, L.: Synchronization of the behavior within the nests of the ant *Leptothorax acervorum*-I. Discovering the phenomenon and its relation to the level of starvation. *Bull. Math. Biol.* **52** (1990) 597–612.

16. Franks, N.R., Wilby, A., Silverman, B.W., Tofts, C.: Self-organizing nest construction in ants: Sophisticated building by blind bulldozing. *Anim. Behav.* **44** (1992) 357–375.
17. Gordon, D.M., Goodwin, B.C., Trainor L.E.H.: A parallel distributed model of the behaviour of ant colonies. *J. Theor. Biol.* **156** (1992) 293–307
18. Goss, S., Deneubourg, J.-L.: Autocatalysis as a source of synchronised rhythmical activity in social insects. *Insect. Soc.* **35** (1988) 310–315.
19. Hatcher, M.J., Tofts, C., Franks, N.R.: Mutual exclusion as a mechanism for information exchange within ant nests. *Naturwissenschaften* **79** (1992) 32–34.
20. Herbers, J.M.: Social organization in Leptothorax ants: Within- and between-species patterns. *Psyche* **90** (1983) 361–386.
21. Hölldobler, B., Wilson, E.O.: *The Ants.* Berlin: Springer-Verlag (1990)
22. Miramontes, O.: *Complexity and Behaviour in Leptothorax Ants*, Master Thesis, Open University, U.K. (1992)
23. Robinson, G.E.: Regulation of division of labor in insect societies. *Ann. Rev. Entomol.* **37** (1992) 637–702.
24. Solé, R.V., Miramontes, O.: Information at the edge of chaos in Fluid Neural Networks. *Physica D* **80** (1995) 171–180
25. Solé, R.V., Miramontes, O., Goodwin, B.C.: Oscillations and chaos in ant societies. *J. Theor. Biol.* **161** (1993) 343–357.
26. Solé, R.V., Miramontes, O., Goodwin, B.C.: Emergent behavior in insect societies: Global oscillations, chaos and computation. In *Interdisciplinary Approaches to Nonlinear Complex Systems. Springer Series in Synergetics vol.62* (Haken, H. & Mikhailov, A. eds.) pp.77-88. Berlin: Springer Verlag (1993).
27. Tofts, C.: Algorithms for task allocation in ants (a study of temporal polyethism: Theory). *Bull. Math. Biol.* **55** (1993) 891–918
28. Tofts, C., Hatcher, M., Franks, N.R.: The autosynchronization of the ant *Leptothorax acervorum* (Fabricius): Theory, testability and experiment. *J. Theor. Biol.* **157** (1992) 71–82

Formalizing the Ant Algorithms in Terms of Reinforcement Learning

Ann Nowé* and Katja Verbeeck

asnowe, kverbeec @info.vub.ac.be
Vrije Universiteit Brussel, Pleinlaan 2, 1050 Brussel, Belgium

Abstract. In this paper we place ant algorithms in a reinforcement learning framework. We concentrate on the original Ant System and we briefly discuss Ant Colony system. We show that ant-quantity and ant-density can be considered as TD(0) algorithms which only take into account immediate reinforcement. Whereas ant cycle is basically an on-policy Monte Carlo method. We introduce the notion of decay traces, for modeling the decay of trail.

1 Introduction : Ants and Reinforcement Learning

In this paper we try to formulate the Ant System (AS) algorithms as Reinforcement Learning algorithms. AS was the first system in the Ant Colony Optimization area. This field studies artificial systems that are inspired by real ant colonies behavior. The main observation is that real ants are capable of finding shortest paths thanks to a simple pheromone trail-laying mechanism. A good overview of the state-of-the-art in the field and the definition of a meta-heuristic is given in [2]. In the next section we begin by introducing AS. We then analyze the three algorithms of AS in section 3 and finally we make a comparison of the experimental results of AS with our theoretical findings in the conclusions. Details on the Reinforcement Learning theory can be found in [5]. A more elaborate version of this work is reported in [4].

2 Ant System

The three algorithms of Ant System for solving the TSP and ATSP, presented in [1] are ant-density, ant-quantity and ant-cycle. At a time instance t, every town i has a number of ants, who chooses the next town to go with a probability that is a function of the town distance and of the amount pheromone trail present on the connecting edges. Transitions to already visited towns are disallowed (forced by a list of visited towns which is in the memory of the ant). The algorithms differ in the way how they update the pheromone trail of the ants. Let τ_{ij} be the intensity of the trail on edge (i,j). The trail is updated as follows:

$$\tau_{ij}(t+h) \leftarrow \rho\tau_{ij}(t) + \Delta\tau_{ij}(t,t+h) \tag{1}$$

* Postdoctoral Fellow of the Fund for Scientific Research-Flanders(Belgium)

where ρ is a trail decay coefficient such that $0 \leq \rho \leq 1$ and :

$$\Delta\tau_{ij}(t, t+h) = \Sigma_{k=1}^{m} \Delta\tau_{ij}^{k}(t, t+h)$$

$\Delta\tau_{ij}^{k}(t, t+h)$ is the quantity per unit of length of trail substance laid on edge (i, j) by the k-th ant between time t and $t+h$. For ant-density and ant-quantity, $h = 1$, for ant-cycle $h = n$. The total number of ants in the system is m, the transition probability is:

$$p_{ij}(t) = \begin{cases} \frac{(\tau_{ij}(t))^{\nu}(\eta_{ij}(t))^{\beta}}{\Sigma_{j \in allowed}(\tau_{ij}(t))^{\nu}(\eta_{ij}(t))^{\beta}} & \text{if } j \in \text{allowed} \\ 0 & \text{otherwise} \end{cases} \tag{2}$$

$\eta_{ij} = 1/d_{ij}$ is called the visibility. For the ant-density model we have :

$$\Delta\tau_{ij}^{k}(t, t+1) = \begin{cases} Q_1 & \text{if } k\text{-th ant goes from } i \text{ to } j \text{ between } t \text{ and } t+1 \\ 0 & \text{otherwise} \end{cases} \tag{3}$$

while for the ant-quantity:

$$\Delta\tau_{ij}^{k}(t, t+1) = \begin{cases} \frac{Q_2}{d_{ij}} & \text{if } k\text{-th ant goes from } i \text{ to } j \text{ between } t \text{ and } t+1 \\ 0 & \text{otherwise} \end{cases} \tag{4}$$

and for ant-cycle :

$$\Delta\tau_{ij}^{k}(t, t+n) = \begin{cases} \frac{Q_3}{L^k} & \text{if } k\text{-th ant used edge } (i, j) \text{ in its tour} \\ 0 & \text{otherwise} \end{cases} \tag{5}$$

L^k is the length of the tour done by the k-th ant. Q_1, Q_2 and Q_3 are constants.

2.1 Formalization of Ant-Quantity and Ant-Density

TD(0) View We can view ant-density and ant-quantity (see (1) and (3) or (4)) as a TD(O) algorithm, with discount factor $\gamma = 0$. Let the different towns in the TS problem stand for the states of the environment in which the reinforcement learner, which is here every individual ant, tries to learn. The possible actions an ant can take in a certain state are the possible edges it can follow to end up in the next town/state. The value function one tries to learn, is the trail of pheromone every edge receives. This value belongs to a town i and edge leading to a different town j. So the values τ_{ij} are state-action values. In terminology of reinforcement learning: $Q(s, a) = Q(i, ij) = \tau_{ij}$. As this value only depends on local information, see (3) and (4) and no long term reward is taken into account, we conclude that : $\gamma = 0$. As a result, learning only depends on the immediate reinforcement the learner receives for taken action ij in state i. In this view, (1) $(h = 1)$ closely resembles the update rule of a TD(0) algorithm :

$$Q_{t+1}(i, ij) \leftarrow Q_t(i, ij) + \alpha(r_{t+1} + \gamma Q_t(j, jk) - Q_t(i, ij))$$

Where r_{t+1} is the immediate reinforcement, α the step size parameter and jk the next action considered. When $\gamma = 0$ this becomes:

$$Q_{t+1}(i,ij) \leftarrow Q_t(i,ij) + \alpha[r_{t+1} - Q_t(i,ij)] = (1-\alpha)Q_t(i,ij) + \alpha r_{t+1} \quad (6)$$

Comparing this to update rule (1) shows that step parameter α can be considered as $1-\rho$, and immediate reinforcement r_{t+1} really is $\frac{\Delta\tau_{ij}}{\alpha}$. As a step size parameter is usually taken very small, this equalities seems very natural. The fact that the ants action-selection procedure, given in rule (2) is not genuine ϵ-greedy [1] (cities already visited are explicitly excluded) should not be a problem, because of the combination of several ants walking around, one starting in every town. This is comparable with the exploring start condition in [5]. The difference with the TD(0) algorithm is that state-action pairs in (1) are also updated when this action is not selected. For every time step where there was no ant crossing link ij, the value τ_{ij} is updated according to: $\tau_{ij}(t+1) \leftarrow \rho\tau_{ij}(t)$. What this really means is that an action is punished for not being taken. One could simply see this punishment as coming from the environment.

We also found this behavior in the theory of Learning Automata, [3]. A Learning Automata is really a vector of action probabilities, which can be updated with various learning schemes. For instance in the reward-penalty scheme actions are rewarded when they proved to be good or others proved to be bad. Conversely, actions are punished when they proved to be bad or others proved to be good. Being good in the ant-density and ant-quantity environment can then be interpreted as being visited. However the precise update scheme here seems not to be linear. Since little is known of non-linear update schemes, this should be further analyzed.

Decay eligibility trace Another place where we find this kind of updating of state-action values, without being selected, is in the TD(λ) algorithms, where a trace of eligibility marks states or state-action pairs as being eligible for undergoing learning changes not only by the immediate reinforcement but also by reinforcements received by actions taken in the future, see [5]. However the definition of the trace here would reduces to 1 in the state-action pair currently visited, and to 0 elsewhere, because state-action values only depend on the immediate reinforcement. So we have to introduce another kind of trace for the decay of τ_{ij}, which we will name the decay trace. Let us write such a trace starting from rule (6) :

$$\begin{aligned}\Delta Q_{t,t+1}(i,ij) \leftarrow & -\alpha Q_t(i,ij) + \alpha r_{t+1} = -\alpha(1-\lambda')\Sigma_{l=0}^{\infty}(\lambda')^l Q_t(i,ij) + \alpha r_{t+1} \\ = & -\alpha(1-\lambda')Q_t(i,ij) - \alpha(1-\lambda')\Sigma_{l=1}^{\infty}(\lambda')^l Q_t(i,ij) + \alpha r_{t+1}\end{aligned}$$

This can be separated in two update rules:

$$\Delta Q_{t,t+1}(i,ij) \leftarrow \begin{cases} -\alpha(1-\lambda')Q_t(i,ij)e_t(i,ij) + \alpha r_{t+1} & \textit{ij is visited} \\ -\alpha(1-\lambda')Q_t(i,ij)e_t(i,ij) & \textit{otherwise} \end{cases} \quad (7)$$

[1] In an ϵ-greedy policy, most of the time an action which has maximum estimated value is chosen, but with a non zero probability ϵ a random action is chosen.

where $e_t(i,ij)$ is taken as:

$$e_t(i,ij) = \begin{cases} \lambda' e_{t-1}(i,ij) & \text{if } (i,ij) \neq (i_t, ij_t) \\ 1 & \text{if } (i,ij) = (i_t, ij_t) \end{cases}$$

What happened is that the computation of $Q_t(i,ij)$ was spread over subsequent time steps just as in equation (1). Comparing rule (1) ($h = 1$) with (7), we see that $r_{t+1} = \frac{\Delta\tau_{ij}}{\alpha}$. Note that we also derived this equality in our interpretation of TD(0). Furthermore $(1-\rho)$ can be interpreted as $\alpha(1-\lambda')$. This is our earlier learning factor multiplied with a factor $(1-\lambda')$.

2.2 Modeling the Ant Cycle Algorithm

Monte Carlo View For the ant-cycle algorithm, the obvious choice was to try to characterize it as a Monte Carlo method, see [5]. The general update rule for an every visit MC method is given by:

$$Q_{t+n}(s,a) \leftarrow Q_t + \alpha(R_t - Q_t(s,a)) \tag{8}$$

α is a constant step-time parameter and R_t is the actual return of the episode:

$$R_t = r_{t+1} + \gamma r_{t+2} + \gamma^2 r_{t+3} + \ldots + \gamma^{T-t-1} r_T \tag{9}$$

T is the last time step of the episode. Comparing (8) with (1) and (5) ($h = n$) leads to the following equalities: $\alpha = (1-\rho)$ and $R_t = \frac{\Delta\tau_{ij}}{(1-\rho)}$. We make a similar assumption as in the previous subsection that state-action pairs who were never visited during the current episode are punished by the environment by decaying their value by ρ for the next episode. We can write out R_t as in (9) :

$$R_t = \frac{1}{(1-\rho)} \Sigma_{k=1}^{m'} \frac{Q_3}{L^k} = \frac{1}{(1-\rho)} \Sigma_{k=1}^{m'} \frac{Q_3 L^k}{L^{k2}} = \frac{1}{(1-\rho)} \Sigma_{t=1}^{n} (\Sigma_{k=1}^{m'} \frac{Q_3 d_t^k}{(\Sigma_{t=1}^{n} d_t^k)^2})$$

where m' is the total number of ants who passes link ij in its tour in the current episode, and d_t^k is the distance of the edge the k-th ant crosses at time step t. Therefore: $r_{t+h} = \frac{1}{(1-\rho)} (\Sigma_{k=1}^{m'} \frac{Q_3 d_{t+h}^k}{(\Sigma_{t=1}^{n} d_t^k)^2})$ and thus $\gamma = 1$, meaning that we are considering the reinforcement of the different actions at a later time moment with an equal weight. (there is no preference in the sequence of the towns) This update is actually an off-line TD(λ) method where $\lambda = 1$, see [5]. Furthermore the update is on-policy[2] because the tour length is used in rule (5).

Decay eligibility trace To handle the update of non-visited state-action pairs, we use the idea of eligibility traces again. Since we are considering a TD(λ)

[2] On-policy methods attempt to evaluate or improve the value of a policy while using it for control. In off-policy methods the policy used to generate behavior is not related to the one that is being evaluated or updated.

method with $\lambda = 1$ and $\gamma = 1$ here, we could define the classical eligibility trace. However, since updating can only be done off-line it would not be interesting here, moreover it is not the reinforcement (as is the case in classical TD(λ)) but it is the decay of τ_{ij} which is spread over subsequent periods. As in (7) we have two update rules, but now the decay is over episodes:

$$\Delta Q_{t,t+n}(i,ij) \leftarrow \begin{cases} -\alpha(1-\lambda')Q_t(i,ij)e_{(t,t+n)}(i,ij) + \alpha R_t & \textit{ij was visited} \\ -\alpha(1-\lambda')Q_t(i,ij)e_{(t,t+n)}(i,ij) & \textit{otherwise} \end{cases}$$

$$e_{t+n}(i,ij) = \begin{cases} \lambda' e_t(i,ij) & \textit{ij was visited} \\ 1 & \textit{otherwise} \end{cases}$$

3 Conclusions

In this paper we come to the conclusion that ant-density and ant-quantity can be considered as TD(0) algorithms without longterm reward and with a special case of eligibility trace. Therefore finding a global optimal solution requires a good heuristic to guide the exploration, and to result in a better performance than a random search. Since the ants are exploring the search space starting in different towns, the chance to get stuck in a local minimum reduces. This is confirmed by the experiments in [1] for testing the parameter settings. Best results were achieved by setting the value of β high opposed to the value of ν, thus giving more weight to the heuristic, opposed to the auto-catalytic process. On the other hand ant-cycle is more robust, because it is basically a MC method, which is well-suited for optimizing multi-stage decision problems. In particular it is an on-policy MC method, using a soft action selection mechanism. Again this is visualized by the experiments run by the authors [1] : ant-cycle outperforms the other two. The first two algorithms reduce to a non-interesting case of reinforcement learning, whereas ant-cycle can be classified as being a better learner.

References

1. Dorigo, M., Maniezzo, V., Colorni, A., : The Ant System: Optimization by a Colony of Cooperating Agents. IEEE Transactions on Systems, Man, and Cybernetics-Part B, **26**, 1, (1996) 29-41
2. Dorigo M., Di Caro, G., Gambardella, L.M.,: Ant Algorithms for Discrete Optimization. Artificial Life, accepted for publication. Also available as Tech.Rep.IRIDIA/98-10, Universit Libre de Bruxelles, Belgium.
3. Narendra, K., Thathachar, M., : Learning Automata: An Introduction. Prentice-Hall (1989).
4. Nowé, A., Verbeeck, K.,: Ant Algorithms and Reinforcement Learning. Technical Report Vrije Universiteit Brussel Belgium, May 1999.
5. Sutton, R.S., Barto, A.G. : Reinforcement Learning: An introduction. Cambridge, MA: MIT Press (1998).

Using Competing Ant Colonies to Solve k–way Partitioning Problems with Foraging and Raiding Strategies

A. E. Langham* and P. W. Grant

Department of Computer Science
University of Wales Swansea
Singleton Park
Swansea, SA2 8PP, U.K
csjim,p.w.grant@swan.ac.uk

Abstract. The self organizing properties of ant colonies are employed to tackle the classical combinatorial optimization problem of graph partitioning. The graph is mapped onto an artificial environment in a manner that preserves the structural information. Ants from a number of colonies compete for resources. This leads to a restructuring of the global environment corresponding to a good partition. On the example graphs, this is shown to outperform the current best algorithms which are based on recursive bisection techniques.

1 Introduction

Simulation of fluid flow with the Finite Element Method approximates solution values across a domain using a set of discretized elements. In 2-Dimensional problems elements are planar polygons. Partitioning the domain for parallel processing can be mapped onto a graph partitioning problem where each vertex in the graph represents a node in the mesh, whilst an edge in the graph represents the need for communication between two nodes. The graph must be broken into approximately equal size sub–domains with as little communication between domains as possible.

There are graph partitioning problems which are NP–complete [5] and therefore we are looking for a *near optimal* partition in reasonable time. Most partitioning methods employ recursive bisection which can often provide a partition which is far from optimal [14] as regards minimizing the number of edge cuts. What seems optimal at the top level of recursion may provide a poor partition at lower levels given the benefit of hindsight. Recursive Spectral Bisection has been shown to be highly effective compared to alternative methods [13]. Multi–level implementations [8] are yet more effective especially in combination with local refinement using Kernighan Lin [9]. Recently many methods have been

* Supported by a University of Wales Validation Unit research studentship

generalized to partition a graph into more than two sets at each stage of recursion [7]. It is thought that such methods could potentially produce better partitions. However, the direct computation of a good k–way partitioning is harder than a bisection and hence recursive bisection is most commonly used. Dorigo et al. have applied the trail laying properties of ant colonies to the Travelling Salesman Problem, Quadratic Assignment Problem and Routing in telecommunication networks showing high quality results [3], [4], [1]. Kuntz et al. tackled the graph partitioning problem using a clustering algorithm [10] based on brood sorting in ant colonies [2] and a swarm colonization technique [11]. Such methods have shown good results on small graphs.

2 The Approach

Initially, we consider the bisection case in which two competing colonies of ants are used to split the graph into *two* partitions. Each colony is centred around a fixed cell in a grid which represents the environment in which the ants can navigate. The ants must learn to forage for food, each piece of food on the grid represents a node in the mesh which is being partitioned. The ants must find all the food and place it in the appropriate nest so that the set of nodes represented by the food in $Nest_1$ forms a set V_1 and the set of nodes in $Nest_2$ forms a set V_2. The graph bisection problem for a graph $G = (V, E)$ with vertices V and edges E, seeks a partition $V = V_1 \cup V_2$ such that $V_1 \cap V_2 = \emptyset$, $|V_1| \approx |V_2|$ and the number of cut edges $|E_c|$ is minimized, where:

$$E_c = \{(v_1, v_2) \in E | v_1 \in V_1, v_2 \in V_2\}.$$

In order to cut down the search space we map the nodes onto the grid in a manner that represents the structure of the mesh by mapping the geometrical position of nodes in the mesh onto the equivalent geometrical position on the grid. To encourage the colonies to find a reasonable initial partition we place each nest at the center of an equal number of nodes using a recursive bisection heuristic. To further cut down search and utilize the structural information on the grid we prevent any colony from exceeding the desired number of nodes by more than one and prevent ants from taking food from other colonies until their colony has acquired 90 percent of the required number of nodes. The main body of the loop, as shown below, is executed for each ant in each colony during every time step. The grid is initialized by placing the food and locating the nest for each colony. The body is iterated until there is no reduction in the number of cuts over 500 time steps. The best partition found during this time is taken to be the solution.

```
if (carrying_food) then
        if (in_nest_locus) then (drop_food)
        else (move_to_nest)
else    if (food_here) then (pick_up_food)
        else
```

```
if (food_ahead) then (move_forward)
else  if (in_nest_locus) then (move_to_away_pheromone)
      else  if (help_signal) then (move_to_help)
            else (move_to_strong_forward_pheromone)
```

The actions and functions in this code are detailed below in Table 1. It should be noted that this foraging strategy can be generated using Genetic Programming [12].

Table 1. Actions and Functions for Main Algorithm

Function/Action	Explanation
`if_food_ahead`	Return `True` if food is in the adjacent square.
`if_food_here`	Return `True` if food is in the current square.
`if_in_nest_locus`	Return True if food is within two grid squares of the colony nest.
`if_carrying_food`	Return True if ant is carrying food.
`if_help_signal`	Return True if there is a help signal within a locus of two grid squares.
`move_to_nest`	Move one step towards the nest in direction: **if** $(\lvert Nest.x - x\rvert \geq \lvert Nest.y - y\rvert)$ **then if** $(x \geq Nest.x)$ **then** $dir = WEST$ **else** $dir = EAST$ **else if** $(y \geq Nest.y)$ **then** $dir = SOUTH$ **else** $dir = NORTH$.
`move_to_help`	Move one step towards the nearest help signal.
`move_to_away_pheromone`	Move one step away from the nest in the away direction with the strongest pheromone trail.
`move_to_strong_forward_pheromone`	Move to an adjacent square with probability dependent on the amount of pheromone present.
`pick_up_food ( foraging )`	Food not belonging to a colony is picked up if proportion of cuts created is less than half total possible number. Number of ants needed is related to the proportion of cuts. Help signal sent out if not enough ants.
`pick_up_food ( raiding )`	Food already in a colony is picked up with probability p dependent on the change in the proportion of cuts δp_c where $p = \begin{cases} 1.0 & \text{if } \delta p_c > 0 \\ 0.5 & \text{if } \delta p_c = 0 \\ 0.0 & \text{if } \delta p_c < -0.33 \\ 1/(\omega * (\delta p_c)^2) & \text{if } -0.33 \leq \delta p_c < 0 \end{cases}$ where ω depends on δp_c.
`drop_food`	Place food around colony nest using a clockwise search for an empty cell.

3 Results

The Ant Foraging Strategy (AFS) is compared with Recursive Spectral Bisection (RSB), Recursive Spectral Bisection plus Kernighan Lin (RSB +KL) and also Multilevel Kernighan Lin (ML-KL). Each method is tested on two 2–dimensional meshes over two, four and eight partitions. Recursive bisection (rb) and simultaneous multiple partitioning (k-way) methods are implemented in the Chaco 2.0 package [6]. The number of cuts for the partitions produced on two finite element meshes are given in Table 2. Mesh 1, (114 nodes, 308 edges), is a triangular mesh with uniform node density. Food was placed on an 11 by 11 grid. Mesh 2 (a test case from the Chaco package), (286 nodes, 1046 edges), is a quadrilateral mesh with variable density. The food was placed on a 25 by 25 grid. The total number of ants on the grid is always equal to 45 percent of the total grid area. ML-KL is coarsened to 30 nodes for Mesh 1 and 50 nodes for Mesh 2.

Table 2. Number of Cuts Created by Various Partitioning Methods

Mesh	Method	RSB	RSB	KL+RSB	KL+RSB	ML-KL	ML-KL	AFS
		rb	k-way	rb	k-way	rb	k-way	k-way
1	Bisection	25	25	25	25	25	28	25
	Quadrisection	51	51	51	50	52	55	49
	Octrisection	89	103	87	97	84	95	81
Time (secs)	Octrisection	0.11	0.2	0.16	0.31	0.18	0.53	1.45
2	Bisection	29	29	28	28	28	28	28
	Quadrisection	91	91	90	90	88	90	87
	Octrisection	176	195	173	191	164	187	162
Time (secs)	Octrisection	0.19	0.29	0.27	0.46	0.29	0.49	8.73

4 Discussion and Further Work

The Ant Foraging Strategy shows some good results on the test meshes. It performs better than RSB, RSB+KL and ML-KL when implemented for quadrisection and octrisection. It also performs better than RSB, RSB+KL and ML-KL methods when implemented for recursive bisection because it is not dependent on partitions at higher levels of recursion. AFS is relatively efficient compared to other swarm methods as it maps the the structure onto the environment and places nests to provide a rough initial partition quickly using a small population. It is also reasonably comparable to multilevel octrisection methods for uniform density meshes. However variable density meshes need a relatively larger grid to represent sufficient structural information. Hence a method for mapping food onto the grid which represents structural rather geometrical information is needed to make AFS more efficient for arbitrary meshes.

5 Conclusion

Most approaches to the k-way partitioning problem use recursive bisection and local improvement techniques, except spectral quadrisection and octrisection which perform relatively poorly. AFS provides a novel global method for tackling the problem with built in local improvement which can simultaneously partition into as many sets as required. Results show that swarm-based simultaneous multiple partitioning needs further investigation as it can produce better partitions than recursive bisection techniques which is dependent on the partition found at higher levels of recursion. Furthermore, AFS is relatively efficient compared to other swarm–based methods and scales up well to large meshes [12] when multilevel methods are applied and food is placed on the grid using a BFS technique to provide more structural information.

References

1. G. Di Caro and M. Dorigo. AntNet: distributed stigmergic control for communications networks. *Journal of Artificial Intelligence Research*, 9:317–365, 1998.
2. J.L. Deneubourg, S. Goss, N. Franks, A. Sendova-Franks, C.Detrain, and L. Chretien. The dynamics of collective sorting: Robot like ants and ant-like robots. In *First international conference on Simulation of Adaptive Behaviour*, July 1990.
3. M. Dorigo and L.M. Garbadella. Ant colonies for the travelling salesman problem. *Biosystems*, 43:73–81, 1997.
4. L.M. Gambardella, E. Taillard, and M. Dorigo. Ant colonies for the quadratic assignment problem. *Journal of Operational Research*, 50(2):167–176, 1999.
5. M.R. Garey and D.S. Johnson. *Computers and Intractability*. W.H. Freeman and Company, 1979.
6. B. Hendrickson and R. Leyland. The Chaco user's guide, version 2.0. Technical report, Sandia National Laboratories, 1993.
7. B. Hendrickson and R. Leyland. An improved spectral load babancing method. *Parallel Processing for Scientific Computing*, 1:953–961, 1993.
8. B. Hendrickson and R. Leyland. A multilevel algorithm for partitioning graphs. Technical report, Sandia National Laboratories, 1993.
9. B.W. Kernighan and S. Lin. An efficient heuristic procedure for partitioning graphs. *Bell Systems Technical Journal*, 49:291–308, 1970.
10. P. Kuntz, P. Layzell, and D. Snyers. A colony of ant–like agents for partitioning in VLSI technology. In *Fourth European Conference on Artificial Life*, July 1997.
11. P. Kuntz and D. Snyers. Emergent colonization and graph partitioning. In *Third International Conference on Simulation of adaptive behaviour: From Animals to Animats 3*, July 1994.
12. A.E. Langham and P.W. Grant. A multilevel k–way partitioning algorithm for finite element meshes using competing ant colonies. Technical Report CSR1–99, Dept. of Computer Science, University of Wales Swansea, 1999.
13. H.D. Simon. Partitioning of unstructured problems for parallel processing. *Computer Systems in Engineering*, 2:135–148, 1991.
14. H.D. Simon and S.H. Teng. How good is recursive bisection? Technical report, Systems Division, NASA,Ca, 1993.

On Improving Clustering in Numerical Databases with Artificial Ants

N. Monmarché, M.Slimane, G. Venturini

Laboratoire d'Informatique, Université de Tours,
École d'Ingénieurs en Informatique pour l'Industrie (E3i),
64 av. Jean Portalis, 37200 Tours
monmarche,slimane,venturini@univ-tours.fr
Phone: +33-2-47-36-14-14
Fax: +33-2-47-36-14-22

Abstract. We present in this paper a new hybrid algorithm for data clustering. This algorithm discovers automatically clusters in numerical data without prior knowledge of a possible number of classes, without any initial partition, and without complex parameter settings. It uses the stochastic and exploratory principles of an ant colony with the deterministic and heuristic principles of the K-means algorithm. Ants move on a 2D board and may load or drop objects. Dropping an object on an existing heap of objects depends on the similarity between this object and the heap. The K-means algorithm improves the convergence of the ant colony clustering. We repeat two stochastic/deterministic steps and introduce hierarchical clustering on heaps of objects and not just objects. We also use other refinements such as an heterogeneous population of ants to avoid complex parameters settings, and a local memory in each ant. We have applied this algorithm on standard databases and we get very good results compared to the K-means and ISODATA algorithms.

1 Introduction

Clustering is one of the problems to which artificial ants have been applied with success in their early developments. Deneubourg and his colleagues are probably the pionneers in this domain with their work on robotics (Deneubourg et al. 1990, Goss and Deneubourg 1991) which has been used as a basis for more recent work involving artificial ants. For instance, clustering ants have been used in a real world application in the context of VLSI technology (Kunz and Snyers 1994, Kuntz et al. 1997). The problem here is to find a relevant partitioning of a graph. For this purpose, this partionning problem is turned into a clustering problem which is solved by artificial ants. Ants can pick up/drop objects on a 2D board according to a local density measure of similar objets in the neigborhood of the ant. In (Lumer and Faieta 1994), artificial ants cluster together data from a numerical database. This work is important for us because it introduces the concept of clustering ants in data analysis and knowledge discovery in databases problems.

There are at least two motivations for using an ant-based algorithm in a clustering problem. In data clustering (Jain 1998), many algorithms (like K-means and ISODATA (Ball and Hall 1965) used in the following) require that an initial partition is given as input before the data can be processed. This is one major drawback for these methods, and it is important to notice that ant-based approaches to clustering do not require such an initial partionning. One should also notice that many methods for clustering are entirely based on heuristics, and may run fast but are very often locally optimal. One way to improve those methods is for instance to introduce a stochastic search rather than a deterministic one.

So we propose in this paper a new method called AntClass, which is based initially on Lumer and Faieta's work, but with major extensions such as introducing more robust ant-like heuristics, dealing with "unassigned objects", speeding up convergence with the K-means algorithm, using hierarchical clustering on heaps of objects, testing the resulting algorithm on several real world data sets and providing a successful comparison with the K-means and ISODATA algorithms.

The remaining of this paper is organized as follows: section 2 the ant-based heuristics of AntClass. Sections 3 presents the hybrid part of AntClass, i.e. the use of the K-means algorithm, and the hierarchical clustering technique. Section 4 describes the experiments which have been performed with AntClass on artificial and real world data sets, as well as a comparative study. Section 5 concludes on future work and extensions to AntClass.

2 Ant-based principles of AntClass

2.1 Objects, heaps and the 2D board

We assume that a set $E = \{O_1, ..., O_n\}$ of n data or objects has been collected by the domain expert, where each object is a vector of k numerical values $v_1, ..., v_k$. For measuring the similarity between objects we will use in the following the euclidean distance between two vectors, denoted by D. D_{max} will denote the maximum distance value between two objects of E, i.e.: $D_{max} = \max_{O_i, O_j \in E} D(O_i, O_j)$.

Initially all the objects will be scattered randomly on a 2D board. This board can be considered as a 2D matrix C of $m \times m$ cells. This matrix is also considered as being toroidal in order to let the ants travel from one side to another in one step. One first important point which has not been studied in previous studies is to determine the size m of the board automatically. The number of possible cells in the matrix has to be at least greater than the number of objects ($m^2 \geq n$). But if the board is too large, the ants will waste a lot of time and will need many moves before encountering an object. So we have chosen the following relation between m and n: $m^2 = n \times 4$, and we have also observed experimentally that this method gives good results.

Ants will be able to create, build or destroy heaps of objects. A heap H is considered to be a collection of at least two objects. A heap is located on a

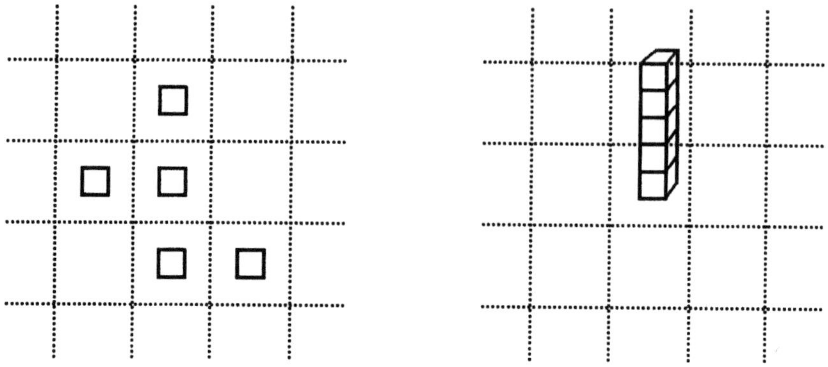

Fig. 1. The definition of a cluster has been changed in AntClass. On the left is represented a cluster as in (Lumer and Faieta 1994). A cluster is thus a spatial pattern, but two different clusters can be in contact and may thus be difficult to identify. On the right is given the new representation of a cluster in AntClass which solves the previous problem and allows us to define more robust heuristics.

given single cell, and is not a spatial pattern, as explained in figure 1. The major advantage of this improvement compared to (Lumer and Faieta 1994) is that a heap or cluster can be easily identified, and it allows us to define more accurate heuristics for dropping or removing objects from a heap, as described extensively in the next section. For instance, ants will be able to remove the most dissimilar object from a heap, or to add a carried object to a heap provided that this object is sufficiently similar to the other objects of the heap.

For this purpose, we have to define the following notations, for a given heap H of n_H objects:

- $D_{max}(H)$ is the maximum distance between two objects of H:

$$D_{max}(H) = \max_{O_i, O_j \in H} D(O_i, O_j)$$

- $O_{center}(H)$ is the center of mass of all objects in H:

$$O_{center}(H) = \frac{1}{n_H} \sum_{O_i \in H} O_i$$

  Objects in this case are considered as vectors of k numerical values. One should notice also that $O_{center}(H)$ does not correspond to a real object in general,
- $O_{dissim}(H)$ is the most dissimilar object in H:

$$O_{dissim}(H) = \arg \max_{O_j \in H} D(O_j, O_{center}(H))$$

1. Initialize randomly the ants positions,
2. Repeat
3. For each ant ant_i Do
   (a) Move ant_i,
   (b) If ant_i does not carry any object Then look at the 8 cells in the neighborhood of ant_i location and possibly pick up an object (see text for explanation),
   (c) Else (ant_i is already carrying an object O) look at the 8 cells around ant_i and possibly drop O (see text for explanation),
4. Until Stopping criterion.

Fig. 2. The general principles of the AntClass algorithm (which are similar to (Lumer and Faieta 1994)).

- $D_{mean}(H)$ is the mean distance between the objects of H and the center of mass $O_{center}(H)$:

$$D_{mean}(H) = \frac{1}{n_H} \sum_{O_i \in H} D(O_i, O_{center}(H))$$

2.2 Ant-based algorithm in AntClass

The algorithm at the heart of AntClass is represented in figure 2. The colony consists of p ants $Ant_1, ..., Ant_p$, and $p = 20$ in the following. Each ant is located on one cell of the board. Initially this position is generated randomly and uniformly, and there is absolutely no central control of the colony.

The ants move in the following way: initially, a given ant ant_i selects a random direction among the 8 possible ones. Then, ant_i has a probability $P_{direction}$ to further continue in this direction when moving next, else it generates randomly a new direction. Each ant also has a *speed* parameter which tells of how many steps it will move in the selected direction before stopping on again. Once it has moved, the ant may possibly pick up or drop an object as described in the next two paragraphs.

When the ant is not carrying any object, it looks for a possible object to pick up by considering the 8 cells around its current position. As soon as one object or one heap is found, then three cases have to be considered:

1. one object alone: the ant has a fixed probability to pick up the object.
2. a heap of two objects: since there are two objects only in the heap, we have the following specific property: $D(O_{dissim}(H), O_{center}(H)) = D_{mean}(H)$. So there is no real heuristic based on distance D to be applied here. This is why we have simply given to the ant a probability $P_{destroy}$ to pick up any of the two objects which results in destroying the heap.
3. a heap of more than two objects: the ant picks up the most dissimilar object in the heap provided that the heap "dissimilarity" is above a given threshold T_{remove}. The dissimilarity is measured by $\frac{D(O_{dissim}(H), O_{center}(H))}{D_{mean}(H)}$.

Parameter	Role	Range
Speed	# amplitude of moves	$[1, 10]$
$P_{direction}$	prob. to move in the same dir.	$[0.5, 1[$
Max_{carry}	object max. carrying time	$[20, 200]$
P_{load}	prob. to pick up a single object	$[0.3, 1[$
$P_{destroy}$	prob. to destroy a heap of 2 objects	$[0, 0.6]$
T_{remove}	min. dissimilarity necessary for removing an object from a heap	$[0.1, 0.2]$
T_{create}	max. dissimilarity permitted for creating a heap of two objects	$[0.05, 0.2]$

Fig. 3. The ants parameters in AntClass. All parameters which are crucial for the results are generated randomly within the indicated bounds.

It is a simple heuristic but which is powerful because it makes the heap more homogeneous.

When the ant is carrying an object, then it looks also at the 8 cells around its current location. For each cell, three cases have to be considered again:

1. the cell is empty: the ant has simply a constant probability to drop the object.
2. the cell contains one object only: the ant will drop the object and will thus create a heap of two objects but provided that the carried object O is sufficiently similar to the one already in the cell (O'), i.e. $\frac{D(O,O')}{D_{max}} < T_{create}$.
3. the cell contains a heap: the ant will add its carried object to the heap provided that it is closer to H's center than the most dissimilar object of H.

In order to avoid that an ant carries an object for a too long time, in the case for instance of a very dissimilar object compared to the others, the ant will drop this object automatically after Max_{carry} iterations on the first empty cell it encounters.

Since real ants have the possibility to memorize several sites in their environment (see for instance (Fresneau 1985)), we have added a memory to each ant in order to speed up the classification. Once an ant has encountered a heap H, it may store the location of H, $O_{center}(H)$ and $D(O_{dissim}(H), O_{center}(H))$ in its memory. Then, the algorithm for dropping an object and also the algorithm that defines the moves performed by the ant are modified as follows: when the ant is carrying an object, it searches in its memory for a heap H on which it could drop the object. If it finds one, then the memory of this heap is activated and the ant will go to H location. If it has not dropped the object on its way to H, then the ant will drop the object on H provided that H is still valid, i.e. it has not been destroyed or too much modified by the other ants. Ants may forget about a heap because they store those heaps in a First In First Out structure.

Furthermore, in order to avoid complex parameter settings and to simplify the use of AntClass by domain experts which may not be computer scientists, we have used an heterogeneous population of ants with different parameters, and thus different behaviors. We have used even more heterogeneous parameters than in (Lumer and Faieta 1994). Initially, the ants parameters will be generated randomly within the bounds represented in figure 3. Always the same values are used in this paper for all the tested data sets. These bounds have been found by several trials on the artificial data sets.

Finally, the stopping criterion of this algorithm is simply the number of iterations.

3 Final AntClass algorithm

3.1 Combining ant-based with heuristic search

The previous algorithm based on ants only has the major advantage of providing a relevant partition of the data without any initial information about the future classification, unlike the K-means or ISODATA algorithms for instance. However, two important problems remain. The first one is due to the fact that some objects are not assigned to any heaps when the ant algorithm stops, what we call in this paper "free" objects. This corresponds for instance to objects which are still carried by the ants or to objects which are alone on the board. The second problem is that if an object has been assigned to a wrong heap then it can take a long time until the object is transported to the right cluster.

So the solution that we propose is to combine two complementary algorithms, i.e. ant based clustering as presented in the previous section, and the K-means algorithm. This heuristic algorithm will use the initial partition provided by the ants as a "starting point". Then, the K-means works in the following way: it computes the center of each cluster, then it computes a new partition by assigning every object to the heap which center is the closest to the object. This cycle is repeated during a given number of iterations until the assignment has not changed during one cycle.

3.2 Hierarchical clustering on heaps of objects

We have noticed experimentally that the two previous steps of AntClass (ants + K-means) usually give good results in terms of misclassification. However the number of classes is always over estimated. We have observed that these two first steps of AntClass generate many small heaps but which are very homogeneous. So the idea which is presented in this section consists in considering those small and homogeneous heaps as objects themselves or "building blocks", and to perform the two same steps (ants + K-means) but on those newly defined objects. Furthermore, another motivation is that hierarchical clustering is a very standard technique in classification, but it has not been used yet with ants.

Let us consider now that the ants+K-means hybrid algorithm has lead to the creation of k heaps of objects. In order to let the ants deal with heaps of objects rather than objects, we have simply adapted the algorithms described previously: ants will be able to carry an entire heap of objects. The algorithm for picking up a heap is globally the same as for objects, and ants will pick up a heap with a probability P_{load}. We have added however another mechanism in order to avoid that ants carry all heaps at the same time, because as will be seen in the next paragraph, the number of heaps will only decrease over time. So if only a few clusters remain on the board, it is important not to carry them all the time because in this case ants would not be able to further cluster them if needed. So once a heap has been dropped, it is marked with a kind of pheromone that prevents other ants from picking up this heap during the next 500 iterations. Ants drop a heap H_1 onto another heap H_2 provided that $\frac{D(O_{center}(H_1),O_{center}(H_2))}{D_{max}} \leq T_{create}$.

When H_1 and H_2 are clustered together, they form only one heap H_3, which can not be separated anymore. This makes the convergence faster.

3.3 AntClass final algorithm

So AntClass consists mainly of four steps: (1) ant-based algorithm for clustering objects, followed by (2) the K-means algorihm using the initial partition provided by the ants, and then (3) ant-based clustering but on heaps previously found, and finally (4) the K-means algorithm once more.

Finally, we should add that all values in the data set are normalized between 0 and 1 in order to avoid any scaling problem between the different attributes.

4 Experimental results

4.1 Experimental settings

We have applied AntClass to the following numerical databases (numbers in brackets indicate respectively, the number of objects, the number of numerical attributes, and the number of classes): Artif. 1 (80, 2, 4), Artif. 2 (270, 2, 9), Artif. 3 (200, 2, 4), Artif. 4 (150, 10, 3), Iris (178, 4, 3), Wine (178, 13, 3), Glass (214, 9, 2-6)[1], Soybean (47, 21, 4), Thyroid (215, 5, 3). "Artif. 1" to "Artif. 4" have been used to evaluate AntClass on databases with know properties where the examples are generated according to gaussian laws (in the same way as Lumer and Faieta). The other real world databases come from the machine learning repository.

In order to evaluate the resulting partition obtained by AntClass (or the K-means and ISODATA algorithms in the comparative study of section 4.3), we have set up the following method. We have used databases for supervised learning, i.e. databases where the true classes are known thanks to a "Class" attribute. But when AntClass is used to partition the data, this "Class" attribute is not given to it. It is only a posteriori that the class information is used to evaluate the results.

We have defined two performance measures to evaluate how close is the obtained partition to the real one. The first measure is a classification error rate. It is computed as follows: for a given cluster H obtained by AntClass, consider

[1] From 2 to 6 classes can be defined for the glass database.

Databases and perf.	1: Ant colony on objects	2: K-means on objects	3: Ant colony on heaps	4: K-means on heaps
Artif. 1: Cl. err.	11.58 %	0.21 %	0.42 %	0.00 %
# of cla.	8.15	7.76	4.24	4
Artif. 2: Cl. err.	17.24 %	0.52 %	2.22 %	0.00 %
# of cla.	22.30	17.07	10.46	9.02
Artif. 3: Cl. err.	20.35 %	6.32 %	6.93 %	4.66 %
# of cla.	15.06	14.98	5.42	4.42
Artif. 4: Cl. err.	22.23 %	3.32 %	2.68 %	1.33 %
# of cla.	5.22	5.18	2.94	2.96

Table 1. Intermediary and final results obtained on each of the four steps of AntClass for the four artificial databases. "Cl. err." stands for "classification error rate" and "# of cla." for "number of classes".

the most represented class among H according to the "Class" attribute. All objects of H that do not belong to this class are considered as being missclassified. The classification error rate is simply the ratio between the total number of misclassified objects for all created clusters and the total number of objects in the database. The second performance measure is simply the number of created clusters.

All runs have been performed on a very standard PC (Pentium 166). One run ends in a few tens of seconds (usually from 10 to 20 seconds). All presented results have been averaged over 50 runs. Finally, one very important point is that the ants parameters in AntClass were always the same for all the databases, i.e. all generated randomly within the same bounds (see figure 3). Ants were simulated during 2000 iterations when clustering objects, and 50000 iterations when clustering heaps. The number of iterations of the K-means algorithm in AntClass was set to 10.

4.2 Results on artificial databases

The results reported in table 1 show the progression of AntClass towards a relevant classification. In the first step of AntClass, the Ant-based algorithm finds an initial partition but with classification errors and really too many clusters. At the end of the second step of AntClass, i.e. the use of the K-means on the initial partition found in the previous step, the classification errors are reduced but the number of clusters is still really too high. This is due to the fact that the K-means algorithm is really sensitive to the initial partition. If this initial partition contains too many clusters, then the final partition is unlikely to be the optimal one. Once the third step has been performed, the ants converge very closely to the right number of classes by working on heaps of objects rather than objects themselves. One can notice that some classification errors remain. At the end of the fourth step, using the K-means once more decreases the classification errors on again. But this time, since the number of classes and an almost optimal

Algo.		AntClass	AntClass	K-means	K-means	ISODATA	ISODATA
Data set	# of cla. (real)	# of cla. (aver.)	cl. err. (aver.)	# of cla. (aver.)	cl. err. (aver.)	# of cla. (aver.)	cl. err. (aver.)
Artif. 1	4	4	0.00 %	5.63	2.15 %	4.53	1.64 %
Artif. 2	9	9.02	0.00 %	9.73	12.78 %	6.38	29.11 %
Artif. 3	4	4.42	4.66 %	7.26	7.30 %	6.58	7.84 %
Artif. 4	3	2.96	1.33 %	9.60	0.00 %	9.53	0.00 %
Iris	3	3.02	15.4 %	6.95	4.63 %	4.59	6.28 %
Wine	3	3.06	5.38 %	8.98	8.57 %	9.09	8.63 %
Glass	2-6	7.7	4.48 %	7.06	50.16 %	2.34	42.98 %
Soybean	4	4.82	0.13 %	7.93	3.89 %	7.94	4.77 %
Thyroid	3	3.28	6.38 %	8.77	8.26 %	1.48	14.72 %

Table 2. Results obtained by AntClass, K-means and ISODATA with artificial and real world databases.

partition have been well determined in the previous step, the K-means really finds optimal or near optimal results.

4.3 Comparative study

We describe now the results obtained with the K-means algorithm and with ISODATA (Ball and Hall 1965). Each algorithm is initialized with 10 classes, and all objects are initially assigned randomly to these classes. The K-means algorithm is used with 10 iterations, and has been described previously in section 3.1. ISODATA is an improved version of the K-means which can (1) delete classes with less than Min_{obj} objects, (2) split one class into two when the deviation in this class is above Max_{dev}, (3) cluster together two classes when their distance to each others is less than Min_{dist}.

In the following, we have used $Min_{obj} = 1$, $Max_{dev} = 1.027$ and $Min_{dist} = 0.117$. These last two values have been determined with the data, which favors ISODATA because this represents an important initial information which has not been given to AntClass. To compute Max_{dev}, the data have been normalized linearly between 0 and 1, and we have computed for each class the deviation of the objects around the center of their real classes. These values have been averaged on all the datasets in order to compute Max_{dev}. Min_{dist} has been computed in the same way by considering the average over each database of the minimum distance between two classes centers.

ISODATA is also run during 10 iterations. We have tried larger numbers of iterations for both algorithms but the results were similar.

Results obtained by the three algorithms (AntClass, K-means and ISODATA) are represented in table 2. As can be seen, AntClass outperforms the two other algorithms, both in terms of classification errors and of correct number of classes. The only exception is for Fisher Iris database. In this database, the Setosa class is completely distinguishable from the two others. The last two classes are more difficult to separate, unless more than three clusters are created, which is what K-means of ISODATA do.

5 Conclusion

We have presented in this paper a new hybrid and ant-based algorithm named AntClass for data clustering in a knowledge discovery context. The main features of this algorithm are the following ones. AntClass deals with numerical databases. It does not require any initial information about the future classification, such as an initial partition or an initial number of classes. AntClass introduces new heuristics for the ant colony, and also an hybridization with the K-means algorithm in order to improve the convergence. We have also introduced in AntClass hierarchical clustering where ants may carry heaps of objects and not just objects. Furthermore, AntClass uses an heterogeneous population of ants in order to avoid complex parameter settings to be performed by the domain expert. Finally, AntClass has been tested with success on several databases, including real world ones.

Future work consists in testing how this model scales with larger databases of several thousands of examples. We are also considering other sources of inspiration from real ants for the clustering problem. For instance, ants that meet on the board could exchange objects.

References

Ball G.H. and Hall D.J., ISODATA, a novel method of data analysis and pattern classification, Technical report, Stanford Research Institute, 1965.

Deneubourg J.-L., Goss S., Franks N., Sendova-Franks A., Detrain C. and Chretien L. (1990), The dynamic of collective sorting robot-like ants and ant-like robots, Proceedings of the first Conference on Simulation of Adaptive Behavior 1990, J.A. Meyer and S.W. Wilson (Eds), MIT Press/Bradford Books, pp 356-363.

Fresneau D. (1985), Individual foraging and path fidelity in a ponerine ant, *Insectes Sociaux, Paris*, 1985, Volume 32, n 2, pp 109-116.

Goss S. and Deneubourg J.-L. (1991), Harvesting by a group of robots, Proceedings of the first European Conference on Artificial Life 1991, F.J. Varela and P. Bourgine (Eds), MIT press/Bradford Books, pp 195-204.

Jain A.K and Dubes R.C. (1988), Algorithms for Clustering Data, Prentice Hall Advanced Reference Series.

Kuntz P. et Snyers D. (1994), Emergent colonization and graph partitioning, Proceedings of the third International Conference on Simulation of Adaptive Behavior: From Animals to Animats 3 (SAB94), D. Cliff, P. Husbands, J.A. Meyer, S.W. Wilson (Eds), MIT-Press, pp 494-500.

Kuntz P., Layzell P. and Snyers D. (1997), A colony of Ant-like agents for partitioning in VLSI technology, Proceedings of the fourth European Conference on Artificial Life 1994, P. Husbands and I. Harvey (Eds), MIT press, pp 417-424.

Lumer E.D. and Faieta B.(1994), Diversity and Adaptation in Populations of Clustering Ants, Simulation of Adaptive Behavior: From Animals to Animats 3 (SAB94), D. Cliff, P. Husbands, J.A. Meyer, S.W. Wilson (Eds), MIT-Press, pp 501-508.

Communication and Language

Ants War with Evolutive Pheromone Style Communication

Hidenori Kawamura, Masahito Yamamoto, Keiji Suzuki and Azuma Ohuchi

Laboratory of Harmonious Systems Eng., Research Group of Complex Systems Eng., Division of Systems and Information Eng., Graduate School of Eng., Hokkaido University
Nishi 8, Kita 13, Kita-ku, Sapporo, Hokkaido, 060-8628, Japan

Abstract. This paper is concerned about the origin of pheromone communication in complex societies, e.g., colonies of real ants and bees. The aim of the work is to study whether pheromone communication among artificial ant agents in a cooperative foraging scenario can arise simply by letting selection favor successfully foraging genotypes. For such aim, we introduce the ants war in which ant agents foraging more food items than the opponent can survive to next generation. In the experiments, we confirmed the emergence of pheromone communication without a specific predefined one.

1 Introduction

In nature, many species of animals exhibit a high level of cooperative behavior and use various means of communication for their lives. Typical examples are societies of real ants and bees, which are known as social insects [7]. In ant and bee colonies, individuals have simple limited abilities but the macro-level behavior is sufficiently complex to enable the ants and bees to adapt to their dynamic environment. Pheromone, which is a simple chemical medium of communication, is known to play an important role in the formation of their complex behavior. For instance, in ant colonies, ants that are discovering food items lay the way to their nest with a specific type of pheromone that attracts other ants. Ants that sense the pheromone follow the pheromone trail to the food. When all the items of food are carried back to the nest, the pheromone evaporates, and the pheromone trail, which is no longer necessary, automatically disappears [1]. Several researchers have studied designed pheromone communications in multi-agent systems [3][5][6].

In this paper, we focused on the fact that these social insects have obtained pheromone communication in their evolutionary process, and the origin of pheromone communication in complex societies. The aim of the work is to study whether pheromone communication among artificial ant agents in a cooperative foraging scenario can arise simply by letting selection favor successfully foraging genotypes. For such aim, we introduce the ants war in which ant agents foraging more food items than the opponent can survive to next generation. This evolutionary process is based on results of the war, and any special fitness

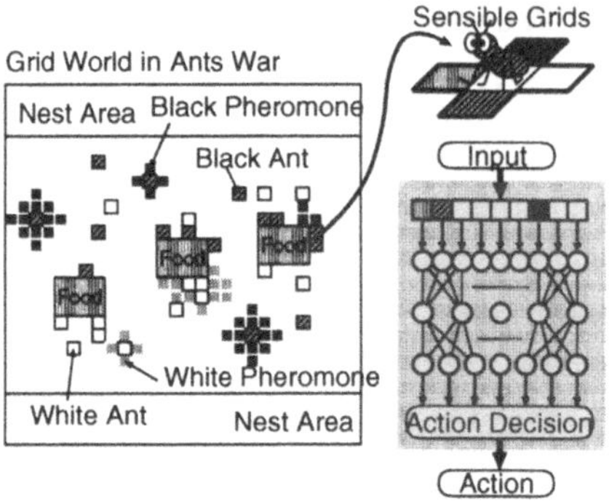

Fig. 1. Ants War Environment and Architecture of the Ant Agent

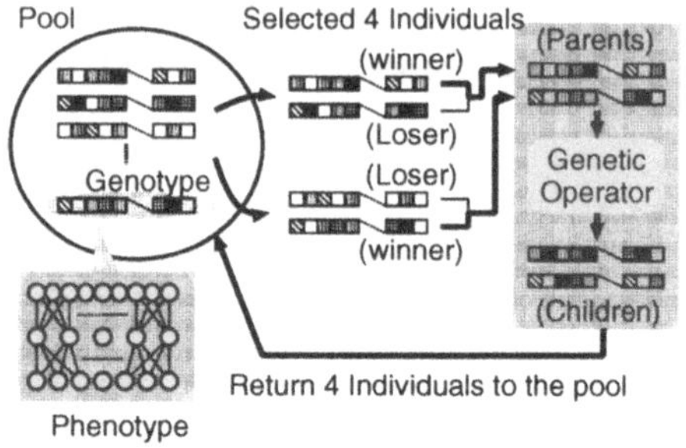

Fig. 2. A series of GA operations in one generation

functions are not implemented. The meanings of pheromone are not defined by the designer, and the ant agents autonomously gain pheromone communication throughout an evolutionary process in Genetic Algorithm (GA). In the computer experiments, we confirmed the emergence of pheromone communication without a specific predefined one although simple genetic operators were implemented.

2 Ants War and Evolutionary Strategy

The ants war is introduced as the stage for emerging of pheromone communication, and it is designed in the grid environment with several kinds of components, i.e., "white ants colony", "black ants colony", "white pheromone", "black pheromone" and "food items" (see Fig. 1). The objective of each colony is to carry back more food items than the opponent. Each ant can sense only information on neighbor grids of its current position, and must decide actions from only sensing information. The information includes whether the sensing ant do adjoin fellows or not, opponents or not, food items or not, and the intensities of own colony pheromone. The ants in a colony have same 3 layer general neural network to decide its action from 7 actions, i.e., "move forward", "move backward", "move to right", "move to left, "force to the food", "stand by", and "put pheromone". The ants of any single colony are assumed to be completely identical for a simplification. The food items are too heavy to be carried back by a few ants, thus the ants are required to cooperate with each other and to allocate their resource for appropriate food items. Namely, the communication among a colony of ants is very important to win this war. Phermone as an only communication medium for ants, which is put and sensed by ants, is non-negative real value on each grid, and it automatically diffuses and evaporates as its characteristics.

For evolutive pheromone communication, GA is applied to genotypes which are encoded according to the weight sets of neural networks directly. At each

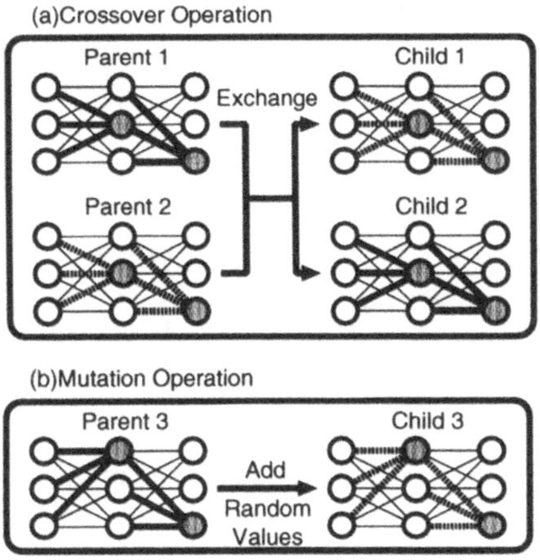

Fig. 3. The crossover and mutation operations used in the GA

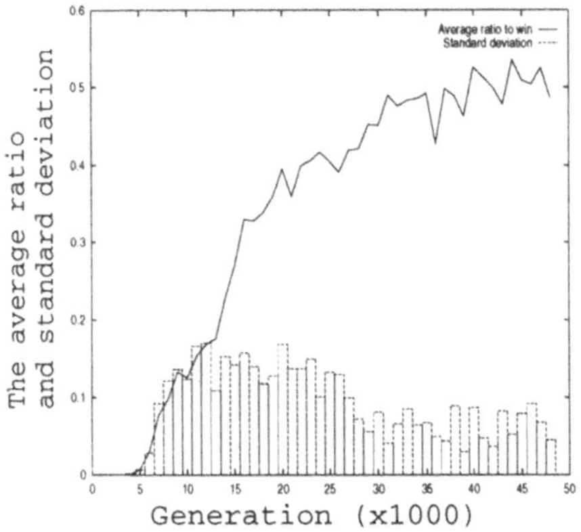

Fig. 4. The graph representing the ratios to win against the criterion and the standard deviation

generation of GA, two pairs of genotypes are randomly selected from the genetic pool, and two colonies of each pair are fought in the ants war. Two winners can become parents and reproduce their offspring applied mutation and crossover genetic operators [4] (see Fig. 3). The number of genotypes in the pool are always kept the constant. In this GA, the selection pressure for genotypes is based on only results of the war, and any special fitness functions are not implemented. We study whether pheromone communication can arise or not without a specific predefined one by using this simple evolutionary mechanism.

3 Experimental Results

We made 7 trials of computer experiment with 20 genotypes population in the pool. The max generation of GA was set to 50,000. At first, it is necessary to show whether the evolution properly occurred or didn't. However, it is difficult because GA doesn't have any global fitness functions to evaluate genotypes. Thus, we measured the progress of the evolution by the average ratios to win against the final population as the criterion population. All individuals in the population of each generation were fought to all ones in the criterion final population. [2]. The number of 400(20 genotypes × 20 genotypes) matches were made for evaluating one population. If the progress was occurred favorably, the average ratios to win against the criterion would be gradually increased from the first population to the final, and be close to 0.5 finally. In Fig. 4, the solid and the dashed lines represent the average ratio to win and the standard deviation in experimental trials respectively. We can confirm the evolution progressed in each trial and the selection pressure certainly worked. Moreover, the final populations as the criteria almost show the best performances in all populations, and

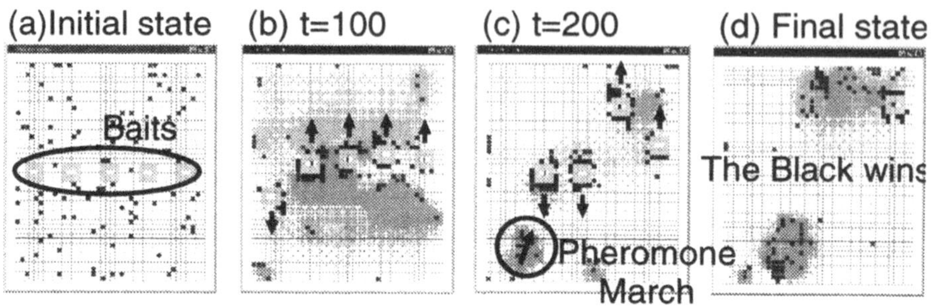

Fig. 5. The screen shots in the ants war with the effective pheromone communication.

the choice of criteria is suitable. If the choice is not appropriate, most populations would win against the criterion or most would lose, and we can't appropriately evaluate each population.

Next, we observe the evolved ant agents in 7 trials. The obtained ant agents are roughly classified into following three types.

Type 1: The drifting ants without pheromone. These ants tend to drift on the grids randomly. If they touch the food or fellow ants in front, take the forcing action. They scarcely put a pheromone in the war, i.e., the emergence of the communication was not occurred.

Type 2: The ants drifting up and down with the simple pheromone communication. These ants drift up and down on the grids, and the ants clinging to the fellows at the left side continuously take the forcing action. Accordingly, They often take the formation like a bar. If no ants touch the food, they fall into the deadlock situation. When the leftest ant in the bar formation doesn't touch the food, this ant occasionally puts the pheromone. Then, other ants sensing the pheromone move from current positions, and they can resolve the deadlock situation. We can say that this is the simple pheromone communication.

Type 3: The marching ants with the effective pheromone communication. The ants sensing little intensity of pheromone move backward, and they take forcing action when be in contact with the food or the fellow at front side. In addition, they indiscriminately put the pheromone. So, the ants which are initially placed at upside area reach appropriate foods efficiently. While, the remainders gather in the downside area, so the intensities of the pheromone in the downside become greater by indiscriminate putting action. The ants sensing strong pheromone begin to march forward with putting the pheromone, and they can reach food items. Especially in the situation that the food is obtained and disappeared, they quickly begin to march forward and reach other foods again. The screen shots of the war with this type of ants are shown in Fig. 5. In the war, most ants were observed to take part in forcing the foods, and they highly cooperated with each other to win the war by using the pheromone communication.

For investigating which types of ants is better than the others, the final populations of each ants were fought with each other as evaluating the populations with 400(20 × 20) matches. The results are as Type 1(372 wins) - Type 2(28 wins), Type 2(8 wins) - Type 3(392 wins) and Type 1(20 wins) - Type 3(380 wins). The reason that type 2 is very weak is that they are poor to find the baits and cooperate for forcing although they use the primitive pheromone style communication. Type 3 is strongest in all 3 types as we expected, and it is said that the good communication as type 3 is effective for the cooperation to win.

4 Conclusion

Throughout computer experiments, we confirmed that pheromone communication emerged without a specific predefined one. The meanings of pheromone were different in experimental cases, and the ants cooperated to survive in GA by using of effective pheoromone communication. We conclude it is possible for artificial complex societies to emerge their communications for their objectives based on simple evolutionary strategies.

References

1. Agosta, W. C.: Chemical Communication - The Language of Pheromone -. W. H. Freeman and Company (1992)
2. Cliff, D., Miller, G. F.: Tracking the Red Queen: Measurements of adaptive progress in co-evolutionary simulations. The proc. of the Third European Conf. of Artificial Life. (1995) 200–218
3. Colorni, A., Dorigo, M., Maniezzo, V.: Distributed Optimization by Ant Colonies. The Proc. of ECAL91- European Conf. on Artificial Life. F.,Varela and P., Bourgine (Eds.), Elsevier Publishing. (1991) 134–142
4. Montana, D. J., Davis, L. D.: Training Feedforward Networks Using Genetic Algorithms. The Proc. of the International Joint Conf. on Artificial Intelligence. (1989) 762–767
5. Nakamura, M., Kurumatani, K.: Formation Mechanism of Pheromone Pattern and Control of Foraging Behavior in an Ant Colony Model. The Proc. of the Fifth International Workshop on the Synthesis and Simulation of Living Systems. C., G., Langton and K., Shimohara (Eds.), MIT Press. (1997) 67–74
6. Suzuki, K., Ohuchi, A.: Reorganizations of Agents with Pheromone Style Communication in Multiple Monky Banana Problem. The Proc. of Intelligent Autonomous Systems 5. Y., Kakazu, M., Wada, T., Sato (Eds.), Ohmsha. (1998) 615–622
7. Williams, G. C.: The Pony Fish's Glow. Orion Publishing Group (1997)

Sexual Signalling in an Artificial Population: When Does the Handicap Principle Work?

Jason Noble

Center for Adaptive Behavior and Cognition
Max Planck Institute for Human Development
Lentzeallee 94, D-14195 Berlin, Germany

Abstract. Males may use sexual displays to signal their quality to females; the handicap principle provides a mechanism that could enforce honesty in such cases. Iwasa et al. [1] model the signalling of inherited male quality, and distinguish between three variants of the handicap principle: pure epistasis, conditional, and revealing. They argue that only the second and third will work. An evolutionary simulation is presented in which all three variants function under certain conditions; the assumptions made by Iwasa et al. are questioned.

1 Sexual Signalling and the Handicap Principle

Sexual selection is a distinct subset of natural selection. The idea is that evolution is an exam with two papers: in order to reproduce, an animal must not only survive to adulthood, but, in a sexual species, it must gain mating opportunities with members of the opposite sex. One of Darwin's insights was that selection for sexual attractiveness and selection for survival could exert opposing evolutionary pressures. If, for some reason, females came to prefer males with elaborate and costly ornaments, such as the peacock's tail, then sexual selection would push towards yet more costly ornaments, because males with longer tails experience greater mating success. At the same time, natural selection would push for less costly ones, because males with longer tails are more vulnerable to predation and less likely to survive to adulthood.

An early explanation for extreme male ornament traits and female preferences was that an initial, random bias led to linkage between trait and preference genes and that a runaway cycle of exaggeration then took place [2]. Another possibility, more recently explored, is that male ornaments function as "indicator mechanisms", i.e., that they are used by males to signal their quality as mating partners to females [3]. This paper describes an evolutionary simulation, based on a population-genetic model by Iwasa et al. [1], that explores the conditions under which this kind of sexual signalling could evolve.

The suggestion that male ornaments are signals of mate-quality leads to a problem in understanding how honest signalling could be maintained. Why should low-quality males ever honestly signal their condition, when by doing so they will make themselves unlikely to be chosen as mates? Why wouldn't all males produce the maximum advertisement, regardless of their true quality—all claiming, in effect, to be the most desirable. Zahavi's handicap principle provides a possible mechanism by which a sexual signalling system could be kept honest.

Zahavi [4] suggested that honesty could only be maintained in a communication system if the signals were costly in some way. He proposed the counter-intuitive idea that signallers sacrifice some of their fitness (i.e., impose a handicap on themselves) in order to produce signals that will be believed by receivers. In the sexual signalling case, perhaps males are signalling their quality (e.g., vigour, viability or territory size) to females, and are being rewarded with a mating episode if they "convince" a female that they are of high quality. Let us suppose that the male signal is tail length. Zahavi realized that if growing a longer tail was cheap, i.e., if it had little deleterious effect on male fitness, then the signalling system would be vulnerable to bluffing: *all* males would come to have long tails, and the female preference for longer-tailed males would no longer be selected for. But if growing a long tail was costly in fitness terms, the communication system would not be corrupted by bluffers: lower quality males would not be able to afford the necessary resources for growing a long tail. Tail length becomes an honest indicator of male quality because "cheating" is prohibitively expensive. Zahavi reasoned that only those communication systems in which the signal happened to be costly would escape collapse due to bluffing. Therefore, the stable systems found in nature are maintained by this mechanism.

When the handicap principle was first introduced, it was generally not accepted by theoretical biologists. Population-genetic models [5, 6] seemed to show that it could not be evolutionarily stable. However, the potential effectiveness of the handicap principle has been validated by several mathematical models in recent years; foremost among these is a model by Grafen [7]. This model establishes that the handicap principle can work, but specifies an important proviso: the unit cost of producing the signal must be greater for a low quality signaller than for a high quality signaller. In other words, the fitness cost of extending one's tail by an extra centimetre must be higher for unhealthy or weak males than for healthy strong ones.

The handicap principle was maligned and misunderstood because Grafen's proviso about differential unit costs was not clear from Zahavi's original formulation, and because several distinct interpretations of the principle are possible. Iwasa et al. [1] attempted to cut through the confusion. They detailed three variant interpretations of the handicap principle, listed below, and suggested that different findings concerning the evolutionary stability of handicapped signalling could be explained by the fact that some authors were modelling one version and others another.

1.1 Pure Epistasis Handicap

In this variant, a particular set of genes determine a male's tail length, and the longer his tail, the less likely he is to survive to reproductive age. However, his survival is also determined by his quality: higher-quality males are more likely to survive, and for any given level of quality, a male is more likely to survive if he has a *shorter* tail. Therefore the males that are most likely to die before reaching reproductive age are those of low quality with long tails. Observing the adult population, one would find a correlation between the genes for quality and tail length. In technical terms, epistatic selection has resulted in linkage disequilibrium; in plainer language, long tails are linked to high quality, because all the long-tailed low-quality males died young. In consequence, a female's preference for mating with long-tailed adult males will mean that she is more likely to achieve her goal of mating with a high-quality male.

1.2 Conditional Handicap

A long tail still reduces a male's chances of survival to reproductive age, and again survival is primarily determined by quality. However, the expression of the gene for tail length is modified by quality: males of lower quality will not realize their full, genetically specified tail length but will grow a proportionately shorter tail. It is assumed that only the highest quality males have the resources to fully realize the tail length encoded in their genes. Because the expression of the tail-length gene is quality-dependent, observable tail length is correlated with quality even before mortality has taken its toll. A female preference for long tails will therefore translate into a preference for high-quality males.

1.3 Revealing Handicap

The expressed tail length of males is determined directly by a gene, as with the pure epistasis handicap. Survival to reproductive age depends on quality modified by tail length, as before. However, when the males reach reproductive age and are competing to be selected by females, only high-quality males succeed in maintaining their tails at their original, genetically specified length. Males of lower quality are less well able to withstand the rigours of their environment, and their tails are shortened due to, for example, attacks by predators or parasites [8]. Low quality males *reveal* their status by tending to have shorter tails as adults. Females preferring to mate with long-tailed males will thus mate with higher quality males on average.

2 Modelling Sexual Signalling of Genetic Quality

Almost all of the models that have recently demonstrated the plausibility of the handicap principle [7, 9, 10] have made a major simplifying assumption: namely, that the underlying male quality of interest to females is environmentally determined. This could mean, for example, that males are advertising their level of nutrition, or the quality of the territory they possess. This assumption misses the interesting subset of cases in which males are believed to be informing females of their *genetic* quality. For example, when sage grouse *Centrocercus urophasianus* mate the males contribute only their sperm, leaving all other aspects of the project of raising offspring to females. Nevertheless, the females choose their mate carefully on the basis of his ornaments, display behaviour, and central position in the mating arena [11, 12]. If the males are advertising anything in this case, it must be their inherited genetic quality.

It is not clear that an honest signalling equilibrium will exist for signals of genetically determined quality. A central problem is that there might not be any residual variation in male quality, and thus nothing to signal about. If the males were honestly advertising their quality, and females were choosing to mate with high-quality males, then after a few generations the males will all be clustered around the optimum quality level. As Maynard Smith [13] has argued, there should be no heritable variation remaining in fitness-related traits at equilibrium. And yet female sage grouse pay the costs of choice (e.g., time costs and predation risk) in order to choose the best male, when the male will contribute only his genes. This is known as the paradox of the lek: why aren't modern sage grouse males all maximally viable, and thus equally attractive to females? The most likely answer is that mutation on fitness-related traits is negatively biased. That is, a single mutation event affecting the genes controlling a fitness-related trait is more likely than not to decrease the value of that trait. There is some empirical evidence for this: at least one component of fitness is mildly heritable in *Drosophila* [14]; this could only occur if mutational load kept fitness-related traits below their optimum value. A similar conclusion was reached in a review of evidence from many avian species [15].

Iwasa et al. [1] constructed a population-genetic model of the evolution of costly male advertisements and female preferences; they incorporated just such a negative mutation bias on the viability trait. Iwasa et al.'s model purports to show that honest signalling of genetically determined male quality can be evolutionarily stable. It is one of the very few models to deal with genetically determined quality, and provides the basis for the simulation described in this paper.

Iwasa et al. reasoned that if females were prepared to pay a cost for their preference, then there must be information worth having in the expressed values of the advertisement trait, and it was therefore an honest indicator of quality. Iwasa et al. derived three conditions for the existence of such a costly-preference equilibrium. The first was that mutation on the viability trait had to be negatively-biased. Otherwise, male viability levels would be clustered around the optimum, and the females would be in a position where random mating was just as likely to result in a high-viability partner as was a costly preference. The second condition was that the genetic correlation between preference and viability had to be greater than the product of the correlations between advertisement and preference and between advertisement and viability. Another way of putting this is that there must be a link between preference and viability that does not come about solely because of their joint relationship with the male advertisement trait. Finally, Grafen's proviso, that the unit costs of signalling must be greater for lower-quality signallers, must also hold.

Iwasa et al.'s second equilibrium condition implies that whereas the conditional and revealing handicaps will work, the pure epistasis handicap will not.

In the conditional and revealing handicaps, the viability trait directly affects the expression of the male's advertisement—viability modifies the expression of the genes for growing an ornament of a particular size, or, in the case of the revealing handicap, low viability means that a large ornament cannot be successfully maintained as an adult. Valuable information for females concerning male viability has thus been built into the expressed male trait. However, in the pure epistasis handicap, the realized size of the male advertisement is only linked to viability indirectly, via differential survival. This means, in turn, that there is no special link between the genes for preference and for viability.

In order to keep their analysis tractable, Iwasa et al. made a number of simplifying assumptions. Most critical was the assumption that the genetic covariances between male advertisement, female preference, and the viability trait could all be treated as positive constants. In the real world, genetic covariances are of course not constant but change as the population evolves over time; even if we suppose that the covariances might start out positive, it is not clear that they would remain so. And it is more likely that the covariances in a plausible initial population would be close to zero. The problem highlights a weakness of the population-genetic approach: despite the name, there is in fact no population, which means that such important variables as genetic variances and covariances must be input into the model as parameters, rather than being measurements that are made with respect to an evolving lineage. The question as to whether Iwasa et al.'s conclusions would hold without assuming constant positive covariances [3] presents an excellent opportunity for an evolutionary simulation.

3 Description of the Model

The work reported here is an implementation of Iwasa et al.'s [1] model as an individual-based evolutionary simulation. The population consists of sexual individuals breeding in discrete, non-overlapping generations. Individual organisms have both a genotype and a phenotype; the genotype consists of real-valued genetic parameters. Each organism carries a gene for the male advertisement or ornament trait (t_{gen}), the female preference trait (p_{gen}), and the general viability trait (v_{gen}). An individual's phenotype consists of two real values: either t_{phen} or p_{phen}, depending on sex, and v_{phen}. Genotypic and phenotypic parameters are always real numbers between zero and one inclusive.

3.1 Development stage

At birth, each individual's sex is chosen at random, and its phenotypic trait values are determined. Normally, each trait is read off the genome, then a random gaussian error term is added ($\mu = 0$, $\sigma = 0.005$), and the resulting value stands as the expressed trait. The phenotypic male advertisement is normally read off the genotypic value of t_{gen}. However, in the conditional and revealing handicaps the male's viability also influences the expressed ornament size—the phenotypic viability is therefore calculated first. For the conditional handicap, the advertisement size that would otherwise be expressed is reduced by an amount proportional to v_{phen}. In other words, $t_{phen} = t_{gen} \times v_{phen}$. Only males with the maximum possible viability actually produce an advertisement that is as big as their genotype specifies.

3.2 Survival stage

Some individuals survive to adult reproductive age, and some die young.[1] An individual's basic probability of survival is equal to its phenotypic viability: less viable animals are less likely to survive. However, both male advertisements and female preferences are costly, and the cost of these characteristics is manifested as a reduction in an individual's probability of survival, according to the degree of the trait's phenotypic expression.

Grafen's proviso, in which the unit costs of advertisement are lower for higher-quality signallers, is enforced at this stage. The basic probability of survival (v_{phen}) is first converted to an odds ratio, and then scaled by $(1 - t_{phen})^{C_{adv}}$, where C_{adv} represents the cost of advertising. If $C_{adv} = 0$ then there is no cost at all to males for growing ornaments; if $t_{phen} = 0$ then a male will pay no costs regardless of how high C_{adv} might be. The scaled odds ratio is then converted back to a probability value. The result of all this manipulation is the following expression for the probability of survival:

$$p_{survival} = \frac{v_{phen}(1 - t_{phen})^{C_{adv}}}{v_{phen}(1 - t_{phen})^{C_{adv}} - v_{phen} + 1} \; . \tag{1}$$

The scaling factor implements Grafen's proviso, because individuals with high phenotypic viability will be best able to bear the costs of advertisement.

The survival costs of female preference are assessed in exactly the same way as the costs of the male trait: p_{phen} is simply substituted for t_{phen}, and C_{pref} for C_{adv}, in (1). Theories of handicap signalling generally do not require that female preference should involve anything other than a simple cost that is independent of viability; however, calculating female costs in the same way as male ones allows the costs borne by each sex to be directly compared.

3.3 Mating stage

Surviving males and females are then able to breed: females get to exercise their preferences, and males may experience the benefits of their costly ornaments. A surviving female is randomly chosen, and she is then presented with a "lek" of eight males, also selected at random. With a probability equal to her preference

[1] To prevent extinctions, in the rare event that no males (or no females) survive to adulthood, one male (or one female) is randomly chosen for resurrection.

value $(p_{phen})^2$, she selects the male with the largest expressed advertisement trait to mate with. If she does not choose this male, she chooses randomly from among the eight males. Thus, high-preference females are likely to end up mating with the male with the biggest ornament, while zero-preference females will mate with anyone. The results of an earlier simulation [16] suggest that this method can be effective in producing sexual-selection effects, and that eight is a reasonable lek size. Note that in Iwasa et al.'s model, female preferences were expressed relative to the population mean of the male advertisement trait. Having a model in which individuals really exist allows us to avoid the dubious assumption that females could know the population mean; instead, females choose a mate from among those males they happen to come into contact with.

The mate selection process continues until sufficient offspring have been produced to stock the next generation. Crossover is simple: newborn individuals inherit the mean of their parents' values for each real-valued genetic parameter. The mutation operator is a random gaussian ($\mu = 0$, $\sigma = 0.03$) added to each gene. The all-important negative mutation bias on viability is implemented by subtracting 0.003 from whatever value a newborn individual's genetic viability would otherwise have been. If the mutated value of any trait would be less than zero or greater than one, it is truncated accordingly.

4 Results

The population consisted of 100 individuals, and evolution proceeded in each run for 5000 generations. Unless otherwise stated, the results summarize a window period over the last 500 generations, and are averaged across 10 repeated runs in each case. The repeated runs in the various conditions were each performed with a different seed for the pseudo-random number generator. The simulations have been conducted over a range of values for the advertising and preference costs C_{adv} and C_{pref}. Earlier work [16] suggests that males will be prepared to bear much higher costs in advertising than females will tolerate in expressing a preference, and the range of cost levels investigated reflects this.

4.1 Pure Epistasis Handicap

Iwasa et al. argue that the key evidence for sexual signalling is the willingness of females to bear a costly preference; Figure 1(a) shows that such preferences do indeed evolve as long as both male and female costs are not excessive. But are females really gaining information about male quality from the advertisement traits they observe? Figure 1(b) shows the correlation, for adult males, between

[2] Phenotypic preference values of less than 0.1 are in fact set equal to zero, i.e., females with sufficiently low preference values mate randomly. This is to avoid a situation in which there is selection pressure for random mating but the mean value of p never quite reaches zero due to recurrent mutation. This would lead in turn to a small female preference being manifested, which might well be enough to push males towards advertising when they would not otherwise have done so.

their expressed advertisement and their underlying quality. It is clear that when the cost of advertising is greater than zero but less than about 5, male ornament size is modestly correlated with viability, and therefore *does* carry information. Signalling of viability does not occur across the full range of cost values, but it certainly appears to be occurring in one region. This contradicts Iwasa et al.'s claim that the honest advertisement of viability cannot be evolutionarily stable given the pure epistasis handicap.

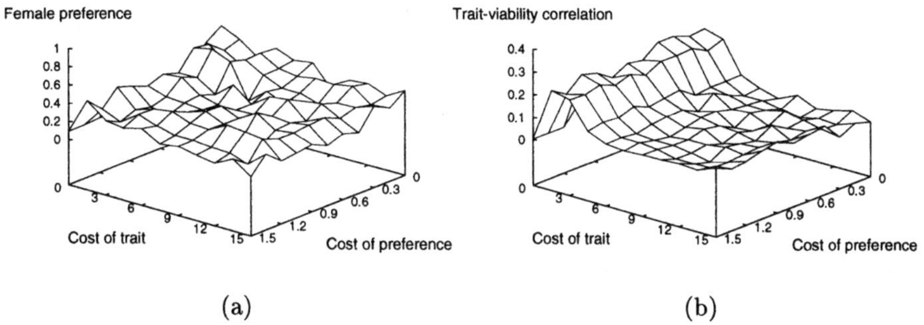

Fig. 1. Results for the pure epistasis handicap condition, by C_{adv} and C_{pref}. (a) Mean female preference values. (b) Correlation between the expressed male advertisement trait and underlying viability.

4.2 Conditional Handicap

Figure 2(a) shows the mean values for female preference: again, females were sometimes prepared to bear a cost in order to choose ornamented males. It is interesting to observe that, although preference falls off somewhat as its cost increases, high preference values are most likely to evolve when the cost of *advertising* is low. Figure 2(b) shows significant correlations between expressed male advertisement and underlying viability, especially when advertising costs are low. The results therefore support Iwasa et al.'s conclusion that an honest-signalling equilibrium could be stable under the terms of the conditional handicap.

4.3 Revealing Handicap

The results for the revealing handicap simulations were very similar to those for the conditional handicap; no additional graphs will be presented.

5 Discussion

Honest signalling of viability occurs in the pure epistasis handicap, despite the fact that Iwasa et al. [1] claimed it could not. In the conditional and revealing

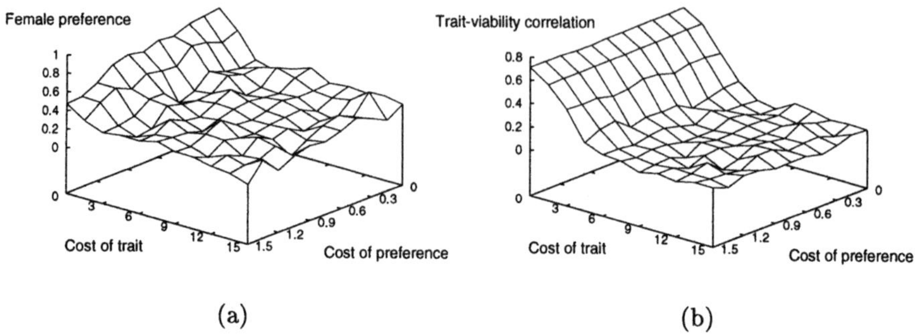

Fig. 2. Results for the conditional handicap condition, by C_{adv} and C_{pref}. (a) Mean female preference values. (b) Correlation between the expressed male advertisement trait and underlying viability.

handicaps, on the other hand, the results presented here are in accordance with Iwasa et al.'s prediction that honest signalling could be evolutionarily stable. Iwasa et al. intend their paper to clarify some of the controversies around the handicap principle. They argue that their findings explain why some earlier papers have concluded that the handicap principle can work while others have concluded that it cannot: different authors have tried to model different versions of the idea. Iwasa et al.'s intended clarification is an admirable goal; however, the results of the simulations presented here suggest that their conclusions must be taken with a grain of salt. Their assumption that genetic covariances could safely be treated as positive constants does not appear to have been a reasonable one.

The conditional and revealing handicaps deserve closer scrutiny. Consider Figure 2(b), which shows the correlation between expressed advertisement and viability in the conditional handicap case; results for the revealing handicap were similar. The graph shows that the highest correlations were achieved when advertising was cost-free. In the pure epistasis handicap, by contrast, we find that "talk is cheap" in these cases: Figure 1(b) shows that male advertisement was never an indicator of quality when the cost of advertising was zero. Why then, in the conditional and revealing handicap conditions, can females trust the advertisement levels of males who, in theory, can choose any advertisement level they like because there is no cost involved? The answer is that the males *cannot* choose any advertisement level that they might like. The stipulation that the expression of the ornament trait is condition-dependent (i.e., modified by viability) builds in an informational link between advertisement and viability in a rather uninteresting way. It seems disingenuous of Iwasa et al. to hold up the existence of costly female preference and honest advertisements as a deep result when the way in which the male trait is expressed itself enforces honesty. Thus we find that the genetic correlation between the advertisement trait and viability remains low in the conditional handicap case, but the correlation between the actual expressed advertisements and the underlying viability of adult males is very much higher: the condition-dependent expression of the ornament means that females automatically get useful information about viability. In addition, it is a little odd to claim that "handicap" signalling is occurring when the cost-free signals are the most reliable. Considered closely, Iwasa et al.'s claims for the conditional and revealing handicaps amount to little more than the uncontroversial observation that females will attend to unfakeable information about male quality.

Finally, it should be noted that all of the simulation results depend on Grafen's proviso that the unit costs of advertisement (and in this case preference as well) should be lower for higher-quality individuals. One's faith in the simulation results must depend on one's faith in Grafen's proviso as a real-world condition.

Acknowledgements

Thanks to Seth Bullock, Andrew Pomiankowski and Steven Siller for comments.

References

1. Iwasa, Y., Pomiankowski, A., Nee, S.: The evolution of costly mate preferences II. The "handicap" principle. Evolution **45** (1991) 1431–1442
2. Fisher, R.A.: The Genetical Theory of Natural Selection. OUP (1930)
3. Andersson, M.: Sexual Selection. Princeton Univ. Press (1994)
4. Zahavi, A.: Mate selection—a selection for a handicap. JTB **53** (1975) 205–214
5. Maynard Smith, J.: Sexual selection and the handicap principle. JTB **57** (1976) 239–242
6. Kirkpatrick, M.: The handicap mechanism of sexual selection does not function. Am. Naturalist **127** (1986) 222–240
7. Grafen, A.: Biological signals as handicaps. JTB **144** (1990) 517–546
8. Hamilton, W.D., Zuk, M.: Heritable true fitness and bright birds: A role for parasites? Science **218** (1982) 384–387
9. Hurd, P.L.: Communication in discrete action-response games. JTB **174** (1995) 217–222
10. Bullock, S.: A continuous evolutionary simulation model of the attainability of honest signalling equilibria. In Adami, C., et al., eds.: Artificial Life VI, MIT Press (1998) 339–348
11. Wiley, R.H.: Territoriality and non-random mating in sage grouse (*Centrocercus urophasianus*). Anim. Behav. Monographs **6** (1973) 85–169
12. Bradbury, J.W., Gibson, R.M., Tsai, I.M.: Hotspots and the evolution of leks. Anim. Behav. **34** (1986) 1694–1709
13. Maynard Smith, J.: The Evolution of Sex. CUP (1978)
14. Partridge, L.: Mate choice increases a component of offspring fitness in fruit flies. Nature **283** (1980) 290–291
15. Pomiankowski, A., Møller, A.P.: A resolution of the lek paradox. Proc. Roy. Soc. Lond. B **260** (1995) 21–29
16. Werner, G.M.: Why the peacock's tail is so short: Limits to sexual selection. In Langton, C.G., Shimohara, K., eds.: Artificial Life V, MIT Press (1996) 85–91

Modeling the Evolution of Communication: From Stimulus Associations to Grounded Symbolic Associations

Angelo Cangelosi

Centre for Neural and Adaptive Systems
University of Plymouth
Plymouth PL4 8AA (UK)
angelo@soc.plym.ac.uk

Abstract. This paper describes a model for the evolution of communication systems using simple syntactic rules, such as word combinations. It also focuses on the distinction between simple word-object associations and symbolic relationships. The simulation method combines the use of neural networks and genetic algorithms. The behavioral task is influenced by Savage-Rumbaugh & Rumbaugh's (1978) ape language experiments. The results show that languages that use combination of words (e.g. "verb-object" rule) can emerge by auto-organization and cultural transmission. Neural networks are tested to see if evolved languages are based on symbol acquisition. The implications of this model for Deacon's (1997) hypothesis on the role of symbolic acquisition for the origin of language are discussed.

1. Symbol acquisition in the evolution of communication

The synthetic approach of Artificial Life has recently been applied to studying the evolution of communication and language (Steels, 1997a). Some models have been used for the simulation of the emergence of simple lexicons in populations of simulated organisms (e.g. Cangelosi & Parisi, 1998; Steels, 1997b) or in small communities of robots (Steel & Vogt, 1997). In these studies organisms evolve shared lexicons for describing entities and relations of the environment. Other models have focused on the evolution of syntax (e.g. Batali, 1994; Kirby, in press). Simulated organisms evolve different syntactic languages starting from a given set of syntactic structures and constraints, and devices for syntax acquisition.

The first type of models, that focus on lexicon emergence, do not make any explicit reference to the role of syntax in language origin. Their aim is to model the early stages of the evolution of (animal) communication. Indeed, in animal communication systems, no syntactic structures have been observed. For example, no animal communication systems have been found that share one of the main properties of human languages, i.e. the combination of words to express different and new meanings. These models of lexicon evolution study communication systems based on simple signal-object associations. Organisms learn and evolve simple stimulus associations between objects in the environment and signals.

In the second type of models the evolution of syntax is simulated. These models, that for example show how syntax can emerge without natural selection (Kirkby, in press), do not explain the possible role of syntactic languages for organisms' adaptation and survival. Moreover, the associations that organisms learn are self-referential symbol-symbol relationships. These models are subject to the symbol grounding problem (Harnad, 1990) since they lack an intrinsic link between their symbols and the entities and relations existing in the organisms' environment. Internal symbols need some form of sensorimotor grounding. Due to the symbol grounding problem, the role of these models for understanding the evolution of cognition is reduced. The Artificial Life methodology used in this model, instead, allows to overcome the symbol grounding problem. Simulated organisms will use symbols whose semantic referents are constituted by categorical representations in the neural network's hidden layer. These semantic representations are activated by the actual presence of their referents in the organism's world.

Recently, Terrence Deacon (1997) proposed an explanation for the fact that animal communication and human language differ. A variety of animal communication systems have been studied (Hauser, 1996), however, there is no apparent continuity between animal communication systems and complex human languages. That is, no "simple languages", using some elementary forms of word combinations or syntax, have been found in the animal kingdom. The existence of simple languages could explain the gap between animal and human communication. Deacon (1997; 1996) believes that this is due to the symbol acquisition problem. In fact the main difference between animals and humans relies on symbolic references. There is a significant difference between the referencing system of simple object-word associations and that of symbolic associations. In animals, simple associations between world entities and words can be explained by mere mechanisms of rote learning and conditional learning. An animal acquires genetically, or learns, that a word's sound is always associated with a specific object. Instead, symbolic associations have double references, one between the word (symbol) and the object, and the second between the symbol itself and other symbols. A language-speaking human knows that a word refers to an object and also that the same word has logical (syntactic) relation with other words. Due to the possible combinatorial interrelationships between words, there can be an exponential growth of reference with each new added word.

The difference between these types of associations, and their relation to the models of language origin, is graphically represented in Figure 1. Figure 1a represents a communication system based on simple associations between objects and words. It refers to the models of the acquisition of lexicons. Figure 1b represents the models of the origin of syntax. It only shows word-word associations, but this system is not based on real symbolic associations as a link is missing between words and objects. Words are self-referencing and they lack a grounding in the external world. Figure 1c shows a system based on grounded symbolic associations. The arrows represent references between words, and references between words and objects. This third type of association system can be simulated through Artificial Life methodology, upon which the present model is based.

It should be noted that the relationship between words and objects, that constitute the grounding of symbols to entities of the real world, is not a direct link between mental symbols and real objects. Instead, it is a link between mental entities (the symbols or words) and other mental entities (such as concepts) that constitute the semantic reference. These categorical representations, that Deacon (1996) calls "indexical", are useful to "sort out" the extensive perceptual variability of objects in the real worlds. The ability of humans and animals to create categories, e.g. through categorical perception, constitutes the "groundwork" of cognition (Harnad, 1987). From this it is possible to build more complex cognitive skills, such as language.

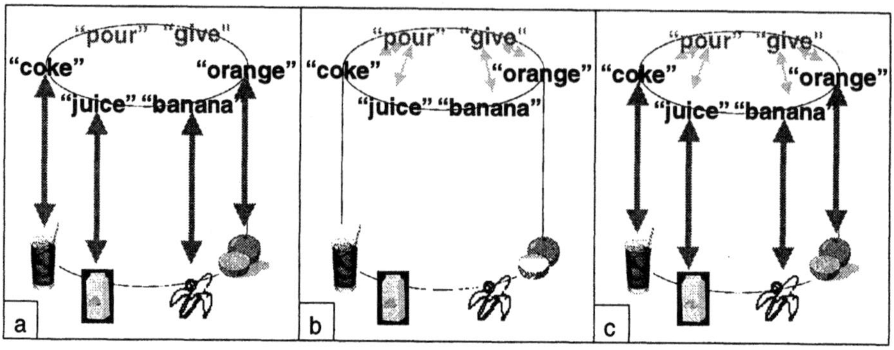

Fig. 1. Associations between objects (pictures) and symbols (words) in language origin models. (a) Simple stimulus associations between objects and words in the models of the origin of lexicons. (b) Self-referential associations between words that lack sensory-motor grounding in the models of the origin of syntax. (c) Grounded symbolic associations. Words have links with objects and logical relationships between themselves. Objects and words were chosen from Savage-Rumbaugh & Rumbaugh's (1978) experiments on ape language.

Deacon's hypothesis on the role of symbolic learning in the evolution of human language is supported by ape language studies and by neuropsychological and neurophysiolgical evidence (Deacon, 1997). For example, experiments on language acquisition in chimpanzees have been used to support the idea that animals tend to learn language using simple word-object associations. However, apes can be taught real symbolic associations under special experimental conditions (Savage-Rumbaugh & Rumbaugh, 1978). Moreover, in these language-speaking animals the spontaneous use of the grammatical rule "verb-object" has also been observed (Greenfield & Savage-Rumbaugh, 1990).

This paper aims to test a model of the origin of communication and language that deals with the evolution of different types of associations. The model should be able to study how different word/object relationships can evolve and also to define the mechanisms that explain the passage from communication based on simple stimulus association to languages based on grounded symbolic references. For this reason languages based on two-word combinations will be evolved. The behavioral task is influenced by ape language studies (Greenfield & Savage-Rumbaugh, 1990).

2. Method

The simulation method combines the use of artificial neural networks and genetic algorithms. It uses the methodology and theoretical framework of Ecological Neural Networks (ECONET: Parisi, Cecconi & Nolfi, 1990). Populations of organisms are evolved according to their behavioral performance in foraging tasks. Organisms' behavior is controlled by neural networks.

In the present simulation, the environment setting for the foraging task consists of a 2D grid of 100x100 cells. About 1200 cells are occupied by randomly placed foods (mushrooms). The foods are grouped into two main functional categories: edible mushrooms (`E`), i.e. foods that need to be collected to increase organisms' fitness, and toadstools (`T`), i.e. mushrooms that must be avoided. The first category of edible mushrooms is then split into three functional subcategories: white (`e1`), yellow (`e2`), and gray (`e3`). These are called functional categories because they require organisms to perform a different task when approached (e.g. white mushrooms `e1` should be picked and cut, whilst other colored mushrooms require different actions). The fitness formula adds one point for each `e1/e2/e3` mushroom that an organism approaches and properly treats according to its color. When a toadstool is collected, the fitness is decreased by one point. The toadstool category does not have any functional subcategory. Even though toadstools are perceptually classifiable into three categories (`t1`, `t2`, `t3`), these are not functional categories because the fitness formula removes one point for each toadstool that the organism reaches, regardless of their appearance.

The organization of the foraging task stimuli into a hierarchy of functional categories was derived from the experimental setting of ape language studies. For example in Savage-Rumbaugh & Rumbaugh (1978) chimpanzees had to learn to use different lexigrams (graphic symbols in a keypad) to name solid foods (e.g. banana, orange) and drinks (coke, milk). Since they receive food from a vending-machine, they also need to learn a lexigram for the verb associated to solid foods ("give") and that for the liquid drinks ("pour"). These stimuli constitute a hierarchy of two high-level functional categories (verbs) followed by four low-level categories (two foods and two drinks). In our model organisms will have to learn a name for each of the three edible subcategories (e.g. "white" for `e1`, "yellow" for `e2`, and "gray" for `e3`), plus a common verb for the whole edible category, e.g. "approach". All toadstools will require the use of a common verb, such as "avoid". The three toadstool subcategories do not require a specific name to be identified, but organisms will be allowed to name them.

The neural networks controlling the organisms' behavior have a 3-layer feedforward architecture (see figure 2). The input layer has 29 units, organized into three groups of sensory units. In the first group there are 3 units, one for each of the 40-degree neighboring visual field. The unit corresponding to the visual field in which the closest mushroom is perceived will be activated. Its activation value id the distance of this mushroom (range 0-1). The second group of input units has 18 nodes that encode some (visual) features of mushrooms. In fact, each mushroom has a set of 18 binary features. The mushrooms of each subcategory share a common binary pattern. For example `e1` mushrooms share the prototype `111***************` and `e2` have `***111************`. An asterix (`*`) represents random bits. The third group of input nodes are localist language units. Each unit is activated whenever the corresponding word is used. The hidden layer has 5 units. The output layer has two groups of unit. The first 3 units encode the actions. Two binary bits control the movement (move one cell forward, turn left, turn right, stand still) and one unit for discriminating between `e1`, `e2`, and `e3` (activation $a<.2$, $.2\geq a>.8$, $a\geq .8$ respectively). The second group has 8 linguist units. These units are organized into two clusters of winner-takes-all units. Only two words at a time will be active (one per cluster). One cluster has 2 units, the second 6 units. The hidden and output units use the sigmoid activation function.

Evolution is organized into two sequential stages. The first stage takes 300 generations and organisms do not communicate at all. They only use the mushroom position and feature information to evolve the proper foraging action. The population consists of 80 individuals. Organisms live in the same environment for 1000 actions (20 epochs of 50 moves each). The 20 organisms with the highest fitness level are selected and each reproduce making 5 offspring. The new organisms' genotypes, i.e. the set of connection weights encoded as real numbers, are then mutated by slightly modifying 10% of the weights.

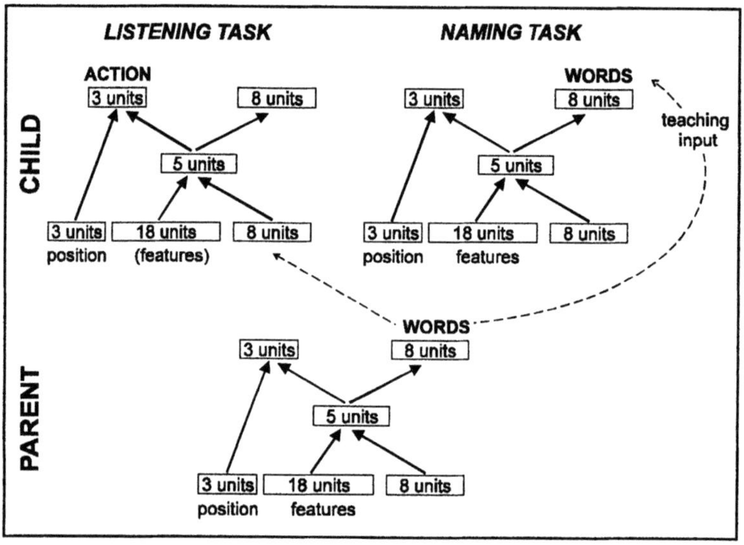

Fig. 2. Neural network architecture and the interaction between the child organism and its parent. The parent uses words to describe the closest mushroom. In the Listening Task the child uses these words to decide which action to take. In the following Naming Task the child uses the parent's words as teaching input for error backpropagation.

In the second stage of evolution, starting at generation 301, communication between organisms is allowed. The 20 parent organisms are kept together with the 80 new siblings. The parent organisms work only as speakers and language teachers. They cannot eat mushrooms and do not replicate. The need for two-stage simulation, in which only the foraging behavior is evolved at first, and subsequently communication is introduced, was suggested by a previous study (Parisi, Denaro & Cangelosi, 1998) in which the simultaneous evolution of foraging and language proved difficult. During each time interval, child organisms perform two actions (figure 2). The first is a Listening Task. To discriminate the type of mushroom children use the words suggested by their parents as input. In fact most of the time children rely solely on the parents' linguistic input because they only perceive the mushroom's features 10% of the time. After the network activation cycle of the Listening Task, children perform a Naming Task. They use the mushroom's 18-bit features to name the food type. An error backpropagation algorithm is then applied. The error is computed using the parent's words as teaching input, so that children learn the same linguistic description given by their parents. Some noise is added to the error between the child's linguistic output and the parents' teaching input. This is to allow variability in the process of cultural transmission of language (Parisi *et al.*, 1998). The same backpropagation algorithm is used for an Imitation task, where the organism's neural network learns the auto-association of the input-output linguistic stimulus. Organisms live for 2000 time steps, a longer lifetime than in the first stage as backpropagation learning requires more stimulus presentations. When the select organisms reproduce, their new offspring inherit the parent's connection weights before backpropagation learning occurred. No Lamarkian inheritance of learned weights is allowed. Darwinian selection will continue evolving the ability to approach/avoid mushrooms. The language learning between parents and offspring permits cultural transmission between consecutive generations.

During the second stage of evolution the interaction between parents and children can result in the emergence and auto-organization of a shared language. As they are only allowed to perceive the mushroom's features 10% of the time, it should facilitate the evolution of a good language that discriminates at least the functional categories **T**, **e1**, **e2**, **e3**. Fitness depends on the correct identification of these four categories.

3. Results

The first stage of evolution, which does not permit communication, was repeated 10 times using different random populations. Nine of these replications resulted in an optimal classification behavior. Organisms evolved the ability to approach edible mushrooms **E** and avoid all toadstools **T**. Moreover, according to the type of edible mushrooms **e1**, **e2**, **e3**, they produced the correct activation in the third node of the output action units. In the sole population where the evolved behavior was poor, organisms were unable to discriminate between **e2** and **e3**. The average fitness for the 9 successful populations is shown in Figure 3 (first 300 generations). At generation 300 the best individual of each population on average collects 90 edible mushrooms (i.e. 4.5 mushrooms per each of the 20 epochs), and avoid all toadstools.

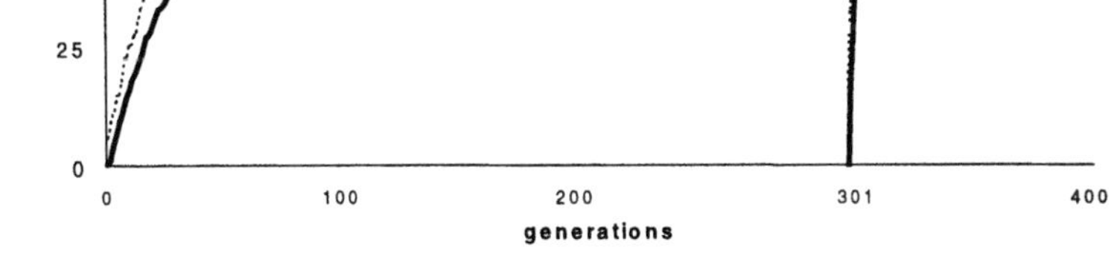

Fig. 3. Fitness of the best individuals and of the groups of 20 selected parents. Simulation without communication (generations 1-300) and during the evolution of communication (generations 301-400). The values of generations 1-300 are averaged over 9 successful simulations. The fitness of generations 301-400 is averaged over 11 populations.

For the second stage of evolution only the nine successful populations were used. The only simulation with unsuccessful fitness growth was not used because language evolution requires a preliminary ability to discriminate behavioral categories (Parisi *et al*, 1998). This stage took 100 generations. For each population, two random starting conditions were executed. In total, 18 replications were performed.

The results of the distribution of evolved languages are shown in Table 1. In 11 of the 18 runs, populations evolved good languages, i.e. the use of at least four words/word-combinations to distinguish the four behavioral categories **T**, **e1**, **e2**, **e3**. These languages emerged through a process of auto-organization of the lexicon, due to the interaction between organisms and the process of cultural transmission. The average fitness for the 11 successful populations is shown in figure 3 (generations 301-400). In the remaining 7 populations the emerged language was poor. That is, some mushroom types were incorrectly labeled due to the lack of a specific symbol, or symbol combination. Therefore the fitness is very low since these mushrooms were incorrectly described and collected.

	Single word	Word combination	Verb-object	TOTAL
Good languages	1 (9%)	3 (27%)	7 (64%)	11
Imperfect languages	1 (14%)	2 (29%)	4 (57%)	7

Table 1. Distribution of language types in the simulations for the evolution of communication.

In the previous section we explained that the linguistic output units are organized into two winner-takes-all clusters. The first cluster is made up of 6 linguistic units (words), and the second has 2 units. The cluster-based structure does not imply that a combination of two words is always necessary to describe a mushroom. In fact the optimal behavior requires the production of only four actions, and therefore four words from the first cluster are enough to name these categories. This is what happened for one of the good language populations. Here organisms used four words of the first cluster to name the four categories **T**, **e1**, **e2**, **e3**. The two words of the second clusters were not systematically associated to any mushroom.

When both clusters are used, there are several possibilities of combining words. However, we are interested in identifying word-combination rules that resemble known syntactical structures. In particular we want to establish if a verb-object rule has emerged. Considering the populations where good communication evolved, ten (91%) evolved languages that use combinations of symbols. Among these, three populations (27%) use various combinations of two words, and seven (64%) use verb-object rules. The way we can identify the verb-object rule is because in the two-word cluster each linguistic unit is systematically associated only to one of the high-order categories **T** and **E**. One "verb" symbol is always used for all toadstools ("avoid") and the other for all edible mushrooms ("approach"). The units in the 6-word cluster are used for distinguishing single "objects" (mushroom types) with which the two verbs systematically couple.

4. Discussion

The aim of this research was to develop a model of the evolution of communication systems based on simple syntactic rules, such as word combination. Moreover we were interested in establishing whether the evolved word-object relationships were based on symbolic learning or mere object-word associations. The resultant description shows that languages that use combination of words (e.g. verb-object rule) can emerge by auto-organization and cultural transmission. During the first stage organisms forage using the 18-bit feature information to discriminate mushrooms. Throughout the second stage, the foraging strongly depends on the evolution of a useful language, as features are rarely available. This condition has resulted in the rapid emergence of a shared language. In fact, within 30 generations the organisms' average fitness is the same as in generation 300, when the mushroom feature information was available all of the time. Moreover, the final fitness at generation 400 is higher (99 for the best organism) than that at generation 300 (91). This could be due to the fact that for neural networks it is easier to process discrete information, such as localist linguistic input, rather then processing the 18-bit feature information.

The percentage of evolved good language is 61%, as 11 out of 18 populations evolved useful languages. The remaining 7 populations (39%) evolved imperfect languages. However the discriminative quality of these languages was relatively good. In the majority of them only one of the four functional categories is incorrectly labeled. Two of the edible mushrooms categories are named by the same word/word-combination. Note that these imperfect languages also tend to use word combinations, and in particular most of them evolve the use of a verb-object rule.

After having shown that it is possible to evolve by auto-organization communication systems based on the combinatory rule "verb-object", we want to analyze the kind of referencing systems that organisms use when they associate words with objects. We are interested in establishing if the evolved languages are based on the use of grounded symbols, i.e. words that have a direct association with objects and that have logical relationships between them. We used a symbol acquisition test consisting of the training of organisms with a perfect combinatory language using the verb-object rule. The test was structured into three stages. In the first stage, organisms learn to name each of the four categories **e1**, **e2**, **t1**, **t2**. The teaching input is not provided by the parent organisms, but directly from the researcher. At this stage verbs are not used, and no names are taught for the two categories **e3** and **t3**. In the second stage, organisms learn to associate the two verbs "approach" and "avoid" with the categories **e1/e2** and **t1/t2** respectively. It is now expected that organisms learn the logical relationship between the names of the two edible mushrooms **e1** and **e2** and the verb "approach". The same symbolic association between the verb "avoid" and the names of toadstools **t1** and **t2** should be learned. In the final stage the learning of the names of categories **e3** and **t3** is finally introduced. The association the two verbs with these new names is not taught. In fact it is expected that after the training the organisms that learned real symbolic relationships between verbs and names will be able to generalize the verb-object relationship to the new mushroom names. If the verb "approach" is not associated with **e3** it means that organisms did not learn any symbolic association between the names of **e1** and **e2** and the verb "approach". They simply learned two independent object-word associations, one between **e1** and its name, and another between the same **e1** and the verb "approach".

This type of symbol acquisition test has been used in ape language studies. In the experiment where chimpanzees learned to associate "pour" with the name of the solid foods banana and orange (Savage-Rumbaugh & Rumbaugh, 1978), animals were then tested with new names of foods. Only those animals that made the correct generalization were considered to have learned symbolic associations.

The symbol acquisition test was repeated with 10 different populations. After the three learning stages, seven populations produced the correct associations **e3**-"approach" and **t3**-"avoid". The success criterion was the production of the correct verb for more than 75% of **e3** and **t3** mushroom types (N=8). In three populations the learning of the names for **e3** and **t3** did not produce the activation of the proper verb. It means that these organisms did not learn any symbolic association. In the seven successful populations, instead, the language is based on logical relationships between the mushrooms' names and the two verbs. The relationships between words and real objects, and between verbs and objects' name, allow neural networks to generalize the association of new names with the correct verb category.

These results show that neural networks can learn simple languages that use symbolic associations. These symbols are grounded in the environment because of the ecological simulation framework that allows a direct link between words and the objects with which organisms interact. However, the network's simple feedforward architecture allows the use of other non-symbolic strategies for language learning. In fact, during some simulations organisms appear to learn languages that do not use symbolic relationships. More complex and biologically-inspired neural network architectures could steer the learning towards symbolic acquisition, rather than simple stimulus associations. As neurophysiological data suggests (Deacon, 1997) the cortico-cortical connections in the human brain could help explain why humans can easily learn symbolic associations, while animals tend to learn conditional stimulus associations (except in controlled experiments as shown in ape language studies). Deacon's analysis of neuropsychological experiments on patients with prefrontal cortex lesions suggests that this area, and its connection with other cortical regions, could play a major role on symbol acquisition. The absence or underdevelopment of the prefrontal cortex in animals would subsequently explain the lack of symbol-based languages in animal communication systems. Our simple neural networks are not meant to represent any real neural systems. However, it is possible to design more articulate neural architectures that are inspired to specific connection patterns observed in the brain. Research is ongoing into understanding which particular neural network architectures allow symbol acquisition in language learning.

The model described in this paper allowed the simulation of the evolution of languages using simple syntactic rules and symbol acquisition. To create a selective pressure for the auto-organization of useful languages the experimental condition made the foraging behavior highly dependent on the parents' linguistic input (as mushroom features are only available 10% of the time). Moreover, the communication between parents and children, and the cultural transmission of language in the Naming Task, allowed the auto-organization of a population-shared language. The potential of this model for the study of the evolution of animal communication and human language is high. In future studies both the experimental setting, defining the availability of input mushroom features and of the linguistic input, and the model parameters controlling the pattern of communication and language learning between organisms, can be systematically changed to test specific hypothesis of the origin of language. For example, this model allowed us to focus on the important distinction between communication systems based on simple object-signal associations and languages based on symbolic relationships. The integration of the model with ongoing studies of the role of neural network architectures for the learning of symbolic representations will help to test Deacon's (1997) hypothesis. It will evaluate the role of symbol reference in language origin and on the co-evolution of brain structures, i.e. the prefrontal cortex, and language and symbolic acquisition.

References

Batali J. (1994). Innate biases and critical periods: Combining evolution and learning in the acquisition of syntax. In R. Brooks & P. Maes (eds), *Artificial Life IV*, Cambridge, MA: MIT Press, 160-171.

Cangelosi A. & Parisi D. (1998). The emergence of a "language" in an evolving population of neural networks. *Connection Science*, **10**(2): 83-97.

Deacon T.W. (1996). Prefrontal cortex and symbol learning: Why a brain capable of language evolved only once. In B.M. Velichkovsky & D.M. Rumbaugh (eds), *Communicating meaning: The evolution and development of language*, Mahwah NJ: LEA Publishers

Deacon T.W. (1997). *The symbolic species: The coevolution of language and human brain*, London: Penguin.

Greenfield P.M. & Savage-Rumbaugh S. (1990). Grammatical combination in *Pan paniscus*: Process of learning and invention in the evolution and development of language. In S.T. Parker & K.R. Gibson (eds), *Language and intelligence in monkeys and apes*, Cambridge University Press, 540-579.

Harnad S. (Ed.) (1987). *Categorical Perception: The groundwork of cognition.* New York: Cambridge University Press.

Harnad S. (1990). The Symbol Grounding Problem. *Physica D* 42: 335-346

Hauser M.D. (1996). *The evolution of communication.* Cambridge, MA: MIT Press.

Kirby S. (in press). Syntax without Natural Selection: How compositionality emerges from vocabulary in a population of learners. In Chris Knight (ed.) *Approaches to the Evolution of Language.*

Kirby S. & Hurford J. (1997). Learning, culture and evolution in the origin of linguistic constraints. In P. Husband & I. Harvey (eds). *Proceedings of the Fourth European Conference on Artificial Life*, London: MIT Press

Parisi D., Cecconi F. & Nolfi S. (1990). ECONETS: Neural networks that learn in an environment. *Network*, **1**: 149-168.

Parisi D., Denaro D., & Cangelosi A. (1998). The evolution of language as an aid to categorization. *Technical Report*, CNR-GRAL, Institute of Psychology. Poster presented at the *2nd International Conference on the Evolution of Language*, London.

Savage-Rumbaugh S. & Rumbaugh D.M. (1978). Symbolization, language, and Chimpanzees: A theoretical reevaluation on Initial language acquisition processes in four Young *Pan troglodytes*. *Brain and Language*, **6**: 265-300.

Steels, L. (1997a) The synthetic modeling of language origins. *Evolution of Communication*, **1**(1).

Steels L. (1997b). Self-organising vocabularies. In C.G. Langton & K. Shimohara (eds), *Proceedings of the ALIFE V conference*, Cambridge MA: MIT Press, 179-184.

Steels L. & Vogt P. (1997). Grounding adaptive language games in robotic agents. In P. Husband & I. Harvey (eds). *Proceedings of the Fourth European Conference on Artificial Life*, London: MIT Press

Acknowledgement. This research was partially supported by a Grant of the Nuffield Foundation (UK) for Newly Appointed Lecturers (SCI-180-97-116).

Emergence of Speech Sounds in Changing Populations

Bart de Boer, Paul Vogt

AI-Lab
Vrije Universiteit Brussel
Pleinlaan 2
1050 Brussel
{bartb,paul}@arti.vub.ac.be

Abstract: This paper shows that realistic and coherent vowel systems can emerge from scratch in a population of agents that imitate each other under human-like constraints of production and perception. The simulation is extended so that populations can change; old agents can be removed, and new agents can be added. In these circumstances vowel systems can also emerge and be preserved. It is shown that sometimes an age structure in the population can improve preservation of the vowel systems.

1 Introduction

The study of human languages is not only concerned with describing the huge variety of phenomena that are encountered in human languages, but also with finding explanations for similar phenomena that recur in languages that are neither related historically nor geographically. Such phenomena are called language *universals,* or, because there are always exceptions, universals tendencies. This paper is concerned with explaining the universal tendencies of human vowel systems. Although the human vocal tract is capable of producing a wide variety of different vowels (at least 45 different basic vowel qualities [10]) one finds that all human languages use only a very small subset of these. Furthermore, these subsets are not chosen randomly, but they exhibit remarkable regularities. When one looks at the vowel systems of widely different languages, such as, for example the 451 languages in the UPSID (UCLA Phonological Segment Inventory Database [12, 13]) one finds that the maximum number of vowel qualities in any language is 15 in Norwegian [10] while the minimum number is 3 (although there are languages that are reported to have only two vowels [10]). One also finds that certain vowels, for example [i], [a] and [u] occur very frequently (in 87%, 87% and 82% of the languages in UPSID, respectively) while others, such as [y], [œ] and [ø] occur only rarely (in 5%, 2% and 3% of the languages, respectively). But one also finds that the structure of vowel systems tends to be regular. For example, if a language contains a front, unrounded vowel of a certain height, for example [ɛ] (which occurs in 41% of the languages in UPSID) one also tends to find the rounded back vowel of the corresponding height, in this case [ɔ], which occurs in 36% of the languages, but in 73% of the languages that also contain an [ɛ]. Many such regularities can be found and typologies of vowel systems have been based on them [2, 16].

A possible explanation of these phenomena is that they occur because humans have an innate, cognitive disposition towards using certain speech sounds and combinations of speech sounds. Innate distinctive features, rules and markedness constraints are proposed. Distinctive features are properties (often binary) of speech sounds that are used to distinguish them from each other. Rules determine in what way abstract distinctive features can be combined into phonemes (the basic abstract speech sounds that can distinguish meanings of words), and how the phonemes are transformed into an actual speech signal. Some features are more marked than others, meaning that they are less likely to occur. There are a number of problems with this theoretical framework, but the most important of these is that it does not actually explain anything at all. The features, rules and markedness constraints are inferred from observations of human sound systems. It would thus be circular to use them to explain the very phenomena from which they have been inferred.

An independent explanation of the structure of vowel systems can be derived from functional criteria, such as acoustic distinctiveness and articulatory ease. Lindblom and Lilencrants [11] have shown that optimization of acoustic distance between the members of a vowel repertoire results in predictions of the most frequently occurring vowel systems in human languages. Subsequent improvements on their model (see e.g. Schwartz *et al.* [15]) have confirmed their results. There are two problems, however. The first problem is that the number of vowels in the system has to be fixed beforehand. The second problem is that the model depends on explicit optimization. It shows that human vowel systems are in general optimized for acoustic distinctiveness, but it does not give an explanation of how the optimization takes place in human languages. It is clear that human language users do not optimize the sound system of their language explicitly. In fact, when learning a language, children imitate their parents as accurately as possible, more accurately in fact than strictly necessary for successful communication. This can be observed from the fact that children do not only learn the language of their parents, but also their dialect. Although people speaking different dialects can often understand each other perfectly, they are nevertheless aware of subtle distinctions in pronunciation. The optimization model is incomplete. It explains *why* vowel systems are the way they are—they are optimized—but not *how* they have become optimized.

The work described in this paper is based on the theory that self-organization in a population of language users drives sound systems towards optimality. It is based on the theories of Luc Steels of language as a complex dynamic system [17, 18, 19] and is also related to other work on the origins and the evolution of language [7, 9]. In the theories of Luc Steels language is considered as much to be a phenomenon of a population as it is knowledge of individuals. This approach does not consider language in terms of abstract ideal knowledge of an individual, nor does it deal with idealized speaker-hearer interactions. Rather, it stresses the fact that language is an open, distributed system, where no member of the population has complete knowledge of the language, let alone control, where interactions can be messy and incomplete, where new words, meanings and constructions can enter the language, and where individual speakers of the language can enter and leave the population. Coherence in the language is maintained through self-organization. Change, which is viewed as an inherent property of language, is driven by speech errors, by contradicting drives to minimize articulatory effort and to maximize communicative success and by conscious creation of new words and expressions. As language is seen as a complex dynamic system, the only way to explore it is with computer simulations.

In the work described in this paper a population of agents that can produce, perceive and learn speech sounds in a human-like way is modeled. The agent's goal in life is to imitate the other agents in the population as well as possible. The work presented here is not the first work that tries to explain the universal tendencies of human vowel systems through the interactions of agents in a population. First Glotin [2] and later Berrah [1] have both built computer simulations of populations of agents that imitate each other's speech sounds. However, in their simulations, the number of vowels in each agent is fixed beforehand and the main process that shapes the vowel systems is still explicit optimization. The interactions between the agents only serve to make the vowel systems in the population coherent, not optimized.

In previous papers, [3, 4] it has been shown that the proposed model really results in coherent and realistic vowel systems. Meanwhile the research has progressed to larger populations and more realistic results (compare, for example the figures in [3] with the figures in this paper.) Many (extremely boring) experiments have been done to test the sensitivity of the system to parameter changes and small changes of algorithm, but the self-organisation has been shown to be remarkably robust. The interested reader is referred to [5]. The main focus of this paper will be on investigating the results of population dynamics. Human languages are spoken in open populations. Old speakers of the language can die and children have to learn the language. With the computer simulations it can be investigated what happens in the case of a changing population. This is a test for the model: will the agents preserve vowel systems, just as happens in human language, and will it still be possible to have vowel systems emerge from scratch? Also socio-linguistic hypotheses that are hard to test in reality can be investigated with this model: what happens with different population replacement rates and is there an advantage in having speakers of different ages learning with different rates? These questions are treated in the section 3.

2 The Simulation

The simulation models the interactions of agents in a population. Each agent is equipped with a realistic articulatory synthesizer for producing vowels, a model of human perception of speech signals for calculating the distance between different signals and an associative memory in which they can store the associations between articulatory prototypes, acoustic prototypes, the number of times the prototypes have been used and the number of times they have been successfully used. Whenever an agent hears an acoustic signal, it calculates the distance between this signal and all its acoustic prototypes and considers the prototype that is closest to the signal as the one that it has recognized. Vowel signals are represented as the frequencies of the first four peaks (formants) of the acoustic spectrum of the vowel. Whenever an agent wants to produce a vowel, it takes the vowel's articulatory prototype and synthesizes its formants with the articulatory synthesizer. Some noise is added (by shifting the formant frequencies randomly) so that an articulatory prototype is never exactly realized the same way acoustically. The distance between a signal and the acoustic prototypes of vowels is calculated in a two-dimensional space with the logarithm of the frequency of the first formant as one dimension and the logarithm of the frequency of the *effective second formant* as the second dimension. The distance is calculated as a weighted Euclidean distance where the effective second formant frequency is weighted as 30% of the first formant frequency.

The effective second formant frequency is a non-linearly weighted sum of the second, third and fourth formant frequencies. It is based on the observation that due to the limited resolution of human perception of high frequencies, signals that have multiple peaks at high frequencies are perceived as identical to signals that have only one peak at high frequencies. The position of this single peak is a function of the position of the peaks in the original signal. In this paper it is calculated by a function that is based on Mantakas *et al.*'s [14] weighting function. It would have been better to calculate the distance in the original 4-dimensional acoustic space, but all available models of human vowel perception were based on the effective second formant.

The interactions between the agents have been called *imitation games*. For each imitation game, two agents are selected from the population at random. One of the agents will play the role of *initiator*, the other will play the role of *imitator*. Basically, the initiator selects an acoustic vowel prototype form its repertoire at random and synthesizes the acoustic signal for this vowel. The imitator listens to this signal and finds its acoustic prototype that is closest to it. It then synthesizes the articulatory prototype that is associated with this acoustic prototype. The initiator listens to this signal and finds *its* closest acoustic prototype. If this is associated with the articulatory prototype it originally produced, the imitation game is successful. If not, it is a failure. This information is communicated non-verbally to the imitator.

In reaction to the imitation game, both the initiator and the imitator update their repertoire of vowels. The use and success (if the game was successful) counts of the vowel prototypes that were produced are increased. Whenever the success/use ratio of a vowel drops below a certain value, (0.7 throughout the paper) and its use count is sufficiently high (5 throughout the paper) it is removed from an agent's repertoire. The imitator can make subsequent changes to its repertoire. If the imitation game was successful, the imitator shifts the articulatory prototype it used so that its acoustic realization matches the signal that was heard more closely. This is done in order to increase the coherence in the population. If the imitation game was a failure, and if the vowel that the imitator used was successful, the imitator adds a new vowel to its repertoire that closely matches the signal that was heard. The articulatory prototype of this vowel is determined by a hill-climbing heuristic: the agent makes a first guess, produces it, listens to itself, makes small articulatory adjustments and iteratively improves its first guess. The vowel is added because the other agent probably had two vowel prototypes at a location where this agent had only one, thus causing confusion. If the success/use ratio of the vowel that was used is low, however, it is moved closer to the perceived signal in an attempt to improve it. Finally, both the imitator and the initiator merge vowels that come too close together in acoustic or articulatory space.

When these interactions are iterated a large number of times in the population, realistic vowel systems emerge. The process of emergence is illustrated in figure 1. In this figure, the acoustic vowel prototypes of all the agents in a population of twenty agents are plotted in the acoustic space formed by the first formant (on the vertical axis) and the effective second formant (on the horizontal axis) in the logarithmic Bark frequency scale. Note that the scales on the axes increase from top to bottom and from right to left, respectively. This has been done in order to get the vowel prototypes in a configuration that corresponds to the way phoneticians usually plot vowels, with front vowels towards the left and high vowels towards the top of the graph. Note that due to

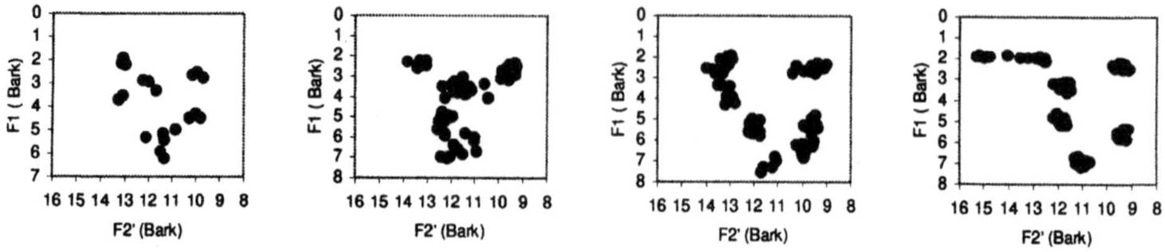

articulatory limitations, the acoustic space that can be reached by the agents is limited to a triangular region with the apex at the bottom of the graph. Whenever there is a cluster in the graph, this means that most agents in the population have an acoustic prototype near this position. The number of clusters in the graph indicates the number of vowel prototypes of each agent.

The frames in figure 1 show the vowel systems in the population after 50, 500, 2500 and 15 000 imitation games from left to right. It can be seen that initially, the vowel prototypes are spread randomly through the available acoustic space. After 500 imitation games, some structure becomes visible. Vowel prototypes cluster together and multiple clusters start to form. After 2500 imitation games, the clusters are dispersed more evenly throughout the available acoustic space and after 15 000 imitation games, the clusters have become compact and more or less evenly dispersed. When this stage has been reached, the agents' vowel systems remain relatively stable. Although they never stop changing completely, the possible changes are relatively small, consisting mostly of small shifts of the clusters and the occasional merging of existing clusters or emergence of a new cluster if space is available. The acoustic noise setting for this graph was 10%.

In order to get an idea of the realism of the emerging vowel systems, several measures can be defined, for example the energy of the vowel systems as defined by Liljencrants and Lindblom [11], the average success of imitation or the average number of vowels per agent. These will be used later in this paper. However, direct inspection and classification of the agents' emerging vowel systems is also a good way to get an idea of the realism of the emerging systems. An example of the classification of emerging vowel systems with five vowels is given in figure 2. This figure was formed by doing 25 000 imitation games in 100 populations of 20 agents. The acoustic noise was set to 15%. Of the emerging vowel systems, 49 consisted of five vowels (the number of vowels in the final vowel systems is not determined beforehand). From each of these populations, one random agent was selected. The vowel systems of these agents were classified according to the relative position of the vowels in the

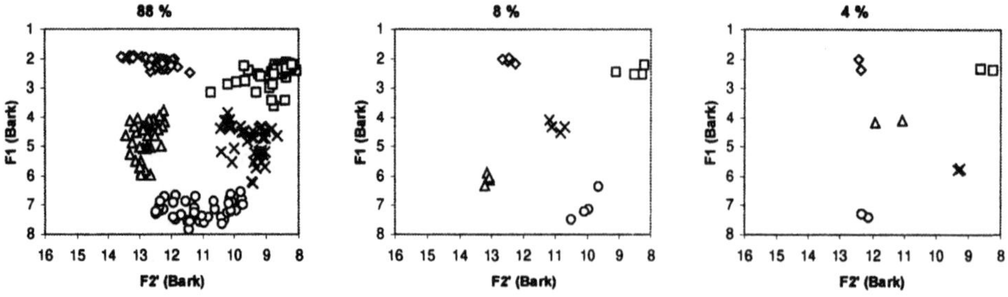

system in the same way as Crothers [2] has classified vowel systems of human languages. It can be seen that the different types occurred (from left to right) in 88%, 8% and 4% of the cases. When the results are compared with those of human languages one finds a very good match. Schwartz *et al.* [15] found that of the languages in UPSID, 89% were of the leftmost type, and the middle and rightmost type both occurred in 5% of the cases. Equally good matches were found for other numbers of vowels, although there were some problems for very low and very high numbers of vowels (see [5, chapter 6] for more details).

Apparently the model makes realistic and coherent vowel systems emerge in a population by no other process than local interactions between the agents under articulatory and acoustic constraints.

3 An Open Population

In the real world, populations of language users are not static. People can die, and children get born. Old speakers continually leave the population and new speakers enter it. This usually does not disrupt the language. Research on the social dynamics of language using traditional methodologies is extremely difficult, as the factors influencing language change are very hard to identify, let alone that they can be controlled. In the artificial setting of the imitation, game, carefully controlled experiments with changing populations can be performed. From this more insight can perhaps be obtained in the social dynamics of real language. A number of experiments are presented that give an idea of the possibilities of the application of computer simulations to the study of sound change. Other parts of language, most notably the emergence of lexicon and semantics, have already been studied by other researchers (see e.g. [20]).

The population is changed by removing and adding agents at random. Making this process random was deemed to be the most realistic (as in human populations birth and death are basically also random processes) and the least likely to introduce artifacts due to regularities in the replacement. In every imitation game in which it participates, an agent has a certain probability of being removed. Also, in every imitation game there is a small probability of an agent being added to the population. Of course, the newly added agent's vowel repertoire is still empty. It will have to learn the sound system of the population to which it is added. The probabilities of adding new agents and removing old ones will have to be equal so that the number of agents in the population remains relatively stable. However, depending on the value of the probabilities, the rate of replacement in the populations can be high or low.

A comparison between different rates of replacement is shown in figure 3. The leftmost frame of this figure shows the original vowel system that was used as the

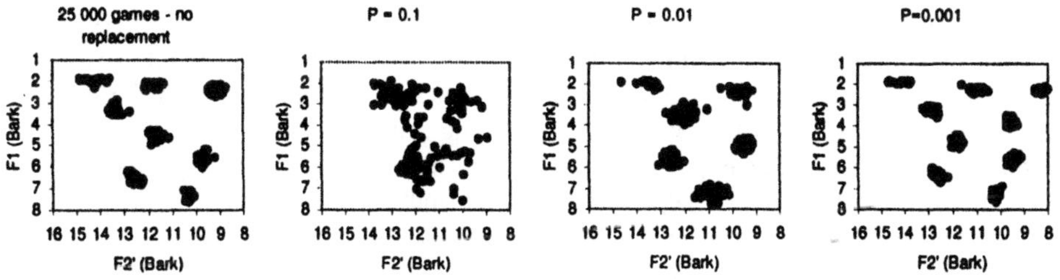

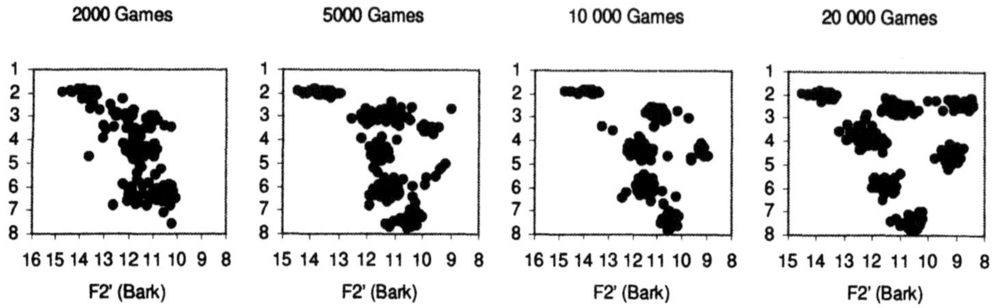

starting configuration for all other experiments presented in this section. This vowel system was obtained from running 25 000 imitation games in a population of 50 agents with an acoustic noise of 10%. The second graph from the left shows the vowel system of a population of agents in which the probability of adding or removing agents was 0.1. It shows the situation after 2500 imitation games. With these parameter settings, it was unlikely that any agent from the original population still remained in the final population. The third graph from the left shows a vowel system from a population with a probability of adding or removing agents of 0.01. As it is unfair to compare populations that have undergone unequal numbers of replacements, this population's vowel system is shown after 25 000 imitation games. The rightmost frame shows the vowel systems of a population with a replacement probability of 0.001 after 250 000 imitation games. From these graphs it is clear that the higher the replacement rate, the worse the agents are able to preserve the original vowel system. This has nothing to do with the absolute number of replacements. These are approximately equal for all graphs shown. Rather, the time in which agents stay in the population (and therefore the time they have to learn the vowel system) determines how accurately a vowel system is preserved. This does not mean that the systems with a high replacement probability do not become stable. After a while they settle in a state in which they contain fewer prototypes and the prototype clusters are more dispersed.

In fact, coherent and realistic vowel systems can even emerge from scratch in an open population. This is illustrated in figure 4. It shows a similar emergence of a vowel system as figure 1. The replacement probability was 0.01 and the acoustic noise was 10%. The emergence is slightly slower than in figure 1, but this is due to the fact that the population is larger—50 agents instead of 20. There are fewer vowel prototypes per agents and the prototype clusters are slightly bigger. Nevertheless, a coherent, successful and realistic vowel system emerges even though at the end of the run, no agent that was present at the beginning is present anymore.

In the experiments shown so far, there is no cost for articulatory effort. If an agent wants to imitate a signal it has heard, it can talk and listen to itself an unlimited number of times in order to perfect its vowel prototypes. In reality, such articulatory effort probably has a certain cost. Therefore an experiment was done in which the number of times an agent can talk to itself for improving a new sound is limited to 10. The interesting finding of this experiment was that it then becomes advantageous to introduce an age structure in the population, so that younger agents can change their articulatory vowel prototypes more quickly than older agents. The older agents thus provide a stable target to which the younger agents can move their vowels. The results are shown graphically in figure 5. The original vowel systems are shown in gray, the

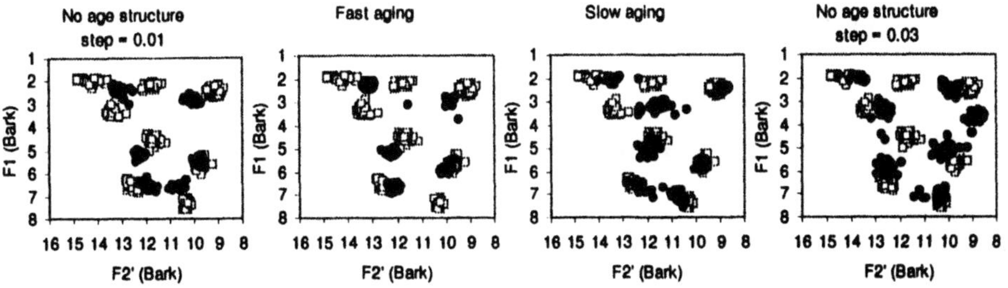

final vowel systems, obtained after 15 000 imitation games are shown in black. In the leftmost and rightmost frames, the populations do not have an age structure. The difference between the two figures is that the step size with which the articulatory vowel prototypes were improved was 0.01 in the leftmost and 0.03 in the rightmost frame (0.03 was the value that was used in all other experiments presented so far). In the two middle frames, young agents used a step size of 0.03 and old agents tended towards step size of 0.03. The agents' step sizes changed in the following way:

$$\varepsilon_t \leftarrow \varepsilon_{t-1} + \alpha(0.01 - \varepsilon_{t-1}) \quad (1)$$

Where ε_t is the step size at time t and α is a parameter that determines the speed with which the agents change their step sizes. The value of α was 0.1 for the left center frame and 0.01 for the right center frame. The difference in performance between the different systems in the figure are a bit hard to judge, therefore table 1 gives the measures of the system. The success is the average over the communicative success of the agents in the population. The energy of the vowel systems is calculated according to [11] as the sum over the reciprocal of the squared distance between all the vowels in an agent's vowel system. The size is the average number of vowels in the agents' vowel systems and the similarity is the communicative success that can be achieved between an agent that has the original vowel system and an agent that has the vowel system that emerges at the end of the simulation run.

Population:	$\varepsilon_0 = 0.01$ $\varepsilon_\infty = 0.01$ $\alpha = 0$	$\varepsilon_0 = 0.03$ $\varepsilon_\infty = 0.01$ $\alpha = 0.1$	$\varepsilon_0 = 0.03$ $\varepsilon_\infty = 0.01$ $\alpha = 0.01$	$\varepsilon_0 = 0.03$ $\varepsilon_\infty = 0.03$ $\alpha = 0$
Success:	0.8531 ± 0.040	0.7701 ± 0.035	0.7930 ± 0.041	0.8041 ± 0.041
Energy:	4.04 ± 0.59	5.55 ± 1.10	5.10 ± 0.95	3.83 ± 0.72
Size:	4.77 ± 0.40	5.66 ± 0.63	5.61 ± 0.58	5.14 ± 0.50
Similarity:	0.7347 ± 0.023	0.8231 ± 0.028	0.8235 ± 0.032	0.7884 ± 0.032

Table 1: Measures of the populations with age structure.

The statistics were obtained from 100 runs of each system. Shown are the averages for every measure and their standard deviations. Using the Kolmogorov-Smirnov test it can be shown that the vowel system sizes and the similarity measures in the populations with age structure are significantly higher (at the 1% level) than in the populations without age structure, indicating that the original vowel system is preserved better. The fact that the success is lower for the two systems with age structure is not important, as success always tends to be a little lower for larger systems.

These experiments show that the fact that old agents learn less well than young agents can sometimes be beneficial.

4 Conclusion and Future Work

The experiments presented here have shown that a) coherent and realistic vowel systems can emerge from scratch in a population of agents that imitate each other under constraints of production and perception and b) that the same simulation can be used to investigate the consequences of population dynamics on the emergence and the preservation of sound systems. The first result is of direct relevance for phonetics. It shows that the agents do not need innate predisposition towards certain structures in order for these structures to emerge. Traditionally it has been assumed that the fact that certain phonetic and phonological structures are found in unrelated human languages is proof for the theory that these structures are innate. This research shows that they can as well be the result of self-organization in a population. The second result is not directly relevant for linguistics, but it does show that the methodology can be used successfully for investigating phenomena as complex as dynamics of populations of language users. It incidentally also shows that the self-organizing emergence is extremely robust and happens even if the population is changing. Finally, the results of the simulations with age structure show that the fact that older language users learn less quickly than younger ones might not just be an unfortunate consequence of aging, but might help to preserve sound systems (and possibly other aspects of language) across the generations.

Of course, the simulation is still very simple. From the point of view of phonetics, one of the main things that needs to be investigated is whether it also works with more complex sounds. If one wants to investigate sound change seriously, one needs to take interactions between different speech sounds into account. Also, most of the more interesting universal tendencies of human sound systems have to do with consonants and combinations of sounds.

From the point of view of population dynamics, more complex experiments with spatial distribution of agents, or with populations with different sound systems that come into contact with each other can be conceived. Some of these experiments have already been done for other aspects of language (see e.g. [20]).

In any case, the use of computer simulations of populations of agents that learn and use realistic speech sounds has proven to yield interesting results. The experiments done thus far are only a beginning, but they show that the technique is promising and can help to shed new light on the complexities of human language.

5 Acknowledgements

This research was done at the artificial intelligence laboratory of the Vrije Universiteit Brussel. It is part of an ongoing research project into the origins of language and intelligence. Funding was provided by the GOA 2 project of the Vrije Universiteit Brussel. Part of the work was done at the Sony computer science laboratory in Paris, France. My thanks also go to an anonymous reviewer for providing useful suggestions for improving the paper.

References

1. Berrah, Ahmed Réda (1998) Évolution Artificielle d'une Société d'Agents de Parole: Un Modèle pour l'Émergence du Code Phonétique, Thèse de l'Institut National Polytechnique de Grenoble, Spécialité Sciences Cognitives.
2. Crothers, John (1978) Typology and Universals of Vowel systems, in: Joseph H. Greenberg, Charles A. Ferguson & Edith A. Moravcsik (eds.) Universals of Human Language, Volume 2 Phonology, Stanford: Stanford University Press pp. 93–152
3. de Boer, Bart (1997) Generating vowel systems in a population of agents. In Phil Husbands and Inman Harvey (eds.) *Fourth European conference on Artificial Life*, Cambidge (MS) MIT Press pp. 503–510
4. de Boer, Bart (1998) Emergence of Sound Systems through Self-Organisation. In Han La Poutre' and Jaap van den Herik (eds.) *NAIC '98 Proceedings*, Amsterdam pp. 37–46
5. de Boer, Bart (*in preparation*) *Self-organisation in vowel systems*, Vrije Universiteit Brussel Ph. D. Thesis (to be defended 4-6-1999)
6. Glotin, Hervé (1995) La Vie Artificielle d'une société de robots parlants: émergence et changement du code phonétique. DEA sciences cognitives-Institut National Polytechnique de Grenoble
7. Hurford, Jim (to appear) Social transmission favours linguistic generalization. In Chris Knight & Jim Hurford (eds.) The evolution of language, Cambridge: Cambridge University Press.
8. Jakobson, Roman & Morris Halle (1956) Fundamentals of Language, the Hague: Mouton & Co.
9. Kirby, Simon (to appear) Syntax without natural selection: how compositionality emerges from vocabulary in a population of learners. In Chris knight & Jim Hurford (eds.) The evolution of language. Cambridge: Cambridge University Press.
10. Ladefoged, Peter & Ian Maddieson (1996) The Sounds of the World's Languages, Oxford: Blackwell.
11. Liljencrants, L. & Björn Lindblom (1972) Numerical simulations of vowel quality systems: The role of perceptual contrast, Language 48 pp. 839–862.
12. Maddieson, Ian (1984) Patterns of sounds, Cambridge University Press.
13. Maddieson, Ian & Kristin Precoda (1990) Updating UPSID, in: UCLA Working Papers in Phonetics 74, pp. 104–111
14. Mantakas, M, J.L. Schwartz & P. Escudier (1986) Modèle de prédiction du 'deuxiéme formant effectif' F2'—application à l'étude de la labialité des voyelles avant du français. In: Proceesings of the 15th journées d'étude sur la parole. Société Française d'Acoustique, pp. 157–161.
15. Schwartz, Jean-Luc, Louis-Jean Boë, Nathalie Vallée & Christian Abry (1997), Major trends in vowel system inventories, Journal of Phonetics 25, pp. 233–253
16. Schwartz, Jean-Luc, Louis-Jean Boë, Nathalie Vallée & Christian Abry (1997), The Dispersion-Focalization Theory of vowel systems, Journal of Phonetics 25, pp. 255-286
17. Steels, Luc (1995) A Self-Organizing Spatial Vocabulary, Artificial Life 2(3), pp. 319–332.
18. Steels, Luc (1997b) The Synthetic Modelling of Language Origins, Evolution of Communication 1(1): pp. 1–34
19. Steels, Luc (1998b) Synthesising the origins of language and meaning using co-evolution, self-organisation and level formation, in: James R. Hurford, Michael Studdert-Kennedy & Chris Knight (eds.) Approaches to the Evolution of Language, Cambridge: Cambridge University Press pp. 384–404
20. Steels, Luc & Frédéric Kaplan (1998) Spontaneous Lexicon Change. In: *Proceedings of COLING-ACL 1998*, Montreal, pp.1243–1249.
21. Vanvik, A (1972) A phonetic-phonemic analysis of Standard Eastern Norwegian, in: Norwegian Journal of Linguistics 26, pp. 119–164

SPECIES: An Evolutionary Model for the Emergence of Phonetic Structures in an Artificial Society of Speech Agents

Ahmed-Reda Berrah[1] and Rafael Laboissière[2]

[1] LORIA, BP 239,
54506 Vandœuvre-lès-Nancy, France
Ahmed-Reda.Berrah@loria.fr
[2] Institut de la Communication Parlée – INPG
46, Av. Félix Viallet, 38031, Grenoble, France
rafael@icp.inpg.fr

Abstract. This paper addresses the emergence of a common phonetic code in a society of communicating speech agents using evolutionary techniques. Predictions for the large vowel systems of the world's languages using the Maximum Use of Available distinctive Features (MUAF) principle are discussed. Simulations of the use of supplementary phonetic features in large vowel systems are presented. These experimental results show how simple local rules of interaction between speaking agents are sufficient to explain some of the universal characteristics of the phonological structure of the world's languages.

1 Introduction

Whatever the language considered, speech production can be considered as a succession of closing and opening gestures of the vocal tract, i.e. syllables made up of consonants *C* and vowels *V* respectively. Maddieson and Precoda [9] gathered phonological descriptions of a representative sample of the world's languages. They obtained an inventory consisting of 920 possible phonemes including 654 consonants, 177 vowels and 89 diphthongs. One of the conclusions of this work is that a majority of languages use five vowels and, roughly, twenty consonants. Although the number of combinations C^{5}_{177} for V and C^{20}_{654} for C are very high, typologies carried out [8, 13] show that phonological systems do not have an arbitrary structure.

To explain these regularities, the generativist linguists estimate that the universal tendencies are axiomatically imposed. For instance, Chomsky [2] thinks that all humans are provided at birth with the capability to recognise a common set of distinctive features. During the first years of language acquisition, this set is filtered in order to keep only the distinctive features of the mother tongue.

We propose however a radically different point of view, which posits that sound system universals emerge from the basic characteristics of the speech production and perception mechanisms, as well as from the speaker/listener interaction (cf. [7]). According to this approach, favoured sound systems are those that allow the best accomplishment of the communication task taking into account the constraints of the speech production and perception systems.

Along these lines, several studies have addressed the universal tendencies of the world's languages, resulting in a series of prediction models [6, 11]. These models allow, on the one hand, the prediction of the structure of phonological systems, and on the other hand the general explanation of why certain sounds imposed themselves in the world's languages. However, neither the evolutionary processes are addressed, nor the aspects related to the interaction between speakers that are at the origin of these universal tendencies.

The central interest of this paper is the study of prediction models of sound systems, in particular those that determine the emergence of phonological inventories from principles of interaction between communicating agents. This approach is innovative and relates to other recent work such as of Steels [12], within the framework of Artificial Life. Our goal is twofold: first, to model the formation of phonetic inventories by investigating the basic mechanisms of the interaction between speakers and listeners and, second, to relate those mechanisms to the emergence of sound systems.

The "SPEech CommunIcating agEnt Society" (SPECIES) model of vowel perceptual exchanges in a society of speech agents is described. In the presented simulations, we explore the role of the principle of "Maximum Use of Available distinctive Features" (MUAF) in structuring phonetic inventories. This principle is invoked to explain why languages with crowded vowel inventories tend to use additional phonetic features.

2 Description of the Emergence Model

Speech communication can be seen as a sequence of perceptual objectives "negotiated" between a speaker and a listener. An important question immediately raises its head: how are the perceptual objectives negotiated? The answer to this question lies in a deep comprehension of the physiological and cognitive mechanisms involved in speaker-listener interactions. Of course, this is a overwhelming task. In the simulations presented here, we implemented a simplified version of local interactions between speech agents, who can communicate using vowel-like, static sounds.

The main aim is to make a "society" of speech agents that can adapt their productions through local interactions, called hereafter *transactions*. For each simulation, the number of distinct phonetic items (i.e. the size of the lexicon) is fixed in advance. This is in contrast with other similar recent studies, in which the number of items can vary during the simulation of a society of communicating agents [3].

Speech agents in our model are simple agents which can produce sustained vowels, represented in a perceptual space, which is composed of subspaces, each one representing a particular vocalic feature. Although some of these subspaces are unidimensional (adequately describing vowel features like duration or nasalisation), we will always consider one of the subspaces as a bidimensional formant space. Formants are the frequency values at which maxima of the acoustic spectrum occur. It is well known that vowels are well characterised by the first two formants (F_1 and F_2) alone [4], which justifies the choice made in the present study. In a former version of our model [5], the articulatory level was also included. We decided not to include it in the present work in order to thoroughly understand both the mechanisms and the implications of the perceptual level alone.

Each agent in the society has a lexicon composed of a fixed number of items initialised randomly. Pairs of agents are randomly chosen to communicate using an item of their lexica. We define an interaction between two agents as a transaction, in which one of the agents takes the place of the speaker and the other that of the listener. The speaker chooses randomly one of its items and produces it. The listener relates the features of the signal sent by the speaker to its own lexicon at the perceptual level. Finally, the listener adjusts its items in order to adapt its lexicon to this new sensed item.

The listener adaptation to a new sensed item constitutes the important aspect of the simulations. Indeed, lexicon convergence, the key to our problem, depends in a great part on this adaptation. We assume that a person who learns a new word (or sound) will try to first repeat this word (or sound). So, a speech agent, which perceives an unknown item, will try to repeat it. This is achieved by moving in the formant space of this agent the nearest item towards the perceived item (*attraction*), and moving the others away in order to avoid confusion risks, and to maximise the perception distinction (*repulsion*). The details are published in [1].

3 Experimental Validation

Despite the striking simplicity of our model, the simulations replicate well-known phenomena observed in phonetic inventories of the world's languages. We will present only the simulation results of predictions for large vowel systems. Simulation results of predictions for vowel systems with small numbers of items are published in [1].

In this set of simulations, we explore what Ohala called the principle of "Maximum Use of Available distinctive Features" (MUAF), which plays a role in the structuring of phonetic inventories [10]. The MUAF principle is invoked to explain why languages with crowded vowel inventories tend to use additional features, like nasalisation and duration, to preserve distinctiveness.

We propose a modified version of our emergent model to reproduce the MUAF principle in vowel systems. The goal is to incorporate the fact that once a feature is used, it is explored systematically before going to the next. The features that provide greater gain in perceptual distinctiveness are explored first. The repulsion of different items in the listener lexicon is done using only one of the two features at a time (namely that for which the gain in distinctiveness is greater).

In order to investigate this idea, an experiment with vowels was conducted. We proposed the addition of a perceptual feature, which makes the perceptual space three-dimensional (two for the formants and one for the additional feature). Fig. 1 shows the results. Right panels represent projections of the 3-D perceptual space (left panels) into the F_1–F_2 plan. Fig. 1(a.1) and Fig. 1(a.2) show a 5-vowel system that is /i,e,a,o,u/ the most widespread system in the UPSID database of the world's languages [8]. It is possible to see that the new feature is not systematically explored in Fig. 1(a.2), contrary to the case shown in Fig. 1(b) which represents a 9-vowel system. One could say that this last system is also a variant of /i,e,a,o,u/, but the interesting point is that, as shown in Fig. 1(b.1), vowels tend to use the additional feature and that is quite clear for /i/, /a/, /e/ and /u/. The same observation can be done with the 12-vowel system represented in Fig. 1(c) where an additional feature is recruited for 5 vowels.

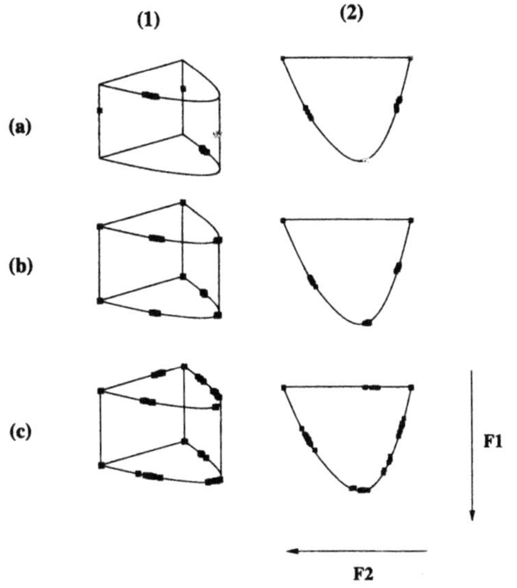

(3)

9 vowels		12 vowels	
type	#	type	#
9+0	17	7+5	8
5+4	4	6+6	4
6+3	2	12+0	2
7+2	2	8+4	1
8+1	1	9+3	1

Fig. 1. Final lexica for simulations in an extended perceptual space, composed of the formant space and an additional feature. (a) 5 vowels, (b) 9 vowels and (c) 12 vowels. The table in (3) summarizes the statistics for the vowel system types with 9 and 12 vowels present in UPSID. $n+m$ means a system with n basic vocalic timbers and m modified (nasalized, longer, etc) vowels. See [1] for details.

These simulation results fit with the data on the world's languages. Indeed, the 9-vowel system shown in Fig. 1(b) is observed in four languages, while 50% of the sixteen 12-vowel system indexed in UPSID use an additional feature for 5 vowels like the system shown in Fig. 1(c) [8].

4 Conclusions and Further Developments

The SPECIES model is a platform for exchanges of vowels within a community of speech agents. The attainment of a common vowel lexicon shared by all the agents is a consequence of the cooperative aspects inherent in the exchanges between the agents. Certainly, the account for the speech interaction in the model is still quite crude (this is a work in progress), but our simulations provide a framework for more realistic studies, which would not be possible with other approaches, like [11]. Simulation experiments related to the prediction of vowel systems were conducted. The predictions obtained are in accordance with the statistics of the world's languages. In spite of its predictive power, this model is not yet fully realistic. In the future, an articulatory model of speech production, as well as more complex linguistic units like syllabic consonant-vowel transitions may be incorporated into SPECIES. Finally, we think that the above results on vowels universals support the hypothesis that language is a phenomenon which is created actively by speakers and listeners, evolving under functional constraints, like exchanges and cooperations based on perceptual distinctiveness.

References

[1] A.-R. Berrah. *Évolution d'une société artificielle d'agents de parole : un modèle pour l'émergence des structures phonétiques.* PhD thesis, Institut National Polytechnique, Grenoble, France, 1998.

[2] N. Chomsky and M. Halle. *The Sound Pattern of English.* MIT Press, Ma., 1968.

[3] B. de Boer. Generating vowel systems in a population of agents. In J. Husbands and I. Harvey, editors, *Proceedings of the Fourth European Conference On Artificial Life*, pages 503–510. MIT Press, 1997.

[4] G. Fant. *Acoustic theory of speech production.* The Hague, Mouton, 1960.

[5] H. Glotin and R. Laboissière. La vie artificielle d'une société de robots parlants : Émergence et changements du code phonétique. In *Du Collectif au Social -Actes de Journées de Rochebrune*, pages 113–125, Paris, 1996.

[6] J. Liljencrants and B. Lindblom. Numerical simulation of vowel quality systems: the role of perceptual contrast. *Language*, 48:839–862, 1972.

[7] B. Lindblom. Systemic constraints and adaptive change in the formation of sound structure. In J. Hurford, editor, *Evolution of Human Language.* Edinburgh Univ. Press., Edinburgh, 1997.

[8] I. Maddieson. *Patterns of Sounds.* Cambridge University Press, UK, 1986.

[9] I. Maddieson and K. Precoda. Updating upsid. *UCLA Working Papers in Phonetics*, 74:104–111, 1990.

[10] J.J. Ohala. Moderator's introduction to symposium on phonetic universals in phonological systems and their explanation. In *IXth International Congress of Phonetic Science*, volume 3, pages 181–185, Copenhague, 1980.

[11] J-L. Schwartz, L-J. Boë, N. Vallée, and C. Abry. The dispersion-focalization theory of vowel systems. *Journal of Phonetics*, 25:255–286, 1997.

[12] L. Steels. Synthesising the origins of language and meaning using co-evolution and self-organisation. In J. Hurford, editor, *Evolution of Human Language.* Edinburgh Univ. Press., Edinburgh, 1997.

[13] N. Vallée. *Systèmes vocaliques: de la typologie aux prédictions.* PhD thesis, Université Stendhal, Grenoble, France, 1994.

Collective Learning and Semiotic Dynamics

Luc Steels[1,2] and Frederic Kaplan[2]

[1]VUB AI Laboratory, Pleinlaan 2, 1050 Brussels,
[2]SONY Computer Science Laboratory, 6 Rue Amyot, Paris
steels@arti.vub.ac.be, kaplan@csl.sony.fr

Abstract. We report on a case study in the emergence of a lexicon in a group of autonomous distributed agents situated and grounded in an open environment. Because the agents are autonomous, grounded, and situated, the possible words and possible meanings are not fixed but continuously change as the agents autonomously evolve their communication system and adapt it to novel situations. The case study shows that a complex semiotic dynamics unfolds and that generalisations present in the language are due to processes outside the agent.

1 Introduction

In recent years it has become clear that the complex adaptive systems approach pioneered by Artificial Life research can fruitfully be applied to the study of the origins and evolution of language [9], particularly to the emergence of shared sound systems [3], the self-organisation of lexicons [7], [11], grounded word meaning [12], and the origins of grammar [4], [1], [5]. In all this research, the same mechanisms for the generation and maintenance of complexity are being used as exploited in other Artificial Life research, and a similar complex dynamics can be seen to emerge.

This paper focuses on *grounded* lexicons as they emerge from the local interactions of a group of distributed autonomous robotic agents, grounded in a real world physical environment through visual sensing. Consequently, and in contrast with other work so far, the meanings of words are no longer given as symbols supplied by a designer, nor is it assumed that hearers have perfect knowledge of what meaning is intended by a speaker. Rather the agents must autonomously infer the possible meanings of unknown words from their visual interpretation of the situations they encounter. Agents never get immediate feedback on whether they had the right meaning, only whether the communication was successful. Grounding introduces two additional difficulties: The ontology must be sufficiently robust to handle variations in the data. Anomalies in perception come from differences in the perception (for example one agent segmenting into different objects than the other). This causes the rise of additional ambiguities which need to be damped by the semiotic dynamics. We show that under these conditions, which are more realistic with respect to human natural language acquisition and use, a very complex semiotic dynamics is generated which nevertheless manages to self-organise into a successful communication system.

2 The Talking Heads Experiment

The robotic setup used for the experiments in this paper consists of a set of 'Talking Heads' connected through the Internet. Each Talking Head features a Sony EVI-D31 camera with controllable pan/tilt motors for horizontal and vertical movement (figure 1), a computer for cognitive processing (perception, categorisation, lexicon lookup, etc.), a screen on which the internal states of the agent currently loaded in the body are shown, a TV-monitor showing the scene as seen through the camera, and devices for audio in- and output. Agents can load themselves in a physical Talking Head and teleport themselves to another Head by travelling through the Internet. By design, an agent can only interact with another one when it is physically instantiated in a body located in a shared physical environment. The experimental infrastructure also features a commentator which reports and comments on dialogs, displays measures of the ontologies and languages of the agents and game statistics, such as average communicative success, lexical coherence, average ontology and lexicon size, etc. For the experiments reported in this paper, the shared environment consists of a magnetic white board on which various shapes are pasted: colored triangles, circles, rectangles, etc.

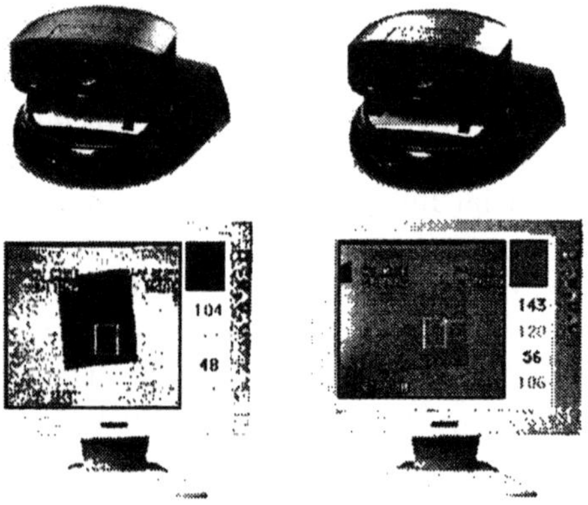

Fig. 1. Two Talking Head cameras and associated monitors showing what each camera perceives.

The guessing game The interaction between the agents consists of a language game, called the guessing game. The guessing game is played between two visually grounded agents. One agent plays the role of *speaker* and the other one then plays the role of *hearer*. Agents take turns playing games so all of them develop the capacity to be speaker or hearer. Agents are capable of segmenting the image perceived through the camera into objects and of collecting various sensory data about each object, such as the color (decomposed in RGB channels), average gray-scale or position. The set of objects and their data constitute a *context*. The speaker chooses one object from this context, further called the *topic*. The other objects form the *background*. The speaker then gives a verbal hint to the hearer.

The verbal hint is an utterance that identifies the topic with respect to the objects in the background. For example, if the context contains [1] a red square, [2] a blue triangle, and [3] a green circle, then the speaker may say something like "the red one" to communicate that [1] is the topic. If the context contains also a red triangle, he has to be more precise and say something like "the red square". Of course, the Talking Heads do not say "the red square" but use their own language and concepts which are never going to be the same as those used in English. For example, they may say "malewina" to mean [UPPER EXTREME-LEFT LOW-REDNESS]. The verbal hint is in this experiment assumed to be transmitted completely accurately.

Based on the verbal hint, the hearer tries to guess what topic the speaker has chosen, and he communicates his choice to the speaker by pointing to the object. A robot points by transmitting in which direction he is looking. The game succeeds if the topic guessed by the hearer is equal to the topic chosen by the speaker. The game fails if the guess was wrong or if the speaker or the hearer failed at some earlier point in the game. In case of a failure, the speaker gives an extra-verbal hint by pointing to the topic he had in mind, and both agents try to repair their internal structures to be more successful in future games.

Agents start with no prior designer-supplied ontology nor lexicon. A shared ontology and lexicon must emerge from scratch in a self-organised process. The agents therefore not only play the game but also expand or adapt their ontology or lexicon to be more successful in future games.

The Conceptualisation Module Meanings are categories that distinguish the topic from the other objects in the context. The categories are organised in discrimination trees where each node contains a discriminator able to filter the set of objects into a subset that satisfies a category and another one that satisfies its opposition. For example, there might be a discriminator based on the horizontal position (HPOS) of the center of an object (scaled between 0.0 and 1.0) sorting the objects in the context in a bin for the category 'left' when HPOS < 0.5, (further labeled as [HPOS-0.0,0.5]) and one for 'right' when HPOS > 0.5 (labeled as [HPOS-0.5,1.0]). Further subcategories are created by restricting the region of each category. For example, the category 'very left' (or [HPOS-0.0,0.25]) applies when an object's HPOS value is in the region [0.0,0.25].

A distinctive category set is found by filtering the objects in the context from the top in each discrimination tree until there is a bin which only contains the topic. This means that only the topic falls within the category associated with that bin, and so this category uniquely filters out the topic from all the other objects in the scene. Often more than one solution is possible, but all solutions are passed on to the lexicon module.

The discrimination trees of each agent are formed using a growth and pruning dynamics coupled to the environment, which creates an ecology of distinctions. Discrimination trees grow randomly by the addition of new categorisers splitting the region of existing categories. Categorisers compete in each guessing game. The use and success of a categoriser is monitored and categorisers that are irrelevant for the environments encountered by the agent are pruned. More details about the discrimination game can be found in [10].

Verbalisation module The lexicon of each agent consists of a two-way association between forms (which are individual words) and meanings (which are single categories). Each association has a score. Words are random combinations of syllables. When a speaker needs to verbalise a category, he looks up all possible words associated with that category, orders them and picks the one with the best score for transmission to the hearer. When a hearer needs to interpret a word, he looks up all possible meanings, tests which meanings are applicable in the present context, i.e. which ones yield a possible single referent, and uses the remaining meaning with the highest score as the winner. The topic guessed by the hearer is the referent of this meaning.

Based on feedback on the outcome of the guessing game, the speaker and the hearer update the scores. When the game has succeeded, they increase the score of the winning association and decrease the competitors, thus implementing lateral inhibition. When the game has failed, they each decrease the score of the association they used. Occasionally new associations are stored. A speaker creates a new word when he does not have a word yet for a meaning he wants to express. A hearer may encounter a new word he has never heard before and then store a new association between this word and the best guess of the possible meaning. This guess is based on first guessing the topic using the extra-verbal hint provided by the speaker, and on performing categorisation using his own discrimination trees as developed thus far. These lexicon bootstrapping mechanisms have been explained and validated extensively in earlier papers [11].

The conceptualisation module proposes several solutions to the verbalisation module which prefers those that have already been lexicalised. Agents monitor success of categories in the total game and use this to target growth and pruning. The language therefore strongly influences the ontologies agents retain. The two modules are structurally coupled and thus get coordinated without a central coordinator.

Semiotic Dynamics We propose the notion of a *semiotic landscape* (which we also call RMF-landscape) to analyse grounded semiotic dynamics. The semiotic landscape is a graph, in which the nodes in the landscape are formed by referents (objects), meanings (categories) and forms (words), and there are links if the items associated with two nodes indeed co-occur (figure 2). The relations are labeled RM for referent to meaning, MR for meaning to referent, RF for referent to form, FR for form to referent, and FM for form to meaning and MF for meaning to form. The RMF-landscape in figure 2 (taken from the experiment to be discussed later) contains an example where the same object $O3$ is designated by two meanings [G-0.25,0.5] and [G-0.375,0.5]. The first meaning, which is more general, is expressed by two words "xu" and "fepi" and the second meaning by the word "pasi". Usually we see much more complex situations and complexity further increases when the same meaning is also used to denote other referents (which is obviously very common and indeed desirable). We track the changes

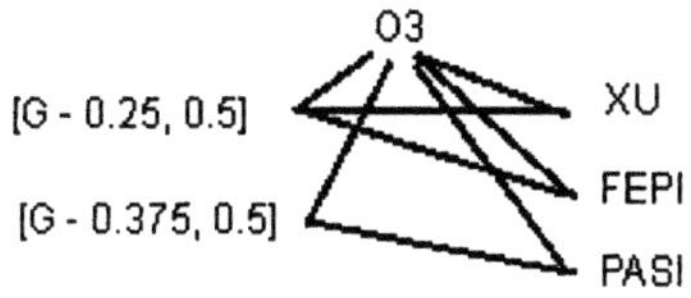

Fig. 2. A semiotic landscape represents the co-occurrences between referents, meanings and forms.

in the semiotic landscape by recording the actual verbal behavior of the agents while they engage in language interactions, more specifically, by collecting data on the co-occurrence of items such as the forms used with a certain referent or the meanings used with a certain form. Frequency of co-occurrence is represented in competition diagrams, such as the RF-diagram in figure 6, which plots the evolution of the frequency of the observed referent-form co-occurrences for a given referent in a series of games. Similar diagrams can be made for the FR, FM, MF, RM and MR relations.

3 Case Study

For real world environments, the set of possible referents is infinite, so the semiotic landscape is infinite. For purposes of analysis, we therefore need to restrict the possible environments and thus the possible referents artificially and then study the semiotic dynamics very precisely. For the present paper, we analyse a test run involving 20 agents and 8 objects, which means 4 x $C_8^4 = 280$ possible situations. The run starts with 4 objects (and hence 4 possible situations) and after every 5000 games a new object was introduced. During the final 15000 games, no new objects were introduced. The overall evolution of the dynamics is shown in figure 3. We see that success in communication or discrimination mounts quickly, drops after a new object is introduced, but increases again as the agents develop new concepts and words. We also see that it takes less and less time to absorb new objects, indicating that the language is not situation-specific. And that the system evolves towards maximum communicative and discriminatory success.

The main goal of this paper is to show how an analysis in terms of the semiotic dynamics aids to understand the evolution of the system. We examine one word, "fepi" which is general in the sense that it is clearly used for more than one object ($O3$ and $O5$) and in many different contexts (figure 4 left)

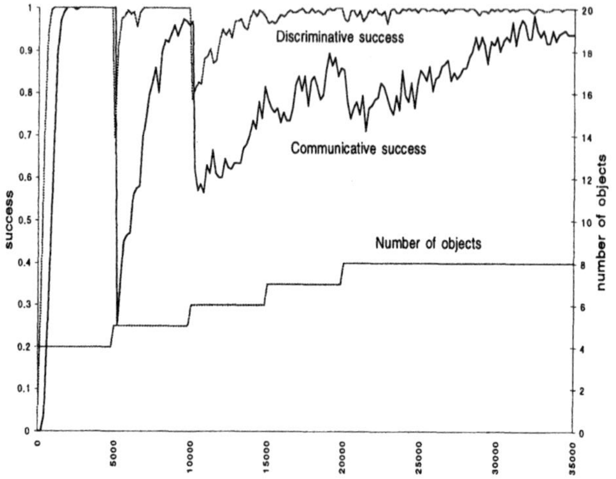

Fig. 3. The graph shows the communicative and discriminatory success for a series of 35000 language games.

Closer examination of the meanings (figure 4 right) reveals that "fepi" 'means' a particular shade of green, categorising a green-intensity [0.25,0.5] on the G channel of the RGB data. The category is labeled [G-0.25,0.5]. When examining the forms used to express this category (figure 5) we see that "fepi" has indeed emerged as dominant for this meaning, but that earlier on the word "xu" was dominant, which raises the question how "fepi" managed to overtake "xu" in the long run.

When inspecting in more detail the game traces, we see that "fepi" is created in game 328 by agent-3, playing the role of speaker, in order to refer to object $O3$ using the meaning [G-0.25,0.5]. Agent-19 acquires at this point the same meaning for [G-0.25,0.5] as hearer. In one sense, we could say that agent-19 learns this meaning of "fepi" from agent-3 but that is not entirely accurate. Agent-19 constructed a possible meaning for "fepi" and this happened to be the same as the one used by agent-3, but this is accidental. This is a first important observation. Agents only indirectly learn the language from others. They construct a language which is compatible with the language used by others in the situation encountered - and in turn influence by their own use the language of other speakers. Compatibility between individual language systems occurs through the positive feedback between language use and communicative success.

As figure 6 (left) shows, "fepi" is not immediately successful. Instead "xu" wins the initial competition against "fepi" and several other words for designating $O3$. Typically, multiple ways to designate the same object develop but this synonymy is damped due to the lateral inhibition between the different forms competing for the same meaning.

"xu" has also the meaning [G-0.25,0.5] although there are some other meanings associated with "xu" as well, which are also distinctive for $O3$. Because

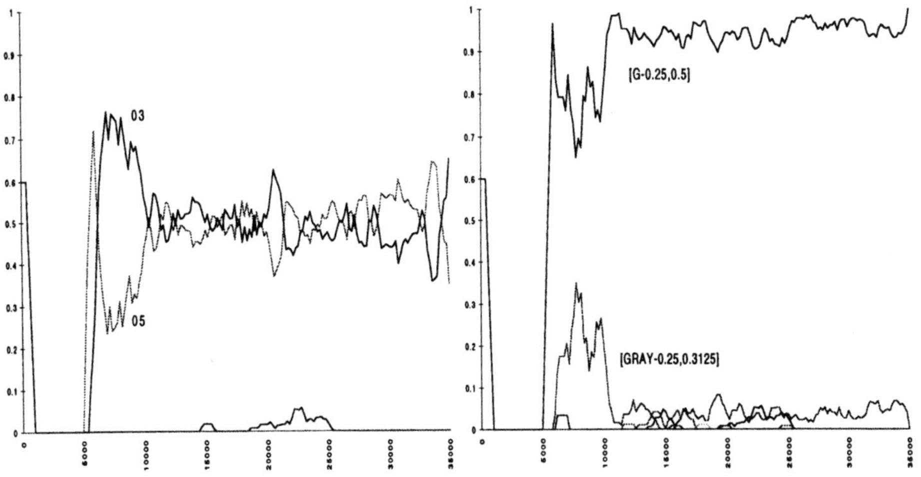

Fig. 4. Left: FR diagram showing the objects referred to by the word "fepi". After 10000 games "fepi" is consistently used for $O3$ and $O5$. Right: FM diagram showing the meanings of "fepi".

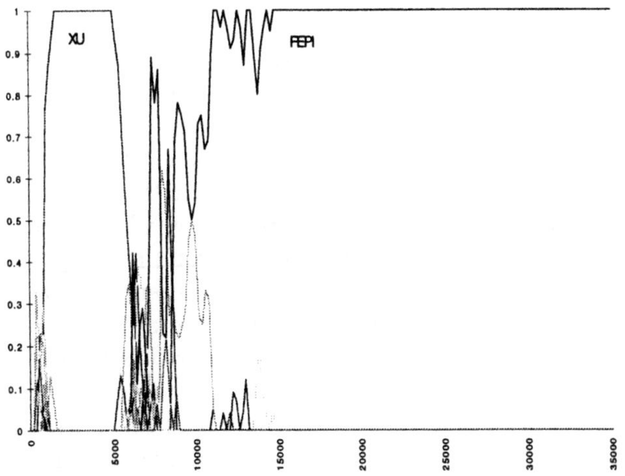

Fig. 5. MF diagram showing the different words circulating in the population for expressing the concept [G-0.25,0.5].

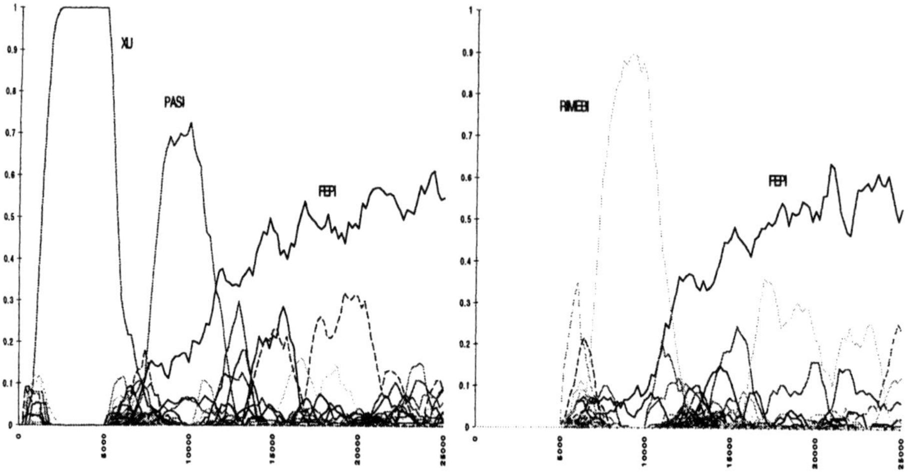

Fig. 6. Left: RF diagram showing the different words being used for identifying $O3$. Right: RF diagram showing the different words used for $O5$.

agents only get feedback about reference and not about meaning, words are polysemous until disambiguated by situations in which its different shades of meaning are incompatible.

In game 5000, the arrival of a new object $O5$ destabilises the association between "xu" and [G-0.25,0.5]. Closer examination reveals that $O5$'s green value is a bit lighter (in the range [0.25,0375]) than that of $O3$ (which is in the range [0.375,0.5]), so that a more refined distinction is necessary if both objects occur in the same context. As seen from the RF-diagram in figure 6 left, "xu" is no longer used for $O3$. Instead the word "pasi" comes to dominate. "pasi" has indeed the more specific meaning [G-0.375,0.5]. At the same time we see from the RF-diagram for $O5$ (figure 6 right) that the word "rimebi" dominates for designating $O5$. As expected "rimebi" has the second more specific meaning [G-0.25,0.375]. The more general word "xu" is still useful in contexts where the refined distinction is not necessary and so we would expect that "xu" continues to exist. However this is not the case. "xu" looses out completely and its role is taken over by "fepi". Why is this so? "xu" looses its strength because (1) the main meaning of "xu" is often not distinctive enough and therefore a game fails, and (2) other meanings competing for "xu" gain (as seen from figure 7) pulling down the original green-based meaning due to lateral inhibition. The weakening of "xu" opens the way to "fepi" which still carries the more general meaning of green and does not have competitors. We see from the RM diagram (figure 7 right) how first a general meaning coupled to "xu" is used for $O3$, then a more specific meaning coupled to "pasi" (after game 5000) and then again a more general meaning coupled to "fepi". The more general meaning increases because due to the arrival of several new objects, the more abstract green category becomes more useful again, even though the more specific meaning is still occasionally needed.

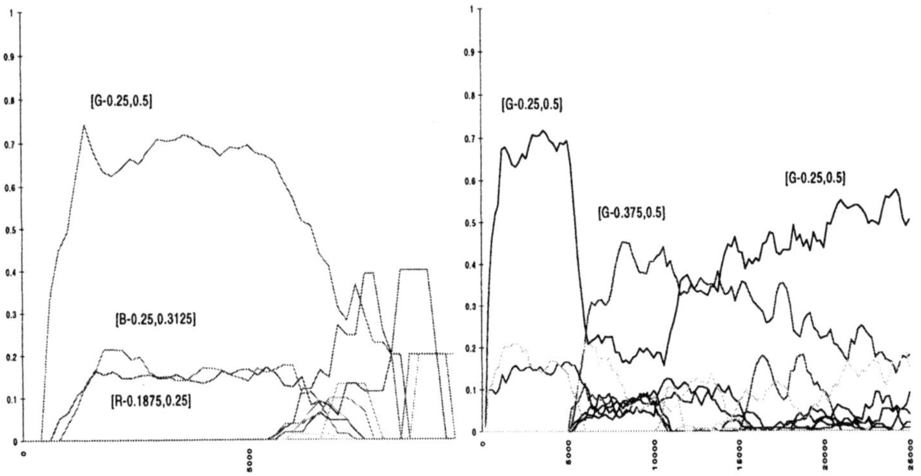

Fig. 7. Left: FM diagram showing the different meanings of XU. After game 5000, the meaning of "xu" becomes unclear and the word falls in disrespute. Right: RM diagram showing the meanings used to identify O3.

A similar picture is seen for the meanings used for $O5$ where "rimebi" and "fepi" are used depending on the degree of distinction required by the context.

4 Conclusion

The main goal of the paper was to analyse a case study of semiotic dynamics based on data recorded from a group of 20 distributed autonomous robotic agents playing language games about real world scenes through visual observations. We see that the word-meaning pairs active in a population show a complex picture. The agents have never exactly the same lexicon but each has their own ideolect. Moreover periods of heavy competition alternate with periods of relative stability. Stability occurs when one word temporarily manages to become dominant in a winner-take-all process. Semiotic dynamics therefore shows similarities to the dynamics exhibited by other types of complex adaptive systems such as punctuated equilibria in species evolution or evolution in adaptive cooperative games [6]. It illustrates the hypothesis that language is an evolving complex dynamical system which self-organises and gets transmitted in a cultural process.

5 Acknowledgement

This work was financed and carried out at the Sony Computer Science Laboratory in Paris. We are indebted to Toshi Doi and Mario Tokoro to be able to work in this superbly productive environment. The BABEL software tool under development by Angus McIntyre at SONY CSL-Paris has proved invaluable for efficiently exploring the issues discussed in this paper. We are also grateful to Angus McIntyre and Joris Van Looveren from the VUB AI Laboratory for contributions to the specific software underlying this experiment.

References

1. Batali, J. *Computational Simulations of the Emergence of Grammar.* In: Hurford, J. et.al. [ed.] Approaches to the Evolution of Language: Edinburgh Univ. Press. Edinburgh, 1998.
2. Belpaeme, T., L. Steels and J. Van Looveren (1998) The construction and acquisition of visual categories. In: Birk, A. (ed.) *Workshop on learning robots*, Brighton. Springer-Verlag, Berlin.
3. De Boer, B. Generating vowel systems in a population of agents. In Phil Husbands and Inman Harvey, eds. Fourth European Conference on Arficial Life, Brighton, 1997. MIT Press, Cambridge, MA
4. Hashimoto, T. et Ikegami, T. (1996) Emergence of net-grammar in communicating agents. BioSystems 38 (1996) 1-14.
5. Kirby, S. Language Evolution without Natural Selection. From vocabulary to syntax in a population of learners. 2nd Evolution of Language Conference, London, 1998.
6. Lindgren, K. and Nordhal, M. Cooperation and Community Structure in Artificial Ecosystems. In: Langton, C. Artificial Life: an overview. MIT Press, Cambridge, MA
7. Oliphant, M. The dilemma of Saussurean communication. Biosystems, 1996, 37 [1-2], pp 31-38.
8. Quine, W (1960). Word and Object. MIT Press, Cambridge, MA.
9. Steels, L. (1997) The synthetic modeling of language origins, Evolution of Communication Journal, 1(1): 1-34.
10. Steels, L. (1997) Constructing and Sharing Perceptual Distinctions. In: van Someren, M. and G. Widmer (eds.) (1997) Proceedings of the European Conference on Machine Learning. Springer-Verlag, Berlin.
11. Steels, L. and Kaplan, F. Stochasticity as a source of innovation in language games. In Adami, C., Belew, R., Kitano, H. and Taylor, C. [eds.] Proceedings of Artificial Life VI, Los Angeles, June 1998, MIT Press, p. 368-376.
12. Yanco, H. and L. Stein (1993) An Adaptive Communication Protocol for Cooperating Mobile Robots. In: Meyer, J-A, H.L. Roitblat, and S. Wilson (1993) *From Animals to Animats 2. Proceedings of the Second International Conference on Simulation of Adaptive Behavior.* The MIT Press, Cambridge Ma. p. 478-485.

Analyzing the Evolution of Communication from a Dynamical System's Perspective *

Edwin D. de Jong

Vrije Universiteit Brussel
Artificial Intelligence Lab
Pleinlaan 2, B-1050 Brussels, Belgium
edwin@arti.vub.ac.be
http://arti.vub.ac.be

Abstract. We study the evolution of communication where concepts are developed individually by agents and relations between concepts and forms (words, signals) are learned through interaction with other agents. By constructing concepts based on experience with the same environment, agents develop similar conceptual systems. Concepts represent situations in the environment. The system of associations between forms and meanings is viewed as a dynamical system. This paper presents first results with investigating the phase space of the system. The analysis contributed to understanding the interaction between association strengths of different agents and of different meanings.

Introduction

In studying the evolution of communication, an important question is that of how stable systems connecting concepts and forms (words, signals) may come about. In this paper, we will be investigating the case where these connections are learned by a group of agents in a shared environment. The concepts themselves are formed by the agents based on experience within the environment using *autonomous concept formation* [de Jong, 1999].

Although the fact that simple local adaptive behavior of agents causes a system of communication to emerge is interesting in itself, a deeper understanding of this process is valuable. Since the complete system of agents, concepts and forms is a complex adaptive system (see e.g. [Steels, 1996a], [Steels, 1997]) containing a large and varying number of changing elements, it is difficult to predict the behavior of the system given the rules which govern it. A viewpoint that has been found to be more appropriate for these types of systems is the dynamic systems approach.

In the literature, several examples of dynamic systems approaches to language exist. A number of researchers have investigated motor behavior during speech,

* An extended version is available as AI-MEMO from the VUB AI Lab

e.g. [Saltzman, 1995], [Browman and Goldstein, 1995], [Tuller and Kelso, 1990]. In [Elman, 1995], an analysis of the activation space of recurrent neural networks is presented as a dynamical view on grammar and embedding. A difference between that work and ours, is that grammar does not play a role here. Instead, the focus here is on the use of language, and concepts are grounded by use in the environment, see [Steels, 1996b].

In section 1 the alarm calls experiment and the association adaptation mechanism are briefly explained. In section 2 the resulting system of communication is described and the phase space of the system is examined. The final section presents conclusions.

1 Environment and Behavior of Individual Agents

In this section we describe the environment used in the experiments and the mechanism that, when used by individual agents in that environment to adapt form-meaning associations and to determine their situation, leads to a stable system of communication.

The Environment The task faced by the agents is to occupy a safe position within a 10 by 3 grid whilst being hunted by predators. At each timestep, each agent in turn selects an action and receives a success value. The action consists of choosing the vertical position (bottom, middle or top row) and a horizontal displacement (1 step to either side, or staying). Three predators exist. When one is present, only a single row is safe (success 1); agents in the other two rows receive success 0. Sensor information consists of an agent's horizontal and vertical position, and a number indicating the predator (none, or predator 1, 2 or 3). The predator indicator is not completely reliable though; this is where the benefit of communication comes in. In 10% of the cases when a new predator arrives, an agent is not able to see it, depending on its horizontal position, and this remains so until it leaves (after 25 timesteps).

Formation of Situation Concepts Based on the relationship between sensor information and success, agents learn to distinguish different *situations*. Knowing in which situation it is allows an agent to choose optimal actions. *Situation concepts* are formed using autonomous concept formation, described in [de Jong, 1999]. Briefly, the method as it is used here identifies regions in the sensor-action space where success is constant. Since concepts are formed based on experience, the number of concepts an agent will develop is not known in advance, and may differ among agents.

Adaptation of Form-Meaning Association Strength Once agents have concepts at their disposal that capture relevant situations in the environment, a basis for developing communication is present. At each timestep, every agent produces a signal corresponding to its perceived situation. When some agent does not receive sensory information that would allow it to determine its situation, this information may be obtained through communication. If the association between a signal received and a situation concept other than the one indicated by the sensors is strong enough, the agent can deduce its actual situation from that signal and act accordingly. This has been shown in [de Jong, 1999]. Associations between meanings (situation concepts) and forms (words, signals) are adapted locally by each agent based on (1) frequency of occurrence and (2) success in determining the situation. The system is open with respect to the signals the agents use, and the number of signals is not fixed.

2 Results

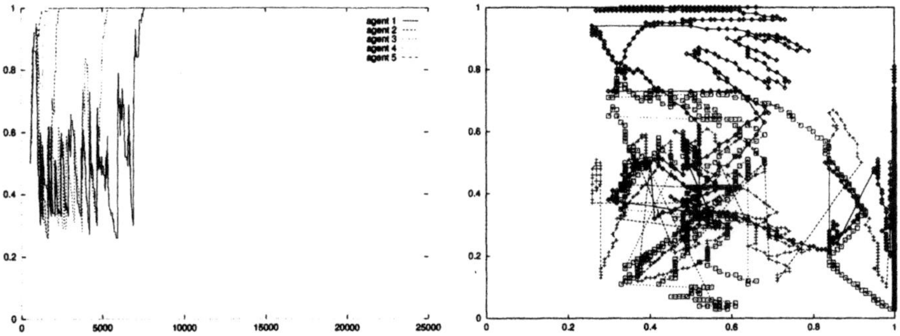

Fig. 1. Left: The evolution over time of the association strength of the word `fi` with each agent's meaning for the first predator (P1). Right: Phase plot for agent 1. Each point represents a pair of association strengths, one between `fi` and the meaning for P1 (always horizontal), and one between `fi` and another meaning (vertical)

This section presents the results of the experiment. In a particular experiment, see figure 1 (left), 4 agents developed strong associations between the word `fi` and the situation where predator 1 is present. For agent 1, this association fluctuated (t = 5000) before it was established (t = 8000). In the phaseplot in figure 1 (right), this final situation is represented by the points at the vertical line to the right (x = 1). The main part of the plot consists of trajectories where the association between `fi` and P1 (P1FI) competed with the other meanings. Of particular interest is the folded curve just right of the middle at the top of the graph. Here, the strength of P1FI repeatedly reaches high values (the horizontal position, around 0.75), but returns to lower values each time. Apparently, the competing meaning (corresponding to the vertical position) was too high to be overcome. As we know from the time series, P1FI did move to 1 later on. This is represented by the series of squares starting just below the line that was mentioned, and describing a curve towards the lower right corner, indicating the meaning gave way to P1FI to become associated. The fact that the system's state never reaches the upper right corner corresponds to the competitive nature of form meaning associations; in the top right corner, a signal would be simultaneously associated with two meanings. It can be deduced from the algorithm that such a state is unlikely to be stable. What analyzing the algorithm doesn't show is *how* the system evolves over time, i.e. through what states it tends to move. That is the result of complex interaction between the association strengths of multiple agents and multiple meanings and words for each of those agents. This is why it can be useful to investigate the phase space of a system, in addition to time series of the relevant data.

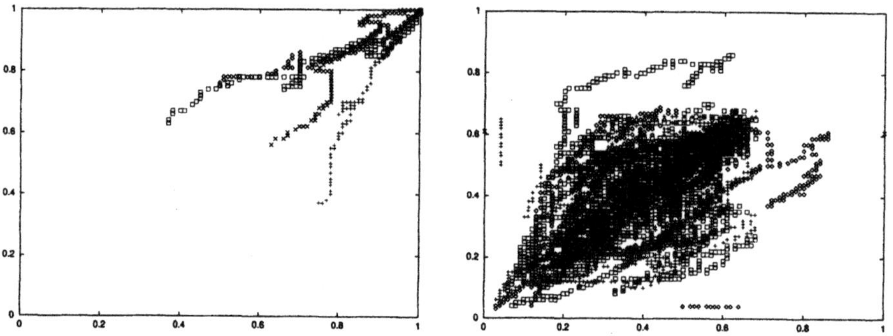

Fig. 2. Left: Phase plot of the association strength of the word `ci` with the first predator for different pairs of agents. Right: Phase plot of the association strength of the word `bo` with the second predator for different pairs of agents

A rather different pattern is seen in figure 2 (left), which shows a phaseplot where the associations between the word `ci` and P1 are plotted for pairs of agents. This word eventually becomes associated to P1 for every agent. What the phaseplot shows us is how the association strengths influence each other. Whereas near the middle of the graph, where association strengths are moderate, variation of the strengths appears rather flexible, this changes when the system reaches the top right corner. In that phase, the association strengths have been observed (also in other graphs) to tend towards each other, resulting in an increased density of points around the diagonal line $y = x$. This is what one would intuitively expect; once most agents use some word for a certain concept, the rest will follow and adopt the word as their preferred word for that concept.

Figure 2 (right) shows the association strength patterns for a word (BO) and a meaning (P2) that did not become associated. Similar to the previous graph, the association strengths of different agents for a particular combination of form and meaning tend towards each other when extreme values are reached, though this time the values are low instead of high. Qualitatively, this can be understood as a word losing its association with a particular concept. When the word is not used anymore by some critical mass of the population, the rest follows and also discards the word.

3 Conclusions

The main conclusion that can be drawn from the experiments is that the evolution of communication, as it was observed in the experiments presented here, can appropriately be viewed as the behavior of a dynamical system. Interacting elements are the vital part of such a system, and the composition of their behavior soon becomes too complex to analyze. By taking a dynamical systems perspective, different ways of analyzing the system become available. The analysis of experiments based on phase plots revealed relationships that would probably not have been found otherwise. Particularly, a tendency of different agents to converge towards the same association strength, both high (1) and low (0) for certain combinations of meanings and signals was found. Furthermore, the behavior of the competition between one form and several meanings of a single agent could be observed. It may be concluded that investigating the phase space of a learned system of communication gave new insights, and is a promising method. An interesting continuation of the analysis would be to investigate relationships between order and control parameters of the system [Kelso, 1995].

References

[Browman and Goldstein, 1995] Browman, C. P. and Goldstein, L. (1995). Dynamics and articulatory phonology. In Port, R. F. and Van Gelder, T., editors, *Mind as Motion: Explorations in the Dynamics of Cognition*, chapter 7, pages 175–193. MIT Press.

[de Jong, 1999] de Jong, E. D. (1999). Autonomous concept formation. In *Proceedings of the Sixteenth International Joint Conference on Artificial Intelligence IJCAI'99.*

[Elman, 1995] Elman, J. L. (1995). Language as a dynamical system. In Port, R. F. and Van Gelder, T., editors, *Mind as Motion: Explorations in the Dynamics of Cognition*, chapter 8, pages 195–225. MIT Press.

[Kelso, 1995] Kelso, J. S. (1995). *Dynamic patterns: the self-organization of brain and behavior.* MIT Press, Cambridge, Ma.

[Saltzman, 1995] Saltzman, E. L. (1995). Dynamics and coordinate systems in skilled sensorimotor activity. In Port, R. F. and Van Gelder, T., editors, *Mind as Motion: Explorations in the Dynamics of Cognition*, chapter 6, pages 149–173. MIT Press.

[Steels, 1996a] Steels, L. (1996a). Emergent adaptive lexicons. In Maes, P., editor, *From Animals to Animats 4: Proceedings of the Fourth International Conference On Simulating Adaptive Behavior*, Cambridge, MA. The MIT Press.

[Steels, 1996b] Steels, L. (1996b). Perceptually grounded meaning creation. In Tokoro, M., editor, *Proceedings of the International Conference on Multi-Agent Systems ICMAS'96. Kyoto, Japan.*

[Steels, 1997] Steels, L. (1997). The synthetic modeling of language origins. *Evolution of Communication*, 1(1):1–34.

[Tuller and Kelso, 1990] Tuller, B. and Kelso, J. (1990). Phase transitions in speech production and their perceptual consequences. In Jeannerod, M., editor, *Attention and Performance*, volume XIII, pages 429–452. Erlbaum, Hillsdale, NJ.

Syntax out of Learning: The Cultural Evolution of Structured Communication in a Population of Induction Algorithms

Simon Kirby

Language Evolution and Computation Research Unit, Department of Linguistics, University of Edinburgh, Scotland
simon@ling.ed.ac.uk
http://www.ling.ed.ac.uk/~simon

Abstract. A new approach to the origins of syntax in human language is presented. Using computational models of populations of learners, it is shown that compositional, recursive mappings are inevitable end-states of a cultural process of linguistic transmission. This is true even if the starting state is no language at all. It is argued that the way that knowledge of language is transmitted through a learning bottleneck profoundly influences its emergent structure. This approach provides a radical alternative to one in which the structure of language is viewed as an innate, biological adaptation to communicative pressures.[1]

1 The origins of syntax

Human language is unique among natural communication systems in having a compositional and recursive mapping between meanings (mental representations to be communicated) and forms (linear strings of, typically phonetic, gestures). It is also extremely unusual in the way it is learnt. Each generation acquires at least some of the meaning-form mapping by observing the use of the previous generation's mapping. It has been argued that this type of learning of mappings is also unique to humans (Oliphant, 1998). In this paper I will explore, using a working model of linguistic transmission, the links between these two features of language. In particular, I aim to explain the origins of syntax in language by looking at general properties of the transmission of learned behaviour.

The explanation put forward here contrasts radically with what is perhaps the dominant approach to the origins of syntax. In many linguists' view, syntax is more or less fully specified by a learning mechanism (the Chomskyan Language Acquisition Device, or LAD) which significantly constrains the language learner with prior knowledge about the nature of language (Chomsky, 1986). Prior biases or constraints on a learner are assumed to be innate, and therefore genetically determined. This leads naturally for many researchers (see Pinker and Bloom,

[1] This work benefited greatly from conversation with and comments from Jim Hurford, Mike Oliphant, and Ted Briscoe and was supported by ESRC grant R000237551.

1990 for example) to the conclusion that the best explanation for the structure of language is one which invokes biological evolution. In other words, the Language Acquisition Device is a biological adaptation that evolved through natural selection in response to the need to communicate "propositional structures over a serial interface" (Pinker and Bloom, 1990, 707).

2 Evolution without natural selection

The standard (biological) evolutionary approach to the origins of syntactic structure typically ignores the dynamics of the social/cultural transmission of language.[2] However, recently there has been more interest in the formal and computational properties of the influence of learning on the dynamic process of language transmission, historically, from one generation to the next (Niyogi and Berwick, 1997; Steels, 1997; Briscoe, 1998; Batali, 1998; Hurford, 1998). This paper explores the hypothesis that this process alone is enough to explain the emergence of the central unique features of syntax even where the prior bias of the learners is not subject to evolutionary change, and is relatively unconstraining.

Linguistic transmission can be viewed as a repeated transformation of information between two domains: the internal or I-domain, and the external or E-domain[3] (Kirby, 1999a). The I-domain contains languages that exist as grammatical knowledge in individuals' brains, whereas the E-domain contains languages that exist as sets of actual utterances. This conception has interesting parallels with the conception of genetic transmission in biology. Figure 1, compares the transformation of information between phenotypic and genotypic domains with the transformation between E- and I-domains.

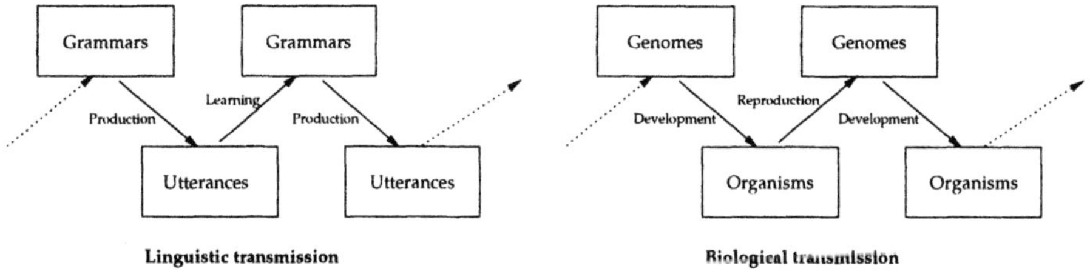

Fig. 1. Transformation of information in linguistic and biological adaptive systems.

In both these adaptive systems, the biological and the linguistic, the transformations that map between the two domains can be seen as "bottlenecks" on the transmission of information over time through the whole system. For example,

[2] See (Briscoe, 1998; Kirby and Hurford, 1997) for counter examples of papers which look at the combined effects of cultural transmission and biological evolution.

[3] These domains are parallel to Chomsky's (1986) I-language and E-language.

a particular piece of genetic information may not persist because the phenotype that it expresses may not survive long enough to reproduce. Similarly, a grammatical regularity may not persist because the utterances that it gives rise to cannot successfully reconstruct it through learning. Just as the bottleneck on transmission of genetic information has implications for the structure of the organisms that emerge (this is the basis of the neo-Darwinian synthesis, after all), we should expect the equivalent bottleneck on the transmission of grammars to impact on the eventual structure of languages.

3 Modelling linguistic transmission

In order to test how the learning bottleneck influences the structure of language, a working model of linguistic transmission must be created. The model consists of a simple population of agents who have a pre-specified "world" of concepts about which they can communicate, and a pre-specified set of symbols with which they can create utterances. The possible communicative behaviour of an agent is determined by that agent's internal grammar which is learnt solely through observation of the behaviour of other agents in the population. The population model is "generational" in that members of the population die and are replaced, but the initial state of the agents is always the same, the agents are not prejudiced or rewarded according to their behaviour in any way, and the only information that flows through the system is in the form of utterances — in other words, there is no natural selection or biological evolution in the simulation.

Semantics The semantic space in the model — the space from which meanings of utterances for the agents is chosen — is made up of combinations of simple atomic concepts. These concepts include, for example:

john heather
loves admires
believes

These atomic elements can be combined into simple predicate-argument propositions, which may have hierarchical structure. For example:

admires(heather,john)
believes(heather,admires(john,heather))

The full set of propositions is infinite in principle, because using predicates such as believes propositions can be limitlessly nested. In the simulations presented here, there are five "object" concepts (such as john, heather etc.), five "action" concepts (such as loves, admires etc.), and five "embedding" concepts (such as believes, knows etc.).

Utterances The utterances of the agents are linear concatenations of the 26 lower-case letters of the alphabet. There is no principled limit to the length of utterance, and the shortest utterance is 1 symbol long. Pairs of strings of symbols and propositions make up the E-domain objects in the simulation. So, if the agents happened to speak English, an E-domain object might look something like:

$$< \texttt{heatherlovesjohn}, \textsf{loves(heather,john)} >$$

Grammars The I-domain objects in the model are a simple form of definite-clause grammars (DCGs). Critically, this choice of formalism does not build-in compositionality or recursion. Consider an agent that could produce the string **heatherlovesjohn** meaning loves(heather,john). Here are two (of many, many possible) grammars that this agent may have internalised:

$$S/\textsf{loves(heather,john)} \rightarrow \texttt{heatherlovesjohn}$$

$$\begin{array}{l} S/p(x,y) \rightarrow N/x \quad V/p \quad N/y \\ V/\textsf{loves} \rightarrow \texttt{loves} \\ N/\textsf{heather} \rightarrow \texttt{heather} \\ N/\textsf{john} \rightarrow \texttt{john} \end{array}$$

The S symbol in these grammars is the start symbol for the grammar, and N and V are arbitrary node labels. The material after the slash on the non terminals is the semantic representation of that node. For left-hand side non-terminals in this formalism, this semantic representation may be a fully specified proposition, an atomic concept, or a partially specified proposition (i.e. one with variables replacing concepts). The right-hand side non-terminals' semantics may only be variables.

The grammar on the right is compositional. The meaning of a sentence is made up by combining two strings of type N with a string of type V, and the semantics for the sentence is built up from the semantics of these individual sub-strings using the variables x, p and y. The grammar on the left, however, is completely non-compositional. In no way is the meaning of the string a function of its parts.

3.1 Learning

Clearly, the central part of the simulation is the agents' ability to take E-domain objects (string/meaning pairs) and build I-domain objects (grammars). In fact, the agents are essentially induction machines designed to store pairs of strings and meanings, and find possible generalisations over these pairs (if such generalisations exist). The algorithm presented here has been designed specifically with simulation tasks in mind — it is extremely simple and efficient. Although no claims are made for its efficacy as a practical learning algorithm for DCGs in general, it does model in a simple way the dual processes of rote learning and search for generalisation that should be the core of any theory of language acquisition.

The algorithm is described in some detail in Kirby (1999b), and outlined informally[4] in figure 2. It works in two stages for each string/meaning pair. Firstly, the individual pair is incorporated into the learner's grammar by addition of the simplest possible rule that will generate that pair. For example, give the string/meaning pair:

$$< \texttt{heatherlovesjohn}, \textsf{loves(heather,john)} >$$

the following rule would be added to the grammar:

$$S/\textsf{loves(heather,john)} \rightarrow \texttt{heatherlovesjohn}$$

The second stage of learning (which takes place after incorporation of each rule) is a process of searching for possible generalisations (essentially subsumptions) over pairs of rules. For example, given the pair of rules:

$$S/\textsf{loves(heather,john)} \rightarrow \texttt{heatherlovesjohn}$$
$$S/\textsf{loves(heather,gavin)} \rightarrow \texttt{heatherlovesgavin}$$

the algorithm will replace these with a more general rule (via the "chunking" step g.3):

$$S/\textsf{loves(heather,}x\textsf{)} \rightarrow \texttt{heatherloves}\ N/x$$

The N category is a newly invented category, with its own rules added to the grammar:

$$N/\textsf{john} \rightarrow \texttt{john}$$
$$N/\textsf{gavin} \rightarrow \texttt{gavin}$$

3.2 Invention

Sometimes, the agents may wish to express meanings for which their grammar does not provide a way of saying. In fact, in the initial stages of the simulation, this will be very frequent, since the population will start with no language at all. The model therefore requires some way for creative language invention to occur. The simplest way to model this is for the agents to produce a string of random symbols whenever they encounter a meaning they do not know how to produce. In the simulations presented here, the strings vary between 1 and 3 symbols in length.[5]

[4] The chunking algorithm is not given here for lack of space (see Kirby, 1999b for details). The example in the text shows a typical application of chunking.

[5] The complete invention algorithm is discussed elsewhere (Kirby, 1999b). The only complication for implementation purposes is what should be done if the agent knows how to say almost all of a particular meaning. The algorithm in effect carries out a top-down production for the meaning as best it can given the grammar, and generates a random string at the point in the tree where the generation fails. This method ensures that invention does not ever introduce new structure. In other words, the invented string will never be more compositional than grammatical strings that already exist' in the agent's language.

Induction algorithm: Given a meaning m, a string s and a grammar g:

i.1 parse s using g, **if** the parse is successful, **then return** g.
i.2 form g', the union of g and $S/m \rightarrow s$.
i.3 apply a generalisation algorithm to g':
  g.1 take a pair of rules $< r_1, r_2 >$ from g'.
  g.2 **if** there is a category label substitution c to c', that would make r_1 identical to r_2, **then** rewrite all c in g' with c', **go to g.5**.
  g.3 **if** r_1 and r_2 could be made identical by "chunking" a substring on either or both their right hand sides into a new rule or rules, **then** create the new rules, and delete the old ones in g', **go to g.5**.
  g.4 **if** r_1's right hand side is a proper substring of r_2's and r_1's semantics is identical to either the top level predicate or one of the arguments of r_2's semantics, **then** rewrite r_2 in g' to refer to r_1, **go to g.5**.
  g.5 delete all duplicate rules in g'.
  g.6 if any rules in g' have changed, **go to g.1**
i.4 **return** g'.

Fig. 2. Outline of the induction algorithm.

3.3 Simulation cycle

For the simulations reported here, the simplest possible population model is used. At any one time there are only two individuals: a learner and a speaker. The speaker is required to produce an utterance for a meaning chosen at random, either by using its grammar or by employing invention if the grammar cannot generate a string (if the speaker invents, then it uses its utterance as input to its own learning to ensure consistency). The learner is given the randomly chosen meaning and the speaker's string, and uses the pair as input to the induction algorithm. This process is repeated a fixed number of times, and then the speaker is removed, the learner becomes a speaker, and a new learner with an empty grammar is introduced.

4 Experiments

The graphs in figure 3 show the average behaviour of ten simulation runs starting with different random seeds. In all the runs, each speaker first tries to produce 50 randomly chosen degree-0 meanings (i.e. those with no embedding), then 50 randomly chosen degree-1 meanings (with one embedding) and finally 50 randomly chosen degree-2 meanings (with two embeddings). The graphs show the proportion of the meanings of each type that the speakers were able to produce without resorting to random invention. The other line on the graphs show a measure of size of the grammars (number of rules divided by 300 to fit it on the same plot).

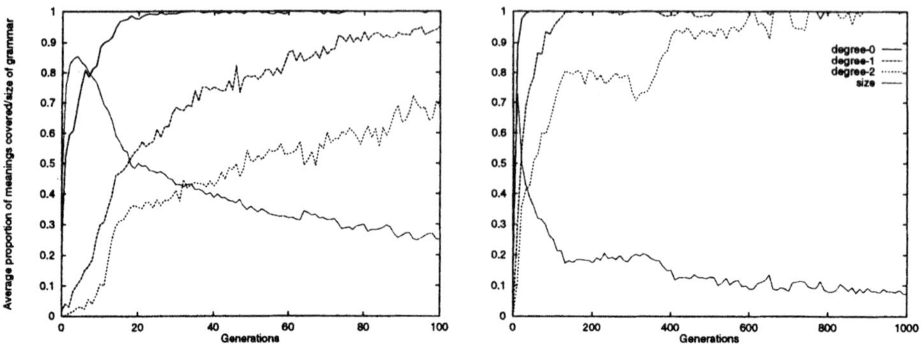

Fig. 3. Average of ten simulation runs, plotted on two different time-scales. Coverage of the different meaning types increases and the size of the grammar decreases over time.

To demonstrate more clearly what the behaviour of the evolving system is, we can examine the grammars of one population over time in some detail:

The emergence of vocabulary In the very early stages of the simulation, the speakers are not able to express many of the meanings without resorting to random invention — this is due to the fact that the population starts with no language at all. Immediately, there is a rapid rise in the size of grammars, as new innovated utterances are acquired by learners. The listing below shows part of an early grammar, which is best thought of as simply a long list of unrelated, idiosyncratic vocabulary items.

S/loves(pete,mary) → axk
S/thinks(john,likes(gavin,heather)) → ew
S/decides(heather,says(mary,detests(john,pete))) → pq
S/admires(pete,heather) → njl
S/decides(john,believes(pete,hates(pete,mary))) → vz
S/admires(gavin,mary) → s
S/knows(mary,decides(mary,loves(gavin,pete))) → c
S/detests(pete,john) → my
S/believes(pete,believes(mary,likes(mary,heather))) → z
S/likes(gavin,mary) → but
S/loves(gavin,pete) → poe
S/thinks(gavin,loves(heather,gavin)) → oeb
S/decides(mary,hates(gavin,john)) → vri
S/detests(heather,gavin) → wb
S/thinks(mary,hates(john,pete)) → y
S/thinks(john,hates(mary,pete)) → pff
S/decides(pete,detests(heather,mary)) → fi
S/believes(gavin,decides(gavin,admires(mary,john))) → go
S/believes(pete,knows(gavin,loves(heather,gavin))) → jt
⋮

The emergence of compositionality Figure 3 shows that after a brief initial rise the size of grammars soon starts to fall, and the proportion of degree-0 meanings that are expressed without invention increases. The grammar fragment below shows how this happens. Here, by the 100th generation, a compositional encoding has emerged, although it is only reliably used for degree-0 meanings. This particular grammar has two categories for arguments (arbitrarily named A and

C), which can be thought of as case-marked nominals, and one verbal category for predicates (named D). (N.B. there are many other idiosyncratic rules for more complex meanings in this grammar that are not shown.)

$S/p(x,y)$ → gj C/y z A/x D/p
A/gavin → dl
A/heather → tej
A/john → n
A/mary → qp
A/pete → h
C/gavin → x
C/heather → ovp
C/john → i
C/mary → h
C/pete → y
D/admires → b
D/detests → gl
D/detests → xe
D/hates → c
D/likes → e
D/loves → m
S/knows(john,decides(pete,loves(mary,john))) →
qixoyia
S/decides(x,thinks(heather,likes(pete,mary))) →
C/x jwyjtejdbznuy
⋮

Here is an example sentence from this language with an English gloss (NOM and ACC stand for nominative and accusative):

(1) gj ovp z qp b
Heather-ACC Mary-NOM admires
"Mary admires Heather"

The emergence of recursion Finally, the percentage of degree-1 and degree-2 meanings that are expressible without invention increases as the size of the grammar continues to decrease. By generation 1000, an extremely concise language has emerged that can completely express the infinite meaning space (the entire grammar is shown below). This is done by using two verbal categories (B and D), one nominal category (A), and two sentence rules, one of which is recursive. In other words, a highly regular, expressive, syntactically structured language has emerged. This language persists without significant change for the rest of the simulation (which was terminated after 10000 generations).

$S/p(x,y)$ → gj A/y f A/x B/p
$S/p(x,q)$ → i A/x D/p S/q
A/gavin → dl
A/heather → tej
A/john → n
A/mary → qp
A/pete → h
B/admires → b
B/detests → wp
B/hates → c
B/likes → e
B/loves → m
D/believes → g
D/decides → u
D/knows → ipr
D/says → p
D/thinks → m

Here are examples from this language, again with glosses in English (notice how little the degree-0 structure has changed, apart from a simplification of the nominal system so that the nominative nouns are now used for accusatives as well):

(2) gj tej f qp b
Heather Mary admires
"Mary admires Heather"

(3) i h ipr gj tej f qp b
Pete knows Heather Mary admires
"Pete knows that Mary admires Heather"

(4) i dl p i h ipr gj tej f qp b
Gavin says Pete knows Heather Mary admires
"Gavin says that Pete knows that Mary admires Heather"

5 Conclusion: evolution through learning

The computational model shows how significant linguistic evolution can occur without any biological evolution. The learners in the simulation are not innately constrained to acquire only syntactic languages, nor are they rewarded in any way for their ability to communicate. The simulation is not seeded with any particular language — in fact, the first speaker has no linguistic knowledge to draw on whatsoever — and the only way information is transmitted over time is by observational learning. And yet, from this apparently simple set up, a linguistic system emerges. First, the language of the community is an impoverished vocabulary-like system, with idiosyncratic signals being assigned to whole complex meanings. Later, a compositional mapping emerges, with a lexicon for atomic concepts and a simple way of putting these together to form sentences. Finally, recursion evolves so that the agents are able, with only a small vocabulary and a couple of rules of combination, to express an infinite range of possible meanings.

The obvious questions that we are left with are "why does the language evolve in this way?" and "how general is this result?" Ultimately, a satisfactory answer to both these questions may well lie in a mathematical analysis of the properties of the transmission of information via observational learning. For the moment, however, it seems that the learning bottleneck that exists between the E-domain and I-domain in figure 1 must be what is driving the system to evolve.

Firstly, consider a non-compositional system such as the early ones in the simulation run described above. Here, the mapping between meanings and strings is such that each rule in the grammar can at best only account for one meaning-string pair. This necessarily means that to learn any particular rule in such a grammar, a learner must be exposed to exactly that pairing. The chance that a particular non-compositional I-language will survive over time are slim in this situation, because every string-meaning pairing in that I-language will have to appear in the input data to the learner. Given that the number of utterances that the learner hears is smaller than the number of meanings that a speaker might be called upon to produce, then a randomly chosen, complete, non-compositional language *cannot* be transmitted through the learning bottleneck. Notice, this must be true for any learning algorithm.

Now, contrast this situation with one in which there is a compositionality in the language. Given an E-language in which there is some pattern to the string-meaning mapping, it is no longer true that a particular pairing can only persist if it is directly observed. A learner can generalise from an observed set of string-meaning pairs to ones that are unseen. This means that a compositional language will be far more transmittable through the bottleneck. A smaller set of example pairings will required in order to reconstruct the language accurately. Again, as long as the learning algorithm is able to exploit pattern, or is biased towards generalisation, then this is true in general.

It seems as if the process of the transmission of learned behaviour has certain general properties whatever our assumptions about the learners. These properties will tend to favour the emergence of systems with patterns that can be exploited by learning. Where the behaviour that is being learned is a mapping between two structured spaces, we can expect the process of transmission to give us mappings that preserve structure. For the case of language — where these spaces are propositional meanings on the one hand, and linear strings on the other — the mappings that will inevitably emerge will be syntactic.

References

Batali, J. (1998). Computational simulations of the emergence of grammar. In J. Hurford, C. Knight, and M. Studdert-Kennedy (Eds.), *Approaches to the Evolution of Language: Social and Cognitive Bases*, Cambridge, pp. 405–426. Cambridge University Press.

Briscoe, E. J. (1998). Language as a complex adaptive system: co-evolution of language and of the language acquisition device. In P. Coppen, H. van Halteren, and L. Teunissen (Eds.), *8th Meeting of Comp. Linguistics in the Netherlands*, Amsterdam, pp. 3–40. Rodopi.

Chomsky, N. (1986). *Knowledge of Language.* Praeger.

Hurford, J. (1998). Social transmission favours linguistic generalisation. In C. Knight, J. Hurford, and M. Studdert-Kennedy (Eds.), *The Emergence of Language.* To appear.

Kirby, S. (1999a). *Function, Selection and Innateness: the Emergence of Language Universals.* Oxford: Oxford University Press.

Kirby, S. (1999b). Learning, bottlenecks, and the evolution of recursive syntax. In E. J. Briscoe (Ed.), *Linguistic evolution through language acquisition: formal and computational models.* Cambridge: Cambridge University Press. Forthcoming.

Kirby, S. and J. Hurford (1997). Learning, culture and evolution in the origin of linguistic constraints. In *Fourth European Conference on Artificial Life*, pp. 493–502. MIT Press.

Niyogi, P. and R. Berwick (1997). Populations of learners: the case of Portugese. Unpublished manuscript, MIT.

Oliphant, M. (1998). Rethinking the language bottleneck: Why don't animals learn to communicate? In C. Knight, J. Hurford, and M. Studdert-Kennedy (Eds.), *The Emergence of Language.* To appear.

Pinker, S. and P. Bloom (1990). Natural language and natural selection. *Behavioral and Brain Sciences 13*, 707–784.

Steels, L. (1997). The synthetic modelling of language origins. *Evolution of Communication 1*(1).

Modelling the Evolution of Linguistic Diversity

Daniel Livingstone and Colin Fyfe

Applied Computational Intelligence Research Unit,
Dept. CIS, University of Paisley, Paisley, PA1 2BE, UK
{livi-ci0,fyfe-ci0}@paisley.ac.uk

Abstract. Some recent Artificial Life models have attempted to explain the origin of linguistic diversity with varying conclusions and explanations. We posit, contrary to some existing Artificial Life work, that linguistic diversity should naturally emerge in spatially organised populations of language learners, and this is supported by our experimental work and by recent literature.

1 Introduction

There exists a rich literature on linguistic diversity, much focussed on tracing the roots and relationships between languages and regional variations in dialect. This division between language and dialect is not clear-cut. Two different languages may be mutually intelligible while two dialects of the one language may possess a low degree of mutual intelligibility [1]. *Dialect continua* make the distinction between languages and dialects harder still. These chains of dialects can cover large areas and cross many national boundaries, each dialect intelligible to speakers of neighbouring dialects. Yet dialects at the ends of the chains may be mutually incomprehensible.

Artificial Life inspired research has shown how populations of agents can negotiate a language and how communication schemes can evolve. Generally, these attempt to negotiate a single language common to all agents. There exists a small body of Artificial Life based work which studies linguistic diversity. These have tended to ignore the simple hypothesis that linguistic diversity may be the non-adaptive result of spatially distributed learning. The model presented here investigates this.

2 Dialect in Models of the Evolution of Language

A small number of Artificial Life papers address the issues of language diversity and innovation directly. Maeda and Sasaki, [2], consider the effects of contact between different linguistic groups, demonstrating the subsequent language re-organisation.

Kirby, [3], studies the evolution of Universal Grammar and shows, implicitly, how spatial organisation can affect the evolution of languages. Hashimoto and Ikegami, [4], show how pressure for greater expressive and parsing ability leads to increasingly complex grammars which are able to generate and interpret wider ranges of signals.

Arita and Taylor [5] argue that spatial distribution can explain the origin of linguistic diversity. Inheritance and mutation are important factors for generating diversity within this model, where learning reduces diversity. Yet in human language it is the cultural, not genetic, factors which determine which language is learned. In Arita and Koyama [6] an innate language is used to study the evolutionary dynamics of vocabulary sharing. One conclusion is that mutation rate is important in defining the class of vocabulary sharing. This is not surprising with a heritable vocabulary, but what this represents linguistically is not explained. Steels and Kaplan [7] demonstrate stochasticity, from errors in language use, as a source of change in language. The formation of multiple, distinct, dialects is not covered in this study in which the agents are not spatially organised. While Nettle and Dunbar [8] cite examples and use simulation to demonstrate that dialect has adaptive value in a social context, we will show that adaptive explanations are not required for evolution of linguistic diversity.

3 A Model for Studying the Evolution of Dialect

We adapt an earlier model [9,10] in which the emergence of dialects was observed but not studied, focussing now on the evolution of dialects. Genetic evolution of agents is removed, and learning modified to take place between successive generations of agents. Our model of language is a greatly simplified one, in which agents learn to map signals sent by others to internal states. These internal states could be said to relate to 'meanings' or to external observations.

A language agent is modelled by a fully connected Artificial Neural Network, (ANN) which has two layers of nodes, with N inputs (the internal state, $\boldsymbol{x}$) and M outputs (the signal, $\boldsymbol{y}$). The output layer has an additional bias node.

The internal state is a sparse bipolar vector (only one node set to +1). This is fed forward to determine the agent's signal for that state, each output being thresholded to a bipolar value, (1). Signals are arbitrary bipolar vectors. For interpretation of signals, competition is applied at the internal state layer, so any signal fed back from the language layer corresponds to only one internal state. For M output neurons, there are 2^M possible signals in the language, and for N input neurons there are N possible states. In experiments detailed here, M = N = 3. Agents have three possible internal states and eight possible signals, providing redundancy in the signalling ability.

$$y_j = \sum_{i=0}^{N-1} x_i w_{ij} \tag{1}$$

$$\text{then } Y_j = 1 \text{ if } y_j \geq 0, \; Y_j = -1 \text{ if } y_j < 0$$

$$\Delta w_{ij} = \eta\left(x_i - x_i'\right) y_j \tag{2}$$

[11] suggests using transmission behaviour to train language reception and reception behaviour to train language production. Accordingly, the signal from a teacher is presented at the output layer of thc learner and fed back to produce an internal state. The error between the teacher's and the learner's internal states is used for learning (2), performed when the learner misclassifies the signal.

The first generation of agents are initialised with random weight values and then provide training for one another for t training rounds, as in the following algorithm:

```
For each agent (in random order)
   Pick another agent to be teacher
   Generate a training signal from teacher
   Train pupil on signal and state
```

When each generation is created, the agents are born with zero valued weights. The new agents learn first from agents in the parent generation for t training rounds, and then further negotiate language amongst themselves for $t/2$ rounds.

4 Experimental Results

Two sets of experiments were run. In the first set, agents inhabit a one-dimensional world. The second set was performed in a dimensionless world, in which every agent is equally likely to interact with every other agent.

4.1 Evolution of Dialects in a 1-D World

Agents are organised linearly and teacher agents are selected according to a normal distribution curve centred on the current learner agent. An agent may not be it's own teacher. To visualise the languages, the three bits of a signal are used to set red, green and blue colour values of pixels. Plotting the pixels for each agent for each state, three columns are drawn each generation. Adding the columns for each generation to those of the previous, three bars are formed. A number of simulations were run, the results of one are shown in Fig. 1. The use of different signals can be seen to spread and recede over time.

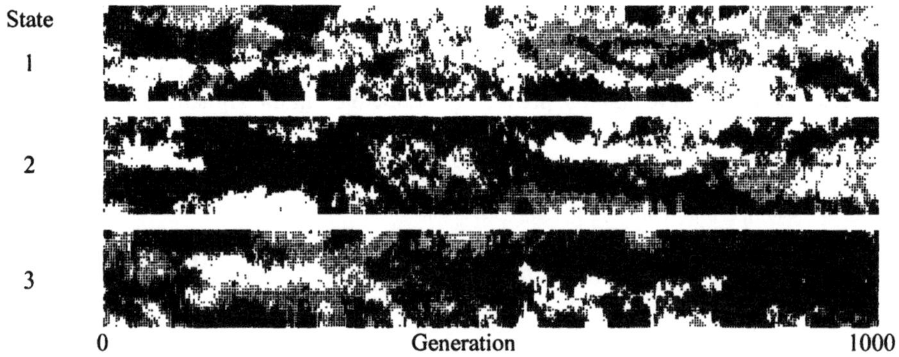

Fig. 1. The 'evolution' of signals representing three states in a spatially distributed population of 120 agents. Each bar shows the evolution of signal use for one of the three internal states

A measure of the difference between signals can be found using the Hamming distance. With a small difference it is possible that both signals are interpreted the same, due to redundancy in the signals. The Hamming distance between signals of neighbours, plotted over the whole population, can indicate language boundaries, and is shown for the final generation of the experiment above (Fig. 2 left). There is a high degree of change over the population. Despite this, signals used by adjacent agents for a particular meaning rarely vary by more than one bit. This could indicate a dialect continuum across the population. In a number of places, the signals used for two or all three of the meanings change together. These might indicate linguistic discontinuities.

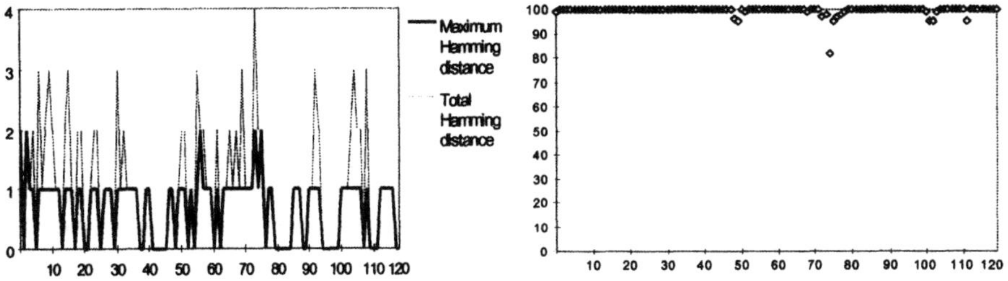

Fig. 2. The maximum (for any one signal) and total (over three signals) Hamming distances between signals used by adjacent agents in a spatially distributed population (left) and the percentage of communicative successes over the spatially distributed population (right)

Fig. 2 (right) plots the communicative success of the agents. Despite the high amount of diversity in the signals used, most agents achieved perfect scores. The lowest scoring agent correctly interpreted 82 percent of signals, showing that a dialect continuum exists from one extreme end of the population to the other.

4.2 Evolution of Dialects without Spatial Organisation

In these experiments agents interact with each other with uniform probability. No distinct dialects are observed, and populations quickly negotiate a global language. Signal redundancy allows the languages to use two or more signals for each state, where each signal is associated with only one state (Fig. 3). Eventually even this diversity disappears, and a one-to-one signal-state mapping established. Using signal uncertainty as a measure, we have determined that there is significantly less diversity under these conditions (mean uncertainty of 0.75 bits). In contrast, where agents are spatially organised no significant diversity loss is observed or measured after 10,000 or even 100,000 generations (mean uncertainties of 2.14 and 2.05 bits respectively).

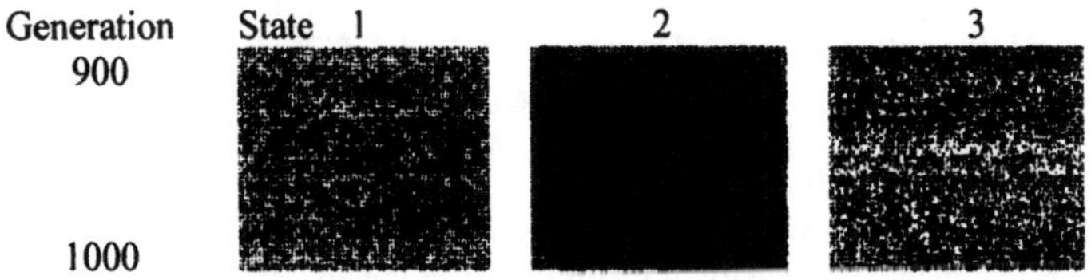

Fig. 3. The evolution of signals without spatial distribution

5 Conclusions

While adaptive explanations of language variation show how dialect may be exploited, our model shows a linguistic system that maintains a high level of diversity without adaptive benefits. Inheritance and mutation are also absent in our model and no errors occur in information transmission, so these are also not required to maintain diversity. Stochasticity occurs in the random selection of agents and in choosing a state to signal, and is not completely removed. The role of stochasticity for linguistic innovation is not disputed, and in our model a population without diversity cannot develop it. There are qualitative similarities between spatial organisation of dialects in our model and the organisation of human dialects. The dialects in our model form a continuum, connecting dialects that are, at the extremes, not mutually intelligible.

We conclude that sustained linguistic diversity can emerge from the interactions of agents, as a consequence of learning and the distribution of agents. Support for this comes from recent work [12] which uses evidence from global linguistic data, suggesting that linguistic diversity is related to *ecological risk*. Ecological factors determine the social networks formed by societies, which in turn determine linguistic diversity. This is the same as our conclusion – that linguistic diversity emerges as a result of constraints on interactions between language learners. We are currently working to determine the effects different parameters have on diversity.

References

1. Crystal, D.: The Cambridge Encyclopedia of Language, Cambridge University Press (1987)
2. Maeda, Y., T. Sasaki, et al.: Self-Reorganizing of Language triggered by "Language Contact". ECAL 97, MIT Press (1997)
3. Kirby, S.: Fitness and the selective adaptation of language. In Approaches to the Evolution of Language, J. R. Hurford, M. Studdert-Kennedy, and C. Knight (eds.), Cambridge University Press (1998) 359-383
4. Hashimoto, T. and Ikegami, T.: Evolution of Symbolic Grammar Systems. ECAL 95 (1995)
5. Arita, T. and Taylor C. E.: A simple model for the evolution of communication. Evolutionary Programming V, MIT Press (1996)
6. Arita, T. and Koyama Y.: Evolution of Linguistic Diversity in a Simple Communication System. Artificial Life IV, UCLA, MIT Press (1998)
7. Steels, L. and Kaplan F.: Stochasticity as a Source of Innovation in Language Games. Artificial Life VI, UCLA, MIT Press (1998)
8. Nettle, D. and Dunbar R. I. M.: Social Markers and the Evolution of Reciprocal Exchange. Current Anthropology **38**(1) (1997) 93-99
9. Livingstone, D. and Fyfe C.: A Computational Model of Language-Physiology Coevolution. Computing and Information Systems, University of Paisley **5**(2) (1998). 55-62.
10. Livingstone, D. and Fyfe C.: Modelling Language-Physiology Coevolution. In The Emergence of Language, J. R. Hurford, M. Studdert-Kennedy, and C. Knight (eds.), Cambridge University Press (in press)
11. Oliphant, M.: Formal Approaches to Innate and Learned Communication: Laying the Foundation for Language, University of California, San Diego (1997)
12. Nettle, D.: Explaining global patterns of language diversity. Journal of Anthropological Archaeology **17** (1998) 354-374.

Compression and Adaptation

Tracy Teal[1], Daniel Albro[2], Edward Stabler[2], and Charles E. Taylor[1]

[1] Department of Organismic Biology, Ecology and Evolution, Box 951606, University of California, Los Angeles, CA 90095, USA {tracyt,taylor}@biology.ucla.edu
[2] Department of Linguistics, Box 951543, University of California, Los Angeles, CA 90095, USA albro@humnet.ucla.edu, stabler@ucla.edu

Abstract. What permits some systems to evolve and adapt more effectively than others? Gell-Mann [3] has stressed the importance of "compression" for adaptive complex systems. Information about the environment is not simply recorded as a look-up table, but is rather compressed in a *theory* or *schema*. Several conjectures are proposed: (I) compression aids in generalization; (II) compression occurs more easily in a "smooth", as opposed to a "rugged", string space; and (III) constraints from compression make it likely that natural languages evolve towards smooth string spaces. We have been examining the role of such compression for learning and evolution of formal languages by artificial agents. Our system does seem to conform generally to these expectations, but the trade-offs between compression and the errors that sometimes accompany it need careful consideration.

1 Introduction

Why are some systems more adaptable than others? A core feature of nearly all successful adaptive systems is the ability to distill experience into schemas, models or theories and then employ those abstracted structures in new circumstances. Information about the environment is not simply recorded as a look-up table, with generalization to new situations happening only at look-up time. Rather, it is plausible that salient features of situations are noted, and then departures from expectations are noted. This is the essence of compression.

Gell-Mann [3] has argued that a compressed form "is usually approximate, sometimes wrong, but it may be adaptive if it can make useful predictions including interpolation and extrapolation and sometimes generalize to situations very different from those previously encountered. In the presence of new information from the environment, the compressed schema unfolds to give predictions or behavior or both." We would like to know more about the role of compression in adaptation. For example, can we identify features of compression that make some forms more or less likely to be successful? What sorts of compression are best? How much is desirable?

This perspective on learning is not new, especially for cultural evolution. It is evident, for example, that there is frequently a practical necessity to simplify things so that we can understand them. Hence the ubiquitous use of simplified models in science. Which models are themselves best, is also a matter of simplicity. William of Ockham, among others, has observed that "it is futile to do with more what can be done with less" [18]. In another domain, Chomsky [2] has placed the desirability of compressed, "minimal" grammar at the heart of his theory of human language. Nonetheless, there have been few attempts to explore, in a systematic way, how compression alone may affect adaptation.

In this paper we first state some heuristics that we conjecture to be generally, if not always, true. We then introduce some elementary definitions and principles about data compression and formal languages. We have chosen to work with formal languages because it is possible to discuss them concretely and because such systems quite clearly show changes in their ability to generalize [14, 6, 11, 10]. We next discuss how compression can occur as an agent learns a language by hearing examples of it. We also describe some experiments where agents learn the languages with compressed grammars, and where the languages evolve as a result. Finally, we discuss some features of compression and evolution in our system in the light of our conjectures. We emphasize that while our discussion will be directed mostly to cultural evolution - scientific theory and language, we believe these heuristics apply to other adaptive complex systems, such as organic evolution, immune systems, and neural networks.

2 Conjectures about compression and adaptation

Conjecture 1. **Compression aids in generalization.**

From a series of observations like "Crow A is black" and "Crow B is black" we compress a look-up table of crows and their colors to the generalization that "All crows are black." The generalization is clearly smaller, more "compressed" than a list of many instances. The precise characterization of the circumstances in which such generalization is appropriate, the problem of induction, is a long-standing philosophical problem.

The history of science is a history of finding generalizations that allow a succinct statement of the facts. Following the invention of the spectroscope, the spectral emissions from various elements, including hydrogen, were cataloged. About 1885 J. J. Balmer discovered formulae that would describe the frequencies for hydrogen emissions both compactly and accurately, though they were simply formulas without a model behind them. In 1913, Niels Bohr published a model for the atom that would compactly describe emissions from hydrogen, and several other atoms, in very compressed and desirable form.

While the history of science can be regarded as a series of successively better compressions, it should also be recognized that the resulting compressions may make predictions that are only approximate or even wrong. Although models can give us insight into systems, the actual model used greatly affects the predictions that can be made and the types of behaviors that can be explained. Many scientific theories, no matter how well they might compress a set of observations, are subsequently proven wrong.

Conjecture 2. **Compression occurs more easily in a "smooth", as opposed to "rugged", string space.**

Related or connected sets of observations form a better basis for generalization than do similar or unrelated ones. We would like to be concrete about the meaning of "related or connected" in this context. Kauffman [8] has explored the use of adaptive landscapes in a variety of contexts, and we build on his example by exploring a string space of languages, below.

Unrelated observations, like Lord Morton's mare for Darwin's theory of inheritance, can make theory formation quite difficult [13]. As a result, it is thought generally better to focus initial scientific study on simple and well-defined systems, where the smoother space is more easily explored, as Mendel did.

Smoothness and ruggedness are, to some extent, a property of the substitution operators. In molecular evolution for example, adding or deleting tandem repeats of 2 or 4 nucleotides is often easier than adding or deleting a single nucleotide [21].

Conjecture 3. **Constraints from compression make it likely that natural languages come to have smooth string spaces.**

Language learners are regarded as systems that aim to identify rule systems that describe the (infinite) language of the community on the basis of finite evidence. This can only be successful in certain circumstances, whether one assumes that success is perfect identification in the limit (the "Gold" paradigm) [4], or that success is feasible convergence to arbitrarily good approximate identification [9, 20]. Since humans clearly learn to speak natural languages that can generate an infinite number of sentences, and do so largely from examples and without conscious awareness of grammatical rules, most linguists believe that, roughly speaking, when a young person is confronted with a new sentence they will "try out" candidate grammars and then, from those grammars that can accommodate it, choose the one that is simplest [1, 5, 19].

Of course, human language learners hear ungrammatical sentences and sentence fragments, but this "noise" apparently does not make communication with the community impossible. We postulate that sufficiently rare complications are typically not incorporated; they will be catalogued as exceptions or simply ignored. As a result, there will be a natural selection for languages that are progressively smoother. This is perhaps analogous to the way that biological systems sometimes "solve" other NP-complete problems. They seem to employ only those parts of the problem space that can be solved simply, even though the general problem might be insoluble [15].

3 Data compression

Data can be compressed only when it contains some regularity that can be exploited. [17]. When there is a regularity, listing it repeatedly makes for an eliminable redundancy.

It is important to recognize that some schemes might provide very efficient coding of data, but would involve lengthy and complicated specifications of the decoder. The decoder could, for example, simply contain a list of any finite set of strings to be encoded. A better measure of compression would recognize both costs. The *minimum description length algorithm (MDL)* we use accomplishes this. In the formal language framework the sum of the *grammar-encoding-length* (the cost of the rule set) and the *data-encoding-length* (the cost of the coding of the data) is minimized [12, 16].

4 Formal languages

A (formal) language is a set of strings of symbols drawn from some alphabet. Here we shall be concerned only with *regular* languages, languages that can be accepted by a *finite automaton.* Such automata can be described by a quintuple, $(Q, \Sigma, \delta, q_0, F)$, where Q is a set of states, Σ is an input alphabet, $q_0 \in Q$ is the initial state, $\delta \subseteq Q \times \Sigma \times Q$ is a transition function, and $F \subseteq Q$ is the set of final states [7]. An automaton is said to be *deterministic* if the transition δ is a function $\delta : Q \times \Sigma \to Q$ with respect to its first two arguments. Associated with each finite state automaton is a *transition diagram*, with an arc connecting states q_1 and q_2 labeled with vocabulary element a if, and only if, $(q_1, a, q_2) \in \delta$. The automaton accepts a string s if, and only if, the string labels a path from the initial state to a final state.

For purposes of illustration, consider languages `1-s` and `1-r`. Language `1-s` consists of the strings `aaaa`, `aaab`, `aaba`, `abaa` and `abba`. Language `1-r` consists of `aaab`, `abaa`, `aaba`, `abbb`, and `bbab`. Their respective "prefix tree" transition diagrams are shown in Figure 1.

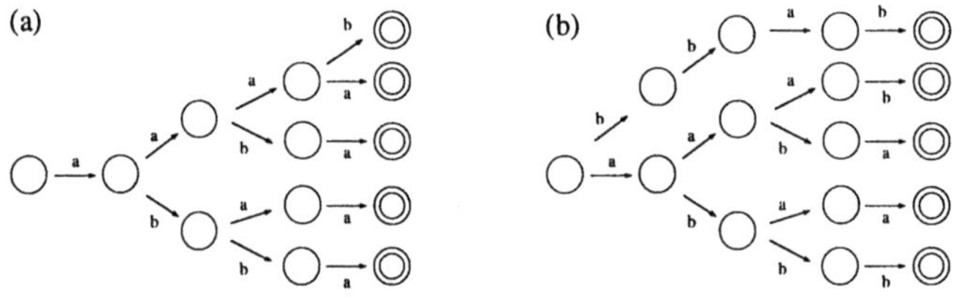

Fig. 1. Transition state diagrams for uncompressed grammars: (a) language 1-s (b) language 1-r.

Since each string can be specified by a path through a deterministic automaton, we calculate one simple approximation of the data-encoding-length in bits given by the formula:

$$\sum_{i=1}^{m} \sum_{j=1}^{|s_i|} \log_2 z_{i,j},$$

where m is the number of sentences in the sequence of strings encoded, $|s_i|$ is the length of the i'th string s_i, and $z_{i,j}$ is the number of ways to leave the state reached on the j'th symbol of sentence s_i. (A more succinct encoding is obviously possible when the probabilities of the transitions are not uniform. The simple approximation suffices for purposes of this preliminary investigation.)

To specify the automaton itself, we must specify all the triples $(q_1, a, q_2) \in \delta$ and we must also specify the final states, so we calculate the grammar encoding length:

$$|\delta|[2(\log_2|Q|) + \log_2|\Sigma|] + |F|[\log_2|Q|],$$

where $|\delta|$ is the number of triples in δ and $|F|$ is the number of final states in F. The MDL of a language is defined as the sum of its grammar-encoding-length and its data-encoding-length. For language `1-s` the MDL is 119.3 and for `1-r` it is 168.0.

We refer to language `1-s` as "smooth" and language `1-r` as "rugged." The reasons for this are evident from Figure 2, which shows how the strings in each set are related by symbol substitution. (More generally, one might use deletion and duplication operators, as well as symbol-substitutions. Duplication and deletion are known to be important for the evolution of DNA and for natural language, but these operators add considerable complication for some of the comparisons we are interested in, so for this paper we shall restrict our attention to operators that are simple symbol substitutions.)

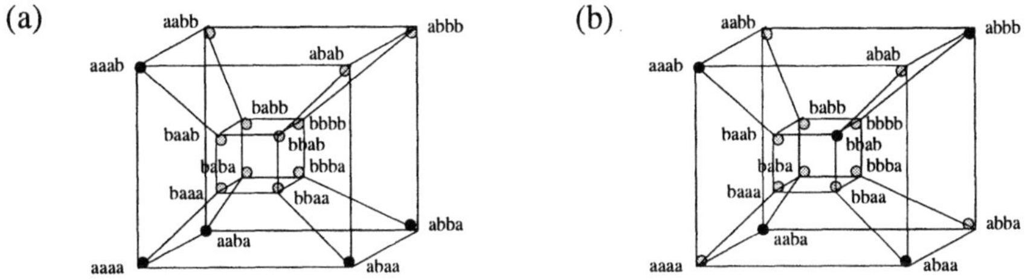

Fig. 2. Hypercube representation of the landscape for strings of length 4. Nodes connected by lines are one symbol-substitution away from each other. Black dots are included in the language. Gray dots are not included. (a) Representation of language `1-s` (b) Representation of language `1-r`.

Each string in `1-s` is one symbol-substitution away from its nearest neighbor. The symbol in only one position is changed from one string to the next, and the strings are very similar or regular, while in `1-r` no string is closer than two substitutions away, making the strings not very related. Further, in the hypercube representation, there is a hyperplane in `1-s` that provides regularity which might be utilized for compression, while such regularity is not evident in `1-r`. This occurs in much the same way that the equation of a line offers a compressed representation of a set of data points in linear regression. These two grammars also differ in their MDL measures: the MDL of 1-s is only 71% that of 1-r.

5 Grammar compression

Our model of language learning will be that of a child, who when listening to an adult speaking a language s/he does not yet understand, tries out different candidate grammars that might accept that language. The minimalist theory supposes that those grammars found suitable are then examined further, and the grammar with the shortest MDL is preferred and tentatively accepted as describing the parent's language.

In our study, an agent is exposed to all the legal sentences in the language – say of 1-s or 1-r. It constructs an automaton with associated grammar as shown in Figure 1, above. The agent then compresses the original grammar by attempting to combine states and transitions in such a way that the outcome is still a deterministic automaton and then combines them iff the MDL of the new automaton is smaller than that of the starting one. Given a machine $A = (Q, \Sigma, \delta, q_0, F)$, the result of merging states $q_i, q_j \in Q$ is the machine A' which has these two states replaced by a new state q_{ij} as follows:

$$A' = ((Q - q_i, q_j) \cup \{q_{ij}\}, \Sigma, \delta', q_0', F')$$

where: $q_0' = q_{ij}$ if q_0 is either q_i or q_j, $F' = (F - q_i, q_j) \cup \{q_{ij}\}$ if either q_i or $q_j \in F$, and δ' is the result of replacing all instances of both q_i and q_j by q_{ij} in all the triples (q_n, a, q_m) that define δ. This is applied recursively in a hill-climbing manner.

The compressed transition diagrams and hypercubes for languages 1-s and 1-r are shown in Figures 3 and 4.

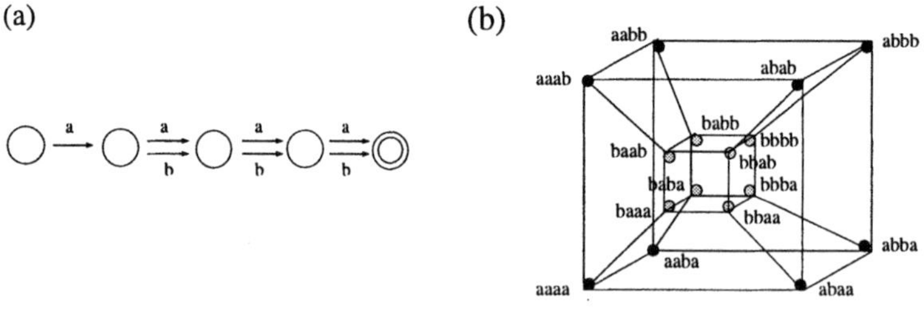

Fig. 3. (a)Transition diagram of the compressed grammar for language 1-s. (b) Hypercube representation of the sentences included in the compressed grammar.

The compressed 1-s is much smaller than that of the original ($MDL = 56.8$ vs. 119.3), due to changes both in the number of states (11 to 5) and in the number of transitions (12 to 7). All of the accepted strings in this example are of

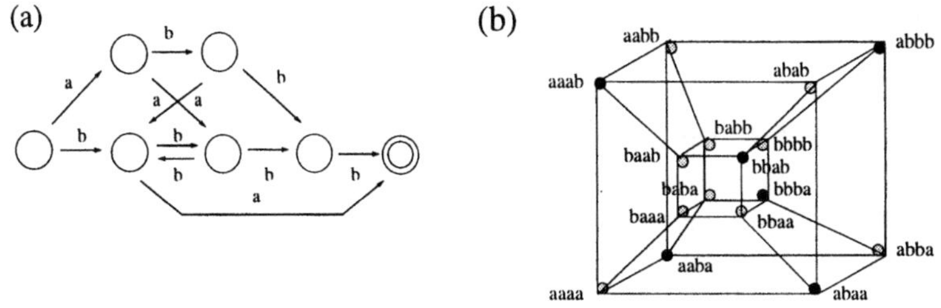

Fig. 4. (a)Transition diagram of the compressed grammar for language `1-r`. (b) Hypercube representation of the sentences included in the compressed grammar.

the correct length, but the compressed version generalizes to include the entire outer cube of sentences in the hypercube. In "real life" such a generalization may or may not be desirable.

The compressed version of language `1-r` is also substantially smaller than the original ($MDL = 92.6$ vs. 168.0). Again this resulted both from a decrease in the number of states (15 to 7) and in the number of transitions (15 to 11). Compression did not lead to new planes or observable regularities, except that the one new sentence of length 4 which was added, `bbba`, was also a distance 2 from its nearest neighbor. Here, however, generalization was such that sentence lengths other than 4 were allowable, ranging from lengths of 2 to infinity, so long as certain regularities, such as embedded tandem repeats of `bb`, are observed.

6 Compression of languages

Compression results from exploiting regularities in the data. We expect that languages with smooth string spaces contain more regularity to be exploited than languages with rugged string spaces. Therefore, we expect more compression from smooth languages. Is this the case?

We created 40 languages with each of several degrees of smoothness. Each consisted of 10 strings, with lengths of 6 symbols drawn from the alphabet $\{a, b\}$. Smoothness was varied as follows: *distance-1, distance-2, distance-3* and *random*. All strings were 1, 2, 3 or random symbol-substitutions, respectively, from their nearest neighbor, drawn successively from a random starting string.

These languages were presented to agents who created the finite state automata grammars for them, then compressed the grammars using the algorithm described above. Compression was measured by MDL of original and compressed grammars. Error was measured by presenting the agents with all possible strings of length 6, then counting the number that were accepted but were not strings in the original language. Recall that the compression algorithm will always accept the original language, so error can also be viewed as the amount of generalization (to strings of length 6) that the grammar achieves; it says nothing about error in recognizing strings of other lengths.

As would be expected, smoother languages could be encoded more economically than the rugged ones. The amount of compression that was achieved is measured by the compressed MDL which was: 126.6 for distance-1, 199.6 for distance-2, 174.8 for distance-3 and 181.4 for the random language.

It is evident that compressed MDL was much less (i.e. more compression was possible) for the distance-1 languages than for the distance-2 ones. This was expected. But the grammars for distance-3 and random languages suddenly became more compressible.

One possibility for the higher compression of distance-3 and random might be that they were unable to extract regularity and simply accepted more strings into the language. The mean numbers of errors was consistent with this explanation: 7.18 for distance-2, 29.37 for distance-3 and 27.42 for random languages. However, the mean number of errors for distance-1 was also large, 23.48, so at best the explanation is still unclear. It is possible that this will change if the agents are presented with more examples, so that the cost of encoding the data is significantly higher for distance-3 and random landscapes. We are attempting to understand this better.

7 Evolution of language

We explored the role of compression in the evolution of language in a simple way. Each language initially began with 10 strings of length 6, drawn from the alphabet $\{a, b\}$ as above. There were 10 replicates of distance-1 languages and 10 of distance-2. The parent in generation n then produced 10 sentences, at random from its language, and presented these to generation $n+1$, who would generate the appropriate grammar. The generation $n+1$ would then compress that grammar and, with the compressed grammar, produce 10 more sentences for generation $n+2$ and so on. This was continued for 10 generations.

There were clear changes observed with all languages. Foremost among the changes were the numbers of strings in the language. While all began with an average 7.7 strings without duplicates, after 10 generations this had changed to an average of 11.8 for distance-1 and to 6.1 for distance-2. This is statistically significant at the .05 level by a paired t-test. Transmission from generation 0 to 10 was lossy, with the average number of strings that were included in the original set of examples lost by generation 10 being 2.3 for distance-1 and 5.3 for distance-2.

The mean smallest distance between strings stayed about the same in both distance-1 (from 1.0 to 1.18) and distance-2 (from 2.0 to 1.72). We had expected that smoothness would increase with time, but this apparently was not the case since there were not significant changes. An increase must be expected from the distance-1 languages, because they began at the lowest value possible, and there is nowhere else to go but up. The mean distance did increase slightly. For the distance-2 languages, however, the mean smallest distance decreased only slightly.

Somewhat paradoxically, the mean MDL for both grammars decreased, in spite of their lack of change in ruggedness. Figure 5 illustrates the change in MDL which occurred, showing a large difference in the early generations, with only modest change later.

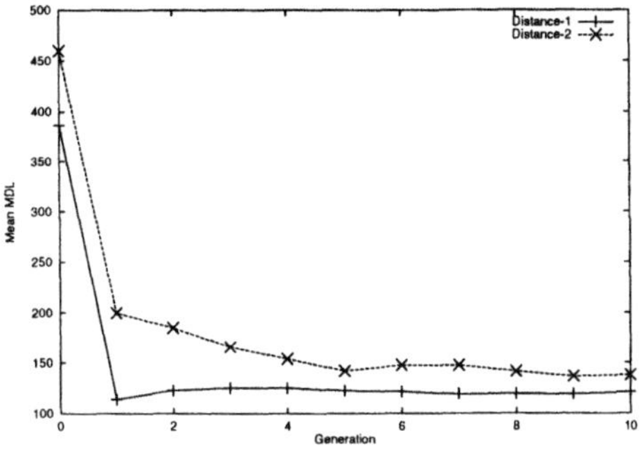

Fig. 5. Mean Minimum Description Lengths for evolving languages of symbol substitution distances 1 and 2 for generations 0 to 10.

It is clear that while the complexity of the grammar was decreased in both, the distance-2 grammars remained more complex, though they admitted a much smaller number of sentences.

8 Discussion

We have examined several conjectures about compression in adaptive complex systems. We employed agents that could learn and evolve one class of formal languages, using one compression algorithm. This learning and evolution was based on syntax alone, without any reference to semantics, which is likely to be important [10]. Our results, while so limited, do seem illuminating.

Conjecture 1: Compression aids in generalization This was examined with systems that learned and compressed languages of varied ruggedness. In all cases we observed a compression of grammar, and in nearly all cases the compressed grammars generalized to admit new strings into the language.

It should be recognized that such generalization may be desirable, because it reduces the complexity of rules, but it can also admit mistakes. To date we have made only a cursory study of the tradeoff between MDL and error in our system, but there clearly are important issues that will warrant further study.

Conjecture 2:Compression occurs more easily in a "smooth", as opposed to "rugged", string space We explored compression in systems where strings were 1-, 2-, or 3- symbol substitutions apart from their nearest neighbor or were randomly placed in the sentence space. We observed that even the uncompressed grammars were smaller for the smooth than for the rugged languages. Compression, as measured by compressed MDL, was clearly greater in in languages where sentences were 2 symbol substitutions apart than if they were only 1 substitution apart. We interpret this to mean that patterns can more easily be identified and exploited by the compression algorithm in the smoother language. However, where smoothness is still less, in the distance-3 and random languages, the compression is also greater than for distance-2 languages. The reason(s) for this remain unclear, but there is a correlation between compression ratio, error rate, and number of examples presented that is important here and needs further exploration.

Conjecture 3: Constraints from compression make it likely that natural languages come to have smooth string spaces

For the purposes of this paper we accept as true the theory that when learning language we (a) compress grammar with simple rules and (b), all else being equal, apply these compressed rules preferentially to new situations. The agents in our study automatically created grammars which admitted all legal sentences. These grammars eliminated some redundancy, but retained logical equivalence. They were already *programmed*, quite literally, to conform to this theory. They were also programmed to compress those grammars, when so directed.

Here it is important to distinguish carefully between the smoothness of a language and the complexity of the grammar needed to describe it. While related, they are not the same. Both distance-1 and distance-2 languages evolved into languages with sentences that were, on average, separated by about the same number of symbol substitutions as when they started. That is to say, they did not become smoother in the sense of becoming more connected or to consist of strings lying together on the same hyperplane. At the same time, the grammars describing them came to have smaller MDLs. That is, both languages came to be described by simpler grammars. When we generated languages for compression, we observed that the smoother languages did have, on average, smaller MDLs, but not invariably so. Clearly there is some subtlety. The suggestion here is that grammatical complexity does, indeed, become simpler, but that this is not the same as saying that the languages become smooth, as smoothness is used in this paper.

In his study of evolving bit strings, albeit with semantic content, Kirby [10] observed two phase transitions in grammatical structure. The first of these occurred when the languages became suddenly more expressive, with a concomitant increase in grammatical complexity. The second phase change occurred after a high degree of expressivity was achieved, then the grammatical complexity suddenly started to become much less. That study, and ours, are in agreement that even in such simple cases as evolving bit strings there are unlikely to be simple rules about changes in smoothness of languages, or grammatical complexity – though in the long run both studies did result in simpler grammars after sufficient time.

In summary, we observed broad agreement with the conjectures made prior to the start of the study. It is evident, however, that even in our simple system there is significant subtlety that must be recognized in the inductive tradeoff of generalization through compression versus error of overgeneralization.

Acknowledgement This work was supported by grant #5BR9720410 from the National Science Foundation. We thank Stuart Kauffman and members of the UCLA Cognitive Science LIS Research Group for helpful discussions.

References

1. N. Chomsky. *Aspects of the Theory of Syntax.* MIT Press, Cambridge, MA, 1965.
2. N. Chomsky. *The Minimalist Program.* MIT Press, Cambridge, MA, 1995.
3. M. Gell-Mann. Talk at Santa Fe Institute, January 9 1990.
4. E. M. Gold. Complexity of automaton identification from given data. *Information and Control*, 37:302–320, 1978.
5. P. Grünwald. A minimum description length approach to grammar inference. In G. Scheler S. Wermter, E. Riloff, editor, *Symbolic, Connectionist and Statistical Approaches to Learning for Natural Language Processing*, LNCS #1040. Springer-Verlag, Berlin, 1996.
6. T. Hashimoto. Usage-based structuralization of relationships between words. In P. Husbands and I. Harvey, editors, *Fourth European Conference on Artificial Life*, pages 483 – 492, Cambridge, MA, 1997. MIT Press.
7. J.E. Hopcroft and J.D. Ullman. *Introduction to Automata theory, Languages and Computation.* Addison Wesley, Reading, MA, 1979.
8. S.A. Kauffman. *The Origins of Order: Self-Organization in Evolution.* Oxford University Press, New York, 1993.
9. M.J. Kearns and U.V. Vazirani. *An Introduction to Computational Learning Theory.* MIT Press, Cambridge, MA, 1994.
10. S. Kirby. Syntax without natural selection: How compositionality emerges from vocabulary in a population of learners. Unpublished ms., 1999.
11. S. Kirby and J. Hurford. Learning, culture and evolution in the origin of linguistic constraints. In P. Husbands and I. Harvey, editors, *Fourth European Conference on Artificial Life*, pages 493 – 502, Cambridge, MA, 1997. MIT Press.
12. M. Li and P. Vitányi. Minimum description length induction, bayesianism and kolmogorov complexity. In *1998 IEEE International Symposium on Information Theory*, MIT, Cambridge, 1998.
13. J.A. Moore. *Science as a Way of Knowing.* Harvard University Press, Cambridge, MA, 1993.
14. P. Niyogi and R.C. Berwick. The logical problem of language change. Technical Report A. I. Memo No. 1516, MIT Artificial Intelligence Laboratory, July 1995.
15. C.H. Papadimtriou and M. Sideri. On the evolution of easy instances. *Unpublished manuscript*, 1998.
16. J. Rissanen and E. Ristad. Language acquisition in the {MDL} framework. In E. Ristad, editor, *Language Computations.* American Mathematical Society, Philadelphia, PA, 1994.
17. K. Sayood. *Introduction to Data Compression.* Morgan Kauffman, San Francisco, CA, 1996.
18. L. B. Slobodkin. *Simplicity and Complexity in Games of the Intellect.* Harvard University Press, Cambridge, MA, 1992.
19. E. Stabler. *Computational Minimalism: Acquiring and Parsing Languages with Movement.* Basil Blackwell, Oxford, 1999.
20. V.N. Vapnik. *Statistical Learning Theory.* Wiley, NY, 1998.
21. J. Weber and C. Wong. Mutation of human short tandem repeats. *Human Molecular Genetics*, 2(8):1123–1128, 1993.

Effective Lexicon Change in the Absence of Population Flux

Christopher Dircks, Scott C. Stoness

School of Computing Science, Simon Fraser University, Burnaby, BC, Canada

Abstract. Multiple agents, equipped with a feature-based phonetic model and a connectionist cognitive model, interact via the *naming game* paradigm, such that lexicon formation and change is an emergent property of this complex adaptive system. Our system converges on a coherent lexicon and effective language change is demonstrated, even in the absence of a changing population, which brings into question claims made in earlier work. We argue that our phonetic and cognitive models tend towards a cognitive validity which was absent from previous work in this area, while maintaining the flexibility of other systems.

1 Introduction

The pace of technological advance and invention in today's world forces the aware individual into intimate contact with language change, especially in the form of lexical creation. It is certainly possible to examine word-formation and word-adaptation processes which occur in response to a terminological void introduced by the creation of a new object or the discovery of a new idea; however, until recently practical investigations into longer-term linguistic change have been severely curtailed by the lack of appropriate experimental subjects. Where historical sources exist, some theoretical conclusions have been drawn, but written records often do not accurately capture phonological processes, and in any case, testing the predictions of such theories might take years, or even generations – assuming, of course, that the design of a controlled experiment is possible in the first place.

Recently, members of the Sony CSL Paris have begun to examine language change as the emergent behaviour of a complex adaptive system. Such a view of linguistic variety enables researchers to apply artificial life and multi-agent techniques to this problem, thus circumventing the difficulties associated with prediction testing by moving the domain of experimentation to the main memory of a computer. In a recent paper produced by this group ([8]), four factors are postulated as necessary for "effective lexical change" to occur: self-organization, stochasticity, agents tolerant of variation, and a changing population.

We perform our investigations into the nature of lexical change within the framework of a multi-agent simulation using the *naming game* (first introduced in [6]) as a paradigm for agent interaction; however, we have augmented our implementation with a more realistic phonetic model than is present in [8], the starting point for our work. The behavioural mechanisms of our agents are based on an internal connectionist model.

2 The Naming Game

Research involving multiple software agents requires an unambiguous specification of the precise details of all possible agent-agent and agent-environment interactions. To this end, [6] introduces the *naming game*, an austere paradigm for interaction tailored especially for the development, transmission, and evolution of a lexicon.

The *naming game* is appropriate for any simulation with a fixed number of agents and objects; an interaction proceeds as follows: (1) Two agents are selected at random out of the population; one is designated as the Speaker, the other as the Hearer. (2) The Speaker chooses an object at random and names the object; that is, accesses an internal representation and produces a linguistic form. (3) The Speaker (virtually) points at the object. (4) The Hearer interprets this combination of linguistic and extra-linguistic information produced by the speaker to be a reference to a particular object.

The game succeeds if the Hearer correctly interprets the information provided by the Speaker; if the two agents do not agree on the object referenced, then the game fails.

Presumably, upon completion of a *naming game* interaction, learning occurs; while the bulk of research uses the adaptive rules outlined in [6], there is no reason to suppose that this framework must be coupled with precisely those learning rules. We demonstrate that the *naming game* provides an excellent paradigm for use with agents possessing different internal mechanisms and learning procedures.

3 The Phonetic Model

We enhanced the original simulation by replacing the random generation of words from character sets which characterizes [8] with a binary feature representation where agents' utterances are consonant-vowel pairs, a somewhat more realistic phonetic model.

Chomsky-Halle features and the cardinal vowel system provide us with a loose basis for our model. There are four consonant features (anterior, coronal, voiced continuant) and three vowel features (closed, mid, back), allowing 16 possible consonants and 8 possible vowels, giving our agents 128 potential abstract CV combinations.

Rather than purely binary values, we allow real-valued features across the interval [0,1], which changes the domain of the utterances of our Speaker agents from idealized phonemes to a naïve abstraction of actual acoustic productions. The cardinal vowel system specifies infinitely variable tongue positions relative to standard points, and voicing delays and tongue positions for consonants also vary across speakers. In real world communication, the hearer takes this fuzzy encoding and interprets it as abstract phonemes; this is a process which our agents model.

4 Agent-Internal Connectionist Architecture

Each agent has two completely separate neural nets; one outputs an utterance given precise object information (the S-Net: e.g. SpeechNet), and one (the H-Net: e.g. HearNet) identifies a particular object given an utterance and appropriate extra-linguistic information.

The S-Net is a two-layer, fully connected feed-forward network with an input node corresponding to each object, and an output node for each phonetic feature. After choosing an object as the topic of the conversation, input node activations are initialized to 0 except for the topic node, which is set to 1.0. Connection weights between the layers range over the interval [0,1], and since only one input node is on, the values of the appropriate weights translate directly as output layer activation levels. These activations can be interpreted as the value of the corresponding phonetic feature in the utterance.

The H-Net interprets the combination of non-linguistic cues and phonetic information from the S-Net, ultimately choosing an object which (hopefully) matches the original topic. There are two sets of input nodes, corresponding to phonetic features and object information. The activation levels from the output nodes of the S-Net are transferred directly to the phonetic nodes, while the object nodes are set up to represent the extra linguistic 'pointing' information from the Speaker. We model the uncertainty inherent in physical indications of location by probabilistically 'smearing' the virtual pointing information. Five of the ten objects are given a weight, thus modelling the viewer's confusion about the target, which could arise because of visual similarity between the objects, their spatial proximity, or other, more complicated issues. The combination of this imperfect physical information and the phonetic communication allows the H-Net to decide upon a topic.

Our simulation achieves an 86% success rate using phonetic information only, but this jumps to 95% when non-linguistic cues are also processed. Object information alone, however, proves too fuzzy, as only a 50% success rate is achieved on this basis.

Following input layer initialization, activation levels are multiplied by the appropriate weights and a sum of products calculation is performed for each output node. Competitive inhibitory links between objects are then implemented as a winner-take-all model; the output node with the highest score is selected as the Hearer's interpretation of the Speaker's actions.

5 The Training Regimen

Learning only occurs when the *naming game* fails; individuals are unlikely to adapt their language model when communication succeeds. When a topic misunderstanding occurs between humans, the two participants do not simply leave; rather, the speaker is likely to repeat the word, or even touch the object to ensure successful communication. Thus it is reasonable to suppose that our Hearer agents, upon failure of the game, would have accurate information about both the Speaker's utterance and the intended topic. If communication fails, the Hearer adapts such that in the future its speech will more closely resemble the Speaker's and it will be more accepting of that Speaker's accent.

Weights in the S-Net are randomly initialized to values between 0 and 1, providing our Speaker with a random initial language. The phonetic weights in the H-Net are initialized on the interval [0,1], but represent the phoneme's contribution to an object's score, rather than explicit phoneme values. H-Net weights between input and output object nodes are not trained. Diagonal weights are given uniformly high values, other weights are given random lower values. This approach simplistically models similarities between objects.

During training each feature of the Hearer's S-Net is examined independently to determine if it would binarize to the same value as the corresponding feature in the Speaker's utterance. If so, its value is reinforced; if not, it is punished. The punishment equation moves values towards 0.5, while the reinforcement function tends towards 1 or 0, depending on the weight's polarity. Random fuzz of 5% is also required, otherwise the punishment function will never move a weight across the fixed point of 0.5.

The only trained weights in the H-Net are those between the phonetic input nodes and the output nodes (object weights remain fixed at their initial values). Again, training occurs only when the Hearer has chosen the wrong object; this weight modification makes the H-Net more likely to settle on the topic given the same phonetic input.

Accordingly, we decrease the score of the false positive, and increase the score of the correct answer, a straightforward procedure which does not complicate the dynamics of the network, and gives weight values in a reasonable range without explicit constraints. Furthermore, there is a certain level of cognitive justification, since humans are more likely to take note of errors and correct responses than to try to correct near-mistakes.

6 The Simulation World

There are 20 agents and 10 objects in our world, numbers chosen to coincide with [8]. However, our simulation successfully scales to at least 50 agents and 20 objects.

There is no attempt to make the agents internally consistent at 'birth', and we never explicitly test for internal consistency – while this could result in the unfortunate situation where an agent cannot understand itself, in practice this does not occur.

In experimental runs, agents are selected to speak in order, with a randomly chosen Hearer. In simulation runs with population flux, every 2000 games a random individual is removed from the population and a new, randomly initialized agent takes its place. Since simulation runs consist of either 100,000 or 400,000 instances of the *naming game*, either 50 or 200 new individuals are created during a run with population change.

7 Experimental Results and Discussion

In our long-term trials, we achieved success rates around 95% when the population was stable, and around 80% with a dynamic population. We ran 9 short and 6 long simulations with a static population, and 9 short and 7 long simulations with population flux. As can be seen from Table 1, we found lexical change in both the dynamic and static populations. Lexical form change is defined as a single form which is spoken by 60% or more of the population for a significant period of time (at least 10% of the simulation length), which is then replaced by a second form which comes to dominance.

Table 1: Percentage Form Change

	Static		Dynamic	
	Short	Long	Short	Long
Form Change	6%	10%	13%	47%

It is the final claim in [8] that four factors are necessary for effective lexicon change to occur. It is difficult to see how one could prove, through simulation only, that these four factors form necessary as well as sufficient conditions for lexical change; indeed, our investigations disprove the fourth factor, showing that while lexical change is much more frequent when population change is introduced, it occurs even in a static population.

One might argue that our definition of language change is too weak; however, a typical example has a form spoken by 65% of the population for 40,000 instances of the *naming game* before it is replaced. By contrast, it often takes as little as 5,000 instances to achieve stability in cases where a single form dominates with >90% of the speakers.

Even a single instance of change refutes the claim that population flux is a necessary condition for lexicon change; we recorded lexicon change in 4 of the 9 short interaction runs. This means that in almost 50% of the runs with static populations, at least one form underwent effective lexicon change. Furthermore, the incidence of lexical change is higher in the longer runs, which indicates that it is not an artifact of the initialization of the system.

Unfortunately, 'effective' change has not yet been defined. Perhaps a form switch after a period of 60% dominance is not 'effective' change. We do not observe forms which are completely (e.g. >90%) dominant falling into disuse, but our results are from only 15 simulation runs; perhaps in 100 or 1000 runs this might occur. Our results are indicative that even given this strong definition, 'effective' language change can occur in a static population, and we have amply demonstrated the extinction of forms spoken by a majority of speakers (13% of all forms in our longer simulations).

9 Conclusions and Future Work

We have shown that the *naming game* is an effective paradigm even when completely divorced from its original implementation, and that similar results to those reported in [7] and [8] can be obtained using a connectionist architecture for the agents' cognitive model, and with a phonetic model which is more linguistically accurate. These results and other recent work indicate that modelling language as the emergent behaviour of a complex adaptive system can be a valuable tool for linguistic investigation.

We would like to model physical constraints of the vocal tract so as to have the agents produce more realistic sound combinations, especially with an expanded feature set, allowing a larger number of phonemes. We hope to introduce an object model in which objects are represented by feature vectors rather than simply atomic nodes, to see if hierarchical concepts might be instantiated as lexical items under these conditions.

The occurrence of lexical change in a static population, which we've observed contrary to the expectations of [8], merits further investigation. [8] grafts stochastic measures onto a deterministic backbone, and uses a rigid, orthogonal phonetic model, training agents on complete words. We use an inherently stochastic connectionist approach, and our real-valued phonetic model increases flexibility. In fact, our H-nets are trained to accept a schema (in the sense of [4]), and the S-Net of the Hearer is trained on the piecewise error. We suspect that these modifications allow our simulation to be permissive enough of variation so as not to require the external prodding via population flux which is necessary to produce lexical change in [8].

References

[1] Carr, Philip. *Phonology*. London: The Macmillan Press, Ltd, 1993.

[2] Caudill, Maureen and Charles Butler. *Naturally Intelligent Systems*. Cambridge, MA: MIT Press, 1991.

[3] Gurney, Kevin. *An Introduction to Neural Networks*. London: UCL Press Ltd., 1997.

[4] Holland, John H. *Adaptation in Natural and Artificial Systems*. Cambridge: MIT Press, 1995.

[5] Rogers, Henry. *Theoretical and Practical Phonetics*. Toronto: Copp Clark Pitman, Ltd., 1991.

[6] Steels, Luc. "Self-organizing vocabularies" In: Langton, C(ed.) *Proceedings of Alife V*, Japan.1996.

[7] Steels, L. and F. Kaplan. "Stochasticity as a source of innovation in language games," *Proceedings of Artificial Life VI*, Adami et al eds. Los Angeles: MIT Press, 1998.

[8] Steels, L. and F. Kaplan. "Spontaneous Lexicon Change," *Proceedings from COLING-ACL 98*. Montreal: Université de Montréal, 1998.

[9] Steels, L and P. Vogt. "Grounding adaptive language games in robotic agents." In Harvey, I. et.al. (Eds.) *Proceedings of ECAL 97*, Brighton UK, July 1997. The MIT Press, Cambridge Ma., 1997.

Self-Organization of a Lexicon in a Structured Society of Agents

Pierre-yves Oudeyer
e-mail : pyoudeye@ens-lyon.fr
Ecole Normale Superieure de Lyon
46, allee d'Italie, 69007 Lyon, France

Abstract. The naming game is a formal mechanism that describes the development of a lexicon in a society of culturally interacting agents. We will here use a cellular automaton version of this game to study the influence of an extra-linguistic structure over the evolution of the lexicon, but also the influence of language over this a priori structure. This extra-linguistic structure will be coded by first giving a location in a 2-D world to agents, and then by allowing them to move in relation to the outcome of the naming games. The results we will present show strong self-organization phenomena, such as the appearance of language and geographical clusters, in addition to the basic properties of the game (high communication success).

1 Introduction

Language as a complex dynamical system has been increasingly studied in the last decade. Formal models have been built to investigate the question of the emergence of language through self-organisation in a society of culturally interacting agents. This work is practically done both in simulations and by using robotic agents ([1]). One of the questions raised is : how can a lexicon shared by many agents emerge ? How can initially random form-meaning associations self-organize through cultural evolution ? The naming game is a formal mechanism that has thus been introduced to tackle this question ([3]), and will be described in section 2. Many results have already been worked out, but most of them concentrated on the language itself and did not consider the possibility of having a priori extra-linguistic structure in the agent society. The first step was done in [2] where the idea of spatialization was introduced : the agent society is given a fixed topological structure (that can be based on a geographical or social distance for instance) that determines with who agents may speak, as we will see in section 2. We will present new results in section 3, based on a purely cellular automaton model of the naming game We will show that this topological structure has an important impact over the lexicon development, and that emergent phenomena such as the appearance of what we will call language clusters. Then, in section 4, we will generalize this model by having a variable topological structure in which agents will move according to the outcome of the naming game they are involved in : the language modifies the structure.

2 The Naming Game : Formal Model

Let us first define what is a lexicon L : it is a set of associations form meaning $L = \{(f_i, m_i, s_i)\}$ where a form f_i is a symbol, a meaning m_i denotes a category/object, and s_i is the score of the association, constrained by $s_i \in [0,1]$. Two operations can be performed over a score : increase by SC or decrease by SC. Because of the constraint $s_i \in [0,1]$, when the result of an operation is greater than 1, it is mapped to 1, and when it is smaller than 0, it is mapped to 0. Moreover, both forms and meanings can appear several times in different associations.
An agent a_i is here a t-uple $a_i = ((x_i, y_i), L_i, w_a, w_c, IS, SC)$ where (x_i, y_i) is its location in a 2-dimentional world, L_i is its current lexicon which is initially empty, w_a the probability of accepting a new association (f_i, m_i, s_i) when needed as explained below, w_c the probability of creating a new association (f_i, m_i, s_i) when needed, IS the initial score of accepted and created associations, and SC the quantity by which a score can be changed (increased or decreased) when needed.
A society of agents is defined as the 3-uple $Soc = (\{a_i\}, loc, \{m_i\})$ where $\{a_i\}$ is the set of agents, *loc* the locality used as explained hereafter, and $\{m_i\}$ is the set of meanings/objects that agents will have to name by associating them one or several forms.
A round of the naming game consists in picking up randomly one agent, called the speaker s, who himself chooses randomly one of its *loc* neighbors (the hearer h), according to the euclidean distance. Then the speaker chooses randomly a topic $m \in \{m_i\}$. Let $E = \{assoc_k = (f_k, m_k, s_k) | (m_k = m) \wedge (assoc_k \in L_s)\}$:

- if $E = \emptyset$, then this means that the speaker has no form for m, and then he creates (with probability w_c) randomly a new form f with probability w_c and add (f, m, IS) to L_s. The round ends and the result is said to be FAILURE.
- if $E \neq \emptyset$, then s chooses randomly $(f, m, s) \in E$ which is an association in E with a maximal score, and will be called the last preferred form for m for agent a_i, or just preferred form for m. Then, s points to m with an extra-linguistic tool (the hearer always understand what is the topic here), and utters f. Let us now consider the predicate $P = (\exists s_k | (f, m, s_k) \in L_h) \wedge (s_k > 0)$:
  - if P is false, then this means that the hearer does not understand and the round ends with the result FAILURE. The speaker decrease the score of its association involving f and m. The association $\{(f, m, IC)\}$ is added to L_h with the probability w_a.
  - if P is true, then this means that the hearer does understand, and the round ends with result SUCCESS. The speaker increase the score of its association involving both f and m, and decrease all associations involving either f or m.

3 Experiments with agents that have a fixed position

The first measure we are going to use is the evolution of SUCCESS/FAILURE with the number of rounds. The second one is what we will call evolution of form-spread for a given meaning m, along with the number of rounds. A form-spread for m in round r is defined as $FS_{m,r} = \{(f_i, p_i) | (\exists a \in Soc | f_i =$ last preferred word of a for $m) \wedge (p_i =$ percentage of agents that have f_i as their preferred form for $m)\}$. The form-spread graphs for m will thus be composed of $max_r|FS_{m,r}|$ curves, each one corresponding to one $f_i \in FS_{m,r}$ and representing the evolution of the associated p_i in function of the number of rounds. This measure allows to follow the evolution of a given preferred form for a given meaning along with the number of rounds and within the population of agents. Other measures will be used and presented when needed.

The experiments in this section deal with agents whose location does not change with time. All agents have the same parameters w_a, w_c, IS and SC. When parameters are not precised, it means that they have default value : 100 agents randomly located in 2-D space, 4 meanings, $loc = 8$, $w_a = 0.1$, $w_c = 0.1$, $IS = 0.1$ and $SC = 0.1$.
The first results yielded by those experiments are that like in the non spatialized naming game (NSNG) ([2]), success reaches more than 85 percent after 1500 games, and this at the same speed. But unlike NSNG, where one form rapidly dominates for a given meaning, several stable forms appear here and go to an equilibrium where each one keeps its percentage of users : Figure 1 shows form-spread for one of the meanings. This is true for nearly all parameters, except when $IS = 1$ where only one form eventually dominates.

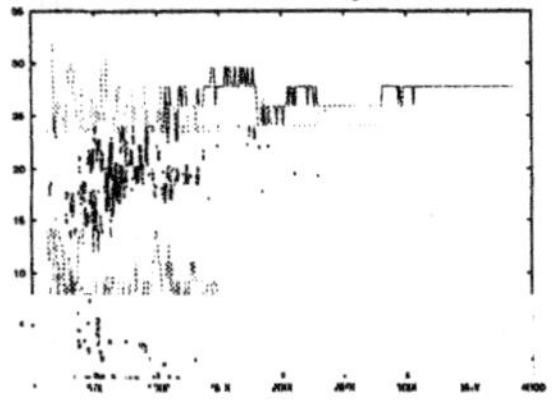

Figure 1 : evolution of form-spread for meaning 1 with 100 agents

Moreover, language clusters for a given meaning m appear : users of a given preferred form for m are geographically neighbors. Moreover, experiments repeated 100 times over 300 randomly located agents yielded the following results about these clusters :

- for a given locality, the number of clusters is very stable ;
- the number of these clusters at convergence state along with locality follows a power law ;
- the occurence of a given size of cluster for a fixed locality follows also a power law (in function of this size).

An interesting phenomenon appears when we introduce agent flux : every n rounds, one agent is randomly choosen and removed while a fresh one (with empty lexicon) is added in a random place. Furthermore, flux is here introduced when the equilibrium has been reached. A low flux (every 30 games in a population of 25 agents) produces no change in communication success, but destabilizes language clusters and surprisingly one preferred form quickly dominates for a given meaning (see Figure 5) , in the same way as in non-spatialized naming game : instead of bringing diversity, agent flux brings unity to the language. Medium flux (every 15 games) shows the same phenomenon, but now success falls to about 50 percent. Finally and logically, when flux is high (every 10 games), language collapses : success falls to 30 percent and no form manages to install itself.

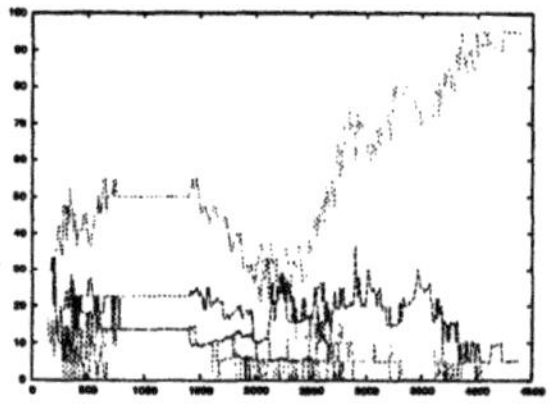

Figure 5 : form-spread for meaning 1 with 25 agents and flux ($n = 30$) introduced after 1400 rounds

4 When Agents Move

Now that we have studied basic properties of the impact of a structure in the society over the development of a lexicon, we will investigate the case in which agents have no more a fixed location but instead move according to the outcome of the rounds : topology and language are bi-directionnaly coupled. Let us first precise how this coupling is defined. At the end of each round, if the result is a SUCCESS, then the speaker moves towards the hearer in the following way : $(x_s, y_s) \Longrightarrow (x_s + C * (x_h - x_s), y_s + C * (y_h - y_s))$ where (x_s, y_s) is the position of the speaker before he moves, (x_h, y_h) is the position of the hearer, and C a constant typically equal to $\frac{max_i(x(a_i)) - min_i(x(a_i))}{|Soc|}$ where $x(a_i)$ stands for the abscisse of agent a_i in the first round : this does that agent displacements are very smooth. When the result of a round is a FAILURE, the speaker performs the same displacement but with $-C$ instead of C. Last, in all experiments agents have a random position in the first round.

Let us now come to what we observe. First, the evolution of success is the same as when agents did not move, contrarily to the a priori idea that it would grow faster because agents try to avoid failure and foster success : the reason is because meanings interact through space. On the other hand, we also reach a similar stable-state in form-spread for a given meaning but the difference is that the number of stable forms is here much smaller and grows very slowly with locality and size of population, and that these stable forms are not frozen for ever : small changes appear, explained by the bumping of clusters as explained below. The most interesting result comes if we now look at the evolution of topology in relation to these stable forms for a given meaning. Indeed, geographical clusters (defined only by the location of agents) quickly emerge as shown on Figure 6.

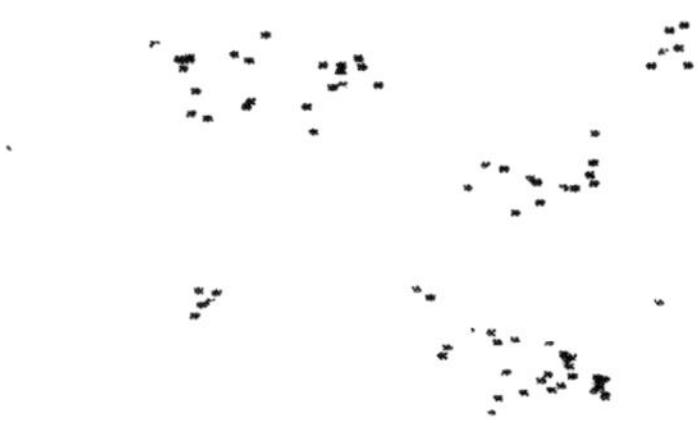

Figure 6

Moreover, most of the time, these geographical clusters are also language clusters for every meaning. This means that in the same geographical cluster, most of agents share the same preferred word for each meaning. Because agents do not stop moving even when they are in a cluster with other agents having the same preferred forms, clusters also move, and this is done in a random walk manner. The consequence is that sometimes two clusters simply bump into each other. Two behaviors may appear : either they bounce and continue independantly their walk, or they melt. The possibility of melting is given when these two clusters have a common form for one or more meanings : this constitutes an attraction force that can hold them together. When they melt, in the case where IS is close to 1, the resulting cluster has a very interesting property : one form soon dominates for each meaning. The important thing is that the origin of these winning forms (one cluster or the other) is random. Thus we observe a real fusion of languages : for instance, cluster c_1 could have form f_1 for m_1 and f_2 for m_2, cluster c_2 have form $f_1^{'}$ for m_1 and $f_2^{'}$ for m_2, and the resulting cluster $c_1 + c_2$ can happen to have form f_1 for m_1 and form $f_2^{'}$ for m_2.

5 Conclusion

We have shown that introducing extra-linguistic structure in a society of agents that use the mechanism of the naming game leads to the emergence of language clusters. Moreover, by coupling the outcome of the rounds to the displacements of agents, we saw that language itself shapes the extra-linguistic structure.

References

1. P. Vogt, 1998, The Evolution of a Lexicon and Meaning in robotic agents through self-organization, In : C. Knight and J.R. Hurford (eds.). The evolution of Language (selected papers from the 2nd International Conference on the Evolution of Language, London, April 6-9).
2. L. Steels, 1997, Language Learning and Language Contact. In : W. Daelemans, A. Van den Bosh and A. Weijters (eds.), Worshop Notes of the ECML/MLnet Familiarization Workshop on Empirical Learning of Natural Language Processing Tasks. Prague. pp 11-24.
3. L. Steels, 1996, Self-oganizing vocabularies. In Langton, C. (ed.) Proceedings of Artificial Life V. Nara, 1996

Modeling Categorization Dynamics through Conversation by Constructive Approach

Takashi Hashimoto

Knowledge Science, Japan Advanced Institute of Science and Technology (JAIST)
1-1, Tatsunokuchi, Ishikawa 923-1292, JAPAN
hash@jaist.ac.jp

Abstract. Categorization dynamics as the clustering of words in word relation is studied by a constructive approach which is suited to inquire evolutionary linguistics with dynamical view on language. Word meaning is represented by relationship among words. Tthe relationship should be derived from usage of language. Being founded on this usage-based view, we define an algorithm to evaluate word relationship. Using the algorithm, cluster structure and its dynamics of words are shown in a model with communicating artificial agents. The relevance of clustering with linguistic categorization is discussed.

1 Dynamical View on Language

There are two ways of viewing language: structurally and dynamically. The structural view is a static one in which language structure, for example, syntax, dictionaries, or pragmatic rules, offers idealized approaches to language. The alternative view is dynamic. It concentrates on the actual use of language rather than abstract notions of how language ought to be. It is possible to better understand the value of the second approach by thinking of metaphor. Metaphoricalexpressions are creative and dynamic precisely because they can "bend" or "break" the rules of conventionally structured language. By producing or understanding metaphorical expressions, especially creative or unique metaphors, our internal models should change. We can not say valid or not valid for such creative expressions, since the expressions are so novel that it is not valid for a conventional language structure. We should consider whether the expressions are to be accepted or not. If we accept them, our internal structure changes and language structure might also come to be modified. In the dynamical view, the whole system of such dynamical processes is considered as 'language.'

Constructive approaches are highly advantageous to understanding complex systems [4]. These approaches are also useful for studying evolutionary linguistics [7]. In contrast to conventional linguistics which attempts to describe various language phenomena, in the constructive approach the emergence of global order as language-like behavior is modeled through interaction among individuals. However, not only emergence but also the dynamics of global order should be observed in constructive models, since language is indeed an ever-changing system.

Perhaps the internal dynamics of individuals should be taken into consideration to study evolutionary language system so as for individuals to change their internal states and relationship to others and circumstances.

2 Modeling – Word Relationship and Conversation

We have proposed *usage-based viewpoint* on meaning [2] which have claimed that meanings of words should be discussed in terms of how language is used [9]. Interrelationship among words can be employed as a representation of meanings of words to some extent. This point of view implies that relationship of one word to other words should be derived from analyzing the usage of the word in the language, not entirely from its indication or reference. Moreover, a word in a sentence is understood from not only relation with only entities mentioned by the sentence but position in the whole system of language.

Based on this viewpoint, we discuss dynamics of categorization by observing how the relationship among words changes through conversations. Building relationship in use of language is a dynamical process performed by language users. We call this process *sense-making* process [1] to emphasize its subjective nature. The sense-making process is modeled by positioning a word in the relationship among all words.

The algorithm to evaluate relationship between words is basically attributed to Karov and Edelman's work [5] with two revisions. The one is to calculate relationship dynamically in the course of conversation, since what interests us is not in the final form of category but in the dynamics of categorization. The other is to consider 'texts' on higher level than sentences[1]. A text is a stream of sentences uttered and accepted. The relationship between words is defined by the linear combination of the terms of usage-similarity and appearance-similarity using a coefficient α^w as[2]

$$R(w_i, w_j) = \alpha^w(\text{usage-similarity}) + (1 - \alpha^w)(\text{appearance-similarity}) \ .$$

The first term is designed to calculate usage similarity of words in sentences by considering the syntagmatic relationships between words, i.e. words used in a sentence are in strong relationship. Since this algorithm is applied iteratively for each sentence in texts, words used not in a sentence but at the same position in different sentences grow their relationship. In other words, this algorithm is able to capture the paradigmatic relationship from the syntagmatic one.

The second term seizes the similarity among patterns of appearance of words in texts. Words with resemblant patterns of appearance among texts, e.g., words used much often in particular texts but not so in other texts, raise their relationship. Conversely, words with different patterns of appearance weaken their relationship. This is realized by calculating the correlation of appearance in texts.

[1] A text is a set of sentences. In our paper, this is applied to a conversation.

[2] As space is limited, for the detail of the algorithm, see [3].

We model a simple conversation process between agents having word relation matrices as their internal structures. Here, we focus on dynamical changes of internal structures of agents through exchanging sentences, the simplest act of using language. A conversation between agents starts with uttering a sentence about a topic displayed to the agents. After the beginning of the conversation, each sentence is not restricted to the topic but there is some relevance with the previous sentence. In this model, to express this relevance, at least one word in the accepted sentence should be used in a reply sentence.

The procedure of conversation in a text is as follows:
1. A speaker agent produces a sentence about a topic.
2. The sentence is modified according to the creativity rate, c, and then uttered to a hearer agent.
3. The speaker's word relation matrix is updated in terms of the uttered sentence.
4. The hearer accepts the uttered sentence if there are less than two unknown words in the sentence[3]. If the sentence is not accepted, the speaker turns to another topic. (go to 1.)
5. The hearer's word relation matrix is updated in terms of the accepted sentence. If there is an unknown word, the matrix is expanded to incorporate the new word.
6. To reply to the utterance, the role of speaker and hearer are exchanged between them. (go to 1.)

When the number of accepted sentences or that of rejected sentences in a text reach some values, the text ends up. Then another pair of agents and a topic are selected for a new text.

3 Summary of Simulation Results

In one conversation, two agents from five are randomly selected as a speaker and a hearer. Sentences are produced artificially by agents by arranging words in which 5 different characters are combined[4]. The number of topics to be displayed to agents is 10. The maximum of accepted sentences in a text is 100 and that of rejected sentences is 5. The parameters are $\alpha^w = 0.4$, $c = 0.1$.

The followings are the major results:

1. Agents develop cluster structure in their own word relation matrices. We observe two characteristic types of clusters. One is flat type in which words have strong relationship with each other. The other is gradual type in which word relationships change gradually. As a result of development, these two types of clusters exist in combination.

2. Relationship among words drastically changes when a new word is used or a word is used in an unusual way. For example, at the 21st text in Fig.1(a) most words with strong relation with a word in new usage weaken their relation value and vice versa.

[3] Note that the criterion for acceptance of uttered sentence by the hearer lays down the limitation of ability to make sense for new words.

[4] The number of words and that of sentences are in principle infinity.

3. The structure of clusters has stability and adaptability. The change of position of words in cluster structure is examplified in Fig.1(b). The words in a new usage, linked with a dashed arrow, moves its belonging cluster. The other words move so coherently in each cluster that the whole structure of clusters is not modified very much.

4. Structure common to agents develops in the course of conversations.

5. Agents also develop structure peculiar to individuals, because they go through diversified experiences of conversations.

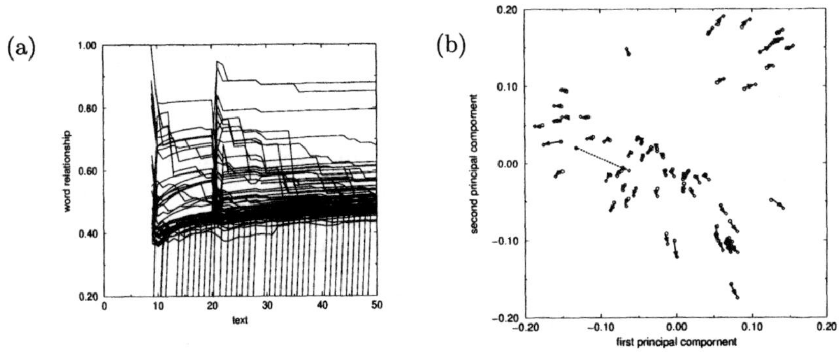

Fig. 1. (a) Transition of word relationship. The x and y axes are the number of texts and the relationship of all words with a word in an agent's word relation matrix, respectively. (b) Dynamics of cluster structure caused by the rapid change shown in (a). This is a scattered diagram from principal component analysis of matrices.

4 Discussions

The clustering can be regarded as categorization through conversations, since words in a cluster have stronger relation with each other and weaker relation with words outside the cluster. Typical clusters are a combination of two types of clusters, flat and gradual, that is, a flat center with gradual expansion into the peripheral. The cluster structure shares some characteristics with the prototype category [6, 8].

In the traditional notion of category, the membership of a category is thought to be defined rigidly like the set notion. In the prototype category theory, in contrast, the membership is matter of gradient and the boundary of a category is fuzzy. The category of liquid containers provides an example. Bottles and glasses are the typical members of the category. Glasses are similar to bowls, bowls are to soup plates, and soup plates are to flat dishes. Although neighboring members are fairly similar, the last one may not be the member of the category, but the boundary which defines the membership of the category is unclear. Another important feature of prototype categories is stability and adaptability with which languages should equip themselves to establish communication and to be flexible about changes. Prototype category and our cluster structure are akin in these traits[5].

Agents develop both commonality and individuality. The structure common to agents implies the emergence of a social system, in which some words are used in the same way by most agents. The words acquire, in the speculative view, virtual references in the society[6]. For a developmental enquiry, we should study how word relationship which reflects relation among prepared entities changes or expands with communication.

The present algorithm shows not a simple convergence but drastic turnovers, which are usually brought by new combinations of words. The turnover behavior locally restructures words in clusters. Such new combinations of words is like metaphorical expressions which often tie different semantic domains by using words from the separated domains in one sentence. And such metaphorical expressions, if they are totally impressive, might modify our internal models, or world view, dramatically. Therefore, it is important for dynamics of linguistic categorization not only to develop clusters but to modify the clusters by a small impact. This is also important for maintaining the dynamics at the global level.

The coefficient parameter α^w controls nonlinearity of the present system. Although the results reported here are seen in the broad area of α^w, the system is likely to fall into fixed and uniform structure at the too large value of α^w. If the creativity rate c is too large, the system has too strong randomness for us to find any significant structure in word relation matrices and their dynamics.

References

1. M. Fukaya and S. Tanaka. *Kotoba no ⟨Imiduke-ron⟩ (A Theory of "Sence Making" of Language)* [in Japanese]. Kinokuniya Shoten, Tokyo, 1996.
2. T. Hashimoto. Usage-based structuralization of relationships between words. In P. Husbands and I. Harvey, editors, *Fourth European Conference on Artificial Life*, pages 483–492. MIT Press, Cambridge, MA, 1997.
3. T. Hashimoto. Dynamics of internal and global structure through linguistic interactions. In Conte Sichman and Gilbert, editors, *Multi-Agent Systems and Agent-Based Simulation*, pages 124–139. Springer-Verlag, Berlin, 1998.
4. K. Kaneko and I. Tsuda. Constructive complexity and artificial reality: an introduction. *Physica*, D75:1–10, 1994.
5. Y. Karov and S. Edelman. Similarity-based word sense disambiguation. *Computational Linguistics*, 24:41–59, 1998.
6. G. Lakoff. *Women, Fire, and Dangerous Things.* The University of Chicago Press, Chicago, 1987.
7. L. Steels. The synthetic modeling of language origin. *Evolution of Communication*, 1(1):1–34, 1997.
8. J. R. Taylor. *Linguistic Categorization – Prototypes in Linguistic Theory.* Oxford University Press, Oxford, 1995.
9. L. Wittgenstein. *Philosophische Untersuchungen*. Basil Blackwell, 1953.

[5] Prototype category has some other attributes[8].

[6] Some abstract notions might be created in our society and internalized in using language.

Author Index

Lecture Notes in Artificial Intelligence (LNAI)

Vol. 1535: S. Ossowski, Co-ordination in Artificial Agent Societies. XVI, 221 pages. 1999.

Vol. 1537: N. Magnenat-Thalmann, D. Thalmann (Eds.), Modelling and Motion Capture Techniques for Virtual Environments. Proceedings, 1998. IX, 273 pages. 1998.

Vol. 1544: C. Zhang, D. Lukose (Eds.), Multi-Agent Systems. Proceedings, 1998. VII, 195 pages. 1998.

Vol. 1545: A. Birk, J. Demiris (Eds.), Learning Robots. Proceedings, 1996. IX, 188 pages. 1998.

Vol. 1555: J.P. Müller, M.P. Singh, A.S. Rao (Eds.), Intelligent Agents V. Proceedings, 1998. XXIV, 455 pages. 1999.

Vol. 1562: C.L. Nehaniv (Ed.), Computation for Metaphors, Analogy, and Agents. X, 389 pages. 1999.

Vol. 1566: A.L. Ralescu, J.G. Shanahan (Eds.), Fuzzy Logic in Artificial Intelliegence. Proceedings, 1997. X, 245 pages. 1999.

Vol. 1570: F. Puppe (Ed.), XPS-99: Knowledge-Based Systems. VIII, 227 pages. 1999.

Vol. 1571: P. Noriega, C. Sierra (Eds.), Agent Mediated Electronic Commerce. Proceedings, 1998. IX, 207 pages. 1999.

Vol. 1572: P. Fischer, H.U. Simon (Eds.), Computational Learning Theory. Proceedings, 1999. X, 301 pages. 1999.

Vol. 1574: N. Zhong, L. Zhou (Eds.), Methodologies for Knowledge Discovery and Data Mining. Proceedings, 1999. XV, 533 pages. 1999.

Vol. 1582: A. Lecomte, F. Lamarche, G. Perrier (Eds.), Logical Aspects of Computational Linguistics. Proceedings, 1997. XI, 251 pages. 1999.

Vol. 1585: B. McKay, X. Yao, C.S. Newton, J.-H. Kim, T. Furuhashi (Eds.), Simulated Evolution and Learning. Proceedings, 1998. XIII, 472 pages. 1999.

Vol. 1599: T. Ishida (Ed.), Multiagent Platforms. Proceedings, 1998. VIII, 187 pages. 1999.

Vol. 1600: M. J. Wooldridge, M. Veloso (Eds.), Artificial Intelligence Today. VIII, 489 pages. 1999.

Vol. 1604: M. Asada, H. Kitano (Eds.), RoboCup-98: Robot Soccer World Cup II. XI, 509 pages. 1999.

Vol. 1609: Z. W. Ras, A. Skowron (Eds.), Foundations of Intelligent Systems. Proceedings, 1999. XII, 676 pages. 1999.

Vol. 1611: I. Imam, Y. Kodratoff, A. El-Dessouki, M. Ali (Eds.), Multiple Approaches to Intelligent Systems. Proceedings, 1999. XIX, 899 pages. 1999.

Vol. 1612: R. Bergmann, S. Breen, M. Göker, M. Manago, S. Wess, Developing Industrial Case-Based Reasoning Applications. XX, 188 pages. 1999.

Vol. 1617: N.V. Murray (Ed.), Automated Reasoning with Analytic Tableaux and Related Methods. Proceedings, 1999. X, 325 pages. 1999.

Vol. 1620: W. Horn, Y. Shahar, G. Lindberg, S. Andreassen, J. Wyatt (Eds.), Artificial Intelligence in Medicine. Proceedings, 1999. XIII, 454 pages. 1999.

Vol. 1621: D. Fensel, R. Studer (Eds.), Knowledge Acquisition Modeling and Management. Proceedings, 1999. XI, 404 pages. 1999.

Vol. 1623: T. Reinartz, Focusing Solutions for Data Mining. XV, 309 pages. 1999.

Vol. 1632: H. Ganzinger (Ed.), Automated Deduction – CADE-16. Proceedings, 1999. XIV, 429 pages. 1999.

Vol. 1634: S. Džeroski, P. Flach (Eds.), Inductive Logic Programming. Proceedings, 1999. VIII, 303 pages. 1999.

Vol. 1637: J.P. Walser, Integer Optimization by Local Search. XIX, 137 pages. 1999.

Vol. 1638: A. Hunter, S. Parsons (Eds.), Symbolic and Quantitative Approaches to Reasoning and Uncertainty. Proceedings, 1999. IX, 397 pages. 1999.

Vol. 1640: W. Tepfenhart, W. Cyre (Eds.), Conceptual Structures: Standards and Practices. Proceedings, 1999. XII, 515 pages. 1999.

Vol. 1647: F.J. Garijo, M. Boman (Eds.), Multi-Agent System Engineering. Proceedings, 1999. X, 233 pages. 1999.

Vol. 1650: K.-D. Althoff, R. Bergmann, L.K. Branting (Eds.), Case-Based Reasoning Research and Development. Proceedings, 1999. XII, 598 pages. 1999.

Vol. 1652: M. Klusch, O.M. Shehory, G. Weiss (Eds.), Cooperative Information Agents III. Proceedings, 1999. XI, 404 pages. 1999.

Vol. 1674: D. Floreano, J.-D. Nicoud, F. Mondada (Eds.), Advances in Artificial Life. Proceedings, 1999. XVI, 737 pages. 1999.

Vol. 1688: P. Bouquet, L. Serafini, P. Brézillon, M Benerecetti, F. Castellani (Eds.), Modeling and Using Context. Proceedings, 1999. XII, 528 pages. 1999.

Vol. 1692: V. Matoušek, P. Mautner, J. Ocelíková, P. Sojka (Eds.), Text, Speech, and Dialogue. Proceedings, 1999. XI, 396 pages. 1999.

Vol. 1701: W. Burgard, T. Christaller, A.B. Cremers (Eds.), KI-99: Advances in Artificial Intelligence. Proceedings, 1999. XI, 311 pages. 1999.

Vol. 1704: Jan M. Żytkow, J. Rauch (Eds.), Principles of Data Mining and Knowledge Discovery. Proceedings, 1999. XIV, 593 pages. 1999.

Vol. 1705: H. Ganzinger, D. McAllester, A. Voronkov (Eds.), Logic for Programming and Automated Reasoning. Proceedings, 1999. XII, 397 pages. 1999.

Lecture Notes in Computer Science

Vol. 1652: M. Klusch, O.M. Shehory, G. Weiss (Eds.), Cooperative Information Agents III. Proceedings, 1999. XI, 404 pages. 1999. (Subseries LNAI).

Vol. 1653: S. Covaci (Ed.), Active Networks. Proceedings, 1999. XIII, 346 pages. 1999.

Vol. 1654: E.R. Hancock, M. Pelillo (Eds.), Energy Minimization Methods in Computer Vision and Pattern Recognition. Proceedings, 1999. IX, 331 pages. 1999.

Vol. 1655: S.-W. Lee, Y. Nakano (Eds.), Document Analysis Systems: Theory and Practice. Proceedings, 1998. XI, 377 pages. 1999.

Vol. 1656: S. Chatterjee, J.F. Prins, L. Carter, J. Ferrante, Z. Li, D. Sehr, P.-C. Yew (Eds.), Languages and Compilers for Parallel Computing. Proceedings, 1998. XI, 384 pages. 1999.

Vol. 1661: C. Freksa, D.M. Mark (Eds.), Spatial Information Theory. Proceedings, 1999. XIII, 477 pages. 1999.

Vol. 1662: V. Malyshkin (Ed.), Parallel Computing Technologies. Proceedings, 1999. XIX, 510 pages. 1999.

Vol. 1663: F. Dehne, A. Gupta. J.-R. Sack, R. Tamassia (Eds.), Algorithms and Data Structures. Proceedings, 1999. IX, 366 pages. 1999.

Vol. 1664: J.C.M. Baeten, S. Mauw (Eds.), CONCUR'99. Concurrency Theory. Proceedings, 1999. XI, 573 pages. 1999.

Vol. 1666: M. Wiener (Ed.), Advances in Cryptology – CRYPTO '99. Proceedings, 1999. XII, 639 pages. 1999.

Vol. 1667: J. Hlavička, E. Maehle, A. Pataricza (Eds.), Dependable Computing – EDCC-3. Proceedings, 1999. XVIII, 455 pages. 1999.

Vol. 1668: J.S. Vitter, C.D. Zaroliagis (Eds.), Algorithm Engineering. Proceedings, 1999. VIII, 361 pages. 1999.

Vol. 1671: D. Hochbaum, K. Jansen, J.D.P. Rolim, A. Sinclair (Eds.), Randomization, Approximation, and Combinatorial Optimization. Proceedings, 1999. IX, 289 pages. 1999.

Vol. 1672: M. Kutylowski, L. Pacholski, T. Wierzbicki (Eds.), Mathematical Foundations of Computer Science 1999. Proceedings, 1999. XII, 455 pages. 1999.

Vol. 1673: P. Lysaght, J. Irvine, R. Hartenstein (Eds.), Field Programmable Logic and Applications. Proceedings, 1999. XI, 541 pages. 1999.

Vol. 1674: D. Floreano, J.-D. Nicoud, F. Mondada (Eds.), Advances in Artificial Life. Proceedings, 1999. XVI, 737 pages. 1999. (Subseries LNAI).

Vol. 1675: J. Estublier (Ed.), System Configuration Management. Proceedings, 1999. VIII, 255 pages. 1999.

Vol. 1976: M. Mohania, A M. Tjoa (Eds.), Data Warehousing and Knowledge Discovery. Proceedings, 1999. XII, 400 pages. 1999.

Vol. 1677: T. Bench-Capon, G. Soda, A M. Tjoa (Eds.), Database and Expert Systems Applications. Proceedings, 1999. XVIII, 1105 pages. 1999.

Vol. 1678: M.H. Böhlen, C.S. Jensen, M.O. Scholl (Eds.), Spatio-Temporal Database Management. Proceedings, 1999. X, 243 pages. 1999.

Vol. 1679: C. Taylor, A. Colchester (Eds.), Medical Image Computing and Computer-Assisted Intervention – MICCAI'99. Proceedings, 1999. XXI, 1240 pages. 1999.

Vol. 1680: D. Dams, R. Gerth, S. Leue, M. Massink (Eds.), Practical Aspects of SPIN Model-Checking. Proceedings, 1999. X, 277 pages. 1999.

Vol. 1682: M. Nielsen, P. Johansen, O.F. Olsen, J. Weickert (Eds.), Scale-Space Theories in Computer Vision. Proceedings, 1999. XII, 532 pages. 1999.

Vol. 1684: G. Ciobanu, G. Păun (Eds.), Fundamentals of Computation Theory. Proceedings, 1999. XI, 570 pages. 1999.

Vol. 1685: P. Amestoy, P. Berger, M. Daydé, I. Duff, V. Frayssé, L. Giraud, D. Ruiz (Eds.), Euro-Par'99. Parallel Processing. Proceedings, 1999. XXXII, 1503 pages. 1999.

Vol. 1688: P. Bouquet, L. Serafini, P. Brézillon, M. Benerecetti, F. Castellani (Eds.), Modeling and Using Context. Proceedings, 1999. XII, 528 pages. 1999. (Subseries LNAI).

Vol. 1689: F. Solina, A. Leonardis (Eds.), Computer Analysis of Images and Patterns. Proceedings, 1999. XIV, 650 pages. 1999.

Vol. 1690: Y. Bertot, G. Dowek, A. Hirschowitz, C. Paulin, L. Théry (Eds.), Theorem Proving in Higher Order Logics. Proceedings, 1999. VIII, 359 pages. 1999.

Vol. 1691: J. Eder, I. Rozman, T. Welzer (Eds.), Advances in Databases and Information Systems. Proceedings, 1999. XIII, 383 pages. 1999.

Vol. 1692: V. Matoušek, P. Mautner, J. Ocelíková, P. Sojka (Eds.), Text, Speech, and Dialogue. Proceedings, 1999. XI, 396 pages. 1999. (Subseries LNAI).

Vol. 1694: A. Cortesi, G. Filé (Eds.), Static Analysis. Proceedings, 1999. VIII, 357 pages. 1999.

Vol. 1701: W. Burgard, T. Christaller, A.B. Cremers (Eds.), KI-99: Advances in Artificial Intelligence. Proceedings, 1999. XI, 311 pages. 1999. (Subseries LNAI).

Vol. 1704: Jan M. Żytkow, J. Rauch (Eds.), Principles of Data Mining and Knowledge Discovery. Proceedings, 1999. XIV, 593 pages. 1999. (Subseries LNAI).

Vol. 1705: H. Ganzinger, D. McAllester, A. Voronkov (Eds.), Logic for Programming and Automated Reasoning. Proceedings, 1999. XII, 397 pages. 1999. (Subseries LNAI).

GPSR Compliance

The European Union's (EU) General Product Safety Regulation (GPSR) is a set of rules that requires consumer products to be safe and our obligations to ensure this.

If you have any concerns about our products, you can contact us on ProductSafety@springernature.com

In case Publisher is established outside the EU, the EU authorized representative is:

Springer Nature Customer Service Center GmbH
Europaplatz 3
69115 Heidelberg, Germany

Batch number: 10723231

Printed by Printforce, the Netherlands